ESTATES GAZETTE LAW REPORTS

Estates Gazette Law Reports

1988

Volume 1

Edited by
J Muir Watt OBE MA
of the Inner Temple, barrister

Ernest Speller
Publishers' editor

THE ESTATES GAZETTE LIMITED
151 WARDOUR STREET, LONDON W1V 4BN

First published 1988
ISBN for complete set of 2 volumes: 0 7282 0123 2
ISBN for this volume: 0 7282 0124 0

ISSN 0951-9289

Typesetting by Digital Graphics, 147-149 Wardour Street, London
Printed and bound at the Bath Press, Avon

CONTENTS

SUPPLEMENT

Cases not reported in "Estates Gazette"

TABLE OF CASES

Titles of cases shown in bold type
Names of parties reversed shown in ordinary type

INDEX OF SUBJECT-MATTER

HOUSING
Improvement grants
"Certified date"
Lengthy interval between physical completion of works and confirmation from local authority departments that they were satisfied in regard to works — Was former date or latter date the "certified date"? — Held former was correct date — It is date of actual completion of works to a standard, namely, the local authority's satisfaction, not date when last department has sent confirmation of its satisfaction — Housing Act 1974, s75(6).
***R* v *Westminster City Council, ex parte Hazan (CA),* 29**

IMPROVEMENT GRANTS — See **Housing**

INTERFERENCE WITH CONTRACTUAL RIGHTS — See **Tort**

LAND REGISTRATION — See **Landlord and tenant** (Overriding interest)

LANDLORD AND TENANT
Business tenancies
Security of tenure — See **Landlord and tenant** (Landlord and Tenant Act 1954 Part II)
Covenant
Apportionment of liability
Collapse of part of warehouse roof due to accumulation of rain-water and blocked outlets — Tenants, whose goods were damaged, claimed on basis of implied contract by landlords, alternatively in nuisance and negligence — Landlords claimed on tenants' covenant to repair — Held it was a problem of apportionment by causation, not under Law Reform (Contributory Negligence) Act 1945, which applied to torts only — Landlords liable in nuisance and negligence, tenants in breach of covenant to repair — Liability apportioned 90% to landlords, 10% to tenants.
***Tennant Radiant Heat Ltd* v *Warrington Development Corporation (CA),* 41**
Forfeiture
Relief
Alteration of parties' position — Discretion to refuse relief in exceptional circumstances for non-payment of rent — "Eleventh-and-a-half hour" application after series of steps to delay giving up possession — After tenants' assurance and court order for possession in undefended proceedings, landlords negotiated with and subsequently granted lease of premises to third party — Tenants nevertheless continued to resist — Judge's refusal of belated summons for relief upheld — Position of parties had altered and rights of third party had intervened.
***Silverman* v *AFCO (UK) Ltd (CA),* 51**
Landlord and Tenant Act 1954 Part II
Agricultural holding — See **Agricultural holdings** (Definition)
Contracting-out with court authority
S 38(4) — Correct procedure — Possible for tenancy to be granted conditionally on court's approval of contracting-out, but tenancy granted unconditionally cannot be brought within s38(4)(a) by subsequent application to, and purported approval by, court — Underlease granted unconditionally but with recital of authorisation and blank spaces left for subsequent entry of county court order — Held purported contracting-out invalid.
***Essexcrest Ltd* v *Evenlex Ltd (CA),* 69**
Landlord's opposition to new tenancy
S 30(1)(*f*) — Building lease — Landlords, freeholders, first proposed to sell subject site outright, but as tenants were unwilling to leave, decided to proceed by a building agreement followed by 99 years' lease — Held landlords' required intention could be shown by grant of building lease and that critical date for establishing intention was date of hearing.
***Spook Erection Ltd* v *British Railways Board (CA),* 76**
S 30(1)(*f*) — Building lease — Landlords had considered different methods, including outright sale, but eventually decided to adopt surveyor's advice to proceed by granting 125 years' building lease — Held landlords had shown requisite intention.
***P F Ahern & Sons Ltd* v *Hunt (CA),* 74**
Ss 30(1)(*f*) and 31A — Shop occupied by tenant as café was one of three owned by landlord — Complete refurbishment of the three proposed, including demolition of walls and shop fronts and making the three units into one open area — Held, subject to s31A, ground under s30(1)(*f*) had been established — S 31A did not apply because works proposed would cause very substantial interference with operation of café.
***Blackburn* v *Hussain (CA),* 77**
New tenancy application
Snack-bar restaurant rent — Decision below preferring landlords' surveyor's figure of £16,500 pa, with interim rent of £13,200, to tenant's surveyor's figure of £9,100 — Tenant criticised reliance on rent of next-door property, where take-away pizza bar run by pizza chain subsidiary — Held, rejecting criticism (as subject property's user clause would allow take-away business) that area had moved up-market, so that rent properly fixed did not allow tenant to make profit.
***Giannoukakis* v *Saltfleet Ltd (CA),* 73**
Notice to terminate tenancy
Validity — Notice served in respect of one of two properties (service stations) demised to one tenant — Whether lease contained provisions sufficient to create two separate tenancies — Held, distinguishing *Dodson Bull Carpet Co Ltd* v *City of London Corporation* and *Southport Old Links Ltd* v *Naylor,* that a clause in the lease clearly created separate tenancies of the two properties.
***Moss* v *Mobil Oil Co Ltd (CA),* 71**
Validity — Premises consisted of offices on 9th floor together with storage accommodation on two other floors and free car-parking space — S 25 notice mentioned only 9th floor office accommodation in description of premises on statutory form — Held notice invalid — Not clear to reasonable tenant reading it what was comprised in notice.
***Herongrove Ltd* v *Wates City of London Properties plc (Ch),* 82**
Withholding of licence or consent
S 53 — Covenant preventing change of user without landlords' consent — Consent refused — County court judge decided landlords had arrived at their decision to refuse in unreasonable way, acting on misconceptions and incorrect information — Held judge in error in basing decision on procedural steps by which landlords arrived at their refusal — Having found landlords' reason, the next step was objective one of deciding whether landlords' ground reasonable.
***Tollbench Ltd* v *Plymouth City Council (CA),* 79**
Lease or licence
Possession claim
Licences or joint tenancy — *Street* v *Mountford* analysed — Four occupants of flat each with separate licence agreement expressly stating there was no right to exclusive possession of any part — Held by majority that, despite differences of periods of occupancy and sums payable, occupants had joint tenancy — Sir George Waller, dissenting, held that none of "four unities" required for a joint tenancy was satisfied.
***A G Securities* v *Vaughan (CA),* 36**
Occupants, a young man and girl friend, each signed identical agreement negativing exclusive possession of flat and providing that licensor or persons authorised by him could use flat in common with licensees — Agreements modelled on *Somma* v *Hazelhurst* — Whether occupants held joint protected tenancy under Rent Act — Held county court judge who believed he was applying *Street* v *Mountford* in holding agreements bad as "artificial transactions designed to evade the Rent Acts" was wrong — Agreements not a sham — *Somma* distinguished — Guidance given on *Street* v *Mountford.*
***Antoniades* v *Villiers (CA),* 59**
Leasehold enfranchisement
"Low rent"
Cottages and garages — Whether rateable value of garages should be aggregated with rateable values of two cottages occupied as one dwelling — Garages built after date when cottages first appeared together in valuation list, which was "appropriate day" for determining whether rent was a "low rent" — On this basis rent more than two-thirds of rateable value of cottages excluding garages — Appeal from Court of Appeal **[1987]** 1 **93** dismissed — Leasehold Reform Act 1967, s4.
***Dixon* v *Allgood (HL),* 85**
Price of freehold
Agency commission on fire insurance premium — Present tribunal consider themselves bound by Lands Tribunal's decision in *Re Castlebeg Investments (Jersey) Ltd's Appeal,* provided sufficient evidence available that purchaser of freehold could obtain commission on single insurance premium — On evidence, held that commission element of claim should properly be admitted.
***Lynch* v *Castlebeg Investments (Jersey) Ltd (LVT),* 223**
Rateable value limits — See **Negligence** (Rating valuation)
Maintenance
Managing agents
Appointment — Landlords a tenants' association with tenants who were also shareholders — Lease provided for appointment of managing agents, to act as experts and not as arbitrators — Association appointed as managing agents a committee of tenant shareholders — Tenant challenged propriety of this appointment on grounds of lack of independence and expertise — Held no formal professional qualification required for appointment — Managers independent and had sufficient expertise — Subsidiary point in regard to thermostatic radiator valves decided in favour of landlords also.
***New Pinehurst Residents Association (Cambridge) Ltd* v *Silow (CA),* 227**
Overriding interest
Registration
Lease of shop in development area — Provisions in agreement giving rise to overriding interest under Land Registration Act 1925 — Undertaking that occupiers would be granted, on completion of new development, "a lease of a shop in a prime position" — Difficulty as to rights of occupiers, there being no contract between them and landlords and their only rights were in respect of their overriding interest, which had not been registered as an estate contract — Rulings as to occupiers' rights in different circumstances — Rights were to prime site in development area, not necessarily to present location.
***Ashburn Anstalt* v *Arnold (No2) (CA),* 64**

LANDLORD AND TENANT — *continued*

Rent Acts

Furnished tenancy

Fair rent decision of rent assessment committee challenged — Whether tenancy a furnished tenancy although bulk of furniture in flat belonged to tenant — Tenant had agreed to change to newly renovated flat, landlords stating they had no objection to her using her own furniture on a daily basis, their furniture to be stored and reinstalled at end of tenancy — Rent officer and committee assessed fair rent on basis of furnished tenancy — "Comparables" of furnished tenancies put forward and discussed — Held committee justified in assessing fair rent on basis of furnished tenancy — Rent Act 1977, s70(1) (*b*).

***R* v *London Rent Assessment Panel, ex parte Mota (QB),* 89**

Long tenancy

Rent less than two-thirds of rateable value — Provisions in 1972 lease — Parties unaware that s80(1) of Housing Act 1969 had altered law by excluding sums payable for services in determining whether rent of tenancy was less than two-thirds of rateable value — Lease contained provisions designed to ensure that amount for services did not bring rent above two-thirds limit — Purpose was to enable premiums to be lawfully demanded on grant or assignment of tenancy — Held that, although badly drafted and based on a misapprehension, clauses could be given intelligible meaning and entitled lessor to claim full service charge.

***Investment & Freehold English Estates Ltd* v *Casement (CA),* 100**

Possession

Agreement by protected tenant with purchaser of dwelling-house to join with landlord (the vendor) in giving vacant possession on completion — Tenant after completion did not give possession — Held, applying *Barton* v *Fincham* and distinguishing *Dudley & District Benefit Society* v *Emerson,* that agreement to give up possession cannot be enforced against protected tenant if entitled under Rent Act to remain in possession and chooses to do so — Rent Act 1977, s98(1).

***Appleton* v *Aspin (CA),* 95**

Res judicata — Room in rooming house purported to be let to defendant by person without authority who cheated her out of her deposit — Defendant did not have tenancy but was in fact a trespasser although she did not realise it — After being locked out, defendant obtained interim injunction restoring her occupation and judgment for damages to be assessed — Later, when possession sought by landlords, she pleaded *res judicata,* contending that previous judgment had determined she was a tenant and protected by Rent Act — Held this contention failed.

***Ozer Properties Ltd* v *Ghaydi (CA),* 91**

Tenant's counterclaim — Allegations of breaches of implied repairing covenants — Only evidence of breaches given at trial related to minor malfunction of hot-water and central-heating system — Judge had refused to regard tenant's solicitors' letter or counterclaim, incorporating surveyor's report, as valid notices of disrepair under *O'Brien* v *Robinson* rule — Held judge had decided these points correctly and had considered all relevant factors on issue of reasonableness in making possession order — Landlord and Tenant Act 1985, s11.

***Al Hassani* v *Merrigan (CA),* 93**

Shorthold tenancy

Grant for term of one year subject to provision that landlords shall give tenant three months' notice in writing prior to expiry date in event of their not wishing to renew agreement for further term — Landlords gave no such notice and tenant remained in occupation — Possession order made — Application to set it aside — Held possession order invalid so far as based on arrears of rent but valid under Case 19 in Schedule 15 to Rent Act 1977 — Housing Act 1980, s52(5).

***Gent* v *De La Mare (CA),* 104**

Succession to statutory tenancy

Meaning of "member of first successor's family" — Claimant, now 70, had resided with a family since 1941 after death of her parents and boy friend — Claim to succeed as second successor on death of daughter of original tenants — Claimant had been treated by these tenants like an adopted daughter and by their daughter like a sister — Held "family" had to be given its ordinary meaning — Although claimant had been treated like a member of the family she was not a member and had not been adopted — Rent Act 1977, Sched 1, para 7.

***Sefton Holdings Ltd* v *Cairns (CA),* 99**

Person cohabiting with deceased tenant — Whether member of tenant's "family" — Respondent had lived with tenant for some years, later with intention of marrying — Judge found relationship, although comparatively short, permanent and stable and concluded that respondent was member of tenant's family and was residing with him at time of, and for six months before, his death — Held no grounds for interfering with judge's conclusion.

***Chios Property Investment Co Ltd* v *Lopez (CA),* 98**

Two-homes tenant

Whether statutory tenancy continued in respect of London flat when tenant had lived in Lancashire cottage since 1978 — Flat had been and still was occupied by tenant's children and furniture had been left there — Tenant had not been in London since 1978 until trial in 1987 — His wife had been in London three times since then for special reasons — Tenant said he would probably return to flat if his wife predeceased him and both might be forced to return by increasing age — Judge found *corpus possessionis* and *animus revertendi* established — Court of Appeal agreed, although a borderline case.

***Brickfield Properties Ltd* v *Hughes (CA),* 106**

Rent provisions

Underlease

Rent of underlease to be fraction $\frac{78}{85}$ of rent payable under superior lease — Effect of surrender of superior lease — Superior lease surrendered in leaseback transaction by which new lease created subject to and with benefit of underlease — Underlessees argued that formula in original superior lease could not be operated — Held, applying principles of *Sudbrook Trading Estate Ltd* v *Eggleton* that, as formula was mere machinery, the court could substitute its own machinery to arrive at fraction mentioned in underlease — Inquiry ordered accordingly.

***R & A Millett (Shops) Ltd* v *Leon Allan International Fashions Ltd (Ch),* 45**

Rent review clause

Construction

Delay of first review date when road works not completed — Date to be delayed "one year for every year or part of a year" until completion of works or notification of abandonment — Whether starting date was date mentioned in clause, March 31 1973, the reason for which was speculative, or July 5 1980, earliest review date provided for in lease — Held some meaning should be given to date, March 31 1973, mentioned in formula — Taking this date as starting point, delayed date for first review was July 5 1988.

***Ladbroke Group plc* v *Bristol City Council (CA),* 126**

Long lease of industrial estate with rent reviews every 21 years — Review rent to equal one half of "estimated aggregate rack rental value of the demised premises" — Majority of units sublet — Held valuer not required to value estate on basis of single hypothetical letting of whole estate — Valuer had to treat estate as divided at material date and substitute for actual rents paid, where appropriate, the rental value of each separate unit — Voids must be dealt with in accordance with normal practice of valuers.

***Leigh* v *Certibilt Investments Ltd (Ch),* 116**

Exclusion of clause

Hypothetical tenancy to be on terms and conditions in actual lease "save as to rent" — Whether hypothetical tenancy should itself include rent review provisions contained in lease — Authorities considered included *British Gas Corporation* v *Universities Superannuation Scheme Ltd* — Held provision ambiguous — In absence of clear words requiring construction that rent review provisions should be disregarded, effect should be given to underlying commercial purpose — Hypothetical tenancy should include rent reviews.

***British Home Stores plc* v *Ranbrook Properties Ltd (Ch),* 121**

Notice and counternotice

Action by landlords for forfeiture for non-payment of rent — Question as to rent payable depended on operation of rent review clause — Clause provided for trigger notice by landlords, agreement as to rent, or counternotice by tenant electing to have rent determined by independent surveyor — Whether letter from tenants saying "Please accept this letter as formal objection and counternotice. We would suggest an early meeting to discuss the matter in detail" was effective election — Held letter constituted effective election.

***Glofield Properties Ltd* v *Morley (QB),* 113**

Time-limits

Whether time of the essence for provision in review clause despite general statement to contrary — A para in review provisions stated generally that time was not to be deemed to be of the essence — A subsequent para, however, provided that if a review not completed within 12 months of commencement of relevant period rent would be increased by 25% of rent currently payable — Held, affirming decision of deputy judge, that time was not of the essence; the statement to this effect applied generally.

***Panavia Air Cargo Ltd* v *Southend-on-Sea Borough Council (Ch) (CA),* 111, 124**

Use

Restrictions in lease and in licence to assign to present tenants — Lease covenant restricted user to offices in connection with lessee's business of mortgage finance and insurance consultants — Licence to assign restricted user to offices for assignees' diplomatic mission — Held provision requiring hypothetical letting to contain similar terms to those of actual lease meant original terms, not personal arrangements in licence affecting current tenants — Assumption that hypothetical tenant would be bound not to use premises save in connection with business of mortgage finance and insurance consultant.

***SI Pension Trustees Ltd* v *Ministerio de Marina de la Republica Peruana (Ch),* 119**

RATES AND RATING — *continued*

Negligent valuation — See **Negligence** (Rating valuation)

Proposal

Validity

Separately rateable hereditaments proposed in place of single hereditament, a showroom and warehouse in SE London — Proposal held invalid by LVC — By tribunal hearing date VO accepted ratepayers' occupation was of one hereditament but contended his deciding on two hereditaments did not invalidate proposal — Held there was fundamental flaw in proposal in seeking to assess as two hereditaments a property comprising only one — Notice of appeal bad in not correctly describing hereditament — No jurisdiction to entertain appeal.

***Aluwihare (VO)* v *MFI Properties Ltd (LT),* 219**

Rateability

Stud premises

Whether hereditaments occupied and used for breeding and rearing of thoroughbred horses qualified for agricultural derating — Decision of Court of Appeal affirmed — Buildings not used in connection with agricultural operations on land — Thoroughbred horses not "livestock" within meaning of s1(3) of Rating Act 1971 — *Per* Lord Goff, breeding and rearing thoroughbred horses not an "agricultural operation" — General Rate Act 1967, s26(3) and (4) — Rating Act 1971, ss1 and 2.

***Whitsbury Farm & Stud Ltd* v *Hemens (VO) (HL),* 159**

Rateable occupation

Exclusive occupation

Warehouse — Company in sole *de facto* occupation during relevant period — Appellants complained magistrates considered only fact of sole occupation — Cases showed other matters should be considered, such as estates or interests in premises — Owners had given appellants right to store goods on premises, receiving payment according to tonnage stored, but owners retained right to permit others to store goods on premises — Owners shared control of access with appellants — Case remitted to magistrates.

***Channel Shipping (Newport) Ltd* v *Newport Borough Council (QB),* 162**

Non-active use

Whether sports ground left unused since purchase to be treated as unoccupied — Furniture and other equipment left behind and part-time caretaker cut grass to maintain pitches and bowling greens — Magistrates decided owners in rateable occupation — Held magistrates approached matter wrongly — Question to be asked was whether, but for presence of plant, machinery or equipment, the hereditament would have been found to be unoccupied — General Rate Act 1967, s46A(1).

***Sheafbank Property Trust plc* v *Sheffield Metropolitan DC (QB),* 164**

Warehouse owners

Whether owners in rateable occupation of warehouse bay which had not in fact been used during material time — Magistrates decided they were as result of their interpretation of estate agents' advertisements for letting of premises and of powers contained in appellant company's memorandum of association — Held that, although court might not have reached same conclusion, magistrates were entitled on the evidence to find appellants in rateable occupation of bay.

***Calmain Properties Ltd* v *Rotherham Metropolitan Borough Council (QB),* 166**

RECEIVER

Remuneration and expenses

Recovery of sums

Appointment of receiver and manager by court to receive rents and to manage block of flats on *Hart* v *Emelkirk* basis — Main litigation concerned repairs to flats — Refusal by Vinelott J **[1987]** 1 **67** to order landlords to meet deficit in moneys coming into receiver's hands for payment of his expenses and remuneration upheld — Correct to apply *Boehm* v *Goodall* principle that receiver not an agent or trustee for parties or under their control — It could be hardship to make party personally liable before rights determined at trial.

***Evans* v *Clayhope Properties Ltd (CA),* 33**

RENT ACTS — See **Landlord and tenant** (Rent Acts)

RENT REVIEW CLAUSE — See **Landlord and tenant** (Rent review clause)

REPAIRING COVENANT — See **Landlord and tenant** (Repairs)

RES JUDICATA — See **Landlord and tenant** (Rent Acts — Possession)

SERVICE CHARGES — See **Landlord and tenant** — (Rent Acts — Long tenancy)(Service charges)

SHORTHOLD TENANCY — See **Landlord and tenant** (Rent Acts)

STAKEHOLDER — See **Vendor and purchaser** (Stakeholder or vendor's agent)

SURVEYORS ARCHITECTURAL SERVICES — See **Tort**

TENANTS' ASSOCIATION — See **Landlord and tenant** (Service charges) (Maintenance)

TORT

Interference with contractual rights

Development condition

Surveyors' action against bank arising out of development project — Plaintiffs employed by developers to render comprehensive service, involving architectural and ancillary work — Defendants, who had provided finance for acquisition of site and had legal charge thereon, decided to carry out development themselves — Condition that plaintiffs should be replaced as architects — Held defendants acted in pursuance of equal or superior rights.

***Edwin Hill & Partners* v *First National Finance Corporation plc (QB),* 15**

TOWN AND COUNTRY PLANNING

Advertisements

Estate agents' boards

Offences — Whether agent commits offence if, after he has lawfully erected sale board, another agent, with or without former's knowledge, erects second board — Appellant agent's convictions upheld — Deemed consent under regulations restricted to one board — When second board added consent ceased to operate — Absence of appellant's knowledge of existence of two boards immaterial — Offence of absolute liability — No *mens rea* required — Town and Country Planning (Control of Advertisements) Regulations 1948.

***Porter* v *Honey (DC),* 202**

Development

Material change of use

Caravan stationed on agricultural land — Whether material change of use — Purpose of caravan was to provide place to store and mix calf feed and as shelter for farmer and wife — Held stationing of caravan was ancillary to use of land for agriculture and therefore not development — Question of material change of use did not arise — Town and Country Planning Act 1971, s22(2)(*e*).

***Wealden District Council* v *Secretary of State for the Environment (CA),* 187**

Enforcement notice

Appeal to Secretary of State

"Planning unit" — Material change of use by installation and use of grain silos — Complaint that silos used for storage of grain not grown on planning unit and not used for feeding of livestock on that unit — Whether error of law to regard planning unit for purpose of enforcement notice as being something smaller than totality of agricultural holding farmed by applicant — Held "planning unit" a matter of fact and degree — No error of law — Town and Country Planning Act 1971, s246(4).

***Fuller* v *Secretary of State for the Environment (CA),* 185**

Issue estoppel

Application of principle — Enforcement notices alleged breach of planning control consisting of change of use from hotel to hostel — Previous inspector's decision in 1982 that use was that of hotel — Held principle of issue estoppel applied to appeals against planning decisions and enforcement notices and did apply to four cases included in appeal — Cases remitted for rehearing — Town and Country Planning Act 1971, ss36 and 88.

***Thrasyvoulou* v *Secretary of State for the Environment (CA),* 192**

Tree preservation order

Compensation

Refusal to fell woodland subject to the order which landowner wished to grub out for grazing — Lands Tribunal held owner entitled to compensation based on diminution in value of land affected, not merely on basis of loss of timber which he was prevented from felling — Held tribunal's approach correct.

***Bell* v *Canterbury City Council (CA),* 205**

Validity

Application to quash order — Development company claimed, applying *Wednesbury* principles, that authority improperly influenced by immaterial consideration — Developers had amended their plan, in applying for approval of matters reserved from outline permission, by avoiding removal of tree — Allegation that authority had taken account of fact that developers had redrawn plan to retain tree and had been influenced by this fact in making order — Court not satisfied that authority improperly influenced by this consideration — Town and Country Planning Act 1971, ss60, 61, 245.

***Bellcross Co Ltd* v *Mid Bedfordshire District Council (QB),* 200**

TREE PRESERVATION ORDER — See **Town and country planning**

UNFAIR CONTRACT TERMS ACT 1977 — See **Negligence** (Mortgage loan report — Disclaimer)

USER COVENANT — See **Landlord and tenant** (Rent review clause)

VALUERS — See **Negligence**

VENDOR AND PURCHASER

Caution

Vacation

Vendors' registration of caution to protect their lien — Motion by purchasers for removal — Vendors contended certain sums payable by purchasers under special conditions in contract of sale were part of purchase price and thus protectable by caution — Held that on analysis the additional sums provided for payments by purchasers if they carried out specified developments, which they were under no obligation to carry out — Sums not part of purchase price — Order for vacation of caution.

***Woolf Project Management Ltd* v *Woodtrek Ltd (Ch),* 179**

Misrepresentation

Drainage

Action by purchasers of house against vendors alleging misrepresentation and against purchasers' surveyors alleging negligence — House represented as having main drainage but in fact drained into cesspool through joint system with adjoining property — Misrepresentation in particulars corrected by vendors' agents in letter to purchasers' solicitors, but letter lost and correction never conveyed to purchasers — Misrepresentation repeated by vendors and their solicitors — Held that purchasers' surveyors, who relied on vendors' assurance as to drainage, were not negligent — Action against vendors failed because knowledge of correction by purchasers' solicitors was imputed to purchasers.

***Strover* v *Harrington (Ch),* 173**

Innocent

Auction of freehold house — Particulars stated incorrectly that first floor let to weekly tenant, whereas in fact it was also subject to 99 years' concurrent lease — Purchasers unaware of this lease before auction, but their solicitors had copy of lease after auction and before completion — Held trial judge in error in holding that purchasers, having completed, could not sue for damages — Purchasers not so precluded from suing.

***Production Technology Consultants Ltd* v *Bartlett (CA),* 182**

Preliminary enquiry as to whether house to vendor's knowledge had been subject to woodworm, rising damp, dry or other rot or infestation — Misrepresentation alleged in reply — Held purchasers had not relied on representation, alternatively that vendor had believed on reasonable ground that facts represented were true — No breach of duty of care proved — Disclaimer clause would not have provided protection for vendor if held to be in default — Misrepresentation Act 1967, s2(1).

***Cooper* v *Tamms (QB),* 257**

Stakeholder or vendor's agent

Solicitor

Money held by vendors' solicitors in sale of hotel — Issues were (1) whether solicitors received money as stakeholders or as vendors' agents and (2) if as stakeholders, had contingency occurred which entitled them to release money to vendors? — Held answer "yes" to both questions — Contingency occurred when lease assigned to purchaser — Solicitors justified in releasing money and action against them dismissed.

***Tudor* v *Hamid (CA),* 251**

AGRICULTURE

A

Court of Appeal

November 12 1987

(Before Lord Justice DILLON and Lord Justice CROOM-JOHNSON)

SHORT v GREEVES

Estates Gazette February 27 1988

[1988] 08 EG 109-112

B **Agricultural holding — Whether a garden centre was an agricultural holding or the subject of a business tenancy under Part II of the Landlord and Tenant Act 1954 — Whether tenancy substantially a tenancy of an agricultural holding — Whether question had to be decided under the Agricultural Holdings Act 1948 or the Agricultural Holdings Act 1986 — Whether substantiality should be determined by proportion of turnover — Whether nature of tenancy had changed from being in its origin clearly a tenancy of an agricultural holding to being a business tenancy — Appeal by landlord from decision of county court judge that the property had remained**
C **an agricultural holding — The tenancy, which began in 1982, was of a horticultural enterprise, a garden centre laid out in a familiar pattern, with the front part of a greenhouse used for retail sales and a display area round it — There were rows of roses, wallflowers, primulas and some conifers and equipment such as garden paving and sheds — Between 1982 and 1985 an increasing number of articles were brought on to the land from elsewhere for sale, such as garden seats, bags of peat, paving, earthenware pots for plants, a much larger number of garden sheds, bird tables, wooden half-tubs, produce such as eggs and potatoes, electric propagators, chemical sprays and applicators, hanging baskets and even**
D **garden gnomes — The county court judge found that the bought-in produce had come to represent more than half the turnover and put the proportions at roughly 40 per cent home-grown produce and 60 per cent bought-in produce — Although the appellant challenged this calculation, the Court of Appeal saw no basis on which they could interfere with it — The history of the litigation was that the appellant had given the respondent tenant a section 25 notice under the Landlord and Tenant Act 1954, dated October 3 1984, purporting to terminate the tenancy on April 8 1985 and indicating that a new grant would not be opposed — The respondent applied**
E **for a new tenancy without prejudice to his contention that the land was an agricultural holding — The relevant date for determining the nature of the tenancy was either October 3 1984 or April 8 1985 (it was immaterial which one it was) — The parties were agreed that the matter fell to be determined under the Agricultural Holdings Act 1986, not the Agricultural Holdings Act 1948, although Dillon LJ, in the light of Schedule 13 to the 1986 Act, had considerable reservations — However, here again, having regard to *Howkins* v *Jardine*, it made no real difference — Held, affirming the decision of the county court judge, that the turnover figures were in no way conclusive — The judge was**
F **justified, looking at the circumstances as a whole, in concluding that this was an agricultural holding at the relevant date — A "not insubstantial part" of the business consisted of horticulture — A court is not lightly to treat a tenancy as having ceased to be within the protection of the agricultural holdings code — There was evidence to support the judge's finding that the character of the tenancy had not changed — Appeal dismissed**

The following cases are referred to in this report.

Howkins v *Jardine* [1951] 1 KB 614; [1951] 1 All ER 320; [1951] 1 TLR 135, CA G
Wetherall v *Smith* [1980] 1 WLR 1290; [1980] 2 All ER 530; (1980) 40 P & CR 205; [1980] EGD 62; 256 EG 163, CA

This was an appeal by the landlord, Mrs Rosemary Joan Greeves, against a decison of Judge Figgis, at Kingston County Court, holding that the tenant, Robert William Short, was the tenant of an agricultural holding at Ditton Hill Garden Centre, Ditton Hill, near Surbiton, Surrey.

Simon Buckhaven (instructed by R H Speechley & Owen) appeared on behalf of the appellant; T Hugh Berry (instructed by Woodhouse Russell & Mitchell) represented the respondent.

Giving judgment, DILLON LJ said: This is an appeal by the H
respondent in the proceedings, Mrs Greeves, against an order of His Honour Judge Figgis made in the Kingston County Court on April 27 1987. The appellant is the landlord and the respondent to this appeal, Mr Short, is the tenant of a property known as the Ditton Hill Garden Centre, Ditton Hill, near Surbiton in Surrey, which is close to the Esher bypass. The question which the judge had to decide was whether that property at the relevant dates — and I will come in a moment to what they were — was an agricultural holding or business premises; in other words, whether it was at the relevant dates subject to the Agricultural Holdings Act or subject to Part II of the Landlord and Tenant Act 1954.

The Landlord and Tenant Act 1954 provides by section 25 that the
landlord may terminate a tenancy to which Part II of the Act applies J
by notice in the prescribed form, and by section 24 that a tenant under a tenancy to which that Part of the Act applies might apply to the court for a new tenancy if the landlord has given notice to terminate the tenancy. But Part II of the 1954 Act provides by section 43 that it does not apply to a tenancy of an agricultural holding, and the definition section provides that "agricultural holding" has the same meaning as in the Agricultural Holdings Act.

In the present case the appellant gave the respondent a notice in the prescribed form under Part II of the 1954 Act, which is dated October 3 1984, purporting to terminate the tenancy on April 8 1985 and stating that an application to the court under Part II of the 1954 Act for the grant of a new tenancy would not be opposed. The tenant then K
on January 9 1985 applied to the court under Part II for the grant of a new tenancy. However, that was without prejudice to the contention that the land was an agricultural holding. The judge had to decide whether it was an agricultural holding or a tenancy of business premises within Part II. In view of the wording of sections 24 and 25, it is plain that the relevant date is October 1984 or January 1985, the dates of the notice to terminate the tenancy and of the tenant's application. There is no distinction on the facts between the two.

Counsel were agreed in the county court, and are agreed in this court, that the question whether in October 1984 or January 1985 the land was an agricultural holding has to be decided under the Agricultural Holdings Act 1986. For my part I have very

A considerable reservations about that approach. Section 98 of the 1986 Act provides that the Act applies in relation to tenancies of agricultural holdings whenever created, agreements whenever made and other things whenever done. But Schedule 13 to the Act provides that nothing in the Act shall apply in relation to a notice to quit an agricultural holding or part of an agricultural holding given before the commencement of the Act. It seems to me, therefore, that the natural approach may well be to say that the question the judge had to decide — whether or not the land was an agricultural holding at the time of the landlord's notice to terminate any business tenancy of the land — fell to be decided under the Agricultural Holdings Act 1948, which was then in force.

Be that as it may, I do not think that it makes any real difference in B substance to the outcome. Under section 1 of the 1986 Act an agricultural holding is defined as meaning:

> the aggregate of the land (whether agricultural land or not) comprised in a contract of tenancy which is a contract for an agricultural tenancy.

Then a contract of tenancy relating to land is said to be a contract for an agricultural tenancy if, having regard to all relevant circumstances,

> the whole of the land comprised in the contract, subject to such exceptions only as do not substantially affect the character of the tenancy, is let for use as agricultural land.

Under the 1948 Act, section 1 provided that an agricultural holding meant the aggregate of the agricultural land comprised in a contract C of tenancy. The courts, interpreting that where part of the land was and part was not used for agriculture, laid down the general approach in a decision of this court in *Howkins* v *Jardine* [1951] 1 KB 614. There Somervell LJ said at p 623, after referring to various authorities:

> I have therefore come to the conclusion, this being in substance an agricultural tenancy, and the definition not providing for severance, that the tenant succeeds.

In that case the question arose because there were certain cottages included in the letting which were not occupied for the time being by agricultural workers, and the claim was that the tenancy could be severed and the agricultural protection could continue with the D agricultural land but not for the cottages. The court held that severance was not possible, and they regarded it as absurd to treat cottages contained in what was in substance an agricultural tenancy as coming in and going out of the Act according to the occupation of the tenants at the moment. Jenkins LJ said at p 628:

> . . . one must look at the substance of the matter and see whether, as a matter of substance, the land comprised in the tenancy, taken as a whole, is an agricultural holding. If it is, then the whole of it is entitled to the protection of the Act. If it is not, then none of it is so entitled.

He rejected as unworkable a literal construction of the word "used" as requiring actual use of every inch of the holding for agriculture or for the purposes of an agricultural trade or business and said that the E letting of the cottages to other persons, because they were not required for agricultural tenants, should not affect the result.

The statutory definition of "agriculture" both in the 1948 Act and in the 1986 Act includes "horticulture" and "use of land as market gardens and nursery grounds".

As a matter of history, the land with which we are concerned, the Ditton Hill Garden Centre, consists of some 6.2 acres, and it was let to a Mr Stone in 1973 on a yearly tenancy. In April 1980 that tenancy was assigned to Mr Short, who paid a price of £4,000 for the plants and stock. He was granted a new tenancy agreement of the property on July 21 1982 by the appellant. That agreement was for a period of five years commencing from April 1 1980. It provided:

> THE Landlord agrees to let as a Garden Centre and the Tenant agrees to take F on the terms and conditions following:
>
> ALL THAT land at Ditton Hill Summerfield Lane Surbiton Surrey heretofore used as a Garden Centre together with the shop or store erected thereon or on some part thereof which for the purpose of identification only is shown edged red on the Plan annexed hereto . . .

It is, of course, not in dispute that a garden centre may be an agricultural holding or may be merely a retail shop which is a business premises subject to Part II of the 1954 Act. One can easily think of garden centres which are plainly within Part II of the 1954 Act and are not agricultural holdings at all and other garden centres which are equally plainly agricultural holdings where retail sales are predominantly sales of plants grown on the same land.

This tenancy agreement included, among many usual covenants, a covenant by the tenant "to manage the property in accordance with G the rules of Good Husbandry so as to preserve the fertility of the soil and to keep the land free from weeds and in good order". That is, of course, a standard covenant in tenancies of agricultural holdings. Then there was an arbitration clause which provided:

> ALL disputes and differences which may hereafter arise between the parties hereto under or in connection with this agreement not being disputes or differences compulsorily referred to Arbitration under the Agricultural Holdings Act 1948 . . . shall be referred in accordance with the Arbitration Act 1950 or any statutory modification or re-enactment thereof for the time being in force to a single Arbitrator to be appointed in default of agreement by a member of [a particular firm]. No award made under this sub-clause shall include any matters compulsorily referred to Arbitration under the Act of 1948.

Mr Short grows roses on parts of the land, also plants such as H wallflowers and primulas and some conifers. We have been supplied with an aerial photograph of the garden centre taken in 1980 or 1982. There is a greenhouse at right angles to the road with a display area round it. That is in the centre of the road frontage of the garden centre. The front part of the greenhouse was used for retail sales. To the east and west ends of the land plants are grown in the open, but near the greenhouse there are plants in containers for sale. One can see relatively near the greenhouse rows of roses; there are also some containerised conifers, relatively small amounts of what looks like garden paving, and four small garden sheds, and a parking area. The conclusion, I apprehend without any hesitation, would be that at the time of the aerial photograph and at the time of the grant of the J tenancy, it was a tenancy of an agricultural holding. The roses sold are now put into containers or pots for sale, but they are still all roses which have been grown by Mr Short on his land.

What has happened since is that parts of the land to the east and west have become somewhat derelict, but they are not really put to any other use except horticulture. Part of the land at the west end is not particularly good land. But, much more significantly, the display area has been devoted to a very much greater extent to the sale of garden sundries, which are brought on to the land from elsewhere for sale. We have a number of photographs which show in 1984 and 1985 garden seats, gnomes, bags of peat, paving, earthenware pots for plants and matters of that sort, and also a very much larger number than before of timber garden sheds which are sold, in effect, by a K concession granted to the makers of the sheds, who pay Mr Short a percentage on each shed sold.

There is a list of bought-in goods for resale among the documents. They include, apart from matters which I have mentioned, turves, bird tables, wooden half-tubs, decorative concrete lions, eggs, potatoes, electric propagators, chemical garden sprays and applicators, hanging baskets and all sorts of matters of that kind, and greenhouses as well as garden sheds.

What is submitted for the landlord (the appellant) is that there is now, on an area which it is agreed is about 1.2 acres, a thriving shop selling predominantly goods which have not been grown on the land — indeed, a large part of which are not even plants at all — and L therefore a substantial part of the land is used for non-agricultural purposes. It is not ancillary to the use of the rest of the land for growing roses and the other plants I have mentioned as grown on the land. Therefore, applying either test, you cannot say that the land as a whole is substantially an agricultural holding.

On the other side, we have been referred to the decision of this court in *Wetherall* v *Smith* [1980] 2 All ER 530, where Sir David Cairns, in giving the leading judgment in this court, pointed out that there was no doubt that land could cease to be an agricultural holding if the agricultural use were abandoned, but he also said (p 537G):

> . . . the cases show that the tenancy is not to be regarded as alternating between being within and outside the 1948 Act as minor changes of user take place, and that, when the tenancy is clearly an agricultural one to start with, M strong evidence is needed to show that agricultural user has been abandoned.

The learned judge had the advantage of a visit to the site, albeit in 1987 rather than at the dates which are relevant for the decision of the case. He also had a certain amount of evidence as to the takings of Mr Short from his business in the garden centre. The judge commented that Mr Short's records and his discovery had been hopeless. The landlord had tried to recalculate the figures. The judge went on to say that the other side's recalculation of the figures he regarded as equally hopeless. He said that he could not accept either of the figures in relation to the proportion which the trade had assumed. He then said:

> I am prepared to find on a balance of probabilities that a greater proportion of

A present trade arises from the sale of bought-in plants and garden requisites and the sale on commission of certain garden sheds and things of that kind. On a balance of probabilities I am prepared to find that that represents more than half the turnover on the premises, but I find that the home-grown produce is a substantial part. If I were to have to give a proportion, I would say that approximately 40% of the turnover is related to home-grown produce and 60% from garden requisites and bought-in produce.

Over the years the garden requisites and bought-in produce has increased in relation to the other.

The appellant has challenged that rough-and-ready computation of 40% of the turnover and has said that it should be something much less, say 20%, attributable to home-grown produce. I see no basis on which this court should interfere with the figure which the judge has given, doing the best he could, as he was compelled to, with the B inadequate figures that have been produced by Mr Short. In fairness to Mr Short, whose accounting system seems to have been rather rudimentary, it may be said that he was not very much concerned, for any of the purposes for which he was keeping accounts, to distinguish between home-grown produce and bought-in produce. He was merely concerned with the overall result.

The question then is whether at the relevant dates the tenancy was still a tenancy of an agricultural holding. The turnover figures, of course, are not in any way conclusive as to that. The holding and all the circumstances have to be regarded as a whole. The judge took the view that it was a pretty borderline case, but he concluded at the end that it was an agricultural holding at the relevant dates. Then he made C the declaration which I have indicated.

We have had some discussion whether the judge misdirected himself. At one point he said that he found that genuinely a part of the business, a not insubstantial part, consisted of horticulture. That might well have been a misdirection if that was all he meant and if he was entirely disregarding what was done in the display area, but it is plain that he was not disregarding that. He was looking at the matter as a whole. He was not putting out of his mind what was done in the display area and considering that it was enough if a not insubstantial part of the business still consisted of horticulture.

Viewing the matter afresh, I can see no basis for differing from the judge's conclusion. This is still the business of a garden centre that it always was. Although the sales of bought-in plants and other goods D have increased, it is still based on the sale also of Mr Short's roses. The court is not lightly to treat a tenancy as having ceased to be within the protection of the Agricultural Holdings Act. With horticulture of this nature, reasons of husbandry may dictate the number of plants that are grown and the amount of land that is used for growing them at any particular time as well as reasons of commerce. I do not doubt that the judge was right in concluding that the bulk of the turnover now comes from bought-in goods, but I would not differ from his view that this was still at the relevant dates a tenancy of an agricultural holding.

Accordingly I would dismiss this appeal.

Agreeing, CROOM-JOHNSON LJ said: It was common ground both E before the learned judge and in this court that the test to be applied, or the definition to be applied, of what is an agricultural holding is that to be found in the Agricultural Holdings Act 1986 and not in the 1948 Act. It was opened to us that this tenancy had originally been an agricultural holding, but that from 1984 onwards it changed radically by selling bought-in goods and what can be described as the ancillary garden items like sheds and things of that kind instead of the home-grown produce with which Mr Short had begun. It was urged that, if the original agricultural use had been wholly or substantially abandoned, that meant that the terms of the tenancy agreement were no longer applicable and that the protection of the Agricultural Holdings Act was no longer available to Mr Short.

The learned judge eventually came to the conclusion, to which F Dillon LJ has referred, that he thought it was a borderline case, and he said: "Some of the land to look at it now as shown in the photographs appears in rather a derelict condition. But none the less I find that genuinely a part of this business, a not insubstantial part consists of horticulture. The greater part of the land is used for this purpose." I interpret that finding, if one eliminates the double negative in the sentence "a not insubstantial part consists of horticulture", that he was in fact finding that there was a substantial part of the use of the land which consisted of horticulture. It was not really advanced to this court that what Mr Short had done was to abandon the horticultural use of the land so much as that the development of the bought-in materials and goods in the shop had changed the character of the tenancy.

One can, of course, radically change the nature of a tenancy by G increasing the bought-in side even if the production and sale of the home-grown plants remains quite a substantial part of the trade which is carried on on the premises. But I did not understand the learned judge to be finding that this is what happened in the present case. On the facts as they were deployed, and on the facts as they were found by the judge, it did not seem to me that this could possibly be described as an abandonment by Mr Short of the agricultural activities which he had carried on before the shop developed. It is quite clear that the shop turnover has overtaken the value of the home-grown sales.

The turnover is not the conclusive matter, however. It is an indication which one can use in order to see if the nature of the tenancy has in fact changed. The learned judge came to the H conclusion that the nature of the tenancy had not altered. In my view, there was evidence to support his finding and, having come to the conclusions which he did, I, too, would dismiss this appeal.

The appeal was dismissed with costs; legal aid taxation of respondent's costs ordered.

Queen's Bench Division

October 5 1987

(Before Mr Justice McCULLOUGH)

J

R v DAIRY PRODUCE QUOTA TRIBUNAL FOR ENGLAND AND WALES, EX PARTE HALL & SONS (DAIRY FARMERS) LTD

Estates Gazette March 19 1988

[1988] 11 EG 79-83

Milk quotas — Reference of disputed construction to the Court of Justice of the European Communities — Issue was as to whether in calculating the applicants' direct sales quota sales of milk which were not derived from their own cows but were bought in from other producers could properly be taken into account — The Dairy Produce Quota Tribunal for K England and Wales, upholding the views of the Minister of Agriculture, Fisheries and Food, decided that sales of milk which was bought in should not be taken into account in the calculation of the applicants' quota — Applicants sought in the present proceedings to have the tribunal's decision quashed — Applicants also complained that the reasons given by the tribunal were inadequate, but McCullough J, although accepting this criticism, held that it would be pointless to quash the decision on this ground alone and require the tribunal to reconsider the matter if the substantive issue could only be decided against the applicants — After considering the Dairy Produce Quota Regulations 1984 and EEC Council L Regulations, in particular article 5c(2) of Council Regulation 804/68 and article 6(1) of Council Regulation 857/84, McCullough J came to the conclusion that milk produced otherwise than on the applicants' own holding should not be taken into account in determining quota — It was clear that if the applicants had been dealing only with milk which had been bought in it would not have been taken into account and in any case milk bought in should have been taken into account in the determination of someone else's quota — However, in view of the importance and complexity of the matter, and a possible ambiguity in article 6(1) of Council Regulation 857/84, it was M desirable to obtain a ruling from the European Court before giving a decision — The judge suggested certain questions for counsel to consider before the order was made in accordance with RSC Ord 114 for reference of the matter to the European Court — Case then to be adjourned pending a reply from that court

The following cases are referred to in this report.

Bulmer (HP) Ltd & Anv. v *Bollinger SA* [1974] Ch 401; [1974] 3 WLR 202; [1974] 2 All ER 1226, CA

CILFIT SrL and Lanificio di Gavarda SpA v *Minister of Health of Italy* (Case 283/81) [1982] ECR 3415; [1983] 1 CMLR 472, European Ct

A This was an application by Hall & Sons (Dairy Farmers) Ltd to quash a decision of the Dairy Produce Quota Tribunal for England and Wales refusing to alter the estimated primary direct sales quota notified to the applicants by the Ministry of Agriculture, Fisheries and Food.

S Isaacs (instructed by Summers & Co, of Beaconsfield, Bucks) appeared on behalf of the applicants; G Pulman (instructed by the Solicitor, Ministry of Agriculture, Fisheries and Food) appeared on behalf of the respondent tribunal.

Giving judgment, McCULLOUGH J said: The appellant company, which I shall call "Halls", seek an order to quash a decision of the Dairy Produce Quota Tribunal for England and Wales dated August 29 1985 that it was unable to alter the estimate of Halls' primary
B direct sales quota of which the Ministry of Agriculture, Fisheries and Food had notified Halls by letter dated January 21 1985.

Halls supply milk and products made from milk both to wholesale and retail customers. Some of this milk comes from their own cows, the rest they buy. Halls contend that to calculate their dairy produce quota all of their sales of milk and milk products should be taken into account. The minister contends that the sales which result from milk which Halls buy in are irrelevant and that only the sales which result from milk produced by their own cows are to be taken into account. The tribunal agreed with the minister. On this issue the outcome of this application depends.

In their notice of motion Halls make the subsidiary complaint that
C the tribunal failed to give adequate and intelligible reasons for its decision and thereby broke rule 11 of the rules of procedure of the Quota Tribunal, August 1984, which requires the tribunal to record the decision and the reasons for its decision. What the tribunal recorded was:

> Decision. The Tribunal directs that it is unable to vary the award of Primary Quota (1323193 litres) notified on 21st January 1985. Reasons. Whilst the Tribunal received and accepted the carefully prepared submission from the Appellant it concluded that the points adduced were not such as could properly be taken into account when assessing this objection. So far as the Tribunal could ascertain the assessment of Primary Quota did take into account the Direct Sales of the business from its own cows.

Read in the context of the application the effect of this was that the
D tribunal rejected Halls' submission on the issue to which I have referred and Halls must have so understood it. However, despite the inclusion of the heading "Reasons", no indication of the tribunal's reasoning was given. I therefore accept that there was a breach of rule 11. But if this was the full extent of the tribunal's failure, it cannot avail Halls. To quash the decision on this ground alone and thus to require the tribunal to reconsider the matter would be pointless if the issue in question could only properly be determined in favour of the minister. To require the tribunal to give its reasons before this court considered the matter further would be equally futile.

The allocation of direct sales quota is governed by regulation 4 of Schedule 1 to the Dairy Produce Quota Regulations 1984 (SI 1984 no 1047). Regulation 4 reads:

E

> Direct sales quota shall be established for any producer in accordance with these regulations and the Community legislation, and Schedule 1 shall apply in respect of the allocation, transfer and surrender of direct sales quota and matters consequential or incidental thereto.

In Schedule 1 para 3 is material and this reads:

> For the purpose of calculating primary direct sales quota under this Schedule the dairy produce sold by direct sale from a holding in a base period shall be deemed to be the dairy produce sold by direct sale during that base period (whether from that holding or elsewhere) by the producer occupying that holding on 2nd April 1984 but not dairy produce sold by direct sale by any other producer.

There is no need to say any more about the word "primary" than
F that primary direct sales quota is the type of direct sales quota with which we are concerned here.

Three things are immediately clear: (i) that direct sales quota is to be established for a producer; (ii) that it is to be established not only in accordance with SI 1984 no 1047 but in accordance with the Community legislation; (iii) that the calculation of his primary direct sales quota is to be by reference to the dairy produce sold by direct sale from a holding in a base period.

Where, as here, there has been no change in the occupation of a holding, these provisions additionally say no more than that for the purpose of calculating a producer's primary direct sales quota the dairy produce sold by direct sale from his holding in a base period shall be deemed to be the dairy produce which he has sold by direct sale during that base period (whether sold from that holding or G
elsewhere). I have inserted the word "sold" in the phrase in brackets because this must necessarily be implied.

Where, as here, no produce was being sold from any place other than the holding, what this boils down to is that X is deemed to be X. One is no further forward in discovering what X is, ie what is meant by the dairy produce sold by a producer by direct sale.

A number of the words used in regulation 4 para 3 of Schedule 1 are defined in regulation 2. Each of those material to the present case leads one, directly or indirectly, to the relevant EEC Regulations.

"Producer" and "holding" have the meanings assigned to them by article 12 of Council Regulation 857/84.

"Direct sale" means a sale referred to in article 12(h) of that regulation. H

"Direct seller" has a corresponding meaning.

"Direct sales quota" means "the quantity of dairy produce which may be sold by direct sale from a holding in a quota year without the direct seller in occupation of that holding being liable to pay levy"; and

"Levy" means "the levy payable under the Community legislation to the competent authority referred to therein, described in article 1 of the Council Regulation 857/84 (which deals with the fixing of the levy)".

Regulation 857/84 refers back to Regulation 856/84, which in turn amends Regulation 804/68. Before looking at the individual definitions it is pertinent to see, if only in the broadest outline, what is J
the scheme of the legislation in so far as it affects milk and milk products, ie milk, cream, butter and cheese.

Regulation 804/68 established, as part of the Common Agricultural Policy, the common organisation of the market in milk and milk products.

The preamble to Regulation 856/84 recited, *inter alia*:

(1) that the market in milk products in the Community was suffering from structural surpluses as the result of an imbalance between supply and demand;

(2) that quantities of milk delivered were increasing at a rate such that the disposal of surpluses was imposing financial burdens and market difficulties which were jeopardising the very future of the common agricultural policy; K

(3) that the most effective method and the one having the least drastic effect on the income of producers was the introduction of an additional levy on quantities of milk delivered beyond a guaranteed threshold; and

(4) that the purpose of the levy, which was provided for by the Regulation, was to regulate and stabilise the market in milk and milk products.

The new article 5c of Regulation 804/68 (which was introduced by article 1 of Regulation 856/84) said, so far as material:

> 1. . . . an additional levy payable by producers or purchasers of cows' milk shall be introduced. The objective of the said levy shall be to curb the increase in milk production . . . The levy system shall be implemented . . . in accordance with one of the following formulas L

Formulas A and B were then set out.

> 2. The levy shall also be payable by every milk producer on the quantities of milk and/or milk equivalent he has sold for direct consumption and which . . . exceed a reference quantity to be determined.

In the United Kingdom legislation this reference quantity is called a "quota".

Article 1 of Regulation 857/84 fixed the levy. Article 6 so far as material read:

> 1. Each producer of milk and milk products referred to in Article 5c(2) of Regulation (EEC) No 804/68 shall be assigned a reference quantity corresponding to the direct sales made by the producer . . .

In article 12 come the definitions. I take first that of "producer". It M
is as producer and only as producer that Halls are entitled to a quota or reference quantity, and as producer that they are liable to pay the levy if their direct sales exceed it. (See article 5c(2) of Regulation 804/68, article 6(1) of Regulation 857/84 and regulation 4 of SI 1984 no 1047.)

"Producer" means "A . . . person . . . farming a holding . . . selling milk or other milk products directly to the consumer and/or supplying the purchaser" (article 12(c)).

"Holding" means "all the production units operated by the producer . . ." (article 12(d)).

"Milk or milk equivalent sold directly to consumption" means "milk or milk products processed into milk equivalent, sold without going through an undertaking treating or processing milk''. (article 12(h)).

Halls are, in my judgment, correct in saying that: (1) all the milk and milk products which they sell are sold without going through an undertaking, treating or processing milk, ie they are all sold directly to consumption; in other words all their sales are direct sales (to use the United Kingdom expression). (2) they operate a production unit and farm a ''holding''. (3) they are, therefore, a ''producer''.

None of this can be gainsaid. The question is whether such direct sales as derive from milk produced other than on their own holding are to be taken into account in the calculation of their reference quantity or quota. There is nothing express in Regulation 856/84 to say whether it should or not. Article 5c(2) merely says that every milk producer's reference quantity is ''to be determined'' and, subject to the requirement that the sum of the reference quantities shall not exceed the maximum figures specified in article 5c(3), Regulation 856/84 leaves the determination of reference quantities to each member state, it being implicit that in making them regard will be had to the objectives of the Community legislation.

A consideration of the objects of the Community legislation and in particular the passages from the recitals to Regulation 856/84, to which I have referred, together with the second sentence of article 5c(1) (''The objective of the said levy shall be to curb the increase in milk production . . .''), point to the conclusion that the sales derived from milk produced other than on the producer's own holding should not be taken into account in determining the reference quantity referred to in article 5c(2).

However, the terms of article 6 of Regulation 857/84 are or may be to the opposite effect. ''Each producer of milk and milk products referred to in article 5c(2)'' — and Halls are one — ''shall be assigned a reference quantity corresponding to the direct sales made by the producer.'' No limitation is placed on the direct sales which are to be taken into account. Is one to be implied? Having regard to the objects of the legislation I would say that it was.

In reality Halls are doing two things: (1) they are farming a holding and selling its products directly to consumers. (2) they are buying in milk and selling its products directly to consumers. I omit reference to any other types of sale. The former activity is within article 5c(2), entitles them to a quota and renders them liable, if the quota is exceeded, to pay the levy. That latter activity is, in my belief, outside it. Had Halls been conducting only the second activity, they would clearly not be within article 5c(2). I can see no reason why, merely because they are in addition conducting the first activity, the second should also come within article 5c(2). The same, in my opinion, is true of para 3 of Schedule 1 to SI 1984 no 1047.

The minister asserts, and it is not disputed, that the milk which Halls buy in is taken into account in the determination of someone else's quota. If Halls' argument were correct the same milk would be taken into account twice. This would hinder the working of the scheme and would make for unfairness between one quota holder and another.

Were Halls right and were they, in consequence, to be allocated a primary direct sales quota of, say, 10m litres on the basis that about 1m litres of sales came from their own milk and about 9m litres from milk bought in, there would be nothing to stop them from doubling their herd, producing 2m litres themselves and, if need be, reducing the amount bought in to 8m litres. It would not prevent the 1m litres which Halls would no longer be buying from being produced. More, not less milk, overall would be produced as a result, yet the objective of the levy is ''to curb the increase in milk production''.

It is not without significance that no other applicant for direct sales quota has contended for the construction advanced by Halls. According to the affidavit of Mr C F Hall, their managing director, there are about 190 of them who both produce milk and buy it in.

As I understand him, Mr Isaacs, counsel for Halls, accepts that these anomalous results would ensue if his argument were correct. His submission is that the wording of the relevant United Kingdom provisions, interpreted as they must be in the light of the relevant EEC Regulations, compels acceptance of the construction for which he contends. I am only able to accept his argument to the extent that Article 6(1) of Regulation 857/84 raises an ambiguity. But in my opinion, for the reasons I have given, it is implicit that the sales in question are not to be taken into account. Left to myself I would so decide.

The final question is whether, before reaching a decision on this application, I should, in accordance with article 177 of the EEC Treaty, request the European Court of Justice to give a ruling. In this connection I have been referred to both *Bulmer (HP) Ltd* v *Bollinger SA* [1974] Ch 401 and *CILFIT SrL and Lanificio di Gavarda SpA* v *Minister of Health of Italy* (Case 283/81) [1982] ECR 3415.

The interpretation of article 6(1) of Regulation 857/84 is necessary to enable me to give judgment. I believe that this, rather than any definition in article 12, is the central provision. If a limit is to be implied on article 6(1) of regulation 857/84, it may be that a corresponding one is to be implied in article 5c(2) itself, for clearly if the sales in question are to be ignored in the determination of the reference quantity, they must likewise be irrelevant to the levy. Be that as it may, to give judgment a decision on a matter falling within article 177 is necessary. Should I decide it myself or ask for a ruling?

There is, so far as I have been told, no previous decision on the point. The applicants want me to refer it; the minister does not. He says that the answer is so clear it would be pointless to refer and one should not overload the court. However, if I do, the matter will be decided once and for all. If I do not, Halls may appeal to the Court of Appeal, who may or may not refer. That way there would be the possibility of an extra hearing and extra expense.

The point, as the minister concedes, is one of importance. Mr Isaacs says that only if the matter is clear beyond reasonable doubt should I not refer (see *CILFIT* p 3430) and he invites my particular attention to paras 16 to 20 on that page. I do not see this as a question which could turn on the comparison of different language versions; nor do I see it as one dependent on terminology peculiar to Community law. No understanding of any legal concept is involved. The question is a pure one of construction in the light of the objectives of the legislation in question. I believe that I have been able to identify the material objectives from the passages to which I have been referred. I cannot think that this is a topic on which the law may evolve. The right answer, whatever it be, will always have been the right answer and will, as far as anyone can see, so remain.

However, there remains this. The provision or provisions in question have to be interpreted in the light of the provisions relating to the Common Agricultural Policy as a whole. The levy provisions in relation to milk and milk products are complex. I have not, for example, been able to see how the determination of reference quantities for producers under formula A or the determination of reference quantities for purchasers under formula B are intended to work or how they and the provisions of article 5c(2) interact. There may well be something in these other provisions which bears on the point in question. Understandably, counsel's submissions have not been able to take account of any feature of possible relevance in the working of the levy system in other member states. I can see consequences which, if Halls' argument is correct, would run counter to the objects of the scheme, but others may see consequences which, if the minister's argument is correct, would be similarly undesirable. My view, clearly held on the material and arguments placed before me though it is, may be wrong. Whether judged by the criteria of *Bulmer* v *Bollinger* or by those of *CILFIT*, I think it is better for a ruling to be requested rather than to decide straight away.

I will not decide what question or questions to ask. I will hear counsel first, but for what it is worth I suggest something along the following lines: (1) Is the reference quantity referred to in article 6 of Regulation 857/84 to be calculated from all of the direct sales made by the producer in the relevant period? (2) Is it only to be calculated from such of his direct sales as derive from the milk produced by the farming of his own holding? (3) In making the calculation are direct sales which derive from milk which he has bought in to be ignored?

In one sense that is to ask the same question three times, but there might be advantages in dividing it into three in that way.

Case referred under RSC Ord 114 to the Court of Justice of the European Communities for the rulings of the court on questions to be settled by counsel and approved by the judge. Case adjourned pending the result of the reference to the court.

AUCTIONS

Court of Appeal

January 27 1988

(Before Lord Justice GLIDEWELL and Lord Justice TAYLOR)

BARNARD MARCUS & CO v ASHRAF

Estates Gazette May 7 1988

[1988] 18 EG 67-71

Auctions — Auctioneers' claim to commission — Two firms instructed at different times and two commissions eventually paid — After auction sales for which the defendant owner of a house had given instructions to one firm had proved abortive, the defendant gave instructions to a second firm, the plaintiffs, without withdrawing instructions from the first firm — The terms arranged with the plaintiffs included a provision that a commission of 2% (reduced from 2.5%) on the purchase price would be payable if a sale of the property, whether arranged by the auctioneers or not, was effected after the acceptance of instructions and before the auction or within 28 days after the auction — It was also provided that if the property was withdrawn between the date of instructions and the auction date a withdrawal fee of £200 and expenses would be payable — In the event the plaintiffs did not put the property in a January auction as planned, but undertook to put it in a sale on February 3 — However, on January 15 the property was sold by the firm who were originally instructed and who still held instructions to sell — The latter firm claimed and were paid their agreed commission of 1.5% on the sale — The plaintiffs also claimed that commission was payable in accordance with their instructions, as mentioned above, and, on the claim being disputed, brought proceedings against the defendant in the county court — The county court judge rejected the claim to commission but held that there had been a "constructive withdrawal" of the property from the February 3 auction — He therefore awarded the plaintiffs the withdrawal fee of £200 and expenses — The plaintiffs appealed — On appeal the defendant's counsel did not attempt to support the judge's conclusion that there had been a withdrawal, but put forward other grounds for resisting the commission claim — His suggestions that the terms of the contract had been subsequently varied so as to exclude a sale otherwise than by auction, or that contractual relations had been terminated by the failure to get the property into the January sale, were rejected by the court as unsupported by evidence — They also rejected, after considering various authorities, the argument that the provision for commission to be paid whether the sale was by the auctioneers or not amounted to a penalty clause which should be struck down — The clause in question had been in use for years without challenge and was to be found as a precedent in the *Encyclopaedia of Forms and Precedents* — Plaintiffs' appeal allowed and defendant's cross-appeal dismissed

The following cases are referred to in this report.

Bernard Thorpe & Partners v *Snook* [1983] EGD 77; (1982) 266 EG 440

Export Credits Guarantee Department v *Universal Oil Products Co* [1983] 1 WLR 399; [1983] 2 All ER 205; [1983] 2 Lloyd's Rep 152, HL

Meacock (John) & Co v *Abrahams Loescher (Third Party)* [1956] 1 WLR 1463; [1956] 3 All ER 660, CA

This was an appeal by the plaintiffs, Barnard Marcus & Co, from the decision of Judge Edwards at West London County Court, rejecting the plaintiffs' claim against the defendant, Mohammad Sharif Ashraf, for commission on the sale of the defendant's house at 84 West Hill, Putney, London SW15, but awarding the plaintiffs a withdrawal fee. There was a cross-appeal by the defendant (respondent).

Jeffrey Burke QC and Colin Wynter (instructed by L B Marks & Co) appeared on behalf of the plaintiffs; R Paul Stewart (instructed by Corsellis Church Rackham) represented the defendant.

Giving the first judgment at the invitation of Glidewell LJ, TAYLOR LJ said: This appeal arises from an auctioneers' claim for commission. The defendant, Mohammad Ashraf, was the owner of a house at 84 West Hill, Putney. He wished to sell it, and in July 1985 he put it up for auction through a firm of estate agents and auctioneers named Edwin Evans. Although a sale was agreed the purchaser failed to complete. Further attempts to sell were unsuccessful. At an auction on December 3 1985 the property failed to reach the reserved price.

By now the defendant was very anxious to effect a sale as he had completed the purchase of another property. Edwin Evans had no imminent plans for a further auction, so without withdrawing instructions from them, the defendant, through his solicitors, approached the plaintiffs, also a firm of auctioneers and estate agents. They agreed on December 5 1985 to use their best endeavours to enter the property in an auction fixed for January 22-23 1986, but they required the defendant to sign their terms of business that very day. The defendant was leaving for a trip abroad, so his wife attended the plaintiffs' office and signed on his behalf. The defendant had previously had discussions with the plaintiffs, earlier in 1985, with a view possibly to putting the property in their hands. He had been told at that time the terms upon which such a contract could be concluded, but nothing further occurred.

The document the defendant's wife signed is in the form of a letter from the vendor to the plaintiffs. After the name of the defendant and the name of the property it reads as follows, so far as relevant:

> I hereby instruct your firm to submit my . . . interest in the above-mentioned property for sale by way of public auction, subject to the following: 1. Your commission will amount to 2.5% of the purchase price (minimum commission £150.00) plus VAT, and will be payable by me if a sale of the property, whether arranged by the Auctioneers or not, is effected after the acceptance of instructions and before the Auction, or within 28 days after the Auction.

Clause 2 provided for contribution towards the auction expenses and I need not read it in full. Clause 3 reads:

> If the property is withdrawn from the market between the date of these instructions and the Auction itself, then a withdrawal fee of £200.00 may be charged plus VAT in addition to the expenses as set out in 2 above.

There then follow a number of other clauses. The only one which needs to be mentioned is clause 15, which provides for the date of the auction, which was left blank. That clause ends with this phrase: ". . . but [I] will allow you to change such date and venue at your discretion."

The only alteration to the printed form was the substitution in longhand of 2% for 2.5% in clause 1. That reduction was negotiated by the defendant's solicitors, and in particular by a Miss Mustafa of that firm. It was confirmed in a letter of December 6 1985 in the following terms:

> Further to our telephone conversation yesterday when you kindly agreed to use your best endeavours to place the above property in your auction sale fixed for January 22 and 23 1986, we now enclose herewith Special Conditions of Sale. We trust you will be inspecting the premises which are presently vacant for the purposes of assessing a reserve price and to take photographs.

The letter then went on to speak of details of sales particulars and the request was made that a board should be put up outside the premises.

The letter concluded with these two paragraphs:

> We also confirm our telephone conversation when you kindly agreed that your commission fees will be reduced to 2% of the sale price in the event of the above property being sold at auction.
>
> We look forward to hearing from you in due course but should you require any further information, would you kindly contact Miss Mustafa of instructing solicitors.

In the event the plaintiffs failed to put the property in the January auction, but wrote on December 10 1985 to the defendant's solicitors as follows:

> Thank you for your letter of December 6 regarding the above property, the contents of which we have noted. We write to inform you that we are unable to place this property into the January Auction Sale but will place the same in the following sale which is on February 3 and is only ten days later, we trust this meets with your approval.

The letter then went on to deal with an acknowledgement of the special conditions of sale being received and a reference to the "For Sale" board and concluded in this way:

> The commission will amount to 2% of the sale price plus VAT in the event of a successful sale. Our mutual client has already been to our office and signed our terms of business and we enclose herewith a copy of the same for your information.

During the first few days of January the plaintiffs sought details from the defendant's solicitors for their catalogue and asked for the keys to the property. However, on January 15 1986 the property was sold for £116,000 by Edwin Evans to a Mr Desaur. On learning of this the plaintiffs rendered an account to the defendant on January 17, claiming commission at 2% plus VAT on "sale of the above property prior to auction in the sum of £116,000.00". In addition they claimed sums in respect of printing and advertising which they had in fact incurred and which are not in dispute. The defendant was of course also faced with an account from Edwin Evans. They claimed their agreed commission of 1.5% on the sale price. They also claimed, and were paid, half the like commission on the abortive sale they had arranged in early 1985. The purchase by Mr Desaur was completed on February 12 1986.

To round off this history — which gives a revealing insight into the economics of auctioneering — I should mention that Mr Desaur himself put the property back within a short time into the hands of the plaintiffs to sell. They sold it for £140,000 and no doubt recovered a commission from Mr Desaur.

The defendant disputed the plaintiffs' claim for 2% commission, amounting in effect to £2,668. The plaintiffs issued proceedings in the West London County Court. The case came before His Honour the late Judge Edwards on April 29 1987. He rejected the defendant's argument that clause 1 of the terms of business, signed on the defendant's behalf, governed only the January auction, and the suggestion that the letters of December 6 and 10 between the plaintiffs' and defendant's solicitors created a new contract. He also rejected the argument that clause 1 amounted to a penalty clause.

However, he held in the defendant's favour on the basis of clause 3, that the property had been withdrawn between the date of the instructions to the plaintiffs and the auction. He put the matter in this way:

> There is evidence that what Barnard Marcus got was news from the new purchaser of the property that the property had been sold. Mr Chopping for Barnard Marcus sought and obtained confirmation from the defendant's solicitors. Can this course of events be called a withdrawal, in which case the minor liabilities ensue, or must it be a sale as per clause 1? Ordinary commission payable in the event of auctioneer selling property — main purpose of commission. I am not bound to look for a construction of clause 1 adverse to the plaintiffs' customers when clause 3 is there to hit the eye as well. What happened here is that there was a sale but also a constructive withdrawal. The defendant had done this a number of days before auction date.
>
> I find that there was a withdrawal and giving priority to clause 3 over clause 1, I find that the plaintiffs' claim for a commission fails. I must say I am happy to be able to do this. To my mind it is a sensible construction. Mr Ashraf having put the plaintiffs to some effort on his behalf is liable to them for the sums contained in clauses 2 and 3.

In other words the learned judge found that the defendant's solicitors (who had confirmed to the plaintiffs before January 17 that the property had been sold) in effect had constructively withdrawn the property from the auction due on February 3. Accordingly he awarded the plaintiffs £200 under clause 3, expenses under clause 2 but no commission under clause 1. It is against that decision that the plaintiffs appeal to this court.

Clearly the learned judge took the view that a vendor ought not to have to pay two commissions in respect of only one sale. However, the basis upon which he decided the case is one which cannot be right. Mr Stewart, appearing on behalf of the defendant, realistically made no attempt to support it. The defendant did not withdraw the property from the market: on the contrary he sold it on the market. That sale was an event specifically provided for in the plaintiffs' terms of business as entitling the plaintiffs to a commission. There was no conflict between clauses 1 and 3. Each provided for a different situation. There was here no withdrawal, constructive or otherwise. There was a sale at a time and in circumstances covered by clause 1.

While making the concession that the judgment cannot be upheld on the judge's grounds, Mr Stewart seeks to argue two points which are canvassed in his cross-notice. First he argued that the terms of the contract were not those contained in the plaintiffs' terms of business and signed by the defendant's wife. He submits that those terms were varied by discussion between the defendant's solicitor, Miss Mustafa, and Mr Chopping on behalf of the plaintiffs.

In her letter of December 6 Miss Mustafa confirmed that discussion by stating that commission of 2% would be paid "in the event of the above property being sold at auction". Mr Stewart concedes that she may have misunderstood Mr Chopping, but once she wrote that letter he contends it was up to Mr Chopping to correct any mistake she had made as to the effect and tenor of that conversation.

In his letter of December 10 Mr Chopping said:

> The commission will amount to 2% of the sale price plus VAT in the event of a successful sale.

Mr Stewart argues he ought to have made clear that "successful sale" was not confined to a sale at an auction but included all sales within clause 1 of the signed terms of business.

In my judgment the very next sentence of Mr Chopping's letter did just that. He wrote:

> Our mutual client has already been to our office and signed our terms of business and we enclose herewith a copy of the same for your information.

That sentence, directed to a solicitor, made it abundantly clear that the terms of business which had been signed and which were the common form of contract entered into by these plaintiffs applied, save for the reduction of 2.5% to 2% which had been orally negotiated.

Mr Stewart also argues that the failure of the plaintiffs to get the property into the January sale in effect brought contractual relations to an end unless they were specifically renewed. He says the plaintiffs themselves accepted that view of the facts in the letter of December 10 by saying, after pointing out that they had been unable to get the property into the January sale and proposed to put it in the February sale 10 days later, "we trust this meets with your approval".

In my judgment that phrase was not an acknowledgement that contractual relations had come to an end. It was a courteous phrase and it may have been meant as no more than a reminder to the solicitors or the defendant himself that he had the option of withdrawing the property at any time that he chose to do so. In any event the course of dealing between that letter from Mr Chopping and the sale on January 15 makes it clear that the defendant was accepting the plaintiffs' terms. The arrangements for information to be provided for the catalogue and for the keys to be provided went ahead, and on January 9 Miss Mustafa wrote a letter to the plaintiffs indicating that all was going proceeding. In my judgment that can only mean it was going ahead on the basis of the terms of business which the defendant had signed, a copy of which had been sent to the solicitors and as to which they had made no objection whatsoever.

Mr Stewart's second argument, which he presented valiantly and persuasively, is that clause 1 amounts to a penalty clause. This point was taken before the learned judge although not fully developed. It had not been raised on the pleadings. Mr Stewart contends that the provision for commission to be paid, whether the sale was by the auctioneers or not, is a disguised form of penalty. It is designed, he submits, to frighten the client off from selling the property otherwise than through the plaintiffs. It is therefore *in terrorem* and should be struck down.

He cited a passage from *Chitty on Contracts*, 25th ed, vol 1 at paras 1724 and 1725, where the learned author discusses the distinction between liquidated damages and penalties. But the more relevant passage in my view is at para 1676 of the same volume. The heading of the paragraph is as follows: "An agreed sum cannot be a penalty unless payable upon breach." The passage goes on as follows: "A sum payable upon performance of the claimant's contractual obligation cannot be a penalty; nor, at common law, can a sum which is payable upon an event other than a breach of contract be a penalty."

In *Export Credits Guarantee Department* v *Universal Oil Products Co* [1983] 1 WLR 399, the House of Lords made it clear that a payment can be a penalty only if it is payable on a breach of contract as opposed to being payable under the contract. Mr Stewart grasps that nettle and seeks to argue that this sale by the defendant, otherwise than at the plaintiffs' auction, was a breach of contract. He submits that apart from paying the auctioneer, the only obligation on the client is to make the property available for the auction. If he sells elsewhere he is therefore in breach of contract. Were that not so, he submits, there is no way the client could be in breach of contract other than by failing to pay.

In my judgment the contract, so far from seeking to prevent the client from selling elsewhere or withdrawing the property, made specific provision for those events. Such permitted conduct cannot be regarded as a breach of contract, simply because there would otherwise be no way of breaching it except for non-payment. As Mr Burke pointed out, there are many contracts in which the only obligation on a party is to pay.

Clause 1 was not, as the learned judge seems to have considered, an unusual term in an auctioneer's contract. Similar terms have appeared regularly in such contracts for many years and have stood the test not only of time but of scrutiny by this court.

We were referred to the decision in *John Meacock & Co* v *Abrahams (Loescher, Third Party)* [1956] 1 WLR 1463. In that case a mortgagee, unable to recover his loan, put up five houses, upon which it was secured, for sale through the plaintiffs, a firm of auctioneers. The plaintiffs' terms of business were substantially the same as in the present case. The properties were duly entered for auction by the plaintiffs, but the day before the auction was due the mortgagor managed to redeem the mortgage and he sold the properties himself to the sitting tenants. The plaintiffs claimed commission. Denning LJ (as he then was) at p 1466 recited the relevant clause and went on as follows:

> The object of this clause is quite plain. It is to cover the case where a client instructs auctioneers to sell a property for him by auction, and then, before the auction actually takes place, the client sells the property himself. In such a case the client must still pay full commission to the auctioneer, for the very good reason that the auctioneers' work in putting up the property, advertising it, and so on, may well have been one of the causes why the client was able to sell the house himself. The clause clearly covers a sale by the client before the auction: and I think it would also cover a sale with his concurrence. But the auctioneers seek here to extend it to a sale by a third person. It covers, they say, a sale by the mortgagor without the consent of the second mortgagee.

A little later, at p 1467, Denning LJ said:

> It must be remembered that it is a clause put in by the auctioneers for their own benefit, in which they stipulate that, although they have not done all the work that was contemplated, nevertheless they are to have full commission. For instance, in this case they have not had to hold an auction at all. I can visualise cases where they may have done nothing beyond accepting instructions. Suppose a client gives them instructions one day and sells himself the next, before they have done anything. Nevertheless they stipulate for full commission irrespective of the amount of work they have done. Such a stipulation should, I think, be given its natural meaning and not be extended so as to be given a wide unnatural meaning.

In the result, in that case, the Court of Appeal ruled that the auctioneer was not able to recover commission, but that was because the sale was not by the auctioneers' client but by a third party. It was common ground in the judgments of all three members of the Court of Appeal that had the sale been by the client himself there would have been commission recoverable. Indeed Morris LJ (as he then was) was prepared to go further and find in the auctioneers' favour even on the facts of that case.

More recently we were referred to the decision of Nolan J in the case of *Bernard Thorpe & Partners* v *Snook* (1982) 266 EG 440. The case concerned an auctioneers' claim for commission on the sale of a farm effected privately by a vendor within three months after the date of auction, the farm having failed to reach the reserve price. The commission clause provided, somewhat similarly to the clause in the present case, that if a sale of the property, whether arranged by the auctioneers or not, was effected within three months after the auction, commission on the price realised should be payable to the auctioneers on the same scale as for a sale by auction.

Nolan J at p 443 said:

> In his second submission Mr Latham rightly said that clear and unequivocal language is required to entitle the plaintiffs to commission in respect of a sale in which they played no part. The language of the critical clause in the present case, argues Mr Latham, does not satisfy that requirement. In order to do so it would need the addition of some such words as "whether the auctioneer is an effective cause of the introduction of the purchaser or not". I am bound to say, however, that the clause seems to me to state clearly and unequivocally that the auctioneer is entitled to commission if a sale is effected within three months even though it is not arranged by the auctioneer. It is not in dispute that a sale has been effected for this purpose when contracts were exchanged. Subject therefore to the third point for decision, those words in my judgment cover the present case.

It does not seem that there has ever been any argument raised as to whether a clause such as appears in the terms of business of these plaintiffs could be regarded as a penalty clause. In my judgment it is most unlikely that if there were any merit in that point it would have escaped all those who have dealt with this clause over the years, including those concerned in the cases just cited.

We have been told that a similar clause is to be found in the *Encyclopaedia of Forms and Precedents*, both in the 4th and 5th editions. In my judgment the clause is effective; it is not a penalty clause. It may seem to a layman to be an injustice and one has sympathy in the present case with the defendant, who found himself in effect paying two and a half commissions before he was able to sell this house, but in my judgment it is not possible to accede to the arguments of Mr Stewart, persuasive though they were, either that the terms of the contract as printed did not apply to the sale or that this was a penalty clause.

Since it is conceded that the learned judge reached his decision on a fallacious basis the effect of this appeal must be that the cross-notice is rejected and the appeal must be allowed.

GLIDEWELL LJ agreed and did not add anything.

The appeal was allowed with costs in the Court of Appeal and below; judgment for plaintiffs for £3,595.75; credit to be given for £416.75 already paid.

COMPULSORY PURCHASE AND COMPENSATION

Queen's Bench Division

November 12 1987

(Before Mr Justice MANN)

EVELYN DE ROTHSCHILD AND ERANDA HERDS LTD v SECRETARY OF STATE FOR TRANSPORT AND ANOTHER

Estates Gazette May 21 1988

[1988] 20 EG 87-88

Compulsory purchase — Highways Act 1980 and Acquisition of Land Act 1981 — Application to quash compulsory purchase provisions — Complaint by applicants in regard to the route proposed by the county council for a bypass road to relieve traffic through the town of Leighton Linslade — Proposed route would cross applicants' land — Applicants had put forward four alternative schemes involving part of their land, which they were prepared to sell at a valuation for the purpose of the bypass — These schemes had been rejected by the county council and the inspector at the public inquiry — The Secretary of State in his decision letter said that he "does not believe that any of the suggested alternatives has sufficient advantages or benefits which would justify its adoption in place of the scheme as proposed by the county council" — It was submitted by the applicants that this showed an erroneous approach in regard to the question of onus — The onus was on the acquiring authority to justify the order depriving the applicants of their land — Held that the applicants' submission was misconceived, being a criticism of the Secretary of State's wording by way of an exercise in semantics — The Secretary of State, although expressing his view in a negative manner, was in substance approving the inspector's conclusion that the route proposed by the council was the best solution in the public interest — The recognised grounds of challenge were those mentioned in *Ashbridge Investments Ltd* v *Minister of Housing and Local Government* and *Associated Provincial Picture Houses Ltd* v *Wednesbury Corporation*, and these were not in issue here — A technical amendment in the notice of motion was allowed to conform with section 23 of the Acquisition of Land Act 1981 — Application dismissed

The following cases are referred to in this report.

Ashbridge Investments Ltd v *Minister of Housing and Local Government* [1965] 1 WLR 1320; [1965] 3 All ER 371; (1965) 63 LGR 400; [1965] EGD 216; 195 EG 205, CA

Associated Provincial Picture Houses Ltd v *Wednesbury Corporation* [1948] 1 KB 223; [1947] 2 All ER 680, (1947) 45 LGR 635, CA

Brown v *Secretary of State for the Environment* (1978) 40 P&CR 285; [1979] JPL 454

Gordondale Investments Ltd v *Secretary of State for the Environment* (1971) 23 P&CR 334; 70 LGR 158; [1972] EGD 72; 221 EG 625, CA

Hanily v *Minister of Local Government and Planning* [1951] 2 KB 917; [1951] 2 All ER 749; (1951) 49 LGR 769; 2 P&CR 161, CA

Prest v *Secretary of State for Wales* (1982) 81 LGR 193; [1983] EGD 282; 266 EG 527, CA

This was an application under section 23 of the Acquisition of Land Act 1981 with a view to quashing the authorisation for the compulsory purchase of land owned and occupied by the applicants contained in the Bedfordshire County Council (Leighton Linslade Southern Bypass) CPO (No 2) 1985. The applicants were Evelyn de Rothschild and Eranda Herds Ltd. The first respondent was the Secretary of State for Transport and the second respondents were Bedfordshire County Council.

Jeremy Sullivan QC and B Ash (instructed by Horwood & James, of Aylesbury, Bucks) appeared on behalf of the applicants; John Laws (instructed by the Treasury Solicitor) represented the Secretary of State; C George (instructed by the county secretary and solicitor, Bedfordshire County Council) represented the council.

Giving judgment, MANN J said: In this case Mr Jeremy Sullivan QC on behalf of Mr Evelyn de Rothschild and Eranda Herds Ltd seeks to quash the authorisation of the compulsory purchase of land owned and occupied by the applicants conferred by the Bedfordshire County Council (Leighton Linslade Southern Bypass) CPO (No 2) 1985. The order was made by the county council in its capacity as a highway authority and in the exercise of powers conferred by sections 239, 240, 246 and 250 of the Highways Act 1980. The confirming authority in regard to that order was the Secretary of State for Transport (see Act of 1980, sections 247(1), (2) and 329(1); also the Acquisition of Land Act 1981, section 2(2)). Objections to this confirmation were made by (among others) the applicants and accordingly a public local inquiry was held by an inspector appointed for the purpose (see Act of 1981, section 13(2)). After considering the inspector's report the Secretary of State on November 26 1986 wrote that he had decided to confirm the order. The present application purports to be made under section 23 of the Act of 1981. The significance of "purports" will appear.

The compulsory purchase order (together with a side roads order and a bridge order) would enable the county council to construct a much-needed southern bypass to the town of Leighton Linslade, which lies on the heavily trafficked Thame-Stevenage corridor. The applicants objected to the route of the bypass as proposed by the county council in so far as it crossed their land at Ascott Farm. The applicants put forward four alternative schemes which they argued were superior on engineering, agricultural and environmental grounds. All four were upon land owned by the applicants which the applicants offered to sell at valuation. The inspector in his report dated July 5 1986 was not attracted by the applicants' case. At para 196 he wrote:

> I am of the opinion that there are no factors, including the offer to sell alternative land at Ascott Farm, which individually or together outweigh the disadvantages of extra cost and exceptional delay which would result from the adoption of the objectors' alternative route and I believe that the Bedfordshire County Council has shown unequivocally that the Order route is the best in the public interest under the circumstances.

This passage was faithfully summarised in the decision letter at para 14 as follows:

> The Inspector is of the opinion that there are no factors which individually or together outweigh the disadvantages of extra cost and delay which would result from the adoption of the objectors' alternative route. He believes the County Council has shown unequivocally that the Order route is the best in the public interest under the circumstances.

It is at para 17 of the decision letter that the issue arises upon which the parties debate. It is in these terms:

> The Secretary of State has carefully considered the report and recommendations of the inspector together with all matters raised by the objectors both at the public inquiries and in writing. He notes that the need for the bypass to relieve traffic in Leighton Linslade and neighbouring villages and to improve access to the industrial area of the town is not in dispute, and that there is a large measure of support from the public both for the bypass and for work to start on it as soon as possible. From the evidence the Secretary of State does not believe that any of the suggested alternatives has sufficient

advantages or benefits which would justify its adoption in place of the scheme as proposed by the County Council.

Mr Sullivan submits that in those last words is to be found a wrongdoing by the Secretary of State in that he has placed on the applicants the onus of showing that their alternatives (or any of them) had advantages over the promoted route whereas the onus is ever on the acquiring authority to show that its route had advantages which were so decisive as to warrant confirmation of the compulsory purchase. Mr Sullivan referred me to two authorities. The first in time was the decision of Forbes J in *Brown* v *Secretary of State for the Environment* (1980) 40 P&CR 285. At p 291 the learned judge said:

It seems to me that there is a very long and respectable tradition for the view that an authority that seeks to dispossess a citizen of his land must do so by showing that it is necessary, in order to exercise the powers for the purposes of the Act under which the compulsory purchase order is made, that the acquiring authority should have authorisation to acquire the land in question. If, in fact, the acquiring authority is itself in possession of other suitable land — other land that is wholly suitable for that purpose — then it seems to me that no reasonable Secretary of State faced with that fact could come to the conclusion that it was necessary for the authority to acquire other land compulsorily for precisely the same purpose.

The second authority was the decision of the Court of Appeal in *Prest* v *Secretary of State for Wales* (1982) 81 LGR 193. At p 198 Lord Denning MR said:

I regard it as a principle of our constitutional law that no citizen is to be deprived of his land by any public authority against his will, unless it is expressly authorised by Parliament and the public interest decisively so demands: and then only on the condition that proper compensation is paid: see *Attorney-General* v *De Keyser's Royal Hotel Ltd* [1920] AC 508. If there is any reasonable doubt on the matter, the balance must be resolved in favour of the citizen. This principle was well applied by Forbes J in *Brown* v *Secretary of State for the Environment* (1978) 40 P&CR 285, where there were alternative sites available to the local authority, including one owned by them.

Watkins LJ said at p 211:

In the sphere of compulsory land acquisition, the onus of showing that a compulsory purchase order has been properly confirmed rests squarely on the acquiring authority and, if he seeks to support his own decision, on the Secretary of State. The taking of a person's land against his will is a serious invasion of his proprietary rights. The use of statutory authority for the destruction of those rights requires to be most carefully scrutinised. The courts must be vigilant to see to it that that authority is not abused. It must not be used unless it is clear that the Secretary of State has allowed those rights to be violated by a decision based upon the right legal principles, adequate evidence and proper consideration of the factor which sways his mind into confirmation of the order sought.

Fox LJ said at p 216:

I can only conclude that, in a case where the Secretary of State decided to confirm the compulsory purchase order primarily on considerations of cost, and where shortly before his decision he was asked to take account of land acquisition costs, he confirmed the order without material as to what the latter costs were. Accordingly, I do not think that he can have given the proper degree of consideration to the overall question of cost. The onus of establishing that a compulsory purchase order has been properly made must be on the acquiring authority. The question of cost was a material issue. One of the elements in the total cost was land acquisition cost. I am not satisfied that the Secretary of State had adequate material to judge the latter cost when he made his decision.

I was also referred to the departmental circular (Department of the Environment 6/85) which gives guidance on compulsory purchase and which in para 7 advised that proposals should not be made "unless there is a compelling case in the public interest".

In the light of the authorities and (for what it is worth) of the circular, Mr Sullivan submits that the Secretary of State's approach in para 17 flaws the decision because of the error as to onus. That the onus was upon the acquiring authority was, he submitted, a legal principle which had here been transgressed. Mr Laws for the Secretary of State (whose submissions were adopted by Mr George for the county council) denied that there was any such principle. The power to confirm or not to confirm was in its exercise to be judged as is any other exercise of a discretionary power and is subject to no special rule. I was referred to *Ashbridge Investments Ltd* v *Minister of Housing and Local Government* [1965] 1 WLR 1320 where, in regard to a challenge to a compulsory purchase order's confirmation, Lord Denning MR said at p 1326F:

The court can only interfere on the ground that the Minister has gone outside the powers of the Act or that any requirement of the Act has not been complied with. Under this section it seems to me that the court can interfere with the Minister's decision if he has acted on no evidence; or if he has come to a conclusion to which on the evidence he could not reasonably come; or if he has given a wrong interpretation to the words of the statute; or if he has taken into consideration matters which he ought not to have taken into account, or vice versa; or has otherwise gone wrong in law. It is identical with the decision of a lower tribunal which has erred in point of law.

Here are to be found the conventional grounds of challenge and in particular the grounds which derive from *Associated Provincial Picture Houses Ltd* v *Wednesbury Corporation* [1948] 1 KB 223. I do not regard the *Prest* case or the *Brown* case as being other than examples of challenges on those grounds. In the *Prest* case there was a failure to take into account a relevant factor, that is to say the cost of acquiring the subject land. Thus, Lord Denning MR at p 202:

. . . he failed to take into account the cost of acquiring the CPO site as against the cost of acquiring the alternative site offered by Sir Brandon. This was a most relevant consideration. It would probably have made a crucial difference because, even at that date in 1978, there was a potential of development for industrial use which would have given a considerable "hope" value to the CPO site. The Secretary of State ought to have had regard to this point — in the public interest — even though it was not canvassed by the parties at the inquiry. In any event he ought to have considered it — after receiving the letter of October 20 1978 — and asked for evidence of values before coming to his decisions. If he had considered it, the only reasonable conclusion would be that the compulsory purchase order would not have been confirmed.

The *Brown* case is an example of what now would be regarded as perversity, for at p 291 I find in the passage I have already recited the words "no reasonable Secretary of State faced with that fact could come to the conclusion".

In the instant case there is no perversity nor is there any suggestion of disregard of a relevant consideration. The inspector concluded in para 196 of his report that the order route was the best in the public interest and at para 20 of the decision letter the Secretary of State expressed his agreement with that (among other) conclusion. The use of the negative in para 17 whereas the positive could readily have been employed is not in my judgment a vitiating factor. Once the true and only grounds of challenge are identified the usage seems to me to be criticisable only by way of an exercise in semantics (see *Gordondale Investments Ltd* v *Secretary of State for the Environment* (1971) 23 P&CR 334 at p 340).

For the reasons given I dismiss this application. However, I must return to my observation that the application purports to be made under section 23 of the Act of 1981. The notice of motion is headed as follows:

IN THE MATTER OF THE HIGHWAYS ACT 1980 THE ACQUISITION OF LAND ACT 1981 . . . BEDFORDSHIRE COUNTY COUNCIL (LEIGHTON LINSLADE SOUTHERN BYPASS) COMPULSORY PURCHASE NUMBER TWO ORDER 1985 . . .

The body reads:

TAKE NOTICE that the High Court of Justice, Queen's Bench Division at the Royal Courts of Justice, Strand, London WC2A 2LL will be moved at the expiration of 35 days from the service upon you of this Notice or so soon thereafter as Counsel may be heard by Counsel on behalf of the above named Applicants for orders that:

1. The decision of the First Respondent contained in a letter dated the 26th November, 1986, confirming the above-mentioned orders and scheme made by the Second Respondent be quashed.

The intention is I think plain, but Mr Laws and Mr George submitted that there was here a defect in that section 23 (which alone is the source of the court's jurisdiction) does not permit of a challenge to a "decision". The complaint could readily be met by moving the words "The decision of the First Respondent contained in a letter dated November 26 1986, confirming" to the grounds of the application so that they would read "The decision of the First Respondent contained in a letter dated November 26 1986 is not within the powers of the Act in that . . ." and then as in the original. Mr Laws (at least initially) and Mr George argued that there was no power to allow an amendment now in that so to do would transgress the rule that an application under section 23 must be made within six weeks from the date on which notice of confirmation is published (section 23(4)(b)). Prior to the expiration of the period it is accepted that there is power to amend (see RSC Ord 20, r 5 and r 7). Does the lapse of time make any difference? In this case I think not. It was, as I have said, plain what the applicants were seeking: they were raising the point that the authorisation was not empowered to be granted by the Act. This they did, albeit infelicitously, within the six-week period. In that circumstance I have jurisdiction to allow the amendment (see *Hanily* v *Minister of Local Government and Planning* [1951] 2 KB 917) and in order that the notice of motion should be in proper form I allow the amendment.

A I should add that although the attack on the order was in general terms it was narrowed during the hearing and in response to an undertaking by the county council to so much of the order as directly affects the land of the applicants. B

The application was dismissed with costs incurred by the Secretary of State only.

ESTATE AGENTS

Queen's Bench Division

July 31 1987

(Before Mr Justice ROSE)

EDWIN HILL & PARTNERS v FIRST NATIONAL FINANCE CORPORATION PLC

Estates Gazette February 20 1988

[1988] 07 EG 75-82

Tort of actionable interference with plaintiffs' contractual rights — Action by firm of chartered surveyors against a bank — Judge's summary of the five essential ingredients of the tort — (1) There must be interference with the plaintiffs' contract, (2) the defendants must know of the existence of the contract, (3) the defendants must intend to interfere with the performance of the contract, (4) the defendants' interference must cause damage to the plaintiffs, and (5) the defendants' conduct must not have been justified by the possession of a right in themselves (ie in the defendants) equal or superior to that of the plaintiffs — After outlining the events which gave rise to the present litigation, Rose J found that there had been direct interference by the defendants with the plaintiffs' contract to be employed to give comprehensive service, involving architectural and ancillary services, in relation to a project for the construction of a new office block to be called Wellington House on a site in Waterloo Road, London SE1 — The judge also found that the defendants knew of the plaintiffs' employment and that they intended to bring the plaintiffs contract to an end — Whether the defendants were justified in their actions depended on whether they were acting in pursuance of an equal or superior right — In deciding this question the factual background had to be considered — The plaintiffs' contract was for employment throughout the whole duration of the project, provided that the developers carried it through — In the difficult years which followed the collapse of the property market in 1974 finance was not easy to find — Various groups showed an interest and then backed away — Eventually the defendant corporation, which had provided finance to enable the Wellington House site to be purchased and had a legal charge on the property, decided to carry out the development themselves, financing it out of their cash flow — They made it a condition, however, that the plaintiffs' services should be dispensed with and other architects and letting agents appointed — At the time when they interfered with the plaintiffs' contract the defendants not only had a legal charge but had provided the finance which had enabled the project to get started during a period of disturbed markets and afterwards to attract other finance — Held that in all the circumstances the defendants did have equal or superior rights compared with those of the plaintiffs and that (to the extent that this is material) they acted reasonably to protect their interests in accordance with those rights — Action dismissed

The following cases are referred to in this report.

Attorney-General v *Newspaper Publishing plc* [1987] 3 WLR 942; [1987] 3 All ER 276, CA

Daily Mirror Newspapers Ltd v *Gardner* [1968] 2 QB 762; [1968] 2 WLR 1239; [1968] 2 All ER 163, CA

Dirassar and James v *Kelly Douglas & Co* (1966) 59 DLR 2nd 452

Emerald Construction Co Ltd v *Lowthian* [1966] 1 WLR 691; [1966] 1 All ER 1013, CA

Greig v *Insole* [1978] 1 WLR 302; [1978] 3 All ER 449

Independent Oil Industries Ltd v *Shell Co of Australia Ltd* (1937) 37 NSW State Reports 394

Knapp v *Penfield* (1932) 256 NY Supplement 41

Lubenham Fidelities & Investment Co Ltd v *South Pembrokeshire District Council* (1986) Construction Law Journal III

Merkur Island Shipping Corporation v *Laughton* [1983] 2 AC 570; [1983] 2 WLR 778; [1983] 2 All ER 189; [1983] 2 Lloyd's Rep 1, HL

Pratt v *BMA* [1919] 1 KB 244

Pritchard v *Briggs* [1980] Ch 338; [1979] 3 WLR 868; [1980] 1 All ER 294; (1979) 40 P&CR 1, CA

R v *Hancock* [1986] AC 455; [1985] 2 WLR 357; [1986] 1 All ER 641, HL

R v *Moloney* [1985] AC 905; [1985] 2 WLR 648; [1985] 1 All ER 1025, HL

Read v *Friendly Society of Operative Stonemasons of England, Ireland and Wales* [1902] 2 KB 88

South Wales Miners Federation v *Glamorgan Coal Co Ltd* [1905] AC 239

Stratford (JT) & Son Ltd v *Lindley* [1965] AC 269; [1964] 3 WLR 541; [1964] 3 All ER 102; [1964] 2 Lloyd's Rep 133, HL

Thomson (DC) & Co Ltd v *Deakin* [1952] Ch 646; [1952] 2 All ER 361; [1952] 2 TLR 105, CA

Torquay Hotel Co Ltd v *Cousins* [1969] 2 Ch 106; [1969] 2 WLR 289; [1969] 1 All ER 522, CA

This was an action for damages by the plaintiffs, Edwin Hill & Partners, chartered surveyors, of London SE1, on the ground of alleged actionable interference by the defendants, First National Finance Corporation, with the plaintiffs' contractual rights. The action arose out of a project for the development of a site in Waterloo Road, London SE1, by the construction of an office block. A previous action arising out of the same project, in which the present plaintiffs were successful in a claim for fees against Leakcliffe Properties Ltd and Mr Alan R Pulver, was reported at (1984) 272 EG 63 and 179.

Igor Judge QC and Richard Davies (instructed by Rowe & Maw) appeared on behalf of the plaintiffs; Colin Smith QC and Andrew Onslow (instructed by Titmuss, Sainer & Webb) represented the defendants.

Giving judgment, ROSE J said: In this action the plaintiffs claim damages for alleged actionable interference with their contractual rights by the defendants. This tort has seen some development in recent years, particularly in the field of trade disputes, and I think it convenient at the outset to set out the framework of law within which I have approached the facts. I express at once my gratitude to counsel on both sides for their researches into relevant authorities and for the clarity of their submissions; these have made my journey in search of appropriate legal principles if not free from difficulty then at least less hazardous than it otherwise would have been.

In *Greig* v *Insole* [1978] 1 WLR 302 Slade J (as he then was) summarised at p 332 the five essential ingredients of the tort in a passage which was adopted by May LJ in giving the judgment of the Court of Appeal in *Lubenham Fidelities & Investment Co Ltd* v *South Pembrokeshire District Council* (1986) Construction Law Journal 111 at p 130. That summary I gratefully adopt and, as will become apparent, it provides the basis for the propositions which seem appropriate to the present case.

These are:

(1) *There must be interference with the plaintiffs' contract.* There are three particular aspects to this which call for comment in the present case. First, in the absence of unlawful means (and none is alleged here) the interference must be direct (see Morris LJ in *Thomson (DC) & Co Ltd* v *Deakin* [1952] Ch 646 at p 702 and Lord

A Denning MR in *Torquay Hotel Co Ltd* v *Cousins* [1969] 2 Ch 106 at p 138E-H). Second, the defendants' conduct in order to amount to interference must go beyond mere advice and amount to inducement, pressure, persuasion or procuration (see *Thomson (DC) & Co Ltd* v *Deakin per* Lord Evershed MR at p 686, Jenkins LJ at p 697 and Morris LJ at p 702). Third, interference with execution of the contract will suffice (see Lord Denning MR in *Torquay Hotel Co Ltd* v *Cousins* at p 138D cited with approval in *Merkur Island Shipping Corporation* v *Laughton* [1983] 2 AC 570 at p 608C-E by Lord Diplock with whose speech the other members of the House of Lords agreed).

(2) *The defendants must know of the existence of the plaintiffs' contract* (see *Thomson (DC) & Co Ltd* v *Deakin* at p 697 *per* Jenkins B LJ and *Merkur Island Shipping Corporation* v *Laughton* at p 608G *per* Lord Diplock). Again there are three particular relevant aspects. First, they need not know the precise terms (see *Emerald Construction Co Ltd* v *Lowthian* [1966] 1 WLR 691 *per* Lord Denning at p 700H to p 701A and *per* Diplock LJ at p 703H to p 704A and also *Greig* v *Insole per* Slade J at p 336 D). But they must have sufficient knowledge of the terms of the contract to know that their interference involves a breach (*Stratford (JT) & Son Ltd* v *Lindley* [1965] AC 269 at p 324A *per* Lord Reid and at p 332F-G *per* Lord Pearce). Second, recklessness as to the terms will suffice (*Emerald Construction Co Ltd* v *Lowthian per* Lord Denning at p 701A and *Daily Mirror Newspapers Ltd* v *Gardner* [1968] 2 QB 762 C at p 781A *per* Lord Denning and at p 784C-G *per* Davies LJ). Third, it is immaterial that the defendants have no knowledge of the contract at the time when they first had inconsistent dealings with the contract breaker: continuance of such dealings after notice of the contract can constitute actionable interference (*Thomson (DC) & Co Ltd* v *Deakin per* Jenkins LJ at p 694).

(3) *The defendants must intend to interfere with performance of the contract* (*Merkur Island Shipping Corporation* v *Laughton* at p 608G *per* Lord Diplock). Here two aspects merit comment. First, malice or spite need not be proved and motive is irrelevant (*South Wales Miners Federation* v *Glamorgan Coal Co Ltd* [1905] AC 239 at p 246 *per* Lord Macnaghten, at p 252 *per* Lord James and at p 255 *per* Lord Lindley). Second, there has been considerable dispute D before me as to the meaning of intention and as to how it is to be decided. For the defendants, Mr Smith submits that intention is not established merely by proof that interference is the natural and probable consequence of the defendants' conduct: he relies on Upjohn J at pp 662-3 in *Thomson (DC) & Co Ltd* v *Deakin*; he draws attention to the word "aim" suggesting a subjective approach which is used by Lord Evershed MR in *Thomson* v *Deakin* at pp 676 and 681; he relies on the formulation of the elements of the tort as including "an intention to inflict harm" by a majority of the Court of Appeal of British Columbia in *Dirassar and James* v *Kelly Douglas & Co* (1966) 59 DLR 2nd 452; he submits that an intention to cause damage is a necessary element in the blacking cases (such as *Merkur*, see p 609B-D); and he draws attention to the discussion of intention E in the criminal law by the House of Lords in *R* v *Moloney* [1985] AC 905 and *R* v *Hancock* [1986] AC 455, which, he submits, establishes a subjective test of intent which he says is as applicable in present tort as in the law of murder.

For the plaintiffs, Mr Judge submits that Lord Evershed's use of the word "aim" finds no echo in the other judgments of Jenkins LJ and Morris LJ in *Thomson* v *Deakin* or anywhere else, that in *Emerald Construction* v *Lowthian* at p 700G Lord Denning and at p 704B Diplock LJ speak of "deliberate" conduct which, he submits, stresses that the tort cannot be committed by accident or (as appears from the *Lubenham Fidelities* case) by negligence; he relies on the words of Lord Halsbury LC in *South Wales Miners*' case at F p 244, followed by Slade J in *Greig* v *Insole* at p 338A — "people are presumed to intend the reasonable consequences of their acts".

In my judgment, although there are cases when the defendant's intention is to inflict harm, no formulation of the ingredients of the tort in the English authorities identifies this as a requirement. In many cases interference will produce harm, and damage is an essential element of the tort. But the intention which is required is, it seems to me, an intention to interfere with the contract. It is trite to say that in relation to the facts, a judge sits as a jury. It is therefore appropriate that I should remind myself, as Mr Smith invites me to, of the considerations relevant to intention and proof of intention discussed in *R* v *Moloney* and *R* v *Hancock* (a course, incidentally, which seems to be in accordance with the judgments of the Master of the Rolls and Lloyd LJ, not as yet fully reported, in *A-G* v *Newspaper Publishing plc and others*, Court of Appeal, July 17 1987*). It seems G to me that those consequences of an act which are "natural and probable" (as described by Lord Scarman in *R* v *Hancock* at p 473A) or "reasonable" (as described by Lord Halsbury in the *South Wales Miners*' case at p 244) are *prima facie* intended but that the inference of intention can be rebutted. In determining whether, in the present case, the defendants intended to interfere with the plaintiffs' contract, I have to look at all the circumstances including what appears in the contemporary documents, what was done and said at the time and what explanations are given in evidence on the defendants' behalf. Actual foresight and probability of consequence are both evidential factors which, among others, help in deciding H what the intention was. I bear in mind, too, that as the defendants are a corporate entity rather than a human being, it is perhaps more readily to be inferred that reasonable, natural and probable consequences were intended. Intention is of course to be distinguished from motive and desire (see *per* Lord Bridge in *R* v *Moloney* at p 926E).

(4) *The defendants' interference must cause damage to the plaintiffs*. There is no issue before me on this aspect. The parties are agreed that the extent of any damage suffered and the damages which are appropriate will, if liability is established before me, be determined by the Official Referee.

(5) *The defendants' conduct must not have been justified*. The first aspect of this matter which caused some difficulty is as to who bears J the burden of proof. Counsel's submissions were notably brief. Mr Judge submitted that the burden is on the defendants and he relies on the words "sought to be justified" in the speech of Lord Halsbury at p 244 in the *South Wales Miners*' case. Mr Smith was inclined to concede this but, at the same time, drew my attention to the way in which Slade J puts the matter in *Greig* v *Insole* at p 332G, namely: "The plaintiff must successfully rebut any defence based on justification." It is, however, to be noted that at p 340D Slade J puts the matter differently: "The defendant may still escape liability if he can in the particular circumstances show sufficient 'justification' in law for what he did." On principle, it might be said that absence of justification is an essential ingredient of the tort which the plaintiffs ought to have to prove. But in *Read* v *Friendly Society of Operative* K *Stonemasons of England, Ireland and Wales* [1902] 2 KB 88 at p 96, Darling J (with whom Channel J agreed) said, in relation to the defendants: "Sufficient justification they may have had, and they may prove it." Furthermore, in the *South Wales Miners*' case Lord Macnaghten at p 246 appears to imply that the burden is on the defendants, Lord James at pp 251-2 refers to the defendants establishing "the existence of good cause and excuse" and concludes that "no justification in law is established by them", and Lord Lindley at p 254 states, in terms: "The defendants have to justify a particular line of conduct which was wrongful." In the light of these authorities, therefore, in so far as Mr Smith made the concession he was right to do so and, in so far as he did not, I find that the burden of L proof is on the defendants to prove justification.

Whether what happened here is capable of being justification is hotly disputed. Mr Smith submits that justification is open to the defendants because in the words of Professor Heydon in the 2nd edition of his book on *Economic Torts* at p 39: "Defendants are justified in protecting their existing contracts, property interests and financial interests." He accepts that justification has been very narrowly circumscribed where there are no prior contractual rights (see, for example, *per* Goff LJ in *Pritchard* v *Briggs* [1980] Ch 338 at p 415H to p 417H), but here, he says, there were existing, inconsistent contractual rights in conformity with which the defendants were entitled to act even if the plaintiffs' contract was interfered with. He cites *Read* v *Friendly Society of Operative* M *Stonemasons* (a decision of the Divisional Court which the Court of Appeal at [1902] 2 QB 732 confirmed). Darling J there at p 96 says:

> Sufficient justification for interference with the plaintiffs' right must be an equal or superior right in themselves

ie in the defendants. Mr Smith seeks support from three cases in other jurisdictions: *Dirassar and James* v *Kelly Douglas & Co* to which I have already referred, *Independent Oil Industries Ltd* v *The Shell Co of Australia Ltd* (1937) 37 NSW State Reports 394 and *Knapp* v *Penfield* (1932) 256 NY Supplement 41.

In *Dirassar* it was held that those who financed a shopping centre

*Editor's note: Now reported at [1987] 3 WLR 942.

A development were not liable for interfering with an architect's contract when they exercised a right of extra judicial foreclosure arising under the agreement entered into with the developer before the date of the architect's contract with the developer. Of all the cases cited to me, these facts are much the closest to the present case. But with great respect to the majority judgments, it is not easy to extract either the ratio of the decision or applicable principles. Justice Maclean found both justification and lack of intent without clearly distinguishing between the two. Justice Branca came close to equating motive with intention and did not appear to contemplate the possibility of more than one intention.

In *Independent Oil* both parties were wholesale vendors of petrol.
B The plaintiffs sold to retailers with a condition that resale would be at a selling price fixed by the plaintiffs. The defendants sold to the same retailers on conditions that they would maintain a prescribed margin between their buying and selling price and would sell all corresponding grades of petrol from other sources at the same price. The defendants increased their prices and refused to supply to retailers who sold the plaintiffs' petrol at the lower price. Certain retailers thereupon, in breach of their agreement with the plaintiffs, sold the plaintiffs' petrol at the higher price. It was held that, although the defendants' action had the effect of causing the retailers to break their contract with the plaintiffs, the defendants were exercising their right to sell on their own conditions and had not knowingly and intentionally procured the breaches. Mr Smith relies
C on three passages in particular in the judgment of Jordan CJ. At p 415 he says:

> The authorities state that there may be lawful justification for the wilful procuring of the breach of a subsisting contract. In this connection it is to be observed that an act which would in itself be wrongful as infringing some legal right of another person may be justified if shown to be no more than reasonably necessary for the protection of some actually existing superior legal right in the doer of the act.

At p 418 he says:

> . . . it is apparent that the primary question in the present case is, what is the proper inference of fact to be drawn from the evidence. Does it appear that the defendant companies have simply been exercising their undoubted right to refuse to sell their own goods except on their own terms and their undoubted
> D right to say that they will sell their own goods only upon their own terms: *Sorrell* v *Smith* [1925] AC 700 at 714, 747; *Ware and De Freville Ltd* v *Motor Trade Association* [1921] 3 KB 40 at 69, 84, or, have they been employing these *prima facie* lawful acts as part of the machinery of a deliberate, and to some extent successful, attempt to procure, without legal justification, the breach of subsisting contracts with the plaintiff company? If the latter is the proper inference, they have undoubtedly been guilty of wrongful acts, notwithstanding that the machinery used, standing by itself would be unquestionably lawful. The question is purely one of fact, ...

At p 420 he says:

> They were simply refusing to sell their own [named] spirit, except at retail rates, to retailers who sold spirit at rates which the plaintiff company prescribed. No doubt this afforded a temptation to retailers to break their
> E contracts with the plaintiff company in order to be able to make profits on the defendants' goods. But I do not think that the defendants were endeavouring to procure such breaches. They were not saying: "We ask you to sell [the named spirit] at the higher price, and if you don't, we will not sell to you except on prohibitive terms." They were saying: "We will not sell to you except on prohibitive terms so long as you choose to sell [the named spirit] at the lower price."

In *Knapp* v *Penfield*, the plaintiff was a professional beauty claiming to be an actress, who was engaged by a producer to play the lead in a musical production financed by a promoter. Her performing qualities were so inadequate as to imperil the show's prospects of success and therefore, at the insistence of the promoter, she was dismissed, though the balance of the salary due under her
F contract was paid to her. She brought an action against the promoter and others. Justice Hammer held that the defendants were acting within their rights and did not wrongfully cause an injury to the plaintiff. He referred at p 44 to "the exercise of a . . . superior right" and said:

> The continuance of the enterprise and of the benefits depended upon the success of the play, without which the [promoter's] investment was lost. The right of the plaintiff was subordinate to that of said defendants, whose contract was paramount to plaintiff's. Without defendants' contract, plaintiff would not have the employment under her contract . . . The very money which paid her agreed compensation arose out of the defendants' prior agreement with [the producer].

He went on: "Such interference was privileged, excusable, and justified." He also said:

> The plaintiff suffered no financial loss, since she received all the money or compensation to which under the contract she was entitled. G

Mr Smith submits that the defendants in the present case, in financing the development themselves, and in requiring the dismissal of the plaintiffs in circumstances which I shall outline later, were doing no more than they were reasonably entitled to do to protect their investment. He further submits that intention and justification are inextricably entwined and that if the defendants have existing rights it is more reasonable to infer an intention to minimise the bank's exposure than to damage the architects. (I comment parenthetically that the flaw in the second part of this argument is that it presupposes that there can be only one intention, which is not, in my judgment, a sustainable proposition.)

Mr Judge submits that lawful justification is not established by H proof of the existence of a contract which gives specific rights to a party who has procured a breach of a contract unless that party is exercising the strict rights given by the contract. He contends that the British Columbia, New South Wales and New York cases illustrate and support this principle. He further relies on the passage in *Read* v *Friendly Society of Operative Stonemasons* at p 97 where Darling J says:

> No one can legally excuse himself to a man, of whose contract he has procured the breach, on the ground that he acted on a wrong understanding of his own rights . . .

this, says Mr Judge, supports the proposition that only the exercise of *actual* rights can provide justification. He relies also on a passage in J the judgment of McCardie J in *Pratt* v *BMA* [1919] 1 KB 244 at pp 265-266 and in particular the words at p 266:

> No justification exists by reason of the fact that the defendants acted either for the advancement of their own trade interests or for the advancement of the interests of those with whom they were associated.

He submits that a defendant who, instead of enforcing his rights under an existing contract, chooses to enter into a new contract which intentionally interferes with the plaintiff's contract cannot claim justification.

With deference to the arguments which have been addressed to me, I do not believe that the principle of law which I must apply in relation to justification is seriously in doubt: it is, as was said in *Read* v K *Friendly Society*, did the defendants have "an equal or superior right in themselves"? It is the application of that principle to the facts of the present case which gives rise to difficulty.

Against this background of legal principle I now turn to the facts.

The oral evidence came from Mr Farino, a partner in the plaintiffs, and Mr Rutter, Mr Otway and Mr Dyer on behalf of the defendants. As part of the support operation to which I will refer later, Mr Rutter, on two years' secondment from Barclays, was a director of the defendants from April 1978, and Mr Dyer, on secondment from Lloyds, was the defendants' managing director from April 1977. Mr Otway was, throughout, the defendants' chief surveyor. The relevant documents were conveniently reduced to a single master bundle, P1. Most of the material events occurred over a period of L about six years which ended over eight years ago. I do not believe that any of the witnesses in this case was other than honourable, honest and doing his best to recall what happened. But, with such a time-scale, recollection is inevitably impaired, and reconstruction, albeit unwitting, is almost inevitable. The contemporary documents are plainly of great importance.

Much of the history is common ground. The plaintiffs are a firm of chartered surveyors. From the late 1950s they provided architectural services at a variety of sites up and down the country. More particularly, from the early 1970s they provided such services for a Mr Pulver, a solicitor and property developer, and for companies which he controlled including Leakcliffe Properties Ltd (to which I M will refer as Leakcliffe) at sites in Birmingham, Sheffield and Basildon. Mr Pulver's finance for these developments came from the defendants, a bank. At no time up to and including trial have the defendants ever questioned the plaintiffs' competence as architects.

In October 1973, shortly before the property market collapsed, Mr Pulver embarked on another project, which came to be known as Wellington House, in Waterloo Road, London SE1. Mr Otway valued the property on October 25 and Mr Pulver agreed to buy on November 1. The defendants provided Leakcliffe with the finance — initially an overdraft facility subject to a first legal charge of £5m. In offering this facility on November 14 1973, the defendants did not reserve any right to nominate the architect, though they were entitled to be satisfied with the drawings before making available a final facility of £1.4m for refurbishment and other works. The defendants knew from October 1974, if not before, that the plaintiffs were acting as architects for this scheme: Mr Pulver appointed them in November 1973. In March 1976 the overdraft facility was continued. In July 1976 Leakcliffe agreed to buy the freehold and in August executed a first legal charge in favour of the defendants, charging the property with all moneys payable to the bank. Clause 3(G)(ii) gave the defendants a power to sell and to appoint a receiver exercisable on default after demand for payment. By clause 4(A) the receiver had power to complete any unfinished building, to borrow money, to obtain planning permission and to employ architects. By clause 4(D) the receiver had power to "commence, continue or complete . . . any work of construction, building or development". The major landmarks of the architectural services provided for Mr Pulver by the plaintiffs (principally through Mr Farino) in connection with this development between 1973 and 1979 are set out in the papers before the court. The period was one of changing and lost fortunes in the London property world. The boom of the early 1970s was followed by the collapse in late 1973, marked on December 17 1973 by an embargo on office development imposed by the Government. A number of secondary banks collapsed or verged on collapse, and some (including the defendants) were supported by the English and Scottish clearing banks by way of a Support Group. Because of the embargo, the Wellington House development could proceed only if an office development permit were obtained. It is common ground that in obtaining one in 1975 the plaintiffs achieved a considerable coup. In September 1977 the plaintiffs obtained planning permission subject to conditions that the ground floor should be a supermarket and the original facade should be retained. Informal notification that permission would be granted was given in early 1977, and from that date Mr Farino and Mr Pulver applied themselves to the difficult problem of financing the development.

In July 1977 the view was canvassed within the defendants that their interests would best be served by carrying out the development themselves. But in August Mr Dyer wrote to Mr Pulver saying the defendants would not put up the money for this. At that time the sum required was £5.7m on top of the existing £6m overdraft, and the site value had fallen considerably, possibly to as low as £2m. Clearly, the defendants could have put in a receiver at this stage but they took the decision to give Mr Pulver more time. Mr Pulver continued to press the defendants to provide development finance themselves, but Mr Dyer continued to refuse — not surprisingly, as the defendants at that time lacked the necessary resources and were being themselves propped up by the Support Group. In January 1978 the overdraft had reached £7m and the defendants threatened to exercise their rights under the mortgage unless, by April, progress had been made. Again the idea of direct building was suggested within the defendants and this time Mr Otway supported it: I find that this meant development directly by the defendants and not (as Mr Otway suggested in his evidence) building after putting in the receiver.

Meanwhile, reports, brochures and costings were being prepared by Mr Pulver with a view to exciting the interest of a backer — several of whom appeared and disappeared during 1978. By way of example, a report was prepared for Philips Electrical Pension Fund which was sent to Mr Rutter in June 1978. A similar brochure was sent to Mr Rutter in October 1978. The report and brochure both, in like terms, name the plaintiffs as architects in "The Professional Team to be employed in connection with the development" and say, "The Professional Team have agreed their fees at ten per cent of the final building cost".

By early 1979 Manufacturers Hanover Property Trust (MHT) were a front runner as potential backers. But by this time the climate of the property market had changed again. There was a shortage of office buildings in central London and the outlook was optimistic. This was reflected in the defendants' approach which, without the knowledge of the plaintiffs or indeed Mr Pulver, markedly changed.

On December 19-20 1978 Mr Rutter and Mr Dyer agreed that they should keep all options open, in particular the MHT offer, and Mr Rutter was prepared also to consider an offer by Samuel Properties Ltd (Samuel). On December 21 the MHT proposal was becoming very specific. On December 29 Mr Rutter and Mr Pulver met Samuel, who expressed a wish to check on the plaintiffs, who were unknown to them but were acceptable to Samuel as joint letting agents. Samuel wanted their in-house team to supervise the project. On January 4 1979 MHT made an offer to finance the development on terms which included their having a first charge. An estimated cost of £5.8m plus professional fees of £600,000, that is about 10%, was discussed. By January 12 Mr Rutter and Mr Dyer had come down firmly in favour of MHT in preference to Samuel. The deficit on build-out was estimated at only £2.6m instead of £6m on a sale: the relevant calculation by the defendants took Mr Pulver's figure of £5.8m and added 10% for professional fees. On January 30 a meeting at which Mr Pulver was not present took place between Mr Rutter and MHT at which final agreement between the defendants and MHT seems to have been very close. But on the next day, as appears from the internal Barclays documents obtained by Mr Rutter and sent by him to Mr Dyer, Sir Richard Pease, who was vice-chairman of Barclays (a member of the Support Group) and a non-executive director of the defendants, suggested that the defendants should themselves finance from their own cash flow. On February 1 this view, including a suggestion that the present team could not "fulfil satisfactorily the function that will be required of them", was strongly supported by Mr Carslake of Barclays. No further relevant documents apparently exist until February 14, when Mr Dyer canvassed, for discussion by Mr Rutter and others, four possible courses, the first of which was that the defendants should themselves finance the development. By February 20 1979, still working on Mr Pulver's figure of £6.4m, Mr Dyer, on behalf of the defendants, prepared a note for the Support Group saying that the MHT terms were not attractive "to us" and saying:

> In view of the considerable possible upside in the development I consider we should provide the finance ourselves out of cash flow. This will be to the eventual benefit of FNFC and the Support Group. Our chief surveyor considers that the correct way to deal with this development is to employ a leading architect and a leading letting agent at the outset. This will ensure as far as possible that a suitable building is constructed which will be attractive to occupiers and to funds seeking property investment. We shall have to start afresh as our customer has used a surveyor with a lesser known agent and this may eventually detract from the value of the investment.

Mr Dyer concludes by proposing:

> 1) We proceed with the detailed work up to tender stage at a cost of about £250,000. 2) If satisfactory figures emerge, and after taking top professional advice, we proceed with the development. 3) We retain sufficient of our cash flow to cover the cost.

That course was favoured by the NatWest and Barclays representatives from the Support Group and on the same day a note was made, apparently by Mr Rutter: "Decision conveyed to Pulver by phone." On February 23 Mr Pulver replied, for him unusually briefly: "Thank you." On February 26 Mr Rutter wrote to MHT saying any deal with them was off.

Events then moved swiftly to a climax. Mr Pulver, following a meeting with the defendants on February 26, saw Mr Farino and told him he had good and bad news. The MHT deal had been discontinued but the defendants had decided to finance the scheme after all. The bad news was that this was conditional on Mr Pulver dispensing with the plaintiffs' services. Mr Farino said, and I accept, that Mr Pulver was very upset "approaching despair" but said he would fight the decision on behalf of them both, because he believed his best chance of success was to retain the plaintiffs. Mr Farino was shattered by the news. On February 28 a meeting took place at the defendants' offices between Mr Farino, Mr Rutter and Mr Otway. Mr Otway's assistant, Hammond, was present but he did not give evidence. Mr Rutter chaired the meeting. Mr Farino took with him documents which were distributed as part of Mr Farino's attempt to persuade the others that the plaintiffs should be retained. There are differences in the recollection of those present as to some of what was said. The differences are of little ultimate significance in relation to the issues in this case, but, so far as is necessary, I prefer and accept Mr Farino's account. I do so for a number of reasons. First, it accords more readily with the reference to litigation in para 2 of Mr Rutter's note following the meeting. Second, having seen and heard Mr Otway, I am satisfied that, to put it no higher, the robustness of his manner (Mr Rutter described him as "forthright") makes probable Mr Farino's account of his reactions and the tone and content of what he said. Third, the importance of this meeting to Mr Farino and its impact on his practice and pride was such as to burn what occurred more indelibly into his mind than into the mind of the defendants' representatives. Fourth, the evidence of Mr Rutter and Mr Otway amounts to no more than "we do not now recall" the matters in dispute.

Mr Otway's response to Mr Farino's presentation of his

A credentials was, "I don't know about all this. What is the tallest building you have built? And what is the largest value of building you've built?" Mr Farino replied that neither question allowed for the complexity of what the plaintiffs had done. When Mr Rutter confirmed that the defendants were concerned about protecting their position, Mr Farino suggested the appointment of a consultant to oversee the plaintiffs' work and drawings. Mr Rutter agreed this was worth consideration. Mr Otway rejected the proposal out of hand whereupon Mr Rutter said, "We must be fair to Mr Farino. You must not allow your personal grudges to interfere with this". It was agreed that the plaintiffs could act in the letting of the building. Mr Farino was very upset that, having spent five years on the project, he was being told he could not build the building. He asked how the B defendants proposed to pay for the work which the plaintiffs had done. Mr Otway said, "That's no concern to us — that's for Pulver". Mr Farino said, "If that's your solution to this problem I'll have to see you in court". There was a further brief conversation on the way to the lift, which left Mr Farino feeling that Mr Rutter was on his side. Mr Rutter's combined note in relation to the meetings on February 26 and 28 is in the papers and it also records: "Pulver advised. 2/3/79".

Mr Pulver wrote to the defendants on March 2 and Mr Dyer replied on March 5. That reply expressly did not seek to deal with the questions raised by Mr Pulver about the position of the plaintiffs. On this aspect Mr Dyer merely said, "We have no obligation to your advisers and have never given them instructions." The letter made C plain the defendants' view that the plaintiffs must go. And I find that the option given to Mr Pulver to find alternative finance within a month was not one which Mr Dyer contemplated was in the least likely to be accepted. I return to this later. On March 6, having received this letter, Mr Pulver saw Mr Farino at his office and told him there was no way he could continue using the plaintiffs: they must recognise that he had no alternative but to get rid of them. A meeting was held between Mr Pulver, Mr Rutter and Mr Dyer on March 7, and on March 8 Mr Pulver was sent by the defendants a list of four architects' names from which to choose: in due course he made the choice. The plaintiffs sought legal advice and a letter before action was written on August 13 1979.

In an action tried by Hutchison J in 1984* the plaintiffs claimed D damages from Leakcliffe for wrongful termination, on March 6 1979, of their contract to provide architectural services at a fee of 10% of the total cost of the development on a *quantum meruit* basis if the development were not carried through. Hutchison J found liability in favour of the plaintiffs, and damages remain to be assessed. Although the present defendants were not a party to that action, Mr Smith, on their behalf, does not seek before me to contest or go behind Hutchison J's finding that there was a contract between the plaintiffs and Leakcliffe for the plaintiffs to act from the beginning to the end of the Wellington House development.

I turn from this summary of largely undisputed events to consider the way in which they fit into the legal framework which I have earlier E attempted to identify. In doing so, I shall have to resolve the more significant factual disputes.

(1) *Interference with the plaintiffs' contract.* I find there was direct interference by the defendants. Indeed, Mr Smith does not contend to the contrary. He does not challenge the existence of a contract between Mr Pulver and the plaintiffs, or Hutchison J's findings that it contained no term entitling Mr Pulver to dispense with the plaintiffs' services and that Mr Pulver, in dismissing the plaintiffs, was in breach of his contract with them. I find that by February 20 1979, if not before, the defendants had made the decision to finance the build-out and it was an integral part of that decision that the plaintiffs should not be the architects; this was clearly inducement, pressure and procuration. The decision was conveyed to Mr Pulver F face to face on February 26; it was repeated to him, probably on the telephone, on March 2 and it was reiterated in the letter of March 5. Put at its lowest, this gave rise to interference by the defendants with the execution of Mr Pulver's contract with the plaintiffs because when, on March 6, Mr Pulver told the plaintiffs that he would have to dispense with their services, he did so because, and only because, of the defendants' insistence.

(2) *Knowledge.* The defence admits that the defendants knew the plaintiffs were employed by Mr Pulver. Mr Smith accepts that they knew the plaintiffs were employed as architects; but he contends that they did not know that the plaintiffs had a binding contract to the end of the development nor did they have sufficient knowledge that their conduct would interfere with the terms of the contract. I am unable to accept this submission. G I find that probably by February 20 and certainly by March 3 (when they received Mr Pulver's letter of March 2) the defendants, who knew that the plaintiffs had been working on the scheme for five years, also knew that Mr Pulver had agreed with the plaintiffs that they would act as architects of the development at a fee of 10% of the overall building costs. I so find having regard to a number of matters.

First, there are the terms of the brochures which Mr Rutter agrees he saw in June and October 1978; no doubt he paid less attention to these than he would have done had the defendants at that time been about to finance the development themselves. But, as I have already indicated, such a proposal had been canvassed within the defendants' organisation in 1977 and again in 1978 (by which time it had Mr Otway's support). H And I am unable to accept Mr Rutter's evidence that the documents were of no importance to him and that he regarded the reference to 10% of final building cost as being no more than an estimate without any significance in relation to the plaintiffs' position. Without financing the development, the defendants in 1978 stood to lose about £6m and I am quite sure, that being so, that no one who, like Mr Rutter, had spent his life in banking, would have failed to keep an eye on details of this kind.

Second, as Mr Rutter said, "the name of the plaintiffs was always cropping up", and if, as he says, he made no assumption about any contractual relationship between them and Mr Pulver, that could only have been because he was closing his eyes to the obvious. This conclusion is to some extent confirmed by what happened at the J meeting with Samuel, attended by Mr Rutter, on December 29: the only inference from Samuel's expressed wish (from which Mr Rutter did not dissent) to accept the plaintiffs as letting agents but check their credentials as architects is that those present at the meeting assumed a live and continuing relationship between Mr Pulver and the plaintiffs.

Third, the reference to "starting afresh" in Mr Dyer's memo for the Support Group is inconsistent with a belief that there was no continuing relationship between Mr Pulver and the plaintiffs.

Fourth, Mr Farino's threat on February 28 to sue, was, as I believe Mr Rutter realised at the time, though he has since forgotten, a reference to the plaintiffs' contract with Mr Pulver.

Fifth, point 2 on Mr Pulver's letter of March 2, particularly when K contrasted with point 1, is, in my judgment, comprehensible only as a reference to a continuing legal obligation by Mr Pulver to the plaintiffs: the evidence of Mr Rutter and Mr Dyer on this was conflicting and unsatisfactory. Mr Rutter said at first that he believed the reference was to a moral obligation; later he made the incredible suggestion that it might refer to something other than Wellington House. Mr Dyer said he believed the reference was to money due for work already done — which would leave point 1 unexplained. And I do not accept Mr Dyer's evidence that he discussed these points with Mr Pulver on March 7: if he had done so, I would have expected a reference to this in the contemporary note.

In my judgment, it is no answer for Mr Dyer and Mr Rutter to say, L "Mr Pulver and Mr Farino never told us there was a contract and we never asked": to put it at its lowest, the defendants were reckless as to whether the terms of the contract between Mr Pulver and the plaintiffs extended to employment of the plaintiffs right through to the building of the development.

(3) *Intention.* Mr Smith submits that even if interference with the plaintiffs' contract was the natural and probable consequence of the defendants' conduct, it was not intended. He relies on the evidence of Mr Dyer and Mr Rutter who both said that they had no intention of causing harm or damage to the plaintiffs. Mr Otway said it never crossed his mind to give any consideration at all to the arrangements Mr Pulver had made with the plaintiffs: his total concern throughout was the defendants' interests. M Mr Dyer said: "I do not think I was putting any pressure on Mr Pulver. He had unattractive alternatives, but so did we all." He also said: "Until March 7 no decision had been made by the defendants to finance the build-out," and, "by February 23 only Mr Pulver's agreement was left before we went ahead." I accept that Mr Dyer believes those answers to be true. But in my judgment they are unacceptable. It is implicit in the findings I have already made in relation to interference that I do not accept that no decision was made until March 7. Clearly the decision was to be implemented in stages and required the continued approval of the Support Group, but I do not accept that Mr Pulver's agreement was necessary for the spending of £250,000 for tender advice. And the suggestion that Mr Pulver was in any sense an equal bargaining

A partner whose decision was genuinely relevant and in doubt is at variance with the contemporary documents. These show that until the defendants had made their decision, Mr Pulver was kept in ignorance of the defendants' change of mind in relation to financing the development. His views were not sought as to whether the plaintiffs should be kept on, and the decision to finance the development and dispense with the plaintiffs was passed on to him as a *fait accompli.* It is not without significance that in relation to the defendants' attitude to MHT Mr Rutter said: "Pulver wouldn't know about this: I didn't think we had to tell him." The defendants plaintiffs. He had failed, despite strenuous efforts over a long period, to obtain finance elsewhere and the defendants were knew that Mr Pulver had no effective choice but to dismiss the B unwilling to surrender their first charge — which unwillingness was clearly a deterrent to other potential lenders. Mr Rutter said: "It never crossed my mind he would not accept." The defendants knew that the plaintiffs' dismissal was not just foreseeable but was the inevitable consequence of their decision that other prestigious architects should be appointed as a condition of their financing the development. Indeed, Mr Dyer said at the end of cross-examination:

> I realised the conditions imposed by the defendants would bring to an end whatever contractual relationship there was between Mr Pulver and the plaintiffs.

Clearly the defendants' primary intention was to guard and further their own best commercial interests. As Mr Rutter put it: "In C financial terms we hoped to do far better from a build-out than exercising our rights under the legal charge." And I accept, despite Mr Rutter's somewhat curious remark to Mr Otway about personal grudges, that on February 28 the defendants were not motivated by malice or spite. But in my judgment the conclusion from all the evidence is, inescapably, that the defendants *did* intend to bring the plaintiffs' contract to an end.

(4) *Justification.* As I have indicated, the question here is: Did the defendants have an equal or superior right? This, in my judgment, is an entirely separate matter from intention. It is clear from the evidence of Mr Rutter, to which I have just referred, that the defendants deliberately chose, for commercial reasons, not to D exercise their rights under the legal charge. Mr Dyer, too, said: "I wouldn't have been very keen on building out with a receiver under the mortgage because we, like Mr Pulver, liked people working alongside us." And it cannot, in my judgment, be said that the defendants believed they were, or indeed gave any thought to the possibility that they might be, exercising rights equivalent to those under the legal charge. But that is not the end of the matter, for the defendants' belief about their rights does not provide the proper test.

What is, in my judgment, material in the present case is not just that the defendants had a legal charge at the time they interfered with the plaintiffs' contract, but that they had agreed to provide Mr Pulver with finance, subject to a first legal charge, at the very outset in November 1973. It is common ground that such a charge would E give the defendants the right to sell the property and appoint a receiver and therefore that such rights were implicit in the defendants' provision of finance from the beginning. Had the defendants not provided finance, Mr Pulver would not have been in a position to retain the plaintiffs' services for this project. He required financial backing in order to purchase the property; and the defendants' backing came at or about the time when the property market started to collapse and only a month before the Government's embargo on office development. If the defendants had not provided finance at the time they did, it is inconceivable that others would have done so shortly afterwards. In my view, this timetable of events points at least to priority in the defendants' rights. Furthermore, both as envisaged in November 1973 and as ultimately F executed, this charge was not for a specific sum. It secured all the moneys due to the defendants on the overdraft. It is not therefore apt to describe the defendants as entering into a new contract when they financed the development. They were making further advances on the basis of a security which already existed.

In these circumstances, I am satisfied that the defendants did have equal or superior rights to those of the plaintiffs and that (to the extent that this is material) they acted reasonably to protect their interests in accordance with those rights.

It follows that, in my judgment, the defendants were legally justified in acting as they did.

Accordingly, the plaintiffs' claim fails.

Queen's Bench Division G

November 5 1987

(Before Sir Douglas FRANK QC, sitting as a deputy judge of the division)

SAVILLS v SCOTT

Estates Gazette March 26 1988

[1988] 12 EG 115-118

Estate agents' commission — Claim for commission against a person who had held himself out as agent for one company as H vendors but who was in fact acting as agent for another company not named — Defendant gave the plaintiff estate agents instructions to put on the market a property known as Neardown Stables and purported to be acting as agent for Neardown Stables Ltd — The asking price for the property was £225,000, but the plaintiffs were successful in obtaining an offer of £235,000 and were authorised to accept it — The plaintiffs then submitted an account for £8,941.75 in respect of their commission —Defendant at that stage asked for a reduction of £500 and subsequently suggested a settlement by instalments — As payment was not made either of the reduced total or by instalments, the plaintiffs' solicitors wrote a letter J before action — In reply the defendant's solicitors claimed that the defendant was the wrong party to sue; the action should have been directed against Neardown Stables Ltd — Subsequently, however, the defence was that the defendant was not personally liable as he was acting as agent for another company, based in Jersey, Monolith Holdings Ltd, which was the freehold owner of Neardown Stables — The judge found as a fact that the defendant was holding himself out as agent, not for Monolith Holdings nor for an undisclosed principal, but for Neardown Stables Ltd — Monolith Holdings had been referred to only once, in the subject heading of a letter, and its K status had not been explained — Held that the defendant was personally liable to the plaintiffs for the agreed commission — "It does not seem to me to need authority for the proposition that where a person holds himself out as an agent for a named person but is in fact acting as agent for an unnamed person then the agent is personally liable"

No cases are referred to in this report.

In this action Savills, estate agents, sued the defendant, R A Scott, for commission in respect of the sale of Neardown Stables, a property in Upper Lambourn, Berkshire. L

R Mawhinney (instructed by Ellis Peirs & Young Jackson) appeared on behalf of the plaintiffs; S Redmayne (instructed by Malkin Cullis & Sumption) represented the defendant.

Giving judgment, SIR DOUGLAS FRANK QC said: This is an action by a well-known firm of estate agents (now a limited company) for commission in respect of the sale of the property known as Neardown Stables, Upper Lambourn, Berkshire. The undisputed facts are:

1. Mr A R Adams-Cairns, a chartered surveyor then employed by the plaintiffs as an associate and now a director, received a telephone call from one David Williams asking if the plaintiffs would be interested in acting as selling agents for the property.
2. On August 30 1983 Mr Adams-Cairns visited the stables where he met the defendant and Mr Williams. He was shown round the M property. A question arose as to planning permission about which he was told to get in touch with a firm of solicitors, namely, Reid Minty & Co.
3. On September 7 Reid Minty wrote to Mr Adams-Cairns about the planning permission and, although they referred to "the present owner", they did not say who that owner was.
4. Within a day or two the plaintiffs wrote to the defendant at his London address confirming that they had been given instructions to place the property on the market at an asking price in the region of £225,000. The plaintiffs sent another letter on the same date to the defendant in which they set out the terms of the agency. They repeatedly referred to "you" and "your", thereby implying that

A it was the defendant who was appointing the plaintiffs as the selling agents. However, in one part of the letter they said "if your clients are not happy with this approach . . .".

5 On October 5 1983 Mr Scott, on notepaper at the head of which was printed "Neardown Stables Limited", wrote to the plaintiffs saying:

> I confirm that I am requested on behalf of the owners of the property to instruct you to sell the property, Neardown Stables at Upper Lambourn as Sole Agents. I enclose our cheque on account as requested for £495 and look forward to hearing from you as soon as possible on intended progress.

In fact the cheque was drawn on the account of Vision Diffusion B Ltd, a company of which the defendant was a director although that was unknown to the plaintiffs.

6 On November 15 1983 the plaintiffs wrote to Mr Scott confirming that they had been given instructions to accept an offer of £235,000.

7 On November 22 1983 Reid Minty & Co wrote to the plaintiffs a letter which was headed:

> Dear Sirs,
> Re: Monolith Holdings Limited
> Neardown Stables, Upper Lambourn, Berks

This was the first occasion on which the name of Monolith Holdings Ltd had appeared in any document. Their status in the C transaction was not stated.

8 On December 13 1983 the plaintiffs sent to Mr Scott a letter saying that they had learned that contracts had been finally exchanged and enclosed a copy of their account for £8,941.75, the amount claimed in this action.

9 In January 1984 the defendant telephoned Mr Adams-Cairns asking for the amount to be reduced, and on January 20 Mr Adams-Cairns replied that provided the account was paid by January 25 they would reduce their account by £500. The account was not paid.

10 On February 9 the defendant visited the plaintiffs' offices and suggested settling the account in three instalments, namely, £5,000 on February 10, £1,500 by February 29 and £1,650 by the D end of March. None of the agreed payments were made and on February 16 the plaintiffs' solicitors sent a letter before action.

11 On March 23 1984 the defendant's solicitors wrote to the plaintiffs' solicitors a letter acknowledging service of the writ and in which they said that the plaintiffs had chosen the wrong defendant. They continued as follows:

> Your client's letter of September 9 1983 is in our view clearly an invitation to treat and it finishes by asking for written instructions to proceed. Those were supplied by letter dated October 5 1983 which although signed by Mr R A Scott were on notepaper of Neardown Stables Limited the actual owners of the property. Your client's own particulars prepared later on clearly set out that the Vendors were Neardown Stables Limited, and there E was no question of Mr Scott becoming personally liable nor indeed is this indicated in the correspondence. In those circumstances your client's action is clearly against Neardown Stables Limited not Mr Scott.

12 Monolith Holdings Ltd is a company registered in Jersey and was in fact the freehold owner of Neardown Stables.

There is one important question of fact which is in dispute and that is whether at the meeting on August 30 1983 either Mr Scott or Mr Williams, or both, mentioned that the owners of the property were Monolith Holdings Ltd as they allege, that is to say, that they told Mr Adams-Cairns that Monolith Holdings were the owners. Mr Adams-Cairns denied that Monolith Holdings Ltd were mentioned. The defendant said that he believed he mentioned their name and that Mr Williams certainly did. If that is so, I do not understand why Mr F Williams was not called to corroborate that assertion. That is a significant and unexplained omission in the evidence for the defendant. Moreover, on or about the day that the plaintiffs were first instructed, Mr Adams-Cairns opened a diary for this transaction in which he appeared to have entered the names of everyone concerned, including the names of the solicitors and the partners, and no mention is made of Monolith Holdings Ltd. The defendant's solicitors could not have asserted so forcibly that the owners were Neardown Stables Ltd unless that is what they had understood from their client. Why, I ask rhetorically, did the defendant not tell his solicitors that he had told Mr Adams-Cairns that the owners were Monolith? As I have said, the first time that company was mentioned in any document was in Reid Minty's letter of November 22 1983. I am satisfied that that was the first mention in any form of that G company to the plaintiffs or Mr Adams-Cairns.

The defence is that the defendant was not contracting personally but was acting as agent for Monolith, which was known to the plaintiffs. My finding of fact on that question in itself defeats that defence. I would go on to say that on a true construction of the documents, and particularly the letter of October 5 1983, the defendant was holding himself out as acting as agent not for Monolith nor for an undisclosed principal but for Neardown Stables Ltd. It does not seem to me to need authority for the proposition that where a person holds himself out as an agent for a named person but is in fact acting as agent for an unnamed person then the agent is personally liable. It is obvious, for example, that although a person may be willing to trade with an English registered company without H further inquiry they might take a different view with a company registered in the Channel Islands. Accordingly, the arguments canvassed and the authorities cited on the question of an agent's liability for an undisclosed principal are in my judgment irrelevant.

It follows from the foregoing that there will be judgment for the plaintiffs for the amount claimed.

Queen's Bench Division

March 4 1988 J

(Before His Honour judge WHITE, sitting as a Judge of the High Court)

BRODIE MARSHALL & CO (HOTEL DIVISION) LTD v SHARER AND ANOTHER

Estates Gazette May 14 1988

[1988] 19 EG 129-144

Estate agents — Commission claim — Sale of hotel — "Sole selling agreement" — Claim rejected by vendors — The subject property was a high-standard country house type of K hotel, set in 45 acres of land at Stanstead Abbots in Hertfordshire, including a nine-hole golf course — The commission claim was based in the alternative on two different clauses in the instructions — The first was a provision that if "during the period of our sole selling agreement" the vendor were to deal with a person not introduced by the plaintiffs or by any other agents, the plaintiffs were to be entitled to the same commission as if they had carried out the introduction — The second clause provided for commission if the plaintiffs introduced directly or indirectly a person who agreed (whether or not subject to contract) to purchase the property at the specified price or L such other price as the vendor was willing to accept — There was some confusing and conflicting evidence, but the judge found that these clauses did form the basis of the agency agreement between the parties and he rejected objections that the wording was uncertain or ambiguous or that the provisions were onerous and unusual and had not been fairly and reasonably brought to the attention of the vendors

Judge White in the course of his judgment distinguishes between three types of agency: (a) the "conventional agent", who must be the effective cause of a sale to be entitled to commission and may be one of several agents instructed by the M vendor, (b) the sole agent, who has the sole right to sell but not to the exclusion of the vendor himself, and (c) the agent who has the sole selling right and will be entitled to commission not only if the purchaser is introduced by another agent but also if the vendor himself sells during the term of the agency — The last type is called a "sole selling agreement", although referred to once (with a risk of confusion) as a "sole selling agency agreement" — This last type is said to be commonly used in the sale of hotels and was used in the present case

In the event the subject property was sold as the result of the introduction by a commercial property consultant who was

A **not an agent in the sense intended by the instructions, namely, an agent instructed by the vendors to sell the property — Judge White held that in consequence of this sale the plaintiffs were entitled to commission under their "sole selling agreement" clause, rejecting a submission that the plaintiffs' instructions had been withdrawn before the sale — It did not therefore matter that the plaintiffs failed under the alternative (introduction) clause, on the ground that the person interested did not "agree to purchase" the hotel — Judgment for plaintiffs**

The following cases are referred to in this report.

B *Interfoto Picture Library Ltd* v *Stiletto Visual Programmes Ltd* [1988] 1 All ER 348
Jaques v *Lloyd D George & Partners* [1968] 1 WLR 625; [1968] 2 All ER 187; [1968] EGD 186; (1968) 205 EG 1207, CA
Parker v *South Eastern Railway Co* (1877) 2 CPD 416
Thornton v *Shoe Lane Parking Ltd* [1971] 2 QB 163; [1971] 2 WLR 585; [1971] 1 All ER 686; [1971] 1 Lloyd's Rep 289, CA

In this case the plaintiff estate agents, Brodie Marshall & Co (Hotel Division) Ltd, who specialised in business concerning the hotel and catering trades, claimed against the defendants, Stuart Arthur Sharer and his wife, Wendy Sharer, commission in respect of the sale of the Briggins House Hotel, Stanstead Abbots, Hertfordshire. The C hotel was sold, but not through the instrumentality of the plaintiffs, to Norfolk Capital for £2.9m.

Gordon Bennett (instructed by Amhurst Brown & Colombotti) appeared on behalf of the plaintiffs; Charles Flint (instructed by M S Marks & Co), represented the defendants.

Giving judgment, JUDGE WHITE said: This is a claim by estate agents specialising in the hotel and catering trades for commission upon the sale of a country house hotel by the defendants in October 1984. Although not instrumental in the introduction of the purchasers, they rely on the terms of their standard "sole selling agreement", upon which they say they were instructed by the first defendant to act as agents for him and his wife (the second defendant) to sell their property and business a year before.

D Under the terms of that agreement they seek a commission upon the sale. Alternatively, they claim commission, again under the terms of that agreement, for the introduction of a would-be purchaser in November 1983 who, they claim, agreed, subject to contract, to purchase the property and business at a price which was acceptable to the defendants. That sale, in the circumstances I shall outline, was never completed, but the plaintiffs argue that they were, nevertheless, entitled to commission for the introduction. If they succeed on the first ground of the claim they would, they say, be entitled to 1.75% of the sale price, which was £2.9m, plus interest. The sum due upon the alternative ground would require an inquiry, as the would-be purchasers agreed to pay £2.4m, plus an additional E amount to be calculated upon the net disposable profits of the hotel in the two-year period that followed the sale. The commission rate would, the plaintiffs claim, be calculated at half that which would have been due had the sale been completed, namely 1.25% of the total sum.

The first defendant denies that any commission is payable. Although he did instruct the plaintiffs to act as his agents in October 1983 for the sale of the hotel, he did so on a very limited basis and not upon the terms of their standard sole selling agreement as they claim. His liability for commission was only to arise, he says, upon the completion of a sale to a purchaser introduced by them. The eventual purchaser was not introduced by them, and in any event he does not accept that by October 1984 they were still acting as his agents. His F instructions to them had been withdrawn, he says, or had lapsed in April or May 1984 when the proposed sale to the would-be purchasers who had been introduced by the plaintiffs fell through. It fell through because the would-be purchasers withdrew after five months of fruitless negotiation. In these circumstances, if he was bound by the plaintiffs' sole selling agency agreement, which he strenuously denies, no commission was payable under its express terms.

Further, the defendants argue that the sole selling agreement which the plaintiffs seek to rely upon was in law unenforceable, either because it was void for uncertainty or because the plaintiffs did not take sufficient steps to bring the crucial terms which are unusual and oppressive in nature to the attention of the first defendant at the time the contract was entered into.

This in outline is the framework of the issues, both of fact and law, G within which evidence has, over five days, been led upon the events leading up to the proceedings and the relationship of the two central characters, Mr Nairn, the chairman and managing director of the plaintiffs, and the first defendant, Mr Sharer. I shall throughout the judgment refer to him simply as Mr Sharer.

I begin by observing that the relationship between the two men clearly had a personal as well as a business dimension, which, quite apart from the matters with which I am concerned, led to tension between them. Exactly what that was, or its relevance to the business relationship, within which there were some features which left the court with an uneasy feeling that they had not been fully explained, is uncertain or unclear. Both parties presented their cases with a curtain H drawn over the reason for any of the personal tensions referred to by the defendant, Mr Sharer. The court proceeds on the evidence that it has.

The property at the centre of the case is a high-standard country house hotel set in 45 acres of land at Stanstead Abbots in Hertfordshire. The main building was a Georgian house and in the grounds there is now a nine-hole golf course. When purchased by Mr Sharer in 1980 it was empty, but he sought and obtained planning permission for its present use. He then converted the property for hotel accommodation and opened the business, employing a general manager to run it, in March 1982. The freehold was purchased with the financial assistance of a sleeping partner, and the hotel business run by a company, Briggins House Hotel Ltd, of which he and his J wife were the sole directors. It is common ground that in the next two years a successfully run four-star country house hotel was brought into being.

It was Mr Sharer's first venture into the hotel trade. He had previously been in the cleaning cloth business, following which he had developed two factories. In about 1977 he began to look to the hotel or leisure business for a development project, and at this point first had contact with Mr Nairn, who was then working for Christie & Co, to whom I shall refer as Christies, who were estate agents specialising in this field. Contact was by telephone, merely indicating his interest in possible properties that might come on to the market. Later that year Mr Nairn bought a stake in the plaintiffs, but the loose contact with Mr Sharer was maintained with occasional particulars K being sent to him and discussed by telephone.

Both Christies and the plaintiffs specialised in the hotel, catering and leisure trade. It is a very specialised field of estate agency work, particularly in the hotel trade, with only a handful of agents operating countrywide. Not only does the understanding and valuation of hotel businesses and property require particular expertise but very often sales need to be handled under conditions of confidentiality to protect a running business from being undermined by a general knowledge, particularly in the staff, of the uncertainties inherent in the situation. Mr Nairn has explained that in this field an agent may become very much more involved in a transaction than an ordinary property agent. At one stage he described himself as being L more of a broker than an estate agent.

Payment for services for the sale of a property are usually by way of commission based on a percentage of the sale price, and in general terms there are three basic types of agency:

(a) the conventional agent, who must be the effective cause of a sale to be entitled to commission and may be one of several agents instructed by the vendor;

(b) the sole agent, who insists on the sole right to sell but not to the exclusion of the vendor; and

(c) the sole selling agent, who has the sole selling right and will be entitled to commission not only if the purchaser is introduced by another agent but also if the vendor himself sells during the term of the agency. I will look at the particular legal implications of these M types of agency in due course. For the moment I merely note by way of background that, on the evidence which I accept, sole selling agreements are, because of the specialised nature and problems of selling hotel businesses, commonly, although not invariably, sought by agents in this field. It was certainly the policy of the plaintiffs, with very few exceptions, to seek such terms when accepting instructions to act, as it was of another firm, Robert Barry, a specialised agency referred to during the hearings. It was not, however, the practice of Humberts, who deal with hotels in their leisure division.

Mr Nairn and Mr Sharer, after loose occasional contact from 1977 onwards, first met in 1980 after the purchase of the Briggins House property. Mr Sharer sought Mr Nairn's professional advice upon the options open to him with the empty house and planning permission yet to be obtained. His position was analysed in Mr Nairn's letter of November 25 1980. The options included not only running a hotel business at the property but immediately marketing it again, either with the benefit of planning permission for office or for hotel use. In either case he quoted a commission fee "payable in the event of sale". There would have been no running business to protect and this would have been a sale as a conventional agent.

In the event Mr Sharer decided to start his own hotel business, running it through Briggins House Hotel Ltd, and the hotel opened in March 1982.

During the next two years Mr Nairn's professional relationship with Mr Sharer continued. He gave advice as to the effect of the planned improvements upon the value of the property and he forwarded particulars of other properties which might be suitable for a similar future project. By 1984 the hunt for a follow-on property appeared to be intensifying, and to this end, in May of that year, Mr Nairn placed a half-page advertisement in the *Country Life* magazine. Further, by then Mr Sharer had begun negotiating with the Harlow Council with a view to obtaining a lease of a site opposite Harlow Station with planning permission for a conference centre and hotel. Again, Mr Nairn was asked to advise on this possible project.

I accept his evidence that in those two years he made at least six visits to the hotel, and although it was not on the market Mr Nairn sensed from the discussions that Mr Sharer, whom he understood to be a developer rather than an hotelier, would soon be selling and moving on to something afresh.

I accept that there was some general talk between the two men in this period as to ways of selling the hotel, with Mr Sharer stressing the importance of confidentiality if this could be achieved. Further, Mr Nairn anticipated, because of his continuing professional involvement with Mr Sharer, that if the hotel were to be sold he would be asked to act as agent. Up to mid-1984, however, no firm decision or plan had been made.

The decision to sell followed the opportunity Mr Sharer had during that year to purchase in his wife's name the sleeping partner's share. Mr Nairn was first asked for a valuation prior to a bank loan, but the resulting financial interest and repayment burden precipitated a decision to sell. The decision was taken by Mr Sharer either just before or during a long telephone conversation with Mr Nairn in the evening of October 22 1984.

Neither party now has a clear recollection of how the telephone call came about, but it is plain that Mr Sharer wanted to discuss the sale of the hotel with Mr Nairn, and in the course of the discussion gave him instructions to act as his agent to sell the property by approaching on the basis of strict confidentiality a few selected possible buyers at offers at around £2.5m plus stock at valuation.

There is a dispute as to what was said or agreed about the terms upon which Mr Nairn was to act and the commission he was to receive, to which I will return later. What is not in dispute is that Mr Sharer's instructions to sell were given orally on this occasion.

The following day Mr Nairn wrote to confirm his instructions. On the back of the letter were standard printed terms and conditions of business for a sole selling agreement, including a scale fee of 3% of the price. Again I will look at the terms in more detail at a later stage of this judgment. For the moment I merely note that in para 2 of the letter Mr Nairn made specific reference to the terms and conditions overleaf:

> You will see that the scale C overleaf is the lowest scale at three per cent, and, in variance with our normal terms and conditions, I would agree that if the owners of

and he gave the names of three possible purchasers

> the Mandeville Hotel or your Nigerian friend (name to be supplied when you can remember who he is), Peter Isles of Norfolk Capital and Mr Peter Harris of the Carnarvon at Ealing, should become the purchasers, that our fees would be one-half the scale set out on the terms and conditions overleaf.

Those three potential purchasers had already made approaches to Mr Sharer, and he had insisted in that telephone conversation that if a sale did take place to one of them the plaintiffs should have only half their fee. The eventual sale was to Norfolk Capital (I will refer to them throughout as "Norfolk") and the plaintiffs do restrict their claim to half the fee which they claim would otherwise have become due.

The scale of fees set out in Mr Nairn's letter of October 23 was subsequently varied. On receiving the letter several days later, Mr Sharer telephoned Mr Nairn complaining of the scale level, and it was agreed that, instead of the blanket 3%, the fee should be 2% if the price was under £2.5m, 3% if it was between £2.5m and £2.6m, and 4% above £2.6m. The call was noted by Mr Nairn as having taken place on October 29 and he confirmed the new scale by letter the following day. He wrote: "The fee scale we have agreed upon in accordance with our terms and conditions is to be revised," and he then sets out the new scale.

There is an important dispute of fact as to whether on this occasion Mr Sharer also, as he claims, rejected the plaintiffs' terms and conditions, insisting that he retained the right to instruct other agents and that commission would be payable only on a sale. That there was such a discussion is disputed by Mr Nairn and I will return to this issue in due course.

There was a further agreed variation of the scale in a discussion just under a fortnight later, and the effect is recorded by Mr Nairn in his letter of November 13 1984. He wrote:

> I take your point regarding the scale of fees and if I say I will now finally set out what has been finally agreed, you will no doubt let me know if it is not what you understand to be the case. Our fee scale is to be as follows: up to £2,499,999.99: 2½ per cent of the total consideration. From £2.5 million to £2,599,999: three per cent: from £2.6 million to £2,999,999: 3½ per cent,

and the commission was to be 5% if the price was over £3m.

The object that Mr Sharer had in mind in insisting upon a graduated scale of this kind was to build in an incentive to the plaintiffs to get the highest price that they could.

By November 13 1984 the terms upon which the plaintiffs had been engaged by Mr Sharer were therefore finally agreed. A central issue in the case is: what, apart from the scale fees, were the agreed terms and conditions?

I pause here to note that unknown to Mr Nairn, Mr Sharer was, at about the same time, separately instructing Christies to act for him as agents to sell — again upon the basis that only a few selected potential customers should be approached to protect confidentiality. Four days after he had instructed Mr Nairn, Mr Sharer had a telephone conversation with Mr Caile of Christies as a result of which that firm was given authority to discreetly market the hotel. In his letter confirming the instruction Mr Caile wrote: "However, I will emphasise to respective buyers that the business is not on the market and that it is only because of the possibility of another centre, are you at all considering a sale at this time." From this I infer that Mr Sharer did not disclose to Christies that Mr Nairn had been separately instructed.

The letter continued: "Should we be instrumental in introducing a purchaser who duly completes a transaction then, as agreed, we shall look to you for our fee in accordance with the scale detailed below." This was up to £2.5m, 2%, up to £3m, 2.5%, and over £3m, 3% — a similar but not identical scale to that which he had agreed the day before with Mr Nairn. No standard forms of contract were introduced by Christies into this arrangement.

Acting upon their instructions, Christies informed a Mr Hazelrigg of Kingsmead Hotels, a potential purchaser, of the hotel by letter dated November 2 1984. It so happened that Mr Nairn also made an approach to him by telephone on November 5 and was of course told that he already knew of the property, but with a price tag of nearer £3m than the £2.6m Mr Nairn was at the time quoting.

According to Mr Nairn, after hearing that Christies had got hold of the property he telephoned Mr Sharer, reminding him that he had sole rights and warning him that he might find himself paying two sets of fees, and expressing concern over the risk to confidentiality. "I distinctly recalled," Mr Nairn said, "him saying that he would get in touch with Christies as they should not be offering it." Mr Sharer went on to explain that he had seen Mr Caile prior to instructing Mr Nairn who had said that he would be willing to sell it at £3m, but Mr Nairn was the sole agent, and it was just a matter of "clearing it up".

If this conversation did take place it clearly was very crucial, being the clearest indication of Mr Sharer's acceptance of the plaintiffs' exclusive agency agreement.

Mr Sharer denies that this conversation ever took place. He does, however, recall being told in November by Mr Caile that he had had a row with Mr Nairn, who had complained of Christies sending the particulars to Mr Hazelrigg, who he claimed was his (Mr Nairn's) client. Mr Sharer's recollection is that when told of this he simply commented that the two agents would have to "sort it out between themselves". He certainly did not speak to Mr Nairn about it or accept that the plaintiffs were sole selling agents for the hotel.

I again leave this important issue of fact until later. Whatever the truth, nothing further was heard from Christies until the following May when, shortly after the protracted negotiations with purchasers introduced by Mr Nairn finally came to an end, Mr Sharer had a telephone conversation with a Mr Cameron, who by then had replaced Mr Caile.

The introduction of a potential purchaser by Mr Nairn followed immediately after Mr Sharer had authorised him to approach potential purchasers. On October 18 1984 agents acting for a father and son, Ralph and Daniel Levy, with a "roll-over tax situation" wrote to the plaintiffs as part of their search for a suitable hotel with vacant possession in London or the Home Counties. There was urgency in their situation as for tax purposes the purchase of any hotel had to be agreed before mid-December.

The Levys were introduced to Mr Sharer by October 30 1984. They visited the premises on November 2 and a written offer quickly followed on November 7 1984. In the urgency of their situation, with no time to have a full accountant's report, the basis of the offer was not a finite sum but an initial consideration of £2.4m plus an additional variable sum based on a projection of profitability for the next two years assessed on the figures available.

I pause to note that the hotel had traded for only two years, that audited accounts for the year ending April 1984 were not yet available, and that full trading had yet to come on stream. The works of improvement and enlargement had still to be completed or put in hand. This was the background to the offer that was made.

There followed two meetings, one at Mr Nairn's office on November 16 and one at the hotel on November 20. Agreement was reached on the footing of the formula suggested by the Levys. There was to be an escalator and a de-escalator provision. Mr Sharer was to be retained for the two years in which the provisions were to operate on a managerial basis at a wage of £25,000 a year, and the purchasers agreed to provide the new trading company which was to be set up to run the business with £500,000 capital to enable Mr Sharer to apply for the necessary planning consents for the extensions which he had had in contemplation, and which could be expected to enhance the hotel's profitability, plus an additional £500,000 for further capital expenditure in the period.

At the two meetings detailed terms were hammered out and agreed as set out in Mr Nairn's letter to Mr Sharer's solicitors of November 21 1984, copies of which were sent to Mr Levy's solicitors. Mr Nairn also drew up a memorandum of sale, although not specifically instructed to do so. Mr Sharer's solicitors acknowledged Mr Nairn's letter and the memorandum, and having seen their client with his accountant raised no objection or query. It was agreed that the Levys' solicitors, Slaughter & May, should draft the agreement.

It is clear that the parties had at this point agreed in general terms, subject to contract, to the sale to the Levys of the hotel and the crucial terms upon which the formal contract was to be drawn. By the law of this country the parties had not yet legally bound themelves, but both recognised that a "deal had been struck". Mr Nairn has described how at the conclusion of the meeting on November 20 1984 "there was a tremendous feeling that a deal had been struck". Mr Daniel Levy added: "In our view, in our accountant's view and in our solicitor's view (who would not have been instructed if there had not been an agreement) there was unquestionably a deal struck."

The effect of that "deal", which, as I shall now indicate, never was embodied in a formal contract, is (in the context of the provisions of the plaintiffs' standard terms) a subject of challenge by the defendants.

In coming to the deal no mention was made by Mr Sharer of his interest in the Harlow site. His interest still was only of a possible opportunity to obtain a lease and planning permission for another hotel development, but it was relevant not only because if it came to fruition it could cause complications with the usual restraint of trade covenant that would be embodied within the contract of sale, but also because in the eyes of the council the link of a potential developer with the Briggins House Hotel which was close by could have been a factor which would affect their ultimate decision whether or not to grant a lease, and in any event the potential at Harlow could be of interest to the new purchasers of the Briggins House Hotel. The business of one could complement the business of the other.

The Harlow project was disclosed to and discussed with the Levys at a further meeting with Mr Sharer, the date of which is not clear but was probably December 10. At that meeting a change of the consideration which had been previously agreed was also broached, as well as Mr Sharer's continuing involvement over the next two years as a manager of the business. I have no doubt that it was Mr Sharer who sought to make the variations from the original agreement, whether or not this was as a result of tax counsel's advice (he had seen tax counsel on December 5) or simply because he had changed his mind — got cold feet about staying on and wanted an outright sum rather than run the risks inherent in the escalator provisions.

In the result negotiations were reopened, and the original deal was varied. The Levys, who were anxious to conclude a deal, now agreed to pay £2.6m for the hotel (a halfway figure between £2.4m and £2.8-9m which they thought it was possible for Mr Sharer to have obtained upon the escalator). Mr Sharer's continuing management involvement was dropped and his interest, if it materialised, in the Harlow site was to be purchased by them for £200,000.

Slaughter & May were reinstructed and with speed produced a draft agreement on the new basis on December 12 1984. There was by then great urgency on the Levy side: contracts were to be signed by December 17, which would have just met their tax deadline.

Their solicitors, however, advised that the Harlow deal should be separately catered for, and the draft contract contained no reference to it. It was referred to in their covering letter. In that letter the Levys' solicitors suggested that it should be discussed later.

Further, to protect their clients against the uncertainties of the accounting position they reintroduced the de-escalation clause — linked to a net disposable profit figure of £200,000. Mr Levy junior has explained that on Mr Sharer's figures that should have caused no problem for him, but it certainly had not been part of the new arrangement.

As a result Mr Sharer's solicitors immediately reacted by telex, complaining that the new draft did not reflect the agreement terms both as to Harlow and as to the escalator. Further negotiations on principle were, they suggested, required, and a new basis for agreement put forward — namely a total consideration of £3.2m which was to include £400,000 to be set aside in a joint account as capital for the improvements at the hotel, reflecting what had been agreed in principle from the beginning on this point. A de-escalator clause would be agreed to only if it were balanced by a similar escalator clause using the same trigger figure. At this point agreement slipped away from the parties.

Negotiations for the sale of Briggins House and the Harlow lease nevertheless continued, but was no longer under the pressure of time on the Levys' side, as their tax deadline had now passed. It is clear that positions on both sides began to shift. At the beginning of January 1985, following a long conference with Mr Nairn, Mr Sharer put forward new proposals. In summary £2.7m was to be paid for the hotel plus stock at valuation. The purchasers would pay an additional £100,000 if and when planning consent was obtained for the additional accommodation at the hotel. This made £2.8m for the hotel alone: if the Harlow interest were required this would cost an additional £200,000, £3m in all.

The Levys' agent reported to his clients at this stage that having spent a long time negotiating with Mr Nairn he was reasonably confident that these were the minimum terms upon which the vendor was prepared to treat. In a letter to Mr Nairn the agent asked for a "solemn undertaking that he would treat on that basis". This reflected a feeling on the Levys' side that Mr Sharer had, as it were, shifted the goal posts.

New draft contracts were then prepared, this time by Mr Sharer's solicitors, with March 1 as the proposed completion date. Solicitors' correspondence on the draft followed, but on January 28 1985 Mr Sharer's solicitors telexed that he wished to retain the two lodge gates. This was the first time that this had been mentioned and was rejected out of hand by the Levys. It was not pursued further, but to the purchasers again it appeared to be an attempt to change the firm basis upon which they had, in this new phase of the negotiations, been prepared to treat.

Apart from this hiccup, it appeared nevertheless that the negotiations were going well. On January 21 Mr Sharer suggested to Mr Nairn that if the Levy deal went through his fee should, instead of being calculated on the scale agreed the previous October, be £50,000 on completion, plus the lease of one of two identified flats in a country manor which he owned or, if they were not available because they had been previously sold to other parties, a further £50,000 payable as to half on the sale of the flat and as to the remainder on the acquisition by Mr Sharer of another property through Mr Nairn or at the end of 18 months, whichever was the sooner. Mr Nairn agreed in principle to the variation and the timing and details were referred to

A in letters from him to Mr Sharer on February 11 and 20. In the event nothing came of the proposed arrangement, as the Levy sale did not go through, and neither the plaintiffs nor the defendant seeks to rely upon it outside the particular circumstances that were envisaged in the discussion. The basic issue in the case remains as to the terms upon which Mr Nairn was originally instructed.

In the middle of February there was a further turn in the negotiations, this time at the instigation of the Levys. Wanting more time, they sought an option arrangement, giving them an option to complete on notice in April: this would have extended the deadline from the beginning of March to the middle of April. The consideration was to be £20,000. It was agreed to by Mr Sharer, who again sought Mr Nairn's advice on the position. He was plainly B becoming disturbed about the level of fees that were mounting, and the resultant risk if the deal did not eventually go through. An option gave him some protection, but no less than three new draft contracts were then prepared, on February 27 and 29 and March 1.

In the meantime inventories were being taken, and further accounting information relating to the trading situation of the hotel business was being pressed for by the Levys. Matters dragged on through March, and by the end of the month Mr Sharer was getting impatient. On April 4 his solicitors telexed to say that he did not wish to proceed further.

He has said, and I accept, this was to bring matters to a head. It did just this, but unhappily led to further offers and counteroffers.

C In response, the Levys offered to proceed immediately, in spite of not having had some of the accounting information that they had earlier requested, at £250,000 below the price previously agreed, the drop obviously being seen as a protection in the absence of the figures that they otherwise were insisting on seeing. There was a further exchange. Mr Nairn discussed in this period the trading figures with Mr Mendoza, the Levys' agent, and shortly afterwards the Levys indicated a willingness to proceed at the original price, so that by April 30 it appeared that the parties were *ad idem* and ready to complete. There was, however, one final condition that the Levys sought to make at the restored price, namely that the accounting information which had been requested earlier should be made available before exchange, not, they assured, for the purpose of D seeking a renegotiation but to give them an opportunity to discuss and comment on any points that arose.

Mr Sharer was not prepared to accept such a condition without a protection against the costs involved, as well as the overall cost for which he was on risk if the Levys backed out. He was seeking a 10% deposit while the information was given, with compensation for delay and expenditure if the Levys pulled out equivalent to 1% of the price. Following this telex sent on May 2 there was a further conversation between Mr Sharer and Mr Daniel Levy at which the transaction was called off.

There is a dispute as to which of the two parties withdrew. I will return to this later. But the fact is that on May 9 the papers were E returned by solicitors. The conversation would have taken place shortly before. Mr Sharer's recollection is that it took place on May 3, but there is no record to support this. But if correct, he by that time was aware that Norfolk were about to make an offer at £3m.

There is a further dispute as to whether Mr Sharer informed Mr Nairn at this time of the breakdown of the negotiations. There is no correspondence or record of any communication between them after March — a time when the negotiations, although hanging fire, were still alive — although Mr Nairn was, as I have indicated, discussing with the Levys' agent the hotel trading figures in April. It is Mr Nairn's evidence that he was never told that the Levy transaction had come to an end. He knew of difficulties, particularly in connection with the new building works, which would have taken time to resolve, F and he said he was not unduly concerned, although he was given no indication of how it was progressing between May and September 1985. During this period he asked Mr Sharer several times about the hotel and in reply Mr Sharer gave a general assurance that everything was all right, but there were still works to be done. At no time did he say the Levy deal had been called off. Mr Sharer's recollection is that after he received planning permission for the new extension — this was on April 24 — he phoned Mr Nairn asking him to withdraw the property from the market, saying he was going to start the building works. This was at the end of April or beginning of May. He further told him that the Levy deal was dead and told Mr Nairn not to introduce further customers. He would, he said, wait until the building works were complete and would then be looking for £3.5m, at which he recalls Mr Nairn commenting that he was "ahead of the G field as usual".

That such a discussion ever took place is strenuously denied by Mr Nairn.

In his affidavit, sworn on July 16 1986, Mr Sharer deposed that he told Mr Nairn in this telephone call that until the building works were complete he would not market or sell the hotel. If this was said, it was at the time plainly untrue, as he was then quite separately about to treat, or in the process of treating, with Norfolk, the eventual buyers of his hotel.

In mid-March, Mr Sharer was told by a commercial property consultant whom he contacted in his search for another property that this company could be interested in Briggins House Hotel. As a result, an appointment for Mr Isles of that company to visit was made H in the last week of April. After that visit, which took place on April 29, an offer in writing dated May 1 was made at a price of £3m, which Mr Sharer accepted subject to contract by letter on May 7.

The Levy transactions had come to an end either just before or just after his letter of acceptance was written. Certainly at the time that they did come to an end Mr Sharer would have been aware of the possibility of a pending offer.

Norfolk's follow-up investigations continued throughout the summer. During this period Christies had re-established contact with Mr Sharer, and on May 8, after a lapse of over six months, were instructed by him to "discreetly market" the hotel on the terms that had been agreed on the previous October, seeking offers in excess of J £3m.

Mr Sharer informed them that he had an interested party — indeed the day before he had, "subject to contract", accepted Norfolk's offer. The Prudential and Thistle Hotels were later introduced by Christies. Both showed some interest during the summer but nothing came of either introduction.

In early October Norfolk completed at £2.9m.

Mr Nairn learnt for the first time of the Norfolk purchase when reading the *Financial Times* in a traffic jam on the M1. He telephoned his secretary from the car and an invoice for his fee under what he contended was a continuing binding agreement with Mr Sharer was sent. This was rejected, and the writ followed in December of that K year.

This in outline was the long and at times tortuous sequence of events leading to these proceedings.

I now turn to the issues of fact and of law which lie along the trail and which were left for the time being to the side.

(1) Upon what terms were the plaintiffs engaged by the defendant to act as agent for the sale of the Briggins House Hotel?

This issue raises questions both of fact and of law.

(i) *The facts*

I deal first with the facts. Before dealing with this particular area of factual dispute I make the following general observations about the two principal witnesses, Mr Nairn and Mr Sharer. Neither was L entirely satisfactory in the witness box. The detailed recollections of both, I have no doubt, have genuinely suffered with the passage of time.

Mr Nairn was frustratingly vague at times, giving sketchy recollections of events or meetings of which detail might have assisted in piecing the evidence together. He did impress, however, as a frank and straightforward witness. I am sure that where there was vagueness it was because of a genuine lack of recall and not simply a device to obscure an unhelpful truth.

Mr Sharer, on the other hand, at times gave the impression of allowing hindsight to distort or to fill gaps in his recollection. Further, the complaints that he was at the very least disingenuous in his affidavits when dealing with the negotiations with Norfolk from M April 1985 onwards, and with the related discovery, were patently justified. Further, his explanations of these matters were unhappily marked by an obvious lack of candour.

I am satisfied that his evidence was not always reliable. Of the two witnesses, where there was a straight conflict I find the evidence of Mr Nairn the more reliable. Nevertheless, each recollection has to be set and tested against the surrounding facts and documents of the time.

At the heart of the factual issues as to the terms upon which the plaintiffs were engaged by the defendant are two telephone conversations — one on October 22 1984 and one which took place shortly after Mr Sharer had received Mr Nairn's letter of October 23 1984. The two men differ as to what was said on each occasion. I

A accept that the call on October 22 followed discussions that had taken place during 1984, in which ways of marketing the hotel had been touched upon in general terms. In the context of the need for secrecy Mr Nairn had explained that, if instructed, he would expect the hotel to be placed exclusively in his hands. Mr Sharer would have been aware that Mr Nairn would proceed on the basis that he was to be the only agent instructed, but the words "sole selling agent" were not used and no detailed discussion as to the terms would have taken place.

In the course of the telephone call on October 22 I accept Mr Nairn's evidence that (1) he indicated that his fees would be 3%; (2) he made it plain that he would be accepting instructions on the basis B that he was to be the only agent; and (3) because of the need for secrecy he was only to approach the six potential purchasers. It was at that point that Mr Sharer indicated that he had already had approaches from three potential purchasers and a variation of the 3% fee was agreed in respect of them. I reject Mr Sharer's evidence that figures were not mentioned.

I am not satisfied that there was any further meaningful discussion about other conditions upon which the plaintiffs would be prepared to act. In particular, Mr Nairn did not attempt to give details of the terms governing liability or fees if Mr Sharer instructed another agent, or sold without the intervention of an agent, or at what point fees would become payable. Further, he did not use the words "sole selling agent". It would nevertheless have been in general terms clear C that, if instructed, Mr Nairn was to have the exclusive rights to sell. This was implied in his explanation of how secrecy was to be protected. Further, I accept that it is probable, as Mr Nairn recalls, that it was left that night that he would be confirming in the usual way — that is by letter — in the morning. Mr Nairn clearly would have relied upon the confirmatory letter to spell out the detailed terms upon which he would be prepared to act. In the conversation the fee was the focus of attention.

The letter he wrote the following day did draw attention in the second paragraph to the plaintiffs' terms and conditions printed overleaf. It has been argued that the reference was only to the fee scale and that the paragraph is ambiguous as to whether the plaintiffs are relying upon the main body of the terms and conditions.

D In my judgment, it was implicit in the second paragraph of the letter, from the heading of the reverse side which read "Brodie Marshall Hotels — terms and conditions of business", and from the wording of the scale C (3%) provision that these were the terms upon which the plaintiffs were prepared to act for the defendant.

On receiving the letter — it was delayed a few days because it was wrongly addressed — Mr Sharer has described how he dealt with it:

> I started to read it, but being the type of businessman I am, I looked to the fee. I didn't give much notice to the other parts — I read parts of it but my focus was on the fee . . . I can't recall reading any of it except focusing my eye on the fee. I obviously read "sole selling agreement" but did not read any terms and conditions on the back.

E Although he was, as I have found, told what the plaintiffs' fee was on October 22 1984, he telephoned Mr Nairn expressing "despondency" at the commission rate. It is common ground that Mr Nairn then agreed to a graduated scale. What is in dispute is whether Mr Sharer also rejected the plaintiffs' terms and conditions. His recollection was: "I basically said I didn't know what he was trying to prove to me . . .. We had not discussed or agreed upon any terms or conditions or fees", and later: "I said I am not interested in any of your terms. I am only interested in your fee if you are successful in selling. I said I did not want to enter into any form of sole selling agreement. I made it plain I reserved the right to instruct other agents."

Mr Nairn denies this formed part of the conversation, and I have F no hesitation in accepting his evidence upon this. He made a note at the time which recorded the change of fee scale but made no mention in that note of a rejection of the standard terms. I do not for a moment think he would neither have made a note of such a crucial matter nor reacted either by seeking to redefine his terms or sending a follow-up letter on the point. I reject Mr Sharer's evidence: I am sure this is an example of time and hindsight distorting a recollection.

What is plain is that Mr Sharer was well aware, on receipt of Mr Nairn's letter of October 23, that the plaintiffs were proposing to act for him upon a "sole selling agency basis", the terms of which were set out on the back of the letter for him to read. It is not entirely clear just how far he did read those terms — he clearly looked at the heading and the provision setting out the scale fee at 3% at the bottom of the page. I think the truth simply is that he glanced at them G but did not bother to read them carefully.

Accepting Mr Nairn's evidence that there was no dissent upon the terms from him when the new fee was negotiated, those terms impliedly became part of the contractual arrangement between the parties.

The position was confirmed by letter, by Mr Nairn, on October 30, in which he wrote, "The fee scale we have agreed upon in accordance with our terms and conditions is to be revised", and he set out the new scale.

That scale was again revised following a further call from Mr Sharer, and that revision, as I have indicated, is set out in Mr Nairn's letter of November 13 1984. I reject the submission that I should H interpret the words "I will now finally set out what has been agreed" as embodying all the other terms and conditions; in other words they were not being embodied. The letter plainly referred solely to the variation of the fee scale.

Linked to the initial basic issues of fact is the dispute as to a telephone call Mr Nairn said he made to Mr Sharer on learning that Christies appeared to be marketing the hotel at the same time. According to his recollection Mr Sharer accepted they should not be acting for him, gave an explanation and said he would talk to them. As a result I am asked to infer Christies' activities did cease. Mr Sharer denies this conversation. He says Christies told him of a row with Mr Nairn and he simply replied that they should sort it out between themselves. It is a curious dispute of fact which is not easy to J resolve, as both witnesses here gave very vivid detailed accounts of the calls, and there are no helpful independent pointers to the truth. Nevertheless, I again accept Mr Nairn's recollection in preference to that of Mr Sharer.

I was asked to infer that the very fact that Mr Sharer instructed Christies at the same time that he was instructing Mr Nairn was inconsistent with his having accepted a "sole selling agency agreement" with the plaintiffs, or of being aware of its terms. I have not overlooked the point, but I conclude that the truth very simply was that on instructing Christies Mr Sharer either overlooked or disregarded his obligations to Mr Nairn.

Those obligations were defined by the plaintiffs' written terms and K conditions of business as set out on the reverse of the letter of October 23 1984 and I now turn to the issues of law that arise upon them.

(ii) *Issues of law*

The defendant argues that if the conditions did form the basis of the agency agreement he is not legally bound by those terms that the plaintiffs seek to rely upon. Either they were void for uncertainty or, being unusual and onerous for this class of agreement, the plaintiffs did not take sufficient care to bring them to his notice at the time of the contract.

The claim in so far as it seeks a commission upon the Norfolk sale in October 1985 relies particularly upon clause 8:

> If during the period of our sole selling agreement the owner or the person L directly instructing us, deals with a person not introduced by us or by any other agents we shall also be entitled to the same commission in the same circumstances as if we had introduced such a person.

The alternative claim is founded upon clause 1, and that reads:

> The owner will pay our commission if we introduce directly or indirectly a person who agrees (whether or not subject to contract) to purchase the property at the price specified overleaf or such other price as the owner is willing to accept. However, no commission will be payable in respect of a proposed sale which is not completed on account of the purchaser withdrawing from the transaction.

Clause 7 provides that the agreement continues until terminated by either party giving not less than 21 days' notice in writing, to expire not earlier than three months from the letter. M

At any time the vendor is, nevertheless, entitled to withdraw the property from the market on payment of a fee of £100 or the plaintiffs' out-of-pocket expenses, whichever is the greater (clause 10). The fee, which is stated to be a reduced fee (as compared with the plaintiffs' fees for straightforward agencies or sole agency agreements) is 3%: in the case of a proposed sale which is not completed half of that rate is payable, provided that the purchaser has not withdrawn from the transaction (clause 6). The commission becomes payable on completion of the sale. If completion does not take place within one calendar month of exchange of contracts the commission becomes due on exchange of contracts or, in the case of a proposed sale which is not completed, the commission becomes

A payable (if the purchaser has not withdrawn from the transaction) when the transaction terminates (clause 3).

It is the defendant's submission that the terms relevant to either head of claim (commission due upon the Norfolk sale or the Levy introduction) are not sufficiently clear to found the contractual rights which the plaintiffs seek to enforce. They are, it is argued, onerous and unusual, and to bind the defendant require not only unambiguous language but also that the effect be fairly and reasonably brought to his attention — brought home to him to ensure that he has agreed to them. It is claimed that the terms were not fairly brought to his attention. In this context I have been referred to three authorities, *Jaques* v *Lloyd D George & Partners Ltd* [1968] 1 WLR 625, *Thornton* v *Shoe Lane Parking Ltd* [1971] 2 QB 163, and
B *Interfoto Picture Library Ltd* v *Stiletto Visual Programmes Ltd* [1988] 1 All ER 348. I refer briefly to passages from judgments in each of these authorities which provide the basis of the arguments which have been put upon the terms of this agreement.

In *Jaques* v *Lloyd D George & Partners Ltd* Lord Denning MR at p 630 said:

> The principles which in my opinion are applicable are these: when an estate agent is employed to find a purchaser for a business or a house, the ordinary understanding of mankind is that the commission is payable out of the purchase price when the matter is concluded. If the agent seeks to depart from that ordinary and well understood term, then he must make it perfectly plain to his client. He must bring it home to him such as to make sure he agrees to it.
> C When his representative produces a printed form and puts it before the client to sign, he should explain its effect to him, making it clear that it goes beyond the usual understanding in these matters. In the absence of such explanation, a client is entitled to assume that the form contains nothing unreasonable or oppressive. If he does not read it and the form is found afterwards to contain a term which is wholly unreasonable and totally uncertain, as this is, then the estate agent cannot enforce it against the innocent vendor.

Megaw LJ in *Thornton* v *Shoe Lane Parking Ltd* at p 172 said:

> . . . it may not be necessary for a defendant to prove more than that the intention to attach *some* conditions has been fairly brought to the notice of the other party. But at least, where the particular condition relied on involves a sort of restriction that is not shown to be usual in that class of contract, a defendant must show that his intention to attach an unusual condition *of that particular nature* was fairly brought to the notice of the other party. How
> D much is required as being, in the words of Mellish LJ in *Parker* v *South Eastern Railway Co* (1877) 2 CPD 416, 424, "reasonably sufficient to give the plaintiff notice of the condition" depends on the nature of the restrictive condition.

Finally Dillon LJ in *Interfoto Picture Library Ltd* v *Stiletto Visual Programmes Ltd* at p 352, after a review of the authorities that followed *Parker* v *South Eastern Railway*, said:

> It is in my judgment a logical development of the common law into modern conditions that it should be held . . . that, if one condition in a set of printed conditions is particularly onerous or unusual, the party seeking to enforce it must show that that particular condition was fairly brought to the attention of the other party.

E There are, therefore, three questions that have to be asked:

1 Are the terms relied upon clear and unambiguous?

2 If they are, are they particularly onerous or unusual? and:

3 If they are, were they brought fairly to the attention of the defendant?

I take clause 8 first. It is argued that clause 8 has an ambiguity or alternatively, if it is clear, it would not, in its literal sense, entitle the plaintiffs to commission on the Norfolk sale.

The focus of the argument is the wording "deals with a person not introduced by us or by any other agents". The problem arises from the facts of the case. Norfolk was introduced by an agent, a Mr Meyers, although he was not an agent who had been instructed to sell
F the property by the defendants. For this purpose he was not the defendants' agent. Mr Meyers, a commercial property consultant, had been asked by Mr Sharer to look for a property that could be developed in a way similar to the Briggins House Hotel in February 1985. He also happened to have been retained by Norfolk to search for suitable hotels for that group. On visiting the property for the purpose of Mr Sharer's business, in the course of conversation Mr Meyers asked if he could make an appointment for his clients (Norfolk) to view, and the first defendant agreed. This led to the ultimate sale.

There was, it is argued, an introduction here by an agent, be it not the defendants', and therefore the transaction was not covered by the words of clause 8. Neither does it appear to be within the relevant clause which makes provision for commission to be paid upon a sale G
to a purchaser introduced by "agents appointed by the [defendant] to sell his property". It is argued that although this appears to be a "loophole" in the drafting, the construction should be *contra proferentem*, the wording is ambiguous and the right to commission on the facts of this case is not established.

I reject this argument. If clause 8 is read in isolation there is some force in it, but in my judgment when read as it should be, together with clause 7, the meaning is clear — namely that "other agents" refers simply to other agents appointed by the vendor to sell the property in question. The words "other agents" clearly have the same meaning in clause 8 as in the penultimate sentence of clause 7: "If this is contravened and a person is introduced by other agents", referring to agents instructed to sell the property as described in the H
first half of that sentence. That meaning is plainly carried on into the following clause and the two should be read together.

In my judgment clause 8 is sufficiently clear and unambiguous to be relied upon in the sense that the plaintiffs claim.

Further, I find on the evidence that "sole selling agency agreements", of which clauses 7 and 8 would form natural and usual terms, are, in that part of the estate agency business concerned with the sale of hotels as running concerns, commonly, although not invariably, used. Practice varies, but agreements of this kind are by no means unusual in this class of transaction for the reasons I have already outlined.

This type of agency is, of course, more onerous for the vendor than J
straightforward or sole agencies, as he may become liable for commission even if he sells on his own, but this is not an unreasonable compensation, provided it is kept to a limited period, for the agent who, to protect secrecy, is restricted from going all out to make a sale.

Of course a sole selling agreement is one of the three basic types of agency agreement and it is clearly necessary for the agent when entering into the contract to bring the nature of the agreement he is insisting upon fairly to the attention of the vendor.

I am satisfied that this did happen in this case. The nature of the agency was brought fairly to the vendor's notice. Mr Nairn's letter of October 23 followed discussions in which the way he would handle a sale to protect confidentiality, including his having the exclusive rights to sell, was made clear to Mr Sharer. The letter directed his K
attention to the terms and conditions on the reverse side where they were clearly and fairly set out in very legible print. Anybody turning the page could not avoid seeing that these were the terms of a "sole selling agreement".

When considering what steps an agent should take, all the circumstances of the particular case have to be considered, and I here bear in mind not only the discussions which preceded the letter but that Mr Sharer had time to consider the terms — this, for example, is a long way from the facts of *Interfoto Ltd* v *Stiletto Ltd*, both as to the circumstances and as to the unreasonableness of the relevant terms. Further, the terms were worded in a straightforward manner.

Indeed, in reality it is not the defendant's case that he was misled at L
the point of contract or was in some way lulled into a false sense of security about the terms: his case is that he did see that these terms were the terms of a sole selling agency but, in the telephone call that followed, expressly refused to be bound by them, insisting on the retention of his right to employ other agents. As I have indicated, I do not accept his evidence about that conversation.

For these reasons I am satisfied that the defendants were bound by the terms of clause 8.

I now turn to clause 1. Again it is argued that this clause is uncertain, and in any event was unusual as well as onerous, requiring particular attention to be drawn to it.

The wording which it is said is ambiguous is:

> if we introduce . . . a person who agrees (whether or not subject to contract) to M
> purchase the property at the price specified overleaf or such other price as the owner is willing to accept [the commission becomes payable].

What, it is asked, is meant by the word "agrees"? There may be various stages of agreement, in outline, in principle, in detail: at which point, if an agreement is to be subject to contract, does the agent become entitled to his commission?

This is a particularly English problem, set as it is upon the background of "offers subject to contract" which the drafting of the clause has sought to provide for. In my judgment the meaning nevertheless is clear. The wording implies that, apart from the need to reduce the terms into a formal contract, the parties had reached full agreement. If it were not for the formality of a written agreement the

A agreement would be sufficiently complete to be binding upon them in law. I accept defendant's counsel's wording of the appropriate test: namely, "a complete and certain agreement which would be legally binding but for the 'subject to contract' condition". Clearly, in the reduction of the terms into a formal contract details may still require negotiation, but these essentially would be of form and not of substance: or using a phrase employed by witnesses in the course of the case, simply dotting the i's and crossing the t's. A central question in this case is whether such an agreement was ever in fact reached, and I return to this issue a little later.

A provision of this kind is, nevertheless, unusual. Commission is usually payable upon the completion of a sale, and I refer again to *Jaques* v *Lloyd D George & Partners Ltd*. I accept that if clause 1 is to
B be relied upon the plaintiffs do have to show that they brought it fairly to the defendant's attention. I pause to note, however, that although more onerous than a "payment on completion" provision it is not more onerous than a term requiring payment upon the introduction of a person who is ready, willing and able to purchase. There must be a full agreement and there is in another clause protection for the vendor if the purchaser withdraws in the last phase of the transaction.

Be that as it may, the attention of the vendor must be fairly drawn to the term, having regard to its particular nature.

Was attention therefore fairly drawn to it? It was not specifically signposted in the terms and conditions, but in all the circumstances I
C conclude on the facts of this case that reasonable steps were taken. The condition is clearly and fairly set out at the beginning of the terms. Mr Sharer had time to consider the terms at his home and I again refer to the general background I have already outlined. I am satisfied that this clause also was binding.

My attention has also been drawn to the term (not less than three months) and the requirement of three weeks' notice in writing to terminate the agreement, my attention being drawn in connection with both clause 8 and clause 1. On the evidence I am satisfied that that term was not unusual or excessive for this class of agreement and that the provision for three weeks' notice in writing was reasonable. It was, in any event, always open to the vendor to withdraw the property from the market without notice on payment of the agent's
D out-of-pocket expenses.

I now turn to the next main issue. Do the plaintiffs prove that under the terms of their agency agreement they are entitled to commission either for the introduction of the Levys or upon the sale to Norfolk?

I begin with the Levys transaction. The plaintiffs have not satisfied me that they are entitled to commission under this head of claim. They have not shown that the Levys "agreed to purchase" the hotel within the meaning of clause 1.

Their case essentially is that there was such an agreement reached by October 30 upon the price the Levys had first offered, which was accepted by the defendant. The parties were then ready to reduce the agreement into a formal contract.

E The difficulty that the plaintiffs face is that, although in the negotiations that took place between the Levys and the defendant this point was indeed reached at that stage, the negotiations were thereafter almost immediately reopened so that the initial intention of the parties was never taken further.

In my judgment the negotiations which took place between October and the following May have to be seen as one continuing transaction or set of negotiations, and it would be wholly artificial to view them, as I am asked to, as two transactions, one of which ended in early December 1984 and the other in early May 1985. I accept that the initiative to reopen the negotiations came from Mr Sharer, but reopened they were and both parties from then on put a series of new offers and counter offers. In the end they were unable to agree. That
F is the reality and on this ground alone I reject this head of claim.

I would also, however, hold that if the deal reached on October 30 1984 was viewed in isolation it was not so complete as to be an agreement within the meaning of clause 1. This was a complex transaction involving not only the sale of the property but the business as well as the employment of the defendant for a two-year period, and the promise of capital for works which it was hoped would be carried out during that time once planning permission was G
obtained. Although the main principles had been agreed, there were still matters outstanding, such as the precise terms upon which the defendant was to stay on, how the capital which was to be provided for the works was to be handled, and the extent of the warranties — as well as other smaller points left open. These were not merely matters of detail, they were substantial issues, not just questions of form. Although there was the euphoric atmosphere at the time that a "deal had been struck", important terms still remained outstanding. The agreement was not complete.

The plaintiffs do not, therefore, succeed under clause 1. Had there been such an agreement I indicate that I would not have held that either in December or May the Levys withdrew from the transaction within the meaning of that phrase in the terms and conditions. On H
both occasions the negotiations were either reopened or called off by mutual consent.

Finally I return to the Norfolk sale.

There is on this part of the claim a further issue of fact to be resolved upon which turns the question: was the agreement between the parties terminated before that sale took place?

As I have noted, there is a dispute on the evidence between the two main witnesses as to whether Mr Sharer called off or suspended the agreement by telephone in late April or early May 1985. It is not suggested that written notice was given, but that Mr Nairn accepted a withdrawal of instructions — mutual consent, it is argued, was to be implied. Alternatively the property was simply withdrawn from the J
market by Mr Sharer.

I have no hesitation in accepting Mr Nairn's denial that there was any such telephone call. It is clear that in April — probably late April — he was still concerned with the transaction. There is, as I have noted, an undated note by him of a discussion with the Levys' agent about trading figures, and I note that in any event Mr Sharer himself could not have known until, at the earliest, May 3 that the transaction was off. I think it is inconceivable that if Mr Nairn had been told it was off by Mr Sharer or that it had been taken off the market, he would not have recorded this or reacted in some clear way to the new situation.

I was troubled during the hearing by Mr Nairn's apparent failure between May and September to press inquiries as to what was the K
state of the transaction, but I accept that in the brief discussions during this period with Mr Sharer he was given assurances that matters were progressing, be it slowly, and that the time factor was not in this context significant.

If Mr Sharer's affidavit sworn in July 1985 was to be believed, he told Mr Nairn in that conversation that he had taken the hotel off the market. As can now be seen following the history of discovery in this matter, upon which I have already commented, that would have been patently untrue and could only have been said at the time to mislead the plaintiffs. I think the truth is that no such telephone conversation took place but that the defendant did in the later brief meetings keep the true position from Mr Nairn. I am quite sure that when sitting in his car in a traffic jam in early October 1985 he, reading the *Financial* L
Times, for the first time became aware of the situation and that the sale to Norfolk took him completely by surprise. I do not for a moment think he would have sent his invoice unless he genuinely thought the agreement was still effective. I repeat, he gave the impression of being straightforward in his evidence and I accept that his response at that point was genuine for the very simple reason that the defendant had never given him notice or led him to believe that the Levy transaction was not still in hand.

The agreement was, therefore, I find still binding on the defendants and commission became payable to the plaintiffs on the sale to Norfolk. Norfolk was one of the three nominated potential purchasers in respect of whom commission was to be paid at 50% of M
the agreed scale and the calculation will be made accordingly. There will therefore be judgment for the plaintiffs in the sum of £58,865.63, plus the contractual interest from the date of completion.

Subject to the award to them of the costs of an interlocutory application, the defendants were ordered to pay five-sixths of the plaintiffs' costs.

HOUSING ACTS

A

Court of Appeal

December 9 1987

(Before Sir John DONALDSON MR, Lord Justice NICHOLLS and Mr Justice CAULFIELD)

R v WESTMINSTER CITY COUNCIL, EX PARTE HAZAN

Estates Gazette March 5 1988

B [1988] 09 EG 57-61

Housing improvement grant — Housing Act 1974, section 75(6) (repealed by Housing (Consequential Provisions) Act 1985, but applicable in the circumstances of this case) — "Certified date" after completion of works — Local authority's appeal from decision of McCowan J in favour of applicant for judicial review — Proceedings by way of judicial review had been taken after a lengthy delay by the authority in paying the final instalment of £1,993 in respect of a grant of £19,928 for improvement works — The "protracted and tortuous history" is summarised in the judgment of Nicholls
C **LJ — The appeal, however, is concerned only with a short point of construction in regard to the meaning of "the certified date" in section 75(6) — This is defined as "the date certified by the local authority by whom the application was approved as the date on which the dwelling first becomes fit for occupation after the completion of the relevant works to the satisfaction of the local authority" — The works in question were physically completed by November 30 1982, but clearance by various departments of the authority, such as environmental health, planning and district surveyor, had not been given before May 12 1983 — It was submitted by the**
D **authority that "the certified date" could not be before May 12 1983 as the earliest date on which the authority were satisfied in regard to the works — Held, affirming the decision of McCowan J, that the correct date was November 30 1982 — The material date is the completion of the relevant works to a particular standard, namely, the satisfaction of the local authority, not the date when the last of the departmental confirmations of this satisfaction is received — In the normal way, it would be hoped, the interval between these two dates will be short — Appeal dismissed**

No cases are referred to in this report.

E This was an appeal by Westminster City Council from the decision of McCowan J granting an application by James Hazan for judicial review following difficulties in regard to payment of an improvement grant by the council.

Eldred Tabachnik QC and Miss Elisabeth Laing (instructed by G M Ives, City Solicitor, Westminster City Council) appeared on behalf of the appellants; the respondent, James Hazan, appeared in person.

Giving the judgment of the court at the invitation of the Master of the Rolls, NICHOLLS LJ said: This is the judgment of the court on an appeal by the City of Westminster against a decision of McCowan J given on June 12 1987 on an application for judicial review made by Mr James Hazan. The application relates to a housing improvement

F

grant, and this appeal raises a short point of construction on a section of the Housing Act 1974.

Mr Hazan bought 9 Thorngate Road, London W9, in July 1981. It consisted of one residential unit, in a state of disrepair. He decided to convert the property into two self-contained units, a basement flat and a separate flat above. He applied for an improvement grant on August 28 1981, and his application was approved by the City of Westminster, to which we shall refer as "the council", by letter dated March 23 1982. The council agreed to pay Mr Hazan by way of grant the sum of £19,928, subject to certain conditions, on completion of the relevant works to the council's satisfaction.

The works were put in hand, and between March and November G
1982 the council paid Mr Hazan five instalments on account of the total grant, which left outstanding as the final instalment the sum of £1,993. On November 4 1982 Mr Lucas, senior grants officer of the council, inspected the property. He recorded that the works were 90% complete.

On the following day the council's director of housing circulated a memorandum to the director of environmental health. He stated that the alterations and improvements to the property were virtually complete, and he inquired whether the works had been completed to the satisfaction and requirements of the director of environmental health under the Public Health Acts. The memorandum sought similar confirmation from the city planning officer, regarding means H
of fire escape, and from the district surveyor, regarding the requirements of the London Building Acts and byelaws. The confirmation sought was forthcoming from the director of environmental health on December 7 1982, from the district surveyor on February 2 1983, and from the city planning officer on May 12 1983. It was not until this latter date that the council was satisfied as to the completion of the relevant works, although in fact the works were all completed by Mr Hazan before the end of November 1982.

This was only the beginning of Mr Hazan's difficulties. For the next four years efforts by Mr Hazan to obtain payment of the final instalment came to nought. In particular the council contended that the grant by Mr Hazan of a long lease of the basement at a premium in J
January 1983 was a breach of one of the grant conditions, so that, far from Mr Hazan being entitled to payment of the final instalment, he was under an obligation to repay all the money already paid to him. Thus, when Mr Hazan brought proceedings against the council in April 1985 in the Westminster County Court for payment of the final instalment of £1,993, he was met with a defence and also with a counterclaim by the council seeking repayment from him of £17,935, with interest amounting to some £4,000.

We need not pursue the rights and wrongs of that particular issue, because subsequently the council abandoned its claim for repayment and accepted that Mr Hazan was entitled to be paid the final instalment of £1,993, with interest. What happened was this. In K
August 1985, at a meeting between the parties, representatives of the council said they preferred to have the matter dealt with by way of judicial review, and they proposed that the county court action should remain stayed pending the outcome of judicial review proceedings. This course was accepted by Mr Hazan. Accordingly, following leave granted by Webster J on November 14 1985, Mr Hazan applied for, first, an order of mandamus requiring the council to nominate a date on or about November 30 1982 as the certified date within section 75(6) of the Housing Act 1974 in respect of the basement residential unit at 9 Thorngate Road; second, a declaration that £1,993 of the improvement grant remained due and owing to him by the council and, third, an order of certiorari quashing the decision of the council to recover grant money already paid to him.

A One week before the application was due to be heard the council wrote to Mr Hazan's solicitors on June 5 1987:

> We are writing to you to make clear our position in relation to the hearing of this matter listed for June 12 1987.
> 1 We are prepared to undertake to consider properly the question of certifying a date for the purpose of section 75(6) of the Housing Act 1974 and to certify such a date according to law, in relation to the basement flat at 9 Thorngate Road, London, W9.
> 2 We are prepared to accept that the remainder of the grant monies, £1,993, offered to the applicant will be paid to him.
> 3 We are prepared to revoke any decision we have made to recover the grant monies already paid to the applicant together with interest thereof.
> Since we are conceding all the relief sought in your form 86A to which you are entitled we hope it will not be necessary to trouble the Court with [a]
> B substantive hearing. So far as costs are concerned we will pay the applicant's costs to be taxed if not agreed down to May 7 1987. We are content that the question of costs thereafter be decided by the learned judge, if that question cannot be agreed between us.

Having regard to the protracted and tortuous history of this matter Mr Hazan was understandably suspicious of the cryptic terms of para 1 of this letter. So he did not accept this proposal, and the application came before McCowan J on June 12. Before the judge there was no dispute on paras 2 and 3 of this letter: the council accepted that Mr Hazan was entitled to payment of the final instalment and that he was not obliged to make any repayment. Issue was joined on para 1, which goes to the relief sought in the application by way of
C mandamus. For Mr Hazan it was submitted, and the judge accepted, that on the evidence the relevant works in the basement flat had been completed by November 30 1982 at the latest. The judge held that it would be wholly perverse for the council to come to any different conclusion. The council submitted that, even so, the "certified date" within section 75(6) could not have been before May 12 1983 as the earliest date on which the council was satisfied regarding the works. This raised a question of construction of that section. The judge rejected the council's submission and held that on the virtually undisputed facts and on his interpretation of the section, November 30 1982 was the date which the council was obliged to certify as the certified date, and accordingly he ordered the council to nominate
D that date as the certified date in respect of the basement flat.

From that order the council has now appealed, regarding this decision as raising an important question of principle. The council is concerned to establish that the certified date, contrary to the judge's construction of the legislation, is not geared to the date of physical completion of the works but is geared to the date on which the council is satisfied as to the completion of the works.

We turn to the legislation. Part VII of the Housing Act 1974, as amended by section 107 of and Schedule 12 to the Housing Act 1980, contained the material provisions at the time of Mr Hazan's application for a housing improvement grant and the council's consideration of that application. Part VII made extensive provision empowering local authorities to make grants for the provision,
E improvement and repair of buildings. Part VII of the 1974 Act has now been repealed by the Housing (Consequential Provisions) Act 1985, but it has been re-enacted, with amendments immaterial for present purposes, in Part XV of the Housing Act 1985.

By section 56 of the 1974 Act, local authorities are empowered to make four different kinds of grant, including improvement grants. Section 61 concerns improvement grants and provides, by subsection (1), that a local authority shall pay an improvement grant if an application for such a grant has been approved and if the conditions for the payment of the grant have been fulfilled. Section 61(2) provides that an application for an improvement grant shall not be approved unless the local authority is satisfied that, on completion of the relevant works, the dwelling to which the application relates will
F attain the required standard. The "relevant works" are the works in respect of which the grant is sought (section 57(2)). The "required standard" is defined in section 61(3). The authority may in certain circumstances reduce the "required standard": see section 61(4) and (5).

"The certified date" has several important functions in the scheme of the legislation. We mention some of them. Section 73, the first section in a fasciculus of sections concerned with conditions attached to a grant, imposes conditions as to the occupation of the dwelling to which a grant relates for a period of five years beginning with the certified date. Section 75 provides for a grant condition to be binding on owners of the dwelling during a period of five years or, in the case of a dwelling in a housing action area, during a period of seven years
beginning, in each case, on the certified date. In the event of the G
breach of a grant condition at a time when it is binding on the owner, the amount of the grant is repayable with compound interest from the certified date: section 76. Again, under section 60 a certificate of future occupation covers a period of five years beginning with the certified date. Further, by virtue of section 81(2), if the applicant for a grant ceases before the certified date to have an interest in the land as required by section 57(3), he ceases to be entitled to a grant and he must repay to the local authority any instalments of grant already paid to him.

Thus the certified date is of importance, both to the applicant for a grant and to the local authority. "The certified date" is defined in section 75(6) as follows:

H

> In this Part of this Act "the certified date", in relation to a dwelling in respect of which an application for a grant has been approved, means the date certified by the local authority by whom the application was approved as the date on which the dwelling first becomes fit for occupation after the completion of the relevant works to the satisfaction of the local authority.

A definition to the same effect appears in section 499 in the Act currently in force, the Housing Act 1985.

For the council Mr Tabachnik submitted that, on the express wording of section 75(6), a duty to certify the certified date cannot arise until the date when the following three conditions are first all satisfied: (a) the dwelling is fit for occupation, (b) the works are complete in fact and (c) the works are complete to the satisfaction of
the local authority. Condition (a) can be fulfilled only after J
conditions (b) and (c) are fulfilled. On the facts in the present case that state of affairs did not exist until May 12 1983 and, accordingly, until then the council cannot have had a duty to certify the certified date.

This submission raises two closely related points. The first is that, on this construction of the definition, the words "after the completion of the relevant works to the satisfaction of the local authority" contain two separate conditions. We cannot accept this. On our reading of them these words predicate one event, not two events. The one event is the completion of the relevant works to a particular *standard*, namely to the satisfaction of the local authority.
That event occurs when the relevant works are physically completed K
to that standard. That the works as completed have attained that standard will become known only when the local authority has inspected the finished works and approved them. But when the authority has made that inspection and approved the finished works, the date of completion of the works to the required standard will be the date on which the applicant ceased from his labours, not the (later) date on which the local authority approved of what the applicant had done.

In the normal course of things the interval between these two dates will, one would like to think, be a short one. One would have hoped that a local authority would be able to inspect the finished work within a very few days after being notified by the applicant of the
completion of the relevant works and express its satisfaction or L
dissatisfaction with those works more or less at once. In which event there will, in most cases, be no significant practical consequence flowing from the use of one date rather than the other in applying the definition of the certified date. But the present case shows, in stark form, that this will not always be so. Here the works were completed by November 30, but the last of the departmental approvals which the council considered it necessary to obtain before it could be satisfied as to the completion of the works was not forthcoming until the following May 12, which was nearly six months later.

This leads us to our second comment. It is implicit, if not explicit, in Mr Tabachnik's submission that the date to be certified by the local authority can never be earlier than the day on which the local
authority is first in a position to give the certificate, namely, the day M
on which the authority reaches its overall conclusion on whether the works have been completed to its satisfaction and the dwelling has become fit for occupation. We are unable to accept that construction of the language of the definition in section 75(6). If the works are physically completed on day 1 and inspected by the local authority on day 2, but it is not until day 180 that the last of the relevant departmental confirmations regarding completion and fitness for occupation is forthcoming, the date which the local authority is then required to certify under section 75(6) is, in our view, day 1, not day 180.

We recognise that with the construction of section 75(6) which we prefer there is the possibility of dispute over when the works were

A physically completed prior to their inspection on behalf of the authority. However, we do not think that this possibility is sufficient to cast serious doubt on the correctness of this construction of the subsection. An applicant for a grant can be expected, in his own interests, to notify the authority of completion of the works at the earliest opportunity. So that if an applicant were to claim that the works had been completed at some considerably earlier date the local authority would normally be justified, in the absence of a cogent explanation for the delay, in declining to accept such a claim at its face value. The alternative construction, contended for by the council, would have the effect that even though, to the knowledge of all concerned, the works were physically completed by a particular date, the date to be certified when all the necessary administrative B steps have been taken will not be that date but will be a date which, if one can judge from what occurred in the present case, may be many months later. We have difficulty in supposing that Parliament can have intended such a result.

Mr Tabachnik placed some reliance on the condition for payment of a grant set out in section 82(5). Under that subsection payment of a grant is conditional upon the works "being executed to the satisfaction of the local authority". It was submitted that the phrase "to the satisfaction of the local authority" has a temporal connotation: it must be looking to a point of time later than completion of the works, because the authority is not obliged to pay the grant until it is satisfied. The language of the crucial words in the C definition in section 75(6) ("completion of the works to the satisfaction of the local authority") is essentiallly the same, and that subsection ought to be construed similarly, because otherwise the certified date might precede the date on which the payment condition D in section 82(5) is fulfilled, which would be an extraordinary position.

In our view the fallacy in this argument lies in the premise. The premise is the proper interpretation of section 82(5). This raises the same point of construction as section 75(6). In our view, as with section 75(6), so with section 82(5): the phrase "to the satisfaction of the local authority" does not have temporal significance. In each case the phrase is stating a standard to which the relevant works must be completed or executed. The payment condition in section 82(5) is fulfilled as soon as the works are physically executed to that standard. The local authority's obligation to pay arises accordingly. But, plainly, a local authority must have a reasonable opportunity to check that the works have been executed to the required standard. E Provided the authority makes the payment with reasonable promptness, there will be no question of the authority being in default or of interest being payable. What is reasonable will depend on all the circumstances (but, we add in parentheses, we find it difficult to think of circumstances where a delay of the order of six months could be reasonable). On this construction of section 82(5) there will be no discord between the certified date and the date on which the payment condition in section 82(5) is satisfied.

For these reasons we are unable to accept the council's submissions on the construction of section 75(6). In our view the judge reached the correct conclusion. We dismiss this appeal.

F *The appeal was dismissed with costs; costs assessed at £150; application for leave to appeal to the House of Lords was refused.*

Court of Appeal

November 18 1987

(Before Lord Justice PURCHAS and Lord Justice NOURSE)

EVANS v CLAYHOPE PROPERTIES LTD

Estates Gazette January 23 1988

[1988] 03 EG 95-96

Landlord and tenant — Receiver and manager appointed by the court to collect rents and manage a block of flats in poor repair on the *Hart* v *Emelkirk* basis — Main litigation proceeding between a tenant of one of the flats and the landlords — Whether the court had power, in an interlocutory application, before the rights of the parties had been determined at the trial, to order the landlords to meet a deficit in the moneys coming into the receiver's hands for the payment of his expenditure or remuneration — The income received by him from the rents paid by long leaseholders and protected tenants was wholly inadequate to meet the expenses of management — Appeal from a decision of Vinelott J rejecting the application and holding that the court had no jurisdiction to indemnify a receiver save to the extent of the assets of which he was put in possession by the order of the court — Held by the Court of Appeal that Vinelott J had correctly applied the principle of *Boehm* v *Goodall* that a receiver and manager appointed by the court is not the agent of the parties nor a trustee for them and they have no control over his expenditure — It could be a hardship to the parties to an action if they were to be made personally liable for expenses incurred by receivers and managers over which they had no control — The court emphasised that the present matter was, however, interlocutory — They declined to express a view as to the trial judge's power to make an order for costs covering the receiver's remuneration or expenditure; that was a question which could only properly be decided when it arose — Appeal dismissed

The following cases are referred to in this report.

Boehm v *Goodall* [1911] 1 Ch 155
Burt, Boulton & Hayward v *Bull* [1895] 1 QB 276
Department of Health and Social Security v *Envoy Farmers Ltd* [1976] 1 WLR 1018; [1976] 2 All ER 173
Flowers & Co, Re [1897] 1 QB 14
Gardner v *London, Chatham & Dover Railway Co* (1867) LR 2 Ch App 201
Hart v *Emelkirk Ltd* [1983] 1 WLR 1289; [1983] 3 All ER 15; (1982) 267 EG 946
R v *Lambeth London Borough, ex parte Clayhope Properties Ltd* [1987] 3 WLR 854; [1987] 3 All ER 545; [1987] 2 EGLR 28; (1987) 283 EG 739, CA

This was an appeal against the decision of Vinelott J refusing an interlocutory application by Richard Denis Collins, the receiver and manager of Dover Mansions, Canterbury Crescent, Brixton, London SW9, for an order under Ord 30 r 3 of the RSC in regard to the payment of his remuneration and expenses. The receiver had been appointed by the court to receive rents and profits under the leases of flats in the block and to manage the properties. The main action was between Rudolph Bayfield Evans, a tenant of a flat in the block, as plaintiff, against the landlords, Clayhope Properties Ltd. The action concerned repairs to the flats. The tenant appealed against Vinelott J's interlocutory decision.

Patrick Ground QC and Anthony Padman (instructed by Zelin Bale) appeared on behalf of the appellant; Roger Cooke (instructed by Bernstein & Co) represented the respondents.

Giving judgment, NOURSE LJ said: This appeal raises a short point of principle as to the court's power, before the rights of the parties have been determined at trial, to order one of them to meet a deficit in the sums coming into the hands of a court appointed receiver and manager for payment of his expenditure or remuneration. Vinelott J, relying on the decision of Warrington J in *Boehm* v *Goodall* [1911] 1 Ch 155, has held *that there is no such power and the question must now be determined by this court.

In *Hart* v *Emelkirk Ltd* [1983] 1 WLR 1289 Goulding J held that the court has power to make an interlocutory order appointing a receiver and manager in a case where (1) a block of flats is held on leases at low rents containing covenants by the landlord to repair, with provisions for the tenants to bear the cost of the repairs by way of service charge, and (2) the landlord, for want of repair, allows the property to deteriorate. That decision, although novel at the time, was an application of the court's jurisdiction to appoint a manager in order to preserve property which is affected by the action. The dispute does not centre, as it usually does, on the ownership of the property but on the liability to preserve it. In that respect the decision ought no doubt to be regarded as an extension of those which had preceded it, although one which is clearly beneficial in the conditions of many modern residential developments.

The order for the appointment of the receiver and manager in the present case was modelled on the order made in *Hart* v *Emelkirk Ltd*, but, as Vinelott J pointed out, there is one very significant difference between the two cases. In *Hart* v *Emelkirk Ltd* the service charge provisions enabled the landlord, and thus the receiver and manager, to recover from the tenants estimated contributions towards the cost of the repairs in advance of their being carried out. In the present case the cost of the repairs and of the attendant expenses is not recoverable from the tenants unless and until it has been incurred in pursuance of the landlord's obligations. Accordingly, the only sums which have come into the hands of the receiver and manager are the rents, negligible in amount, which are wholly insufficient to fund his remuneration and expenses to date, let alone the cost of the necessary repairs. Unless this appeal succeeds, so that the landlord can be ordered to pay what has to be paid, the appointment of the receiver and manager will have been of no practical value.

The facts of the present case are fully stated in the judgment of Vinelott J, which is reported at [1987] 1 WLR 225†, and I do not propose to repeat them. I should emphasise that in its defence the landlord denies that at the date of the issue of the writ on July 12 1983 it was in breach of any express or implied covenant to keep the structure and common parts of the block of flats in good repair, and that its liability in that respect cannot be determined before trial. I should also add, in order to bring matters up to date, that the landlord's appeal to this court in the proceedings for an order by way of judicial review directing the local authority to make repair grants was dismissed on June 11 1987: see *R* v *London Borough of Lambeth, ex p Clayhope Properties Ltd* [1987] 3 WLR 854. Leave to appeal was refused, and a petition for leave has now been dismissed by the Appeal Committee of the House of Lords.

---

*Editor's note: [1987] 1 WLR 225; see also [1987] 1 EGLR 67.

†Editor's note: See also [1987] 1 EGLR 67; (1987) 282 EG 862.

A *Boehm* v *Goodall* was a case of a partnership action in which the court had made, by consent, a final order for dissolution, together with orders for the usual inquiry and accounts, an order for sale of the business as a going concern, and also an order for the appointment of an individual as receiver and manager of the business until sale. Upon completion of the receiver and manager's duties there was a certified balance owing to him on his account, but no assets of the partnership available for its payment. The receiver and manager issued a summons in the action asking that the partners (other than one who had become bankrupt) might be ordered to pay to him the balance thus owing.

The principal argument advanced in support of the claim was that, since the order for his appointment had been made by consent, the
B receiver and manager must be taken to have undertaken the management of the partnership business and to have made expenditure and incurred liabilities at the request of the partners and under an implied promise on their part to repay him. His position was said to be analogous to that of a trustee, who was not confined to his indemnity out of the trust estate, but was entitled to be indemnified by the beneficiaries personally. In rejecting that argument, Warrington J referred to authority, in particular to passages from the judgments of Cairns LJ in *Gardner* v *London, Chatham & Dover Railway Co* (1867) LR 2 Ch App 201 at p 211 and Lord Esher MR in *Burt, Boulton & Hayward* v *Bull* [1895] 1 QB 276 at pp 279 and 280. (He might also have referred to the subsequent observations of Lord
C Esher MR to the same effect in *Re Flowers & Co* [1897] 1 QB 14 at p 15.) What Cairns LJ said was this:

> When the court appoints a manager of a business or undertaking, it in effect assumes the management into its own hands; for the manager is the servant or officer of the court, and upon any question arising as to the character or details of the management, it is the court that must direct and decide. The circumstance that in this particular case the persons appointed were previously the managers employed by the company is immaterial. When appointed by the court they are responsible to the court, and no orders of the company, or of the directors, can interfere with this responsibility.

Warrington J in *Boehm* v *Goodall* said that he could not come to the conclusion that the principles of the cases with reference to trustees
D applied to the case of a receiver and manager appointed by the court without running counter to the decisions in all the cases relating to those acting in the latter capacity. At [1911] 1 Ch 161 he continued:

> Such a receiver and manager is not the agent of the parties, he is not a trustee for them, and they cannot control him. He may, as far as they are concerned, incur expenses or liabilities without their having a say in the matter. I think it is of the utmost importance that receivers and managers in this position should know that they must look for their indemnity to the assets which are under the control of the court. The court itself cannot indemnify receivers but it can, and will, do so out of the assets, so far as they extend, for expenses properly incurred; but it cannot go further. It would be an extreme hardship in most cases to parties to an action if they were to be held personally liable for expenses incurred by receivers and managers over which they have no control.

E He then went on to say that it made no difference that the order there was made by consent and also to reject an alternative argument based on subrogation.

Boehm v *Goodall* was a decision based on statements of principle of high authority. In my judgment it was correctly decided and it applies to this case. Moreover, the decision is one of jurisdiction and, unless any part of the receiver and manager's remuneration and expenditure can be treated as "costs" (see below), no question of discretion arises. Mr Ground, for the plaintiff, has sought to explain or distinguish it in various ways, but I agree with Vinelott J that none of his arguments can avail him, at any rate at this stage of the proceedings.

F The most formidable of Mr Ground's arguments was to this effect. Relying on sections 37 and 51 of the Supreme Court Act 1981 and Ord 30 rr 3 and 8 and Ord 62 r 1 of the Rules of the Supreme Court, he submitted that the court has power in an appropriate case to order a party to pay the costs of a receiver and manager. He relied in particular on the power in section 51(1) to award costs "of and incidental to all proceedings" and the definition of "costs" in Ord 62 r 1(4) as including "fees, charges, disbursements, expenses and remuneration". He said that that definition must, at the least, cover the remuneration of a receiver and manager appointed by the court. He argued that his remuneration is incidental to the proceedings in the sense that his duties are performed in "subordinate conjunction with" the proceedings, and he referred us to the observations of Jupp J in *Department of Health and Social Security* v *Envoy Farmers Ltd* G
[1976] 1 WLR 1018 at p 1021. Finally, Mr Ground pointed to the fact that this avenue of approach was not traversed in *Boehm* v *Goodall* and, indeed, that it is not at all clear that any of the outstanding balance of the account in that case consisted of unpaid remuneration.

When this argument was put before Vinelott J on that occasion by counsel for the receiver and manager, he said at [1987] 1 WLR 225, 230D:

> . . . even if that proposition is well founded, I can see no justification for making an interim order for costs. Whether when the action is heard the court will have power to order the landlord to pay the remuneration of the receiver, either as part of the cost of complying with its repairing obligation or as part of the costs of the action, is a question on which I express no opinion. The use of the court's power to appoint a receiver to enforce a landlord's obligation to H
> repair property, is a new development which poses many novel questions. They will have to be answered in time. I think that it is undesirable that I should attempt to do more than answer the precise question raised by the application now before me.

As to that, I agree that there is no justification for making an interim order for costs of this nature before the issues in the action have been determined. Even if there were power to do so, the judge's decision was well within his discretion and could not be interfered with by this court. I have, however, given careful consideration to the question whether this court ought now to express a view as to the trial judge's power to make an order for costs covering the remuneration or expenditure. Having reflected on the matter, I have come to the conclusion that the judge's approach is as correct for us as it was for J
him. I think that this question can only properly be decided if and when it arises.

For these reasons, which are in all material respects those of Vinelott J, I am of the opinion that he came to a correct decision in this case and I would dismiss the appeal accordingly.

PURCHAS LJ agreed and did not add anything.

The appeal was dismissed with costs, not to be enforced without leave of the court. Legal aid taxation was ordered.

K

Queen's Bench Division

October 21 1987

(Before Mr Justice SCHIEMANN)

R v LONDON RENT ASSESSMENT PANEL, EX PARTE TRUSTEES OF HENRY SMITH'S CHARITY ESTATE

Estates Gazette January 23 1988

[1988] 03 EG 96-98

Landlord and Tenant Act 1985, section 29 — "Recognised tenants' association" — Provision for certificate by a member L
of a local rent assessment panel recognising an association of tenants for the purposes of the legislation relating to service charges — Application by landlords for judicial review of a certificate given by a member (in fact the president) of the London Rent Assessment Panel — A recognised tenants' association (being "an association of tenants of flats in a building") is given a role in the system of protection under the Act for tenants against unreasonable service charges — Dispute as to meaning of "building" in section 29 — The problem in the present case arose because a row of 8 terraced houses had been converted so as to form 5 blocks of flats — M
There was, however, no identity either in terms of services supplied or charges payable between the blocks — The tenants' association which had been formed covered tenants from all 5 blocks — The president of the panel treated the blocks as constituting one "building" and granted a certificate of recognition to the association — He took the view that the legislation could not have intended the creation of a number of separate associations in cases where the tenants of the individual blocks were receiving many of the same services — The landlords submitted that each of the blocks was a separate "building" for the purpose of section 29 —

They contended that the intention was that recognised tenants' associations should be formed for each block of flats with its own service charge regime — It was, in their view, not the intention that one association should operate in respect of different service charge regimes — The judge accepted the landlords' submissions — He considered that the purpose was to enable a number of tenants to exercise their rights through the medium of an association; it was consistent with this that the association should represent one service charge regime — In the present case the president should not have given the certificate and should not have refused to cancel it — The judge declined to give any ruling on the position, to which the president had drawn attention in his affidavit, where there was "some interrelation, if not identity" in the service charge regimes; that was not this case — Landlords' application granted in substance; the certificate would be cancelled by the panel under section 29(3)

No cases are referred to in this report.

This was an application for judicial review by the Trustees of Henry Smith's Charity Estate seeking to quash a certificate issued by the president of the London Rent Assessment Panel giving recognition to a tenants' association covering five blocks of flats formed by conversion of a row of terraced houses, 1 to 8 Onslow Gardens, London SW7. The applicants were the landlords, the freehold owners of the properties. The certificate had been given under section 29(1)(b) of the Landlord and Tenant Act 1985.

Robert Bailey King (instructed by Denton Hall, Burgin & Warrens) appeared on behalf of the applicants; Robert Jay (instructed by the Treasury Solicitor) represented the respondent panel.

Giving judgment, SCHIEMANN J said: Sections 18 to 30 of the Landlord and Tenant Act 1985 contain a code which regulates the power of the landlord to recover service charges in respect of a building consisting of a number of flats.

The physical situation in the present case is set out in para 3 of the affidavit of Mr S H Scott-Barrett [partner of Cluttons, chartered surveyors]. He says this:

> The Applicants are the freehold owners of a row of eight 19th century terraced houses situate at and formerly known as Nos 1, 2, 3, 4, 5, 6, 7 and 8 Onslow Gardens, London SW7.
>
> The above-mentioned houses are no longer houses as such. That is to say they are no longer divided vertically from each other nor are they in separate occupation. The reason for this is that they have been converted into flats on [various dates]. The 8 houses in question have been so converted as to form 5 "blocks" of flats. The division is as follows: (1) 1 and 2 Onslow Gardens now form the building known as "1 and 2 Onslow Gardens"; (2) 3 and 4 Onslow Gardens now form a separate building known as "3 and 4 Onslow Gardens"; (3) 5 Onslow Gardens is a separate building known as "5 Onslow Gardens"; (4) 6 and 7 Onslow Gardens now form a building known as "6 and 7 Onslow Gardens"; and (5) 8 Onslow Gardens is a separate building known as "8 Onslow Gardens".

I shall refer in consequence to the five blocks of flats but bear in mind that they have been created out of a row of houses. The situation must be typical of much of London, and no doubt elsewhere in the country. So this case is of some importance not merely to the parties but to the wider world.

Section 20 of the Act provides that where relevant costs incurred by a landlord in the carrying out of works on a building exceed a certain limit, the excess shall not be taken into account in determining the amount of the service charge, unless the tenants have been given an opportunity of making representations.

The landlord has to serve a notice describing the works to be carried out on the building, accompanied by a copy of the estimates and the notice must invite observations. The landlord must have regard to any observations received in pursuance of a notice. That notice has to be given to each of the landlord's tenants of flats within the building by whom a service charge is paid to which the costs of the proposed works are relevant.

Section 20(3)(b) also provides that:

> . . . if there is a recognised tenants' association for the building, the notice . . . shall also be given to the secretary of the association.

Section 29(1) says:

> A recognised tenants' association is an association of tenants of flats in a building which is recognised for the purposes of the provisions of this Act relating to service charges either — (a) by notice in writing given by the landlord to the secretary of the association, or (b) by a certificate of a member of the local rent assessment committee panel.

The tenants' association with which I am concerned has members from flats in each of the five blocks.

In the present case a member of the panel — indeed, I think the president — has given the appropriate certificate. He has been asked pursuant to section 29(3) to cancel the certificate on the basis that there was no power to issue the certificate in the first place, because the tenants' association concerned has members from several blocks. The request for cancellation has been refused. It is that refusal which is the subject of this application.

The legal relationships between landlord and tenant are contained in 34 leases with obligations in relation to service charges, which differ from block to block.

There is a potential difficulty peculiar to this case in respect of 5 Onslow Gardens, two of whose four tenants are required to contribute to expenditure in respect of 3 and 4. But apart from that, there is no obligation on any tenant of a flat in one block to pay a service charge in respect of work done on another one. I have not heard submissions on this little potential difficulty and I propose to disregard it in this judgment.

It is clear from the legislation that no tenant in one block has the right to be consulted in respect of any proposed charges in any other block. However, in respect of his own block he has various rights.

In section 21(1) it is provided that:

> A tenant may require the landlord in writing to supply him with a written summary of the costs incurred — (a) if the relevant accounts are made up for periods of twelve months, in the last such period ending not later than the date of the request, or (b) if the accounts are not so made up, in the period of twelve months ending with the date of the request, and which are relevant costs in relation to the service charges payable or demanded as payable in that or any other period.

So here is a provision entitling the tenant to require the landlord to give him details.

Section 22(1) says:

> This section applies where a tenant . . . has obtained such a summary as is referred to in section 21(1) . . . whether in pursuance of that section or otherwise.
>
> (2) The tenant . . . may within six months of obtaining the summary require the landlord in writing to afford him reasonable facilities — (a) for inspecting the accounts, receipts and other documents supporting the summary, and (b) for taking copies or extracts from them.

Those are the tenants' rights.

The secretary of a recognised tenants' association also has these rights, but — and this is noticeable — only if a tenant who himself has such rights consents. That appears from section 21(2) and from section 22(1) and (2).

The basic argument for the landlord is that each of these blocks is a separate building for the purposes of the code, and in particular for the purposes of section 29. The essential contention on behalf of the president of the rent panel is that there is only one building, namely nos 1 to 8.

In my judgment, the contention of the landlord is right for these reasons. First, the purpose of the provisions relating to associations is, in my judgment, to permit a number of tenants, each of whom is given rights under the provisions, to exercise those rights through the medium of an association, which can exercise those rights on behalf of the tenants concerned. Second, the building to which reference is made in section 20 is the building upon which it is proposed to carry out works. At the time when a certificate under section 20(1) is applied for, it may well not be known what works will be necessary during the next six months on what building. It seems to me to accord with the scheme of the legislation for recognised tenants' associations to be formed for each block of flats with its own service charge regime. It is really those words which are important, though of course there is nothing to prevent such associations from federating, or otherwise co-operating, one with the other.

The president of the panel recognises that there is little to be said for members of an association who are not concerned with the payment of a particular service charge having a vote on such matters. In consequence he requires in the constitution of a tenants' association which wishes to be recognised a provision limiting the right of voting on the estimates to those tenants concerned with the payment of the service charge. This mitigates the disadvantage of having in a tenants' association those operating under differing

A service charge regimes but does not help me to construe the Act or the proper approach to applying it. It becomes necessary only because of the construction put upon it by the panel president.

He says in para 4 of his affidavit:

> Since 1980, 108 certificates of recognition have been issued by the Panel. Seven of those certificates have been issued where an estate consisted of a number of separate and independent blocks for which one association had been formed. The view has been taken that it could not be the intention of the legislation to restrict the word "building" in this context to mean one single structure, because such a strict construction could lead to the absurd situation of separate tenants' associations having to be formed for each block of a number within one estate, although they all may receive many of the same services. Examples of such common services would include provision of a
> B caretaker or porter, refuse collection, hot water provided from a central plant, lighting, cleaning, and maintenance of gardens or common areas. Bearing in mind the purpose of the legislation and the limited relevance of a tenants' association, a flexible approach has been adopted to the interpretation of the word "building" in this context.

That paragraph is dealing with a situation where there is some interrelation, if not identity, of the service charge regime applicable to the tenant members of the association concerned. That situation is not before me, and I do not rule on it. In the present case there is no such interrelation, still less of service regimes, either in terms of services supplied or charges payable. In such a case it is, in my judgment, not permissible to give the certificate under section 29(1)(b) in the way it has been done, and the president of the panel
C has, in my judgment, erred in not reconsidering his refusal to cancel.

It is right that I should conclude by expressing my sympathy with him. I have not found the point easy, and I do not suppose he did.

The relief sought by the applicants was granted in substance. No formal order quashing the certificate was necessary in view of the undertaking of the president of the panel to cancel it under section 29(3) of the Landlord and Tenant Act 1985.

Court of Appeal

December 21 1987

D (Before Lord Justice FOX, Lord Justice MUSTILL and Sir George WALLER)

A G SECURITIES v VAUGHAN AND OTHERS

Estates Gazette February 13 1988

[1988] 06 EG 112-122

Landlord and tenant — Individual non-exclusive licences or joint tenancy — *Street* v *Mountford* further considered — Court of Appeal divided — Leave to appeal to House of Lords — The present appeal was by occupiers of a flat in Hornsey
E Lane, London N6, from the decision of the county court judge that they were licensees, not tenants, of the flat — The flat consisted of four bedrooms, a sitting-room, and a bathroom and wc — The four individual occupants each had separate short-term licence agreements granting a right to use the flat in common with others with a like right and expressly negativing a right to exclusive possession of any part of the flat — The licence agreements began at different dates and there were some variations in the rents charged — The four occupants concerned in the present proceedings had replaced earlier occupants when vacancies had occurred — The time came
F when the leaseholders who had granted the licences wished to let the flats on long subleases and for this reason gave the occupants of the subject flat notices to quit — This prompted the local authority, acting under section 68 of the Rent Act 1977, to apply to the rent officer for the determination of a fair rent on the basis that the occupants were in law joint tenants of the flat — The leaseholders challenged the validity of this application and sought an injunction to restrain the authority from proceeding with it — As mentioned above, the county court judge accepted the leaseholders' submission and declared that the occupants were licensees, not joint tenants — Three of the four occupants appealed to the Court of Appeal

The majority decision of the Court of Appeal is contained in G
the judgment of Fox LJ (with which Mustill LJ agreed) — After analysing *Street* v *Mountford* and the cases where the arrangements were treated there as shams, the majority applied to the facts of the present case the test that the hallmarks of a tenancy are exclusive possession at a rent for a term — The four occupants had together exclusive possession of the flat — Despite differences in the periods of occupancy, it could be said that there was a joint right of occupancy for a "term" — As regards rent (which in any case was not an essential constituent of a tenancy) the aggregate amounts payable, although different and variable, could be regarded as H
constituting a "rent" — As regards the traditional "four unities" required for the creation of a joint tenancy (possession, title, time and interest), the majority considered that they were satisfied in the present case — Accordingly, when the notices to quit were given the occupants were jointly entitled to exclusive possession of the flat for a term at a rent — Sir George Waller, in a dissenting judgment, drew attention to the considerable difficulties involved in regarding the transactions as constituting a joint tenancy — "A joint tenancy is one where the tenancy commences on the same day for all, where the term is the same for all, where the rent should not be altered without due notice to all and possibly where all J
are jointly liable for the rent" — He entertained serious doubts as to whether any one of the "four unities" had been satisfied — He would have dismissed the appeal — In accordance with the majority view the appeal was allowed, but leave to appeal to the House of Lords was granted

The following cases are referred to in this report.

Addiscombe Garden Estates Ltd v *Crabbe* [1958] 1 QB 513; [1957] 3 WLR 980; [1957] 3 All ER 563, CA
Aldrington Garages Ltd v *Fielder* (1978) 37 P&CR 461; [1978] EGD 347; 247 EG 557, CA
Ashburn Anstalt v *Arnold* [1987] 2 EGLR 71; (1987) 284 EG 1375, CA K
Errington v *Errington and Woods* [1952] 1 KB 290; [1952] 1 All ER 149, CA
Facchini v *Bryson* [1952] 1 TLR 1386, CA
Hadjiloucas v *Crean* [1987] 3 All ER 1008; [1987] 2 EGLR 60; (1987) 284 EG 927, CA
Radaich v *Smith* [1959] 101 CLR 209
Somma v *Hazelhurst* [1978] 1 WLR 1014; [1978] 2 All ER 1011; (1978) 37 P&CR 391; [1978] EGD 269; 246 EG 311, CA
Street v *Mountford* [1985] AC 809; [1985] 2 WLR 877; [1985] 2 All ER 289; [1985] 1 EGLR 128; (1985) 274 EG 821, HL
Sturolson & Co v *Weniz* [1984] EGD 524; (1984) 272 EG 326, CA

This was an appeal by Roderick Lyons, Simon Russell and Christopher Cook (the first defendant, Nigel Vaughan, did not appeal) from the decision of Judge Aron Owen, at Clerkenwell L
County Court, holding that they were licensees and not tenants of Flat 25, Linden Mansions, Hornsey Lane, London N6. Together with Nigel Vaughan and the London Borough of Haringey, they had been the defendants in an action by A G Securities (an unlimited company), the leaseholders of Linden Mansions, as plaintiffs (present respondents).

Stephen Sedley QC and David Watkinson (instructed by Bindman & Partners) appeared on behalf of the appellants; John Furber (instructed by I A Landy Laufer) represented the respondents.

Giving judgment, FOX LJ said: This is an appeal by the second to fourth defendants from an order of Judge Aron Owen declaring that the first to fourth defendants were licensees and not tenants of certain premises at Linden Mansions, Hornsey Lane, London. Linden M
Mansions is a block of residential flats. The present case is concerned with Flat 25. The flat, which is furnished, consists of six rooms. Of those rooms, four are bedrooms, one is a sitting-room and the other is a bathroom and wc. The flat, so the judge found, can, without difficulty, provide residential accommodation for four persons.

Prior to 1974, the plaintiffs, who are leaseholders of Linden Mansions, let the flats on short-term furnished tenancies. Such tenancies did not at that time have full security of tenure under the existing legislation. By the Rent Act 1974, however, residential furnished tenancies were brought into full control. The judge's finding as to the consequences of that were as follows:

> The plaintiffs took counsel's opinion and were advised to grant short-term

A licence agreements separately to individual occupiers. The plaintiffs admit frankly that the purpose of such agreements was to avoid granting tenancies falling within the Rent Acts. Counsel settled the form of agreement which he advised the plaintiffs to use and from that time the plaintiffs did use such [a form of] agreement.

The form of agreement which the plaintiffs have used and which they used in the present case is as follows:

THIS AGREEMENT is made the day of BETWEEN A G SECURITIES of 116 Stanmore Hill, Stanmore, Middlesex (hereinafter called "the Owner") of the one part and (hereinafter called "the Licensee") of the other part.

WHEREBY IT IS AGREED as follows:

1 THE Owner grants to the Licensee the right to use in common with others B who have or may from time to time be granted the like right the flat known as 25 Linden Mansions, Hornsey Lane, N6, but without the right to exclusive possession of any part of the said flat

TOGETHER with the fixtures furniture furnishings and effects now in the said Flat for six months from the day of 19 and thereafter until determined by either party giving to the other one month's notice in writing to take effect at any time.

2. THE LICENSEE agrees with the Owner as follows:

(1) To pay the sum of £ per month for the right to share in the use of the said flat such sum to be payable by equal monthly instalments on the first day of each month the first payment to be made on the day of

(2) Not to damage or cause any damage to the walls or floors of the said Flat or to the fixtures furniture furnishings and effects therein.

(3) To share the use of the said Flat peaceably with and not to impede the use of C the said Flat by such other persons not exceeding 3 in number at any one time to whom the Owner has granted or shall from time to time grant Licence to use the said Flat in common with the Licensee and not to impede the use by such other persons of the gas electricity and telephone services supplied to the Flat PROVIDED that each shares the cost of such services.

(4) If at any time there shall be less than 3 persons authorised by the Owner to use the said Flat in common with the Licensee upon reasonable notice given by the Owner to meet with any prospective licensee nominated by the Owner at the Flat to provide an opportunity to such prospective licensee to agree terms for sharing the cost of services in accordance with Clause 2(3).

(5) Not to assign this Agreement nor permit any other person except as licensed by the Owner to sleep or reside in or share occupation of the said Flat or any part of it at any time.

(6) Not to do or suffer to be done any act or thing which may be a nuisance or D cause damage or annoyance to the Owner or to the other users of the Flat or to the tenants or occupiers of any adjoining property or which may vitiate any insurance of the said building against fire or otherwise or increase the ordinary premium thereon.

3. PROVIDED as follows:

(1) If the said sum or any part thereof shall be in arrear or unpaid for at least 14 days after the same shall have become due or

(2) in the event of any breach of the Licensee of the agreements therein contained

then the Owner may by notice forthwith determine this Agreement but without prejudice to any other remedies of the owner hereunder.

4. IF the Flat becomes uninhabitable as a result of fire then the Licensee shall be entitled to terminate this Agreement forthwith.

5. THE OWNER AGREES with the Licensee to pay all general and water rates E payable in respect of the Flat but not any charges for the supply of gas or electricity or for the use of the telephone.

6. THE Licensee shall pay the sum of £ which shall be retained by the Owner until the termination of this Agreement as a deposit to secure to the Owner the performance by the Licensee of the agreements on the part of the Licensee herein contained but without prejudice to the Owner's rights under and in respect of the Agreement. At the termination of this Agreement and upon the Licensee's vacating the said Flat and subject to proper performance by the Licensee of the Licensee's obligations herein contained the Owner will refund the said deposit to the Licensee.

7. NOTWITHSTANDING Clause 1 the Licensee shall have the right at any time after the expiration of three months from the date in Clause 1 to terminate this Agreement by giving the Owner one month's notice in writing.

8. THE Owner may give notice hereunder to the Licensee by leaving a Notice addressed to the Licensee at the said Flat.

F AS WITNESS the hands of the parties hereto the day and year first before written.

Each of the defendants in the present case signed such an agreement. They signed them separately and at different times. The dates of the agreements with the various defendants and the monthly amounts paid by them were as follows:

Defendant	*Date*	*Amount*
Vaughan (1st defendant)	August 1 1982	£86.66
Russell (2nd defendant)	August 1 1984	£125.00
Lyons (3rd defendant)	March 2 1984	£99.00
Cook (4th defendant)	January 28 1985	£104.00

There was great demand for the flats which came from young single people — secretaries, nurses, students and persons working in various capacities in the City or the West End of London. G

The normal length of an agreement was six months. But it often happened that the occupier would stay on for a further six months or on a month-to-month basis. The turnover was large. From 1980 to 1985, Flat 25 was occupied as follows:

1978	*1980-81*	*1982*	*1983*	*1984*	*1985*
A Abbas			C Bolt	R Lyons	
M Clarke			M Ramsey	S Russell	
D Clarke		N Vaughan			
S Walters	J Adams				C Cook

Thus, in 1980-81 Adams replaced Walters; Vaughan (first defendant) replaced D Clarke in 1982. In 1983 Bolt replaced Abbas and Ramsey replaced M Clarke. In 1984 Lyons (second defendant) replaced Bolt and Russell (third defendant) replaced Ramsey. In 1985 H Cook (fourth defendant) replaced Adams.

The judge found the following facts regarding the system of letting operated by the plaintiffs:

(1) Each person who was accepted by the plaintiffs for a particular flat had to sign a separate agreement and pay a deposit as provided in the agreement. And it would be explained to such person that he was entering into a separate agreement with separate obligations.

(2) The flat No 25 . . . has four of its rooms used as bedrooms and the agreements relating to the residential user of this flat strictly limit the total number of occupants to four.

So this flat would be fully occupied within the terms of the agreement when it contained four residents.

(3) What happened if, for example, one of the residents left creating a J vacancy? In these circumstances Mr Saffrin [a solicitor and director of the plaintiffs] would enquire of the remaining residents whether they had someone whom they would like to fill the vacancy. Sometimes they did. In that case, the person put forward as a prospective new resident would be asked to contact Mr Saffrin who would refer that person to his co-director, Mr Laufer, for an interview.

(4) Mr Laufer, in conducting the interview, would enquire about the occupation of the applicant and, if a student, what grant was being received. He would inform the applicant about the deposit and the amount of the monthly payment. Such an applicant to fill a vacancy may have known the approximate monthly payments from having been told by the current residents. But the sums paid varied because of the ebb and flow of residents and the monthly payments were from time to time increased by modest sums to provide for rate increases and inflation. Later residents accordingly paid K more than earlier ones.

(5) Where no candidate for a vacancy was proffered by the existing occupiers, the plaintiffs would themselves find a person to fill the vacancy . . .

(6) In the case of a candidate found by the plaintiffs, Mr Saffrin would get in touch with the current residents of the flat and ask them to see this candidate who had applied to fill the vacancy.

The plaintiffs had the final decision as to what person would fill the vacancy. But, and this was obviously good estate management, the plaintiffs would not, in order to fill a vacancy, foist on the existing residents of the flat any person to whom they, the existing residents, objected.

(7) What often happened in practice, when a vacancy occurred, was that an advertisement was inserted in the London "Evening Standard" under a column in that newspaper headed "Flat Sharing".

And, as a matter of convenience, the plaintiffs would ask the existing L residents to insert the advertisement, for which the plaintiffs would pay . . .

As a result of such an advertisement, applicants would go directly to the flat, they would see the accommodation offered and would see, and be seen by, the other residents in the flat.

I was told by one of the defendants, Mr Nigel Vaughan, that the response to one such advertisement in the "Evening Standard" resulted in the phone at the flat "ringing for a week" producing, for one advertised vacancy, a hundred applicants from whom the residents took 60 names and conducted a number of interviews before choosing one particular person.

(8) The plaintiffs were anxious to ensure that the prospective incoming person should be acceptable to those already in residence and for that reason allowed them to choose the one they, the residents, thought to be the most suitable. But that person also had to be interviewed by the plaintiffs and thereafter only if the plaintiffs approved would that person be accepted and required to pay M the deposit and sign the agreement.

(9) At no time did the plaintiffs tell the incoming resident which particular bedroom he was to occupy. The internal arrangements were left entirely to the residents themselves. What usually happened in practice, when a vacancy occurred and a newcomer subsequently entered to fill the vacancy, was this: The current occupiers, or some of them, switched rooms so that the newcomer got the smallest and least desirable bedroom. The "pecking order", if I can call it that, appears to have been determined by seniority: the longest-residing occupant got the first choice of whatever particular bedroom had been vacated. The plaintiffs never concerned themselves with these internal arrangements.

I should add that the judge also found that, when vacancies occurred so that there were less than four persons living in the flat, the plaintiffs never required the remaining residents to make up the total of the payments which the plaintiffs would have received if the flat had been fully occupied. Nor did any of the residents ever expect that they would have to do so.

In April 1985, the plaintiffs decided that they wished to let all the flats on long subleases to reduce, so we were told, the burden of management.

In May 1985 the fifth defendant (the Borough of Haringey), in consequence of the service by the plaintiffs of notices to quit upon the first to fourth defendants, made an application to the rent officer under section 68 of the Rent Act 1977 for the determination of a fair rent of the flat on the basis that the first to fourth defendants were tenants of the flat and held such tenancy jointly. The plaintiffs contend that the application was unlawful and of no effect because there was no tenancy and they sought an injunction restraining the borough from proceeding with the application. The plaintiffs also sought judgments for arrears of rent against the first to fourth defendants. The judge held that the first to fourth defendants were licensees and not tenants. He made a declaration accordingly. The appeal is by the second to fourth defendants only — Mr Vaughan does not appeal.

The case turns, in the main, upon the application of principles stated by the House of Lords in *Street* v *Mountford* [1985] AC 809. In that case a written agreement between Street and Mountford granted to Mountford the right to occupy two rooms for £37 per week subject to determination on 14 days' notice. The agreement was entitled "licence agreement" and contained a declaration by Mountford that she understood that the agreement did not give her a tenancy protected by the Rent Act. Mountford and her husband went into occupation of the rooms of which they had exclusive possession. After a few months, Street sought a declaration that the occupancy was under a licence. The House of Lords, reversing the decision of the Court of Appeal, held that Mountford was a tenant.

The principles stated in *Street* v *Mountford* so far as relevant to the present case are as follows:

(1) Exclusive possession is of the first importance in deciding whether an occupier is a tenant (p 823D).

(2) Exclusive possession is not decisive because there are circumstances in which an occupier who has exclusive possession is not a tenant. Thus, he may be an owner in fee simple, a trespasser, a mortgagee in possession, an object of charity or a service occupier (p 618E).

(3) The House, as I understand it, approved (p 826F) the view of Denning LJ in *Facchini* v *Bryson* (1952) 1 TLR 1386 at p 1389 that:

> In all the cases where an occupier has been held to be a licensee there has been something in the circumstances, such as a family arrangement, an act of friendship or generosity, or such like, to negative any intention to create a tenancy.

The occupier there referred to I take to be an occupier for a term or other ascertainable period.

(4) *Street* v *Mountford* was a letting of residential accommodation at a rent for a term. Lord Templeman said at p 826E:

> Unless these three hallmarks are decisive, it really becomes impossible to distinguish a contractual tenancy from a contractual licence save by reference to the professed intention of the parties or by the judge awarding marks for drafting.

(5) The House expressly approved (p 827B-E) the language and logic of Windeyer J sitting in the High Court of Australia in *Radaich* v *Smith* (1959) 101 CLR 209 at p 222 where he said:

> What then is the fundamental right which a tenant has that distinguishes his position from that of a licensee? It is an interest in land as distinct from a personal permission to enter the land and use it for some stipulated purpose or purposes. And how is it to be ascertained whether such an interest in land has been given? By seeing whether the grantee was given a legal right of exclusive possession of the land for a term or from year to year or for a life or lives. If he was, he is a tenant. And he cannot be other than a tenant, because a legal right of exclusive possession is a tenancy and the creation of such a right is a demise. To say that a man who has, by agreement with a landlord, a right of exclusive possession of land for a term is not a tenant is simply to contradict the first proposition by the second. A right of exclusive possession is secured by the right of a lessee to maintain ejectment and, after his entry, trespass. A reservation to the landlord, either by contract or statute, of a limited right of entry, as for example to view or repair, is, of course, not inconsistent with the grant of exclusive possession. Subject to such reservations, a tenant for a term or from year to year or for a life or lives can exclude his landlord as well as strangers from the demised premises. All this is long established law: see *Cole on Ejectment* (1857) pp 72, 73, 287, 458.

(6) While the Rent Acts must not be allowed to alter or influence the construction of an agreement, the courts are required to "be astute to detect and frustrate sham devices and artificial transactions whose only object is to disguise the grant of a tenancy and to evade the Rent Acts" (p 825H).

(7) The decisions in *Somma* v *Hazelhurst* [1978] 1 WLR 1014, *Aldrington Garages Ltd* v *Fielder* (1978) 37 P&CR 461 and *Sturolson & Co* v *Weniz* (1984) 272 EG 326 (pp 825H-826A) were disapproved.

In *Somma* v *Hazelhurst* a young unmarried couple, Hazelhurst and Miss Savelli, occupied a bed-sitting room (containing two beds) for which they paid weekly sums. The landlord did not provide any services. Savelli and Hazelhurst had exclusive possession of the room. The owner required them to sign separate agreements, in the same terms. Under the agreements the room was not let to either Hazelhurst or Miss Savelli nor to Hazelhurst and Miss Savelli. Instead, the agreements required each to share the room with such persons as the landlord might from time to time nominate. The Court of Appeal held that Hazelhurst and Savelli were licensees. In *Street* v *Mountford* the House of Lords was of the opinion that they were tenants and that the form of the agreements was a sham. "The room", said Lord Templeman at p 825F, "was let and taken as residential accommodation with exclusive possession in order that [Hazelhurst] and [Miss Savelli] might live together in undisturbed quasi-connubial bliss making weekly payments. The agreements signed by [Hazelhurst] and [Miss Savelli] constituted the grant to [them] jointly for exclusive possession at a rent for a term" (p 825G).

In *Aldrington Garages* v *Fielder* the respondent entered into occupation of a flat in Highgate with a woman with whom he was living. The premises consisted of two rooms together with a kitchen, bathroom and wc. Each of the occupiers signed a separate agreement with the landlord. The agreements provided that each occupant might use the flat "in common with others who may have been granted the like right". Each agreement provided for the payment by the occupant signing it of £54.17 per month. Subsequently the woman exercised her right under her agreement to determine her "licence" and went out of occupation. The Court of Appeal held that the case was not, in substance, different from *Somma* v *Hazelhurst* and that no tenancy was created. As I read the decision in *Street* v *Mountford* (p 825H), the House of Lords regarded *Aldrington Garages* as another case of a sham transaction which was, in truth, a tenancy.

The same applies to *Sturolson & Co* v *Weniz*. In that case the defendant, his wife and a Mr May together inspected a flat which they were contemplating sharing. Mr Weniz (the defendant) and Mr May, after the visit, each signed an agreement in identical terms on the same day. The agreement provided (*inter alia*) by clause 1:

(a) that the Sharer (ie the signatory) should have the right to share the flat with such other persons as might be approved or nominated by the landord;

(b) that the Sharer should have no right to exclusive occupation of any part of the flat;

(c) that the Sharer should pay a monthly fee of £100 in consideration of the grant of the right of occupancy.

The evidence was that the parties, prior to the signature of the agreements, had been told by the agents for the landlords that all that the landlords were concerned about was the payment of £200 per month and that they could forget about clause 1, which was merely designed to get around the Rent Acts, and that they could be assured that the landlords would not put anybody into the flat. The landlords eventually served notice to quit and sought possession of the premises and arrears of payments due. By the time of the proceedings Mr May had left. It seems that a Mr Strudwick came in at some time, but the only defendant to the proceedings was Mr Weniz. In this case also the Court of Appeal held that the transaction was a licence. The House of Lords in *Street* v *Mountford* considered that that was wrong and that a tenancy was created.

I come now to the application of the law to the present case. Mr Sedley QC, for the second to fourth defendants, submits that those defendants are jointly entitled to a lease of the flat. Mr Sedley expressly disclaims any argument:

(1) that the agreements entered into by the occupiers were shams *or*

(2) that any of the defendants is or was individually entitled to exclusive possession of any part of the flat; accordingly he does not suggest that any of the defendants was entitled to exclusive possession of his bedroom with a right in common with the other three to use the common parts.

I will consider in turn the questions of exclusive possession, term and rent. First, the question of exclusive possession. Each licensee is, by clause 1 of the agreement, granted the right to use the flat in common with others who have, or may from time to time be granted, the like right. Under the provisions of clause 2(3) the number of other persons who may be authorised to use the flat is three. Each of the four occupants could thus have one room.

If one looks at the position as it was when the first to fourth defendants were all in occupation of the flat (so that it was fully occupied) it seems to me that they had, between them, exclusive possession of the flat. They were entitled to exclude anybody; the plaintiffs had no right of entry. That right to exclusive possession belonged, in my opinion, to the occupants as joint tenants. I do not see any other means of giving effect to it. I appreciate that clause 1 of the agreement expressly excludes the creation of any right in the licensee "to exclusive possession of any part of the said flat". But a joint right to exclusive possession of the *whole* flat gives no right to the individual licensee to exclusive possession of *part*. On the other hand, clause 1 confers upon the "licensee" the right to use the flat in common with the other authorised occupiers. The combination of those rights of all four occupiers constitutes the joint right to exclusive possession of the whole flat. I will return later to the question of the constituents of a joint tenancy. I do not think that this interpretation of the arrangement between the parties involves giving any forced or artificial meaning to it at all. I think it simply represents the reality of the position. Strangers have, on any view, no right to interfere with the occupants. And the agreement confers no right on the plaintiffs (nor was any right ever subsequently conferred on them) which allows the plaintiffs to interfere with the full enjoyment of the flat by the four occupants so long as the provisions of the agreement are observed. I conclude, therefore, that the element of exclusive possession which the House of Lords regarded as being of the "first importance" in deciding whether occupiers are tenants is present here.

The next matter is the term. A lease must be for some ascertainable period which, for the purposes of the present case, must be either for a fixed term or for a period continuing from month to month (or other period) until determined under some power of determination. The period of occupancy of the defendants differed. The initial six months' "term" of the first, second and third defendants had expired when the notices to quit were given. The six months' period of the fourth defendant was still running, though it may be (I am not certain as to the dates) that he could give one month's notice under clause 7.

In my view, as from the date when the fourth defendant's agreement took effect, the joint right to exclusive possession would continue until the right of occupancy of any one of the four occupants was determined under the provisions of the agreement. There would be no difficulty in deciding when that date occurred. The first to third defendants had monthly occupancies (ie continuing from month to month on either side). The fourth defendant had a six months' occupancy from January 28 1985 (determinable if he wished after three months) and subject thereto on a monthly basis.

As regards rent, the feature of this case which was not present in either *Street* v *Mountford*, *Somma* v *Hazelhurst*, *Aldrington Garages* v *Fielder* or *Sturolson & Co* v *Weniz* is that each of the occupants was paying a different monthly amount. That is certainly not a common form of rent, but even if it is not rent at all I do not think that prevents a lease coming into existence. A rent is not, in my view, essential to the creation of a valid lease. It is true that in *Street* v *Mountford* Lord Templeman refers on a number of occasions to the fact that exclusive possession of property at a rent for a term will constitute a lease. That is certainly the law. But I do not read Lord Templeman as saying that a rent is essential to the existence of a lease. Section 205(1)(xxvii) of the Law of Property Act 1925 defines "term of years absolute" as "a term of years (taking effect either in possession or in reversion whether or not at a rent) . . .". And Windeyer J in the passage in *Radaich* v *Smith* (*supra*) expressly approved by the House of Lords in *Street* v *Mountford* says:

> What then is the fundamental right which a tenant has that distinguishes his position from that of a licensee? It is an interest in land as distinct from a personal permission to enter the land and use it . . . And how is it to be ascertained whether such an interest in land has been given? By seeing whether the grantee was given a legal right of exclusive possession of the land for a term or from year to year or for a life or lives. If he was, he is a tenant.

For these reasons this court in *Ashburn Anstalt* v *J Arnold & Co* (reported in *The Independent* on October 29 1987)* concluded that a rent is not an essential constituent of a lease. No doubt the existence of consideration may be material to the determination of the question whether the parties intended to enter into legal relations at all. That, however, is not in issue here. The parties were plainly entering into commercial transactions and the occupants were paying a substantial monetary consideration. In these circumstances, I see no reason why the absence of a joint liability to pay a single rent should prevent a lease from coming into existence. A lease is an estate in the land for the duration of the lease. The parties are free to create it for such considerations as they choose.

That, however, is not the end of the matter. It seems to me that the payments in the present case do constitute a rent. If there was a grant of exclusive possession for a term to the four occupants jointly and they are each making periodic payments in respect of their occupation, it seems unreal to regard those payments as other than rent. *Halsbury's Laws of England*, 4th ed, vol 27 at para 211 states: "Rent is the recompense paid by the tenant to the landlord for the exclusive possession of corporeal hereditaments". That seems to me to describe the nature of the aggregate monthly payments made by the four occupants in this case. And not the less so because the amounts payable by the individual joint tenants vary. If a landlord chooses to grant a lease to joint tenants in consideration of sums of differing amounts paid by them individually, it seems to me that there is still a tenancy and still a rent. It would, no doubt, look better if there were a provision for the payment of a single rent with a proviso limiting the amounts to be recovered from the individuals, but that is a drafting device.

I should add that the Law of Property Act 1925, section 205(1)(xxiii) defines "Rent" as including "a rent service or a rentcharge, or other rent . . . or annual or periodical payment in money or money's worth, reserved or issuing out of or charged upon land . . ." The aggregate payments of the occupants would, I think, come within that wording also.

The conclusion which I have reached does not mean that the monetary obligation of the individual tenants is increased. The fact that the payments constitute a rent does not, in my view, alter the nature of the payments or turn the obligation to pay them into joint liabilities.

The result, in my opinion, is that when the notices to quit were given the defendants were jointly entitled to exclusive possession of the flat for a term at a rent. Following the statement of the law in *Street* v *Mountford* and by Windeyer J in *Radaich* v *Smith* (*supra*), to which I have referred, that constituted a tenancy. This is not a case where the landlord provided any services; the defendants were plainly not lodgers. Further, the case is not within any of the exceptions referred to by Lord Templeman (eg mortgagees in possession or service occupiers etc). Nor does the case come within the sort of transaction referred to by Denning LJ in *Facchini* v *Bryson* (*supra*) such as family arrangements and acts of generosity. The arrangements between the plaintiffs and each of the defendants were transactions of a business nature for full consideration.

I should add that the matter does not depend upon the intention of the parties. It is quite clear that the plaintiffs, upon advice, intended to create a relationship which was outside the operation of the Rent Acts, and the defendants may well have been content to accept that in order to get accommodation. The question, however, is not, "What did the parties intend?" The question is what is the effect in law of the rights which they actually created. As I have indicated, it seems to me that they created a joint right of exclusive possession of land for a term and that constitutes a lease.

The judge referred to the four unities which must be present for the creation of a joint tenancy, namely the unities of possession, interest, title and time (see *Megarry and Wade*: *The Law of Real Property*, p 419 *et seq*). I deal with these in turn.

Possession. Each co-owner is as much entitled to any part of the land as the others. This does not, in my opinion, conflict with the provision in the agreements that the licensee is not to have the right to exclusive possession of any part of the flat. None of them does. Each has to share it with the others.

Unity of title. Each joint tenant must claim his title to the land under the same act or document. In my opinion, the joint tenancy was created by a single implied agreement for the grant of exclusive possession to the defendants between the defendants and the

*Editor's note: Now reported in [1987] 2 EGLR 71; (1987) 284 EG 1375.

A plaintiffs at the time when the fourth defendant's agreement was signed. The individual agreements are, no doubt, evidence of the transactions, but it seems to me that the single implied agreement is what finally brings the relationship into being.

Unity of time. The interest of each tenant must vest at the same time. That is satisfied in the present case. The joint tenancy arose when the fourth defendant's agreement was signed.

Unity of interest. The interest of each joint tenant must be the same in extent, interest and duration. Thus one joint tenant cannot act by himself (eg to surrender a lease or give a notice), for by himself he does not have the whole estate. But it is, I think, necessary to distinguish the joint tenancy from the property on which the joint tenancy exists. What we have here is a joint tenancy of a lease. As I
B have indicated, it seems to me that the term of the lease is until any individual occupancy is determined under the provisions of the relevant agreement. Once the six-month period has expired, that can be done upon one month's notice either by the occupant himself or by the plaintiffs. It does not seem to me that the existence of the joint tenancy prevents that or that the giving of such a notice by the occupant interferes with the principle of unity of interest. The joint tenancy of the term cannot alter the term for which the lease exists. The defendants, as joint tenants, had unity of interest in the property of which they were joint tenants.

I conclude, therefore, that there is no objection in principle to the existence of such a joint tenancy as I have indicated.

C It is not in doubt that this flat is property which is within the control established by the Rent Act 1977 and amending legislation. For the reasons that I have indicated, it seems to me that at the material times the defendants were lessees of the flat at a rent and not licensees.

We were referred to *Hadjiloucas* v *Crean* [1987] 3 All ER 1008*. I do not think that the conclusions which I have reached are in conflict with anything decided in that case. In the present case, we have a much fuller knowledge of the facts than had the court in *Hadjiloucas* and, as I understand from Mustill LJ, the argument has taken a different course.

I should add that I have concentrated on the position as it exists in the present case. If, after a tenancy has arisen, in the manner which I have indicated, in consequence of four persons occupying the flat,
D one or more of them goes out of occupation and a new occupant (or new occupants) is or are admitted under the form of agreement used in this case, I think that by implied surrender and regrant a new tenancy will arise.

The present appeal should, in my view, be allowed.

MUSTILL LJ agreed with the judgment of Fox LJ and did not add anything.

Dissenting, SIR GEORGE WALLER said: I do not need to repeat the facts which have been fully set out in the judgment of Fox LJ.

Two possibilities arise in this case. If the agreements can be accepted as stating the true legal position and if in those
E circumstances it can properly be said that no one has exclusive possession, there is no problem and the arrangements will work out with no rewriting of the contract. On the other hand, if the four persons have to be looked at together as one, then as a group they have exclusive possession, they will be joint tenants, they may have taken on obligations of which they have no notice, and the consequences of any change among the occupants may be serious. There are a number of cases where the courts have had to consider similar problems, in particular *Street* v *Mountford* [1985] AC 809, where Lord Templeman, with whom the other members of the House of Lords agreed, gave guidance on what is a very difficult problem. In none of these cases, however, were the facts similar to the present. In this case, unlike the others, each of the occupants started his or her
F occupation on a different day with a different monthly payment.

I will therefore first consider the agreements. Although they were drafted so that the Rent Acts would not apply, it was not submitted to us that this could be said to be a sham device or an artificial transaction. It clearly was not. I can therefore bear in mind Lord Templeman's words that the Rent Acts must not be allowed to alter or influence the construction of the agreement. *Prima facie*, therefore, at the time we are considering there were four separate agreements, one running from August 1 1982, one running from August 1 1984, one running from March 2 1984 and one from January 28 1985. The sums payable were £86.66, £125, £99 and £104 monthly. The agreements were all in the same form and granted the G
right to share the use of the flat and not to impede the use by not more than three others. There were four bedrooms and these were each occupied by one of the "licensees". Who had which bedroom was a matter of mutual arrangement and when one became vacant there might be a change among those left behind. The cost of services had to be shared. There were changes from time to time and that is why each agreement began on a different day.

The House of Lords in *Street* v *Mountford*, where it was admitted that there was exclusive possession, considered the principles which have to be applied in residential cases and Lord Templeman in his speech considered a number of cases where the question which had to be decided was whether there was a licence or a tenancy. In all the cases where there was more than one person in occupation, the H
agreements, if there were more than one, related to an occupation which was to start on the same day, for the same term and for the same rent. In *Somma* v *Hazelhurst* [1978] 1 WLR 1014, which was disapproved by the House of Lords, a young unmarried couple occupied a double bed-sitting room by virtue of two identical agreements. The trial judge held that the agreement was a sham but the Court of Appeal reversed that finding. Lord Templeman in *Street* v *Mountford*, having expressed the view that the agreement was a sham, said in the course of his speech: "The agreements signed by H and S constituted the grant to H and S jointly of exclusive possession at a rent for a term for the purposes for which the room was taken and the agreement therefore created a tenancy." J

Immediately after dealing with *Somma* Lord Templeman said that he disapproved of the decision in *Aldrington Garages Ltd* v *Fielder* (1978) 37 P&CR 461 and *Sturolson & Co* v *Weniz* (1984) 272 EG 326. In *Aldrington*'s case, although there were two agreements, they were identical, they were both signed at the same time and the two parties moved in at the same time. The trial judge found exclusive possession and held that the true intention of the parties was a joint tenancy. The Court of Appeal reversed this decision following the decision in *Somma*'s case. In *Sturolson*'s case the defendant, his wife and a friend looked at a flat, the defendant and the friend each signed a sharing agreement and they were in effect being given exclusive possession. When the landlord sought to obtain possession the trial judge held that the agreement was a sham and found that they had K
exclusive possession and held that they were joint tenants. The Court of Appeal, mentioning *Somma*'s case, rejected the suggestion of sham and reversed the judge's decision. Since in both cases there was a finding of exclusive possession I have no doubt that Lord Templeman would have upheld the trial judge's decision for that reason as in *Somma*'s case.

In *Street* v *Mountford* the fact that exclusive possession was agreed was decisive because, unless the case could be brought within the exceptional cases of service tenancy or family relationship or those mentioned in *Errington* v *Errington* [1952] 1 KB 290 at p 297 by Denning LJ, exclusive possession would result in a tenancy. And, as I have already mentioned, where there was multiple occupation L
which began on the same day and was for the same term and the same rent there would be exclusive possession and therefore a joint tenancy would result (see *Somma, Aldrington* and *Sturolson*). In the present case the only way in which the defendants can be said to be tenants is as joint tenants with exclusive possession, because if each is looked at separately he or she does not have exclusive possession. The suggestion was made before us that each of the occupants might be said to have exclusive possession of a bedroom. In my opinion this is not so. The licence was for the whole flat and the internal arrangements made by the occupants which were varied from time to time cannot affect the limits of the licence. Furthermore, although each has a bedroom which is personal and private, possession of it cannot be said to be exclusive; it is merely an arrangement with the M
other occupants.

If these proceedings had been taken at a time when the original four were there it might well have been possible to argue that the agreements were a sham and that the four were joint tenants, but subsequent events showed clearly (as is agreed) that the agreements represent what is actually happening and are not a sham. While the first four were there it might have been possible to argue that there was unity of possession, interest, title and time. But by the time that Vaughan arrived the presence of these unities was, to say the least, doubtful.

Vaughan, who was the earliest of the appellants to go into the flat, did not have exclusive possession because he had to share with three

* Editor's note: Also reported at [1987] 2 EGLR 60; (1987) 284 EG 927.

A others. Although Lord Templeman disapproved of the decision in *Aldrington*'s case I do not think that he would disapprove of the test set out in the judgment of Geoffrey Lane LJ (at p 471):

> In order to succeed, the defendant must show that the two agreements must be read together, and that, being read together, are effective to create a joint tenancy between himself and [the second occupier], on the one hand, and the owners on the other . . . If the two agreements are separate then there is no exclusive possession and no tenancy.

A joint tenancy is one where the tenancy commences on the same day for all, where the term is the same for all, where the rent should not be altered without due notice to all and possibly where all are jointly liable for the rent.

B Fox LJ has referred to the four unities set out in *Megarry and Wade* as the requirements for a joint tenancy and I will consider them.

Unity of interest. The extent, nature and duration of each occupant's interest is different. Each one started on a different day and as a result the extent was different because the duration was six months in each case and then monthly. Although each one became monthly after six months, there is no "act" to create a joint tenancy as I shall refer to under unity of title.

Unity of title. The unity of title for a joint tenancy has to be the same act or document. *Megarry and Wade* were repeating that which was contained in *Blackstone's Commentaries* (vol 2, Kerr's ed, p 155):

> Joint-Tenancy cannot arise by descent or act of law: but merely by purchase,
> C or acquisition by the act of the party: and unless that act be one and the same the two tenants would have different titles.

Each occupant started with a different document on a different date. Can it be said that it was the same act? In my opinion a new joint tenancy could not arise each time a new person came because the new occupant had six months to run whereas every other occupant was monthly, and it could not arise at the expiration of six months because there was then no "act" which could create a joint tenancy.

Unity of time. The interest of each must vest at the same time. If we were considering the first four occupants it would follow that their interest vested at the same time, but as each new occupant arrived the interest of the others had already vested and at the time when the notices to quit were delivered each occupant's interest had vested on
D different days.

Unity of possession. This must be possession of the whole. Each occupant was given the right to use the flat and so it could be said that there was unity of right to use the flat. But was that possession of the whole? In *Addiscombe Garden Estates Ltd* v *Crabbe* [1958] 1 QB 513, a case where the "licence" contained a clause entitling the landlord to enter the land, Jenkins LJ said:

> . . . The importance of that is that it shows that the right to occupy the premises conferred on the grantees was intended as an exclusive right of occupation, in that it was thought necessary to give a special and express power to the grantors to enter. The exclusive character of the occupation granted by a document such as this has always been regarded, if not as a
> E decisive indication, at all events as a very important indication to the effect that a tenancy, as distinct from a licence, is the real subject-matter of [the] document.

The indication is in the opposite direction in this case, but the plaintiff never claimed the right.

I am prepared to accept that there may be an exceptional case where there might be a special arrangement about one of these characteristics, but in my opinion there cannot be a joint tenancy where there are serious doubts about each of the four unities.

This agreement represents the true situation. To try to make it into a joint tenancy would be to create a wholly artificial situation. Even if there were the strongest reasons to try to make the Rent Acts apply it would be a joint tenancy that departed from all the principles of joint
F tenancies. All occupants would have to be parties to the same agreement. All would have to start on the same day. Whenever one changed there would have to be a new agreement signed by all. All would be liable for the rent. But there is no good reason at all for making the Rent Acts apply. This is a flexible arrangement, generally fairly short term, for which there is a substantial demand evidenced in this case by the large number (100) who answered Vaughan's advertisement. These agreements accurately represent the relationships between the parties.

I regret that I am in disagreement with Fox LJ but I would dismiss this appeal for the reasons I have given.

The appeal was allowed and declaration made that the first, G *second, third and fourth defendants were joint tenants at date of commencement of proceedings. Leave to appeal to the House of Lords was granted.*

Court of Appeal

December 16 1987

(Before Lord Justice DILLON, Lord Justice CROOM-JOHNSON and Mr Justice CAULFIELD)

TENNANT RADIANT HEAT LTD v WARRINGTON DEVELOPMENT CORPORATION

H

Estates Gazette March 19 1988

[1988] 11 EG 71-79

Landlord and tenant — Apportionment of liability — Claims by tenants and counterclaims by landlords following collapse of part of warehouse roof — Collapse due to accumulation of rain-water because of blocked outlets — Appeal from decision of Judge Fitzhugh QC, sitting as a judge of the High Court — The blockage had been caused by failure to clear bird droppings, bird feathers and silt — The tenants, whose goods were damaged, claimed on the basis of an implied covenant by the landlords to keep the roof and outlets in good repair and J **condition; alternatively, they claimed on grounds of nuisance and negligence — The landlords counterclaimed under the tenants' repairing covenants in the lease and also, alternatively, in nuisance and negligence — The judge below held, in favour of the tenants, that the roof over the tenants' unit was not included in the demise and that the landlords were under an implied contract to keep it and its outlets in good repair and condition; and he dismissed the landlords' counterclaim — Held by the Court of Appeal that part of the roof over the tenants' unit was included in the demise and that there was no implied contract by the landlords to keep the roof and its outlets in good repair and condition — The part of the** K **roof included in the demise was subject to the tenants' repairing covenant and the tenants were to that extent in breach of covenant — On the other hand, the landlords, although not liable under an implied contract to repair, were liable in nuisance and negligence save in so far as they were absolved by the tenants' breach of covenant — They were in control of the majority of outlets and they had been expressly warned by architects of the danger of blockages caused by the droppings and feathers of seagulls**

The crux of the appeal was, therefore, the interplay between the liabilities of the landlords and tenants — The L **apportionment provisions of the Law Reform (Contributory Negligence) Act 1945 did not apply because, like the old common law, they did not apply to claims in contract — In the present case the landlords' crossclaim was a claim in contract, their alternative claims in nuisance and negligence having been rejected — The problem here was a problem of causation, and apportionment on that basis was permissible — The court on a broad assessment apportioned 90% liability to the landlords and 10% to the tenants — The judgment below would be varied and an inquiry as to damages ordered on the counterclaim as well as the claim and the landlords** M **ordered to pay 90% of the tenants' damages and the tenants to pay 10% of the landlords' damages — Appeal on that basis allowed**

The following cases are referred to in this report.

Barnes v *City of London Real Property Co Ltd* [1918] 2 Ch 18; (1918) 119 LT 293; 34 TLR 361
Edmonton Corporation v *W M Knowles & Son Ltd* (1962) 60 LGR 124
Forskringsaktielselskapet Vesta v *Butcher*, October 30 1987, unreported
Grant v *Sun Shipping Co Ltd* [1948] AC 549; [1948] 2 All ER 238, HL
Hargroves Aronson & Co v *Hartopp* [1905] 1 KB 472
Leakey v *National Trust for Places of Historic Interest and Natural Beauty*

A [1980] QB 485; [1980] 2 WLR 65; [1980] 1 All ER 17; (1979) 78 LGR 100, CA

Sedleigh-Denfield v *O'Callaghan* [1940] AC 880; [1940] 3 All ER 349; (1940) 164 LT 72; 56 TLR 887

Sleafer v *Lambeth Borough Council* [1960] 1 QB 43; [1959] 3 WLR 485; [1959] 3 All ER 378; (1959) 57 LGR 212, CA

This was an appeal by the defendants below, Warrington Development Corporation, the landlords, against the decision of Judge Fitzhugh QC, sitting as a judge of the High Court, in favour of the respondent tenants, plaintiffs below, Tennant Radiant Heat Ltd. Judge Fitzhugh upheld the respondent tenants' claim and dismissed the appellants' counterclaim.

John Mowbray QC and Nigel Howarth (instructed by A W Mawer B & Co, of Manchester) appeared on behalf of the appellants; Leolin Price QC and Brendan Hegarty (instructed by Pannone Blackburn, of Manchester) represented the respondents.

Giving judgment, DILLON LJ said: The appellant, a new town development corporation ("the corporation") is the owner of a warehouse building at Chetham Court, Winwick Quay in Warrington which was constructed in the late 1970s. The respondent ("the lessee") is the lessee of a part of that building known as unit no 6. On October 22 1980 part of the roof of unit 6 collapsed under the weight of rain-water accumulated above it. The accumulated water poured through the hole made by the collapse of the roof and the lessee's goods stored in unit 6 were damaged. The lessee therefore C started this action, claiming damages against the corporation for the damage to its goods. The claim was put on the ground that on the true construction of the lease of unit 6 and in order to give it business efficacy there is to be implied in the lease a covenant by the corporation to keep the roof of the building, and the rain-water outlets in the roof, in good repair and condition, or alternatively on the grounds of nuisance or negligence. The corporation denied liability and counterclaimed against the lessee, under the repairing covenants in the lessee's lease of unit 6, for the cost of repairing the roof.

The action and counterclaim came before Judge Fitzhugh QC, sitting in Manchester as a judge of the High Court in the Chancery Division, for trial on liability only. By his decision of December 19 D 1985 he held that a covenant by the corporation to keep the roof, the building and the rain-water outlets in the roof in good repair and condition was indeed to be implied in the lessee's lease of unit 6. Accordingly, as it was common ground that if there was such a covenant the corporation was in breach of it, he directed an inquiry as to damages on the lessee's claim, and he dismissed the corporation's counterclaim. The corporation now appeals to this court against that decision. By a respondent's notice the lessee has brought before this court its alternative grounds of claim in nuisance or negligence. The basic facts have never been in dispute, and the issues on the appeal and under the respondent's notice are predominantly issues of law.

The building which includes unit 6 is an extensive single-storey E steel-framed building divided into 22 units which are not all of the same size. The building is approximately 111 m long and 75 m wide — the size of a rugby pitch — and 6 m high. The units are self-contained; each is divided off from the adjoining units by blockwork walls which are built into the steel frame to the full height from floor to roof. Unit 6 is a small unit; it has a depth of 18 m and a width of 10 m, and was described in argument as being roughly 1/49th of the whole building in area.

The roof is an important factor. It is a single continuous deck covering all 22 units in the building, with a shallow fall from the centre to the front, covering half the building, and a similar fall from the centre to the back covering the other half. The roof consists of a specialised metal decking only 9mm thick, topped by a 50mm layer of F foam insulation, with, on top of that, several layers of felt with a reflective surface. The roof is surrounded by a fairly high parapet on all sides, and access to the roof can be obtained only by placing a ladder against the outside wall of the building, climbing the ladder, and clambering over the parapet; there is no access to the roof from inside any of the units. Rain-water drainage from the roof is provided by 24 rain-water outlets, 12 in the front half and 12 in the back half of the roof, each just inside the parapet which I have mentioned. Each outlet is covered by a removable plastic grille. Below each grille is a pipe leading down within the building to normal underground storm water sewers.

On the pleadings it was common ground between the parties that one of these down pipes came down within unit 6, and one of the outlets, with its grille, was in the roof over unit 6. At the trial the G number of outlets in the roof over unit 6 was the subject of controversy. There was evidence from one witness that there were two outlets in the roof over unit 6, but there was other evidence that there was only one, and we see no reason to interfere with the judge's finding of fact that there was only one rain-water outlet in the roof over unit 6.

It seems that at the time of the collapse of the roof of unit 6 in October 1980 only a couple of the units in the building had been let apart from unit 6; all the rest were unlet, unoccupied and still in the possession of the corporation. One of the two others which had been let, however, was one of the units immediately adjoining unit 6.

The accumulation of rain-water on the roof which caused the collapse of part of the roof of unit 6 came about because all the rain- H water outlets on the roof which I have mentioned had been blocked up by bird droppings, bird feathers and silt, so that the rain-water had no means of getting away. This was a danger which had been brought fairly and squarely to the corporation's attention by their architects a year before in a letter of September 6 1979, where the architects said:

> During the recent inspection of the building . . . prior to issue of certificate of practical completion, a significant point emerged concerning the maintenance of the roof finish.
>
> Evidently, a large flock of seagulls located in the area are using the roof as a resting place. In addition to the mess they are creating with droppings, they are leaving behind large quantities of feathers which, during the course of rain accumulate around the roof outlets and obstruct the proper draining of water off the roof. J
>
> I note from your comments on the maintenance check list that the question of keeping the roof clear is an item which you consider the responsibility of the occupants, but in view of this particular problem we strongly suggest that this item is carried out on a regular basis by WNTDC and taken out of the hands of the tenants. I am sure you will agree that it is very unlikely tenants will ever clean their respective areas on a regular basis — in fact tenants have really no way of knowing which areas refer to their units anyway, so the idea is really quite impractical. It is important that the roof must be regularly inspected and kept clean if it is to function properly, otherwise problems will inevitably arise . . .

The judge dealt with the case on his construction of the lessee's lease of unit 6, and, as I have said, he held that there was to be implied in the lease a covenant by the corporation to keep the roof of the K building and the rain-water outlets in the roof in good repair and condition.

The lease is dated November 1 1979 and is for a term of five years from that date at a yearly rent of £2,925, together with an insurance rent. The property demised is described as being "all those premises situate and known as 6 Chetham Court, Winwick Quay, Warrington, in the . . . County of Cheshire and for the purposes of identification only shown edged red on the plan annexed hereto". The property demised was granted, as is almost common form in such circumstances, together with

> the right to free passage of surface storm soil and effluent drainage gas water electricity steam and telephone or any other service or supply from or to the L demised premises through the sewers drains water courses conduits pipes wires and cables then or thereafter during the term in or over under or upon the other buildings and land of the Corporation and its lessee adjoining or near to the demised premises

and subject to a corresponding exception and reservation.

The lessee's covenants in the lease include a full repairing covenant in clause (5)(a) whereby the lessee covenanted:

> to repair and keep the demised premises and all present and future buildings and erections thereon and on every part thereof . . . maintained repaired and cleansed and generally in good and substantial repair and condition.

There is an exception in the event of destruction by fire or special perils in respect of which the corporation had insured, but it is common ground that the corporation's insurance did not cover what M actually happened.

There are further covenants by the lessee as follows:

> 2(5)(b): to pay a fair proportion (to be conclusively determined by the Corporation) of the expenses payable in respect of repairing renewing maintaining rebuilding and cleansing all private sewers drains roads pavements car parking spaces landscaped areas and other things the use of which is common to the demised premises and other premises within the area shown edged blue on the plan annexed hereto and known as Chetham Court, Winwick Quay, Warrington aforesaid (damage by coal mining subsidence excepted) and to keep the corporation indemnified against such proportion of such costs and expenses as aforesaid (damage by coal mining subsidence excepted) . . .
>
> 2(13): not to exhibit on the outer wall or roofs of the demised premises or of

A any building or structure thereon any sign flag or advertisement except such as may previously have been approved by the Corporation and in default the Corporation may enter and remove the same at the Lessee's cost Provided however that the Lessee may erect a sign in such position and of such form colour or design as may be first approved by the Corporation (such approval not to be unreasonably withheld) displaying the name and business of the Lessee.

2(14): not to suspend any weight from the roof or roof members or use the roof of the demised premises for the storage of goods or to place or permit or suffer to be placed any weight thereon or to permit any person or persons to enter thereon save with a view to the execution of necessary repairs and then only in such manner as to subject the roof and roof members to the least possible strain.

B The lessee's covenants run to some 27 in all and are obviously drawn with considerable care and in great detail. The corporation's covenants as lessor are merely the usual qualified covenant for quiet enjoyment and a detailed covenant to insure against destruction or damage by fire or special perils, which it is not necessary to set out.

There are cases, such as *Barnes* v *City of London Real Property Co Ltd* [1918] 2 Ch 18 and *Edmonton Corporation* v *W M Knowles & Son Ltd* (1961) 60 LGR 124 in which it has been held that where in a lease there is a covenant by the lessee to make payments to the lessor for the provision of certain services or for the carrying out of certain works, then a covenant by the lessor to provide the services or carry out the works may be implied. But in the present case the lessee seeks to go very much further than that, and I cannot see that it is entitled to C do so. Where a landlord of a building grants a lease of a part only and is in a position to insist on the lessee taking a lease in a common form of the landlord's choice, it is not at all unusual to find that the lease does not contain any covenant by the landlord to do repairs to parts of the building, however important, and whether included in the demise or not. Yet it has never been held in any general way that it is necessary to imply a repairing covenant on the part of the landlord to give business efficacy to the contract: see *Sleafer* v *Lambeth Borough Council* [1960] 1 QB 43, and see also *Woodfall on Landlord and Tenant*, 28th ed vol 1, para 1-1465. It may well be objectively sensible, or reasonable, that there should be such a landlord's covenant, with a corresponding covenant by each lessee to contribute a proportionate part of the expense, but that is not enough to warrant D implying such covenants.

In the present case there are references to the roof where that is intended, as in the lessee's covenants (13) and (14) cited above. It is impossible, therefore, to construe the lessee's covenant (2)(5)(b) as a covenant to contribute to the cost of repairs to the roof, which is nowhere there mentioned. Although the lease has obviously been carefully prepared, there is no other clause which points in any way towards the lessee's contributing to the costs of repairs to be carried out by the corporation. But without a covenant by the lessee to reimburse the costs, there is no commercial sense, so far as the corporation is concerned, in the corporation's binding itself by covenant to repair the roof of the building; that would be wholly out E of character for this type of commercial lease. Moreover, the natural reading of the parcels in this lease is that they include the roof in so far as the roof is over unit 6.

Accordingly, for my part I would respectfully but without hesitation reject the learned judge's conclusions (a) that the roof over unit 6 was not included in the demise of unit 6 and (b) that a covenant by the corporation to keep the roof and the rain-water outlets in the roof in good repair and condition is to be implied in the lease. On the contrary, I would hold that the part of the roof which is over unit 6 and the rain-water outlet which is in that part of the roof were included in the parcels demised to the lessee by the lease and are therefore covered by the lessee's repairing covenant in clause 2(5)(a) of the lease. It follows that the grounds on which the judge gave F judgment for the lessee and dismissed the corporation's counterclaim do not support his conclusion. It follows also that the lessee was, when the roof collapsed, in breach of its repairing covenant (5)(a) in that it had failed to keep the part of the roof over unit 6 and the rain-water outlet in it maintained, repaired and cleansed and generally in good and substantial condition.

I turn to the lessee's claim against the corporation in nuisance and negligence. On the facts the corporation had possession of all the unlet units and of the parts of the roof above them, including all the rain-water outlets in those parts of the roof — by far and away the majority of the rain-water outlets. The corporation had been clearly warned by their architects of the need to keep the outlets clear and the roof clean, and clearly warned of the foreseeable consequences if that were not done. Yet the outlets were allowed to get, and remain, G blocked up, and the foreseeable consequences followed. On the general principles laid down in such cases as *Sedleigh-Denfield* v *O'Callaghan* [1940] AC 880 and *Leakey* v *National Trust* [1980] QB 485 I have no doubt that the corporation is liable to the lessee in nuisance and negligence save if and in so far as the corporation is absolved from liability by the lessee's own breach of its own repairing covenant in failing to keep the one outlet above unit 6 clear.

The crux of this appeal is therefore the interplay of the two liabilities.

On the facts, the evidence established that by mathematical calculation if the one outlet above unit 6 had been unblocked while all the rest remained blocked and if all the water accumulated on the roof before it collapsed had been free from debris, it would have taken H four hours for the whole of that volume of water to be cleared from the roof by the one outlet assumed to be unblocked. In truth, of course, the water was not free from debris. If the one outlet only had been unblocked at the outset, the water running away through that outlet would have carried down to it further bird feathers and silt, and it would speedily have been blocked up again. It would have had to be regularly unblocked again and again for all the water to get away. Realistically, what was required to get rid of the water and to avoid the collapse of the roof from the weight of the water was that a good majority of the outlets in that half of the roof, not necessarily including the outlet over unit 6, should have been unblocked. If that be right, I would hold that the fault of the corporation in failing to J keep the outlets under its control clear was the major cause of the collapse of the roof over unit 6; but the fault of the lessee in failing to clear the outlet in the roof over unit 6 was not so trivial that it can on the facts be disregarded as a contributory cause of the disaster.

What then is the effect in law? At this juncture it is natural to think of apportionment for contributory negligence under the Law Reform (Contributory Negligence) Act 1945, since (a) apportionment of liability in tort because of contributory negligence is the context in which questions of apportionment of liability usually arise and (b) the decision of this court in the case of *Forsikringsaktieselskapet Vesta* v *Butcher* (October 30 1987) was briefly reported in *The Times* newspaper under the heading "Power to apportion blame in contract" only a few days before the hearing of K this appeal. It is to be remembered, however, that the 1945 Act has a very limited function; it was enacted merely to override the archaic rule of the common law, which had long been recognised as unjust, that if a plaintiff sued in tort for damages for negligence his claim would wholly fail if it were shown that he had been guilty of any degree of contributory negligence, however slight, which was not *de minimis*. In the *Vesta* case this court held that the 1945 Act applied, with the result that liability fell to be apportioned under it, where a plaintiff, who had been guilty of contributory negligence, chose to formulate in contract a claim against a defendant which was in substance a claim in tort and could equally have been formulated as a claim in negligence. That is not this case. But I find the *Vesta* case L helpful in this case for the following passage in the judgment of Sir Roger Ormrod, which I would respectfully adopt:

> . . . I remain quite unconvinced that contributory negligence, as such, at common law had any relevance in a claim in contract. Apart from one case, the word "contributory" is not to be found in relation to contract. It is true, however, that in some cases the rule of strict liability, eg in the case of common carriers or innkeepers, was modified to take into account fault by the passenger or guest but on the basis of causation. Had contributory negligence been a defence at common law to a claim for damages for breach of contract the reports and the textbooks prior to 1945 would have been full of references to it.

The context of the 1945 Act, and the language of section 1, to my mind make it clear that the Act is concerned only with tortious liability and the power to apportion only arises where the defendant is liable in tort and M concurrent liability in contract, if any, "is immaterial": see the passage in the judgment of Pritchard J in *Rowe* v *Turner Hopkins & Partners* [1980] 2 NZLR 550, cited in the judgment of O'Connor LJ at pp 18 and 19. In 1945 "contributory negligence" was one of the most notorious blots on the common law in relation to claims for damages for negligence along with "common employment", both of which were targets for law reform after the war. The Act only applies to cases in which, but for the Act, the claim for damages would have been "defeated by reason of the fault of the person suffering the damage".

In the present case, the claim of the lessee against the corporation is a claim in tort — nuisance or negligence, involving failure to take reasonable care to avoid a foreseeable danger, but the corporation's crossclaim against the lessee is not a claim in tort at all; it is a claim in

A contract — breach of an absolute repairing covenant in the lease — which could not be formulated as a claim in tort. Therefore, in my judgment, the 1945 Act has no application to the present case, since the breach of covenant on the part of the lessee does not fall within the statutory definition of "fault" in section 4 of the 1945 Act, namely:

> negligence, breach of statutory duty or any other act or omission which gives rise to a liability in tort or would, apart from this Act, give rise to the defence of contributory negligence; . . .

More importantly, however, for present purposes, the archaic and draconian rule of the common law which the 1945 Act was passed to override also has no application to the present case, since that rule had no relevance to a claim in, or in my judgment to a breach of, B contract which was not also itself tortious.

The problem which this court faces, on claim and counterclaim alike, is in my judgment a problem of causation of damage. On the claim, the question is how far the damage to its goods which the lessee has suffered was caused by the corporation's negligence notwithstanding the lessee's own breach of covenant. On the counterclaim, the question is how far the damage to the corporation's building which the corporation has suffered was caused by the lessee's breach of covenant, notwithstanding the corporation's own negligence. The effect is that on each question, apportionment is permissible. This is the same result as the 1945 Act would produce, but it is not reached through the Act, because the obstacle which the 1945 Act was passed to override is not there on C either claim or counterclaim in the present case.

In making the apportionment, I regard as particularly significant the number of outlets in the relevant half of the roof for which the corporation and the lessee were respectively responsible, and on a broad assessment I would apportion liability as to 90% to the corporation and as to 10% to the lessee. Accordingly, I would allow this appeal and vary the order of the judge by directing an inquiry as to damages on the counterclaim in addition to the inquiry as to damages on the claim which the judge ordered. I would then order the corporation to pay 90% — and not the whole as ordered by the judge — of the lessee's damages the subject of the claim. I would order the lessee to pay 10% of the corporation's damages the subject of the counterclaim.

D

Agreeing with the order proposed, CROOM-JOHNSON LJ said: I agree that the part of the roof above unit 6 was included in the demise to the plaintiff. Counsel for the plaintiff tried hard to identify the horizontal line in the roof where unit 6 ended and the general roof of the building began, but owing to the peculiar construction of the building there was no such place. Moreover, the clauses in the lease cannot be contradicted. The express covenant to repair by the tenant makes it impossible to imply the covenant which the learned judge felt able to imply. Accordingly the plaintiff is liable to the defendant for its breach of covenant to repair but only to the extent of that part of the roof to which its covenants applied, namely, the roof of unit 6 E and the drainpipe and grille which led from it. The calculation of the damages required consideration later.

I agree with the analysis of the evidence in the judgment of Dillon LJ relating to the effect of the seagull debris over the whole of the roof and the blocking of all the grilles and drainpipes. Save for two units, the whole of the rest of the roof (especially that half of it which included the roof of unit 6) and all the other grilles and drains but one were in the occupation and under the control of the landlord. The result was that the débris-contaminated water from the landlord's part of the roof all flowed down to unit 6 and helped to cause the collapse of that part of the roof. The landlord knew well, from the letter sent to it by its architects, that in those circumstances there was a risk of trouble if it did not keep the roof and drains clean. The F defendant had no answer to the plaintiff's alternative claim in negligence: see *Hargroves, Aronson & Co* v *Hartopp* [1905] 1 KB 472. In that case there was a building let off in flats, of which the plaintiffs were tenants of the top flat. The landlords retained possession of the roof and of a gutter. The landlords knew that the gutter was blocked, but did not clear it. When the top flat was damaged by water flooding, the landlords were held liable in negligence to the plaintiffs. In the present case, in so far as the defendant was in control of most of the roof, the condition of which caused the flooding of unit 6, it is liable in tort for the damages to the plaintiff's goods.

How are the two crossclaims to be treated? The Law Reform (Contributory Negligence) Act 1945 would solve the problem if both claims were in tort. After evaluating each claim, the court would G apportion the liability to pay damages to the extent which it thinks just and equitable having regard to each claimant's share in the responsibility (section 1(1)). But the Act has no application if one of the claims is in a contract which does not depend for its application on the discharge of the duty of care. Section 4, the interpretation section, excludes from the scope of the Act any act or omission which would not, before 1945, have given rise to the defence of contributory negligence. Breach of a strict duty under a contract has never given rise to the defence of contributory negligence. Therefore the 1945 Act has no application, and the rights of the parties here must be dealt with accordingly under the old law.

I should mention that the defendant, in its counterclaim, alleged not only the breaches of the covenants in the lease; it also put its claim H in negligence and nuisance. The learned judge did not deal with these alternative ways of advancing the defendant's claims. They can be dealt with very shortly. The plaintiff was not guilty of negligence, because it would not have been reasonably foreseeable that if its one grille and drainpipe became obstructed on its own the result would have been the flooding of the whole roof. As for the claim in nuisance, it did not (as was pleaded) cause the water to escape "into other parts of the building".

If the 1945 Act has no application, what is the position on the two claims? The evidence clearly indicates that the damage to the plaintiff's premises was attributable to two concurrent causes, both operating contemporaneously. J One was the defendant's negligence and the other was the plaintiff's breach of covenant. I agree with the finding of Dillon LJ that simply as a matter of causation, based upon the amount of water which was wrongfully on the roof, the plaintiff's breach of covenant was a factor of one-tenth of that united cause, and the defendant's negligence a factor of nine-tenths. If the collapse had been caused by the combined negligence of the defendant and some third party, the plaintiff would have been entitled to recover all its damages from the defendant, leaving it to get contribution from its fellow tortfeasor: see *Grant* v *Sun Shipping Co Ltd* [1948] AC 549 at p 563 *per* Lord du Parcq. But that does not apply when the other cause comes from the plaintiff himself.

Take the landlord's claim in contract for the damage to its roof. Could it rely on the covenant in the lease as entitling it to recover in K full? For the same reason that would produce an unjust result.

A tenant who is in breach of his repairing covenant must pay for all the costs of repairs or of the damage to the reversion. But the position is different if the covenantee is the selfsame landlord who has caused nine-tenths of the damage to the roof.

It is not possible, therefore, to allow both the claim and the crossclaim in full and to set off the two awards of damages. Where one is dealing with two contemporaneous causes, each springing from the breach of a legal duty but operating in unequal proportions, the solution should be to assess the recoverable damages for both on the basis of causation. The landlord will pay for the nine-tenths of the damage which its negligence caused. L The tenant will pay for the one-tenth caused by his breaches of covenant.

I agree with the order proposed by Dillon LJ.

CAULFIELD J agreed with the reasoning and conclusions of Dillon and Croom-Johnson LJJ and did not add anything.

The appeal was allowed and the order below varied by directing an inquiry as to damages on counterclaim in addition to inquiry as to damages on claims as ordered by the judge. The defendant was ordered to pay 90% of the plaintiff's damages, the subject of the claim, and the plaintiff ordered to pay 10% of the defendant's damages, the subject of the counterclaim. The order as to costs below to remain, appellant to have costs of appeal. M *Leave to appeal to House of Lords was refused.*

Chancery Division

December 2 1987

(Before Mr Michael WHEELER QC, sitting as a deputy judge of the division)

R & A MILLETT (SHOPS) LTD v LEON ALLAN INTERNATIONAL FASHIONS LTD

Estates Gazette April 2 1988

[1988] 13 EG 50-56

Landlord and tenant — Construction of underlease — Rent formula in underlease fixed the rent payable as the fraction of $\frac{78}{85}$ of the rent of the superior lease, which provided for rent reviews — The explanation of the fraction was that the underlease did not demise the whole of the property comprised in the superior lease — The event which gave rise to the present problem of construction was the surrender of the superior lease in pursuance of a leaseback arrangement by which the property comprised in the superior lease together with certain other properties were included in a new lease — This new superior lease was expressly stated to be subject to and with the benefit of the underlease — It was argued by the underlessees that the effect of the surrender was to put an end to the operation of the rent formula; as the original superior lease had ceased to exist the formula could no longer be invoked to quantify the rent of the underlease — The superior lessees argued that the formula could still apply as a matter of construction notwithstanding the disappearance of the original superior lease — Alternatively they submitted that the rent formula was merely machinery and if for some reason it broke down the court could substitute its own machinery — Held, accepting this submission and applying the principles of *Sudbrook Trading Estate Ltd* v *Eggleton*, that the court would substitute its own machinery to supply the gap — The court would therefore order an inquiry with appropriate machinery to arrive at the proportion of $\frac{78}{85}$ which would have been payable as the rent of the underlease if the original superior lease had not been surrendered

The following cases are referred to in this report.

Lorien Textiles (UK) Ltd v *S I Pension Trustees Ltd* (1981) 259 EG 771
Sudbrook Trading Estate Ltd v *Eggleton* [1983] 1 AC 444; [1982] 3 WLR 315; [1982] 3 All ER 1; (1982) 44 P&CR 153; [1982] EGD 392; 265 EG 215, HL
United Scientific Holdings Ltd v *Burnley Borough Council* [1978] AC 904; [1977] 2 WLR 806; [1977] 2 All ER 62; (1977) 33 P&CR 220; [1977] EGD 195; (1977) 243 EG 43 & 127, HL

This was an originating summons by which the plaintiffs, R & A Millett (Shops) Ltd, lessees of premises at 39 New Street, Birmingham 2, sought a determination of the true construction of rent provisions in an underlease held by the defendants, Leon Allan International Fashions Ltd.

Robert Pryor QC and Paul Rees (instructed by Philip Hodges & Co) appeared on behalf of the plaintiffs; D McEvoy QC and Peter Griffiths (instructed by Martyn W Amey & Co) represented the defendants.

Giving judgment, MR MICHAEL WHEELER QC said: This case turns on the true construction of clause 5 of an underlease dated March 10 1970 ("the underlease") of part of premises at 39 New Street, Birmingham 2. It will be convenient if I first outline the steps by which the parties acquired their interests in the underlease.

1 By a lease dated June 23 1969 ("the original headlease") 39 New Street was demised by Raventop Developments Ltd to Suede Circle Ltd ("Suede Circle") for a term of 21 years from March 25 1969 to March 25 1990 at an initial rent of £8,500 pa, subject to one review at the end of the 14th year, namely on March 25 1983.

2 On March 10 1970 Suede Circle assigned its interest under the original headlease to C & J Clark Retail Ltd ("Clark Retail") and on the same day Clark Retail underlet to Suede Circle most (*but not all*) of the premises demised by the original headlease, retaining part of the ground floor for itself. This is the underlease as above defined. It was for a term of 20 years and one quarter, less one day, from December 25 1969. It would thus terminate on March 24 1990, ie one day before the original headlease.

3 On October 27 1971, Clark Retail assigned, *inter alia*, its interest in the original headlease to Wakefields Stores (Midlands) Ltd ("Wakefields").

4 On December 18 1972, by which time the reversion to the headlease was vested in Ravenseft Properties Ltd ("Ravenseft") and the underlessee under the underlease was still Suede Circle, Ravenseft entered into an agreement with Wakefields for the surrender, *inter alia*, of the original headlease and a leaseback to Wakefields of the property comprised in the original headlease, together with certain other properties, after various alterations had been carried out. This agreement was completed on April 9 1973, when Ravenseft let the new combined premises to Wakefields for a term of 32 years from December 25 1972. This lease ("the 1973 headlease") was expressly stated to be subject to and with the benefit of the underlease. There is no evidence before me as to whether or not Suede Circle was aware of this surrender and leaseback.

5 On August 15 1978, by an assignment of that date, Suede Circle assigned the underlease to the defendant in this action.

6 By a transfer dated September 2 1985, the unexpired residue of (*inter alia*) the 1973 headlease was transferred by Wakefields to the plaintiff in this action.

So much for the parties' chain of title. The rent payable under the underlease was £7,800 per annum "or such other amount as shall become payable under the provisions of Clause 5 hereof"; that clause provided as follows:

> 5. The rent payable by the Lessee as from the 25th day of March 1983 shall be seventy-eight eighty-fifth parts of the rent payable by the Lessor as from such date in respect of the principal premises as fixed in the manner provided by the said Superior Lease or the rent of £7,800 per annum (whichever shall be the greater).

It is common ground that in the underlease "the lessor" means the person for the time being entitled in reversion immediately expectant upon the term thereby granted; "the principal premises" means that part of 39 New Street which had been demised by Raventop Developments Ltd to Suede Circle Ltd by the headlease; and that the "superior lease" means the original headlease.

Basically, the question which I have to decide is as to the rent payable as from March 25 1983 under clause 5 of the underlease, having regard to the fact that in December 1972 the plaintiff's predecessor in title, Wakefields, had surrendered the original headlease and in April 1973 had obtained a leaseback in the form of the 1973 headlease.

Before leaving the underlease, I should also mention clause 8, which provides as follows:

> 8. It is hereby agreed and declared without prejudice but in addition to Lessee's rights under the Landlord and Tenant Act 1954 (as amended) in the event of the Lessor being granted a new Lease of the premises comprised in the said Superior Lease on or before the expiration of the term hereby granted the Lessor will within 14 days of being granted such new Lease give to the Lessee written notice thereof and the Lessee shall have the option to be exercised within one calendar month from the date of the receipt of the aforesaid notice to renew this Underlease for the same term (less one day) as shall be granted to the Lessor by such new Lease at a rent equal to seventy-eight eighty-fifth parts of such rent as shall be payable by the Lessor under such new Lease and containing such covenants provisions and conditions as may be appropriate in the light of those contained in such new Lease and the relationship of the covenants provisions and conditions herein contained to those contained in the said Superior Lease.

It is common ground that if, which the plaintiff denies, clause 8 had any application to a case where the original headlease was surrendered and the underlessor was granted a new headlease, no such notice as the clause contemplates was in fact given or sought.

In view of the provisions in clause 5 of the underlease regarding the rent payable thereunder as from March 25 1983, I must refer to the provisions in the original headlease regarding the manner of fixing the rent thereunder. As I have already mentioned, the basic rent as from May 13 1969 was £8,500 "(or such other amount as shall become payable under the provisions of clause 5 hereof)". To avoid confusion, I must stress that this reference was to clause 5 of *the original headlease*, which provided as follows:

> 5(a) Upon the request of the Lessors made not earlier than twelve months nor later than six months immediately preceding the expiration of the fourteenth year of the said term each party shall within fourteen days appoint a duly

qualified Valuer to negotiate and endeavour to reach agreement with the Valuer of the other party as to the fair market rack rent of the premises at that time for a Lease for the residue of the said term without taking a premium and disregarding the effect on rent of any alterations carried out by the Lessees (other than in pursuance of any obligation to the Lessor) or any change of use consequent thereon or any goodwill attached to the premises by reason of the lessee's occupation thereof on the same terms and conditions (save for this proviso) as are herein contained. And in the event of such Valuers not having reached such agreement by the expiry of the period of six weeks they shall agree upon an umpire to settle the question whose decision shall be final and binding upon both parties;

(b) If within the aforementioned period of fourteen days either party shall not have appointed a Valuer as aforesaid or if within the period of two weeks next following the said period of six weeks the said two Valuers shall not have agreed upon such umpire or such an umpire shall not have been appointed then in the first contingency the other party and in the second contingency either party may request the President for the time being of the Royal Institution of Chartered Surveyors to appoint a member of that body (but not the Valuer appointed as aforesaid by either party) to settle the question and the parties hereto shall be bound as aforesaid by the decision of the person so appointed and;

(c) The rent payable under the Lease from the expiration of the fourteenth year of the said term shall be the rent fixed in manner aforesaid or the rent of £8,500 per annum (whichever shall be the greater).

I should here add, although only for completeness, that the 1973 headlease also contained (once again in a clause 5) elaborate machinery for a series of rent reviews; the first rent review date was December 25 1982, when the initial rent of £27,000 per annum was increased to £70,000. But I must stress that the 1973 headlease demised considerably more property than had been comprised in the original headlease. It is no part of the plaintiff's case that the rent review provisions in the 1973 headlease play any part in the computation of the rent payable under the underlease.

Looking at the original headlease and the underlease together, certain points seem to me to be reasonably clear:

(i) the references to 78/85ths of the rent payable under the original headlease reflect the fact that at the time when Clark Retail entered into the underlease it retained for itself part of the property comprised in the original headlease, and the remaining seven eighty-fifths of the rent which it would have to pay under the original headlease and which it could not recover under the headlease reflected this retention.

(ii) Clause 5 of the original headlease clearly contemplates a review of the rent payable thereunder within the period thus specified by qualified valuers, with a view to their agreeing the then current "fair market rack rent", with provisions for an umpire or, if necessary, a reference to the president of the Royal Institution of Chartered Surveyors in default of agreement. This clause 5, which the lessor under the original headlease was entitled to invoke, bears all the hall-marks of a formula which could quite readily be used, if it became necessary, to quantify the rent which would become payable under the underlease as from March 25 1983.

(iii) Clause 5 of the underlease was clearly intended to protect the underlessor in the event of the rent payable by the underlessor as lessee under the original headlease being increased by reason of the machinery provided for in clause 5 of the original headlease.

(iv) The operation of this rent review machinery did not result in any new contract between the parties to the underlease. They remained underlessor and underlessee respectively (see *per* Lord Diplock in *United Scientific Holdings Ltd* v *Burnley Borough Council* [1978] AC 904 at p930).

For the defendant, Mr McEvoy contends that with the surrender of the original headlease, clause 5 of the underlease in effect ceased to operate so as to be capable of quantifying an increased rent: and in support of this argument he points to, and relies on, the language of clause 5 itself. His main points, I think, are that there was no longer any "rent payable by the Lessor . . . in respect of the principal premises as fixed in the manner provided by the said Superior Lease" because the "superior lease" had disappeared as a result of the surrender. He also relies on clause 8 of the underlease as showing that the parties had contemplated — and provided for — the possibility of the underlease expiring early; that this must have been intended to include the possibility of the surrender of the original headlease and its replacement by a new headlease; and that clause 8 contained a complete statement of the underlessee's rights in those circumstances, rights which it is admitted the underlessee never sought to exercise. Mr McEvoy also relies on the fact that the surrender of the original headlease was brought about by the action of the plaintiff's predecessor in title, so that it does not lie in the plaintiff's mouth to complain now of possible adverse consequences of that surrender.

For the plaintiff, Mr Pryor has two main points. First, that clause 5 of the underlease is merely a formula which operates by reference to the more detailed formula set out in clause 5 of the original headlease and remains capable of being operated purely as a matter of construction and is therefore totally unaffected either by the surrender of the original headlease or by such questions as the rent under the original headlease not being duly paid: payment of that rent, says Mr Pryor, is in no sense a condition precedent to the operation of clause 5 of the underlease. Second, and as an alternative to his first point, Mr Pryor relies on clause 5 as mere machinery and claims that if a term, properly construed, is mere machinery and is not an essential term of the contract, and if that machinery for some reason breaks down, the court will substitute its own machinery. For this, Mr Pryor relies heavily on the recent decision in *Sudbrook Trading Estate Ltd* v *Eggleton* [1983] 1 AC 444 at pp 464 *et seq*: in that case the House of Lords held, in relation to leases each of which contained an option to purchase the freehold reversion, that (and I quote from the headnote at p 445):

as the price was to be ascertained by machinery which, on the true construction of the agreement was a subsidiary and non-essential part of the contract, the court would, if the machinery broke down for any reason, substitute its own machinery, to ascertain a fair and reasonable price.

In the present case, as I have already indicated, I regard clause 5 of the underlease as merely providing machinery for the rent increase by reference to the machinery set out in greater detail in clause 5 of the original headlease. That being so, I respectfully adopt and apply what Lord Fraser said in *Sudbrook* at p 484, which seems to me to be directly in point of the present case:

Where, as here, the machinery consists of valuers and an umpire, none of whom is named or identified, it is in my opinion unrealistic to regard it as an essential term. If it breaks down there is no reason why the court should not substitute other machinery to carry out the main purpose of ascertaining the price in order that the agreement may be carried out.

As to Mr McEvoy's point that the difficulty in the present case stems from the seemingly voluntary act of one of the plaintiff's predecessors in title agreeing to the surrender of the original headlease, I would only say this. In *Sudbrook* the machinery had broken down because the lessors refused to appoint their valuer. On this aspect, Lord Fraser, again at p 484, said:

In that sense the breakdown has been caused by [the lessors'] fault, in failing to implement an implied obligation to co-operate in making the machinery work. The case might be distinguishable in that respect from cases where the breakdown has occurred for some cause outside the control of either party, such as the death of an umpire, or his failure to complete the valuation by a stipulated date. But I do not rely on any such distinction. I prefer to rest my decision on the general principle that, where the machinery is not essential, if it breaks down for any reason the court will substitute its own machinery.

In the present case, I know virtually nothing of the circumstances in which the original headlease came to be surrendered and I must also assume, I think, that the defendant's predecessor in title, Suede Circle, was not asked to agree to the surrender. But in my judgment, Wakefields' apparent failure to consider the possible legal effect of the surrender on clause 5 of the underlease cannot fairly be regarded as anything more than an oversight and is certainly not sufficient to preclude the court from substituting its own machinery, as Lord Fraser had contemplated.

I should also add that there is in my judgment nothing in the point which Mr McEvoy sought to raise on clause 8 of the underlease.

In an amended statement of claim the primary relief for which the plaintiff asks is formulated as follows:

(i) A declaration that on a true construction of clause 5 of [the underlease] the Plaintiff is entitled to have the rent payable by the Lessor as from the 25th March 1983 assessed as between the Plaintiff and the Defendant in the manner provided by [the original headlease];

(ii) Alternatively, an enquiry as to what is 78/85ths of the rent payable by the Lessor as from the 25th March 1983 in respect of the principal premises.

In para 3a of the body of the amended statement of claim, the plaintiff contended that the effect of clause 5 of the underlease was "to incorporate into the underlease the terms of [the original headlease] so far as might be necessary to ascertain the amount of rent payable in respect of the principal premises". In response to a

A question from me as to precisely what would thus be incorporated, Mr Pryor's junior, Mr Rees, in effect incorporated virtually the whole of clause 5 of the original headlease with one or two relatively minor and fairly obvious changes or adaptations.

As to this, Mr Griffiths, who appeared with Mr McEvoy, raised three points:

(1) that it would be necessary to exclude alterations carried out by the underlessor even if in pursuance of an obligation to the headlessor;
(2) that this incorporation would require the underlessee to incur the costs of appointing a valuer, something which the underlessee had never previously agreed to;
(3) that looking at the scheme of the underlease and the original headlease as a whole the underlessor was not to make any profit by B reason of any rent increase under the underlease, and this situation must be preserved.

Mr Pryor accepts point (1). As to point (2), he said that this was something which could be covered as a matter of machinery, but it was not in any way fatal to the substitution by the court of its own machinery. He contended that point (3) was not a good point and that the existence of a profit element was irrelevant, since the machinery was merely designed to ascertain 78/85ths of the fair market rent of the premises in question.

I think there is possibly sufficient in points (1) and (2) (I agree with Mr Pryor on point (3)) not to incorporate specific provisions of clause 5 of the original headlease, even with further relatively minor C changes or additions, although I would have been prepared to treat those changes or additions as *de minimis* if the only solution open to me in order to preserve clause 5 as an effective provision had been to proceed on the proper construction of clause 5 itself. But since, on the case as a whole, I have no doubt that justice requires that clause 5 of the underlease should not be emasculated in the manner for which the defendant contends, and that machinery for implementing what was clearly the original intention of the parties to the underlease should be provided by the court, I propose to proceed on the alternative basis for which the plaintiff contends and to order an inquiry with appropriate machinery so as to establish the 78/85ths proportion of the rent which should have been paid under the underlease as from March 25 1983 in respect of "the principal premises".

D I will hear further argument as to the precise terms of this order at the conclusion of this judgment.

It may have come as something of a surprise to those who argued this case so ably to discover that I have reached my decision without apparent reference to the judgment of Mr Julian Jeffs QC, sitting (as I do) as a deputy judge in the Chancery Division in 1981 in a case called *Lorien Textiles (UK) Ltd* v *S I Pension Trustees Ltd* reported in (1981) 259 EG 771. But in actual fact I have obtained very considerable assistance and encouragement from that decision. It was a case of an underlease and a subunderlease and the effect of the surrender of the former on the rent review provisions in the latter where, in certain events, the new rent to be paid under the E subunderlease was to be fixed in accordance with the covenant and conditions contained in the underlease in question. In broad outline, therefore, *Lorien* reflected many of the elements in the case before me. There were of course — inevitably — a number of differences between the two cases, but none which, in my judgment, go to the heart of the matter. *Lorien* was, of course, decided some six years before the House of Lords had (if I may be excused the expression) changed the rules of the game in *Sudbrook*. It comes as no surprise, therefore, to find Mr Jeffs upholding the proposition that "if the contract provides machinery for determining the rent and the machinery breaks down or becomes inoperable, the court cannot interface or substitute fresh machinery". That was clearly right as matters then stood and it was thus not open to Mr Jeffs to decide the F case before him on the basis on which I prefer to proceed in the present case. He decided *Lorien* on the construction of what was then also clause 5 of the subunderlease. I do not propose to examine in detail the differences between *Lorien* and the present case. Suffice to say that Mr Jeffs found, on the facts before him, (i) that the revised rent in the subunderlease fell to be determined "on the basis set out" in the underlease; (ii) that it mattered not that no rent was payable under the underlease; and (iii) that notwithstanding the determination by surrender of the estate granted by the underlease, it was still possible and proper to fix a yearly rent under it within the meaning of clause 5 of the subunderlease.

While I yield to no one in my admiration and respect for the late Lord Russell of Killowen, I cannot help feeling that the majority decision of his fellow law lords in *Sudbrook* blew a welcome breath of G fresh air into a part of the law which over a century and a half had tended to become somewhat inappropriate and, on occasion, unfair. In the words of Lord Scarman (at p 488):

> I challenge the relevance of ancient authority when construing the terms of the formula used in these leases. I agree with the comments of my noble and learned friends on the injustice which can arise by applying old cases when what has to be considered is a modern contract.

Mr Jeffs, of course, reached his conclusion without the aid of *Sudbrook*. While I do not for one moment suggest that anything I have decided in the present case is new law, I like to think that the result which I have arrived at is at least consistent with the reasoning which underlay the approach of the majority in *Sudbrook*. H

Court of Appeal

November 19 1987

(Before Lord Justice SLADE, Lord Justice BALCOMBE and Lord Justice STOCKER)

MORRIS v LIVERPOOL CITY COUNCIL

Estates Gazette April 9 1988

[1988] 14 EG 59-64

J

Landlord and tenant — Implied covenant to repair under section 32 of Housing Act 1961 (since replaced by section 11 of Landlord and Tenant Act 1985) — Appeal by tenant from decision of county court judge holding that tenant's claim under section 36(1)(a) of the 1961 Act on the ground of breach of covenant to repair failed — The facts were that the appellant was the tenant of a flat of which the local authority were the landlords and it was common ground that the tenancy agreement incorporated the covenants implied by section 32, in particular the covenant "to keep in repair the structure and exterior of the dwelling-house" — One early morning when K **the tenant was away from home there was a fire in the block in which the flat was situated — In the course of their duties the fire brigade, finding that they could get no reply from the appellant's flat, broke down its door (the only access) —The local authority as an emergency measure boarded up the door with thin plywood (the judge found it to be about a quarter of an inch thick) — The tenant later that day removed the sheeting temporarily to inspect the flat but replaced it, and nothing turns on this — About a week later the tenant (who had again been absent) discovered that the boarding had been removed and the flat had been burgled, a number of his possessions having been stolen — The tenant instituted** L **proceedings against the authority — His claim was rejected by the county court judge, apparently on the ground that the authority were not guilty of any breach of duty but, as was pointed out by Slade LJ, the reasons for the judge's conclusions were not altogether clear — The tenant appealed —Although the tenant's claim was originally pleaded in both contract and tort, it was presented in the Court of Appeal as based entirely on a breach of the implied covenant under section 32(1)(a) of the 1961 Act — The tenant claimed (1) that there was a contractual obligation on the authority to repair the door and door frame within a reasonable time after** M **receiving notice of the damage done by the fire brigade, (2) that the authority were in breach of that contractual obligation and (3) that the losses suffered by the tenant through the criminal action of the burglars were the foreseeable result of that breach and were not too remote — Held that there was clearly such a contractual obligation on the authority, but that there had been no finding by the judge, and no evidence on which the Court of Appeal itself could find, that the delay in replacing the door was unreasonable — Accordingly, the tenant had not discharged the onus of proving that the authority had failed to carry out repairs**

A **within a reasonable time — The court expressed the view *obiter* that, if the tenant had established the breach, the loss caused by the burglary would not have been considered too remote — Appeal dismissed**

The following cases are referred to in this report.

Calabar Properties Ltd v *Stitcher* [1984] 1 WLR 287; [1983] 3 All ER 759; [1983] EGD 578; (1983) 268 EG 697, CA
Donoghue v *Stevenson* [1932] AC 562, HL
King v *Liverpool City Council* [1986] 1 WLR 890; [1986] 3 All ER 544; (1986) 84 LGR 871, CA
Smith v *Littlewoods Organisation Ltd* [1987] AC 241; [1987] 2 WLR 480; [1987] 1 All ER 710, HL
B *Stansbie* v *Troman* [1948] 2 KB 48; [1948] 1 All ER 599; (1948) 64 TLR 226, CA
Weld-Blundell v *Stephens* [1920] AC 956

This was an appeal by the plaintiff, John Morris, from a decision of Judge Naylor, at Liverpool County Court, holding that the plaintiff had not established his claim that the defendants, Liverpool City Council, the present respondents, were liable as landlords for losses due to the burglary of the plaintiff's flat at 17A Cam Grove, Liverpool, after a fire.

Eric Somerset Jones QC and Norman Wright (instructed by Yaffe Jackson & Ostrin) appeared on behalf of the appellant; John Kay QC and Miss Tania Griffiths (instructed by the City Solicitor, Liverpool
C City Council) represented the respondents.

Giving judgment, SLADE LJ said: This is an appeal by Mr John Morris from a judgment of His Honour Judge Naylor given on June 17 1987 in the Liverpool County Court. Mr Morris was the plaintiff in the proceedings. The defendant, and respondent to this appeal, is the Liverpool City Council ("the council"). I will take the facts of the case largely from the judge's judgment and otherwise from the plaintiff's evidence, which the judge accepted in its entirety.

At all material times the plaintiff was the tenant and the council was his landlord of a flat known as 17A Cam Grove, Liverpool. It is common ground that the tenancy agreement incorporated the repairing covenants implied by section 32 of the Housing Act 1961. In January 1984, though the plaintiff was a tenant of the flat, he spent
D very little time there because his daughter had been unwell and he spent time looking after her home and her children. However, the judge found that he paid regular, probably daily, visits to his own flat.

In or about the early morning of Sunday January 15 1984 there was a fire in the block where the flat was situated. In the course of dealing with the fire, the fire brigade, finding that they could get no reply from the plaintiff's flat, broke down its door. He was not there. The judge found as a fact that the entry of the fire brigade had effectively ruined both the door and its frame, rendering it insecure. The door was the only means of access into the flat. The council then put emergency measures into effect by boarding up the door with
E sheeting which, as the judge found, was done by means of very thin plywood about a quarter of an inch thick, fastened by a few inadequate nails.

Later on the Sunday, the plaintiff heard of the fire, went to the flat and found that his door had been sheeted up. He came back with his son-in-law. Together they removed the sheeting to inspect the flat's contents. They found at that stage that his property was still intact. Having first tried to mend the door and found this to be impossible, they replaced the sheeting, supplementing the number of nails and not using the same holes in the plywood. They "clenched" the nails over to prevent them pulling through the wood if efforts were made to prise it away.

The next day, Monday January 16 1984, the plaintiff telephoned
F the council's office, asking them to come and mend the door so that he could have access to the flat. A man came round at about 4pm, but he did not repair the door and it remained sheeted up as before.

The following weekend, on the Saturday or the Sunday (January 21 or 22 1984), the plaintiff, on calling at the flat, discovered that the boarding on the door had been taken off and that the flat had been burgled. He found that he had lost a portable television set, an electric fire, a dining-room suite, various bedding, blankets and sheets, a washing machine, a kitchen cabinet and a dinner service. He reported the loss to the police. The council came that night and boarded up the door space with plywood nearly one inch thick, using on this occasion six-inch nails and doing what the plaintiff in evidence called "a decent and solid job". The door and frame, as I understand, were eventually replaced *in toto* on February 5. During G the intervening period, of course, the plaintiff had no way of getting into the flat.

On April 10 1985 the plaintiff instituted proceedings against the council. In his particulars of claim he pleaded that it was an implied term of the tenancy that the council would keep in repair the structure and exterior of the flat, and alleged:

> In breach of the said term and/or negligently the Defendants failed to repair properly or at all damage to the front door to the said flat.

The pleading was thus based both on breach of contract and on tort and was founded simply on the council's failure to repair the damage to the front door.

The council by their defence did not admit that the plaintiff was H tenant of the flat, but this is now accepted. They admitted that section 32 of the Housing Act 1961 applied to the tenancy but denied any liability. In the alternative they submitted that the plaintiff had not properly mitigated his damage or had contributed to it. These latter submissions have not been pursued in this court; they were rejected by the judge.

At the trial the judge heard evidence from two witnesses only, namely the plaintiff himself, whose evidence he accepted in its entirety, and Mr Peter Harrison, a housing assistant employed by the council. Mr Harrison, who had been responsible for the area for about 16 months, told him that burglary in the area was quite common and, where property of the council was left unoccupied without their being told, it was soon vandalised, sometimes within a J day. His evidence was that the plywood first put on the door to the flat by the council would more probably have been about one inch thick rather than one quarter inch as suggested by the plaintiff. However, on this point the judge preferred the plaintiff's evidence. As I have already indicated, he rejected the plea that the plaintiff had failed properly to mitigate his loss or had caused or contributed to it. However, in the event, he dismissed the plaintiff's claim on the grounds which he stated as follows in the last two pages of his judgment:

> Thus, the real question in this case, is whether the Corporation are to be fixed with liability for the acts of third parties, namely the burglars, in the circumstances of this case, it not being in dispute that the door formed part of K the exterior of the flat.
>
> In the course of her submissions, Miss Griffiths, for the defendants referred me to the cases of *Stansbie* v *Troman, King, Perl* and *Lamb* and she anticipated Mr Wright's submissions on the Scots case of *Evans* v *Glasgow District Council* reported in 1978 SLT at p 17. Mr Wright dealt additionally with the *Dorset Yacht* case. Additionally, I have derived what help I can from the case of *Smith* v *Littlewoods* [1987] 1 All ER 710. What Miss Griffiths' submissions come to is this: if there is a duty on the defendants, either in contract or in tort (and she by no means admits that there is) then their duty is simply to act reasonably, and this she contends they did. Additionally, she asserts that if there was a breach of duty, then such breach was not causative of the plaintiff's losses.
>
> Mr Wright in reply says that it is clear that the tenancy here creates a contract, an implied term of which was that the defendants would keep the structure and exterior in repair. He says I should determine whether here the premises L were kept in repair, conceding that there had to be notice of disrepair. As the defendants had notice — because they effected a temporary repair — I should find the temporary repair was not reasonably adequate. As to his claim in tort, Mr Wright sought to distinguish the various authorities, urging me to find that none deals with the situation in this case.
>
> I have thought long and hard about this case, and about the authorities to which I referred earlier. Some might think I have thought far too long, not least of all the plaintiff, whom I have perhaps already indicated I found to be a thoroughly decent, honest man, who, at his time of life deserves some recompense from the courts for the activities of those who do so much to give this city a bad name. Nevertheless, I come back, time and time again, to the words of the headnote to this case of *King* in the All ER "In the circumstances, and having regard to the fact that it was not possible for the council to take M effective steps to defeat the actions of trespassing vandals, the council did not owe a duty of care to the plaintiff in respect of the damage caused by the actions of the vandals". I am, regrettably, unable to say that that which the council here did, namely to use only one quarter inch ply was unreasonable, given what I might call their "workload".
>
> In short, I am not satisfied on the current state of the law that the defendants are liable for the losses *caused by the burglars*

the judge's emphasis

> and if I were so satisfied I would not have been able to say that the defendant's actions were causative of those losses.

With respect to the learned judge I think that the reasons for his conclusion as expressed in the final passages of his judgment which I have read are rather hard to follow. First, as appears from his citation

A from the headnote of *King* v *Liverpool City Council* [1986] 3 All ER 544, that decision was based on the absence of a duty of care owed by the Liverpool City Council to the plaintiff (in that case to prevent damage caused by the action of vandals). However, in holding that the use by the council of one-quarter-inch plywood was not unreasonable, the learned judge appeared to be accepting that the council in the present case did owe a duty to the plaintiff, but to be finding that the council were not in breach of that duty. Whether or not that finding as to the absence of a breach of duty was correct, it does not seem to me to follow from his reference to the *King* decision.

Second, I am puzzled by the judge's phrase "if I were so satisfied" in the last sentence of his judgment which I have already quoted. I do not see how the judge could have been satisfied on the current state of B the law that the council were liable for the losses caused by the burglars unless the council's actions were causative of those losses.

It therefore seems to me that in this court, albeit on the basis of the judge's clear and helpful findings of fact, we have to examine the legal position more or less entirely afresh. It may well be that, with the assistance of leading counsel on both sides in this court, the course of the argument may have been somewhat different from that which it followed in the court below.

The recent decision of the House of Lords in *Smith* v *Littlewoods Organisation Ltd* [1987] 1 All ER 710 well illustrates that the law does not recognise a general duty of care at common law to prevent others from suffering loss or damage caused by the deliberate wrongdoing C of third parties. Lord Goff explained the reasons thus (at p 729C):

The fundamental reason is that the common law does not impose liability for what are called pure omissions.

He referred (at p 729J) to the statement of principle by Lord Sumner in *Weld-Blundell* v *Stephens* [1920] AC 956 at p 986:

In general . . . even though A is in fault, he is not responsible for injury to C which B, a stranger to him, deliberately chooses to do.

However, Lord Goff proceeded (at p 730B) to say:

That there are special circumstances in which a defender may be held responsible in law for injuries suffered by the pursuer through a third party's deliberate wrongdoing is not in doubt. For example, a duty of care may arise from a relationship between the parties which gives rise to an imposition or D assumption of responsibility on or by the defender, as in *Stansbie* v *Troman* . . . where such responsibility was held to arise from a contract.

I turn to *Stansbie* v *Troman* [1948] 2 KB 48, where the brief facts as stated in the headnote were:

A contractor carrying out decorations in a house was to his knowledge left alone on the premises by the householder's wife. During her absence, he left the house to obtain wall-paper. He fastened back by its catch the latch of the yale lock on the front door and closed the door behind him. That door was accordingly then held shut only by its mortise lock, and could be opened by a mere turn of the handle. During the decorator's absence a thief entered the house and stole property, the value of which the householder claimed from the decorator.

The county court judge, after referring to *Donoghue* v *Stevenson* E [1932] AC 562, had held that:

when the decorator was left alone in the house, he was in a position to exercise control over the access into the house and this put him in such a relation to the householder that, in exercising such control, he ought to have had in mind the safety of the householder's goods which were in the house,

and had therefore given judgment for the householder.

It was argued for the decorator that no duty was owed, that the duty must arise from the relationship between the parties, and that it must be a duty within the scope of the contractual relationship existing between them. I agree that the duty must be within the scope of the contractual relationship between these two persons, but I think that that contractual relationship did impose a duty on the plaintiff decorator to take reasonable care with regard to the state of the premises if he left them during the performance of his work. That, I think, F was the measure of the duty.

Tucker LJ then proceeded to consider whether there was a breach of that duty and he said (at p 51):

If I am right as to the existence of the duty, I think that there can be no question but that there was a breach of it, for I do not think that it was acting reasonably, that it was taking reasonable care, to leave this empty house for a period of two hours with the front door in that condition.

Tucker LJ then proceeded to consider whether the negligence of the decorator was causatively responsible for the householder's loss. He quoted the dictum of Lord Sumner from *Weld-Blundell* v *Stephens* [1920] AC 956 (which I have already quoted) and said (at pp 51-52):

I do not think that Lord Sumner would have intended that very general statement to apply to the facts of a case such as the present where, as the judge points out, the act of negligence itself consisted in the failure to take G reasonable care to guard against the very thing that in fact happened. The reason why the decorator owed a duty to the householder to leave the premises in a reasonably secure state was because otherwise thieves or dishonest persons might gain access to them; and it seems to me that if the decorator was, as I think he was, negligent in leaving the house in this condition, it was as a direct result of his negligence that the thief entered by the front door, which was left unlocked, and stole these valuable goods. Except that I would have phrased the nature of the duty somewhat differently from the way in which the county court judge put it, I am in entire agreement with his judgment, and in my view the appeal fails.

On the basis of the decision of this court in *Stansbie* v *Troman* it might conceivably have been arguable on behalf of the plaintiff that, having regard to the contractual relationship existing between him H and the council, the council owed him a general duty to make the flat reasonably secure against intruders as soon as they were aware of the breaking down of the door by the fire brigade. The plaintiff's case, however, was not pleaded in this way and it has not been presented in this way in this court. His case as argued in this court by Mr Somerset Jones is founded exclusively on three clear and simple propositions, which are substantially as follows.

First, there was a contractual obligation on the council to repair the door and door frame within a reasonable time after receiving notice of the injury to them done by the fire brigade; second, the council were in breach of that contractual obligation; and, third, the losses suffered by the plaintiff were the foreseeable result of that breach and were not too remote in law to be recoverable. J

The plaintiff to succeed on this appeal thus has to surmount each of these three hurdles. Mr Somerset Jones accepted that, if he could not succeed in contract, *a fortiori* he could not succeed in tort and therefore, if I may say realistically, did not pursue the alternative claim in tort.

As to the first of these propositions, section 32 of the Housing Act 1961, which was the statutory provision in force at the relevant time, though it has since been replaced by section 11 of the Landlord and Tenant Act 1985, provides so far as material as follows:

(1) In any lease of a dwelling-house, being a lease to which this section applies, there shall be implied a covenant by the lessor —
(a) to keep in repair the structure and exterior of the dwelling-house . . .

It is common ground that the door and frame of the flat were part K of the structure and exterior of the flat within the meaning of this implied statutory covenant. It is also common ground that the effect of this implied obligation is to impose on the landlord the obligation to effect any necessary repair within a reasonable time after receiving notice of the want of repair: see *Calabar Properties Ltd* v *Stitcher* [1984] 1 WLR 287 (at p 298G) *per* Griffiths LJ (as he then was). The plaintiff therefore surmounts the first of the three hurdles which I have mentioned.

It is the second of them which, in my judgment, presents him with far more formidable difficulties. Did the council in breach of their obligation fail to repair the door and frame within a reasonable time after receiving notice of the damage done to them? In contending that L the council were in breach of this obligation Mr Somerset Jones has not submitted that they were in breach on the grounds that the temporary plywood barrier erected by them was inadequate as a temporary barrier. His submission has been that the only way in which the covenant to repair could be complied with was by replacement of the whole door and frame and that a reasonable time for effecting this repair had elapsed between the day when the council first had notice of the want of repair and the following weekend, when the burglary occurred without the repair yet having been done. He stressed that this was a case where the door constituted the sole means of access to the flat and that the plaintiff could not get into it until the door had been replaced. In those circumstances, in his submission, a reasonable period for the doing of the work would M have been at most 48 hours.

However, as Mr Kay pointed out on behalf of the council, the answer to these submissions lies in the findings of the judge and in the state of the evidence. We have been told that counsel appearing on behalf of the plaintiff submitted to the judge both that it was unreasonable for the council to use one-quarter-inch plywood in erecting the temporary barrier and that delay in replacing the door was itself unreasonable. It is not entirely clear to me whether the judge was intending to reject merely the first or both of these submissions as a matter of fact when he said in the passage which I have already read:

I am, regrettably, unable to say that that which the council here did, namely to

A use only one quarter inch ply was unreasonable, given what I might call their "workload".

What is clear is that he made no finding at all that the *delay* in replacing the door was unreasonable. This is a pure question of fact and on this point it therefore becomes of crucial importance to look at the evidence which was before the judge.

No evidence was called on behalf of the plaintiff to establish that the period of delay was in itself unreasonable. In the notes of evidence which the judge has approved, the only relevant passage is to be found in the notes of Mr Harrison's evidence in chief in which he is recorded as having said:

> Call out service should have reported to someone they'd done an emergency B sheeting. Can't say if did. Housing Office maintenance dept react. Take approx 1 week for door and frame (to be fitted). Same, just a frame because of financial problems. Sheeting is emergency top priority. No real deterrent been to a forceable (*sic*) entry. We at one time tried welding a frame but it didn't work.

Though we were told that there was some cross-examination of Mr Harrison, we have no notes of this cross-examination. The notes of evidence are thus somewhat sparse, but we have to do the best we can with them. I read Mr Harrison's evidence, as recorded in these notes, as being to the effect that the council could not arrange for the replacement of the door and frame within less time than about a week. So far as I can tell, the matter was not explored further in evidence before the learned judge.

C Though recognising the scantiness of the evidence on this material point of fact, Mr Somerset Jones submitted to us that the facts more or less speak for themselves. In his submission, if the council could not arrange for the replacement of this door, which afforded the only means of access to one of their tenants' flats, within less than a week, this shows on the face of it that their arrangements were inefficient and inadequate. In his submission, the period of one week was on the face of it unreasonably long, particularly in an area where burglary was known to be rife.

I cannot, for my part, accept this submission. It has to be stressed that the onus throughout this case was on the plaintiff to prove unreasonable delay on the part of the council, not on the council to D prove the converse. If the matter had been explored further in evidence, it may be that the plaintiff could have established the point. However, it seems to me that, if it had been explored, the reasonableness or otherwise of the delay would have depended on evidence as to a number of imponderable matters such as the difficulty or otherwise for the council in obtaining the proper size and quality of door and frame and also on what the judge referred to as the council's "workload". Furthermore, in assessing the reasonableness or otherwise of the period, it seems to me that the court might have been entitled to take into account that the plaintiff was not living in the flat at the time and that the council had already taken at least some steps in an effort temporarily to secure the property.

E In my judgment, on the limited evidence before the judge and before us, as recorded in the notes of evidence, it simply is not open to this court to make a finding, not made by the learned judge, to the effect that the delay of a week or less which elapsed between the council's knowledge of the damage to the door and the burglary was in all the circumstances unreasonable. Accordingly, in my judgment, it has not been proved that, as at the date of the burglary, the council were in breach of contract. Whether they became in breach of contract because of the delay that ensued thereafter is not an issue in the present proceedings. It follows that, in my judgment, this appeal must fail because the alleged breach of contract on which it is founded has not been proved.

F In deference to Mr Somerset Jones' argument, I would add that, if I had found that a breach of contract had been proved, I would, for my part, have held that the loss suffered by the plaintiff through the burglary was not too remote to be recoverable. This conclusion seems to me to be supported not only by the decision in *Stansbie* v *Troman* but also by the comments on that decision in *Smith* v *Littlewoods Organisation Ltd* made by Lord Mackay at 724 of the report of that case. It seems to me that if by the time of the burglary, contrary to my view, the council had already been in breach of their contractual duty to replace the door, they ought, in all the circumstances of this case, reasonably to have foreseen that unauthorised intrusion and theft might well result. As the judge found, burglary in this area was, unhappily, "rife" or "commonplace".

As things are, while I have some sympathy with the plaintiff in his G ill fortune, I can see no grounds in law for holding the council responsible for his losses and I would dismiss this appeal.

Agreeing, BALCOMBE LJ said: The plaintiff has based his claim in contract and relies on the defendant landlord's obligations under section 32 of the Housing Act 1961, which was the statute in force at the relevant date. That section implies a covenant by the lessor to keep in repair the structure and exterior of the dwelling-house. The plaintiff further accepts that the landlord's obligation in this respect is not absolute. As Griffiths LJ (as he then was) said in *Calabar Properties Ltd* v *Stitcher* [1984] 1 WLR 287 at p 298:

> . . . a landlord is not in breach of his covenant to repair until he has been given notice of the want of repair and a reasonable time has elapsed in which the H repair could have been carried out.

The plaintiff gave notice to the landlord of the damage to his front door on January 16 1984, so the only question on this appeal is whether, before the burglary on January 21-22 (which is some five to six days later), a reasonable time for carrying out the repair — ie replacing the front door and the door frame — had elapsed.

Unfortunately, although we are told that this point was argued before the learned judge, it is not expressly dealt with in his judgment. However, it seems to me improbable that the judge, who had expressed great sympathy with the plaintiff, could have decided the case in the way he did unless he was satisfied at the very least that the plaintiff had failed to establish a breach of covenant by the landlord, J that is a failure to repair within a reasonable time. Further, the notes of evidence on this point are sparse in the extreme. Slade LJ has already quoted the passages from the evidence of Mr Harrison, the defendants' housing assistant, and no note of any cross-examination of Mr Harrison exists. I am quite unable to say that the learned judge was wrong on this aspect of the case and, in my judgment, the plaintiff failed to establish a breach of the covenant to repair. If he had established such a breach, for my part, in reliance on *Stansbie* v *Troman* [1948] 2 KB 48, as explained in *Smith* v *Littlewoods Organisation* [1987] 1 All ER 710, I too would have been prepared to hold that the plaintiff's loss caused by the burglary was not too remote.

K Mr Somerset Jones, quite rightly, accepted that under this head of a failure to repair, an alternative claim in the tort of negligence did not add anything. No claim was pleaded by the plaintiff on the basis that the defendants owed him a duty of care to make the premises secure pending the necessary repairs being effected with the consequential liability in tort based on a breach of that duty. So I make no comment on what might have been the result if such a claim had been made, although I can see the difficulties there might well have been in the plaintiff's way of making such a claim.

I agree that this appeal should be dismissed.

Also agreeing, STOCKER LJ said: The appeal before this court has proceeded solely on the basis that the loss sustained by the plaintiff L appellant was caused by the breach by the defendants of the repair obligations implied into the tenancy under section 32(1)(a) of the Housing Act 1961; that is to say that the repair to the door was not effected within a reasonable time before the burglary which took place some six days later.

The facts have been fully stated by my lords. In my view, this appeal fails on that issue, since there was no real evidence on the question of what was a reasonable time for effecting the repairs in question. Mr Harrison, the housing assistant employed by the defendants, said: "It would take approximately one week for a door and frame to be fitted" and "Same, just a frame because of financial problems". No note of any cross-examination of this witness exists. M The plaintiff himself did not call evidence on this issue and his own evidence is critical of the temporary protective sheeting rather than the time taken to effect the permanent repair. There was, therefore, in my view, no evidence to establish that the defendants were in breach of the repair covenant implied into the tenancy. Had a breach of contract been established I also agree that the losses which were in fact sustained were probably recoverable for the reasons given by my lords.

Accordingly, I agree that this appeal fails and must be dismissed.

The appeal was dismissed. No order was made for costs save legal aid taxation of the plaintiff's costs.

Court of Appeal

January 11 1988

(Before Lord Justice SLADE and Lord Justice STOCKER)

SILVERMAN AND OTHERS v AFCO (UK) LTD AND OTHERS

Estates Gazette April 9 1988

[1988] 14 EG 67-70

Landlord and tenant — Forfeiture — Relief — Alteration of parties' position — Discretion to refuse relief from forfeiture for non-payment of rent in exceptional circumstances — Appeal from decision of Turner J refusing relief — "Eleventh-and-a-half hour" application by tenants after a series of steps to delay giving up possession — Tenants at first said that they would not contest landlords' proceedings for possession, payment of rent arrears and mesne profits but would seek a stay in respect of any money judgment — Landlords had, on the basis of tenants' assurance and the order for possession in proceedings not defended by tenants, entered into negotiations for the grant of a tenancy of the premises to a third party — There then followed an application by the tenants, which was dismissed, for a stay of execution on the money judgment; a summons for relief against forfeiture and for a stay of all further proceedings, including the execution of the order for possession; an agreement by the landlords not to execute a writ of possession and fieri facias; a hearing before a master at which the tenants offered only a post-dated cheque, the summons for relief being then dismissed; after the execution by the landlords of the lease of the premises to the third party, a notice of appeal to the judge in chambers from the master's dismissal of the summons for relief; and, two days before the hearing by the judge, the proffer of a cheque for the amount due, properly drawn by a partner in the tenants' firm of solicitors on the firm's client account, but held by Turner J to be too late — Held, on appeal, that the judge had properly exercised his discretion and there was no misdirection — The judge had correctly applied the principles in the case of *Gill* v *Lewis* and the earlier case of *Stanhope* v *Haworth* — The present was a case where the position of the parties had altered and the rights of a third party had intervened — In these circumstances the court can properly refuse relief — Appeal dismissed

The following cases are referred to in this report.

Chandless-Chandless v *Nicholson* [1942] 2 KB 321
Gill v *Lewis* [1956] 2 QB 1; [1956] 2 WLR 962; [1956] 1 All ER 844, CA
Stanhope v *Haworth* (1886) 3 TLR 34

This was an appeal by AFCO (UK) Ltd, Dildar Aziz Khan and Tayyadas Iqbal Fareed, defendants, from the decision of Turner J refusing them relief from forfeiture in proceedings for possession, arrears of rent and mesne profits brought by the plaintiffs (the present respondents), Joyce Silverman, Geoffrey Mayer Silverman and Peter Douglas Ziwan. The premises concerned were the back suite of offices on the mezzanine floor of 92 Regent Street, Westminster.

Leslie Michaelson (instructed by Kennedys) appeared on behalf of the appellants; Laurence Marsh (instructed by Silvermans, of Barnet, Herts) represented the respondents.

Giving judgment, SLADE LJ said: On December 17 1987 Turner J made an order refusing the defendants in these proceedings relief from forfeiture of certain office premises of which they were tenants consisting of the back suite of offices on the mezzanine floor in the building known as 91 Regent Street in the City of Westminster ("the premises"). In view of the urgency of the matter an appeal by the defendants from his decision was heard by us on the last day of the Michaelmas Term, December 21 1987. We dismissed the appeal and said that we would give the reasons for our decision at a later date. This we now do.

The history of the matter is as follows. By a lease dated July 2 1987 and made between the plaintiffs of the first part, the first defendant of the second part and the second and third defendants as guarantors of the third part the premises were demised by the plaintiffs to the first defendant for a term of 10 years from December 23 1985 at an annual rent of £6,250 (subject to variation as provided by the lease) and further rents payable in respect of service and insurance charges. The lease contained a proviso entitling the plaintiffs to forfeit it and to re-enter the premises in the event of the rent or any part thereof being in arrears for 21 days after becoming due.

On August 26 1987 the plaintiffs issued a writ endorsed with a statement of claim asserting that there was a balance of £7,394.07 rent due and unpaid under the lease and claiming possession of the premises, payment of the last-mentioned sum plus interest and mesne profits plus interest.

On September 14 1987 (as we have been told in an agreed chronology) solicitors acting on behalf of the first defendant filed an acknowledgement of service stating that *it did not intend to contest the proceedings* but intended to seek a stay of execution in respect of any judgment entered in respect of the debt or liquidated demand.

On September 28 1987, no notice of intention to defend having been given by any of the defendants, an order was made that the defendants should

(i) give the plaintiffs possession of the premises;
(ii) pay the plaintiffs £7,499.41 and £86.90 interest making a total of £7,586.31;
(iii) pay the plaintiffs mesne profits at the rate of £18.84 per day from September 29 1987 to the time of obtaining possession.

Following the making of that order the plaintiffs entered into negotiations for the grant of a lease of the premises for three years to a chartered quantity surveyor, Mr C S Parsons, with a view to his going into occupation on December 1 1987. In affidavits sworn in these proceedings Mr Parsons and the plaintiffs' solicitor, Mr Gary Chambers, stated that an agreement for the grant of a lease to Mr Parsons was concluded in October 1987. Mr Chambers exhibited copies of two letters passing between the parties respectively dated October 9 and 13. These letters do not by themselves evidence a concluded agreement and there is no particularised evidence before us to show when a legally binding agreement (if any) for the grant of a lease to Mr Parsons was finally reached. However, the correspondence shows that it was the parties' intention that the term of the new lease should begin on December 1 1987 and that by October 13 they had at least reached an advanced state of negotiation. Mr Parsons himself closed his previous office premises in Somerset on October 15 1987.

On October 20 1987 an application by the defendants for a stay of execution on the money judgment was dismissed. On that same day they issued a summons seeking relief from the forfeiture and a stay of all further proceedings in the action, including execution of the order for possession. The summons was returnable before the master on December 1. We have been told that it was simply a coincidence that this was also the date on which the plaintiffs contemplated giving possession of the premises to Mr Parsons. So far as the evidence shows, the issue of this summons was the first intimation by the defendants that they would or might seek to retain possession of the premises.

On October 27 1987, pursuant to the leave of Master Turner, the plaintiffs issued a writ of possession and *fieri facias*. However, their solicitors agreed with the defendants' solicitors that it should not be executed pending the hearing on December 1 1987 before Master Turner.

The defendants apparently took few steps to prepare for that hearing. They filed no affidavit in support of the summons. They made no tender of payment. Effectively all they did at the hearing on December 1 was to make an offer of a post-dated cheque of their own in settlement of the arrears (an offer which they repeated after the hearing). Not surprisingly, Master Turner considered this offer inadequate and dismissed the defendants' summons for relief from forfeiture. The correctness of the master's decision, in the circumstances as they were before him, was not challenged at the subsequent hearing before Turner J. At the hearing before the master the defendants did not seek a stay of execution or mention any intention to appeal.

Following this decision, the plaintiffs took the view that there was nothing to prevent their implementing the arrangements with Mr Parsons. On December 1 1987 they informed him that they would be able to enter into a lease of the premises and arranged to meet him on

A the following day. On Wednesday December 2 at about 11 am he attended their solicitors' offices and shortly thereafter a lease of the premises bearing that date was executed in his favour. The lease was substantially in the same form as the lease to the defendants save that the annual rent was £6,500, the term was three years and there were no guarantors.

At 12.21 pm, *after* execution of this new lease, there arrived at the plaintiffs' solicitors' offices a "faxed" letter from the defendants' solicitors stating that they would probably be instructed to appeal and seek a stay of execution in the meantime.

On December 3 1987 the defendants issued a notice of appeal to the judge in chambers from the master's decision of December 1. On December 10 the defendants' solicitors obtained a stay of execution
B of the writ of possession until December 17 when the appeal was due to be heard by the judge in chambers.

On December 15 1987 Mr Chambers swore an affidavit on behalf of the plaintiffs stating that the total amount due to them was £6,152.84. On December 16 1987 Mr Olympitis, a partner in the firm of the defendants' solicitors, made an offer to the plaintiffs' solicitors by telephone of a cheque for this amount, and on December 17 1987 sent them a cheque drawn on his firm's client account, which was available for examination at the hearing before Turner J. It is common ground that this offer of payment was properly and sufficiently made.

On December 17 1987 Turner J dismissed the defendants' appeal
C from Master Turner's order. The short note of his judgment available to us reads as follows:

> On December 1 1987 Master Turner refused an application by the defendant to be relieved from forfeiture. It was accepted before me that Master Turner's decision was unexceptionable — I would say unappealable except that the defendants have the right of appeal.
>
> The defendants say they are entitled to relief. The circumstances have changed in that at the eleventh-and-a-half hour they have tendered a cheque for the amount that is owing drawn on their solicitors client account.
>
> That would have saved the day but the plaintiffs, not unreasonably given the history of this matter, on December 2 had granted a fresh lease of the premises. There is binding and persuasive authority to which I have been referred — *Gill* v *Lewis* (1956) 2 QB 1 at 10 *per* Jenkins LJ. The plaintiffs are not to be embarrassed by this eleventh-and-a-half hour application. It would
> D be quite wrong.
>
> Appeal dismissed with costs.
>
> Stay of execution of writ of possession pending appeal to the Court of Appeal refused.

On December 18 1987 Parker LJ granted a stay of execution over the hearing of the appeal on December 21 1987. On that day, as we have said, we dismissed the appeal.

The right of a tenant to apply for relief from forfeiture for non-payment of rent is conferred by the combined effect of section 38(1) of the Supreme Court Act 1981 (replacing section 46 of the Supreme Court of Judicature (Consolidation) Act 1925) and section 210 of the Common Law Procedure Act 1852.

These statutory provisions by their terms confer on the court a
E discretionary power to grant relief from forfeiture. As Mr Michaelson rightly stressed on behalf of the defendants, however, it is a well-established general principle that:

> The court, in exercising its jurisdiction to grant relief in cases of non-payment of rent is, of course, proceeding on the old principles of the court of equity which always regarded the condition of re-entry as being merely security for payment of the rent and gave relief if the landlord could get his rent.

(*Chandless-Chandless* v *Nicholson* [1942] 2 KB 321 at p323 *per* Lord Greene MR).

Consequently, as Mr Michaelson also rightly submitted, it is an equally well-established principle that, save in exceptional circumstances, a court which is asked to grant relief from forfeiture for non-payment of rent should do so if by the time of judgment all
F that is due for rent and costs has been paid up or tendered: see, for example, *Gill* v *Lewis* [1956] 2 QB 1 at p13 *per* Jenkins LJ.

However, as Hodson LJ pointed out in the same case (at p17) the so-called right to relief is a right to relief in equity and equity reserves to itself the right to refuse relief in an appropriate case. In the passage in his judgment in *Gill* v *Lewis* to which Turner J clearly intended to refer in the present case, Jenkins LJ, after referring to the decision of this court in *Stanhope* v *Haworth* (1886) 3 TLR 34, said at p10:

> So that case shows that where parties have altered their position in the meantime, and in particular where the rights of third parties have intervened, relief ought not to be granted where the effect of it would be to defeat the new rights of third parties or be unfair to the landlord having regard to the way in which he has altered his position.

Those are some of the exceptional circumstances in which the court G
may properly refuse relief.

In *Stanhope* v *Haworth* an order had been made in favour of a landlord for possession of colliery premises. Relief was applied for towards the end of the six-month period allowed by section 210 of the Common Law Procedure Act 1852. Meanwhile, the landlord had let other people into possession and he and they had spent substantial sums in working, or purchasing plant to work, the colliery on the footing that the lease was at an end. This court agreed with the Divisional Court that the tenant should not be given relief. Lindley LJ said at p35:

> The tenant, he said, had delayed his application for relief until just before the end of the six months, and in the meantime the position of the plaintiff was H
> altered. The plaintiff had in the meantime kept the colliery up, and had entered into an arrangement with other parties to let the colliery to them. There was no explanation of the delay by the tenant except that he had not had the money. That was no reason for allowing him now to have relief to the prejudice of the other parties, and it would be unreasonable to give him such relief.

Lopes LJ (at p35) said that he quite agreed with the Divisional Court:

> that it would be "monstrous" now to give the tenant such relief as he asked. Relief ought not to be granted if the landlord and other parties interested could not be put in the same position as before, and here that was impossible, and the relief asked for would be unjust and inequitable.

Mr Michaelson submitted that there are a number of points which render the present case distinguishable from *Stanhope* v *Haworth*. J
First, he pointed out (and we accept) that there is no clear evidence that the plaintiffs had actually committed themselves to any legally binding agreement to grant a lease to Mr Parsons before they did so on December 2 1987. In these circumstances, he submitted, the grant of the lease to him should be regarded as a purely voluntary undertaking. Second, at the time when the plaintiffs executed that lease they knew of the defendants' wish to remain in the premises, a wish which they had demonstrated before the master, even though they had not been able to support it by a tender of the outstanding rent. Furthermore, at that time they knew or should have known that, by virtue of RSC Ord 58, r1, the defendants had the right to appeal to the judge in chambers from the master's order within five days from its date. Third, so Mr Michaelson submitted, at the time K
when they executed the lease in favour of Mr Parsons, the plaintiffs were aware that the defendants would shortly be in funds because they had offered the plaintiffs a post-dated cheque. Fourth, in his submission, the plaintiffs, in granting the lease to Mr Parsons, should have made it clear to him, if they did not do so, that the defendants were still in possession of the premises and had the right to appeal from the master's order within the next four days. The plaintiffs and Mr Parsons, in proceeding to execute the lease as they did on December 2, were, in Mr Michaelson's submission, doing so at their own risk. The grant of relief to the defendants, he contended, would, in all the circumstances, cause no unfair prejudice to either the plaintiffs or to Mr Parsons. The prejudice to Mr Parsons, it was L
said, would in any event be purely financial, since he was not yet in possession of the premises and would be paying a rack-rent for them.

We see the force of at least some of these submissions and accept that the grounds for refusing relief in the present case are certainly less compelling than they appear to have been on the facts of *Stanhope* v *Haworth*. Nevertheless, the jurisdiction to grant relief for non-payment of rent is of a discretionary nature and in the end, as Mr Marsh submitted on behalf of the plaintiffs, the question for this court was whether sufficient grounds had been shown for us to interfere with the exercise of Turner J's discretion. In our judgment there were not.

In our judgment, a crucial question is whether or not the learned judge was justified in finding as he did that the plaintiffs had acted M
"not unreasonably" in executing a lease of the premises in favour of Mr Parsons on December 2 1987. It is not surprising that, effectively, by far the greater part of the argument before us was directed to this issue. We have already summarised the contentions put forward on behalf of the defendants by Mr Michaelson, who submitted that the grant of the lease on that day, if not a manoeuvre calculated to defeat the defendants' rights to apply for relief on appeal to the judge in chambers, was at least a precipitous act done in what should have been the knowledge that an appeal might follow and that relief might be granted by the judge. Mr Marsh on behalf of the plaintiffs, however, submitted with no less force that, having regard to the history of the matter, there was no sufficient reason why the

A plaintiffs should have been expected to wait beyond December 2 before executing the new lease.

The learned judge accepted the latter submission and we are not disposed to differ from his view, in particular for the following reasons.

On the available evidence, we are, as we have already indicated, willing to assume in favour of the defendants that no agreement for the grant of a lease by the plaintiffs to Mr Parsons had been concluded before the lease in his favour was actually executed. However, as we have also indicated, the evidence does show that by October 13 the parties were in an advanced state of negotiation and that already by that date it was the plaintiffs' stated intention to give him possession of the premises on December 1 1987 if possible.

B Furthermore, in our opinion, as at October 13 the plaintiffs would have been fully justified not only in entering into arrangements for the grant of a lease of the premises to Mr Parsons but in doing so in the belief that they would be able to give him possession at least by December 1. The first defendant in its acknowledgement of service had specifically intimated that it did not intend to defend the claim for possession. No notices of intention to defend the proceedings had been given by any of the defendants. The summons seeking relief from forfeiture had not yet been issued. As at mid-October there was no reason for the plaintiffs to suppose, when making their arrangements with Mr Parsons, that they would not be able to give him vacant possession of the premises as their new tenant in due C course. He himself closed his previous office in Somerset on October 15 1987.

The first intimation from the defendants that they might not be willing voluntarily to vacate the premises came when they issued their summons for relief from forfeiture on October 20 1987. The plaintiffs' solicitor, having issued a writ of possession and fieri facias, subsequently, sensibly, agreed with the defendants' solicitors that it should not be executed pending the hearing of the summons for relief returnable on December 1 1987. The defendants, however, must be taken to have known that the plaintiffs were willing to show forbearance only until that date. The plaintiffs, for their part, in our judgment, were entitled reasonably to assume that, if the defendants had both the intention and the ability to pursue an effective D application for relief from forfeiture, they would make a tender of all the outstanding moneys due to the plaintiffs before the hearing on December 1 or at very latest at the hearing.

The defendants, who had had several more weeks in which to collect the money since the issue of their summons on October 20 1987, did no such thing. All they did at the hearing before the master on December 1 1987 was to make an offer of a post-dated cheque without (so far as the evidence shows) any guarantee that the cheque would be met or, it would seem, any explanation as to why the landlord's further forbearance was sought in this manner.

When Master Turner, surely to no one's surprise, rejected the defendants' application for relief, they did not ask him for any E further stay of execution. Nor did they on December 1 express their intention to appeal from his order.

When their offices opened on December 2 1987 the plaintiffs' legal advisers were presumably aware that the defendants had the right under the rules to lodge an appeal to the judge in chambers from the master's order at any time within the next four days. However, as Mr Marsh asked rhetorically during the course of argument: "How many bites of the cherry are tenants to be allowed in circumstances such as these?" The plaintiffs and their legal advisers could not, in our opinion, be said to have been unreasonable in taking the view that the defendants had come to the end of the road and that there was no sufficient reason why they should defer any longer the grant of the proposed lease to Mr Parsons, who had left his own previous business F premises on October 15 and to whom they had held out the hope and expectation of completion on December 1.

On this crucially important premise we can see no sufficient ground for interfering with the exercise of the learned judge's discretion in refusing the defendants relief from forfeiture in accordance with the principles stated by Jenkins LJ in *Gill* v *Lewis* (*supra* at p10).

At the time when he accepted the lease neither Mr Parsons nor the plaintiffs themselves would, in our judgment, have been unreasonable in thinking that the plaintiffs were in a position to confer on him a good title to the premises which they were purporting to demise. By the time when the defendants made their first tender of the sums due to the plaintiffs, very shortly before the hearing, on December 17 1987, both the plaintiffs and Mr Parsons had G substantially altered their positions by the grant and acceptance of the new lease. Mr Michaelson criticised the learned judge's description of the relevant passage from Jenkins LJ's judgment in *Gill* v *Lewis* at p10 as "binding and persuasive authority". In our judgment, however, that passage and the decision in *Stanhope* v *Haworth* on which it is based supported the proposition that the court may, in the exercise of its discretion, properly refuse relief from forfeiture even to a tenant who belatedly tenders the full amount of outstanding rent and costs if, during the interim period, the landlord has, not unreasonably or precipitously, granted rights in the premises to third parties, on the footing that the original lease is at an end, and the court considers that, in all the circumstances, the grant of relief to the original tenants would cause injustice to the landlord or the third H parties or both.

Such was the learned judge's view of the present case. Because, and only because, of the rights which had been created by the plaintiffs in favour of Mr Parsons during the long period between the judgment of September 28 1987 and the first tender of the outstanding rent on or about December 16, he thought it would be "quite wrong" that the plaintiffs should be embarrassed by this "eleventh-and-a-half hour" application for relief. We do not think that in any respect he misdirected himself in principle. Nor do we regard his decision as plainly wrong. It was a decision to which he was, in our judgment, entitled to come in the exercise of his discretion.

For these reasons, we dismiss this appeal. J

The appeal was dismissed with costs. Leave to appeal to House of Lords was refused. A stay was granted during the period within which the appellants had the right to petition the House of Lords for leave to appeal under an undertaking by the appellants to indemnify the respondents against any damages which they might suffer through the stay having been granted in the event of the petition being refused or the appeal being unsuccessful.

Court of Appeal

November 2 1987 K

(Before Lord Justice DILLON and Lord Justice BINGHAM)

MINCHBURN LTD v PECK

Estates Gazette April 16 1988

[1988] 15 EG 97-99

Landlord and tenant — Landlords' repairing covenant — Tenant's claim in respect of breaches — Landlords' appeal from county court decision — Procedural difficulties caused by pressure of business in the county court — There were two actions by the landlords which were heard together, one claiming sums due under the tenant's mortgage of his L leasehold interest (a lease of 99 years from 1980) and the other claiming relief in respect of arrears of rent and service charge under the lease — There was a counterclaim in both actions by the tenant seeking damages for breach of the landlords' repairing covenant — The order made by the assistant recorder (at the end of a trial in which evidence and argument had to be sandwiched in, over three days, between other causes in his list) included an order for specific performance by the landlords of certain works in a Scott Schedule which it was agreed had not been carried out; an order for the tenant to pay arrears of rent and service charge; a declaration as to the M amounts owing on the mortgage; and an order for the landlords to pay the tenant £1,050 damages — There was also a suspended order for possession against the tenant — In fact all the arrears under the lease and mortgage had been paid by the date of the present appeal and the tenant remained in occupation of the flat — The appeal was directed only to the tenant's counterclaim, apart from a complaint (which the Court of Appeal rejected) against the refusal of the assistant recorder to admit new evidence at a late stage of the trial — The grounds of disrepair relied on by the tenant were two — The first was that he had suffered from dampness as a result of

A **the disrepair of a roof which leaked — The second was that, as a result of settlement and subsidence, cracks which had developed in the main party wall had caused consequential damage to the internal decorations in the flat — It was agreed that the internal redecoration required under this second head should be carried out by the plaintiffs, in lieu of damages, together with the works for which specific performance had been ordered — The Court of Appeal rejected a submission that the decorative disrepair had been the result of subsidence, not of the landlords' failure to repair the main structure under their covenant, as proper notice of evidence to this effect had** B **not been given at the trial — On a question of mitigation of damage, the court made a small reduction in the amount awarded by the judge for discomfort and dampness because the tenant had not given earlier notice of the condition — Judge's order varied to that extent**

No cases are referred to in this report.

This was an appeal by the plaintiff landlords, Minchburn Ltd, from a decision of Mr Assistant Recorder Langan QC, at Willesden County Court, in actions by the plaintiffs against Ernest John Peck, the present respondent, in respect of a top-floor flat at 73 Fordwych Road, London NW2.

Ian Partridge (instructed by A W Kemp & Co) appeared on behalf C of the appellants; Edward Cousins (instructed by Emmott & Levan) represented the respondent.

Giving judgment, DILLON LJ said: This is an appeal by the plaintiffs in the proceedings, Minchburn Ltd, against part of an order of Mr Assistant Recorder Langan QC made in the Willesden County Court on July 10 1987. The defendant in the proceedings, Mr Peck, who is respondent to this appeal, is the occupier of a flat on the top floor of 73 Fordwych Road, London NW2. By a lease of February 18 1980 he was granted a 99-year lease of the flat, of which he had previously been the sitting tenant at a low rent, in consideration of a premium of £10,000: £8,000 of that premium he raised by mortgaging his lease back to the then landlord, a Mr Galler.

D The present plaintiffs, Minchburn Ltd, came on the scene in August 1980, when the freehold reversion in the premises was transferred to them together with the benefit of the mortgage but subject to the defendant's lease. There were two actions. One was commenced in November 1980 and by that the plaintiffs claimed payment of outstanding sums due under the mortgage and other mortgage relief. In 1984, after an order for possession had been made which was set aside, the defendant counterclaimed in the mortgage action claiming damages for breach of the landlord's repairing covenant in the lease. The action under the lease was commenced in 1986 and again there was a counterclaim claiming damages for breach of the repairing covenant. The repairing covenant, clause 3, subclause 4, provides:

E That save insofar as the same fall to be maintained repaired and redecorated and renewed by the Lessee under the covenants on his part herein contained or by the Lessees of the other flats or parts of the Building under the covenants on their part contained in the respective leases thereof and subject to contribution and payment by the Lessee as herein provided the Lessor will maintain repair and redecorate and renew the main structure and foundations the boundary walls fences and areas of the Building and the roof and external parts and other retained parts of the Building and all tanks cisterns sewers drains pipes ducts and conduits serving the Building or any part thereof;

The defendant's obligation as lessee was merely to repair internally.

There was a pre-trial review at which the usual directions were given limiting the number of expert witnesses, and it was provided that each party's experts' reports must be disclosed to the other party F at least 28 days prior to the hearing. The parties had estimated two days for the trial of the two actions. They came on for trial before Mr Assistant Recorder Langan to start on June 11 1987, June 11 and 12 having been allocated as the two days. What happened then is all too typical of what happens to actions of some complexity which fall to be tried in the county court. It was no fault whatsoever of the learned assistant recorder, but it must have made his task, as it also made the task of the parties' counsel, very much more difficult.

On the first day, June 11, they did not have a clear day as there were other matters listed ahead. However, the case started and was opened, and a director of the plaintiffs gave his evidence, Then, however, it was arranged between counsel that, as the defendant's surveyor was not in court that day, the plaintiffs' surveyor's evidence should be held over to the following day, and instead the defendant G himself gave his evidence. So closed the first day.

When the parties arrived at court for the second day, June 12, both surveyors were in attendance, but they found that Mr Langan had a full list of other cases listed ahead of this action then part-heard. What the parties did to avoid wasting time was to send both surveyors off to the premises to have a good look at them and consider what of the works listed in the Scott Schedule, which had been prepared, required to be done. The surveyors came back to the court about the time of the lunchtime adjournment, having reached complete agreement as to the works that required to be done as of that time, but they were not in agreement on the pricing of some of those works. The learned assistant recorder was told that when he came back into court, and he was told that this did not in fact dispose of the issue of H liability.

The defendant called his surveyor to deal with one matter, namely the estimated cost of repair to the interior of the defendant's flat. Evidence on that was given by the defendant's surveyor and he was cross-examined. The plaintiff's surveyor also gave some evidence on the cost of repairs on which he was cross-examined. That evidence was interposed, however, in the course of one of the prior cases which was still part-heard.

Certain new points had been raised between the plaintiffs' surveyor and the plaintiffs' counsel for the first time that morning. They were concerned with the possibility that some of the disrepair had occurred before the grant of the lease in 1980, if that was a J relevant date, and also with whether it might have occurred after the service of the counterclaim in 1984, though attaching weight to that date overlooked the fact that a further counterclaim had been served much later in the second action. Accordingly, counsel for the plaintiffs said that he wished to cross-examine the defendant's surveyor more generally, and he was told by the assistant recorder that there was insufficient time available that afternoon. At any rate, the understanding of counsel for the plaintiffs was that he might reserve his right to ask for a more general cross-examination until the next occasion. Similarly, besides wanting to cross-examine the defendant's surveyor, counsel for the plaintiffs wanted to adduce further oral evidence from his own surveyor on these points, which had not been the subject of any written report which had been K submitted. There had, indeed, been an updating of the reports in writing on the first day of the hearing.

The effect was that there was no time available for an effective hearing to conclude the case on the second day, and it was adjourned at the end of that day to a date to be fixed. A date was ultimately fixed for the conclusion of the hearing roughly a month later on July 10 1987. No order was made or sought requiring the attendance of the defendant's surveyor on that date.

There may well have been some element of confusion in everyone's mind in the very confusing circumstances which were prevailing on that day. In the upshot, on July 10 the defendant's surveyor did not come to court. The plaintiffs' surveyor did. The plaintiffs' counsel L renewed his application to allow the plaintiffs' surveyor to give evidence on the points which I have briefly mentioned which were not the subject of his reports. The judge thought that the expert evidence had already been concluded, but in the upshot, when it became clear that the application to call the plaintiffs' surveyor was being pursued, he ruled against it having regard to the direction that the surveyors' evidence should be contained in reports which should previously have been submitted by each party to the other side.

In this court Mr Partridge for the plaintiffs submits that, in exercising his discretion in that way, the learned assistant recorder was wholly wrong and that injustice had resulted because the plaintiffs have been prevented from raising a good point, namely that the date of subsidence damage, which is part of the damage M complained of, had occurred before the lease was granted. I have considerable difficulty, in view of the breadth of the landlord's repairing clause, in seeing that the point that it occurred before the grant of the lease has any relevance. But, that aside, in my judgment, whether or not to admit fresh evidence at that stage in the trial of the action after there had been the direction for the exchange of experts' reports and when on July 10 the defendant's surveyor was not present in court seems to me an entirely correct exercise of the learned assistant recorder's discretion. It was a perfectly possible view for him to take and accordingly, in so far as the notice of appeal seeks a new trial on the ground of the refusal to admit evidence, I would reject that ground.

A The actual order which the learned assistant recorder made on July 10 was a rather complicated one. He ordered, first, that the plaintiffs do specifically perform the works mentioned in the schedule to his order, being certain works in the revised Scott Schedule which it was agreed had not yet been done. He made certain orders in respect of costs. He ordered the defendant to pay the plaintiffs' arrears of rent and arrears of service charge under the lease, and he ordered the plaintiffs to pay the defendant the sum of £1,050 damages. He also declared the amount owing by way of arrears of mortgage instalment at the date of the judgment to be £7,545.60, and the amount owing upon the mortgage to be £11,623. He ordered the defendant to deliver up possession of the property but suspended that order provided that
B the defendant paid the arrears of £7,545.60 by the expiration of 28 days and thereafter punctually paid the instalments due under the mortgage. We are told that in fact all the arrears have been paid, both under the mortgage and under the lease, and the defendant remains in occupation of the flat.

The appeal is therefore directed only at the counterclaim — primarily the award of damages, but also in one respect the order for specific performance. Apart from the point I have mentioned in relation to the judge's refusal to allow oral evidence to be called, two points are taken on the appeal. The grounds of disrepair relied on by the defendant were two-fold. He said firstly that he had suffered damp from about 1981 onwards as a result of the disrepair of the roof which leaked. He said also that, as a result of settlement and
C subsidence of the building as a whole, cracks had developed in the main party wall with the result that consequential damage was suffered to the internal decorations in his flat. Those were the two matters in respect of which he claimed, under various headings, damages for breach of the repairing covenant. As a matter of practical convenience, however, it was agreed that the internal decorative work required as a consequence of the settlement cracks in the main wall should be carried out by the plaintiffs when they were doing other work, instead of the plaintiffs paying the cost of those works to the defendants by way of additional damages. Therefore the works for which specific performance was agreed included the making good of those internal decorative repairs.

One point that is taken by the plaintiffs on this appeal is that they
D say that they can only be liable to the defendant for failure to repair the structure of the premises and, if the subsidence immediately caused the internal decorative disrepair, that disrepair was caused by subsidence over which the plaintiffs had no control and was not referable to their failure to repair the structure. They accept, however, that, as this is a covenant to repair the structure and the common parts of the premises which were under their control, their covenant to repair is not to be construed as a covenant to repair upon notice. They are liable if they fail to repair, and damage consequently is suffered by the defendant even if they had not been given notice by him of disrepair.

The plaintiffs' counsel has referred us to a number of authorities as
E to the meaning of the word "maintain" in various contexts, for instance in the context of maintaining hoists and lifts in industrial premises. He has submitted that there was no continuing obligation on the landlords to maintain this property in the sense that an employer is under an obligation to maintain machinery. In my judgment, however, the covenant to maintain, repair, redecorate and renew the main structure and foundations of the building and the roof and external parts and other retained parts is as wide a covenant as there could be imposed on the lessor. If the plaintiffs wished to show against the background of that covenant, and the course the case had taken, that the decorative disrepair was not caused by the disrepair of the main structure, it was for them to do so on proper notice of their surveyor's evidence to that effect. It did not happen
F that way at the trial. Accordingly, I take the view that this point cannot succeed in this court. There was a continuing obligation to maintain, repair and renew the main structure. They did not do so. Damage was suffered as a result of the process of settlement.

I should add that the evidence is that the crack appeared on the interior of the flat in 1981. When the surveyors inspected in 1984 they found that there had been substantial settlement. When, however, they inspected again later they found that the settlement had stabilised itself some time not later than the date of the inspection in 1984 and, accordingly, drastic work by way of underpinning was no longer necessary. But that does not lead, in my judgment, to the conclusion that the disrepair was work about which the landlords could have done nothing before the consequences appeared in the decorative repairs within the defendant's flat, nor does it absolve G
them from liability for the making good of that consequential damage.

The other point taken is a question of mitigation of damages. It is said, with particular reference to the roof not having been watertight, that the assistant recorder has awarded damages to the defendant for his discomfort and unpleasant living conditions over a period of four or five years but the defendant should have mitigated his damage by giving earlier notice to the plaintiffs of the bad condition of the property so that they could repair it earlier. So far as the settlement damage is concerned, it appears that they had known about that from the Camden Council from a date in 1982 at any rate, but had done nothing about it. So far as the damp through the damaged roof is H
concerned, the defendant asserted that he had frequently told the plaintiffs that the roof was leaking. But that the judge did not accept, and he found that the defendant had told the plaintiffs of the condition of the roof, and the fact that it was causing leaking and damp in his flat, only in March or early April 1984, at the time that a counterclaim to that effect was introduced into the pleadings in the action.

It appears from invoices produced by the plaintiffs that they spent £165 on the roof in April 1984 and an invoice was rendered by a roofing contractor in that sum for refixing slipped slates and supplying and fitting 40 secondhand good-quality slates, leaving the roof watertight. In the surveyors' reports, on which they were not cross-examined, the plaintiffs' surveyor said that at his inspection J
shortly after that date the roof was all right. The defendant's surveyor said that in an inspection at the end of 1984 the roof was still not watertight. Be that as it may, in April 1986 the plaintiffs employed another roofing contractor to carry out repairs to the roof, to supply 100 good-quality secondhand Welsh slates, and refix all slipped slates. This was done for a sum of £240. This 1986 repair seems to have been satisfactory for rendering the roof watertight.

The judge dealt with the claim that damages should have been mitigated by the defendant by giving notice earlier by expressing the view that the doctrine of mitigation of damage, while applicable in ordinary commercial circumstances, really did not apply at all as between landlord and tenant because, if it were brought in in relation to a repairing covenant in a lease, it would be in effect turning the K
landlord's responsibility into a responsibility to repair upon notice.

I am bound to say that I do not, for my part, agree with that way of putting it, because it does not turn the landlord's responsibility into a responsibility to repair on notice. He remains liable to do the repairs whether or not he has been given notice. It is only a question not of liability but of how far some sorts of damage can be claimed and, in particular, damage for discomfort and inconvenience when the person claiming to have suffered the discomfort and inconvenience could have improved his position by giving notice to the landlord.

It is, however, implicit in that that the landlord would have done something if given notice. The history of the state of repair of this property is not very encouraging in that field. The settlement damage L
was not made good. The roof damage was not adequately made good on the first round, though it was cured two years later. There is, however, a possibility that if the defendant had badgered at a yet earlier date, the final result might have been achieved a bit sooner. On the whole, I am of the view that some small allowance ought to be made by way of mitigation and, accordingly, I would for my part reduce the sum of £800, which the judge awarded as compensation for discomfort and dampness and so forth, to a sum of £700.

Save for varying the judge's order in that respect, I would dismiss this appeal.

Agreeing, BINGHAM LJ said: I agree with what Dillon LJ has said as to the substance of this matter. In particular, I agree with what he M
has said about the mitigation of damage. A landlord and tenant are contracting parties even though their contract is of a special kind. There is, however, nothing in the nature of that contract which excludes the ordinary principle of the law of contract that an injured party cannot recover damage which he could without expense to himself and by straightforward means have avoided. Accordingly, I agree that the appeal should be allowed, but only to the limited extent that Dillon LJ has indicated.

I do wish to add a brief word on the procedural side of this matter. The parties gave, and the court accepted, an accurate estimate of two days as the time needed for the hearing. As it was, the case was tried over three days, June 11 and 12 and July 10 of this year. The first day

A was not a full day of hearing, as Dillon LJ has pointed out, and on the second day, for reasons he has given, there was very little time available to these parties, although a little argument and evidence were sandwiched in between other causes in the assistant recorder's list. The third day of the hearing thus took place on a date over a month later.

This was not an easy or straightforward case. There were involved issues of fact, as there almost always are in actions concerned with disputes about building works and the condition of premises. There were arguments of law both as to the construction of this lease and as to the application of general principles. It was certainly not a case beyond the capacity of this learned assistant recorder, but it was a
B case which would make exacting demands even on an experienced judge. That being so, it was the more desirable that there should be uninterrupted and continuous hearings. No judge would, I think, wish to find himself presented with the schedule that faced this assistant recorder. In saying that, I am not wishing to criticise anyone, but I do wish to emphasise the desirability, as the county court limit rises, that arrangements be made for substantial cases of this kind to be heard in such a way that a judge can concentrate his mind and his energy on resolving the case without unnecessary distractions and without intermissions, which always increase the difficulty and usually increase the length of what remains. It is a tribute to this assistant recorder that he handled the case so effectively and, subject to one relatively very minor point, so
C correctly. I am, however, of opinion that he should never have found himself in this position at all.

The appeal was allowed in part, as mentioned at the end of Dillon LJ's judgment. The respondent was allowed nine-tenths of the costs: legal aid taxation of respondent's costs ordered.

Court of Appeal

January 22 1988

(Before Lord Justice GLIDEWELL and Lord Justice TAYLOR)

SUTTON (HASTOE) HOUSING ASSOCIATION v
D WILLIAMS AND ANOTHER

Estates Gazette April 23 1988

[1988] 16 EG 75-80

**Landlord and tenant — Housing Act 1980 — Acquisition by tenants of flat of long lease of 125 years under "right to buy" provisions in Part I of 1980 Act — Landlords a housing association — Dispute by new long leaseholders (the present appellants) in regard to inclusion in service charge of payments by instalments for replacement of windows of flat — Wet rot in the woodwork of windows of flats in the estate
E had caused leaks and had required constant repairs — Housing association decided to replace the old single-glazed wooden windows by double-glazed UPVC windows — Appellants objected and refused to pay consequent increases in the service charge — Housing association brought proceedings against them and succeeded in the county court — Lease provided for a service charge payable by the leaseholders, covering contributions to the costs of the lessors in carrying out repairs under their covenants and also such additional works and services as the lessors in their discretion considered necessary — In conformity with Schedule 2 to the
F Housing Act 1980 the lease excluded payments by the lessees for making good structural defects unless their existence had been notified to the lessees before the grant of the lease — Appellant lessees criticised the decision of the county court judge on the grounds that she had not distinguished between repairs and improvements; that Schedule 19 to the 1980 Act did not provide for a service charge to cover improvements; and that there was no finding that the cost of repairs was reasonable — Held by the Court of Appeal, rejecting criticisms of the judge's decision, that there was no need for a sharp distinction between repairs and improvements in this case and that some of the works involved both repairs and improvement; that the lease complied with Schedules 2 and 19 G
to the 1980 Act; that, in so far as the service charge related to the making good of structural defects, the lease had given express warning of the need to replace the old wooden windows; that Schedule 19 did not prohibit a service charge from covering improvements if the parties agreed to it; that the lease gave the lessors discretion to carry out additional works and provide additional services; and that the judge had been satisfied on the question of reasonableness — Appeal dismissed**

The following case is referred to in this report.

H

Mullaney v *Maybourne Grange (Croydon) Management Co Ltd* [1986] 1 EGLR 70; (1986) 277 EG 1350

This was an appeal by Leslie Henry Williams and his wife, Deirdre Olive Williams, the leaseholders of a flat at 14 Priory Crescent, Beulah Hill, London SE19, from the decision of Her Honour Judge Graham Hall, at Croydon County Court, in favour of the plaintiffs (the present respondents), Sutton (Hastoe) Housing Association.

T Brudenell (instructed by Cowles & Co) appeared on behalf of the appellants; Miss C Hutton (instructed by Debenham & Co) represented the respondents.

Giving judgment, GLIDEWELL LJ said: The respondent housing association was until 1982-83 the owner of the whole of a residential J
estate at Beulah Hill, Norwood, comprising blocks of houses, maisonettes and flats. The defendants (the present appellants), Mr and Mrs Williams, had been tenants of and occupied one of those flats, 14 Priory Crescent, since 1967.

Mr and Mrs Williams and a substantial number of other tenants who had been in occupation for long periods of the flats or maisonettes, exercised their right to buy their homes under the provisions of the Housing Act 1980. The flat, 14 Priory Crescent, was conveyed to the Williams on January 31 1983. They were granted a lease for 125 years from December 25 1981 at an annual rent of £10. The price was £12,882, which was a substantial discount from the market price under the terms of the Act. The lease provided for the payment of a service charge, to which I shall have to refer in more K
detail later.

The window frames in the flats had been constructed of wood. Over the years — in general terms, without for the moment speaking specifically about 14 Priory Crescent — there had been trouble with wet rot in the woodwork of quite a number of the windows, which resulted in leaks and the necessity for constant repairs.

As a result of this, in 1982 the housing association engaged an architect, a Mr Linden, to advise them about the woodwork and the exterior of the Beulah Hill Estate. In June 1982 Mr Linden reported, recommending that double-glazed UPVC windows should be installed as an alternative to continuing with joinery repairs as and when defects arose. The association therefore decided that, in the L
first place, they would install the double-glazed windows in one of the tower blocks in what are called Menlo Gardens. The intention was that, if that proved to be successful, they would start a phased programme of replacements of the windows in the other blocks of the estate. That work to Menlo Gardens was duly done in the latter part of 1982.

On November 17 1982 the association made an application to the Housing Corporation for a grant for the replacement of the windows to Menlo Gardens and also to some of the flats in Priory Crescent, not at that time including the block in which no 14 is contained.

The Housing Corporation suggested that it might be overall more economic if all the blocks on the estate had their windows renewed in the same contract. Accordingly, in March 1983, by which time the M
Williams had purchased their flat, the association made a fresh application to the Housing Corporation for a grant for double-glazed windows for virtually the whole of the estate. The application included a grant for the block in which 14 Priory Crescent is contained.

In the event the Housing Corporation decided that a grant was not payable for the replacement of windows where the flats had been sold to long leaseholders, but before they reached that decision they made it clear that they had doubts about the matter.

In March 1983 the association wrote to both long leaseholders and those who were negotiating for long leases, telling them of the application they had made to the Housing Corporation for a grant but warning them that the grant might not be payable on the leasehold flats. The Housing Corporation made their decision that the grant would not be payable for leasehold flats on December 20 1983. In the following months the leaseholders were told by the association that the grant was available only for flats held on short leases but nevertheless that the association intended to go ahead with the work of replacing the windows. And so they did.

As the long leaseholders were not going to be assisted by a grant, the association considered how best to ensure that the cost to them which the association intended to pass on to them could be met. They suggested to the long leaseholders that there were two possible methods of achieving this. One, since the effect of the work would be to improve the flats and thus they hoped to increase the value, would be for leaseholders to take out an additional loan on mortgage. But the second alternative which the association were willing to entertain was that they themselves would do the work and pay for it and that they would recoup the cost from the long leaseholders over the period of three years by adding each quarter to the service charge levied.

Many of the new long leaseholders, including Mr and Mrs Williams, took the view that it was not necessary to replace the windows to their flats and they objected, and objected strongly, to the work being done. Nevertheless, the association, after obtaining four estimates for the doing of the work, placed a contract for it, and the windows in the various blocks, including those to 14 Priory Crescent, were replaced during the second half of the year 1984.

In accordance with the offer which they had made, the association started adding to the service charge demanded of the long leaseholders in December 1983 an amount calculated as being (I take it) one-twelfth of the total sum due from each leaseholder; that is to say, a sum to be charged once a quarter over three years. Mr and Mrs Williams refused to pay the service charges so demanded. By October 30 1985 their account was in arrear to the extent of £2,882.11. On that date the association issued proceedings in the Croydon County Court claiming that sum together with interest, and also claiming possession of the flat and forfeiture of the lease. In the event, by the time the matter came to trial, the total sum claimed by the association including interest came to £3,688.80. Of this it was conceded by the defendants that £1,184.06 was due as being nothing whatsoever to do with the replacement of the windows but arising out of other matters, and the defendants submitted to judgment for that sum. There was, however, a contest as to the balance of £2,504.74, which was the cost including interest of the work of replacing the windows. On May 12 1987 Her Honour Judge Graham Hall gave judgment for the plaintiffs for that amount. Mr and Mrs Williams now appeal against that judgment.

It is necessary to consider some of the terms of the lease under which the Williams hold their flat. The first provision relating to the service charge is in clause 3(1) under which:

> The Lessee HEREBY FURTHER COVENANTS with the Lessor and with other relevant Lessees To pay to the Lessor (subject to Clause 7 hereof) an annual service charge (in this Lease called "the Service Charge") of an amount determined in accordance with the provisions of and at the time and in the manner specified in Clause 4 hereof but excluding (except as mentioned in Clause 7(E) hereof) such amounts in respect of the repair of structural defects as the Lessor is precluded from recovering from the Lessee under the provisions of paragraphs 13 to 17 of the Second Schedule of the Act and of insuring against risks involving such repairs or the making good of such defects.

I will return later to say something about clause 7(E) and also about paras 13 to 17 of the Second Schedule to the Act. By clause 4(1):

> THE Service Charge shall be of an amount determined as hereinafter provided and payable at the times and in the manner hereinafter mentioned.

By clause 4(3)(a):

> The amount of the Service Charge for each financial year shall be the aggregate of the following sums:
> (a) 8.76% of the costs of the Lessor in performing the Lessor's covenants set out in clause 5(1)(a) hereof and the costs and expenses incurred by the Lessor in respect of the expenditure described in Part I of the Fifth Schedule during such financial year as and herein called "Block Expenditure".

The Fifth Schedule is not relevant to the circumstances of the present case. By clause 5(1)(a):

> The Lessor HEREBY COVENANTS with the Lessee as follows:
> That the Lessor will:
> (a) Provide and maintain the following services (herein called "the Block Services") namely:
> (i) maintaining repairing renewing cleaning redecorating and in all ways keeping in good condition the Block and in particular the roof foundations and walls thereof . . .

By subclause (viii) of clause 5(1)(a) the covenant also required the:

> carrying out such additional works and providing such additional services as may be considered necessary by the Lessor in its absolute discretion from time to time.

Clause 7(E) is headed "Condition of the Block" and reads:

> The Schedule of Conditions contained in Part III of the Sixth Schedule hereto is a true and accurate Schedule of the condition of the Block at the date hereof and the Lessee hereby acknowledges notification by the Lessor of the structural defects specified therein for the purpose of paragraphs 13 to 17 of the Second Schedule of the Act and clause 3(1) hereof.

Finally, in the Sixth Schedule there is contained in Part III what is described as "Schedule of repairs to the structure anticipated during next 10 years", of which item 2 is:

> Joinery: External timber elements will need to be renewed during the next ten years. This includes windows, timber cladding to fascias and garage doors.

The case for the association can be summarised as follows. The replacement of the windows was a work of repair within the lessor's repairing covenant, clause 5(1)(a)(i). As such, by clauses 4(1) and 4(3)(a) the cost of that work was properly recoverable from the lessee as part of the service charge. If, however, replacement of the former single-glazed wooden windows by double-glazed windows constituted in part an improvement in addition to repair, then that improvement was "additional works . . . considered necessary by the Lessor in its absolute discretion". Thus that fell within the lessee's repairing covenant under clause 5(1)(a)(viii). In so far, therefore, as part of the cost was attributable to improvement, it was still recoverable by clause 4(1) and (3) as part of the service charge.

Before I come to the counterargument of Mr Brudenell for Mr and Mrs Williams, I should refer to the relevant terms of the Housing Act 1980. Section 17 of the Act provides:

> A conveyance of the freehold executed in pursuance of this Chapter shall conform with Parts I and II of Schedule 2 to this Act and a grant of a lease so executed with Parts I and III of that Schedule . . .

Turning to Schedule 2, Part III, para 11(1) starts with the words:

> The lease shall be for a term of not less than 125 years at a rent not exceeding £10 per annum, and the following provisions shall have effect with respect to the other terms of the lease . . .

Obviously the first two parts of that provision were reflected in this lease. Para 13(1)(a) reads:

> There shall be implied, by virtue of this Schedule, covenants by the landlord —
> (a) to keep in repair the structure and exterior of the dwelling-house and of the building in which it is situated (including drains, gutters and external pipes) and to make good any defect affecting that structure.

That covenant was, of course, expressed in the lease. Para 14 deals with covenants by the tenant and does not raise anything relevant in this case. Paras 15, 16 and 17, however, do. So far as is relevant, para 15 provides:

> Any provision of the lease or of any agreement collateral to it shall be void in so far as it purports
> (a) . . .
> (b) to enable the landlord to recover from the tenant any part of the costs incurred by the landlord in discharging or insuring against his obligations under paragraph 13(1)(a) or 12(1)(b) above, or
> (c) . . .
> but subject to section 19 of this Act and paragraph 16 below.

Para 16:

> A provision is not void by virtue of paragraph 15 above in so far as it requires the tenant to bear a reasonable part of the costs of carrying out repairs not amounting to the making good of structural defects or of the costs of making good any structural defects falling within paragraph 17 below or of insuring against risks involving such repairs or the making good of such defects.

By para 17:

> A structural defect falls within this paragraph if —
> (a) the landlord has notified the tenant of its existence before the lease was granted . . .

The effect of those provisions in paras 15, 16 and 17 can be expressed as follows. A provision in a lease which requires a lessee to pay to the lessor the cost of carrying out his (the lessor's) repairing covenant with regard either to the dwelling-house (that is to say, the flat) or to the whole building is permitted by the Schedule and is not void provided that either (a) it requires the lessee to bear a reasonable part of the costs of repairs which do not amount to making good structural defects or (b) in so far as the repairs are to make good structural defects, the lessor notified the lessee of their existence before the lease was granted.

Other provisions in the Act which are relevant to this appeal are section 136, which brings into effect Schedule 19 to the Act and that Schedule. By para 1(1) of Schedule 19:

> For the purposes of this Schedule, a service charge is an amount payable by the tenant of a flat as part of or in addition to the rent —
> (a) which is payable, directly or indirectly, for services, repairs, maintenance or insurance or the landlord's costs of management; and
> (b) the whole or part of which varies or may vary according to the relevant costs;
> and the relevant costs are the costs . . . incurred . . . by or on behalf of the landlord . . . in connection with the matters for which the service charge is payable.

By para 2, which is headed "Limitation of service charge":

> The extent to which relevant costs are taken into account in determining the amount of a service charge payable for any period shall be limited in accordance with the following paragraphs, and the amount payable shall be limited accordingly; and where the service charge is payable before the relevant costs are incurred —
> (a) no greater amount shall be so payable than is reasonable . . .

And by the first words of para 3:

> Costs are to be taken into account only to the extent that they are reasonably incurred . . .

Mr Brudenell's forceful argument goes as follows. Work of the nature of that carried out to the windows in this case may be either repairs or improvement, but it cannot be both. To put that another way, if work constitutes an improvement, it is not work of repair. Mr Brudenell refers us to the decision of Mr Julian Jeffs QC, sitting as a deputy High Court judge, in *Mullaney* v *Maybourne Grange (Croydon) Management Co Ltd* [1986] 1 EGLR 70. That was a case on somewhat similar — indeed, in many ways, very similar — facts to the present. It was a case in which the windows to a flat let on a long leasehold were replaced by the lessors. There was a lessor's covenant to repair in the lease, and the lessee covenanted to pay 2½% of the costs of the lessor carrying out his repairing covenant. There was, however, no clause in the terms of clause 5(1)(a)(viii) of this lease. There was a clause in different terms which related to the word "amenity", and the part of the decision which relates to the argument about the word "amenity" is of no relevance to this case. In relation to the other matter which arose, however, in the course of his judgment Mr Jeffs said:

> The new windows . . . may well have been a very desirable improvement, but I am quite satisfied that they fell into the category of improvements and not simply of maintenance.

He then quoted a passage from vol 1 of *Woodfall's Landlord and Tenant*, para 1-1436, part of which read:

> If a defect in design has given rise to dry rot, the elimination of that existing rot and the replacement of affected timber is, within the principle of renewal of subordinate parts which is dealt with below, an obligation imposed by a covenant to repair, but there is no obligation to undertake the structural alteration which is needed to prevent a recurrence of the rot, for that is improvement and not repair.

Mr Jeffs commented:

> Thus it appears to me in the context of the replacement of the windows that what was being done here went beyond what was necessary for the purpose of effecting a repair and was by way of a long-term improvement. It was looking to the future not to the present.
> Accordingly, I am satisfied that the plaintiff has made out his case on the relief he sought in paras 1 and 2 of the summons which I have just read out.

In the context of facts such as those in *Mullaney's* case I would respectfully not disagree with what fell from the learned judge. But it is noticeable that, because of the terms of that lease, the question whether the work could be both repair and improvement did not arise in that case. Therefore, to that extent, with respect to Mr Brudenell, I do not find that decision of any assistance on the question.

Mr Brudenell's argument continues as follows. Schedule 19, para 1, to the 1980 Act allows a service charge to be made only if it is to recover the cost of items described in that paragraph; that is to say, services, repairs and maintenance or insurance of the landlord's costs of management. Thus the Act does not permit a service charge to be made to recover the cost of an improvement. The third proposition is that the judge did not distinguish, as she should have done, between repairs and improvements. She made no finding as to which category these works fell into, and thus her decision was defective in that respect. Fourth, and as an alternative, even if the service charge can cover improvements as well as repairs, the effect of Schedule 19, paras 2 and 3, is that the cost of repairs must be reasonable, and there is no finding in the judgment to that effect.

So far I have not referred to the judgment and, without intending any disrespect, I do not need to read very much of it. However, the concluding passage, having recited Mr Brudenell's argument that an improvement is not a matter for which the lessor could charge the lessee, reads as follows:

> I find that the replacement of the windows whether or not this amounts to a repair or has an element of improvement as commonly understood in landlord and tenant law is in fact covered by clause 5(1)(a) of the lease, whether under subsection (i) or (viii). The defendants are bound to pay 8.76% of the costs of the block expenditure. The plaintiffs set out to repair the windows and the reference in the lease to structural repairs conforms with the provisions of Schedule 2 to the Housing Act 1980 in that the plaintiffs followed the procedure of that Schedule in paragraphs 15, 16 and 17. That Schedule and Schedule 19 do not make the lease void either expressly or by implication.
> I find that the plaintiffs set out to repair the windows and they conformed with their obligations under the lease and conformed also with all the relevant provisions of the Housing Act 1980 in so doing by following the correct procedure set down herein.
> Judgment for the plaintiffs for the additional sum of £3,506.74.

Miss Hutton for the association, in a clear and very concise argument, submits as follows. First, Schedule 2 to the Act contains provisions which require terms to be implied in leases granted under the right to buy provisions and regulate the terms of such leases. This lease complies with para 1(1) of the Schedule and it expresses the covenants which would otherwise have to be implied by clause 13(1). Moreover, the lease is not void under paras 15 to 17 of the Schedule because where in this lease the lessee is required to pay the costs of repairs, they are repairs within the lessee's repairing covenant and, in so far as they relate to works to make good structural defects, the lease itself in Schedule 6 expressly informed the lessee of the need to renew the wooden windows.

Second, she submits that works of repair can — indeed, I imagine she would say, not infrequently do — involve also improvements. The two are not mutually exclusive, and in *Mullaney* it was not suggested that they were.

Third, Schedule 19 does not say that a service charge can only be levied to pay for the items described in para 1(1) of that Schedule. The effect of Schedule 19 is to limit and impose restrictions on service charges imposed in order to pay for the items described in para 1(1), but the Schedule does not prohibit and has no impact upon a service charge imposed in order to pay for some other item, including improvements. So, in so far as the cost here involves an improvement, and I think that she accepts that it very probably does, then that cost can be recouped by way of service charge. She reminds us of course, although strictly this is not necessary to her argument, that in so far as the work is an improvement, that benefits Mr and Mrs Williams as long leaseholders.

Finally, Miss Hutton argues that in so far as the cost sought to be recovered by way of service charge is the cost of repairs, she accepts that it must be reasonably incurred and that the amount must be reasonable — that is the combined effect of Schedule 19, paras 2(a) and 3. She says that, although the judge did not expressly find that the cost was reasonable, there was during the course of the hearing much argument about whether it was or was not reasonable. There was evidence about it and there was argument about it from both learned counsel. Thus, Miss Hutton argues, when the learned judge said in the penultimate sentence of her judgment "I find that the plaintiffs . . . conformed also with all the relevant provisions of the Housing Act 1980", that must include a finding that both the works themselves and the cost were reasonable.

Having summarised Miss Hutton's submissions, I do not need to repeat them because I agree with all of them.

For those reasons I do not accept Mr Brudenell's arguments and I would therefore dismiss this appeal.

Agreeing, TAYLOR LJ said: *Prima facie* the parties to a lease are entitled to agree such terms as they choose. Mr Brudenell concedes that but for the operation of statutory control, the terms of clauses 3, 4, 5(1)(a)(i) and 5(1)(a)(viii) of this lease would be valid and would entitle the landlords to recover the costs they sought. One therefore has to look to see whether the statute prohibits such recovery.

The landlords rely on two Schedules to the Housing Act 1980. Schedule 2 applies only to conveyances pursuant to the right to buy.

Paras 13 to 17 are relied upon. Para 13(1)(a) imposed an implied covenant on the landlord to keep in repair the structure and exterior of the dwelling-house and of the building in which it is situated and to make good any defects affecting that structure. The effect of paras 15(b), 16 and 17(a) is that a provision enabling a landlord to recover his para 13 costs is void save in so far as it requires the tenant to bear a reasonable part of the costs of carrying out repairs other than making good structural defects or of the costs of making good any structural defects of which notice has been given before the lease was granted.

Complaint is made that the learned judge should have decided, but did not, whether the replacement of these timber-framed windows with double-glazed windows was a repair or an improvement. In my judgment she did not need to do so. If and in so far as the work was a repair, then, subject to the question of reasonableness to which I shall return, the landlord was entitled to recover under para 16. If and in so far as the works were an improvement falling short of making good a structural defect, they were not caught by the Schedule. If and in so far as the landlord was making good a structural defect, then he was entitled under paras 16 and 17(a) to recover a reasonable part of the costs, having given the required notice in Part III of the Sixth Schedule to the lease.

Schedule 19 applies to all leases. Para 1(1) defines the phrase "service charge", but only for the purpose of the Schedule. Mr Brudenell argues that any costs sought to be recovered by way of service charge must fall within that definition or be irrecoverable. Thus, if the replacement of these windows was, as he argues, an improvement rather than a repair, the landlord cannot recover any part of the costs from the tenant.

Miss Hutton submits that Schedule 19 applies only to a service charge as defined for its purposes. If the relevant works do not fall within para 1(1)(a), as improvements would not, then they are not caught by the provisions of the Schedule. But they are not prohibited. The Schedule is not an avoiding provision. In my judgment, Miss Hutton is right. Thus, in so far as the works were repairs, Schedule 19 applies and, subject to reasonableness as specified in paras 2 and 3, the landlords have complied with the requirements of the Schedule as the learned judge found. In so far as the works may be regarded as improvements, they are not caught by the Schedule and the provisions of the lease are effective.

The issue which has concerned me under both Schedule 2 and Schedule 19 is as to reasonableness. The appellants contend that the works were unnecessary. The windows of 14 Priory Crescent had, they said, been recently repaired. On this ground and on grounds of expense they were opposed to the double-glazing being installed. They were required to pay 8.76% of the costs, but, since each tenant paid 8.76%, the tenants as a body were paying the total costs and the landlord was paying nothing. Mr Brudenell argues that para 16 of Schedule 2 permits only that the tenants should bear "a reasonable part of the costs". The word "part", he says, even if qualified by the word "reasonable", implies less than the whole. In effect, however, the appellants were paying the whole cost of the double-glazing of their own flat.

In my judgment, the phrase "a reasonable part of the costs" must be read in the context of para 13, which required the landlord to keep in repair and make good the structure not only of the dwelling-house but of the building in which it is situate. It is a reasonable part of the costs of doing that which he is entitled to recover. For a tenant to have to pay for his own share of the total costs could be regarded, in my judgment, as paying a reasonable part of the whole.

However, Mr Brudenell further argues that the learned judge did not adequately deal with the question of reasonableness either under Schedule 2 or under paras 2 and 3 of Schedule 19. It is true that she did not specifically state in her judgment that she found the costs charged to be a reasonable part, that the amount was reasonable, or that such costs were reasonably incurred. However, it is common ground that these issues were fully argued before the learned judge. She had heard evidence on them; she summarised that evidence. Her judgment ended with the passage which Glidewell LJ has quoted, and in my judgment that passage must import that the provisions as to reasonableness in the Act have been complied with and that she was satisfied about them.

Accordingly I agree with Glidewell LJ that the appeal must be dismissed.

The appeal was dismissed with costs, not to be enforced without the leave of the court. Legal aid taxation of the appellants' costs was ordered. Leave to appeal to the House of Lords was refused.

Court of Appeal

March 17 1988

(Before Lord Justice BINGHAM and Lord Justice MANN)

ANTONIADES v VILLIERS AND ANOTHER

Estates Gazette April 30 1988

[1988] 17 EG 122-129

Landlord and tenant — Licence or tenancy — *Street* v *Mountford* again considered and other cases reviewed — Appeal by owner of house against decision of county court judge in possession action holding that respondent occupants were tenants entitled to the protection of the Rent Act — The subject flat, of which the appellant was the freehold owner, comprised four rooms, a bedroom, a bed-sitting room, a kitchen and a bathroom — Appellant made a practice of granting what he described as licences with the clear intention of avoiding the difficulties attendant on letting properties on tenancies, including the application of the Rent Act — The present occupants, respondents to the appeal, were a young man and his girl friend who wished to share the flat and live as husband and wife — They each signed an identical agreement, of an extremely detailed nature, which provided that it conferred a personal licence only and expressly negatived the grant of exclusive possession — It also provided that the licensor would be entitled to use the rooms in common with the licensees and would be entitled to permit other persons to do so — The appellant made it plain that the flat was intended for single persons sharing and that if the respondents got married they would have to go — He also explained that the Rent Act protection would not apply, although it was evident that the respondents were ignorant of the Act and did not understand the expression "exclusive possession" — Despite the express provision in the agreement the appellant made no attempt to put other occupiers into the flat or to occupy it himself — After a time some difficulties arose in regard to the rent and the respondents were given notices to quit — They took advice and then claimed to be protected tenants — A rent officer accepted this claim and registered a rent which was lower than the sum of the two separate payments of £87 each per calendar month to which the occupants had agreed — The agreements, which were made in February 1985, were clearly modelled on those in *Somma* v *Hazelhurst*

The county court judge in the present case, believing that he was following the guidance of Lord Templeman's speech in *Street* v *Mountford*, which had disapproved *Somma* v *Hazelhurst*, concluded that the present agreements were bad as "artificial transactions designed to evade the Rent Acts" — He therefore found that the respondents had a protected tenancy

Held by the Court of Appeal that the judge had fallen into error in his application of *Street* v *Mountford* — The court formulated a new "broad approach" in five propositions which are summarised briefly as follows — 1. The court must be astute to frustrate sham devices and artificial transactions designed to evade the Rent Act — 2. A "sham" is as defined in *Snook* v *London & West Riding Investments Ltd* or, more shortly, where parties say one thing while really intending another — 3. It is not a crime or contrary to public policy for a property owner to license occupiers to occupy property on terms which do not give rise to a tenancy — 4. The Rent Acts must not be allowed to alter or influence the construction of an agreement — 5. Where an agreement is held to be a sham the task of the court is to give effect to the true bargain; where it is not held to be a sham the task is to construe it and give fair effect to its terms in the context of all relevant surrounding circumstances

In the present case the written agreements could not

A **possibly be construed as conferring exclusive possession and the respondents could succeed only by showing that they were a sham — The respondents argued that they were a sham and that the reality was the grant of a joint tenancy, with each occupant jointly and severally liable for the whole monthly payment of £174 — The facts, however, did not support this allegation and the judge was wrong in having accepted it — The agreements set out the true bargain — *Somma* v *Hazelhurst* was distinguishable from the present case; apart from other points, *Somma* concerned a single room with two beds, so that the right reserved by the licensor to use the room**
B **herself (in addition to the two licensees) could well be regarded as physically impracticable, while here there was a bed-sitting room in addition to the bedroom used by the occupants — Appeal allowed and possession ordered**

The following cases are referred to in this report.

Aldrington Garages Ltd v *Fielder* (1978) 37 P&CR 461; [1978] EGD 347; 247 EG 557, CA
Brooker Settled Estates Ltd v *Ayers* [1987] 1 EGLR 50; (1987) 282 EG 325, CA
Donald v *Baldwyn* [1953] NZLR 313
Hadjiloucas v *Crean* [1987] 3 All ER 1008; [1987] 2 EGLR 60; (1987) 284 EG 927, CA
Shell-Mex & BP Ltd v *Manchester Garages Ltd* [1971] 1 WLR 612; [1971] 1 All
C ER 841, CA
Snook v *London & West Riding Investments Ltd* [1967] 2 QB 786; [1967] 2 WLR 1020; [1967] 1 All ER 518, CA
Somma v *Hazelhurst* [1978] 1 WLR 1014; [1978] 2 All ER 1011; (1978) 37 P&CR 391; [1978] EGD 269; 246 EG 311, CA
Street v *Mountford* [1985] AC 809; [1985] 2 WLR 877; [1985] 2 All ER 289; [1985] 1 EGLR 128; (1985) 274 EG 821, HL

This was an appeal by the owner of the top flat at 6 Whiteley Road, Upper Norwood, London SE19, Agis Antoniades, against the decision of Judge Macnair, at Lambeth County Court, rejecting the owner's application for possession against the occupants of the flat, William Robert Villiers and Sharon Bridger.

The appellant, Agis Antoniades, appeared in person; James
D Harris (instructed by A L Hughes & Co) represented the respondents.

Giving judgment, BINGHAM LJ said: This appeal raises the familiar question whether residential accommodation was let to occupiers as tenants or licensees.

Mr Antoniades is the freehold owner of a house at 6 Whiteley Road, Upper Norwood, London SE19. He let Mr Villiers and Miss Bridger, a young man and a young woman, into occupation of the top flat in the house. In due course he gave each of them notice to vacate the flat. Neither did so, and he sought an order for possession in the Lambeth County Court. Mr Villiers and Miss Bridger ("the occupants") resisted the claim, contending that they were tenants and so entitled to protection under the Rent Acts. Mr Antoniades
E ("the landlord") contended that they were licensees, not entitled to protection. His Honour Judge Macnair upheld the occupants' contention. The landlord appeals.

The top flat comprises four rooms: a bedroom, a room described as a bed-sitting room, a kitchen and a bathroom. The facts found by the learned judge are these.

The landlord had considerable experience in the management of properties. He was fully aware of the disadvantages of letting out property on tenancies and so had made a practice of granting licences. He was approached by Mr Villiers in about November 1984. The landlord told Mr Villiers that he dealt only in licences. The landlord in the course of a telephone call asked Mr Villiers whether he wanted a double bed or a single and Mr Villiers said a double. Before
F an agreement was signed, Mr Villiers made it clear that the flat was for him and his girlfriend. The landlord appreciated (so the judge inferred) that the occupants would share the flat and live as husband and wife. They looked at the flat and approved it in December 1984. They gave references showing that each of them was in employment.

On February 9 1985 the landlord with his wife and a daughter (who is a solicitor) attended at the flat. The occupants waited outside the house and were then let in and each of them signed an identical agreement. The occupants read the terms of the agreement and the landlord explained that these were licences, that the Rent Acts did not apply, that he had the right to put other people in and that no exclusive possession was given to the occupants. No finding is made whether the occupants read the agreement and the landlord gave his explanation before the occupants signed or after, but it appears to G
have been common ground in the evidence that the occupants read the agreement before signing and the landlord's evidence suggests that the explanation was also given before signing. The occupants knew nothing about the Rent Acts but signed because they were thankful to find somewhere to live. They did not understand the term "exclusive possession". The landlord made it plain that they had to go if they got married and that the flat was only for single people sharing. On July 7 1986 (after difficulties over the monthly payments) the landlord gave each of the occupants one month's notice to vacate. No effort was made by the landlord to put other occupiers into the flat or to occupy it himself. The occupants asked the landlord if a friend could stay with them in the flat and he agreed. H
The friend slept on a sofa bed for a period in cramped conditions. From time to time the landlord came and knocked on the door. On one occasion he entered without knocking to show people round the flat. After receiving notice the occupants took advice. They then claimed to be tenants. The rent officer upheld that claim and registered a rent substantially below the total of the payments the occupants had agreed to make.

The agreement signed by Mr Villiers on February 9 1985 was in these terms:

This licence is made the 9th day of February 1985 One thousand nine hundred and eighty five between A Antoniades of 152 Croydon Road, London SE20 (herein-after referred to as the "licensor") of the one part and William Roberts Villiers of 30 Dynewood Close, Bexley Kent (hereinafter referred to J
as the "licensee") of the other part.

Whereas the licensor is not willing to grant the licensee exclusive possession of any part of the rooms hereinafter referred to.

And whereas the licensee is anxious to secure the use of the rooms notwithstanding that such use be in common with the licensor and such other licensees or invitees as the licensor may permit from time to time to use the said rooms.

And whereas this licence is entered into by the licensor and the licensee solely upon the above basis.

By this licence the licensor licenses the licensee to use (but not exclusively) all those rooms (hereinafter referred to as "the rooms") on the Top flat (One bedroom one bedsitting room, kitchen and bathroom) of the building known as and situate at 6 Whiteley Road SE19 (hereinafter referred to as "the building") together with the use of the entrance hall and the staircase outer K
door and vestibule, of the building (shared with the other occupants in the house) and together with the use of the furniture fixtures and effects now in the rooms (more particularly set out in the schedule of contents annexed hereto) from 14/2/1985 for the sum of £87 per calendar month on the following terms and conditions:

1) The licensee agrees to pay the said sum of £87 (on the 14th of each month) monthly in advance.
2) The licensee shall be responsible for the payment of all gas electric light and power which shall be consumed or supplied in or to the rooms and shared parts of the house used by the licensee during the licensee's occupation thereof.
3) The licensee shall use his best endeavours amicably and peaceably to share the use of the rooms with the licensor and with such other licensees or invitees whom the licensor shall from time to time permit to use the rooms L
and shall not interfere with or otherwise obstruct such shared occupation in any way whatsoever.
4) The licensee shall keep the interior of the rooms and all fixtures and fittings therein and all other shared parts of the house used by the licensee in good and clean condition and complete repair (fair wear and tear and damage by accidental fire only excepted) and immediately replace all broken glass.
5) The licensee shall preserve the furniture and effects in the said rooms from being destroyed or damaged and make good pay for the repair of or replace with articles of a similar kind and of equal value such of the furniture and effects as may be destroyed lost broken or damaged (fair wear and tear thereof only excepted).
6) The licensee shall leave the furniture and effects at the expiration or sooner determination of this licence in the rooms or places in which they were at the commencement thereof. M
7) On notice in writing being given to the licensee by the licensor of all wants and repairs cleansings amendments and restorations to the interior of the rooms and of all such destruction loss breakage or damage of or to the furniture and effects as the licensee shall be bound to make good found therein the licensee shall repair cleanse amend restore or make good the same within three weeks of the giving of such notice.
8) The licensee shall not remove any furniture and effects from the rooms without the previous consent in writing of the licensor and neither will move in other furniture.
9) The licensee shall not carry on or permit to be carried on in the rooms any profession trade or business whatsoever.
10) The licensee shall not do or suffer to be done in the rooms any act or thing which may be a nuisance cause of damage or annoyance to the licensor

A and the other occupiers or users of the rooms or the building or of any adjoining premises or which may vitiate any insurance of the building against fire or otherwise or increase the ordinary premiums thereon.

11) The licensee shall not hang or allow to be hung any clothes or other articles on the outside of the rooms or the building but can install a washing line properly supported for this purpose in the back garden.

12) The licensee shall not keep any combustible or offensive goods provisions or materials in the rooms and will not use the rooms in any illegal or immoral way.

13) The licensee shall not cause or permit any damage spoil or destruction to the rooms or to the building (including gardens, fences and gates).

14) The licensee shall not pull down alter or add to or in any way interefere with the locks construction or arrangements of the rooms and the house without the previous consent in writing of the licensor.

B 15) The licensee shall not keep any animals or birds in the rooms nor shall the licensee permit any child or children to reside or stay in the rooms.

16) The licensor shall be entitled at any time to use the rooms together with the licensee and permit other persons to use all of the rooms together with the licensee.

17) This licence is personal to the licensee and shall not permit the use of the rooms by any person whatsoever and only the licensor will have the right to use or permit the use of the rooms as described in clause 16. The licensee under no circumstances will have the right to allow any other people of his choice to use the rooms in any way.

18) The deposit held of £30 (Thirty pounds) will be returned on vacation if no damages or losses occur and if the flat and shared parts are left clean and no such dispute about the deposit will entitle the licensee to remain on the premises.

C 19) The licensee will inform the licensor immediately by recorded delivery in case of any damage or losses to the rooms buildings and contents.

20) The licensee having inspected the rooms and the shared parts of the building and the furniture and fittings declares that he found them in good order and satisfactory.

21) Upon the licensee being in breach of any of the conditions referred to above this licence shall immediately determine without prejudice to any other remedies of the licensor and the licensee shall immediately cease his use of the rooms and the building as permitted hereunder.

22) The licensee (occupier) declares that he is over 18 years old and understood this licence and found flat and shared parts, furniture and fittings all in good state of repair, tidy, clean and satisfactory.

23) The real intention of the parties in all surrounding circumstances is to create this licence which is not coming under the Rent Acts and is binding as written.

D 24) This licence represents the entire agreement of the parties and no oral or other agreements were made and no different explanations or representations were made and only agreements in writing will be legally binding.

25) The licensee read and understood this licence and received copy and the licensee understands that all rooms and all parts of the dwelling will be shared and no exclusive possession of any part or the whole will be allowed to the licensees by the licensor under any circumstances.

Schedule of furniture:

1 Wardrobe, 1 Dressing table, 1 Double-bed with mattress and mattress cover, 1 Tablebed, 1 Electric Cooker, 1 Kitchen cabinet, 1 Fridge, 1 Sink unit, 1 Sinkheater, 1 Table, 4 chairs, 1 Bed settee, 1 Armchair, 1 Sideboard, 2 Double-bar Electric fires, 1 Bathroom suite, 1 Bath heater E electric, Carpets, Lampshades, Bulbs, Curtains, Net curtains everywhere, 2 keys.

26) Subject to clause 21 this licence may be terminated by one months notice in writing given by either party at any time and the licensor reserves the right of eviction without court order.

There was an addendum, which Mr Villiers signed on the same date and at the same time:

The undersigned Mr Villiers hereby agrees that the licence signed on the 9 2 1985 does not come under the Rent Acts and the flat is for single people sharing and if Mr Villiers marries any occupier of the flat then Mr Villiers will give notice and vacate the flat at 6 Whiteley Road, London SE19. The owner Mr Antoniades did not promise any other accommodation in any way. No person will have exclusive possession of above flat as agreed.

F The documents signed by Miss Bridger were, save for her name, in identical terms.

The judge held that the main agreement was drawn up with the Court of Appeal decision in *Somma* v *Hazelhurst* [1978] 1 WLR 1014 in mind.* That is plainly so. When the terms of the present agreement are compared with those of the *Somma* agreement (see pp 1017 E — 1019 F of the report of that decision), it is plain that most of the terms of the present agreement have been closely modelled (subject to some modifications, additions and omissions) on the *Somma* agreement. The reason for this is obvious. The Court of Appeal held that the *Somma* agreements created licences not tenancies. That decision stood as good law when these agreements were made in February 1985. That was the result the landlord wanted to achieve. So he based G his form of agreement on the *Somma* agreement to gain the protection of the Court of Appeal ruling, no doubt taking advantage of his daughter's legal knowledge.

*Editor's note: Also reported at (1978) 246 EG 311.

A month after the landlord and the occupants signed these agreements, *Street* v *Mountford* [1985] AC 809 was argued before the House of Lords. In his leading speech delivered in May 1985, with which the other members of the House concurred, Lord Templeman stressed that the letting of residential accommodation with exclusive possession for a term at a rent ordinarily gives rise to a tenancy.

In *Street* v *Mountford* it was accepted by the landlord (Mr Street) that Mrs Mountford was entitled under her agreement to enjoy exclusive possession of the premises and the case was not complicated H by the presence of other occupiers. But in the course of reviewing the authorities Lord Templeman said:

In *Somma* v *Hazelhurst* [1978] 1 WLR 1014, a young unmarried couple H and S occupied a double bedsitting room for which they paid a weekly rent. The landlord did not provide services or attendance and the couple were not lodgers but tenants enjoying exclusive possession. But the Court of Appeal did not ask themselves whether H and S were lodgers or tenants and did not draw the correct conclusion from the fact that H and S enjoyed exclusive possession. The Court of Appeal were diverted from the correct inquiries by the fact that the landlord obliged H and S to enter into separate agreements and reserved power to determine each agreement separately. The landlord also insisted that the room should not in form be let to either H or S or to both H and S but that each should sign an agreement to share the room in common J with such other persons as the landlord might from time to time nominate. The sham nature of this obligation would have been only slightly more obvious if H and S had been married or if the room had been furnished with a double bed instead of two single beds. If the landlord had served notice on H to leave and had required S to share the room with a strange man, the notice would only have been a disguised notice to quit on both H and S. The room was let and taken as residential accommodation with exclusive possession in order that H and S might live together in undisturbed quasi-connubial bliss making weekly payments. The agreements signed by H and S constituted the grant to H and S jointly of exclusive possession at a rent for a term for the purposes for which the room was taken and the agreement therefore created a tenancy. Although the Rent Acts must not be allowed to alter or influence the construction of an agreement, the court should, in my opinion, be astute to detect and frustrate sham devices and artificial transactions whose only object K is to disguise the grant of a tenancy and to evade the Rent Acts. I would disapprove of the decision in this case that H and S were only licensees and for the same reason would disapprove of the decision in *Aldrington Garages Ltd* v *Fielder* (1978) 37 P&CR 461 and *Sturolson & Co* v *Weniz* (1984) 72 EG 326.

Against that background of authority the learned judge considered the present case. He asked himself whether in truth this was a licence or a tenancy. He acknowledged *Street* v *Mountford* as the clear authority on the question whether an occupation is by virtue of a tenancy or a licence. He said that one of the essential features of a tenancy is exclusive possession. He referred to Lord Templeman as making it clear that the question of exclusive possession is a question of fact not necessarily to be determined by what is in an agreement between the parties. Then, according to the approved note of the L judgment, which is no doubt a truncated version of what was actually said, the judge concluded:

Mr Antoniades insists he has made a Licence. House of Lords tell me to ignore the assertions in the document that it is a licence. What difference does it make here that this is a sharing agreement. It seems clear to me that apart from a few visits by Mr Antoniades no one shared with them during period of occupation, they had exclusive possession (*Street* v *Mountford* comments) (reads re *Somma* v *Hazelhurst*). Hold that these licences are artificial transactions designed to evade the Rent Acts. Finds that there is a tenancy — claim for possession dismissed.

Making (I hope) fair allowance for the abbreviated note of the judgment and for the circumstances in which the judgment was probably given, I none the less conclude that the judge has in this M passage fallen into error:

(1) The House of Lords has not, I think, held that assertions in a document that it is a licence should be ignored. It has held that the true legal nature of a transaction is not to be altered by the description the parties choose to give it. A cat does not become a dog because the parties have agreed to call it a dog. But in deciding whether an animal is a cat or a dog the parties' agreement that it is a dog may not be entirely irrelevant.

(2) Court of Appeal authority since *Street* v *Mountford* shows that sharing agreements do raise rather different questions: see, for example, *Hadjiloucas* v *Crean* (1987) 284 EG 927.*

*Editor's note: Also reported at [1987] 3 All ER 1008; [1987] 2 EGLR 60.

A (3) The judge has, I think, overlooked the fundamental rule of construction that it is not legitimate to use as an aid in the construction of the contract anything which the parties said or did after it was made. Subsequent actions are therefore inadmissible except as evidence of a new agreement or as the basis of an estoppel, neither of which is suggested here: see *Chitty on Contracts*, 25th ed, vol 1, p 447, para 825.

(4) To stigmatise these agreements as "artificial transactions designed to evade the Rent Acts" (unless this amounts to a finding that the agreements were a sham) is to leave unanswered the question whether they succeed in their object. The task of the court, where this issue arises, is to decide whether the owner has succeeded or not.

I do not, therefore, think that the judge's reasons are enough to B support his decision. But that does not mean that his decision is wrong, still less that the contrary decision is right.

The broad approach to be adopted in resolving a question of this kind is, I think, well settled:

(1) The court should be astute to detect and frustrate sham devices and artificial transactions whose only object is to disguise the grant of a tenancy and to evade the Rent Acts (*Street* v *Mountford* at p 825H). The court has to be especially wary and especially careful to see that things like premiums are not being used to conceal payments of rent and so on (*Aldrington Garages* v *Fielder* (1978) 37 P&CR 461*at p 499 *per* Geoffrey Lane LJ as he then was).

(2) A written agreement is a sham where it incorporates clauses by C which neither party intends to be bound and which is obviously a smokescreen to cover the real intentions of both contracting parties (*Hadjiloucas supra* at p 929 *per* Purchas LJ). The accepted definition of a sham is that given by Diplock LJ, as he then was, in *Snook* v *London & West Riding Investments Ltd* [1967] 2 QB 786 at p 802C:

> As regards the contention of the plaintiff that the transactions between himself, Auto Finance and the defendants were a "sham", it is, I think, necessary to consider what if, any, legal concept is involved in the use of this popular and pejorative word. I apprehend that, if it has any meaning in law, it means acts done or documents executed by the parties to the "sham" which are intended by them to give to third parties or to the Court the appearance of creating between the parties legal rights and obligations different from the actual legal rights and obligations (if any) which the parties intend to create. But one thing, I think, is clear in legal principle, morality and the authorities D (see *Yorkshire Railway Wagon Co* v *Maclure* and *Stoneleigh Finance Ltd* v *Phillips*), that for acts or documents to be a "sham", with whatever legal consequences follow from this, all the parties thereto must have a common intention that the acts or documents are not to create the legal rights and obligations which they give the appearance of creating.

Put more shortly, a sham exists where the parties say one thing while really intending another (*Donald* v *Baldwyn* [1953] NZLR 313, at p 321 *per* F B Adams J).

(3) "There is no reason why, if it is possible and properly done, agreements should not be entered into which do not fall within the Rent Acts, and the mere fact that those agreements may result in enhanced profits for the owners does not necessarily mean that the E agreements should be construed as tenancies rather than as licences" (*Aldrington Garages, supra, per* Geoffrey Lane LJ at p 468); ". . . persons are entitled to arrange their affairs to their best advantage so long as the law allows it. That has long been the position in tax cases, and equally long been the position in landlord and tenant and Rent Acts cases" (*ibid, per* Roskill LJ, as he then was, at p 473). It is not a crime nor is it contrary to public policy for a property owner to license occupiers to occupy property on terms which do not give rise to a tenancy.

(4) The Rent Acts must not be allowed to alter or influence the construction of an agreement (*Street* v *Mountford, supra* at p 825G). As Buckley LJ said in *Shell-Mex & BP Ltd* v *Manchester Garages Ltd* [1971] 1 WLR 612 at p 619:

> F It may be that this is a device which has been adopted by the plaintiff company to avoid possible consequences of the Landlord and Tenant Act 1954, which would have affected a transaction being one of landlord and tenant; but, in my judgment, one cannot take that into account in the process of construing such a document to find out what the true nature of the transaction is.

(5) Where a written agreement made between the parties is held to be a sham, the task of the court is to identify and give effect to the true bargain between the parties which the written agreement was intended to conceal. Where a written agreement is not held to be a sham, the task of the court, as with any other agreement, is to construe it and give fair effect to its terms in the context of all relevant surrounding circumstances (*Shellmex & BP, supra*, at p 619; G *Aldrington Garages, supra,* at pp 463, 467 and 473).

I am not quite sure whether the learned judge in this case intended to hold that the agreements were a sham. He describes them as "artificial transactions designed to evade the Rent Acts", thus adopting part of Lord Templeman's language in *Street* v *Mountford* when he referred to "sham devices and artificial transactions whose only object is to disguise the grant of a tenancy and to evade the Rent Acts". I do not know whether the words omitted are significant or whether they simply represent a defect in the note of the judgment. What the judge said was undoubtedly true. The agreements certainly were the produce of considerable artifice. The landlord's object certainly was to give the occupants the rights of licensees and not the rights of tenants in order that the transaction should not be subject to H the Rent Acts. But the words which the judge omitted — if he did omit them — are very significant because they involve a finding that what had in truth been granted was a tenancy, and such a finding is essential to the occupants unless the agreements themselves, properly construed, create a tenancy.

This last point can be disposed of shortly. The written agreements cannot possibly be construed as giving the occupants (jointly or severally) exclusive possession of the flat or any part of it. They stipulate with reiterated emphasis that the occupants shall not have exclusive possession. The lack of exclusive possession is fatal to any claim of tenancy. The occupiers can therefore succeed only by showing that the agreements were a sham. It may be that this is what J the judge held, even though he did not use the word. Certainly it was the case expressly pleaded by the occupants.

Before us counsel for the occupants urged that the written agreements were a sham and that there was a joint tenancy granted to the occupants, each of whom was jointly and severally liable for the whole monthly payment of £174. He relied in particular on:

(1) the conversation about the bed. (The landlord told us that there was already a double bed in the flat so that he had not had to put one in specially. There is no finding to this effect. We have no reason to disbelieve him. But I do not think it much matters.);

(2) the fact that the occupants, a young man and a young woman, approached the landlord together, making it plain they intended to occupy the accommodation jointly; K

(3) the fact that the occupants attended to sign the documents together and that they signed identical documents.

If the issue had been whether the landlord appreciated that the occupants proposed to live together as husband and wife, no possible fault could be found with the judge's conclusion that he did. But that was not the issue. The issue was whether (the written agreements apart) the landlord agreed to give the occupants, jointly or severally, exclusive possession of the flat or any part of it. He plainly did not so agree in express terms. It is not suggested that he did. I do not think the facts relied on could be held to give rise to an implied agreement, particularly having regard to the facts (a) that the flat contained two rooms (apart from the kitchen and bathroom), (b) that there were L two beds in the flat (apart from the double bed), and (c) that the landlord orally insisted on his right to put other people in.

The occupants' contention can be tested by considering two hypothetical situations:

(1) Suppose that the occupants had quarrelled and one had left the flat. Suppose further that the landlord had sought payment of the full monthly rent of £174 from the survivor and that that claim had been resisted. It seems to me that any court would have been bound to uphold the survivor's plea that he or she had agreed to pay £87 monthly in advance and not a penny more.

(2) Suppose that the landlord had proposed to introduce a third occupant into the flat to occupy a bed in the bed-sitting room. Suppose further that the occupants had moved for an injunction to M restrain him on the ground that such conduct would be a breach of the agreement between them. Would they have succeeded? In my view they plainly would not. The written agreements gave him that right. I do not see how the occupants could begin to show any contrary agreement on his part.

If the written agreements are to be discarded as a sham, it must be shown not only that the occupants intended to enjoy a right to exclusive possession but also that the landlord shared that intention. In my view, he plainly did not. He was determined that they should not enjoy that right. Doubtless his determination was conditioned by his desire that the relationship between himself and the occupants should not be governed by the Rent Acts, but that consideration must

*Also reported at (1978) 247 EG 557.

A be understood as fortifying rather than undermining his intention that the occupants should have no right to exclusive possession.

If, as the House of Lords held in *Street* v *Mountford* (at p 825F), and the Court of Appeal in *Hadjiloucas* (at p 929) agreed, the agreements in *Somma* v *Hazelhurst* were an obvious sham, I was at first inclined to think that the same result must follow here. But there are significant differences. *Somma* concerned a single room with two beds in it. In that situation the right reserved by Miss Somma to use the room herself (in addition to the two licensees) could well be regarded as physically impracticable. Here there was a bed-sitting room and two beds in addition to the double bed and the bedroom which the occupants used. The introduction of an additional sharer would not have been physically impracticable, as the sojourn of the B occupants' guest (however uncomfortably) shows. The landlord in argument drew attention to various differences in the agreements: for example, he pointed out that his clause 16 (unlike *Somma* clause 19) contained no limit on the number of licensees; he relied on his clause 24, which had no parallel in the *Somma* agreement; and he relied on his clause 25, which had no exact parallel in the *Somma* agreement. He also relied on the marriage agreement (the public policy implications of which we were not asked to consider). I do not, however, think that these textual points are of major significance. But the facts found by the judge did not, in my judgment, justify him in finding, if he did find, that these agreements were a sham designed to disguise or conceal the true bargain between the parties.

C It follows that, in my judgment, the landlord should have obtained an order for possession. I would accordingly allow the appeal, set aside the judge's order and make an order that the occupants give up possession of the flat in 28 days from today.

This order will affect the sum payable by the occupants to the landlord as at the date of the judge's order. I invite the parties to tell us what sum was properly payable in the light of this judgment.

The landlord complained that the learned judge had not made an order in his favour for payment of interest on the sum in court which was ordered to be paid out to him. I do not know the reason for the judge's decision, but I am not prepared to assume that he exercised his discretion wrongly and I would not therefore disturb his order. The landlord may address us concerning interest on any additional D sum which may now be payable to him.

The landlord complained of the judge's order concerning costs, but this was not raised in the notice of appeal and I can see nothing whatever wrong with the order made in the light of the judge's decision as it stood.

Agreeing that the appeal should be allowed, MANN LJ said: I gratefully adopt and endorse the statement of the facts which are relevant to this appeal contained in the judgment of Bingham LJ.

The issue in the appeal is simply stated. The appellant contends that the respondents are each licensees of the top flat at 6 Whiteley Road, London SE19. The respondents contend that they are joint (and protected) tenants of that flat with a joint and several liability to E pay the rent. His Honour Judge Macnair decided in favour of the respondents' contention. The approved note of his judgment leaves me in doubt as to the basis or bases of his decision. The relevant passage is as follows:

> House of Lords tell me to ignore the assertions in the document that it is a licence. What difference does it make here that this is a sharing agreement. It seems clear to me that apart from a few visits by Mr Antoniades no one shared with them during period of occupation, they had exclusive possession . . . Hold that these licences are artificial transactions designed to evade the Rent Acts. Finds that there is a tenancy . . .

If the consequence in law of a transaction is the avoidance of the application of the Rent Acts, then that is not a ground on which the F transaction can be flawed (see *Shell-Mex & BP Ltd* v *Manchester Garages Ltd* [1971] 1 WLR 612 at p 619). Accordingly, if, and I emphasise "if", the learned judge based his decision on a design to avoid the application of the Rent Acts, then that basis could not be sustained. A licence to occupy is as legitimate as the grant of exclusive possession.

The critical question in any case must be what is the transaction? The judge had before him the two licences of February 9 1985 and regarded the House of Lords' decision in *Street* v *Mountford* [1985] AC 809 as directing him to ignore the assertions in the documents that they were licences. He seems to have regarded himself as subject to such a direction because the documents were modelled upon the documents which this court in *Somma* v *Hazelhurst* [1987] 1 WLR 1014 had decided to be licences but which decision was disapproved G in *Street* v *Mountford* at p 825H. The disapproval was because the arrangements between the owner of the single room and the unmarried but cohabiting couple, whereby each licensee was obliged to share with such other person as the owner might nominate, demonstrated that the arrangement was a sham to disguise the conferment of exclusive possession upon the licensees "in order that H and S might live together in undisturbed quasi-connubial bliss" ([1985] AC at p 825G).

Mr James Harris for the respondents submitted that the transactions here were a sham to disguise the conferment of a joint tenancy (ie of a right of exclusive possession), and that the learned judge had so found. There are two important passages in the authorities which deal with "sham". H In *Snook* v *London & West Riding Investments Ltd* [1967] 2 QB 786 at p 802C, Diplock LJ, as he then was, said:

> As regards the contentions of the plaintiff that the transactions between himself, Auto Finance and the defendants were a "sham", it is, I think, necessary to consider what, if any, legal concept is involved in the use of this popular and pejorative word. I apprehend that, if it has any meaning in law, it means acts done or documents executed by the parties to the "sham" which are intended by them to give to third parties or to the court the appearance of creating between the parties legal rights and obligations different from the actual legal rights and obligations (if any) which the parties intend to create. But one thing, I think, is clear in legal principle, morality and the authorities (see *Yorkshire Railway Wagon Co* v *Maclure* and *Stoneleigh Finance Ltd* v *Phillips*) that for acts or documents to be a "sham", with whatever legal J consequences follow from this, all the parties thereto must have a common intention that the acts or documents are not to create the legal rights and obligations which they give the appearance of creating.

In *Hadjiloucas* v *Crean* (1987) 284 EG 927 at p 931* Mustill LJ said:

> By way of preface it is necessary to distinguish between three situations in which, aside from any question of rectification, the court may take an agreement otherwise than at its face value. The first exists where the surrounding circumstances show that the arrangement between the parties was never intended to create any legally enforceable obligation. The second is the case of the "sham", in the sense in which that word has been used in numerous cases, including *Snook* v *London & West Riding Investments Ltd* [1967] 2 QB 786, 802. K Correctly employed, this term denotes an agreement or series of agreements which deliberately are so expressed with the object of deceiving third parties as to the true nature and effect of the legal relations between the parties. The third situation is one in which the document does precisely reflect the true agreement between the parties but where the language of the document (and in particular its title or description) superficially indicates that it falls into one legal category, whereas when properly analysed in the light of the surrounding circumstances it can be seen to fall into another.

I suspect that Mr Harris would have suggested, had he not been economic in the citation of authority, that the instant case was an example of Mustill LJ's second situation. I suspect that he would probably say the same of *Somma* v *Hazelhurst* and *Street* v *Mountford*. If my suspicions are correct, then I would agree with him. L Whether or not a case is a second situation is a question of fact. In *Street* v *Mountford* there was a concession that exclusive possession had been conferred; in *Somma* v *Hazelhurst* the inference was irresistible from the surrounding circumstances at the time of the arrangements. That the matter is one of fact is to be derived from the speech of Lord Templeman in *Street* v *Mountford* at p 826H and from the judgment of O'Connor LJ in *Brooker Settled Estates Ltd* v *Ayers* (1987) 19 HLR 246† at p 252 and of Purchas and Mustill LJJ in *Hadjiloucas* v *Crean* at pp 930 and 932.

I regret that I do not think the learned judge addressed himself to the factual question of whether the apparent corresponded to the real. He seems rather to have assumed that the condemnation of the M arrangements in *Somma* v *Hazelhurst* was fatal to the appellant's cause regardless of any difference in the factual matrix. He also seems to have had regard to the fact that the respondents' enjoyment of the flat was not in the event disturbed by the appellant or by any other licensee. However, subsequent conduct, or its absence, cannot assist in the identification of the terms of a transaction.

I cannot, with respect to him, sustain the learned judge's decision on any basis which I can identify. What then is to be done? To remit the matter for a fresh hearing would not reveal any novelty, for none was suggested. In my judgment, on the evidence which we have

* Editor's note: Also at [1987] 2 EGLR 60 at p 65.
†Editor's note: Also at [1987] 1 EGLR 50; (1987) 282 EG 325.

A before us, there was here no sham. The real transaction corresponded with the apparent transaction which the parties voluntarily entered into. There was here neither concession of exclusive occupation nor inference to be drawn from the grant of one twin-bedded room. Here was a flat with beds in separate rooms; a table, bed and a bed-settee could have been occupied without embarrassment to the two licensees in their separate double-bedded room. Indeed a friend of the respondent ("Angela") did sleep on the bed-settee for a time, albeit in a cramped condition but with the consent of the appellant.

I would allow this appeal and hold that the respondents were each licensees of the premises and that their licences were each effectively determined by notices in writing dated July 7 1986. I agree with the orders proposed by Bingham LJ.

B Mr Antoniades asserts that he has a claim in respect of the general rate and water rate. No such claim is before the court and I say nothing about it.

The appeal was allowed with costs, not to be enforced without leave of the court. Possession to be given in 28 days, but stay of execution ordered pending application for leave to appeal to the House of Lords. Legal aid taxation of the defendants' costs ordered.

Court of Appeal

February 25 1988

C (Before Lord Justice FOX, Lord Justice NEILL and Lord Justice BINGHAM)

ASHBURN ANSTALT v ARNOLD AND ANOTHER (No 2)

Estates Gazette June 11 1988

[1988] 23 EG 128-132

Landlord and tenant — Lease of shop in development area — Rulings in regard to matters consequential on previous Court of Appeal decision between the same parties — The previous D decision established that an agreement in 1973 created a lease, and not a mere licence, in favour of the present respondents and that certain provisions in clause 6 of that agreement gave rise to an overriding interest under the Land Registration Act 1925 in respect of a property occupied by the respondent company (126 Gloucester Road) — These provisions were never registered as an estate contract (which would have been enforceable against the appellants) with the result that the respondents had no enforceable rights in respect of clause 6 except such as were derived from their overriding interest — There was no contractual relationship between the appellants (assignees of the reversion) and the respondents — The E present proceedings were to determine the effect of the undertaking in clause 6 that, on the completion of a proposed development, the respondents would be granted "a lease of a shop in a prime position at the development" — The development in question included properties other than that occupied by the respondents — The difficulties arose from the limited relief available to the respondents, despite their overriding interest, owing to the absence of a contractual relationship with the appellants and the respondents' failure to register the provisions of clause 6 as an estate contract — The main rulings now given by the Court of Appeal were the F following — (1) The overriding interest merely protected existing rights in the occupied land; it did not extend or otherwise alter them — (2) The respondents had no rights at all over that part of the development site to which their overriding interest did not extend — (3) Clause 6 did not give the respondents a right to require a shop to be built on the 126 Gloucester Road site which was the subject of the overriding interest; it merely gave a right to have the grant of a lease of a shop in a prime position in the development area — (4) There might, however, be circumstances in which the only way to implement the undertaking in clause 6 would be to grant the respondents the lease of a shop on the 126 Gloucester Road site; the judgment of Fox LJ discussed some complex G situations in which this would be so — (5) It was difficult to see in what circumstances an award of damages in lieu of specific performance could arise — Declarations accordingly

No cases are referred to in this report.

The Court of Appeal in a lengthy judgment of the court given on October 27 1987* decided a number of matters in dismissing an appeal from the decision of Mr E C Evans-Lombe QC, sitting as a deputy judge of the Chancery Division. Some consequential matters were left for decision after further argument and are now dealt with in the judgment of Fox LJ.

William Goodhart QC and Peter Cowell (instructed by Fox & H Gibbons) appeared on behalf of the appellants, Ashburn Anstalt; Robert Pryor QC and Miss Erica Foggin (instructed by Pritchard Englefield & Tobin) represented the respondents, Walter John Arnold and W J Arnold & Co Ltd.

Giving judgment, FOX LJ said: We gave judgment on a number of matters in this case on October 27 1987. It was then agreed that we should hear further argument upon consequential matters.

The issues arise in this way. Arnold & Co and their predecessor, Mr W J Arnold, have for many years carried on business at shop premises in Gloucester Road, London. I will refer to them as 126 Gloucester Road. Immediately prior to the agreement of February 28 1973, to which I will refer in more detail later, Mr Arnold owned a headlease of the premises for a term of $52\frac{1}{2}$ years granted on J December 31 1945 and Arnold & Co owned a sublease expiring on September 29 1973.

In the agreement of February 28 1973 (which was made between Arnold & Co of the one part and Matlodge Ltd of the other part) Mr Arnold sold the headlease to Matlodge for £205,000. By a further agreement of the same date (which I will call the 1973 agreement) Arnold & Co sold the sublease to Matlodge.

Clauses 5 and 6 of the 1973 agreement provided as follows:

> 5 From and after completion Arnold shall be at liberty to remain at the property as Licensee and to trade therefrom until September 29 1973 without payment of rent or any other fee to Matlodge save that Arnold shall pay all outgoings so long as it is in occupation of the property. From and after K September 29 1973 Arnold shall be entitled as Licensee to remain at the property and trade therefrom in the like terms save that it can be required by Matlodge Ltd to give possession on not less than one quarter's notice in writing upon Matlodge certifying that it is ready at the expiration of such notice forthwith to proceed with the development of the property and the neighbouring property involving inter alia the demolition of the property.
>
> 6 Matlodge hereby warrant that it intends either itself or its successor in title to redevelop the property and the neighbouring property by the erection inter alia of shops and hereby undertakes that in consideration of the disturbance to be suffered by Arnold it or its successor in title will grant to Arnold on completion of the development a lease of a shop in a prime position at the development with an area available for trading of approximately 1,000 square feet and with carpark facilities within development for a term of twenty one years and the rent in respect of such lease shall for the first seven years be 25 L per cent less than the market rent and thereafter shall be at the market rent subject only to the rent being reviewed to the then market rate at the expiration of the seventh and fourteenth years from the date of its grant. Arnold shall within fifty six days of being offered the grant of such lease elect in writing to Matlodge or its successor whether or not it will take up the lease and if it shall not so elect or if it shall elect not to take up the lease then the provisions of this clause shall lapse. In the event of Arnold taking up the lease then the lease shall not be capable of assignment for seven years from the commencement of the term except at the full market rent. The market rent shall be such rent as may be agreed between the parties in accordance with the criteria laid down in the Landlord and Tenant Act (as amended) and in the event of the parties failing to agree such market rent then either party may upon giving notice to the other have the question of the market rent determined by a chartered surveyor acting as an expert and not as an arbitrator and to be appointed on the application of either party by the President for the time being of the Royal M Institution of Chartered Surveyors whose decision shall be final and binding on the parties and whose fees shall be borne as directed by such expert. This provision for fixing the market rent shall apply to the initial market rent and to the market rent at the seventh and fourteenth years.

On February 24 1973 the benefit of the 1973 agreement was assigned by Matlodge to Cavendish Land Co Ltd ("Cavendish"), which was the owner of the freehold. In 1976 the provisions of clauses 5 and 6 of the 1973 agreement were novated between Cavendish and Arnold & Co. Cavendish subsequently transferred the freehold to the Legal & General Insurance Society, who transferred it to the plaintiffs, Ashburn Anstalt ("Ashburn") in 1985.

*Editor's note: Reported at [1987] 2 EGLR 71; (1987) 284 EG 1375.

In our previous judgment we held, *inter alia*, (a) that clause 5 of the 1973 agreement created a lease and not a mere licence and (b) that the provisions of clause 6 constitute an "overriding interest" under the Land Registration Act 1925 in respect of 126 Gloucester Road (ie the property actually in the occupation of Arnold & Co when the freehold was sold to Ashburn). The provisions of clause 6 were never registered by Arnold & Co as an estate contract; accordingly, Arnold & Co have no enforceable rights in respect of clause 6 against Ashburn (with whom they have no contractual relationship) except such as they may have by reason of the overriding interest.

Ashburn owns, under several different registered titles, a block of property at Gloucester Road, part of which is 126 Gloucester Road. They intend to develop the whole block. As regards planning permission, we were informed that Ashburn's plans for development have been approved but that there is an outstanding question as to a listed building consent. We have not seen any documents regarding planning permission.

We were also informed that in September 1987 Ashburn had entered into an agreement with the local authority under section 52 of the Town and Country Planning Act 1971 by which most of the site occupied by Arnold & Co's premises will be used as a public way.

In January 1988 Arnold & Co commenced proceedings in the West London County Court for the grant of a new lease at 126 Gloucester Road under the Landlord and Tenant Act 1954. The proceedings are based on the decision of this court that clause 5 of the 1973 agreement created a lease.

The questions now put before us are as follows:

Whether in the absence of an undertaking by Ashburn to offer to Arnold & Co a lease of a shop elsewhere on the development, Arnold & Co would be entitled to any of the following relief, namely:

(a) to restrain Ashburn from proceeding with any development which did not provide suitable shop premises on the land now subject to the overriding interest;

(b) to compel Ashburn to offer Arnold & Co a lease of a shop on the land now subject to the overriding interest if shop premises were in fact to be constructed thereon *or*

(c) damages in lieu of specific performance.

These questions relate to the extent of Arnold & Co's rights against Ashburn in respect of the provisions of clause 6 of the agreement. That clause constituted an estate contract. If Arnold & Co had registered that estate contract under the Land Registration Act it would have been enforceable against Ashburn. However, Arnold & Co failed to register it. The result is that clause 6 is void as against Ashburn save to the extent that its provisions constitute an overriding interest. In our previous judgment we held that there was only an overriding interest in respect of the land occupied by Arnold & Co which Ashburn purchased, that is to say 126 Gloucester Road (to the site of which I will refer as "the Gloucester Road site"). Ashburn was never under any contractual liability to Arnold & Co in respect of the provisions of clause 6; the clause was part of a contract between Arnold & Co and Matlodge and, by novation, with Cavendish; Ashburn was never a party to that contract.

The following propositions appear to me to be correct:

(1) Arnold & Co cannot have any different rights against Ashburn than Arnold & Co had, as a matter of contract, against Cavendish. The overriding interest merely protects existing rights in respect of occupied land; it does not extend or otherwise alter them. Thus suppose, for example, that Arnold & Co had contracted for a lease of the whole of the development site but had only an overriding interest in respect of a small part of it. Arnold & Co could not insist upon a lease of that part at an apportioned rent (though it may be, as Mr Goodhart was willing to concede, that it could insist upon a lease of that part — at any rate if it was a substantial part — at the full rent).

(2) Arnold & Co has no enforceable rights at all under clause 6 over or in respect of that part of the development site to which its overriding interest does not extend. Accordingly it has no rights in respect of any land other than the Gloucester Road site.

(3) Clause 6 conferred no rights upon Arnold & Co to require a shop to be built upon the Gloucester Road site. The only right conferred by clause 6 was on completion of the development to have a grant of a lease of a shop in a prime position, with certain special characteristics, for the term and at the rents and subject to the provisions stated in clause 6. It is entirely a matter for Ashburn to decide where it sites shops on the development. It is asserted, as I understand it, that there never was an agreement to grant a lease but merely a warranty sounding in damages. I do not agree with that. I think that the "undertaking" given in clause 6 was a contract. Accordingly, to compel Ashburn to build such a shop would be imposing upon Ashburn a different liability to that for which it contracted in clause 6.

(4) If Ashburn build a shop completely on the Gloucester Road site and do not offer to Arnold & Co any other shop (within the description in clause 6) Arnold & Co are entitled to insist upon the grant of a lease of the shop on the Gloucester Road site. Mr Goodhart was, I think, disposed to accept that, but in any event it seems to me to be correct. What is protected by the overriding interest is a contract to grant a lease. The question is in what circumstances could Cavendish, consistently with the provisions of clause 6, have been compelled to grant a lease to Arnold & Co of the Gloucester Road site. In that connection, Ashburn cannot be in any worse position than Cavendish. For present purposes, the circumstances, I think, are two. First, if the only shop premises built upon the development are upon the Gloucester Road site, the grant of a lease of that shop would be the only way of giving effect to clause 6; second, if shop premises are built on the Gloucester Road site and other shop premises on other parts of the development but Cavendish refused to offer a lease of any shop at all. In those circumstances, I apprehend the court would permit Arnold & Co to make a choice itself and would enforce the choice — which could be the Gloucester Road site. In the present circumstances, Arnold & Co cannot insist upon a lease of a shop outside the Gloucester Road site but, if one is built upon that site and Ashburn refuses to grant a lease of any other, I do not see why Arnold & Co should not be able to insist upon a lease of the shop on the Gloucester Road site. Arnold & Co would have no better rights and Ashburn no worse rights than Arnold & Co and Cavendish respectively would have had if Cavendish had remained the owners of the freehold.

I now come to Mr Pryor's contention on behalf of Arnold & Co. He asserts that, if Ashburn's development is such that a shop is built partly on the Gloucester Road site and partly on land belonging to Ashburn, then Ashburn is bound to offer a lease of the shop to Arnold & Co. That, it is said, is the only way to give effect to the overriding interest. I do not feel able to accept that. Arnold & Co have no interest which they can enforce over any part of the development other than the Gloucester Road site. Any other conclusion would involve, in effect, treating the overriding interest as applying to land to which it certainly does not apply. There are no circumstances in which Arnold & Co could claim a lease of a shop built wholly outside the Gloucester Road site. That is simply because they had no overriding interest in respect of such land. For the same reason, Arnold & Co can have no rights in a shop built partly on the Gloucester Road site and partly outside it. What clause 6 is dealing with is the lease of an entire shop premises. The fact that Arnold & Co has an overriding interest in part of the premises is not a reason for conferring upon Arnold & Co what is, in effect, the same right in respect of other land. The only reservation I make to that is that, if the overlap on the land outside the Gloucester Road site were trivial, it might be possible for the court to disregard it, but I express no concluded view on the point, since its determination would require a consideration of the precise facts of a situation which does not at present exist. If there is any hardship in that result it is essentially the consequence of Arnold & Co's failure to register its clause 6 rights as an estate contract. From what we know of the facts the point seems likely to be academic, since it seems probable that no shop will be built on any part of the Gloucester Road site (either because of the section 52 agreement or because Ashburn do not wish to build there).

The reason, in my view, is that the answer to question (a) is "No" and the answer to question (b) is "Yes". We can consider the precise terms of any declaration with counsel.

As regards any award of damages in lieu of specific performance, it is difficult to see in what circumstances that could arise. At the completion of the development, either Arnold & Co is entitled as of right to a lease of some part of the development or it is not. In any event, the award of damages in lieu of specific performance is a discretionary matter and I do not think we can usefully speculate upon the circumstances in which it might be appropriate. It was argued by Mr Pryor for Arnold & Co that it would be better if this court left the questions considered above to the judge in the Landlord

A and Tenant Act proceedings. I see no need for that course and it seems better that we should deal with them ourselves.

NEILL and BINGHAM LJJ agreed and did not add anything.

Declarations accordingly. Plaintiffs were awarded costs of the further argument.

Court of Appeal

(January 25 1988)

(Before Lord Justice GLIDEWELL and Lord Justice TAYLOR)

B STUROLSON & CO v MAUROUX

Estates Gazette June 18 1988

[1988] 24 EG 102-108

**Landlord and tenant — Landlords' claim for possession and arrears of rent — Tenant's counterclaim for damages for breaches of landlords' covenants to repair and provide services — Appeal from decision of county court judge, who dismissed possession claim, assessed rent arrears due to landlords at £2,988.77 and awarded tenant £5,895 together with interest of £1,250 on his counterclaim — Landlords'
C appeal related only to amount of damages awarded on counterclaim, there being no appeal on liability — The judge had found the tenant's complaints of breaches of repairing and service obligations substantially justified; they were due mainly to the present landlords' predecessors in title and most of the repairs had been completed by the date of the hearing — The larger part of the damages awarded by the judge was in respect of inconvenience, discomfort and injury to health — The award was attacked by the landlords on two grounds — First, it was submitted that the fair rent registered in respect of the property by the rent officer took account, under section 70 of the Rent Act 1977, of the state of disrepair due to the
D landlords' breaches and that the tenant would be twice compensated if he received damages for breach of the covenant to repair — Second, it was alleged that the tenant (and his wife, of whom he was the statutory successor) had failed to take reasonable steps to mitigate their loss — The Court of Appeal rejected both submissions — As regards the first, the rent officer had taken account of the bad state of repair but he had also to place a value on the landlords' repairing covenant, as it could not be assumed that the covenant would not be carried out in future — "The failure thereafter to carry out that covenant sounds in damages" —
E As regards mitigation of loss, in view of all the circumstances, the age of the tenant and the age and state of health of his wife before she died, and their financial straits, the judge was justified in holding that they had done all that they reasonably could — Appeal dismissed**

No cases are referred to in this report.

This was an appeal from a decision of Judge Vick QC, at Wandsworth County Court. The appellants, plaintiffs in the county court, were the landlords of a flat at 31 Thurleigh Court, Nightingale Lane, London SW12, and claimed possession and rent arrears against the defendant tenant, the present respondent, who
F counterclaimed for damages for breach of covenants. The judge dismissed the possession claim but made an award in respect of rent arrears and on the tenant's counterclaim. The appeal by the landlords was only in respect of the latter.

J Ferris (instructed by Rochman Landau) appeared on behalf of the appellants; John Hodgson (instructed by Corsellis Church Rackham) represented the respondent.

Giving judgment, GLIDEWELL LJ said: This is an appeal against an order of His Honour Judge Vick QC in Wandsworth County Court on June 4 1987 when the learned judge ordered as follows: he entered judgment for the plaintiffs, Sturolson & Co, for the sum of £2,988.77 for rent arrears; he dismissed the plaintiffs' claim for possession of 31 Thurleigh Court, Nightingale Lane, London SW12; he ordered that judgment be entered for the defendant on the counterclaim for G
£5,895 together with £1,250 interest; and he ordered the plaintiffs to pay the defendant's costs of the claim and counterclaim.

The plaintiffs appeal to this court in relation to the amount of damages awarded on the defendant's counterclaim. There is no appeal on liability.

The facts as agreed or as found by the learned judge can be summarised as follows. The late Mrs Mauroux became tenant of flat 31 Thurleigh Court, Nightingale Lane, Clapham, in 1963. She and her husband had previously occupied another flat in another block owned by the same landlords, who were then a company called Keene Investments Ltd. At that time Mr Mauroux was 52. Now simple mathematics show he is either 76 or 77.

By 1979, and for many years before that, Keene Investments Ltd H
failed to comply, first, with an express covenant in the tenancy agreement, or a number of express covenants of which the most important was to supply constant hot water both for heating and for domestic purposes, and, second, with the covenant implied by section 32 of the Housing Act 1961 that they should repair the structure and the exterior of the premises. Mr Mauroux, on his wife's behalf, complained continually about the landlords' failure to carry out their covenants and gave written notice from time to time of the various defects. But the landlords did nothing. Indeed, their agent told him in effect that they did not intend to do anything because the land on which the building stood was worth more than the building itself. So after a time Mrs Mauroux withheld her rent. J

In an action commenced in 1979 Keene Investments Ltd sued Mrs Mauroux for arrears of rent, then amounting to £742.76 and they claimed forfeiture of the lease and possession of the premises. By a defence and counterclaim Mrs Mauroux alleged the breaches of covenant to which I have referred and claimed to set off damages for those breaches and, indeed, to counterclaim.

I should have said that, sadly, during the time that she lived at these premises Mrs Mauroux increasingly lost her sight and in the latter years of her life she was, as I understand it, totally blind. She died on June 2 1982. Mr Mauroux remained in possession of the premises as successor to her statutory tenancy under the Rent Act. The state of disrepair continued. Meanwhile, at a date of which I am not certain but which must have been before 1983, the present plaintiffs, K
Sturolson & Co, bought the block of flats from Keene Investments Ltd, subject, of course, to their existing state of disrepair and to the landlords' liability under the repairing covenants and under the express covenants.

On July 1 1983 Sturolson & Co started a second action against Mr Mauroux, again claiming forfeiture and possession for non-payment of rent, which by now amounted to £1,020. However, they also alleged, by amendment, a further breach of covenant by him in that he had for many years acted, I think in a relatively modest way, as agent for wine shippers and this they alleged constituted a breach of covenant not to use the premises for the purposes of a business. Mr Mauroux, as his wife had done previously, pleaded the landlords' L
breaches of covenant as a defence and set off and counterclaimed damages for those breaches.

In all the circumstances, the learned judge refused to order possession, as I have already said.

The period for which Mr Mauroux, both in his own capacity and representing his wife's estate for the years in which she had been tenant, could claim damages for breach of covenant was limited by the Limitation Act, that is to say to the period starting some six years before the first action.

It is fair to say that, unlike their predecessors, the plaintiff company were conscious of their obligation to carry out their covenants, and by the time of the hearing they had completed most if not all of the required repairs. Indeed, they had done the major part M
of the work in the summer of 1983.

Now as I said, there is no appeal by the plaintiffs with regard to liability. So far as damages are concerned, the judge assessed damages under two broad heads. First, he awarded damages for the diminution in the value to the Maurouxes of their tenancy, including in this both damages for the state of disrepair of the premises and for the failure to supply services, including the hot water. Second, he awarded damages for discomfort, inconvenience and injury to health.

Mr Mauroux' complaints as to disrepair and breach of covenant were set out in his counterclaim in the second action in the following terms (para 11):

Particulars

(a) The Lessors have failed to keep the structure and exterior of the premises in repair so that damp has penetrated the external walls of the said flat to the extent that fungus grows on the inside of the walls;
(b) The Lessors have failed to provide hot water to the premises throughout the day or a sufficient supply of hot water to heat the central heating radiators in the flat;
(c) The Lessors have failed to provide lighting to the corridors of the building during the daytime so that the said corridors are in darkness in the daytime and have failed to keep in repair the external lighting of the building;
(d) The Lessors have failed to keep the electrical wiring within the said flat in repair so that the same has been unusable for long periods;
(e) The Lessors have failed to prevent water penetrating into the said flat from a defective bath in the flat above;
(f) The Lessors have failed to keep in repair the window frames of the said flat so that water penetrates into the flat through the said frames and have left broken windows unrepaired for years;
(g) The Lessors have permitted cockroaches and other insects to infest the boiler room below the said flat from whence they have repeatedly penetrated into the said flat and caused a nuisance to the Defendant;
(h) The Lessors failed to repair a defective bath water closet and bathroom hand basin in the said flat for three years;
(i) The Lessors have failed to keep the external drain outside the flat clear so that the said drain repeatedly became blocked so as to be a nuisance to the Defendant.

The judge specifically found for the plaintiffs — he found that he was not satisfied that the defendant's complaints were made out — on two of those items, namely the penetration of water from the bath above and the infestation by cockroaches and other insects. But that apart, he found for Mr Mauroux. He said (p 13):

> The weight of evidence adduced before me has satisfied me that Keene Investments were in deliberate and continuing breach of their statutory obligations to repair from at least 1965 onwards until they were replaced as landlords by Sturolson & Co in late 1982, and there is every indication from all the documents and other evidence that by the 1970s at the latest the state of disrepair had become appalling and had nearly reached its peak, although it continued until repaired by Sturolson & Co in late 1982 and 1983.
>
> Having seen and heard Mr Mauroux give his evidence and studied the documents carefully I am satisfied that the defects complained of by him existed, that he complained regularly about them, and that nothing was done about his complaints by Keene Investments save to a minimal extent.

The judge assessed the damages, not merely under the two heads to which I have referred but separately in relation to each of four periods; the first is the period leading up to the first action, that is to say to October 9 1978; the second is a period from then until June 2 1982, the date of Mrs Mauroux' death; the third is a period from that date until the summer of 1983 when the present plaintiffs had completed the major works of repair; and the fourth is the period to the date of the hearing from the summer of 1983 onwards.

The judge assessed the damages under head 1 for diminution in value for each of those periods as a percentage of the total rent for that period. He did not always choose the same percentage. Mr Ferris, for the plaintiffs, had not objected to the judge adopting that as a method of assessment. The total awarded under this head, to cover both the breach of the covenant to keep in repair and the breach of the covenant to provide services, comes to £1,345. The judge did not specify how much of that was for breach of the covenant to repair and how much was for the breach to provide services.

The balance of the award of damages under the counterclaim is under the second head, inconvenience, discomfort and injury to health. Mr Ferris, who has argued the plaintiffs' case in detail and with skill, has put forward two quite separate grounds of attack against the judgment.

First, he says — and it is the case — that since June 1973 these premises have been let at a rent which is a registered fair rent. That rent has been revised from time to time; indeed, there have been five separate assessments. On June 5 1973 the rent was assessed at £414.72 and, without going through the matter in detail, it was reassessed at February 1977, November 1981 and April 1984, and on August 22 1986 it was fixed at £1,776.

By the Rent Act 1977, section 70, the rent officer (whose task it is to determine a fair rent) is given instructions as to matters which he is to take into account. Section 70(1) provides:

> In determining, for the purposes of this Part of this Act, what rent is or would be a fair rent under a regulated tenancy of a dwelling-house, regard shall be had to all the circumstances (other than personal circumstances) and in particular to —
> (*a*) the age, character, locality and state of repair of the dwelling-house,

(*b*) does not matter and subsection (2) does not matter.

Subsection (3):

> There shall be disregarded —
> (*a*) any disrepair or other defects attributable to a failure by the tenant under the regulated tenancy or any predecessor in title of his to comply with any terms thereof.

So Mr Ferris argues that the rent officer is obliged to take into account, and it must be assumed did take into account, the state of disrepair to which the landlords' failure to carry out their repairing obligation had led. He must, therefore, have fixed the rent at a figure lower than it would have been if the tenancy had not been in that state of disrepair. Thus, Mr Ferris argues, the rent officer has, in effect, already compensated the defendant for the state of disrepair of the premises. He accepts that he has not compensated the tenant for the failure to provide services because, in respect of that, the rent officer is not concerned with the question whether the services are actually provided, he is concerned only with the contractual position; that is to say, he treats the matter as if the landlord was carrying out his obligation under the contract to provide the services.

So Mr Ferris argues that so far as disrepair is concerned, if the tenant is awarded damages for the breach of the covenant to repair, he is in effect being awarded damages for that for which he has already been compensated in the fixing of a lower rent. Therefore, the award of damages of £1,345 must be excessive.

In his judgment the learned judge dealt with this matter, shortly, in the following terms (on p 20):

> Of course I have to have regard to the fair rent assessed and the dates of the assessments and the intervals of time between assessments, and to the considerations to which the rent officer and rent assessment committee are required to direct their attention in arriving at their assessments. These include the nature of both the landlords' and tenants' respective contractual and statutory obligations and the condition of the dwelling-house when they inspect it, bearing in mind that those inspections normally only take place at biannual intervals if not longer. In my view that does not preclude me from determining on an objective basis whether the landlords have been proved to have failed to perform their statutory or contractual obligations to a measurable degree over and above that reflected in the fair rent assessment.

So the judge specifically had it in mind that section 70 of the 1977 Rent Act did require the rent officer to take the state of disrepair into account but concluded in that last phrase that there might well be something which might be the subject of compensation for disrepair over and above that which the rent officer had taken into account.

Mr Hodgson, for Mr Mauroux, argues that this is indeed so. He submits that the rent officer was obliged to value having regard both to the condition of the premises and also taking account of the terms of the contract which included the landlord's implied obligation to repair. Obviously, he submits, that obligation was of value to the tenant. The rent officer was not entitled to assume it would not be carried out, even though in the end it was not. Obviously the rent of a property in disrepair is less than that of a property in repair, even with a repairing covenant in the lease. But the rent for a property in disrepair but with a landlord's repairing covenant will be greater than such a property where there is no repairing covenant. Therefore, he says, the rent officer has not wholly taken disrepair into account because he must have valued the value of the repairing covenant and the failure thereafter to carry out that covenant sounds in damages.

With that argument I agree, and I would therefore reject the appellants' argument under this head.

The second argument for the plaintiffs is that the Maurouxes, on ordinary principles, were only entitled to claim damages under any head if they had taken reasonable steps to mitigate their loss. They could, it is argued, have done this by carrying out at least some of the repairs themselves, of Mr Mauroux carrying them out, and then by withholding the cost of doing that from their rent — to which it is suggested there would have been no answer. By not carrying out any repairs — even the most modest or those requiring the least skill — themselves they have failed to make out a case. Thus to an extent the award of damages under both heads should be reduced.

Now by an unfortunate error (which the county court has explained) the passage in the judgment which was intended to deal with this matter was omitted from the draft of the judgment that was originally circulated. I should have said that the learned judge dictated his judgment, a draft of it was circulated to the parties by post, and thereafter the parties' representatives attended at court on June 4 1982, when the judgment was formally given although I imagine was not actually read out. However, the omission having been pointed out to the judge, he added a passage which was what was originally intended to be inserted, which reads as follows:

Mr Ferris has urged strongly the alleged failure to mitigate by the tenants. The principles to apply are expressed clearly in *Chitty on Contracts* (25th ed) at paras 1714 and 1715 thus: "The Plaintiff cannot recover for loss consequent upon the Defendant's breach of contract where the Plaintiff could have avoided the loss by taking reasonable steps" and "the onus of proof is on the Defendant who must show that the Plaintiff ought as a reasonable man to have taken certain steps to mitigate his loss". The test is that of reasonableness and on the evidence adduced before me I have formed the clear judgment that there was nothing which Mr and Mrs Mauroux have done or failed to do as reasonable ordinary statutory tenants to mitigate their damage or loss beyond what they did having regard to the clear and deliberate policy of Keene Investments to let the block fall into disrepair. Faced with this landlord they did all that they could reasonably be expected to do, in my view, to make their flat habitable.

Mr Ferris argues that, on the evidence, the Maurouxes did almost nothing. They could at least, he says, have cleaned out the drains, or replaced the lights in the corridors, or replaced window panes from time to time. And, as I have said, they could if necessary, if that involved them in some cost, have withheld the rent. Their failure to do so was so unreasonable that no judge could properly find that they behaved reasonably.

With this also I disagree. The parts of the works of repair which it is alleged the Maurouxes could have done were, of course, all relatively minor and it may be that they could have done them. It may be that some other persons differently situated, without a wife who had gone blind and the worry that that entailed, and without the financial stringency that the judge found the Maurouxes were under certainly for most of the relevant period, would have carried out such works. There were many parts of the disrepair which, of course, they could not reasonably have been expected to do — curing the damp coming through the walls where the fungus was growing through is an obvious example.

My view does not depend upon a finding of any fact by this court; I regard it as being a matter singularly within the province of the learned judge. He heard the evidence. He was able to decide, in a way we cannot, what his view was of the relative conduct of the parties, and his view of the conduct of Keene Investments shines through from the passage I have already read and, indeed, one or two other passages in the judgment. They behaved quite disgracefully. The conclusion that he reached that it was not reasonable to expect Mr and Mrs Mauroux to have done anything more than they did is one with which, in my view, we cannot and should not interfere.

For these reasons, I would dismiss this appeal.

Agreeing, TAYLOR LJ said: The appellants' main argument is based upon section 70 of the Rent Act 1977. That requires that in fixing a fair rent for a regulated tenancy regard shall be had "to all the circumstances (other than personal circumstances) and in particular to — (*a*) the age, character, locality and state of repair of the dwelling-house". Thus, any disrepair is to be taken into account in reduction of the rent. Mr Ferris argues that if the tenant thereafter seeks damages for breach of the landlord's covenants, such damages cannot include any sum for disrepair; otherwise, he says, the landlord would suffer a double penalty for the same default.

I cannot agree. The rent officer has to have regard to all the circumstances of the case. Those, in my judgment, include the existence of the landlord's repairing covenant. Mr Hodgson puts it another way: the fair rent must charge the tenant, *inter alia*, for the value of the landlord's obligation to repair. It is true that the rent officer must take account of any disrepair at the time of his assessment, but he is not to assume that any present disrepair will continue indefinitely because of the failure of the landlord to discharge his obligation.

Take two cases for comparison. First, a landlord whose rent is fixed on the basis of present disrepair thereafter rectifies the defects. The tenant continues to pay only the fixed fair rent. Second, another landlord, such as the plaintiffs here, does no repairs at all and indeed the disrepair may gradually progress. The tenant still has to pay the same fixed rent. In those circumstances, I see no reason why a judge faced with a claim for damages for breach of covenant should not award a sum to the tenant in respect of the failure of the landlord to carry out his obligations. He has, of course, to have regard to the fact that the rent is fixed on the basis of existing disrepair, but that fact does not, in my view, exclude the award of an appropriate amount to take account of the continuation or deterioration of that disrepair and of the landlord's failure to discharge that repairing obligation for which the rent includes a charge.

I consider the approach of the learned judge in this case was wholly justified and I agree with Glidewell LJ that that argument addressed by Mr Ferris must fail. I do not wish to add anything to what Glidewell LJ has said in respect of the argument as to mitigation of loss. Accordingly, I agree that this appeal must be dismissed.

The appeal was dismissed with costs: legal aid taxation of defendant's costs until January 7 1988 ordered.

For a further case on this subject see p 227

LANDLORD AND TENANT

BUSINESS TENANCIES

A

Court of Appeal

October 2 1987

(Before Lord Justice DILLON and Lord Justice LLOYD)

ESSEXCREST LTD v EVENLEX LTD

Estates Gazette January 9 1988

[1988] 01 EG 56-60

**Landlord and Tenant Act 1954, section 38(4) — Contracting-
B out with authority of the court — Correct procedure not followed by parties and attempt to contract-out ineffective — Wording of section 38(4)(a) makes it clear by reference to "persons who will be the landlord and tenant" and "a tenancy to be granted" that the court's authorisation must precede the grant of the tenancy — It is possible to grant tenancies conditionally on the court's giving authority, but a tenancy granted unconditionally cannot be brought within section 38(4)(a) by a subsequent application and purported approval — In the present case an underlease was granted unconditionally but including the recital of an authorisation
C under section 38(4), with blank spaces left for the name of the county court and the date of the order to be filled in when the authorisation was granted — Subsequently an application by the parties was made to the court and the registrar made an order expressed to be pursuant to section 38(4) — The underlease was endorsed by the underlessees' solicitors with a memorandum of the order, but there was no re-execution of the underlease or any joint endorsement of agreement excluding the tenancy from sections 24 to 28 of the 1954 Act — Eventually the underlessees applied for a new tenancy, contending that there had been no effective exclusion of rights under sections 24 to 28 of the Act — The county court judge
D accepted this contention and the underlessors appealed — Their appeal was rejected by the Court of Appeal after determining a procedural question with some relevance to the possibility of future litigation, namely, whether the proceedings below were merely an application to strike out for no arguable cause of action or were the hearing of a preliminary application — The court decided in favour of the latter view and upheld the judge's conclusion that there was no basis for treating the grant of the underlease as conditional — It was an effective grant, made in the mistaken belief that authorisation would be forthcoming and that all would be well
E — Unfortunately the parties had gone about the matter in the wrong way — Appeal dismissed**

The following case is referred to in this report.

Joseph v *Joseph* [1967] Ch 78; [1966] 3 WLR 631; [1966] 3 All ER 486, CA

This was an appeal by landlords, Essexcrest Ltd, against the decision of Judge Hammond at Bolton County Court holding, on a preliminary issue, that the application of sections 24 to 28 of the Landlord and Tenant Act 1954 had not been effectively excluded from the grant of an underlease of business premises at 2 Mawdsley Street, Bolton.

F

Paul Chaisty (instructed by Betesh Fox & Co, of Bolton) appeared on behalf of the appellants; Nicholas Dowding (instructed by Wells & Hind, of Nottingham) represented the respondents.

Giving judgment, DILLON LJ said: This is an appeal by landlords against a decision of His Honour Judge Hammond given in the Bolton County Court on February 13 1987. It is concerned with the statutory provisions for the protection of business tenants.

One of the more important of those provisions, contained in Part II of the Landlord and Tenant Act 1954, is the right of the tenant to apply for a new tenancy on the expiration of his existing tenancy. It was held by this court in the case of *Joseph* v *Joseph* [1967] Ch 78, and indeed it is provided by section 38 (1) of the 1954 Act, that a tenant G
cannot contract-out of the rights conferred on him by Part II of the Act; otherwise there would be too much danger of landlords imposing on prospective tenants by refusing to grant tenancies unless the tenants contracted-out of their statutory rights.

However, as a result of *Joseph* v *Joseph* a limited contracting-out, subject to the approval of the county court, was authorised by a new subsection (4) of section 38 of the 1954 Act which was introduced by the Law of Property Act 1969. This provides as follows:

> (4) The court may — (*a*) on the joint application of the persons who will be the landlord and the tenant in relation to a tenancy to be granted for a term of years certain which will be a tenancy to which this Part of this Act applies, authorise an agreement excluding in relation to that tenancy the provisions of sections 24 to 28 of this Act; and (*b*) on the joint application of the persons H
> who are the landlord and the tenant in relation to a tenancy to which this Part of this Act applies, authorise an agreement for the surrender of the tenancy on such date or in such circumstances as may be specified in the agreement and on such terms (if any) as may be so specified; if the agreement is contained in or endorsed on the instrument creating the tenancy or such other instrument as the court may specify; and an agreement contained in or endorsed on an instrument in pursuance of an authorisation given under this subsection shall be valid notwithstanding anything in the preceding provisions of this section.

The difference in tense between subsection (a), which refers to the persons who "will be" the landlord and the tenant, and subsection (b), which refers to the persons who "are" the landlord and tenant, is striking. Moreover, what the court is empowered to do under subsection (a) is to authorise an agreement excluding the revelant J
provisions of the 1954 Act in relation to a tenancy which, on the wording of subsection (a), is a future tenancy — a tenancy "to be granted" for a term of years certain which "will be" a tenancy to which Part II of the Act applies. That power under section 38(4) has been fairly widely exercised and found useful. A practice has also developed (as we have been told) of having conditional agreements for leases to be granted if the court grants authority under section 38(4) but not otherwise. Obviously it is in the interests of commercial parties to have their terms binding, subject to the court's approval, when they come before the court and not to find that once the court has given its approval the matter is thrown back into the melting pot of negotiation once again.

In the present case there were negotiations between the landlords K
and the tenants, acting by their respective solicitors, in December 1982 for the grant by the landlords to the tenants of a tenancy for business purposes of some premises at 2 Mawdsley Street, Bolton. There was then what has been referred to as a completion on December 29 1982. On that occasion an underlease, which is dated December 29 1982 — the figures "29th" and the word "December" being inserted in ink and the words "One thousand nine hundred and eighty-two" being typed — was executed under seal by three parties — the landlords, the tenants and a company acting as surety. Clause 1 of the lease provided that in consideration of a premium of £10,000 paid by the tenant to the landlord (receipt of which is said to be acknowledged) the landlord demised the premises to the tenant from

A

that date to January 31 1983 and thereafter for a term of four years. There is provision for a rent payable yearly and proportionately for any fraction of a year at £6,000 a year until February 1 1986 and for the residue of the term at £15,000 a year with an insurance rent in addition.

The lease contains a number of tenants' covenants which I need not mention, and then at the end, in clause 7(2), there is the following:

> Having been authorised to do so by an order of (blank) County Court made on (blank) One thousand nine hundred and eighty-three under the provisions of section 38(4) of the Landlord & Tenant Act 1954 the parties hereto agree that the provisions of sections 24 to 28 (inclusive) of that Act shall be excluded in relation to the tenancy hereby created.

B

It is not apparent when the premium which I have mentioned of £10,000 was paid. It is quite likely that it was paid on December 29 1982. It is clear that on December 29 1982 the tenant took possession of the premises for business purposes and rent in accordance with the terms of the lease has been paid from that date. It may be that an apportioned sum in respect of rent was paid on December 29 1982.

On December 30 the tenants' solicitors wrote to the landlords' solicitors and said this:

> Thank you for your help yesterday. I am dealing with the stamping of the lease. I look forward to hearing from you when you have had the formal licences from the landlord both for the assignment and for the underletting and also as to the court order.

C

It was plainly understood that the court order would be applied for after December 29.

On December 21 the landlords' solicitors wrote enclosing "joint application under section 38 of the . . . Act for signature by" the tenants or on behalf of the tenants.

On January 10 1983 the landlords' solicitors wrote again to the tenants' solicitors saying that they looked forward to receiving the section 38 agreement duly signed by the solicitors on behalf of their clients, and they added:

> We shall also be grateful if you will provide us with a copy of the duly executed original underlease in order that we may stamp the duplicate underlease.

D

On January 12 1983 the tenants' solicitors returned the joint application under section 38 duly signed. That application was accordingly made. It came before Mr Registrar Taylor in the Bolton County Court on January 24 1983 and he made an order, expressed to be on the joint application of the landlord, the tenant and the surety and expressed to be pursuant to section 38(4) of the 1954 Act, that the three applicants

> be at liberty to enter into an agreement excluding the provisions of sections 24 to 28 inclusive of the said Landlord and Tenant Act 1954 in relation to a tenancy granted by the first applicant to the second applicant such agreement to be endorsed on the said underlease a copy whereof is annexed hereto.

E

It is possible that what was annexed was merely a copy of the draft, but it rather looks as if the document which had been executed by the three parties on December 29 was itself annexed and that the registrar, like the solicitors for the parties, supposed that it would be sufficient if a memorandum of the registrar's order was endorsed on that document.

The landlords' solicitors then wrote to the tenants' solicitors on February 21 1983. They enclosed a copy of the order of the county court of January 24 and said:

> We shall be pleased to receive confirmation that you have no objection to endorsing the underlease to your clients with a memorandum of the order of January 24 1983. We appreciate that reference is made in the underlease to the order but in view of the fact that the underlease is dated December 29 1982 the date of the order cannot really be inserted in the appropriate gaps and it is for this reason that we suggest a memorandum should be placed on the deed.

F

The tenants' solicitors replied on April 6 1983 confirming that they had endorsed upon the underlease a memorandum of what they call "the order of January 4 1983". I take that to be a reference to the 24th. The actual terms of that endorsement are not before us, but plainly there was no re-execution of the underlease by the parties nor was there any joint endorsement of agreement excluding from the tenancy the provisions of sections 24 to 28 of the Act signed by both solicitors on behalf of their respective clients. It was merely a unilateral endorsement that the order which I have mentioned had been made.

On October 2 1986 the tenants made an application in the county court for the grant of a new tenancy under Part II of the 1954 Act. It is their contention that there has been no effective exclusion of their rights under that Act under section 38(4). The landlords' response to

G

that was to issue a summons before the registrar seeking an order that the application for the grant of a new tenancy be struck out on two grounds. The first was that the procedure available to a tenant of business premises in sections 24 to 28 of the 1954 Act was not in this case available to the applicant for the reason that the said sections 24 to 28 were excluded by the order of Mr Registrar Taylor of January 24 1983, pursuant to a joint application of the landlord and tenant. The second as given is:

> . . . that if which is denied the procedure referred to above is available to the applicant the applicant ought not to be granted a new tenancy for the reason that the respondent intends on the termination of the current tenancy to substantially reconstruct the premises comprised in the tenancy and could not reasonably do so without obtaining possession of the premises.

H

In support of the first part of that application Mr Cornthwaite, the solicitor for the landlords, swore an affidavit in which he set out the negotiations of December 1982 and said that it was agreed between the parties that, pursuant to section 38(4) of the 1954 Act, the parties should be at liberty to enter into an agreement excluding the provisions of sections 24 to 28 inclusive of the Act. He then sets out that the joint application was made, and in para 6 he asserts:

> It is to my knowledge and belief that when the parties entered into the underlease dated December 29 1982 it was a condition precedent that a Court Order would be made excluding the provisions of sections 24 to 28 of the . . . 1954 Act as amended and an order by consent was duly made.

J

The assertion that this was a condition precedent is not in that paragraph an assertion of any specific agreement but a contention, as I understand it, that that is the true inference to be drawn from the documents which passed.

On the application which I have mentioned by the landlords, Mrs Registrar Wilby, on December 1 1986, made an order that the application on behalf of the landlords be granted and the application for a new tenancy be dismissed on the ground that the procedure available to a tenant of business premises under sections 24 to 28 of the 1954 Act was not in this case available to the tenant because those sections were excluded by the order of the Bolton County Court of Mr Registrar Taylor of January 24 1983.

K

The tenants appealed against that decision of Mrs Registrar Wilby. It was that appeal which was heard by Judge Hammond. His judgment given on February 13 1987 was, it would seem, a reserved judgment. He allowed the appeal with some reluctance, holding that the order of Mr Registrar Taylor was ineffective as the lease had already been granted and could not be held to have been only conditional on the making of the court order.

By leave of Judge Hammond the landlords now appeal to this court against that decision of the judge.

L

There has been some discussion in the course of the argument of the appeal as to the nature of the proceedings in the court below: was it merely an application to strike out on the ground that there was no arguable cause of action — an application which the judge should only accede to in the clearest of all possible cases? Or was it the hearing of a preliminary issue?

M

The actual form of the summons is not entirely helpful on that, because although the summons asks that the application to grant the new tenancy be struck out, the second ground on which that is sought — that the landlord intends, on the termination of the current tenancy, to substantially reconstruct the premises, and so forth — is plainly a matter appropriate to a preliminary issue rather than to what is more ordinarily understood by striking out. The particular significance of whether it was a striking-out application or a preliminary issue lies in whether it is open to the landlords to litigate the matter again. The basic evidence which was before Judge Hammond was affidavit evidence from solicitors exhibiting contemporary documents which did not give rise to any controversy, except as to the true interpretation of the documents taken as a whole. In a third affidavit, however, Mr Cornthwaite had included a sentence asserting that it had been telephonically agreed, on a date which he does not specify, between himself and a solicitor from the tenants' solicitors who had the conduct of the matter that the lease was subject to the condition that a court order would be obtained. That is a categorical averment of a specific oral agreement, but it was not proceeded with before Judge Hammond because the tenants' counsel had been unable to get instructions on that allegation. It came in an affidavit served at a very late stage indeed, at a time when the solicitor who had had the conduct of the matter was abroad and could not be contacted. It was, therefore, agreed that that sentence should not be taken into account by Judge Hammond, as otherwise the proceedings would have had to be adjourned so that instructions could be taken. The landlords submit that if the judge was only dealing with a striking-out application and not deciding a preliminary issue, they can reintroduce that evidence at a later stage.

Viewing the procedural state of things as a whole, it seems to me that the conclusion we must reach is that the judge heard a preliminary issue by consent of counsel and evidence by affidavit, that particular statement being, by consent of counsel, excluded. Therefore this is not to be regarded as a striking-out application but the hearing of a preliminary issue.

The question for this court then is whether the judge was right in his conclusion on that preliminary issue. After carefully summarising the evidence, he said:

> I have studied the evidence, which consists entirely of affidavits and the exhibits thereto. I can find nowhere any evidence that this was a conditional grant or that there was a suspensory condition.

I entirely agree with that conclusion. It is quite impossible to say, as has been argued for the appellants, that the only possible conclusion from the documents is that the lease was granted conditionally on the court's granting authority under section 38(4) to exclude the relevant provisions of the Act. Everything indicates that the lease was granted as an effective lease in the belief that the necessary consent would be forthcoming from the court and that all would then be well. Even so, when Registrar Taylor made his order all would, as it seems to me, have been well if the parties had re-executed the lease or had endorsed on it some form of agreement which excluded the provisions of the Act in terms which would have constituted a new tenancy thenceforth in respect of the premises. But I cannot think that the unilateral endorsement of the memorandum of the court's order is enough to constitute the agreement authorised by the court excluding, in relation to the tenancy referred to, the provisions of the sections of the Act. The whole of the relevant wording of the Act relates to the future and the way these parties have gone about things unfortunately does not achieve the objective which they had in mind, in my judgment.

Accordingly, albeit with reluctance, I would dismiss this appeal.

Agreeing, LLOYD LJ said: Despite the attractive argument advanced by counsel for the appellants, this appeal must be dismissed for the reasons which Dillon LJ has given.

The appeal was dismissed with costs.

Court of Appeal

November 17 1987

(Before Lord Justice KERR and Mr Justice EASTHAM)

MOSS v MOBIL OIL CO LTD

Estates Gazette February 13 1988

[1988] 06 EG 109-112

Landlord and tenant — Landlord and Tenant Act 1954 — Whether a lease which demised two properties to one tenant contained provisions which were sufficient to create two separate tenancies — Whether, if so, a notice by the landlords under section 25 of the 1954 Act purporting to apply to one only of the properties was valid — Appeal by tenant from decision of county court judge holding that the section 25 notice was valid and dismissing tenant's application for a new tenancy — The lease, which related to two service stations, included in a schedule a provision for apportioning rents as between the two stations — It also included a clause providing that "minimum load" should apply separately to each of the demised premises and that all other covenants, agreements and conditions in the lease "shall be read and construed as if separate and independent leases had been entered into in respect of each of the demised premises and not as one demise" — Held, affirming the decision of the county court judge and distinguishing the cases of *Dodson Bull Carpet Co Ltd* v *City of London Corporation* and *Southport Old Links Ltd* v *Naylor*, that the section 25 notice was valid — The court did not express a final view on the effect of the apportionment of the rents (which the judge did not regard as sufficient to effect a severance) but agreed that the clause quoted above was sufficient to create two separate tenancies — "Within the four corners, so to speak, of one document in the form of one lease the parties have agreed to create two separate demises", thereby achieving what had not been achieved in the *Dodson Bull* or *Southport* cases — Appeal dismissed

The following cases are referred to in this report.

Dodson Bull Carpet Co Ltd v *City of London Corporation* [1975] 1 WLR 781; [1975] 2 All ER 497; (1974) 29 P&CR 311; 236 EG 484
Southport Old Links Ltd v *Naylor* [1985] 1 EGLR 66; (1984) 273 EG 767, CA

This was an appeal by the tenant, Gordon Moss, from the decision of Judge Crabtree, at Rotherham County Court, in favour of the landlords, Mobil Oil Co Ltd, in respect of a notice under section 25 of the Landlord and Tenant Act 1954 concerning the Fitzwilliam Road Service Station at Rotherham. The other service station comprised in the lease was the Maltby Service Station at Tickhill Road, Maltby.

Wayne Clark (instructed by Penningtons Ward Bowie, agents for Broomheads, of Sheffield) appeared on behalf of the appellant; W Poulton (instructed by Metson Cross & Co) represented the respondents. Mr Poulton was not called on.

Giving judgment, KERR L J said: This appeal raises a new point in the context of section 25 of the Landlord and Tenant Act 1954. The question is whether a lease which demised two service stations to one tenant effected a sufficient severance of the demise by an express provision to create two demises or tenancies with the result that a section 25 notice which purported to apply to only one of the service stations had been valid. His Honour Judge Crabtree upheld the section 25 notice in a brief judgment in the Rotherham County Court given on July 31 1987. This is an appeal by the tenant.

The issue arises out of a lease dated September 1 1983 which was headed on the frontispiece:

> LEASE OF
> Fitzwilliam Road Service Station, Fitzwilliam Street, Rotherham, South Yorkshire and Maltby Service Station, Tickhill Road, Maltby, Rotherham, South Yorkshire.

Until one comes to clause 13(a), this was indeed on its face one demise of two service stations as the judge recognised. I must briefly refer to some of the provisions of the lease which are in common form.

Clause 1 contained various definitions including approved opening hours for the Fitzwilliam and Maltby Service Stations respectively. It also defined "Minimum Load" as 27,000 litres "or such other quantity as shall from time to time be specified by Mobil in writing".

Clause 2, headed "Demise", was in the following terms:

> In consideration of the rents reserved hereby, and of the Dealer's covenants contained in this Lease, Mobil hereby demises unto the Dealer:
> First of all those premises known as "Fitzwilliam Service Station"

whose address I have already given

> and "Maltby Service Station"

whose address I have also stated

> more particularly described in the First Schedule hereto (hereinafter called "the Premises"). And secondly the equipment, furniture and other chattels belonging to Mobil in and about the Premises, and listed in the Second Schedule hereto (hereinafter called "the Effects").

The habendum was as follows:

> To hold unto the Dealer for the term of three years commencing the First day of September 1983 (hereinafter called "the Term"), terminable as hereinafter provided, Yielding and paying during the Term the rents specified in the Sixth Schedule hereto at the times and in the manner set out therein.

Clause 7 was a printed proviso for re-entry in common form.

Before turning to clause 13(a) it is convenient to refer to some of the schedules. There was a schedule which identified and described each of the service stations together with its effects. There was another containing dealer's covenants in printed form, to which I need not refer. The Fourth Schedule comprised the landlords' covenants. The Fifth Schedule dealt with the supply of motor fuel and so forth and the Sixth Schedule dealt with rents. That provided different rents for the first, second and third rental years respectively and then apportioned each of those rentals differently and separately as between the Fitzwilliam Service Station and the Maltby Service Station.

That is all I need to mention in the main body of the lease. I agree, as the judge accepted, that, despite the severance of the rents (though it is unnecessary to express a final view on this), on the authorities everything up to this point makes this one demise only of both service stations. Accordingly a section 25 notice would be valid only if it purported to apply to both of them.

However, one then comes to clause 13(a). This was typed and preceded by "Take in attached Rider". The rider itself was in the following terms:

In this Lease Minimum Load shall apply separately to each of the demised premises and all other covenants agreements and conditions contained herein and in the Schedules hereto shall be read and construed as if separate and independent Leases had been entered into in respect of each of the demised premises and not as one demise.

On July 24 1986 a notice to terminate was served which purported to expire on January 29 1987. Accordingly, so far as the time was concerned, this was a good notice for the purposes of section 25. The notice applied only to the Fitzwilliam Service Station. A counter-notice was served on August 18 1986 asking for a new tenancy. But the application for this was not made until December 12 1986, which was out of time, being more than four months after service of the section 25 notice. The time for the application had expired on November 24 1986.

It is somewhat ironic that by the time the section 25 notice expired on January 29 1987 it had been agreed that the Maltby Service Station would be sold by the landlords to the tenant dealer. It is therefore not surprising, since the negotiations for that sale had presumably already started, that the section 25 notice had related only to the Fitzwilliam Service Station. That, however, is neither here nor there. The fact that by the time when the notice expired the Maltby Service Station had already been transferred, or had been agreed to be transferred, forms the subject-matter of a cross-appeal with which we have found it unnecessary to deal.

Mr Clark argued the issues with great clarity on behalf of the tenant, but we found it unnecessary to call on Mr Poulton on behalf of the landlord.

What happened, inevitably in the light of the history, was that having appreciated that the application for a new tenancy was out of time, the tenant took the point that this had been a bad section 25 notice because it applied only to the Fitzwilliam Service Station and not to both of them. Accordingly, he amended his notice of application on March 30 1987 and the question whether the section 25 notice was good was treated as a preliminary issue. In effect, of course, if the issue raised on behalf of the tenant was bad in law, that was the end of the matter, as it proved to be. In the light of clause 13(a) the judge took the view that this section 25 notice was good. Accordingly, he dismissed the application for the new tenancy. Although he subsequently gave leave to appeal it seems clear that no leave was necessary, since the order had been final. The question is whether the judge's conclusion was correct.

We were referred to two cases. Both rest on the principle, which is common ground between the parties and settled law, that a section 25 notice which purports to apply only to part of the demised premises is invalid. They are *Dodson Bull Carpet Co Ltd* v *City of London Corporation* [1975] 1 WLR 781, a decision of Goff J as he then was, which has frequently been followed; and the most recent decision in this area of the law in *Southport Old Links Ltd* v *Naylor* [1985] 1 EGLR 66 (CA).*

In the *Dodson Bull* case Goff J considered at p 784, but without reaching any final conclusion, whether it might be possible, because of sufficiently clear terms, in particular a severance of rents, for one lease or demise to comprise what would effectively be two or more demises for the purposes of section 25.

This point arose in a more acute form in the *Southport* case. One lease, albeit with one unapportioned rent, comprised a golf course (the red land) on the one hand and the clubhouse (the blue land) on the other, with an express provision giving the landlord the right to give 12 months' notice so far as the golf course was concerned, but not so far as the clubhouse was concerned. In those circumstances the landlord gave a contractual notice as well as a section 25 notice, as is frequently done, which were both applicable only to the golf course. The issue was whether, on the construction of that lease, it was possible to treat it as two demises. That was accepted by the county court judge but rejected by this court, partly because there was only one rent. And of course there was nothing equivalent to clause 13(a) as in the present case.

Oliver LJ, as he then was, put the issue as follows on p 68:

Since the section 25 notice in this case relates, and could only relate, to the red land, the argument before the learned judge was based, as it had to be, on the contention that it was possible to spell out of the demise two separate and distinct tenancies, one of the blue land and one of the red land, and thus to validate the notice given only in relation to the red land as a notice given in respect of the whole of the land comprised in a relevant tenancy.

Then he considered certain provisions of the lease and said:

So that unless one can find some separate tenancy of the red land to which, as a whole, a section 25 notice can relate, one is up against the difficulty that it is, on its face, a notice purporting to terminate the tenancy only of part of the holding.

He went on to examine the lease and was unable to overcome that difficulty.

In the present case, however, one does have a provision, in the form of clause 13(a), which has the effect which was absent both in *Dodson Bull* and in the *Southport* case. The judge put it quite shortly as follows:

On the face of it, this was one Lease of two separate premises. The fact that rent was apportioned in Schedule 6 would not make it two separate Leases; and apart from Clause 13(a) everything else points to this being one Lease; but in my view page 15 of the bundle which includes the rider 13(a) and the Lease is the crucial point. Clause 13(a) is incorporated into the Lease and it splits this one Lease into two. This is a contractual clause which says in the plainest possible English that these are two separate Leases. It is my view that this is the plainest possible language. It is putting all these covenants and conditions and so on into one document for convenience's sake, almost for ease of typing one might say as the terms of the two Leases are identical but these are clearly two Leases.

The only point that can be made on that language is that he is not quite accurate when he says that the terms of the two leases are identical, since he himself had rightly mentioned earlier that there were separate rents which were not identical. But subject to that I entirely agree with this construction of clause 13(a).

Mr Clark sought to overcome that point by saying that when one considers the wording of clause 13(a) it does not achieve this purpose. As I followed his argument, he put it on two grounds. First, he appeared to be saying that the words "covenants, agreements and conditions" did not apply to every provision of this lease. In particular I think he was saying that it did not apply to the proviso for re-entry and — of greater importance — to clause 2, which defines the demise.

I do not accept that construction. It seems to me, exactly as the judge said, that no language could express more plainly that this was intended to split this lease into two. Every provision of it was to be read distributively, as applicable to each of the demised service stations severally.

The second and more substantial point on which Mr Clark relied — and in this connection he referred to a later part of the judgment of Goff J in the *Dodson Bull* case at pp 785 and 786 — was that there is nothing in clause 13(a) which enables that clause to be applicable, as he put it, for the purposes of a section 25 notice. He said that, even assuming that it applies to everything in the lease — to every single provision or term of the lease — nevertheless it does not apply to section 25, or to a notice given under section 25, which is not in the lease but arises by operation of law. In my view, that submission is misconceived. The question is not whether the wording of clause 13(a) applies, expressly or by implication, to a notice under section 25 but simply whether or not it has the effect of splitting what, apart from that clause, would be one demise of two service stations into two demises of one service station each. On my view, on its true construction it has precisely that effect, and that was of course the effect intended.

That being the construction of the clause, it must follow that in this case there has been achieved what was not achieved in either *Dodson Bull* or the *Southport* case. Within the four corners, so to speak, of one document in the form of one lease the parties have agreed to create two separate demises. The consequence is that it was open to the landlord to give a valid notice in relation to only one service station pursuant to section 25.

In these circumstances I consider that the judge came to the correct conclusion in deciding the preliminary issue against the tenant and in dismissing the application for a new tenancy. It follows that it is unnecessary to consider what would have been the effect of this notice in relation to the cross-notice, bearing in mind that by the time

* Editor's note: Also reported at (1984) 273 EG 767.

when it expired, though not when it had been given, there was only one service station left in this lease, the other one having been sold to the tenant.

For these reasons I would dismiss this appeal.

EASTHAM J agreed and did not add anything.

The appeal was dismissed with costs.

Court of Appeal

December 8 1987

(Before Lord Justice DILLON and Lord Justice NEILL)

GIANNOUKAKIS v SALTFLEET LTD

Estates Gazette April 30 1988

[1988] 17 EG 121

Landlord and Tenant Act 1954, Part II — Application for new tenancy of snack-bar restaurant — Terms agreed except as to rent — Appeal from decision of assistant recorder fixing rent at £16,500 per annum, with an interim rent of £13,200 per annum — Assistant recorder preferred evidence of landlords' surveyor, who put forward the figure of £16,500, to that of the tenant's surveyor, whose figure was £9,100 — Appellant tenant had been a tenant of the subject premises for 23 years and could not afford to pay the higher rent from the profits of his restaurant business — The judge had considered the rents of various properties but was particularly influenced by the one next door, which was let as a take-away pizza bar run by a subsidiary of Pizzaland International — The reliance on this property was criticised on appeal, on the ground that a take-away business was not really comparable, but the court pointed out that the user clause for the subject premises was wide enough to cover take-away food — Held that there was no justification for interfering with the assistant recorder's decision, which was essentially on a question of fact and based on the evidence — The court had sympathy with the appellant — The case was one of those where the area had become more fashionable and moved up-market, so that the potential take of the premises had become higher than the old-established businesses would yield — Thus a rent properly fixed in accordance with the 1954 Act could be more than the tenant's business could afford — Appeal dismissed

No cases are referred to in this report.

This was an appeal by the plaintiff tenant, Christos Giannoukakis, from the decision of Mr Assistant Recorder Simpson, at Wandsworth County Court, fixing the rent of premises at 244 Lavender Hill, London SW11, where the appellant carried on a snack-bar restaurant. The defendant landlords, the present respondents, were Saltfleet Ltd.

John Speed (instructed by Leon Kaye & Collin) appeared on behalf of the appellant; David Hodge (instructed by Travers Smith Braithwaite) represented the respondents.

Giving judgment, DILLON LJ said: This is an appeal by the plaintiff, Mr Christos Giannoukakis, against an order made by Mr Assistant Recorder Simpson, in the Wandsworth County Court, on July 17 1987. The appellant has been the tenant for some 23 years of premises known as 244 Lavender Hill, London SW11, which he runs as a snack-bar restaurant, with the assistance of his family; and we were told that he has an old-established clientele of people who know and like his restaurant.

The proceeding which came before the learned assistant recorder was a proceeding to fix the new rent for those premises on an application for a new tenancy under Part II of the Landlord and Tenant Act 1954. The landlords had duly given notice in January 1986 terminating the previous tenancy as at July 18 1986, but intimated that they would not oppose an application to the court under Part II of the Act for the grant of a new tenancy. The application was, therefore, duly made, and substantially the terms were agreed, except for the rent. The assistant recorder fixed the rent at £16,500 per year, with an interim rent to be paid from July 18 1986 at the rate of £13,200 per year. The notice of appeal asks that the new lease should be granted at a rent of £9,100 per year, and that the interim rent should be decreased accordingly.

Both parties called evidence at the trial. On the one side there was a Mr Peter B Gaston, a surveyor, who was very familiar with the neighbourhood; he had prepared a report, he gave evidence for the present appellant and came out at the figure of £9,100 a year. On the other side there was a Mr D H Adamson [FRICS], a chartered surveyor and a partner in the firm of Allsop & Co; he came out at the figure which I have mentioned, and his evidence was the evidence preferred by the learned judge.

It is put in the notice of appeal and also in the skeleton argument for the appellant that the learned judge misdirected himself in law. I cannot for my part see that he misdirected himself in law at all. What happened was that he was given evidence of various comparables; he considered them, and he guided himself in particular by the landlords' comparable, which happened to be the premises immediately next door, 242 Lavender Hill, with which Mr Adamson had himself been previously concerned.

No 242 had been recently let as a take-away pizza bar to a subsidiary of Pizzaland International. It is suggested that that was not really comparable to the user of no 244 as a snack-bar restaurant, because no 242 was being used for take-away food only. But the user clause for no 244 was amply wide enough to include use for take-away food and I cannot myself see any reason why no 242 should have been disregarded by the judge. He did not regard no 242 only; he looked at the other comparables to which Mr Gaston had referred; they were a little further away in the same general area and part of the same area. They were essentially retail premises, and the judge's view was that they were not as valuable *qua* user as the premises with which he was concerned, no 244. It has been suggested that the letting of no 242 was a one-off which must have been due to special factors, but no evidence to that effect was put before the judge at the trial, nor was there any application for an adjournment to enable someone to attend and give such evidence. It seems to me that the judge's judgment is based generally on the evidence which he had heard on a question which is essentially a question of fact, namely what is the correct rent for these particular premises in accordance with the formula under the Act.

It is pointed out to us that it is very sad indeed for the appellant, who has been a good tenant of the premises for a very long time, that a rent should be fixed which is higher than he can afford for the sort of business that he has been running in the premises for so many years. This, unfortunately, is capable of happening with business premises. If you have long-established premises where the business has a fairly low profit element it may well be that a rent fixed in accordance with the formula of the Act will be more than the tenant can afford. Fortunately, it does not happen all that often, but I do not think there is anything in the Act which enables the rent to be fixed at a lower rate because the tenant cannot afford a rent which the formula in the Act would produce. It is perhaps particularly likely to happen where an area has, as it were, become more fashionable and moved up-market, where the potential take of premises in the area has become much higher than the old-established businesses would yield. But I see no means, in accordance with the Act, under which that can be alleviated.

We have been referred to a factor that the premises no 242, were better premises than no 244, because they had a return frontage down one side, and it is suggested that the surveyor for the landlords, Mr Adamson, though saying that he had taken a 15% uplift in respect of no 242 because of that benefit, has not clearly indicated how he has done that. Therefore, it is suggested, possibly the rental ought to have been further reduced to no 244. The difficulty about dealing with that in this court is that, first, Mr Hodge, for the landlords, has produced an explanation of how Mr Adamson has arrived at his figures; and, second, that if the figures produced to the judge, and which the judge accepted, were not satisfactory, they ought to have been further probed in cross-examination to make Mr Adamson explain more clearly and in simpler terms exactly what he has done with each figure so as to produce his computation for no 244.

I feel great sympathy for the appellant in the situation in which he now finds himself after so long, but I am unable to see, despite all the help that Mr Speed has been able to give us on the appellant's behalf, that there is any way in which this court can interfere with the decision of the learned assistant recorder. Accordingly for my part I would dismiss this appeal.

A Agreeing, NEILL LJ said: Like Dillon LJ, I feel great sympathy for the appellant and his family. He has been in these premises for nearly a quarter of a century, but I can find no ground whatever on which it is possible to find fault with the decision of the learned judge. Mr Speed has put before us all the considerations which can properly be put forward on behalf of the appellant, but, for the reasons Dillon LJ has given, I, too, would dismiss this appeal.

The appeal was dismissed with costs.

Court of Appeal

B February 24 1988

(Before Lord Justice BINGHAM and Lord Justice MANN)

P F AHERN & SONS LTD v HUNT AND OTHERS

Estates Gazette May 28 1988

[1988] 21 EG 69-73

Landlord and Tenant Act 1954, Part II, section 30(1)(*f*) — Appeal by tenants from decision of county court judge dismissing application for a new tenancy — Issue as to whether landlords had established their intention to carry out C works within the meaning of para (*f*) as explained (although in respect of previous legislation) by Asquith LJ in *Cunliffe* v *Goodman* — It was common ground that a landlord could not satisfy para (*f*) if his intention was to sell the premises outright in order to enable the purchaser to carry out the works, but that it was sufficient if the intention was that the work should be done by the mechanism of a building lease — The county court judge held that in the present case the landlords did intend to carry out the proposed redevelopment by the creation of a building lease and thus satisfied para (*f*) — On appeal the tenants submitted that the evidence showed that the D landlords had not formed a settled intention as to the method of redevelopment — The evidence was reviewed by the Court of Appeal and it appeared that there had been consideration given to different methods, including an outright sale, but that it had been decided to follow the advice of the landlords' professional adviser, a chartered surveyor, who advised that the method should be by way of a 125-year building lease of the site — There was thus evidence on which the judge below was justified in finding that the requisite intention mentioned in section 30(1)(*f*) was satisfied — The facts of the present case were contrasted with those in *Reohorn* v *Barry Corporation* — Appeal dismissed

E The following cases are referred to in this report.

Cunliffe v *Goodman* [1950] 2 KB 237; [1950] 1 All ER 720, CA
Reohorn v *Barry Corporation* [1956] 1 WLR 845; [1956] 2 All ER 742; [1956] EGD 243; (1956) 167 EG 604, CA

This was an appeal by the tenants, P F Ahern & Sons Ltd, from the decision of Judge Paynter-Reece at Romford County Court dismissing their application under the Landlord and Tenant Act 1954 for a new tenancy of a plot of land occupied by the appellants on a site at 238 Crow Lane, Romford, Essex. The landlords, Agnes Hunt, Valerie Smith and Pamela Cosburn, who owned an estate which included the site in question, had outline planning permission for F residential development on the site. The appellant tenants occupied the plot for the purpose of their business, which was that of rubbish disposal contracts.

Alan Steynor (instructed by Hamlin Slowe) appeared on behalf of the appellants; John Norman (instructed by Wortley Byers & Co) represented the respondents.

Giving the first judgment at the invitation of Bingham L J, MANN L J said: There is before the court an appeal against a decision of His Honour Judge Paynter-Reece given at Romford County Court on September 17 1987. The appellant is P F Ahern & Sons Ltd, which carries on business as industrial rubbish disposal contractors. The respondents are respectively the widow and daughters of the late Mr Cyril Hunt. They are also trustees of his estate.

Among the assets of that estate is a site at 238 Crow Lane, G Romford. In area it is 1.13 acres. It has the benefit of an outline planning permission dated June 4 1987. That permission permitted the erection of 28 flats and two houses. If it is implementable it appears that the site value of the plot would be in the region of £600,000.

However, 15,000 sq ft of the plot are occupied by the appellant for the purpose of its business. It is common ground that the occupation is by virtue of a periodic tenancy. It is common ground that the development scheme cannot be implemented unless vacant possession of the entire 1.13 acres is secured.

Accordingly, on April 30 1986 there was served upon the appellant a landlord's notice to terminate under section 25 of the Landlord and Tenant Act 1954. That provoked an application by the appellant H dated July 29 1986 for a new tenancy of the 15,000 sq ft. That application was opposed by the respondents by reference to section 30(1)(*f*) of the 1954 Act. To that I shall advert in one moment.

It was that application which was before His Honour Judge Paynter-Reece on September 17 1987 and which resulted in his upholding the objection and dismissing the application for a new tenancy.

Section 30 of the Act of 1954, so far as is material, provides:

> (1) The grounds on which a landlord may oppose an application . . . are such of the following grounds as may be stated in the landlord's notice under section twenty-five of this Act or, as the case may be, under subsection (6) of section twenty-six thereof, that is to say . . . (*f*) that on the termination of the J current tenancy the landlord intends to demolish or reconstruct the premises comprised in the holding or a substantial part of those premises or to carry out substantial work of construction on the holding or part thereof and that he could not reasonably do so without obtaining possession of the holding.

It is common ground that if, and I quote from *Hill & Redman's Law of Landlord and Tenant* (17th ed) p 767, "the landlord intends merely to sell the premises for the purchaser to carry out the works he cannot rely on ground (*f*)". Why that should be so I do not, for my part, readily understand, but the parliamentary intention is clear.

The issue here is as to the respondent's intention. It is convenient to go at once to the judgment of Asquith LJ (as he then was) in *Cunliffe* v *Goodman* [1950] 2 KB 237. The judgment is at p 253. It is a judgment manifestly not bearing upon the Act of 1954 but which has K ever been held to be applicable to the ascertainment of intention for the purposes of that Act.

Asquith LJ said:

> An "intention" to my mind connotes a state of affairs which the party "intending" — I will call him X — does more than merely contemplate: it connotes a state of affairs which, on the contrary, he decides, so far as in him lies, to bring about, and which, in point of possibility, he has a reasonable prospect of being able to bring about, by his own act of volition.
>
> X cannot, with any due regard to the English language, be said to "intend" a result which is wholly beyond the control of his will. He cannot "intend" that it shall be a fine day tomorrow: at most he can hope or desire or pray that it will. Nor, short of this, can X be said to "intend" a particular result if its occurrence, though it may be not wholly uninfluenced by X's will, is L dependent on so many other influences, accidents and cross-currents of circumstance that, nor merely is it quite likely not to be achieved at all, but, if it is achieved, X's volition will have been no more than a minor agency collaborating with, or not thwarted by, the factors which predominantly determine its occurrence. If there is a sufficiently formidable succession of fences to be surmounted before the result at which X aims can be achieved, it may well be unmeaning to say that X "intended" that result.

I omit a paragraph and continue.

> This leads me to the second point bearing on the existence in this case of "intention" as opposed to mere contemplation. Not merely is the term "intention" unsatisfied if the person professing it has too many hurdles to overcome, or too little control of events: it is equally inappropriate if at the material date that person is in effect not deciding to proceed but feeling his M way and reserving his decision until he shall be in possession of financial data sufficient to enable him to determine whether the project will be commercially worth while.

As I have said, the learned judge dismissed the application, upholding the ground of objection. He said:

> I have come to the conclusion that on the evidence the landlords have established their intention to carry out the redevelopment, and that they intend to carry out the redevelopment by a building lease. Mr Simpson tells me that that is what he will be advising them to do. They have the intention which para (*f*) requires.
>
> It is true no specific lease has been put before me, nor any scheme of joint venture, but it is a highly desirable venture for a company of developers to take on as a lease. Mr Simpson says that in getting planning permission six household names amongst developers made enquiries — he expects many

A more. There would be no difficulty in the trustees disposing of this site by way of a building lease. Advancing on the basis that they can make as much money as possible, the only way to do it is by such a lease, and accordingly the landlords are entitled to rely on section 30(1)(*f*).

The Mr Simpson there referred to is the chartered surveyor who has been advising the respondents.

The learned judge's decision is challenged on a number of grounds in the notice of appeal. The central challenge is, I think, contained in the first ground, which I read:

> The learned judge erred in law in holding that it was the intention of the landlord trustees to redevelop the premises comprised in the holding on the termination of the current tenancy. Particulars. (A) The learned judge should have found on the evidence that at the date of the hearing the landlords had
> B not formed a settled intention with regard to the method of development of the site and therefore he could not be satisfied that they intended to participate in the redevelopment in such a way as to bring themselves within section 30(1)(*f*) of the 1954 Act.
> (B) In the alternative the learned judge should have found on the evidence that the intention of the landlords was to sell the premises freehold with vacant possession to a developer who would carry out the development of the site.

Mr Steynor, who appears for the appellant, has taken us with care through the evidence which was before the learned judge. There are in fact three series of notes.

The appellant itself called no evidence. It could not do so: there was nothing it could contribute to the discussion.

On behalf of the respondents four witnesses were called. The first
C was Mr Terence Hunt, who is the brother of Valerie Smith and Pamela Cosburn. The second was Mrs Cosburn, who is one of the trustees. The third was a Mr Gordon Hunt, another brother. Finally there was Mr A D Simpson, a chartered surveyor with the firm of BE Chartered Surveyors. It is he who was advising upon the site. The learned judge accepted that all of the witnesses were witnesses of truth.

It is transparent that the trustees, with the assistance of Mr Terence Hunt, had decided that they wished to realise the development value of the site. There is no challenge as to their decision. The challenge is on the basis that they were merely saying they would do whatever was necessary to realise the development value and had no intent, as Asquith LJ would have it, to proceed in one way (outright sale) rather
D than another (building lease).

In my judgment it is plain that the trustees and their brothers relied as to the actual method of realisation wholly upon the advice of Mr Simpson. The trustees and their brothers were innocent of such matters. Thus I find a note of Mr Terence Hunt saying "we therefore have sought advice from BE Chartered Surveyors, notably Mr Simpson. He discussed whether the various means of selling the land and the matter was, apart from this, being left totally in his hands. There were other methods apart from outright sale that were discussed but I do not understand them".

Mrs Cosburn, the only trustee called, said: "I am prepared to accept the advice given to me by my brother/Mr Simpson. I do not
E understand the two methods of sale though I do understand that they both involve development of the site."

That leads to the evidence of Mr Simpson. It appears that he had considered three methods by which the development value could be achieved. The first, chronologically first, outright sale of the freehold. The second, a joint venture agreement with a developer. The third, a grant of a building lease at a premium. The second possibility can be dismissed because the learned judge held that the trustees had no power to engage in such an adventure. That is not disputed now.

The sale possibility is of course the most obvious and was first in point of time. It was pursued through 1986 and had been previously pursued. It appears to have been thought that there would be no
F difficulty in securing vacant possession of the site. However, during 1986 it became apparent that not only had the present appellant a tenancy, unlike others on the 1.13 acres, who had licences, but it was unwilling to give vacant possession.

In his evidence Mr Simpson said of the earlier period in 1986: "No advice taken as to risks of not getting the possession, but the opinion given was that the possession would be recovered." That I apprehend is a reference to an opinion given by the solicitors then acting for the trustees. I continue: "More detailed advice has been taken recently; it is clear that there are now two options to pursue — a building lease or a joint venture. I have experience of these." As I have said, the joint venture agreement is not pursuable.

Mr Simpson was cross-examined as to what it was he was going to advise in order that the case would fall within section 30(1)(*f*). His G
critical answer is as follows:

> (Q) As soon as you have vacant possession, you intend to sell the freehold with vacant possession? (A) No, the current circumstances are that we will sell with a 125-year lease — we will not get any less money and bidders will not be reduced. There will be no point in preparing a joint venture agreement only to be told that the freeholders could not dispose of the property. (Q) In practice, property is going to be disposed of freehold? (A) No, by 125-year lease.

So we have a situation where, first, the freeholders had decided to realise the development value by a disposition; second, where the freeholders would achieve whatever disposition their adviser recommended; third, where the adviser is at the date of the hearing firm on a 125-year building lease.

There is a planning permission; finance is no difficulty. Bidders for H
a lease, according to Mr Simpson, would be numerous and no challenge is made to that. It is within the powers of these respondents to grant such a lease.

The matter appears to me to be strikingly different from the case of *Reohorn* v *Barry Corporation* [1956] 2 All ER 742. At p 745 letter A, Denning LJ (as he then was) said:

> There is no building lease, no lessee in existence at the moment, no plans with any details, no knowledge of any financial backing. There is so much still to be explored and discussed, there are so many factors outside the control of the landlords altogether, that I do not think that there is here that "intention not likely to be changed" which is required before the tenant is to be deprived of a new lease. I can well see that in the course of the next few months difficulties J
> may arise which might lead to the abandonment of this plan, or, at all events, to its suspension for an indefinite period. It would not be right that the tenant should be turned out by the landlords on such an uncertain and unsettled intention.

The learned lord justice was in effect saying that the scheme was inchoate in all respects. That is not the case here. There is no perceivable obstacle to the scheme coming to fruition. I would hold that there was ample evidence on which the learned judge could find in the way in which he did that the requisite intention was here established. I would accordingly dismiss this appeal.

I add this: that I am not sorry so to do because the consequence of coming to a different conclusion would mean a rerun of the proceedings in perhaps no more than nine months' time. K

Agreeing, BINGHAM LJ said: Three things have, I think, given rise to difficulty in this case. The first is that there was in the first instance an intention on the part of the trustees and their chartered surveyor that the freehold of the property should be sold outright. That was a decision taken in ignorance of the existence of any business tenancies to create obstacles to that course of conduct.

The second difficulty, as I think, is the lack of legal advice given to the respondents themselves and, indeed, as it would appear, to their surveyor, until a relatively late stage.

The third source of difficulty — but of course one makes no criticism of this — is that the three trustees themselves and their brother Terence appear to have had some difficulty in grasping the L
technical issues raised by the case.

The result of those three things has been that the evidence presented on the trustees' behalf has lacked the precision and unambiguity which better preparation might have ensured. On the other hand, it is quite plain on all the evidence that this is a site which is crying out for development. It is on the edge of a completed development. It is in an area of considerable demand for housing, and the evidence suggests that the property market has at all material times been buoyant. Outline planning permission has been obtained and several potential developers have already expressed interest.

The trustees' position is, and quite understandably is, that they have always been determined to obtain the full development value of the site. There is no reason in the evidence to suggest that there is any M
difficulty in achieving that end. In order to achieve it they have always been willing to follow the advice of Mr Simpson, their professional adviser. His evidence has already been summarised by Mann LJ and culminated in a perfectly clear and unequivocal assertion that in his judgment the way to proceed was by way of a 125-year building lease of the site.

The position therefore is this. The trustees intend to do what their surveyor advises. The surveyor advises a building lease and that is a course which does achieve the trustees' intention of realising the full development value of the site. Because of the effect of the 1954 Act, there is no other course which does achieve that end.

An intention to make an outright sale of the freehold would not

A satisfy section 30(1)(*f*). Therefore the trustees cannot make an outright sale of the freehold with vacant possession of the whole site, and therefore they cannot realise by that means the full development potential.

On the other hand, an intention to grant a building lease, which is what the surveyor asserts that he will advise, and what the trustees say they will follow, does satisfy section 30(1)(*f*). It will permit a lease of the whole site to be given and that will realise the full development potential of the site, which is what the trustees wish.

In my judgment, therefore, they do, as the judge held and for the additional reasons which Mann LJ has given, intend, and did on September 19 (which is the relevant date) to pursue the
B redevelopment of the site by means of granting a building lease to the developers for 125 years. That being so, I agree that the appeal should be dismissed.

The appeal was dismissed with costs.

Court of Appeal

February 29 1988

(Before Lord Justice BINGHAM and Lord Justice MANN)

C SPOOK ERECTION LTD v BRITISH RAILWAYS BOARD

Estates Gazette May 28 1988

[1988] EG 73-74

Landlord and Tenant Act 1954, Part II, section 30(1)(*f*) — Appeal by tenants from decision of county court judge refusing their application for a new tenancy, the landlords having opposed the application on the ground set out in section 30(1)(*f*) — The landlords wished to redevelop the site which was the subject of the tenancy for the purpose of a
D **supermarket and had obtained the necessary planning permission — The issue was as to whether the landlords had established the necessary intention required by para (*f*) — The landlords, who were freeholders, had at first proposed to sell the site outright to well-known supermarket operators, but it became clear that the appellant tenants were not willing to quit — It was therefore decided to enter into a building agreement with the operators, the agreement providing for a 99-year lease to be granted on completion of the buildings — Pending completion a licence fee was payable and the lease would be subject to a premium and a ground rent — From a valuation point of view the total of the sums payable would probably be**
E **equivalent to the value of the freehold, which had been agreed at £650,000**

It was clear law since *Gilmour Caterers Ltd* v *St Bartholomew's Hospital Governors* that the intention required by section 30(1)(*f*) could be satisfied by the landlords' grant of a building lease, but the appellants raised two points of objection — The first was that the present transaction looked at as a whole was tantamount to an outright disposal — The Court of Appeal rejected this suggestion, holding that in principle the case was indistinguishable from *Gilmour* — The other objection was that the landlords' intention at the date of
F **service of the notice opposing the grant of a new tenancy was different from that which was held to be established at the date of the hearing (the earlier idea having been an outright sale) — That objection also failed; the critical date was the date of the hearing — Held that there was abundant material on which the judge was entitled to hold that the ground of objection under section 30(1)(*f*) was established — Appeal dismissed**

The following cases are referred to in this report.

Betty's Cafés Ltd v *Phillips Furnishing Stores Ltd* [1959] AC 20; [1958] 2 WLR 513; [1958] 1 All ER 607; [1958] EGD 92; (1958) 171 EG 319, HL

Gilmour Caterers Ltd v *St Bartholomew's Hospital Governors* [1956] 1 QB 387; [1956] 2 WLR 419; [1956] 1 All ER 314, CA

This was an appeal by the tenants, Spook Erection Ltd, who G
carried on business as market owners, operators and erectors, from the decision of Judge Lee at Evesham County Court, rejecting the appellants' application for a new tenancy of premises at Moreton-in-Marsh, Gloucestershire, of which the landlords were British Railways Board.

John West (instructed by Whiteman, Bretherton & Co, of Gloucester) appeared on behalf of the appellants; P M Kremen (instructed by Simon K Osborne, solicitors' department, British Railways Board, Melbury House, London NW1) represented the respondents.

Giving the first judgment at the invitation of Bingham LJ, MANN LJ said: There is before the court an appeal from His Honour Judge Lee sitting at Evesham County Court in respect of a decision given on H
October 30 1987. By that decision the learned judge refused an application dated August 12 1986 by the present appellants for a new tenancy of premises at Moreton-in-Marsh. The application for the new tenancy was mounted by reference to the business tenancy provisions of Part II of the Landlord and Tenant Act 1954. The particular business carried on by the appellants is described as being that of market owners, operators and erectors.

The freeholder of the premises is the British Railways Board. The board desires to see the site which is the subject of the tenancy redeveloped. The board obtained planning permission for a supermarket and listed building consent for the demolition of the listed buildings upon the site. The board opposed the application for J
a new tenancy on September 18 1986. The grounds of opposition were as follows:

> that on the termination of the current tenancy the Respondents intend to demolish or reconstruct the premises comprised in the holding or a substantial part of those premises or to carry out substantial work of construction on the holding or part thereof and that they could not reasonably do so without obtaining possession of the holding.

That ground of opposition is a reflection of the ground of opposition contained in section 30(1)(*f*) of the Act of 1954.

The facts fall within short compass. In pursuit of its desire to secure the redevelopment of the site, which was presumably surplus to its operational requirements, the board had in mind a sale of its freehold to Budgens. That company is a well-known firm of supermarket K
operators. However, it became plain that the appellants would not leave the site. Accordingly, in order to sustain a ground of opposition under section 30(1)(*f*), the board entered into a building agreement with Budgens. It was a building agreement which contained a term to the effect that a 99-year lease would be granted on completion of the buildings. Pending completion of the buildings there was a licence fee payable. There was a premium payable on the lease and there was a ground rent. As a valuation exercise the sums payable under the heads I have described would probably equate with the value of the freehold, which had been agreed at £650,000.

It is plain law that the necessary intent for the purposes of section 30(1)(*f*) must be an intent extant at the time of the hearing of the application. The intent is not necessarily an intent to reconstruct or L
rebuild with one's own hands. It is sufficient that, for example, the intention should be to effect the rebuilding or the reconstruction by the mechanism of a building lease. That has been clear for upwards of 32 years. It was established in *Gilmour Caterers Ltd* v *St Bartholomew's Hospital Governors* [1956] 1 QB 387 where at p 391 Denning LJ (as he then was) said:

> The landlord intends to demolish or reconstruct the premises even though he does it through the hands of a building lessee. In a way, the grant of a building lease is a means of paying for the work. Just as when you employ a building contractor you pay him in money, so you pay the building lessee by granting him a period of years of occupation. It seems to me that, whether the work is done directly by a building contractor or less directly through a building lessee, the landlord intends to demolish or reconstruct the premises. He M
> intends to have it demolished and reconstructed, and that is sufficient.

Morris LJ (as he then was) said at p 392:

> Mr Heathcote-Williams concedes that, if the landlords in the present case had employed building contractors to demolish and reconstruct and had then granted a lease, the provisions of the Act would have been satisfied.
>
> In my judgment the effect and result is exactly the same under the arrangement which the landlords have in fact made . . . . The scheme, therefore, is that, when the building is completed, there will be a lease, though that lease is to be dated from September 29, 1955, and there is a provision for a rent in money to take the place of a peppercorn rent, which is to apply for the earlier period.
>
> It seems to me that on the facts here the landlords are intending to demolish and reconstruct the premises.

A The decision in *Gilmour* has never been criticised. It must have been followed numerous times in the county courts. The textbooks do not venture adverse comment. Mr West, who presented the case for the appellants with moderation and with an appreciation of the difficulties in his path, said that the *Gilmour* case was distinguishable because here there was a transaction which was tantamount to an outright disposal. He did not and, I think, would not have suggested that the transaction was a sham. The reality of the transaction and its appearance corresponded. The reason for entering into the transaction is, in my view, immaterial. Reality and appearance corresponding, I can see no reason for distinguishing the decision in *Gilmour*.

B The other ground upon which Mr West sought to impugn the learned judge's decision was that at the time of the service of the notice on September 18 1986 the board's intention was different from that which had been sustained at the hearing before the judge. Mr West drew our attention to a passage in the speech of Lord Denning in *Betty's Cafés Ltd* v *Phillips Furnishing Stores Ltd* [1959] AC 20 at p 52. At that page Lord Denning said:

> If the notice had been a dishonest notice in which the landlords had fraudulently misrepresented their intention — or, I would add, if there had been a material misrepresentation in it — I should have thought it would be a bad notice.

There is here no suggestion of fraud or material misrepresentation. Undoubtedly, the mechanism of redevelopment was the subject of
C change in the mind of the board. The hearing date is, however, the critical date and, in my judgment, there was abundant material on which as at that date His Honour Judge Lee could hold that the ground of objection under section 30(1)(*f*) was established. Accordingly, I would, for my part, dismiss this appeal.

BINGHAM LJ agreed and did not add anything.

The appeal was dismissed with costs to be taxed if not agreed. Leave to appeal to the House of Lords was refused.

Court of Appeal

D February 24 1988

(Before Lord Justice PARKER and Lord Justice TAYLOR)

BLACKBURN v HUSSAIN

Estates Gazette June 4 1988

[1988] 22 EG 78-82

Landlord and Tenant Act 1954, Part II, section 30(1)(*f*) and section 31A — Landlord's appeal from decision of county court judge holding that landlord had not established his
E ground of opposition under section 30(1)(*f*) as qualified by section 31A to the tenant's application for a new tenancy of a café — Appellant was the freehold reversioner of the subject premises and of certain adjoining premises — Shop occupied by tenant was one of three in a group and landlord's proposals involved the demolition of the walls of two of the shops, on each side of a central passage, the demolition of the shop fronts, the removal of lavatories and the provision of a staircase leading out of the tenant's café — The proposals in fact involved the complete refurbishment of the three units, making them into one open area — In the view of the Court of Appeal, apart from the possible application of section 31A, it
F was clear that the ground specified in section 30(1)(*f*) was made out — It was common ground that for a minimum period of some eight weeks it would be impossible during the proposed works to run the café in anything like a reasonable way, if at all, particularly as for part of the period the landlord would require the use of the whole of the premises — For section 31A to apply not only must the tenant agree to giving access and other facilities to the landlord for carrying out the work but it had to be shown that the landlord could carry it out without interfering to a substantial extent or for a substantial time with the use of the premises for the operation of a café — In the opinion of the Court of Appeal the interference would, on the contrary, be very substantial — Section 31A was not, G therefore, applicable — The appeal would therefore be allowed

Parker LJ expressed the view obiter that there was great force in a submission that section 31A did not contemplate an agreement which involved the destruction of the subject-matter of the original holding

Two jurisdictional points which had been mentioned were held not to require determination

The following case is referred to in this report.

Norwich Corporation v *Norwich Electric Tramways Co Ltd* [1906] 2 KB 129 H

This was an appeal by the landlord, Ashfaq Ahmed Hussain, from the decision of Judge Crabtree, at Sheffield County Court, holding that the appellant had not established his ground of opposition under section 30(1)(*f*) of the Landlord and Tenant Act 1954 to the application of the tenant, Ernest Blackburn, for a new tenancy of premises used as a café at 682 Chesterfield Road, Sheffield.

Christopher Semken (instructed by Hall Clark, agents for Irwin Mitchell, of Sheffield), appeared on behalf of the appellant; Michael Coles (instructed by Chester & Simmerson, of Sheffield), represented the respondent.

Giving judgment, PARKER LJ said: The respondent in this appeal is the tenant of premises known as 682 Chesterfield Road, Sheffield. J
He became such tenant in 1979 by way of assignment of a lease for a period of five years from June 21 1978. That lease therefore expired by effluxion of time on June 20 1983, but the tenancy, being a business tenancy, continued by virtue of the Landlord and Tenant Act 1954.

The business carried on by the tenant was that of running a café. The appellant is the freehold reversioner of the premises, which are situate in the ground floor of no 682. The appellant is also the freehold reversioner of adjoining premises, nos 684 and 686, both as to the ground floor and the first floor.

On March 21 1984 the appellant gave notice under section 25 of the 1954 Act to terminate the tenancy. Such a notice is required to be in the prescribed form, and section 25(6) provides: K

> A notice under this section shall not have effect unless it states whether the landlord would oppose an application to the court under this Part of this Act for the grant of a new tenancy and, if so, also states on which of the grounds mentioned in section thirty of this Act he would do so.

The only relevant ground in section 30 is that set out in subpara (*f*), which reads as follows:

> that on the termination of the current tenancy the landlord intends to demolish or reconstruct the premises comprised in the holding or a substantial part of those premises or to carry out substantial work of construction on the holding or part thereof and that he could not reasonably do so without obtaining possession of the holding.

The notice to terminate given by the landlord states as follows: L

> I would oppose an application to the court . . . under Part II of the Act for the grant of a new tenancy on the ground . . . that the Landlord requires the premises in view of reconstruction of his own premises and cannot reasonably do so without obtaining possession of the premises.

That is not in terms or by reference the ground specified in para (*f*) of section 30(1).

The tenant's response to that notice was a letter dated March 26 1984 [from his solicitors to the landlord's solicitors] in the following terms:

> Our Client, Ernest Blackburn, telephoned us at approximately noon, on Friday the 23rd . . . to tell us that he received, by Recorded Delivery, from your firm, what seems to amount to a Notice to Quit. We always thought it was Courtesy for the Landlord's Solicitors to send a copy of such Notice and M covering letter, written direct to [the] Tenant, to the Tenant's Solicitors. You are aware that we do act for Mr Blackburn, and we should be pleased to receive your explanation why you have not observed the usual courtesy to date.
>
> We have asked our Client to let us have the communication he has received from you, and we will then comment on the matter in detail, but we gather from our Client that the document you have sent states that your Client will oppose the grant of a new Lease to our Client, for the reason that he requires possession of our Client's property, because he wishes to reconstruct his own premises.
>
> Obviously, we are very sceptical of this ground, and should be pleased to receive copies of your Client's plan for reconstruction, together with, of course, copies of the relevant Planning and Building Regulation Permissions, which he has obtained.

The remainder of that letter is not presently relevant. It provoked no reply, and on June 19 the tenant's solicitors reverted to the matter in the following terms:

> We refer to our letter to you dated March 26 last and are somewhat surprised not to have heard from you, in the past, nearly three months. Please now have the courtesy to reply, giving us an outline, supported by Planning Permissions, plans etc, etc, of your client's proposals for reconstruction of the premises. Our Client, is, obviously, not prepared to vacate these premises at the end of September, and we are in the course of preparing his application to the Court for the Grant of a new Tenancy, because, honestly, it is considered by us that your Client's proposals are not really relevant and do not justify the refusal of the grant of a new Lease.

Again I need not read the remainder.

I should mention that in his notice the landlord required possession to be given on September 29 and not the end of September.

Thereafter the tenant applied for a new tenancy under section 24 of the Act, and in his application he stated as follows:

> On March 26 1984 I served on the Respondent a counternotice dated March 26 1984 stating that I would not be willing to give up possession of the premises on the date of termination.

That statement was not challenged at any time in the proceedings until the opening of this appeal.

The tenant having made his application for a new tenancy, there were directions, discovery, an answer was filed, and on August 17 1987, some very considerable time later, a preliminary issue was ordered. The preliminary issue is specified in these terms:

> The preliminary issue is limited to the sole question of the Landlord's intention to redevelop.

On the face of it, that simply goes to the question whether the landlord's statement of intention was or was not valid, but it has been accepted throughout, and is still accepted, that the intention in the preliminary issue was to determine the question whether or not the landlord had made out the ground for opposition in section 30(1)(*f*) as qualified by section 31A.

The preliminary issue was heard in the Sheffield County Court by His Honour Judge Crabtree on August 26 1987 and he concluded that the landlord had not made out the ground specified in section 30(1)(*f*). The landlord now appeals.

At a late stage a question has been raised by the landlord as to the jurisdiction, it being contended that the tenant had not served a counternotice and that as a result, and by virtue of section 29(2) of the Act, the court had no jurisdiction to entertain his application. It is clear that a point on jurisdiction can be taken as late as a hearing before this court, albeit it has not been taken at an earlier stage in the proceedings. That stems from the decision in *Norwich Corporation* v *Norwich Electric Tramways Co Ltd* [1906] 2 KB 129. There also arises, incidentally and as a result of the jurisdictional point having been raised by the landlord, a further possible jurisdictional point, namely, whether the landlord's notice to terminate the tenancy was itself invalid. I will leave both jurisdictional points until the end of this judgment and deal first with the question of the merits. By way of preliminary, it is necessary shortly to describe both the premises themselves, the adjoining premises and the landlord's proposals.

The premises consist of a group of three shops in an old building in Chesterfield Street. If one faces the premises, the tenant's premises are on the right and consist of a café called "The Cherry Pie". They are enclosed premises and between them and the next-door premises, which are of a similar size and also self-contained, there is a corridor or passage which leads to a third shop at the rear of the building. The three shops, consisting of the tenant's shop, the premises at the rear and the premises to the left of the passage, are at present separate premises, separately numbered and separately entered. There are apparently no toilet facilities in the café which the tenant occupies, but it is common ground that he had a right under his lease, imported probably by virtue of the Law of the Property Act, to the use of certain lavatories which lie behind the neighbouring shop.

The proposals which the landlord has made and which the judge has accepted he does intend to carry out involve the demolition of the walls which contain the two shops on either side of the central passage and the making of the three premises into one open area. They involve also the demolition and replacement of the shop fronts, the removal of the lavatories and the provision of a staircase which, as originally planned, would not have led out of the tenant's café, but which did, as the plans stood at the date of the hearing of the preliminary issue, lead out of the café or was intended to do so.

There are other works involved in the landlord's proposals, but they need not be mentioned other than to say that they come under a general description of the complete refurbishment of the three premises as a whole.

Apart from the qualifications in section 31A, it is abundantly clear that the ground specified in section 30(1)(*f*) is made out. The proposals of the landlord are so extensive that in ordinary parlance he could not have reasonably carried them out without obtaining possession of the holding. To suggest that, without having legal possession of the tenant's holding, he could demolish his containing walls, remove his lavatories and insert a staircase into the premises appears to me to be wholly unreal, and indeed the judge proceeded on the basis that it was by virtue of the provisions of section 31A that the landlord failed to make out the section 30(1)(*f*) ground. It is therefore necessary to examine the terms of section 31A(1). That reads as follows:

> Where the landlord opposes an application under section 24(1) of this Act on the ground specified in paragraph (*f*) of section 30(1) of this Act the court shall not hold that the landlord could not reasonably carry out the demolition, reconstruction or work of construction intended without obtaining possession of the holding if —
>
> (*a*) the tenant agrees to the inclusion in the terms of the new tenancy of terms giving the landlord access and other facilities for carrying out the work intended and, given that access and those facilities, the landlord could reasonably carry out the work without obtaining possession of the holding and without interfering to a substantial extent or for a substantial time with the use of the holding for the purposes of the business carried on by the tenant;

There are thus the following conditions: first, that the tenant agrees to the provision of access and facilities; second, that, given those facilities, the landlord could reasonably carry out the work without, first, obtaining possession and without, second, interfering to a substantial extent or for a substantial time with the use of the holding. The tenant in this case indicated that he would be prepared to afford the landlord all facilities and access which he required for carrying out the work which he intended to do. I will assume for the moment that agreement to the extent required to enable the landlord to carry out these works falls within the terms of section 31A(1)(*a*); but that is of course not enough, because it has to be shown that, given the facilities, the landlord could carry out the work without interfering to a substantial extent or for a substantial time with the use of the premises for the operation of a café.

It is common ground that for a minimum period of some 12 weeks it would be impossible to run this café in anything like a reasonable way, if at all. It is common ground that for part of the period the landlord would require the use of the whole of the premises, albeit it might be possible, indeed probably would be possible, for the café equipment to be shifted around and not actually moved out of the premises themselves.

It is rightly pointed out on behalf of the tenant that the question is a question of fact and degree, and accordingly that this court should be slow to interfere with the decision of the learned judge in the court below. It is, however, clear that this court can and must consider the question whether on the admitted facts the interference can be regarded as being otherwise than substantial within the meaning of the section. For my part, I have no hesitation in saying that, with all respect to the learned judge, I cannot come to the conclusion that interference for this time and to this extent within the business of running a café is otherwise than substantial, and indeed in my view very substantial.

That is sufficient to dispose of the issue arising on this appeal, but it was also submitted on behalf of the appellant that section 31A did not really arise at all, because it cannot have contemplated an agreement which involved the destruction of the subject-matter of the original holding. Were it otherwise it would involve a tenant's being able to say: "I agree to the destruction of my holding, but I demand that, it having been destroyed, I am granted a tenancy, not of my holding, but of some entirely different entity."

Without finally deciding the matter, I should say that I accept that there is great force in that contention and that for my part I do not, as presently advised, accept that it is possible for a tenant by any such agreement to avoid the plain meaning of section 30(1)(*f*).

There remains to be considered the two jurisdictional points. I say "remains to be considered", for in truth, in view of what I have already said, they do not any longer arise. The point taken by the appellant does not arise because he succeeds in his appeal on the merits. The other point, which was adumbrated by the court, is not relied upon by the respondent, save to the extent that the court might

A have come to the conclusion that the other jurisdictional point was *prima facie* valid. Since it has not been necessary to deal with the one, it remains unnecessary to deal with the other.

I would add that I regard the jurisdictional point taken on behalf of the appellant as being wholly without merit and that, were it necessary to go into it, I would have been inclined to conclude, either (a) that the letter of March 26 should be construed as a notification of unwillingness or (b) that the subsequent letter, albeit out of time, had been the subject of a time waiver. It is accepted that it is a procedural matter. The suggestions which were made to avoid the time waiver appeared to me to be not only without merit but without substance. As to the other jurisdictional point, it was raised only by the court, and no doubt it was a matter of wisdom on the part of the respondent
B not to leap upon it with joy, for it might have had consequences which would have been undesirable.

Accordingly, I would allow this appeal.

Agreeing, TAYLOR LJ said: It seems to me that the learned judge allowed himself to be influenced in two respects by what might have been, rather than confining himself to the facts before him. First, he dealt with the case on the basis that, had the landlord so wished, he could have reconstructed the premises in such a way as to enable the tenant to resume his café business on the floor area where it was before, albeit without walls. Second, he was influenced by evidence that the landlord had discusssed with the tenant the possibility of the
C café being relocated upstairs. These matters were not relevant.

The landlord put before the court evidence of what he actually intended to do. There was no suggestion that he did not have a genuine intention to carry out the works as described by his expert and shown on his plans. In those circumstances the issue was whether in regard to that work section 31A applied. That issue involved two questions. First, could the landlord reasonably carry out that work without obtaining legal possession of the holding? Second, could he carry out that work without interfering to a substantial extent or for a substantial time with the use of the holding by the tenant as a café? In my judgment, for the reasons which have been given by Parker LJ, the evidence admitted only of a negative answer to each of those questions. Accordingly, I agree that this appeal should be allowed.

D *The appeal was allowed with costs in the Court of Appeal and below (below on Scale 2).*

Court of Appeal

March 3 1988

(Before Lord Justice MAY and Mr Justice WAITE)

TOLLBENCH LTD v PLYMOUTH CITY COUNCIL

Estates Gazette June 11 1988

[1988] 23 EG 132-139

E **Landlord and tenant — Landlord and Tenant Act 1954, section 53 — Covenant preventing change of user without landlords' consent — Whether landlords' refusal of consent reasonable — Application by tenants for relief under section 53 — Relevance of procedure by which landlords reached decision to refuse consent — Correct approach to consideration of reasonableness — Lease in present case contained a covenant by tenants not without landlords' previous consent in writing to use the demised premises for any purpose other than as a retail shop and not to use them for residential or office purposes — The existing use was as a**
F **restaurant, but trade had dwindled and tenants wished to change use to that of a self-service snack restaurant with about half the area given over to gaming machines or amusement with prizes machines — Landlords in present case were also the planning and the licensing authority — They granted planning permission for the proposed change of use and a gaming licence, but *qua* landlords refused consent to the change — Tenants applied to county court under section 53 and judge made a declaration in their favour — Landlords appealed — Although the county court judge agreed with fears expressed that the proposed change of user would lead to the attraction of undesirable elements, he found in favour of**
the tenants — He accepted a submission that because the landlord council had reached their decision in an unreasonable way, on the basis of misconceptions and incorrect information put before them, the decision itself was unreasonable G

Held by the Court of Appeal that the judge had fallen into error — In determining whether a refusal of consent is reasonable two questions have to be considered — The first, which might be called a subjective inquiry (although these terms are not very helpful), is what was the landlords' reason for refusal — The onus is on the tenant to show the reason and
that it is unreasonable — If the landlord gives no express reason the court will more readily infer that the refusal was unreasonable — Having found what the landlord's reason was, the next inquiry (which might, subject to the same comment, be called objective) is whether that reason in the landlord's mind was reasonable or unreasonable — The reason must not be thought up after the decision was made or be a reason arising *ex post facto* — The judge was wrong in considering the procedural steps taken by the landlords in arriving at their decision and inquiring whether they had acted on information that was accurate and complete — If he had not adopted this wrong approach he would probably have H
found that the refusal of consent was reasonable — Appeal allowed J

The following cases are referred to in this report.

Associated Provincial Picture Houses Ltd v *Wednesbury Corporation* [1948] 1 KB 223; [1947] 2 All ER 680, (1947) 45 LGR 635, CA
Bromley Park Garden Estates Ltd v *Moss* [1982] 1 WLR 1019; [1982] 2 All ER 890; (1982) 44 P&CR 266; [1983] EGD 492; 266 EG 1189, CA
Fredk Berry Ltd v *Royal Bank of Scotland* [1949] 1 KB 619
International Drilling Fluids Ltd v *Louisville Investments (Uxbridge) Ltd* [1986] Ch 513; [1986] 2 WLR 581; [1986] 1 All ER 321; (1985) 51 P&CR 187; [1986] 1 EGLR 39; (1985) 277 EG 62, CA
Lovelock v *Margo* [1963] 2 QB 786; [1963] 2 WLR 794; [1963] 2 All ER 13, CA

This was an appeal by the landlords, Plymouth City Council, from the decision of Judge Sir Jonathan Clarke QC, at Plymouth County Court, in which he granted a declaration in favour of the tenants, Tollbench Ltd, that the landlords' refusal of consent to a proposed change of user was unreasonable. The tenants had applied for relief under section 53 of the Landlord and Tenant Act 1954 in relation to premises at 93 Cornwall Street, Plymouth. K

Michael Beloff QC and Clive Newbury (instructed by Sharpe Pritchard & Co, agents for A Forbes Watson, town clerk, Plymouth City Council) appeared on behalf of the appellants; Anthony Donne (instructed by Goldbergs, of Plymouth) represented the respondents.

Giving judgment, MAY LJ said: This is an appeal from a judgment and declaration of His Honour Judge Sir Jonathan Clarke in the Plymouth County Court on November 10 1987. L

The learned judge had before him on that occasion an originating application by a tenant, the respondent in this court, under section 53 of the Landlord and Tenant Act 1954. The relief claimed was not entirely in accordance with the wording of the statute, but the situation was that the tenant had applied to its landlords, the city council, who were the appellants before us, for their consent to a change of user of premises at 93 Cornwall Street, Plymouth, of which the respondent was the tenant. The appellants had withheld their consent. The respondent contended that in doing so they were unreasonable and sought a declaration from the court to that effect.

The relevant statutory provision is, as I have said, section 53 of the Landlord and Tenant Act 1954. I need only read subsection (1)(*c*) and subsections (2) and (4): M

53-(1) Where a landlord withholds his licence or consent — . . .
(*c*) to a change in the use of the demised property or any part thereof, or to the making of a specified use of that property,
and the High Court has jurisdiction to make a declaration that the licence or consent was unreasonably withheld, then without prejudice to the jurisdiction of the High Court the county court shall have the like jurisdiction whatever the value of the demised property or the rent payable under the tenancy* and notwithstanding that the tenant does not seek any relief other than the declaration.

*Editor's note: The above wording has been to some extent amended: see now County Courts Act 1984, Sched 2, para 23.

A (2) Where on the making of an application to the county court for such a declaration the court is satisfied that the licence or consent was unreasonably withheld, the court shall make a declaration accordingly.

(4) Nothing in this section shall be construed as conferring jurisdiction on the county court to grant any relief other than such a declaration as aforesaid.

The brief facts of the case are these. As I have said, the respondent was the tenant of the appellant of premises under a lease containing not unusual tenant's covenants, among them being subclause 2(10)(ii) and (12), which read in this way:

(ii) Not without the previous consent in writing of the Corporation (which consent shall not unreasonably be withheld) at any time during the said term to use the premises for any purpose other than for the purposes of a retail shop and not to use the premises or any part thereof for residential accommodation
B or for the purpose of offices except offices ancillary only to some other non-prohibited purpose.

(12) Not to use the premises or any part thereof nor knowingly to permit or suffer the same to be used for any illegal or immoral purpose nor for the carrying on of any noxious offensive noisy or dangerous trade or process nor knowingly to permit or suffer to be done in or upon the premises or any part thereof anything which may be or become a nuisance or annoyance to the Corporation their lessees or tenants or occupiers of any adjoining or neighbouring property . . .

Until recently, under an earlier change of user, the premises were used as a restaurant. It seems that the restaurant trade had been badly affected by a substantial amount of pedestrianisation in the area. The respondent consequently conceived the idea to change the user to
C what was originally described as a restaurant with amusement facilities. The learned judge in his judgment described that in this way:

This . . . involves the provision of an attractive frontage . . . a table and seating area with a servery from which on a self-service basis will be sold food for consumption on the premises . . . mainly jacket potatoes . . . pies and puddings . . . In approximately one half of the premises immediately to the rear of the seating area will be machines variously described as gaming machines or amusement with prizes machines. These . . . will not be of the video game type . . . nor of the jackpot type and will be machines which when played can deliver a token or tokens to a maximum of £4 which in turn can be exchanged for an equivalent prize.

However, the learned judge was understandably sceptical about
D the proposed change and, after hearing the evidence, indicated in his judgment that he had no hesitation in finding that the proposal could not properly be described as a restaurant with amusement facilities, but more properly as an amusement hall with ancillary snack refreshments provided.

Be that as it may, the proposed change of user by the respondent required, in addition to the landlord's consent, planning permission from the local planning authority, which happened to be the appellants wearing another hat, and an appropriate gaming licence. Again the application had to be made to the appellants, but in yet another capacity.

Planning permission was granted on July 17 1987. On July 24 the appellants' local economy committee referred the question of
E granting consent as landlord to the full council with a recommendation that the council should refuse it. On August 17 1987 the appellants granted the respondent an appropriate gaming licence. However, despite the success in the planning and gaming fields, on August 26 1987 the respondent was refused landlord's consent under the terms of the lease by the council.

That refusal led to the application by the tenant to the county court, which was the one upon which the learned judge adjudicated. It is right to comment that the evidence led on behalf of the appellants as to the reasons for the local economy committee and the council refusing consent was not substantial. Nevertheless, on what he had the learned judge held, first:

F I am though in no doubt and so find that there was a consensus majority view held at both of these meetings which was articulated by the council officers whom I have heard that, if granted, these premises and the amusement machines in particular could and probably would attract an undesirable element to the area. This in turn could and probably would lead to vandalism and disorder generally and would result in an undermining of the economic fabric of this part of Cornwall Street.

He then held, second:

. . . if one looks at the matter objectively I have no doubt that the fears of the council officers and the views of the councillors as I infer them to be are soundly based and justified and that this proposal could and probably would lead to an attraction to the premises of an undesirable element present in the city. This in turn could lead to the true middle aged and respectable shopper tending to shun not only the premises, but the immediate area surrounding them. This could in turn lead to an undermining of the economic fabric of the G
immediate area.

Having found in those terms, one might have been forgiven for thinking that the learned judge would then swiftly go on to have held the refusal of consent to have been reasonable.

However, he turned in the course of his judgment to consider the second way in which the matter was put before him, and before us, on behalf of the tenant. Briefly, it was to the effect that the council's decision had been reached in an unreasonable way and was therefore necessarily unreasonable itself.

What was argued was that the local economy committee and/or the council had acted on various misconceptions, had taken into account objections which were themselves based on misconceptions, H
had failed to be told that the planning permission granted had been conditional and that so also could the grant of a gaming licence be conditional, with a view in each case to minimise any anticipated trouble that there might be from the new proposals, and that the council and the committee had acted on substantially incorrect memoranda from two council officers. Had the council, it was submitted, not been misled in that way, they could not have reached the conclusion to refuse consent which they did. It was argued, therefore, that any decision so reached in that unreasonable way must itself be unreasonable. The learned judge accepted that both the committee and the council acted in that sense in an unreasonable way, but did not accept that "looked at objectively, this necessarily means that their decision was unreasonable. It could, coincidentally, J
be the correct view for them to take had all the true facts then available to them been actually placed before them". He then asked rhetorically "Where does this leave the application?" He considered whether the test which had to be applied was objective or subjective. He was shown some authorities and concluded that:

it seems to me that the proper test to apply is not a wholly objective test, but one in which it is necessary to have regard to the landlord's state of mind insofar as that can be discerned from a council of some 60 members. That is to be done at the time the decision was made. The council it seems to me can only be permitted to rely upon what actually influenced their minds at the time the decision was made.

He then went on to hold that on that test so stated:

the way in which the council had approached this decision, the misconceived K
facts and the partial picture before them rules out any possibility of the court saying that their decision, for it is their decision which falls to be scrutinised and which is to be tested, could possibly be a reasonable one. It is not saved by saying that they could have acted differently and could by a reasonable process have reached the same conclusion. They acted, in the sense in which I have indicated, unreasonably and their decision was in the event an unreasonable one for them to take.

That finding of the learned judge effectively concluded the matter as it was before him, but he went on to consider a third point which has been raised by the tenant, namely that the landlords' decision to refuse consent had in truth had nothing to do with the relationship of landlord and tenant because no evidence had been adduced before L
the court that the grant or refusal of consent to a change of user had any relevance to any of the council's own proprietary interests. The learned judge expressed unease about the inadequacy of the evidence that there was before him, but on such evidence as he had, which comprised that from only one witness, he felt able to conclude that whatever the precise position may be, the city council are closely concerned either as landlords or ground landlords with at least the majority of the city centre and so concerned with a significant number of properties which would be directly affected by their decision in respect of 93 Cornwall Street.

Nevertheless, as I have said, he held that as the appellants had acted unreasonably in the sense indicated, the refusal of consent was itself unreasonable. He therefore found for the tenant and made the M
declaration accordingly.

The city council now appeal against that order and declaration, asking that it should be set side. There is a respondent's notice seeking to uphold it on grounds additional to those relied on by the learned judge.

The argument for the appellants before us was, first, that the basic principles to be applied in these cases were recently restated by the decision of this court in *International Drilling Fluids Ltd* v *Louisville Investments (Uxbridge) Ltd* [1986] Ch 513,* particularly the passage in Balcombe LJ's judgment from pp 519H to 520E. I respectfully

*Editor's note: Also reported at [1986] 1 EGLR 39; (1986) 277 EG 62.

agree. I do not think it is necessary to refer in detail to the principles which the learned lord justice there set out.

Referring to section 53(2), the argument for the appellants proceeded that the exercise contemplated by the subsection involved two stages: first, finding what were the grounds for withholding consent and, second, determining whether those were or were not unreasonable within the principles set out in the *International Drilling Fluids* case to which I have referred.

Then it was argued that, if the grounds actually relied on were objectively reasonable, any error in the process whereby that decision had been reached was nothing to the point; procedural flaws could only be material, it was contended, if they led to an objectively unreasonable decision. Accordingly, in the light of the basic findings of the learned judge, to which I have already referred, it was submitted that the learned judge ought to have been bound to find that the refusal of consent was reasonable.

Alternatively if, contrary to that submission, the learned judge had been entitled to take into account procedural as distinct from substantial flaws, he could not make a final declaration, as he did, having a final and permanent effect, but only so as to give the appellants the opportunity to consider the request for consent properly. On the findings the result would have been the same and therefore there should be no declaration in any event. Finally, the procedural improprieties upon which reliance was placed, it was submitted somewhat comprehensively, were irrelevant, insignificant and, if any, were matters which affected the local economy committee and not the council. For present purposes I think it unnecessary to elaborate further on that particular point.

The respondent tenant, in seeking to uphold the learned judge's decision, relies on the procedural point and submitted that the learned judge was correct in the approach that he adopted in relation to it. Counsel supported the argument by reference, first, to a passage from the judgment of Slade LJ in *Bromley Park Garden Estates Ltd* v *Moss* [1982] 1 WLR 1019 at p 1034E, which reads:

> The decision of this court in *Lovelock* v *Margo* . . . clearly establishes that in cases such as the present the court has to have regard to the landlord's actual state of mind at the relevant time. The test is not a purely objective one, though no doubt inferences may be drawn as to his state of mind from his words and actions and all the other circumstances of the case. It is therefore necessary to consider what were the factors which actually influenced the plaintiff landlords in the present case, as at [the relevant date].

And also a short quotation from the judgment of Lord Denning MR in *Lovelock* v *Margo* [1963] 2 QB 786 at p 789:

> I am quite clearly of opinion that it is not right to say that this is an objective question, as counsel said. This matter cannot be considered without regard to the state of mind of the landlord herself as to her reasons for refusing consent. How otherwise can a lessee hope to see whether he can assign unless he knows the landlord's reasons for objection?

As to what was unreasonable, counsel referred us to the well-known judgment of Lord Greene MR in *Associated Provincial Picture Houses Ltd* v *Wednesbury Corporation* [1948] 1 KB 223. I say straightaway that in that case what was under consideration was reasonableness in a public law statutory duty context and, with respect to counsel's argument, cogent though it was, I do not think we are assisted by what Lord Greene said in the *Wednesbury* case.

In the alternative, if he were wrong on his main submission, counsel for the respondent tenant submitted that there was in truth no sufficient evidence before the learned judge below to entitle him to find that any grant or refusal of consent to change of user would have affected any proprietary interest of the appellants as landlords and that accordingly the learned judge was right again.

As is so often the case, I do not find an approach which divides consideration of a case into a subjective inquiry on the one hand and an objective inquiry on the other very helpful.

To determine whether a landlord's refusal of consent is reasonable, accepting the submission on behalf of the appellants, requires a consideration of two questions. First, what reason did the landlord actually have for refusing consent, that is to say upon what reason did he act in refusing that consent? That is a subjective inquiry, if one needs so to describe it, in the sense that one has to discover what was in the mind of the landlord at the time when he was refusing his consent. The onus is on the tenant to show that the reason the landlord had was unreasonable and the onus therefore is on the tenant to show the reason and that that reason was unreasonable. One may have, in whole or in part, to infer in certain cases what was in the landlord's mind at the relevant time. In particular, I comment that if the landlord gives no express reasons for withholding his consent, then the court will more readily imply that that withholding was unreasonable, as Lord Goddard CJ pointed out in *Fredk Berry Ltd* v *Royal Bank of Scotland* [1949] 1 KB 619, at p 623. Having then found what reason the landlord did have in mind at the relevant time, one turns to what can perhaps properly be described as an objective inquiry, whether that reason in the landlord's mind was reasonable or unreasonable. The legal guidelines under which one should make that inquiry were laid down, as I have said, in the *International Drilling Fluids* case referred to earlier in this judgment.

I do not think that on its proper construction section 53 of the Landlord and Tenant Act 1954, and in particular the subsections which I have quoted, were intended to cover, or permit the court to consider, the procedure by which the landlord made up his mind whether or not to refuse consent. The procedure may be material on the issue of the genuineness of the stated reason, that is to say on the bona fides of the landlord in any given case, or indeed on what the actual reason was. But *qua* procedure it is, in my judgment, not intended to be covered by the statutory provisions.

I do not understand Slade LJ in the *Bromley* case or Lord Denning MR in *Lovelock* v *Margo* to be saying any more than that one must consider the reasonableness or otherwise of the landlord's refusal at the date he so refused and the reason he then had for refusing consent. If he gave none, as I have said, it may be more readily possible to imply that he was unreasonable. But the essence and true effect of the judgments of Slade LJ and Lord Denning is, I think, that a landlord is not entitled to rely on a reason, for instance, thought up after he has made his decision to refuse consent or a reason, for instance, arising *ex post facto*.

In my opinion, therefore, in this case the learned judge erred in considering what he thought was the procedural unreasonableness of the landlords in the sense to which he referred. Had he not done so, and subject to the evidential point to which I have referred, I have no doubt that the learned judge would have found that the refusal of consent in this case had been reasonable.

On the evidential point, as I have already indicated, I do not think there was over much evidence before the judge. Indeed I am surprised, for instance, by the absence from the court of any one of the 60 councillors who took the relevant decision when this matter came before the Plymouth County Court and that the appellants were not better supported by the evidence of witnesses. As I have also indicated, the learned judge was clearly troubled by the evidential point, but in the end decided that there had been enough for him to make the appropriate decision. Despite the paucity of the evidence, I am in no way persuaded that the learned judge was wrong in the conclusion to which he came on this particular issue.

For these reasons, therefore, I would allow this appeal and I would set aside the declaration made by the learned judge below.

Agreeing, WAITE J said: I will only add a few supporting words in tribute to the helpful arguments that have been addressed to us and in tribute also to the careful attention given to the case by the learned judge, including the one issue on which we have found it necessary to differ from him.

Once it had become established to the judge's satisfaction, first, what the grounds were of the head lessors' objection to the proposed change of user and, second, that those grounds were reasonable, it became unnecessary and indeed irrelevant to inquire whether or not the head lessors, in making the decision to grant or refuse consent, had acted upon information that was accurate and complete in every particular. The judge's secondary finding — that there were shortcomings in the head lessors' processes of consideration, which might have put them at theoretical risk of reaching an erroneous decision on incomplete or inaccurate information — ceased therefore to have any relevance to, and should not have been allowed to affect, the judge's primary finding that the grounds of refusal were reasonable.

I, too, would allow the appeal.

The appeal was allowed with costs.

Chancery Division

December 10 1987

(Before Mr Justice HARMAN)

HERONGROVE LTD v WATES CITY OF LONDON PROPERTIES PLC

Estates Gazette June 18 1988

[1988] 24 EG 108-110

Landlord and Tenant Act 1954, section 25 — Trial of preliminary issue treated by consent as trial of the action — Issue concerned the validity of the notice under section 25 of the 1954 Act purporting to determine the plaintiffs' tenancy of commercial premises — The premises in question, at Winchester House, London Wall, included in the lease consisted of offices on the ninth floor and storage accommodation on two other floors together with free car-parking space — The section 25 notice, however, served by the landlords mentioned as a description of the property only the ninth-floor accommodation — The tenants, relying on such cases as *Kaiser Engineers & Constructors Incorporated* v *E R Squibb & Sons*, submitted that the notice was defective and void — It was argued by the landlords that the tenant receiving the notice must be regarded as a reasonable tenant, with a copy of his lease available, so that he would know what was comprised in his tenancy and would not be misled — Harman J rejected this submission — He agreed that the test was "Is the notice quite clear to a reasonable tenant reading it? Is it plain that he cannot be misled by it?" — But he did not agree that a reasonable tenant in the present case would necessarily be free from doubt as to the scope of the notice — He might think that the landlords' intention was to take back the office floor for some redevelopment but not the other accommodation — The notice did not pass the test applied in the authorities — Notice accordingly declared invalid

The following cases are referred to in this report.

Carradine Properties Ltd v *Aslam* [1976] 1 WLR 442; [1976] 1 All ER 573; (1975) 32 P&CR 12
Hankey v *Clavering* [1942] 2 KB 326
Kaiser Engineers & Constructors Inc v *E R Squibb & Sons Ltd* [1971] EGD 553; (1971) 218 EG 1731; CA unreported
Safeway Food Stores Ltd v *Morris* (1980) 254 EG 1091

This was a preliminary issue directed by a master to be tried in proceedings by originating summons brought by plaintiff tenants, Herongrove Ltd, to determine whether a notice under section 25 of the Landlord and Tenant Act 1954 served by the landlords, Wates City of London Properties plc, was effective. The matter arose out of a lease for 21 years from May 1 1966 of premises at Winchester House, London Wall, London EC.

D A Hochberg (instructed by Slaughter & May) appeared on behalf of the plaintiffs; Nicholas Dowding (instructed by Norton Rose Botterell & Roche) represented the defendants.

Giving judgment, HARMAN J said: I have before me a preliminary issue directed by Master Gowers in May. The issue is that the points raised by para 1 of the originating summons be tried as a preliminary issue. The originating summons was issued by Herongrove Ltd as plaintiff on January 7 1987. The defendant was Wates City of London Properties plc. The originating summons, by para 1, sought a declaration that a notice dated September 8 1986, served by the defendant on the plaintiff, purporting to determine the plaintiffs' tenancy on May 1 1987 under section 25 of the Landlord and Tenant Act 1954, is invalid and of no effect.

The matter arises out of a lease granted on March 31 1966 between a predecessor in title of the defendant and predecessors in title of the plaintiff, whereby a term of 21 years from May 1 1966 was granted. The premises demised were "all those premises more particularly described in the first, second, third and fourth parts of the first schedule hereto, hereinafter referred to as the said premises, forming part of the building known as Winchester House". The first schedule to the lease has four parts to it.

Part I: "all those office premises situate on the 9th floor of phase I of the building".

Part II: "all those storage premises on the lower ground floor of phase I of the building".

Part III: "all those storage premises situate on the basement floor of phase II of the building", and

Part IV: "all those free car parking spaces on level B of the building".

It is quite plain, and has never been disputed, that that is one demise of one set of premises, although in separate parts of the schedule.

The matter has been very well argued on both sides. The notice itself is in the statutory form under section 25 of the Act of 1954 — a Landlord's Notice to Terminate Business Tenancy. It is in the form which has always seemed to me quite extraordinarily inappropriate and inapt, all drafted as if it were between two individuals when, as is notorious, 99% of all leases of business premises are granted by landlord companies to tenant companies. Given that difficulty, the notice is fairly inadequate in its descriptions. The statutory form provides, in relation to clause 2 of the notice, that it shall give a description of the property. In this case the notice to the tenant, with the tenant's postal address, states — clause 2 — that it relates to ninth-floor premises at Winchester House, London Wall.

The point taken is at once, I think, obvious. The premises demised by the lease were the ninth floor and two sets of storage premises and free car-parking spaces. The storage premises, or some of them, and certainly the car-parking spaces, are physically substantially removed in distance from the office premises on the ninth floor.

Mr Hochberg for the tenants cited to me *Kaiser Engineers & Constructors Incorporated* v *E R Squibb & Sons Ltd*, a decision of the late Ungoed-Thomas J in March 1971. The report cited was [1971] *Estates Gazette Digest* 553. In that case there was an underlease of premises which included an office suite on the ninth floor of a building, and a separate underlease of an office suite on the fourth floor of the building, together with a flat on the ninth floor of the building. The notice given was "Tenant of premises known as part of the fourth floor". It was held that the notice was not a good notice because, although premises known as something may well include buildings or land which is not strictly part of that description, but which is generally referred to by that description, yet it was impossible to hold that that notice was clear enough to tell the tenant what he was being asked to give up. It is in my judgment clear, and there was no intention here of arguing the contrary, that no section 25 notice can be given as to part only of premises comprised in a single demise, any more than a common law notice to quit can be given as to part only of premises comprised in a single demise.

That case went to the Court of Appeal, where Ungoed-Thomas J's decision was upheld by Russell and Phillimore LJJ, Cairns LJ dissenting.

There was also cited to me a decision of Walton J in *Safeway Food Stores* v *Morris* (1980) 254 EG 1091, where a set of premises, formerly being five numbers in Kensington High Street and a garage at the back thereof, had been converted into one single supermarket and let under one lease. The notice there referred to premises known as the ground floor and basements of the five old numbers in Kensington High Street. The question before Walton J was whether the garage at the back of the premises could be said to be included within the description "premises known as". He held in that case, and on the facts there, that it plainly was included, and he observed that he would have come to the same conclusion as Ungoed-Thomas J in the *Kaiser Engineering* case without any hesitation or difficulty.

The test which I have to apply is to my mind best expressed by my former master in the law Goulding J in *Carradine Properties Ltd* v *Aslam* [1976] 1 WLR 442, and in particular where that most learned lawyer at p 444 between G and H says:

> I would put the test generally applicable as being this: "Is the notice quite clear to a reasonable tenant reading it? Is it plain that he cannot be misled by it?".

That test he derived from a series of authorities going back to 1796, and he applied it to a notice which had given, when served in September 1974, a termination date of September 1973. He held that it was quite plain that any reasonable tenant reading it must have understood that 1973 was a typing error and must have deduced that 1975 was what was intended, and therefore the notice was good. He also referred to the well-known decision in *Hankey* v *Clavering* [1942] 2 KB 326 and at p 445 of the report of the case concerning

A *Carradine Properties Ltd* at letter C cited Lord Greene, where that learned Master of the Rolls said:

> By his letter

in *Hankey* v *Clavering*

> . . . the plaintiff, on the face of it, was purporting to determine the lease by notice on December 21 1941.

I pause to observe that the lease could only have been determined at Christmas Day 1941.

> The whole thing was obviously a slip on his part, and there is a natural temptation to put a strained construction on language in aid of people who have been unfortunate enough to make slips. That, however, is a temptation which must be resisted, because documents are not to be strained and B principles of construction are not to be outraged in order to do what may appear to be fair in an individual case . . .

Having so cited the authorities, Goulding J decided that the notice in that case was a valid notice and that no reasonable tenant reading it could have been misled.

Mr Dowding, in a most able argument, submitted to me that it was always a test of what the reasonable tenant would understand that was applicable to all points on the construction of any notice, and I consider that that is sound. As it seems to me, it is not only where there are blanks, or where there are words which are inapt so as to amount to *falsa demonstratio* — which in the old words *non nocet*, did no harm — that the principle applies but also where there is an C ambiguity or an inadequacy in a notice. He further submitted that the tenant must be regarded as the reasonable tenant in the position of the actual tenant, that is that he should be expected to have a copy of the lease available, so that he knows what his tenancy is of, and that he must know what the actual position on the ground is.

Mr Dowding made those observations in answer to Mr Hochberg's contention that this lease, granting to the tenant the office premises on the 9th floor and the wholly physically separate car-parking spaces in the level B, which is somewhere in the bowels of the earth, gave to the tenant power to separate the occupations of the two sets of property. Mr Dowding observed that that was hypothetical. It was, of course, accurate as a proposition of possibility, but the tenant receiving the notice would know that it was not in fact the case on the ground. In my view, that again is sound, and one must take the tenant D as standing in his shoes, in his own premises, knowing what he has and with access to his own lease.

Having said all that, I return to the notice; would such a tenant, being a businessman or a business company, receiving a notice which says that it relates to ninth-floor premises at Winchester House, know what was meant by it? Is it "quite clear" — borrowing Goulding J's words — "that it includes not only the ninth floor but also the car-parking on level 'B' "? To my mind it is not. To my mind, any reasonable tenant knowing his position and knowing that he had other premises on the level B in the same building, might well say, "I wonder if they are meaning to take back these offices but not to take back the car-parking and the storage spaces. I know that there is a great deal of talk about redevelopment of office premises. E Perhaps all they want is to redevelop this floor as a whole of which I have only a part, and that they are going perhaps to offer me some other premises in the building and allow me to keep my car-parking spaces."

Of course there is no evidence, and there could not be any evidence, that the tenant actually thought that. There is in fact a simple, straightforward counternotice by the tenant, as one would expect, saying that it will not give up the premises comprised in its tenancy, as is required by the Act and following the form of the Act. Such being the facts, it is pure speculation as to what result this notice had upon the tenant, but I have never had any doubt since I first looked at it that in fact this notice does not tell a reasonable tenant, knowing the F position on the ground, that he is expected to give up the whole of his premises. It might leave him in genuine doubt as to whether he is to give up part, that is the ninth floor, or not, and a notice that has an ambiguity of that sort about it seems to me to be fatally flawed.

In my judgment, therefore, for the reasons given by Goulding J in the *Carradine* case citing Lord Greene, as I have already done, this notice was of no effect.

Declaration given that the notice under section 25 of the Landlord and Tenant Act 1954 purporting to terminate the plaintiff's tenancy was invalid and of no effect. By consent these proceedings on the preliminary issue were treated as the trial of the action. Plaintiff awarded costs.

LANDLORD AND TENANT
LEASEHOLD REFORM

A

House of Lords

November 26 1987

(Before Lord KEITH OF KINKEL, Lord BRANDON OF OAKBROOK, Lord BRIGHTMAN, Lord TEMPLEMAN and Lord ACKNER)

DIXON v ALLGOOD

Estates Gazette January 16 1988

B [1988] 02 EG 59-62

Leasehold Reform Act 1967 — Application by tenant to acquire freehold — Appeal from decision of Court of Appeal who had allowed an appeal from the trial judge — Issue as to whether rent payable by tenant was less than two-thirds of the rateable value on the appropriate day — Question as to what was "the appropriate day" — Subject to a decision as to whether the tenancy was at the requisite low rent on the appropriate day, the tenant fulfilled the qualifying conditions for the acquisition of the freehold — His lease was for 51 years C from 1964 and his rent at the material time was £51.44 a year — The lease was originally of two derelict cottages and 6 acres of land — After reconstruction the cottages were shown in the valuation list with rateable values of £42 and £34 on May 9 1966 and February 6 1967 respectively — The tenant subsequently constructed on an adjoining forecourt a garage building first rated on December 22 1971 with a rateable value of £24 — Having obtained possession of one of the cottages which had been sublet, the tenant occupied both cottages as a single dwelling-house — The cottages and garages constituted the premises of which the tenant gave notice under the 1967 D Act on February 18 1981 to purchase the freehold — The county court judge held that, as the tenant's rent of £51.44 was less than two-thirds of the combined rateable values of the cottages and garages (£100), the rent was a low rent and the tenant was therefore entitled to acquire the freehold — The Court of Appeal reversed this decision on the ground that £51.44 was more than two-thirds of the relevant rateable value, which was £76, the aggregate rateable value of the cottages (excluding the garages) — Held by the House of Lords, affirming the decision of the Court of Appeal, that by virtue of section 4(1)(a) of the 1967 Act the appropriate day in relation to "the house in question" fell to be determined under E section 25(3) of the Rent Act 1977 — The "house in question" consisted of the two cottages which appeared for the first time together in the valuation list on February 6 1967 — The "appropriate day" was not later than that date and the aggregate rateable value was then £76 — It was clearly provided by section 4(1)(a) of the 1967 Act that for the purpose of determining the appropriate day the only question was when the house was first rated either as a single hereditament or two or more hereditaments — If the garages had been built before February 6 1967 their rateable value would have had to be taken into account under section 4(2) — Appeal dismissed

F

No cases are referred to in this report.

This was an appeal by James Dixon, the tenant, from the decision of the Court of Appeal allowing an appeal by the landlord, Lancelot Guy Allgood, from the decision of Judge H G Hall, at Hexham County Court, in favour of the tenant's claim to be entitled to purchase the house and premises known as Riverside Cottages at Acomb, Hexham, Northumberland, under the provisions of the Leasehold Reform Act 1967. The decision of the Court of Appeal is reported at [1987] 1 EGLR 93.

Michael J Gadd (instructed by Park Nelson, agents for Septimas G Ward & Rose, of Newcastle upon Tyne) appeared on behalf of the G appellant; Nigel Hague QC and John H Fryer Spedding (instructed by Hyde Mahon & Bridges Sawtell, agents for Wilkinson Maughan, of Newcastle upon Tyne) represented the respondent.

In his speech, LORD TEMPLEMAN said: The Leasehold Reform Act 1967 enables a tenant occupying a dwelling-house as his residence under a long lease at a low rent to acquire the freehold of the house and premises. A low rent is less than two-thirds of the rateable value.

By a lease dated December 31 1964 the appellant, the tenant, became the tenant of two derelict cottages and 6 acres of land at Acomb, Hexham in the county of Northumberland, for a term of 51 years. The rent was £52 a year. The tenant reconstructed one of the cottages, which he named "Riverside Cottage," for which a rateable value was first shown in the valuation list on May 9 1966; the rateable H value was £42. The tenant occupied Riverside Cottage as his residence. The second cottage was then reconstructed by the tenant, named "Riverside Cottage East" and sublet. A rateable value for Riverside Cottage East was first shown in the valuation list on February 6 1967; the rateable value was £34. The two cottages formed a pair of semi-detached houses divided by vertical internal walls. Adjoining the cottages was a forecourt. The tenant constructed on the forecourt a single building with five doors which he used to garage cars and as a domestic store-room. The new building was separately rated on December 22 1971 under the description "five garages adjoining Riverside Cottage"; the rateable value was £24. In 1977 the tenant obtained vacant possession of Riverside Cottage East, J inserted doors in the vertical walls dividing Riverside Cottage from Riverside Cottage East and, the trial judge held, thereby converted both cottages into one dwelling-house which he calls "Riverside Cottages" and which he occupies as his residence. A small part of the land comprised in the lease was compulsorily acquired and the tenant's rent payable under the lease for the remaining $5\frac{1}{2}$ acres or thereabouts, including Riverside Cottages, was reduced to £51.44.

On February 18 1981 the tenant served notice on his landlord, the respondent, claiming to purchase under the Act of 1967 the freehold of the house and premises "Riverside Cottages". The notice as amended on June 22 1981 claims that "the rateable value of the cottages and premises at the 'appropriate day' was £100". It is conceded that "the premises" of Riverside Cottages includes an area K of about half-an-acre comprising the driveway, garden and forecourt with the five garages erected thereon used for the purposes of the cottages.

The trial judge, Judge H G Hall, held that the tenant was entitled to acquire the freehold of the cottages and premises because his rent of £51.44 was less than two-thirds of £100, the aggregate rateable value of the cottages including the garages. The Court of Appeal (Slade LJ and Waite J) reversed the trial judge because the rent of £51.44 was more than two-thirds of £76, the aggregate rateable value of the cottages excluding the garages. The tenant appeals.

The Act of 1967, as amended, and so far as material, provides as follows:

1(1) . . . this Act shall have effect to confer on a tenant of a leasehold house, occupying the house as his residence, a right to acquire on fair terms the freehold . . . of the house and premises where — . . . (*b*) at the relevant time (that is to say, at the time when he gives notice in accordance with this Act of his desire to have the freehold . . .) he has been tenant of the house under a long tenancy at a low rent, and occupying it as his residence, for the last three years . . .

The tenant fulfilled all these conditions if he was a tenant at a "low rent".

2(1) . . . "house" includes any building designed or adapted for living in and reasonably so called . . .

The trial judge found that though the cottages had originally been two houses they had become one "house" as defined by the Act by the relevant date when the tenant gave notice to purchase under the Act.

2(3) . . . premises [in relation to a house let to and occupied by a tenant means] any garage, outhouse, garden, yard and appurtenances which at the relevant time are let to him with the house and are occupied with and used for the purposes of the house . . .

The five garages and forecourt were part of the premises of the cottages at the relevant time, namely February 18 1981 when the tenant gave notice to purchase under the Act.

If the tenant is entitled to purchase the cottages, the purchase will by virtue of the Act of 1967 include "the premises", ie the five garages, the courtyard and any other outhouses and appurtenances of the cottages, "at the relevant time," ie February 18 1981 when the tenant served notice under the Act. The price payable by the tenant will be the amount which on February 18 1981 the cottages and premises if sold for an estate in fee simple subject to the tenant's tenancy might have been expected to realise in the open market.

3(1) . . . "long tenancy" means . . . a tenancy granted for a term of years certain exceeding 21 years . . .

The tenant's lease was for 51 years.

4(1) . . . a tenancy of any property is a tenancy at a low rent at any time when rent is not payable under the tenancy in respect of the property at a yearly rate equal to or more than two-thirds of the rateable value of the property on the appropriate day . . .

The rent of £51.44 payable by the tenant was more than two-thirds of the rateable value of the cottages, £76, until the garages were built and rated on December 22 1971. If the "appropriate day" falls before December 22 1971, the tenant is not entitled to purchase the cottages and premises under the Act.

4(1)(*a*) "appropriate day" means the 23rd March 1965 or such later day as by virtue of section 25(3) of the Rent Act 1977 would be the appropriate day for purposes of that Act in relation to a dwelling-house consisting of the house in question;

The date March 23 1965 was the date on which the Bill for the Rent Act 1965 was introduced. That Act contained, in section 43(3), provisions similar to section 25(3) of the Rent Act 1977 and by the First Schedule para 1(3) directed that a tenancy was to be exempt from the restrictions imposed by the Rent Act 1965 if the rent payable was less than two-thirds of the rateable value "on the appropriate day".

The cottages were not rated on March 23 1965. The "appropriate day" is by virtue of section 4(1)(*a*) to be determined under section 25(3) of the Act of 1977 in relation to "the house in question", ie the cottages excluding the garages.

Section 25 of the Rent Act 1977 provides:

(1) . . . the rateable value on any day of a dwelling-house shall be ascertained for the purposes of this Act as follows: — (*a*) if the dwelling-house is a hereditament for which a rateable value is then shown in the valuation list, it shall be that rateable value; (*b*) if the dwelling-house . . . consists of . . . more than one such hereditament, its rateable value shall be taken to be such value as is found by . . . aggregation of the rateable . . . values so shown. . . . (3) . . . "the appropriate day" — (*a*) in relation to any dwelling-house which, on 23rd March 1965, was or formed part of a hereditament for which a rateable value was shown in the valuation list then in force, or consisted or formed part of more than one such hereditament, means that date, and (*b*) in relation to any other dwelling-house, means the date on which such a value is or was first shown in the valuation list. (4) Where, after the date which is the appropriate day in relation to any dwelling-house, the valuation list is altered so as to vary the rateable value of the hereditament of which the dwelling-house consists or forms part and the alteration has effect from a date not later than the appropriate day, the rateable value of the dwelling-house on the appropriate day shall be ascertained as if the value shown in the valuation list on the appropriate day had been the value shown in the list as altered . . .

By section 37(6) of the Act of 1967, as amended, section 25(1) (2) and (4) of the Act of 1977 shall apply to the ascertainment for purposes of the Act of 1967 of the rateable value of a house and premises or any other property as they apply to the ascertainment of that of a dwelling-house for the purposes of the Rent Act.

The dwelling-house, Riverside Cottages, which now embraces both the former Riverside Cottage and the former Riverside Cottage East, consists of two hereditaments, namely, the former Riverside Cottage and Riverside Cottage East both of which appeared in the valuation list on February 6 1967. The appropriate day for the purposes of determining the rateable value of the dwelling-house now occupied by the tenant is not later than February 6 1967 and the aggregate rateable value of the two hereditaments which now constitute the tenant's dwelling-house is £76. The rent payable by the tenant is £51.44, which is more than two-thirds of the rateable value of £76 on the appropriate day, and it follows that the tenant is not entitled to purchase the freehold reversion of Riverside Cottages and premises.

Slade LJ delivering judgment in the Court of Appeal in the present case pointed out that there is a clear and unequivocal difference between the provisions of the Act of 1967 which entitle a tenant to purchase "the house and premises" and the provisions of section 4(1)(*a*) which require the appropriate day to be determined in relation to a dwelling-house "consisting of the house in question". The reason for this distinction is not far to seek. A hereditament for rateable purposes is not limited to buildings:

where two or more properties are within the same curtilage or contiguous to one another, and are in the same occupation . . . they are, as a general rule, to be treated for rating purposes as if they formed parts of a single hereditament. There are, however, exceptional cases, where for some special reason they may be treated as two or more hereditaments. That may happen, for instance, when they are situate in different rating areas, or because they were valued at different times . . . : or because they were at one time in different occupations . . . ; or because one part is used for an entirely different purpose . . .: *per* Denning LJ in *Gilbert (Valuation Officer)* v *S Hickinbottom & Sons Ltd* [1956] 2 QB 40, 48.

The Act of 1967 was inspired by the plight of a large number of lessees of houses whose building leases granted in the 19th century were coming to an end. A tenant of a house under such a lease held a wasting asset which was difficult or impossible to sell; he was faced with the prospect of large dilapidation claims and was not entitled to security of tenure under the Rent Acts. By fixing March 25 1965 as the appropriate day for houses rated before that day, the Act of 1967 indicated clearly that events taking place after March 25 1965 resulting in an increase or decrease of rateable value were to be ignored. Changes in rateable value could take place as a result of a quinquennial or other general revaluation or as a consequence of changes taking place to a particular hereditament which included a house. The rateable value of such a hereditament might be reduced as a result of deterioration of the house or the neighbourhood. The rateable value of the hereditament could be increased by an extension or improvement to the house itself or by the erection of a building in the curtilage of the house. All these changes must be ignored. For example, if in the present case Riverside Cottages had been rated before March 25 1965, the hereditament comprising Riverside Cottages would have included the courtyard. If after March 25 1965 the tenant had erected five garages in the curtilage of Riverside Cottages, the rating authority might have increased the rateable value of Riverside Cottages or might have rated the five garages as a separate hereditament. Whether the garages were rated with the cottages or as a separate hereditament, the rateable value attributable to Riverside Cottages and the premises including the five garages for the purposes of the Act of 1967 would remain the rateable value as fixed on March 25 1965. A tenant of a house rated on March 25 1965 cannot improve his position under the Act of 1967 by building in the curtilage of the house garages or other buildings which become rated as a separate hereditament or lead to an increase in the rateable value of the house. Under the Act of 1967 the tenant is entitled to purchase both the house and the garages which are included in the curtilage as premises, provided that the rent payable for the house and premises is less than two-thirds of the rateable value of the house on March 25 1965. Similarly, in the present case the tenant could not improve his position under the Act of 1967 by building five garages in the curtilage of the cottages after the appropriate day. It is significant that section 25(4) of the Act of 1977 does permit an increase in rateable value after the appropriate day in limited circumstances. By section 25(4) such a variation is taken into account but only where the

A variation "has effect from a date not later than the appropriate day". In other words if a garage is built in the courtyard of a house before the appropriate day but is not rated until after the appropriate day or is not taken into account in the rateable value of the house until after the appropriate day, nevertheless effect can be given to the variation for the purposes of the Act of 1967 if the variation when it takes place "has effect from a date not later than the appropriate day". But unless the conditions specified in section 25(4) are satisfied, events which take place after the appropriate day must be disregarded, and the appropriate day itself is by section 4(1) of the Act of 1967 judged by reference to the first day upon which the house as a single hereditament or as two or more hereditaments first appears in the valuation list.

B Mr Gadd, who appeared for the tenant, relied on section 4(2) of the Act of 1967. That subsection provides:

> Where . . . a question arises under section 1(1) above whether [the tenant's] tenancy of the house is or was at any time a tenancy at a low rent, the question shall be determined by reference to the rent and rateable value of the house and premises as a whole, and in relation to a time before the relevant time shall be so determined whether or not the property then occupied with the house or any part of it was the same in all respects as that comprised in the house and premises for purposes of the claim; . . .

If the five garages had been built before February 6 1967, section 4(2) would have required the garages to be taken into account although they were separately rated, because of the requirement that C the question shall be determined by reference to the rateable value of the house and premises as a whole. Section 4(2) also requires that Riverside Cottage East shall be taken into account notwithstanding that Riverside Cottage East was sublet and did not form part of "the house" on February 6 1967. But section 4(2) does not in my opinion alter or cast any doubt on the clear direction in section 4(1)(*a*) that for the purpose of determining the appropriate day the only question is when the house was first rated either as a single hereditament or as two or more hereditaments. In the present case, Riverside Cottages, the house, was rated as two hereditaments on February 6 1967, which is therefore the appropriate day.

Mr Gadd also submitted that "the house" did not come into D existence until 1977 when the two cottages were converted into one house and that 1977 was, therefore, the appropriate day and that the house then consisted of three hereditaments comprised of the two cottages and the garages. But section 4(1)(*a*) of the Act of 1967, read in conjunction with section 25(1) of the Act of 1977, requires the appropriate day to be the day when "the house," consisting of two cottages and no more, was first rated and that day was February 6 1967. A tenant of two semi-detached houses, each rated at £50 on March 25 1965, could not by inserting communicating doors between the two houses and converting them into one house, rated in 1987 at £150, alter the appropriate day or increase the rateable value for the purposes of the Act of 1967. In the present case, "the house" created by the tenant in 1977 consisted of two hereditaments rated for the E first time by February 6 1967.

Mr Gadd, who did not appear before the trial judge or the Court of Appeal, also sought to raise the argument that "the house and premises" did not consist of the whole 5 acres or thereabouts remaining subject to the lease, but was limited to the area of about half-an-acre consisting of Riverside Cottages, the garages, forecourt and garden, occupied and used for the purposes of the cottages. Therefore the sum of £51.44 ought to be apportioned between the cottages and premises on the one hand and the remainder of the 5 acres on the other hand. Such an apportionment might well result in the rent apportioned to the cottages and premises being less than two-thirds of the rateable value of £76. This point was not raised prior to F the appeal to this House. Moreover, it involves questions of fact and evidence and would necessitate a retrial. Your Lordships declined to allow Mr Gadd, in these circumstances, to take the point at this late stage. In the result, in agreement with the reasons expressed by the Court of Appeal, I would dismiss this appeal.

LORDS KEITH OF KINKEL, BRANDON OF OAKBROOK, BRIGHTMAN and ACKNER agreed with the speech of Lord Templeman and did not add anything.

Appeal dismissed.

For a further case on this subject see p 223

LANDLORD AND TENANT

RENT ACTS

Queen's Bench Division

November 3 1987

(Before Mr Justice SCHIEMANN)

Rv LONDON RENT ASSESSMENT PANEL, EX PARTE MOTA

Estates Gazette January 16 1988

[1988] 02 EG 66-70

Rent Act 1977, section 70(1)(b) — Application by tenant for judicial review, challenging decision of rent assessment committee — Whether tenancy was a furnished tenancy, the furniture being "provided for use under the tenancy" within the meaning of section 70(1)(b), despite the fact that the furniture physically *in situ* belonged to the tenant — Tenant of a furnished flat agreed with the landlords that during renovations she would be given temporary accommodation and would then be put back into a renovated flat, although not the one which she had originally occupied — The arrangement, set out in a written agreement, confirmed that she would have a protected tenancy of the new flat, the fair rent of which would be fixed by the rent officer — The agreement stated that the landlords would have no objection to the tenant using on a daily basis her own furniture, in lieu of that supplied by the landlords, which would be removed and retained by them and reinstalled at the end of the tenancy — After she had moved into the new flat a fair rent of £4,500 pa was registered by the rent officer on the basis of a furnished tenancy, adopting an inventory of furniture sent in by the landlords, which the latter confirmed in their application form — The tenant objected to the rent registered and appealed to the rent assessment committee, who reduced the rent to £4,250, but still on the basis of a furnished tenancy — In their reasons the committee accepted that the bulk of the furniture in the flat belonged to the tenant, but said that they were "satisfied that the landlords had provided a full complement of furniture and equipment for the flat and concluded that it was let as furnished accommodation and should be assessed on that basis" — They noted that the tenant had previously had a furnished tenancy from the same landlords — In an affidavit the chairman of the committee observed that, although at the hearing the tenant's representative had pointed out that the tenant was using her own furniture, he had not argued that the tenancy was unfurnished and the only comparable rents put forward were those of furnished tenancies — Held, rejecting the application, that furniture can be "provided for use" within the meaning of section 70(1)(b) although at the tenant's request the furniture is not physically on the premises — In the present case the committee were justified on the evidence in taking the view that the tenant was entitled (a) initially to have a flat furnished, (b) to use her own furniture and have some or all of the landlords' furniture removed and retained by the landlords, and (c) to have the landlords' furniture replaced not merely at the end of the tenancy but also during its currency — The committee were not in error in having regard to the fact that the tenant's original tenancy was furnished — It was not the case, as alleged in the grounds for the application, that the tenant *should* provide her own furniture; the landlords had merely agreed that she *could* do so — Application dismissed

No cases are referred to in this report.

This was an application for judicial review by Mrs Teresa Mota, seeking to quash a decision of a committee of the London Rent Assessment Panel assessing a fair rent for a first-floor front flat at 63 Greencroft Gardens, London NW6. The landlords were Tropicacre Ltd.

Miss Caroline Hunter (instructed by Daniel & Harris) appeared on behalf of the applicant; D P Pannick (instructed by the Treasury Solicitor) represented the respondent committee.

Giving judgment, SCHIEMANN J said: The Rent Act 1977 contains provisions for the registration of fair rents. Section 67(1) provides:

> An application for the registration of a rent for a dwelling-house may be made to the rent officer by the landlord or the tenant, or jointly by the landlord and the tenant, under a regulated tenancy of the dwelling-house.

Subsection (7) of that section provides:

> . . . The provisions of Part I of Schedule 11 to this Act . . . shall have effect with respect to the procedure to be followed on applications for the registration of rents.

I shall come to that schedule in a moment.

Section 70(1) provides:

> In determining, for the purposes of this Part of this Act, what rent is or would be a fair rent under a regulated tenancy of a dwelling-house, regard shall be had to all the circumstances (other than personal circumstances) and in particular to — (a) the age, character, locality and state of repair of the dwelling-house, and (b) if any furniture is provided for use under the tenancy, the quantity, quality and condition of the furniture. (c) Any premium . . . which has been or may be lawfully required or received on the grant, renewal, continuance or assignment of the tenancy.

The relevant provisions of the 11th Schedule are contained in paras 6, 8 and 9. Para 6 provides that once a rent officer has fixed a fair rent the person affected can object. Para 8 provides that if either landlord or tenant requests to make oral representations a committee known as the rent assessment committee shall give him an opportunity to be heard either in person or by a person authorised by him in that behalf. Para 9 provides:

> (1) The committee shall make such inquiry, if any, as they think fit and consider any information supplied or representation made to them in pursuance of paragraph 7 or paragraph 8 above and — (a) if it appears to them that the rent registered or confirmed by the rent officer is a fair rent, they shall confirm that rent; (b) if it does not appear to them that that rent is a fair rent, they shall determine a fair rent for the dwelling-house.

This is an application to quash the decision of such a rent assessment committee. The factual background is that Mrs Mota was a tenant of some property. The landlords wished to renovate it and to decant her while the renovation was going on, and then to put her back in a renovated flat, albeit a different one from that which she had originally occupied.

On April 10 1984 an agreement was reached between the landlords and the tenant which, so far as presently relevant, runs as follows:

> We the Directors of Tropicacre Ltd, the landlords of Mrs Mota confirm: that she is a protected tenant of the furnished Flat 2, of 63 Greencroft Gardens, London NW6, comprising of two rooms, kitchen and bathroom; that her

> A move to Flat 5, 61 Greencroft Gardens, London NW6, is only a temporary move pending works at 63 Greencroft Gardens . . .; that Mrs Mota's tenancy of Flat 2, 63 Greencroft Gardens NW6 continues during the works, and that the rent will continue to be collected for Flat 2, 63 Greencroft Gardens, while the works are under completion; that when the works at 63 Greencroft Gardens are completed . . . Mrs Mota will be entitled to move to the first-floor front flat at 63 Greencroft Gardens, under a protected tenancy, within the meaning of the Rent Act 1977, paying an exclusive fair rent to be registered by the rent officer; that subject to the plans and specifications Mrs Mota's intended flat at first-floor front, 63 Greencroft Gardens, will be a furnished converted self-contained dwelling consisting of two bedrooms, living room, a kitchen and a bathroom, for her own exclusive use with shared use of communal garden; that we are to be responsible for the removal costs and other necessary expenses incurred by the tenants in connection with their
> B transfer to and from the temporary accommodation; that we have no objection to the tenant using on a day to day basis their own furniture, in lieu of that supplied by the landlord, which will be removed and retained by the landlord, and reinstalled at the end of the said tenancy, with the exception of the gas cooker.

That has been signed by Mrs Mota and also by the landlords.

The landlords in due course provided to the rent officer service of the Camden Council an inventory of the contents of the first-floor front flat at 63 Greencroft Gardens. That includes a fair amount of furniture, crockery, glasses, and so on.

In February 1985 the applicant moved into her new flat. It was not the flat which she had been occupying, but rather one on the first floor. It is agreed that the landlords provided the carpets for the flat
C and the cooker. The landlords applied on August 9 1985 for a registration of the rent. In their application form they were asked: "Is any furniture provided under the tenancy? If Yes, give details." They wrote against that: "See the Rent Officer's file" — and no doubt the reference was to the list of contents to which I have referred.

On October 31 1985 the rent was registered and the furniture provided by the landlords was stated to be: "As per inventory, a copy of which is in the Rent Officer's possession." The rent register contains the remark: "Fully furnished to a high standard."

On November 15 1985 an objection by the tenant was duly received. On January 23 1986 that objection was referred to the rent assessment committee. On March 12 1986 there was a hearing before
D the rent assessment committee. The tenant was represented but the landlords were not.

The rent assessment committee reduced the assessment of fair rent from £4,500, which had been the rent officer's figure, to £4,250. They gave their reasons as follows:

> The committee were concerned with assessing the rents of three furnished flats in a substantial semi-detached house located in a good residential area of West Hampstead. On inspection they ascertained that the structure of the building appeared recently to have been overhauled and that major works of decoration had been carried out both to the exterior and the common parts. Internally the committee found that the units had been newly converted and comprehensively refurbished.

E Then they make various findings in relation to that.

They say that the tenant's representative presented evidence to the committee of registered rents for furnished accommodation within the locality. They comment that unfortunately, as in all furnished accommodation, these registrations were of little assistance as they gave no indication of the extent and quality of the furnishings. However, the committee took note of the general rent levels and inspected externally some of the comparables. They found that neither of those properties appeared to have benefited from substantial modernisation works, and that both their situation and character were less favourable than those of the subject premises.

Then comes this paragraph: "The committee took careful note of the differences between the units." In relation to Mrs Mota's flat
F they say they accepted that the bulk of the furniture currently in use belonged to the tenant, Mrs Mota. They were, however, satisfied that the landlords had provided a full complement of furniture and equipment for the flat and concluded that it was let as furnished accommodation and should be assessed on that basis. They noted that the tenant had previously had a furnished tenancy from the same landlords. They assessed the fair rent on the evidence that the landlords were responsible for all repairs and decorations.

There is an affidavit in front of me sworn by the chairman of the rent assessment committee which contains the following:

> The committee concluded that neither the letter dated April 10 1984 nor any other evidence put forward on behalf of Mrs Mota demonstrated any intention to create anything other than a furnished letting. Although [her representative] pointed out that Mrs Mota was using her own furniture, he did G
> not put forward any argument to the effect that the tenancy was unfurnished, and the only comparable rents which were in evidence were of furnished tenancies. He was unable to put forward any satisfactory evidence as to repairing liabilities, or differentiate between the liabilities of the two tenants whom he represented. The committee accepted the landlords' written assertion on the application form, adopted by the rent officer, that they were responsible for all repairs and decorations, and found this to be consistent with a furnished letting.

That decision is challenged on grounds which I permitted to be amended at the beginning of the hearing in front of me, and which, as amended, read as follows:

> The decision of the Respondents erred in law in that: (i) no reasonable Rent Assessment Panel would have assessed the rent on the basis that the H
> accommodation was furnished; (ii) the Respondents took into account an irrelevant consideration, namely that the tenancy of a flat previously let to the Applicant in the same house had originally been furnished; (iii) the Respondents failed to take into account a relevant consideration, namely that the landlords had not provided any furniture and that the landlords had agreed that the Applicant should provide her own furniture; (iv) they misconstrued section 70(1) of the Rent Act 1977, by taking into account furniture which the Applicant had no contractual right to use or alternatively only had to call for the future use of, a right which she had not exercised.

In my judgment furniture can be provided for use under the tenancy within the meaning of section 70(1)(b) notwithstanding that at the tenant's request the furniture is not physically on the premises. Further, one of the circumstances to which the committee would be entitled to have regard is any obligation by the landlord to provide J
furniture on the premises if asked by the tenant so to do.

In my judgment, the committee were entitled to take the view that the tenant was entitled (a) initially to have the flat furnished, (b) to use her own furniture and have some or all of the landlords' furniture removed and retained by the landlords, and (c) to have the landlords' furniture replaced not merely at the end of the tenancy but also during its currency.

My reasons for that view are, first, that she wished for a furnished flat. That is quite clear from the wording of the agreement of April 10 1984, even though at the time that she entered into the agreement she had no furniture in the old flat. I consider that the words at the very conclusion of the April agreement, namely: K

> We have no objection to the tenant using on a day to day basis their own furniture, in lieu of that supplied by the landlord, which will be removed and retained by the landlord, and reinstalled at the end of the said tenancy, with the exception of the gas cooker,

are words which are not apt to cut down the tenant's right to have a furnished dwelling if she so wished. Second, that conclusion is reinforced by the wording "on a day to day basis", and the phraseology that the landlords have no objection to the tenant using their own furniture on a day-to-day basis. Third, I bear in mind that this appears to have been the view of the representative who argued the whole case for the tenant on the basis that she was a tenant of furnished accommodation. L

I consider that the committee were entitled to bear in mind that in fact at the time of their hearing all the furniture on the premises, with the exception of the carpets and the cooker (if these items are properly described as furniture) had been provided by the tenant. It is, however, by no means clear that the committee did not have this in mind in making their assessment of the fair rent. They certainly knew it as a fact, because they referred to it.

The committee were, in my judgment, entitled to have regard to the fact, undisputed before them, that the tenant had in respect of the premises which she had originally occupied been a tenant under a furnished tenancy. This was a background factor against which the arrangements set out in the letter of April 10 1984 had to be seen and a factor, moreover, which had been drawn to the committee's M
attention by the advocate appearing for the tenant.

With that in mind, looking at the grounds of challenge, I reject the challenges, because looking at the first one, I consider that it was perfectly reasonable for the rent assessment committee in these circumstances to assess the rent on the basis that the accommodation was furnished. Second, I do not regard the earlier history as irrelevant. Third, I am not persuaded that the respondent had failed to take into account the fact that the flat was filled with the tenant's furniture. It is not the case, as is alleged in these grounds, that the landlords had agreed that the applicant *should* provide her own furniture. All that the landlords had agreed was that she *could* provide her own furniture.

So far as the assertion in the grounds that the landlords had not provided any furniture is concerned, I am not prepared to take a different view of the factual background from that which was found by the committee after hearing representations on behalf of the tenant. So far as the last ground is concerned, it proceeds from the assumption that the tenant had no contractual right to use furniture provided by the landlords. That factual assumption is one which, in my judgment, the committee were not obliged to make on the evidence before them, which was both that of the written agreement and the further evidence which they received on behalf of the tenant. So far as the tenant's having only a right to call for the future use of the furniture is concerned, and that this was a right which she had not exercised, I am not persuaded that the committee took any view of the facts which is inconsistent with that. In those circumstances, this application to quash the decision of the committee fails.

The application was dismissed with costs.

Court of Appeal

September 29 1987

(Before Sir John DONALDSON MR and Lord Justice STOCKER)

OZER PROPERTIES LTD AND ANOTHER v GHAYDI

Estates Gazette January 23 1988

[1988] 03 EG 87-88

Rent Act 1977 — Possession action — Defendant's plea of *res judicata* — Defendant was in occupation of a room in a rooming house consisting of more than 20 rooms — It had been owned by a succession of landlords and rents had been collected by a succession of persons, some of them authorised agents, but at least one of them a fraudulent person who deceived the present defendant — This person took a deposit of £100 and promised a rent book, but then disappeared — In fact the defendant never had a tenancy and, although she did not realise it, was a trespasser — Subsequently the defendant was locked out of her room but obtained an interim injunction ordering the then landlords (not the present plaintiffs) to allow her to return — In subsequent proceedings as plaintiff against these landlords the present defendant obtained a judgment "that the plaintiff do recover against the defendant damages to be assessed and costs" — This judgment and a subsequent assessment of damages were given in default of service of defence by these landlords — These facts gave rise to the issue in the present appeal by the defendant from the judgment of Judge Hayman granting possession to the present landlords — It was contended that the judgment which the defendant had obtained in the previous proceedings for damages to be assessed created a situation of *res judicata*, determining in her favour that she was a tenant and, it was suggested, that she had a tenancy protected by the Rent Act — Held, rejecting this submission, that in order for a judgment in default to create such a situation there must be an issue which, when formulated, necessarily and with complete precision determined the rights of the parties — It was not, however, necessary in order to support the judgment to postulate that the present defendant had a tenancy of any kind — She was entitled to damages for eviction although she was not in fact a tenant and had not the benefit of the Rent Act — Appeal dismissed

No cases are referred to in this report.

This was an appeal by Miss Zara Ghaydi from a decision of Judge Hayman at Bloomsbury County Court granting to the plaintiffs (the present respondents), Ozer Properties Ltd and Esther Halpern, possession of room 17 at 93-95 Moscow Road, London W2.

The appellant, Miss Zara Ghaydi, did not attend and was not represented; M G Warwick (instructed by Grangewoods) represented the respondents.

Giving the court's ruling on an application for adjournment, SIR JOHN DONALDSON MR said: This appeal has been listed for hearing today. As I have already indicated, we have received a medical certificate which does not say that Miss Ghaydi is unable to attend — it says that she is suffering from depression and a rheumatological complaint — and a letter from a social worker who says that her depression is made worse by the stress of the court case. It does not seem to me that an adjournment would relieve her depression — it would tend to make it worse. This application for an adjournment has been opposed by the respondents. Undoubtedly it does cause great hardship to the respondents if the appeal is adjourned. In those circumstances our view is that we ought to deal with the matter. We have had the advantage of reading it. I will ask Stocker LJ to deliver the first judgment.

Giving judgment, STOCKER LJ said: This is an appeal from a judgment of His Honour Judge Hayman given on December 11 1986 whereby it was ordered that the plaintiffs have possession of room 17 in a building called 93-95 Moscow Road, London W2. The appeal by the appellant is that that order should be set aside. There is also a respondents' notice justifying the judgment on grounds other than those which appear in it.

The short facts are these. The premises in question are a rooming house which contained something in excess of 20 rooms, each having some form of cooking facilities but all sharing bathroom accommodation. Room 17 was occupied by the appellant, Miss Zara Ghaydi. There was a parallel proceedings heard at the same time before the same judge in respect of a claim for possession of room 23 occupied by a man called Sanford. This building was, from early 1980, owned by a company called Gate Investments Ltd. That was a company registered in the Channel Islands, the controlling shareholder apparently being a Libyan resident in Libya. It follows therefore that somebody had to carry on the day-to-day administration of the building and the various occupied rooms in it.

That function was carried on by a man called Shahumi until about 1981, when it seems that he was not accounting properly for the rent he collected or producing to anybody the necessary books. Accordingly, on May 12 1981, a letter was written to him by the owners' solicitors telling him to hand over all books of account and informing him that managing agents would be appointed to take over from him. Until the events of 1984 occurred that was the last contact which it seems Shahumi had with those premises at all. It is quite clear that he ceased to have any authority in the building or to be the landlords' agent from that time onwards. A letter in May 1981 was sent or handed to all the tenants in the various rooms in the premises in effect informing them of the fact that Shahumi was no longer authorised to collect the rents.

The owners then had to make some arrangements for the administration of the building. As an interim matter they appointed their solicitors to do so. In due course two successive firms of estate agents took over the management of the premises and were responsible for the collection of rent. Over that period also one of those managing agents was concerned with trying to negotiate a sale of the freehold. A sale of the freehold was in fact arranged and the contract of sale was entered into on May 25 1984, that being a contract for sale of the premises to a firm called Selous Investments Ltd. By that same contract Selous Investments were authorised by the vendors to collect the rents in the meantime pending completion.

In fact there never was a completion to Selous Investments Ltd. The sale was completed on February 12 1985 to the plaintiffs. Though it would seem that it does not bear on the issues to be decided in this court, there was a trust deed regarding the equity which would be resultant upon the contract of sale itself.

It is now necessary to deal with the letting to Miss Ghaydi and the circumstances in which this claim for possession comes about. On March 21 1984, according to her affidavit — nearly three years after the man Shahumi ceased to have anything to do with the premises — she was in a café with a friend of hers, a Miss Morfid. They were discussing between themselves the difficulties of obtaining accommodation in London. It seems that Shahumi overheard their conversation and indicated that he could perhaps provide rooms for them. He then took them to 93 Moscow Road, having a key to the door, and showed them three rooms. Miss Ghaydi took one of the rooms, which is the room the subject of this claim, no 17, and her friend Miss Morfid took another room. They paid a deposit of £100, rent for the rooms being £20 per week for each room. They were given to understand by this man, who was apparently Shahumi, that he would receipt a rent book for them in due course. In fact he never did so and they never saw him again. They each moved into possession on March 22 1984.

It may have some bearing on the subsequent events that the defendant in the parallel action, Mr Sanford, had an almost exactly similar experience. He was visiting a friend in the premises of 93 Moscow Road and saw the man who purported to be Shahumi, who let to him three rooms in the premises, one for Mr Sanford himself and two for two of his friends. He also took a deposit of £100. Mr Sanford went out and purchased three rent books and had the appropriate copy keys made, having been told by Shahumi that he would receipt the rent books when they were produced. In fact when he returned 20 minutes later Shahumi had disappeared and was never seen again. Thus the events which have been recited seem to have been a common experience for both the two defendants to the action.

I might say at this point that it seems to me perfectly clear that the appellant, and indeed the other persons involved as well, had not got a tenancy at all. Shahumi had no authority whatever to let rooms to them or to take any sums by way of deposit or for any other purpose from them. Accordingly Miss Ghaydi was, though she did not realise it, in my view undoubtedly a trespasser from that moment onwards.

It is now necessary to recite shortly the events which gave rise to an action in which she herself was plaintiff. They arise out of a dispossession which occurred in June 1984. On June 12 Miss Ghaydi returned to the premises in the morning with her friend, having been out for a short visit elsewhere, to find that her keys did not fit the locks. She was confronted by a man called Edwards who told her that in future she would have to ring at the front door in order to gain admittance. That happened for a day or two. On June 15, however, when she returned to the premises, she was confronted, without the door even being opened, we understand, by this man Edwards, who told her that "they were out". He refused her admittance and refused her any access to her room where all her personal belongings were stored. She took action very promptly and applied to the court for an injunction.

On June 15, the very day of her dispossession, His Honour Judge Curtis-Raleigh made an interim injunction in these terms. He ordered that the defendants, who were then Selous Investments Ltd, should:

1. Allow the Plaintiff forthwith to enter and occupy Room 17, 93 Moscow Road, London W2.
2. Should not interfere in any way with the Plaintiff's quiet enjoyment thereof.
3. Should give the Plaintiff forthwith a key to the said premises.

That was an interim injunction. A return date for an *inter partes* hearing was fixed for June 21. On June 21 the same judge dealt with the matter. Undertakings were given in these terms: each of the defendants (ie Selous Investments and Mr Edwards) undertook "not to interfere in any way with the plaintiff's quiet enjoyment of her premises at 93 Moscow Road, London W2," that is room 17. Having made that order, he adjourned generally Miss Ghaydi's action with liberty to restore. Miss Ghaydi had undertaken to issue formal proceedings by Monday June 18. She did so. In accordance with that undertaking particulars of claim were filed on her behalf. It is necessary, in the light of the issue which arises on this appeal, to see the form in which that claim was presented. Having set out the facts relating to her alleged tenancy and to the circumstances in which she was dispossessed, she claimed: "A Declaration that she holds the premises at Room 17 as a protected tenant under the Rent Act 1977". "An injunction" in the terms which had already been granted and in respect of which an undertaking was given. She also claimed damages.

On June 21, as has been recited, Judge Curtis-Raleigh heard the matter *inter partes*, when the undertakings were given to which I have referred. Next, on August 16, the registrar gave orders for the hearing of the action, ordered that it should be consolidated with the similar actions by other people and ordered defences to be filed within 28 days.

In fact the defendants to that action, Selous Investments and Edwards, did not comply with that order, and no defences were in fact ever filed. On February 12, as has already been indicated, there was a conveyance of the premises to the present plaintiffs. On May 7 1985 the present plaintiffs commenced the action with which this appeal is concerned.

The defendants not having filed any defence in accordance with the orders made by the registrar, on May 24 an order was made that they should file their defences within seven days or be disbarred from defending. The solicitors then acting for them applied to come off the record. A date was fixed for an assessment of damages, and on July 24 judgment was given in favour of Miss Ghaydi in her action. It is important to look at the precise terms of that judgment. It is dated July 24 1985 and reads:

IT IS ADJUDGED that the plaintiff do recover against the defendant damages to be assessed and costs.

In fact there was a hearing in respect of the assessment of damages on August 12 1985 before a different judge, when His Honour Judge Honig assessed the damages at the sum of £525. The defendants did not attend either on the assessment of damages or on the occasion when formal judgment was entered in favour of Miss Ghaydi.

Those facts give rise to the issue which arises on this appeal. It was contended before the learned judge that the judgment entered in default of service of the defence created a situation of *res judicata* on the issue as to whether or not Miss Ghaydi was a tenant protected by the Rent Acts. Before the learned judge extensive argument took place, most of which was concerned with the proposition that the present plaintiffs, who were the people to whom in fact the premises had been conveyed, were not privies to Selous Investments and Edwards, who were the defendants in Miss Ghaydi's action and against whom the judgment in default had been given.

For my part I find it unnecessary to consider that aspect of the matter any further, for it seems to me plain from the terms of the order itself which I have just recited that no case of *res judicata* could arise. In order for a judgment in default to create such a situation there must be, as it seems to me, an issue which, when formulated, necessarily and with complete precision determined the rights of the parties. In my view it is really not possible to say that that judgment, which was a judgment for damages to be assessed, in any way determines the issue as to whether or not Miss Ghaydi was a tenant and whether or not that tenancy had the protection of the Rent Act 1977. The judgment was a judgment for damages to be assessed, and those damages were in fact assessed. It is not necessary in order to support the judgment to postulate that Miss Ghaydi had a tenancy of any kind. She would have been entitled to damages having regard to the events of her eviction even if she had not been a tenant or had not had the benefit of the Rent Acts.

Accordingly, in my view the result of the hearing before the learned judge, His Honour Judge Hayman, could be fully justified on the basis that there was no *res judicata* in this case at all without embarking upon the rather complicated investigation as to whether or not the plaintiffs in this action were privies to the judgment given against Selous Investments and Edwards.

Accordingly, although not raised specifically in her notice of appeal, it seems to me manifest for the reasons that I have already given that Miss Ghaydi was not in fact a tenant, she never had a tenancy and was a trespasser at all times. Accordingly, subject to the plea of *res judicata*, the judgment to the effect that the plaintiffs should have an order for possession was really incontrovertible.

For the reasons which I have shortly endeavoured to give it seems to me that this is not a case in which a plea of *res judicata* could succeed either. Indeed that is the matter raised by the respondents' notice in this appeal.

One is bound to feel a great deal of sympathy for Miss Ghaydi. She was an entirely innocent dupe of a fraudulent person, who obtained from her a sum of money which I do not doubt she could ill afford to pay. She has been the victim of a mean and dishonest trick. Nobody could blame her in the least for that, and anybody would have the greatest sympathy for her. However, we have gone into this matter with great care, and accordingly, although she is not present to put forward any arguments in support of her appeal, for my part I have not the slightest doubt that her appeal was bound to fail.

Accordingly and for these reasons I would dismiss this appeal.

SIR JOHN DONALDSON agreed and did not add anything.

The appeal was dismissed with costs, costs up until June 19 1987 not to be enforced without the leave of the court.

A
Court of Appeal

October 28 1987

(Before Lord Justice PARKER and Sir George WALLER)

AL HASSANI v MERRIGAN

Estates Gazette January 23 1988

[1988] 03 EG 88-95

Landlord and tenant — Rent Act 1977 and section 11 of Landlord and Tenant Act 1985 — Possession — Appeal by
B **statutory tenant from decision of county court judge ordering possession, payment of rent and other charges and mesne profits, and dismissing tenant's counterclaim under section 11 of 1985 Act — Rule in *O'Brien* v *Robinson* as to requirement of notice to landlord of want of repair under implied covenants — At the trial the proceedings took an unusual course — Although the defendant tenant's counterclaim alleged breaches of the implied covenants under section 11, particulars were not provided before trial and the only evidence given at the trial related to a malfunction of the hot water and central heating system, which had in fact been cured**
C **after a short delay — The judge refused an application on behalf of the defendant to call the author of a surveyor's report, mentioned in the defendant's pleadings, which had been made available to the plaintiff shortly before the trial — As particulars of damage had not been supplied the judge decided to try the case on the footing that no special damage was alleged — In upholding the decision of the county court judge, the Court of Appeal held as follows — (1) A letter sent by the tenant's solicitors did not constitute notice of defects within section 11 as it promised to submit estimates and state what was necessary in due course, so that the landlord could either do the work or have the cost deducted from the rent —**
D **(2) The judge was correct in refusing to admit the evidence of the surveyor's report or to have its author called in support of the tenant's counterclaim; if the counterclaim amounted to notice by its incorporation of the report, there was in fact no breach at that time — (3) The main question in the appeal was whether the judge erred in making a possession order; he could do so only if he were satisfied that in all the circumstances such an order was reasonable — Although a number of criticisms were made, it was apparent to the court that the judge had given consideration to all the relevant factors going to reasonableness — Appeal dismissed**

E The following cases are referred to in this report.

Griffin v *Pillet* [1926] 1 KB 17
O'Brien v *Robinson* [1973] AC 912; [1973] 2 WLR 393; [1973] 1 All ER 583; (1973 25 P&CR 239; [1973] EGD 296; (1973) 226 EG 297, HL

This was an appeal by the tenant, Kevin Merrigan, from a decision of Judge Quentin Edwards QC, at Bloomsbury County Court, in favour of the landlord, Asaad Mohammed Al Hassani, in the latter's proceedings for possession and other relief in respect of Flat 28, Leith Mansions, Grantully Road, London W9.

Paul Staddon (instructed by Oliver O Fisher & Co) appeared on behalf of the appellant; Wayne Clark (instructed by Canter &
F Martin) represented the respondent.

Giving judgment, PARKER LJ said: The defendant appeals from the order and judgment of His Honour Judge Quentin Edwards QC made in the Bloomsbury County Court on May 11 1987 whereby it was ordered (i) that the defendant deliver up possession of Flat 28, Leith Mansions, Grantully Road, London W9, within two months; (ii) that the defendant do pay the plaintiff the sum of £8,006.07 on the claim; (iii) that the defendant's counterclaim be dismissed; (iv) that the defendant pay the plaintiff's costs of claim and counterclaim on scale 3; and (v) that the defendant pay to the plaintiff by way of mesne profits £78.38 per week from May 9 1987 to delivery of possession.

On July 8 1987 a stay of execution pending appeal was ordered on certain terms with which the defendant has complied.

The present position is that the defendant is still in possession of G
the flat, which he occupies together with his wife and child, a boy aged 6, who unfortunately suffers from a form of cerebral palsy known as spastic diplegia. He has paid into court the full amount of the judgment and has also paid the mesne profits ordered in accordance with the judgment.

It is necessary to set out the facts in some detail. At all material times, the plaintiff acted through an agent, a Mr Lake. On November 20 1983, Mr Lake, on behalf of the plaintiff and the defendant, entered into an agreement which, although described as a licence, operated as a tenancy agreement and created a protected tenancy. This is common ground. The weekly rental was £60. The circumstances in which this agreement came to be made are of H
importance. The flat had been let in July 1983 at a rental of £150 per week, but it had become vacant and the defendant's mother, who lived nearby, informed him that it was available. He then asked Mr Lake if he could have it for a short period while moving from one flat to another. Mr Lake told him that the flat was in a very bad state so he was asking only £60. The defendant went in knowing it was out of repair and that repairs needed doing.

The agreement was expressed to be operative from November 21 1983 to December 19 1983, but it was thereafter extended by agreement and the rent was increased to £78.38.

The rent was duly paid until May 1985. It then ceased to be paid in the following circumstances. On April 15 Mr Lake gave what was described as formal notice to quit the premises on May 13. It is, J
however, common ground that it was ineffective as a notice. On May 13 Mr Lake visited the premises in order to inspect them but was refused entry. A cheque for the next rent was offered, but this was later refused. Thereafter, apart from two payments of £1,000 and £500, no rent was paid nor did the defendant pay for gas and electricity consumed.

On August 8 1985 the plaintiff by his solicitors gave a valid notice to quit on September 9 1985, and thereafter the defendant became the statutory tenant of the premises.

Between September 9 1985 and the commencement of proceedings for possession on January 28 1987, it is, for present purposes, only necessary to mention one event. On March 7 1986 the defendant's solicitors wrote to the plaintiff's solicitors a letter which included the K
following paragraph:

> There are considerable repairs which require immediate attention. Your client's agent has been notified of these matters on countless occasions. The electricity needs immediate attention, the bath has been unusable for a considerable period of time, the pipes off the back of the washing machine have become clogged up. Our clients will obtain estimates from reputable firms and then submit them to your client. Your client will have the opportunity of either arranging for the firms to carry out the work himself or our clients will carry out the work and then deduct it from any rent due.

The plaintiff's claim as pleaded was a simple one for possession on the ground of non-payment of rent and for recovery of arrears together with sums in respect of gas, electricity and telephone use. L
There has been no dispute with respect to the arrears, gas or electricity claims. The claim in respect of the telephone was misconceived and rejected by the judge.

On March 6 1987, the defendant delivered his defence and counterclaim. Much of the defence was misconceived, but it is necessary to set out certain parts thereof as follows:

> 5. (iii) The defendant will aver that he was entitled to withhold rental payments by way of set-off against the said BJ Lake (or if he be found to be the defendant's landlord, the plaintiff) by reason of the matters which are the subject to the counterclaim herein below . . .
>
> 7. It is denied (in so far as it is averred) that the plaintiff is entitled to possession of the premises by reason of the matters alleged or otherwise. The defendant will contend that it is not reasonable for an order for possession to be made against him in the circumstances (and in particular without prejudice M
> to the generality of the foregoing) the matters counterclaimed.

The counterclaim was based on alleged breach of the covenants implied in the contract by section 11(1) of the Landlord and Tenant Act 1985 (formerly section 32 of the Housing Act 1961). It is common ground that the section was applicable. The covenants thereby implied are:

> (a) to keep in repair the structure and exterior of the dwelling-house (including drains, gutters and external pipes), (b) to keep in repair and proper working order the installations in the dwelling-house for the supply of water, gas and electricity and for sanitation (including basins, sinks, baths and sanitary conveniences, but not other fixtures, fittings and appliances for making use of the supply of water, gas or electricity, and (c) to keep in repair and proper

A working order the installations in the dwelling-house for space heating and heating water.

The pleading then alleged particulars of breaches:

> PARTICULARS The plaintiff relies upon the report prepared by the Scrase Hewlitt partnership (surveyors and valuers) dated April 4 1986 which reveals defects of repair of the structure and exterior of the premises (including but without prejudice to the generality of the foregoing, dry rot and defective window sashes), and inadequate and defective installations for the supply of water heating and hot water have not worked for a very long time and sanitation (the WC is in a very poor condition).

The report referred to was no doubt obtained by the defendant's solicitors following the letter of March 7 1986, but it had not been sent to the plaintiff.

B The defendant averred that by the breaches pleaded, he had suffered damage:

> PARTICULARS OF DAMAGE (i) Loss of amenity at the rate of £... per week (ii) Reduction in the value to the plaintiff (and his family) of the premises caused by the landlord's failure to perform the covenant (iii) General damages for vexation, anxiety and distress (iv) Miscellaneous costs. Full particulars will be provided by way of a Schedule before the trial of this action.

Full particulars were not in fact provided before trial of the action.

The plaintiff's reply dated March 10 consisted, so far as now material, in a denial of every allegation in the defence and counterclaim and a plea that if the flat was in a state of disrepair it was due to the acts or defaults of the defendant, his servants or agents and C or the defendant's refusal to permit Mr Lake to have access for inspection and the carrying out of any work for which the defendant might be liable.

Prior to delivery of his full defence, the defendant's solicitors had on February 26 sent a short form of defence both to the court and to the plaintiff's solicitors and had informed both that expert evidence would be called on behalf of the defendant and that, in consequence, the case might last at least half a day. They therefore asked that the hearing date be vacated.

After some dispute, this was done by agreement and on March 13 the defendant's solicitors sent the plaintiff's solicitors a copy of the report referred to in the pleadings. Facilities for inspection had by then been requested by the plaintiff's solicitors, but it was not until D March 26, by which time the hearing had been fixed for May 11, that any facilities were offered. On that day, the defendant's solicitors offered inspection at 11 am on April 8. Inspection finally took place on April 14, and on April 29 the plaintiff's surveyors sent to his solicitors their comments on the report of April 4 and on the defence and counterclaim.

When, on May 11, the action came on for hearing, proceedings took a somewhat unusual course. Both sides were represented by counsel. In opening the case for the plaintiff, counsel observed that the promised particulars of damage had not been given, and the judge indicated that he would try the case on the basis that no special damage was alleged.

E Mr Lake then gave evidence, in the course of which the judge pointed out that the only notice of defects pleaded by the defence was the report of April 4 1986 which had not been supplied until March of 1987. This was no doubt done in fairness to the defendant because it is established law that the obligation to repair under the implied covenants imposed by section 11 does not arise until the landlord has notice of the want of repair. See *O'Brien* v *Robinson* [1973] AC 912. The defendant thereupon applied for and was given leave to amend to allege: (i) Oral notice of hot water not working given in November 1984, but not rectified until April 1985. (ii) Notice by the solicitor's letter of March 7 1986 already quoted and (iii) Service of the report of April 4 1986 in March 1987.

The amendment having been made, Mr Lake gave evidence that he F had been told of the hot water not working in March of 1985, that he had visited on April 15 1985, that he had found that the ball valve on a header tank was stuck, and that he had cured the defect by wiggling it up and down. According to the judge's notice of evidence, he said nothing else in chief about any alleged lack of repairs and none of the matters mentioned in either the solicitor's letter of March 7 1986 or the report of April 1986 were put to him in cross-examination.

No further evidence was called for the plaintiff, which, no matters of disrepair having been put to Mr Lake other than the hot water, is not surprising.

The defendant's counsel then called first the defendant's wife and then the defendant. Both spoke generally of the bad state of the premises but neither gave any evidence, according to the judge's notes, of any want of repair which might fall within section 11 with G the exception of the hot water and central heating, which had been dealt with, albeit after a small delay.

In the course of the evidence, Mr Merrigan offered to pay £3,000 immediately and the balance within a month. Both stated that they had not paid the rent on the advice of solicitors.

The defendant and his wife having given evidence, the defendant's counsel sought to call the author of the report of April 4 1986, but was met with an objection by the plaintiff's counsel which was upheld by the learned judge. It is the judge's decision not to admit this evidence which is the core of the defendant's appeal to this court.

The objection was put on three grounds: (i) The evidence sought to be given was expert evidence which could be admitted only by leave of H the court, since there had been no direction. See RSC, Ord 38 r 36. (ii) Leave should not be given, since no evidence in chief had been given and no want of repair had been put to Mr Lake in cross-examination. (iii) Additionally, no notice of want of repairs specified in the notice of repair had been given until after service of the counterclaim.

The defendant's counsel answered this with the contention that albeit leave was required, it ought to be given because: 1. The defendant's advisers had been led to believe that no objection would be taken. 2. The counterclaim itself constituted notice. 3. (After a question by the judge as to what issue on the pleadings the proposed evidence would go), the evidence went to damages for loss of amenity from the date of the counterclaim, ie from March 6 1987 to May 11 J 1987.

In his notes, the judge records that he declined to hear the evidence on the grounds (1) that it could not go to any issue he had to determine in the light of the fact that no notice was given until after the counterclaim and (2) "in any event the sequence of events".

In his judgment, the judge, in dealing with the counterclaim, said:

> The matter comes before this court with a substantial counterclaim. By amendment oral notice of defects to the hot water system between November 1984 and April 1985 and the letter of March 7 1986 have been pleaded. The pleading also relies upon a report served after the counterclaim had been made. The surveyor's report cannot be noticed as it was served after the counterclaim, and even if I am wrong, it was not served until March 13 1987. The landlord's surveyors went to visit in mid-April and reported on April 24 K 1987. I do not regard the letter of March 7 1986 as notice. By its very terms, it is not.
>
> The only point left on the counterclaim is the question of hot water between November 1984 and April 1985. I find it very difficult to see how the Merrigans could have gone all through the winter without hot water and not written a letter about it. It may be that some sort of oral complaint was made, but all that needed doing was for the ball valve in the header tank to be pushed down. I really do think that a man such as Mr Merrigan who works with cars could do a minor work of that kind and was under an obligation at common law to do it, and that he could and should have done it. I don't think that a tenant can sit back and do nothing. If he had looked around, he could have discovered the cause. I do not think that the counterclaim sounds in damages. Accordingly, I give judgment for the plaintiff on the counterclaim.

L The first ground of appeal is that the judge erred in holding that the letter of March 7 1986 did not amount to notice. I can deal with this contention very shortly. I accept that a notice of want of repair need not specify the precise nature or degree of want of repair. See *Griffin* v *Pillet* [1926] 1 KB 17. If, however, one reads the relevant paragraph of the letter the only matters which could possibly come within the section 11 covenants are the electrical installation needing attention and the bath being unusable.

As to these matters, if they were sufficient to put the plaintiff on inquiry, and I think they were, the plaintiff's obligation would normally have been to attend and inspect and thereafter carry out such repairs as were necessary. The letter, however, expressly stated that the defendant would be ascertaining what was necessary and M submit estimates. This being so, I am unable to see, as was the judge, that the letter was a notice sufficient to bring into being the plaintiff's obligations under section 11. He was being specifically told that he would be told what was necessary in due course and that he could then either do it or have the cost deducted from rent.

The next ground of appeal is that the evidence of the author of the report ought to have been admitted because service of the report amounted to notice of the matters complained of therein and "that the defendant is accordingly entitled to damages for breach of covenant for a period after service of the report (a short period of time to allow for inspection and repair of the premises) until the date of the hearing".

A This ground is, in my view, without substance. The report was not served until March 13 and, despite the provisions of section 11(6) of the Act, the plaintiff was not permitted to inspect until April 14. His surveyors reported within 10 days, which it is not suggested involved any unreasonable delay. At the most, therefore, there could only have been a breach if, on getting his surveyor's report, the plaintiff should have carried out the repairs at some time prior to the trial on May 11. This was not, and could not in my view have been, seriously suggested. Moreover, so far as the counterclaim is concerned, I accept the contention that the matter crystallised at the date of service. If the counterclaim amounted to notice by its incorporation of the report, there was therefore no breach at that time. I accordingly conclude that the judge was right in refusing to admit the B evidence in support of the counterclaim.

There remains what in my view is the only real question on the appeal, namely whether the judge erred in making a possession order. He could do so only if he were satisfied that, in all the circumstances, such an order was reasonable. It is submitted on behalf of the appellant that even if the evidence were rightly excluded as to the counterclaim, evidence as to the current condition of the premises would have been relevant to the issue of reasonableness and should therefore have been admitted on that ground. It is also submitted that the judge (1) had failed to take into account or pay sufficient weight to: (i) the offer of payment made in evidence (ii) the medical condition of the defendant's son (iii) the fact that rent had C been withheld on legal advice (iv) a valid counterclaim; (2) had considered only whether it was reasonable "to expect the landlord to put up with the tenant" instead of "all the circumstances".

As to (1)(iv) above, it follows from what I have already said that this point must fail. As to (1)(ii), the learned judge specifically mentioned the child's condition and took it into account. I cannot say that he did so insufficiently. He did not, it is true, mention the offer of payment made at the trial, but he was not bound to do so. He was entitled to, and no doubt did, regard it as being without significance in view of its lateness.

There is, however, some force in the submissions that: (1) he did not pay any regard to the fact that rent had been withheld on legal advice; (2) he appeared to regard the test as being whether it was D reasonable to require the landlord to put up with a tenant such as the defendant, and (3) evidence as to the condition must have been relevant on the issue of reasonableness and was part of all the circumstances to be taken into consideration.

I deal with these points in turn. As to legal advice, it appears from the evidence that the defendant's solicitors took the view in March 1986 that if the plaintiff did not carry out repairs the defendant could do them and deduct the cost from the rent. There is, however, no evidence that they were advised not to pay gas and electricity bills or that they were advised that there was any basis for withholding rent amounting to many times more than could conceivably be justified on the ground of any non-repair of which they were entitled to E complain. In these circumstances, the failure to mention legal advice or regard it as of significance was, in my view, justified.

As to the contention that the judge applied the wrong test, I reject it because it is apparent from the judgment that the judge considered all the factors going to reasonableness. He plainly did not regard the plaintiff's position only.

There is finally the matter of the excluded evidence. In the light of what the judge described as the sequence of events, I consider the exclusion was justified. The premises were taken at a low rent specifically because they were in a bad state of repair. Although initially taken for two months, the defendant lived in them until 1985 without, on the evidence, any material complaint until the hot water incident in 1984-85. Thereafter, there was no complaint until the F solicitor's letter of March 1986, and that was not pursued in any way. Had the evidence been admitted, it could not in my view have affected the matter in the slightest.

Like the learned judge, I have great sympathy with the small boy, but I do not consider that the judge erred in excluding the evidence or in any way in which he dealt with the question of possession.

I should perhaps add that I have not overlooked arguments which were addressed to us that it was not necessary to plead notice or knowledge or that knowledge of want of repair, albeit no notice had been given by the tenant, was sufficient to give rise to the obligation to repair. I am prepared to accept, without deciding, that both contentions are correct, but neither avails the defendant. Apart from the hot water incident, there was no evidence of knowledge of want of repair falling within section 11 and the absence of necessity to plead G notice or knowledge was in the circumstances of the case irrelevant.

I would dismiss this appeal, but we should, I consider, hear argument on the question of the time which should be allowed to the defendant before the possession order can be enforced.

SIR GEORGE WALLER agreed and did not add anything.

The appeal was dismissed and possession ordered on agreed terms as to enforcement. Directions were given as to payment of mesne profits. The award of costs below was allowed to stand and an order nisi against the Law Society was made in respect of the costs of appeal.

H

Court of Appeal

November 19 1987

(Before Lord Justice PURCHAS, Lord Justice NOURSE and Lord Justice LLOYD)

APPLETON v ASPIN AND OTHERS

Estates Gazette January 30 1988

[1988] 04 EG 123-127

Rent Act 1977, section 98(1) — Whether a court is prohibited J from making a possession order against a protected tenant who has agreed with the purchaser of the subject dwelling-house to join with the landlord, the vendor, in giving vacant possession on completion — Appeal from decision of Millett J, who held that a court was not so prohibited — The agreement by the protected tenant not to enforce any right of possession or occupation against the purchaser was contained in a special condition in the contract of sale, which the tenant signed (in addition to the other two parties) in order to bind herself to the purchaser — The tenant did not receive any monetary consideration — The tenant did not give possession K to the purchaser after completion and the latter brought proceedings for specific performance — The tenant claimed that, as she was a protected tenant, the special condition was rendered inoperative by section 98(1) of the 1977 Act — Held by the Court of Appeal, reversing the decision of Millett J, that the case was covered by the general principle of *Barton* v *Fincham* that the agreement of a protected tenant to give up possession cannot be enforced against her if she is entitled under the Act to remain in possession and chooses to do so — Where the legislature has restricted the jurisdiction of the court, as section 98(1) does, the parties cannot by agreement L give the court a jurisdiction which the statute says that it is not to have — The court distinguished the decision in *Dudley and District Benefit Society* v *Emerson*, upon which Millett J had relied, and rejected an alternative argument based on the suggestion that the tenant could be treated as having assigned her tenancy without monetary consideration to the purchaser, a transaction recognised by section 120 of the 1977 Act — Appeal by tenant allowed

The following cases are referred to in this report.

Barton v *Fincham* [1921] 2 KB 291 M

Dudley & District Benefit Building Society v *Emerson* [1949] Ch 707; [1949] 2 All ER 252; (1949) 65 TLR 444, CA

Jenkin R Lewis & Son Ltd v *Kerman* [1971] Ch 477; [1970] 3 WLR 673; [1970] 3 All ER 414; (1970) 21 P&CR 941, CA

R v *Bloomsbury and Marylebone County Court, ex parte Blackburne* [1985] 2 EGLR 157; (1985) 275 EG 1273, CA

Stirrup's Contract, Re [1961] 1 WLR 449; [1961] 1 All ER 805

This was an appeal by Mrs Ivy Charlotte Plane, the protected tenant of a freehold house at 7 Finch Way, Brundall, Norwich, from a decision of Millett J granting possession of the house to the purchaser, John Appleton, the plaintiff in an action for specific performance, in which the defendants were the vendor, Mrs Norma

Ivy Maud Aspin, her husband, David Patrick George Aspin, and the tenant, Mrs Plane.

J M Stevenson (instructed by Rogers & Norton, of Norwich) appeared on behalf of the appellant; David Iwi (instructed by Bazeley White & Co) represented the respondent (plaintiff).

Giving the first judgment at the invitation of Purchas LJ, NOURSE LJ said: This is an appeal against a decision of Millett J given on April 15 1987 in a purchaser's action for specific performance. The question in issue is whether section 98(1) of the Rent Act 1977 prohibits the court from making an order for possession against a protected tenant who has agreed with the purchaser to join with his landlord, the vendor, in giving vacant possession on completion. Millett J answered that question in the negative, and the tenant now appeals to this court.

The numerous issues before the learned judge, including those raised in third party proceedings by the tenant against her former solicitor, made it necessary for him to state the facts at some length. For the purpose of determining the one question which is still in dispute, a short although possibly important question of law, they can be stated comparatively briefly.

On September 23 1982 Mrs Ivy Charlotte Plane, then a widow of about 65 years, conveyed to her daughter, Mrs Norma Ivy Maude Aspin, a freehold house known as 7 Finch Way, Brundall, Norwich, Norfolk. On the same day the two ladies entered into what has been called an occupation agreement, by clause 5 of which Mrs Aspin undertook that Mrs Plane might for the rest of her life reside in the house subject to payment of a yearly rent of £725, if demanded.

On June 11 1984 Mrs Aspin entered into an agreement for the sale of the house to the plaintiff, Mr John Appleton, the contractual date for completion being September 1 1984. By special condition B it was provided that the sale was with vacant possession on completion. Special condition M, which was addressed to the plaintiff, was in these terms:

> In consideration of your entering into a contract with my Daughter Norma Ivy Maude Aspin for the purchase of the property hereby contracted to be sold with vacant possession I Ivy Charlotte Plane undertake not to seek to enforce against you any right of possession or occupation whether arising by statute or by reason of any equitable interest or otherwise of which I may be or become entitled.

The contract was signed by Mrs Plane as well as by Mrs Aspin, but it is clear that the former signed only in order to bind herself to the plaintiff in the terms of special condition M.

The contractual date for completion passed without possession being given to the plaintiff, and on October 26 1984 he commenced an action for specific performance against Mrs Aspin and Mrs Plane, and also against Mrs Aspin's husband, Mr David Patrick George Aspin. The action related also to another neighbouring property which was vested in Mr and Mrs Aspin, and a decree of specific performance was made in respect of that property before the commencement of the trial in regard to 7 Finch Way.

At the trial the only substantive claim made in the action was for an order for vacant possession against Mrs Plane. To that claim the effective defence was (1) that the occupation agreement of September 23 1983 created a tenancy protected by the Rent Acts; and (2) that the agreement constituted by special condition M of the agreement for sale of June 11 1984 was rendered inoperative by section 98(1) of the Rent Act 1977.

As to that defence, Millett J held, first, that the occupation agreement was not a sham and accordingly that it created a protected tenancy. That holding has not been questioned on this appeal. Second, he held, as a matter of construction, that special condition M, although negative in form, contained an implicit, positive undertaking by Mrs Plane to join with Mrs Aspin in giving vacant possession to the plaintiff on completion. In the court below, Mr Stevenson, for Mrs Plane, advanced a contrary argument of what the judge described as subtlety and ingenuity, but we have heard almost nothing of it in this court and the judge was clearly right to reject it. Third, and for present purposes most importantly, Millett J held that the agreement as he had construed it was not rendered inoperative by section 98(1) of the 1977 Act. It is to that question I now turn.

I start by analysing the effect, so far as material, of the agreement for sale dated June 11 1984. It is accepted by Mr Stevenson, in my view correctly, that the particular agreement constituted by special condition M was an agreement between Mrs Plane and the plaintiff alone. It was not, and cannot be treated as having been, an agreement to which Mrs Aspin was also a party. On the other hand, it is impossible to ignore the fact that Mrs Aspin was a party to the agreement for sale, and that its provisions as a whole became operative between all three parties at one and the same time. Thus it is not in my view correct to say, as did Millett J, that at the date of the agreement with Mrs Plane the plaintiff was only an intending purchaser of Mrs Aspin's interest. He was the actual purchaser. At the moment of time when Mrs Plane agreed with the plaintiff to join with Mrs Aspin in giving vacant possession to him on completion the plaintiff became contractually bound and entitled to acquire the freehold from Mrs Aspin. The equitable interest passed to him. In the eye of equity he became her landlord. He knew of course that she was in occupation.

Section 98(1) is in these terms:

> Subject to this Part of this Act, a court shall not make an order for possession of a dwelling-house which is for the time being let on a protected tenancy or subject to a statutory tenancy unless the court considers it reasonable to make such an order and either —
>
> (a) the court is satisfied that suitable alternative accommodation is available for the tenant or will be available for him when the order in question takes effect, or
>
> (b) the circumstances are as specified in any of the Cases in Part I of Schedule 15 to this Act.

It is agreed that this case does not fall within either of paras (a) and (b). And so the question is whether it falls within the general prohibition or not. On the wording of the subsection alone it appears clear that it does. No 7 Finch Road is a dwelling-house which either is let on a protected tenancy or is subject to a statutory tenancy, it matters not which. It would therefore appear that the court is prohibited from making an order for possession in relation to it. However, the question is not free from authority which shows that this simple approach is not necessarily correct.

In *Barton* v *Fincham* [1921] 2 KB 291, which was decided on section 5(1) of the Increase of Rent and Mortgage Interest (Restrictions) Act 1920, it was held by this court that if the conditions upon which alone an order for possession could be made were not fulfilled, an order could not be made against the opposition of the tenant, even if he had freely and for good consideration agreed with the landlord that he would give up possession. In that case the agreement had been made so that the landlord could sell with vacant possession, although there was no evidence that he had actually entered into a contract for sale. It was emphasised that the statutory fetter had been placed not upon the landlord's action but upon the court's: see p 295, *per* Bankes LJ. At p 299 Atkin LJ said:

> The section appears to me to limit definitely the jurisdiction of the Courts in making ejectment orders in the case of premises to which the Act applies. Parties cannot by agreement give the Courts jurisdiction which the Legislature has enacted they are not to have.

That decision, the authority of which has recently been recognised by this court in *R* v *Bloomsbury and Marylebone County Court, ex parte Blackburne* (1985) 275 EG 1273*, undoubtedly established a general rule that the court is prohibited from giving effect to an executory agreement between a protected tenant and his landlord for the former to give up possession to or at the direction of the latter. And it would have applied to this case if the agreement had been between Mrs Plane and Mrs Aspin. But in *Dudley and District Benefit Building Society* v *Emerson* [1949] Ch 707, another decision of this court, it was shown that the prohibition is not as far-reaching as the wording of section 98(1) would by itself suggest.

The material facts of the *Dudley* case, which I take from pp 708-709 of the report, were these. In 1946 the first defendant, Emerson, agreed to purchase a dwelling-house which was subject to the provisions of the Rent Acts. Immediately on completion of the purchase he charged the property by way of legal mortgage in favour of the Dudley and District Benefit Building Society. That was a mortgage under which principal and interest were repayable by instalments. It contained a provision that the mortgagor's statutory power of leasing under section 99(1) of the Law of Property Act 1925 should not apply. It was alleged that the mortgage had been granted on the understanding that Emerson would personally occupy the premises, but there was no express provision to that effect. In fact Emerson never went into possession of the property and, unknown to the building society, he immediately granted a weekly protected tenancy to the second defendant, Goodlad, who went into possession and had remained in possession thereafter. Emerson got into arrear with the instalments payable under the mortgage. Its statutory power

* Editor's note: Also reported at [1985] 2 EGLR 157.

A of sale having thus arisen, the building society took the familiar step of seeking an order for possession so that it could sell the property out of court. Emerson did not enter an appearance, but Goodlad defended the proceedings, claiming to be a tenant entitled to the protection of the Rent Acts.

The leading judgment was given by Sir Raymond Evershed MR. Having considered the effect of the exclusion of section 99(1), he said that he was content to assume that Goodlad, the tenant, had a precarious title, good as between himself and the mortgagor Emerson, but liable to be defeated on the assertion by the building society of its paramount title: see p 714. He observed that apart from the Rent Acts there would have been no real defence to the claim for
B possession. Then he turned to the material provisions of the Rent and Mortgage Interest Restrictions (Amendment) Act 1933, in particular to section 3(1), which was for present purposes to the same effect as section 98(1) of the 1977 Act, Schedule I to the 1933 Act being the equivalent of Part I of Schedule 15 to the 1977 Act. At the bottom of p 715 the Master of the Rolls, having observed that section 3(1) was quite general in its terms, said:

> . . . But when you examine the terms of Sch I, I think it is quite plain that what the section and the schedule, which is read into it, must be taken to contemplate is a proceeding for possession against the tenant by somebody who may be fairly described — and I deliberately use that phrase for the moment — as a landlord. In other words, it appears to me that the general conception (which I of course entirely accept) that the Acts are designed to
> C protect occupants of dwelling-houses within the Acts against eviction must be qualified to this extent — against eviction at the suit of persons who may fairly and properly be described as landlords of the occupants.

In the next paragraph he buttressed those views by further reference to Schedule I to the 1933 Act and to the definition of "landlord" in section 12 of the 1920 Act, and continued:

> Pausing for a moment at that stage, and turning to the facts of this case, it will at once be seen that the description cannot be fairly, as I think, applied to the mortgagees in their claim against Goodlad. They are not his landlords; they never have been. The evidence makes it, I think, tolerably clear that they have never accepted his tenancy as one which binds them, and it is quite clear that there is no contractual relationship between Goodlad and the plaintiffs, either imported by the statute or otherwise.

D Essentially for the reasons stated in those two passages, the learned Master of the Rolls held that the general prohibition in section 3(1) of the 1933 Act did not apply to proceedings by the mortgagee and that an order for possession should therefore be made against Goodlad. Somervell LJ delivered a reasoned judgment agreeing with the Master of the Rolls. His views can be taken as having been summarised in a sentence at p 721:

> It therefore seems to me that the Act is primarily one which enables tenants to assert certain rights, and to remain in possession, as against landlords who are seeking to obtain possession, and who would, but for the provisions of these Acts, be entitled to possession.

Jenkins LJ also agreed, but without adding any further reasons of his
E own.

In the present case Millett J, having read a passage from the judgment of Sir Raymond Evershed which included the second passage above quoted, said, at p 18E of the transcript:

> In my judgment, that applies equally to the present case. The plaintiff was not Mrs Plane's landlord at the date when she signed the contract, and he never intended to become her landlord. The only contractual relationship between them was that contained in clause M of the contract, and its whole purpose was to prevent the relationship of landlord and tenant coming into existence between them.

He then dealt with Mr Stevenson's arguments for distinguishing the *Dudley* case and, having rejected them, made an order for possession against Mrs Plane. In the main Mr Iwi, for the plaintiff, has adopted
F the judge's reasoning on the appeal to this court.

In deciding whether the decision in the *Dudley* case applies to the present case, I start by observing that the facts there were very striking. A mortgagee had advanced money on the security of property which, if it was not actually to be occupied by the mortgagor personally, was subject to an agreement that he should not have the mortgagor's statutory power of granting leases. Without the mortgagee's knowlege or approval the mortgagor created a protected tenancy. The mortgagor then defaulted on the mortgage repayments and the statutory power of sale arose. He did not defend the mortgagee's action for possession, but the tenant did.

On those facts it would have been alarming, perhaps worse, if the mortgagee had been unable to obtain possession against the tenant in order to realise its security. This court clearly thought, in my G
respectful view correctly, that that cannot have been the intention of Parliament in framing the predecessors to section 98(1). But, in order so to conclude, they had to follow a route which may have led them into expressing views which were not only wider than was necessary for the decision of the case before them, but which should not be taken to be of literal application to other cases depending on other facts.

The general question which arises is whether, before the prohibition can apply, the person seeking possession must be someone who, in the words of Sir Raymond Evershed, can fairly and properly be described as the landlord of the occupant. For a reason which will soon appear, I do not find it necessary to express a view on that general question in this case. I would merely say that I can think H
of circumstances, some of which have been canvassed in argument, where it might seem that the spirit of the decision in the *Dudley* case ought to prevail, even though the person seeking possession does not satisfy the description there propounded. However that may be, I think that here the plaintiff does satisfy the description and, moreover, that we are bound to apply it in this case.

In the *Dudley* case importance was attached to the other provisions of the legislation, including the definition of "landlord". By section 152(1) of the 1977 Act the expression "landlord" is defined to include "any person from time to time deriving title under the original landlord" and also, in relation to any dwelling-house, "any person other than the tenant who is, or but for Part VII of this Act would be, J
entitled to possession of the dwelling-house". Those are two alternatives to which full effect ought, so far as possible, to be given.

I would agree that at the time of the signing of the agreement of June 11 1984 the plaintiff in the present case could not properly be described as someone who, but for section 98(1), would have been entitled to possession of 7 Finch Road. Contractually, he was not entitled to possession until completion of the purchase. On the other hand, it seems to me that different considerations apply to the question whether the plaintiff was a person deriving title under the original landlord.

Millett J dealt with Mr Stevenson's submission to the effect that the plaintiff was such a person at p 19H of the transcript:

> . . . At the date of the agreement with Mrs Plane the plaintiff was only an intending purchaser of the landlord's interest; he was not prepared to enter into a contract of purchase without first having Mrs Plane's agreement to clause M. He never acquired any interest, even equitable, in the property without Mrs Plane's agreement to vacate it; and although on completion he will derive title under Mrs Aspin, that is not the capacity in which he claims against Mrs Plane. K

As I have indicated, that is an analysis with which in two vital respects I respectfully disagree. First, the plaintiff was not an intending purchaser. He was an actual purchaser. Second, it is not, in my view, correct to say that he would only become a person deriving title under Mrs Aspin on completion. That is certainly true so far as the legal estate is concerned. But the equitable interest passed to the L
plaintiff under the contract at one and the same time as Mrs Plane agreed to join in giving him vacant possession. He had derived the equitable title from Mrs Aspin. I therefore conclude that at the date of the agreement the plaintiff was a person deriving title under the original landlord. Whether on that ground alone or on a general view of the relationship between the parties, I think that he was a person who may fairly and properly be described as Mrs Plane's landlord within the views expressed in the *Dudley* case. In the end the correct view may well be that the fact that Mrs Plane's agreement to give up possession was made with the plaintiff and not with Mrs Aspin was a mere technicality which ought not to take it outside the general rule established by *Barton* v *Fincham*.

It remains for me to mention an alternative argument which has M
been advanced to us by Mr Iwi on behalf of the plaintiff. That argument was not, I think, put quite in the same way to Millett J as it has now been put to us. If it was, I would infer that the judge did not think much of it. Mr Iwi relies on section 120 of the 1977 Act, which recognises that an assignment of a protected tenancy is not in itself objectionable. It is objectionable only if there is a condition for the payment of a premium or the making of a loan. Mr Iwi accordingly submits that Mrs Plane could, had she wished, have assigned her tenancy, without monetary consideration, to the plaintiff before entering into the contract for sale or, at any rate, before completion of the purchase. He relies on the fact that Mrs Plane did not receive any monetary consideration from the plaintiff. In those

A circumstances, submits Mr Iwi, the transaction could not have been impugned.

Then, in reliance on certain observations made by Russell LJ when delivering the judgment of this court in *Jenkin R Lewis & Son Ltd* v *Kerman* [1971] Ch 477, at p 496, and by Wilberforce J in *Re Stirrup's Contract* [1961] 1 WLR 449, at pp 452-453, Mr Iwi submits that the court can properly treat the transaction as having been an agreement to assign the tenancy and not to surrender or extinguish it. I think that the observations relied on by Mr Iwi are of no assistance to him at all. Those of Russell LJ were directed only to a surrender by operation of law. The decision of Wilberforce J was to the effect that where a document under seal, which is expressed to be an assent, discloses a manifest intention to pass a legal estate in land, the estate will pass
B even though the document ought properly to have been expressed as a conveyance. That principle, which is no doubt very beneficial, has nothing at all to do with a case where a party has agreed to surrender or extinguish her tenancy and cannot therefore be treated as if she had entered into a completely different agreement, namely to assign her tenancy, being a transaction by which the tenancy is not surrendered or extinguished but preserved. Accordingly, I have no hesitation in rejecting Mr Iwi's alternative argument.

I would, therefore, distinguish the actual decision in the *Dudley* case and hold that section 98(1) applies to the agreement constituted by special condition M. I recognise that this may well be thought to work a hardship on the plaintiff, but I regret that that is not
C something which can be taken into account in defining the extent of the application of a provision such as section 98(1).

For these reasons I would allow this appeal, dismiss the action as against Mrs Plane and grant her a declaration on her counterclaim in terms which can be discussed with counsel.

PURCHAS and LLOYD LJJ agreed and did not add anything.

The appeal was allowed with the costs of the appeal, no order being made as to costs below. Leave to appeal to the House of Lords was refused.

D
Court of Appeal

October 28 1987

(Before Lord Justice PARKER and Sir George WALLER)

CHIOS PROPERTY INVESTMENT CO LTD v LOPEZ

Estates Gazette February 6 1988

[1988] 05 EG 57

**Rent Act 1977, Schedule 1 — Statutory tenant by succession — Whether a person cohabiting with deceased tenant was a member of the deceased tenant's family — Whether test of
E permanence and stability of relationship was satisfied — The respondent, defendant in a possession action in the court below, had lived with the deceased tenant for some two years, initially with the intention of cohabiting but subsequently with the intention of marrying him when financial circumstances permitted — The judge found that both regarded their relationship as permanent and he held that the respondent was a member of the deceased's "family", residing with him at the time of, and for the period of six months immediately before, his death — The judge referred to *Dyson Holdings Ltd* v *Fox, Helby* v *Rafferty* and *Watson* v
F *Lucas* and formulated the test as being that the claimant has to show "that a sufficient state of permanence and stability has been reached in the relationship with the deceased protected or statutory tenant for it to be said that in all the circumstances the occupant was a member of the original tenant's family" — The judge held that this test had been satisfied and that the respondent was entitled to possession as a statutory tenant by succession — Held on appeal that there were no grounds for interfering with the judge's conclusion — He had heard the evidence and had applied the correct test — Criticisms by the landlords founded on the absence of children, the comparative shortness of the relationship and the fact that the
G appellant kept her maiden name could not invalidate the judge's finding based on the evidence and the authorities — The court, however, gave a warning — This was an exceptional case and should not be regarded as a precedent for drawing a similar inference from a short period of cohabitation — Landlords' appeal dismissed**

The following cases are referred to in this report.

Dyson Holdings Ltd v *Fox* [1976] QB 503; [1975] 3 WLR 744; [1975] 3 All ER 1030; (1975) 31 P&CR 229; [1976] EGD 309; 239 EG 39, CA

Helby v *Rafferty* [1979] 1 WLR 13; [1978] 3 All ER 1016; (1978) 37 P&CR 376; [1978] EGD 365; 247 EG 729, CA

H *Watson* v *Lucas* [1980] 1 WLR 1493; [1980] 3 All ER 647; (1980) 40 P&CR 531; [1980] EGD 443; (1980) 256 EG 1171, CA

This was an appeal by the landlord company, Chios Property Investment Co Ltd, from a decision by Mr Assistant Recorder Mann QC, at Bloomsbury County Court, in favour of the defendant, the present respondent, Miss A Arranz Lopez, in an action concerning the possession of Flat 5, 24 Pembridge Square, London W2.

Mark Piercy (instructed by Wegg-Prosser & Farmer) appeared on behalf of the appellants; S Dougall (instructed by Nathoo & Co) represented the respondent.

Giving the first judgment at the invitation of Parker LJ, SIR GEORGE WALLER said: This is an appeal from a decision of Mr
J Assistant Recorder Mann QC, sitting at Bloomsbury County Court. He held that the respondent, Arranz Lopez, is the statutory tenant of Flat 5, 24 Pembridge Square, London W2, in succession to Mr El-Din. The judge held that she was a member of Mr El-Din's family residing with him at the time of, and for a period of six months immediately before, his death.

The background facts were that Mr El-Din was an Egyptian. Miss Lopez, who is Spanish, came to England in 1980 or 1981 and met Mr El-Din in March 1981. They saw each other regularly for the next two years and on February 10 1983 she moved to Mr El-Din's flat with the intention of marrying him as soon as financial circumstances permitted. The judge found that she moved

K with the intention of cohabiting with Mr El-Din initially as his fiancée and thereafter as his wife when financial circumstances should permit, that by July 1985 or thereabouts, when it became apparent to Miss Lopez and Mr El-Din that they were likely to become better off imminently, they resolved to marry and for that purpose to secure the blessing both of his family in Cairo and her family in Spain and that they both regarded their relationship at all material times as being permanent albeit that marriage was to be conditional on improved financial circumstances.

The judge referred to the three recent authorities, namely *Dyson Holdings Ltd* v *[1976] QB 503, Helby* v *Rafferty* [1979] 1 WLR 13 and *Watson* v *Lucas* [1980] 1 WLR 1493, and then set out the test as follows:

L What the claimant has to show is that a sufficient state of permanence and stability has been reached in the relationship with the deceased protected or statutory tenant for it to be said that in all the circumstances the occupant was a member of the original tenant's family.

The judge went on:

Hence the question is one of fact and degree, and it is for consideration now whether the facts of this case warrant the inference that Miss Lopez's relationship with Mr El-Din was of sufficient permanence and stability and endured for the period of six months immediately before Mr El-Din's death.

The judge then summarised the evidence and said:

In my judgment, although the case is not an easy one, Miss Lopez was indeed a member of Mr El-Din's family residing with him at the time of and for the period of six months immediately before the death in the sense of para 3 of Schedule 1 to the Rent Act. In my judgment, Miss Lopez has proved that
M permanence and stability in the relationship which is necessary in order to satisfy the statutory provisions. I have no doubt that had either she or Mr El-Din been asked whether they regarded their relationship as permanent, albeit not yet sealed by marriage, they would have answered in the affirmative sense.

Before this court it has been submitted that in applying the test which I have already quoted, the judge erred and should have asked himself whether the common man would have regarded them as a family. This was an expression used in *Dyson*. In my judgment, the test which the judge posed went to the heart of the relationship and the evidence of the witnesses which the judge accepted, including the evidence of Miss Lopez's reception in Egypt by his family after his death, all supported this.

It was also argued that the absence of children and the shortness of

A the relationship should have resulted in a different conclusion. Counsel cited the very much longer periods in the reported cases referred to above. In my judgment, there can be no rule about length. Of course, the longer the relationship, the easier it will be to infer permanence, but there can be no rule about length. For a relationship of only two years to be regarded as permanent must be rare, but the judge for reasons which he gave found this was permanent and I see no reason for disagreeing. The appellants also submitted that there were periods of being apart, eg when the deceased was in Egypt and at holiday time, but I do not think that these made any difference.

Finally, it was submitted that the fact that she still used her maiden name indicated that they did not constitute a family. This I found the most persuasive of all the appellants' arguments, but at the end of the B day the judge had to make a finding of fact bearing in mind the authorities I have quoted. He heard the evidence, he applied the right test and he came to the conclusion that this was a permanent relationship, so recognised by others, and that Miss Lopez was "a member of the tenant's family". I can see no reason for interfering with this conclusion, and I would dismiss this appeal.

I would only add that this was a most exceptional case, and it should not be regarded as a precedent entitling courts to draw a similar inference from a similar short period of time unless there are wholly exceptional surrounding circumstances.

PARKER LJ agreed and did not add anything.

C *The appeal was dismissed with costs.*

Court of Appeal

November 2 1987

(Before Lord Justice LLOYD and Sir Roualeyn CUMMING-BRUCE)

SEFTON HOLDINGS LTD v CAIRNS

Estates Gazette April 9 1988

D **[1988] 14 EG 58-59**

Rent Act 1977 — Second succession to tenancy on death — Meaning of "member of the first successor's family" in Sched 1, para 7 — Claim by lady, now aged 70, who had lived with the family since 1941 when she was 23, rejected; and appeal by landlords against decision of county court judge in favour of claim allowed — Respondent claimant had come to live with the family, then consisting of a father, mother and daughter, in 1941 after both her own parents had died and her boy friend had been killed in the war — The daughter of the family became the first statutory successor to the tenancy on her E father's death in 1965 — When the daughter died in 1986 the landlords sought possession of the dwelling-house and the respondent claimed to be entitled to remain as second statutory successor — She had continued to live in the house for some 45 years, being treated by the parents as a daughter and by the first successor as a sister — The question was whether she was a member of the first successor's family within the meaning of the statutory provision as the county court judge had held — It was pointed out by the Court of Appeal that "family" had to be given its ordinary everyday meaning — There was a distinction between being a member of the family and being a member of the household — There F was also a distinction between being a member of the family and living as, or being treated as, a member of the family — The principles applicable had been stated in *Rose* v *Collins* and had been applied in *Joram Developments Ltd* v *Sharratt* — The suggestion that the respondent became a member of the family by adoption could not be sustained — Adoption must be *de jure* or *de facto* and applies to a minor, not an adult; in any case in the present context "family" meant the family of the first successor — Held, although the court sympathised with the respondent's position, that the judge had been in error in deciding that she was a member of the first successor's family — Appeal by landlords allowed

G The following cases are referred to in this report.

Brock v *Wollams* [1949] 2 KB 388; [1949] 1 All ER 715, CA
Joram Developments Ltd v *Sharratt* [1979] 1 WLR 928; [1979] 2 All ER 1084; *sub nom Carega Properties SA* v *Sharratt* (1979) 39 P & CR 76; [1979] EGD 514; 252 EG 163, HL
Ross v *Collins* [1964] 1 WLR 425; [1964] 1 All ER 861, CA

This was an appeal by Sefton Holdings Ltd, the landlords, from the decision of Judge Downey, at Liverpool County Court, holding that the defendant (the present respondent), Miss Florence Cairns, was the statutory tenant of 49 Cherry Avenue, Liverpool, by virtue of para 7 of Sched 1 to the Rent Act 1977.

Miss Bernadette Goodman (instructed by Redferns, agents for R H Vyner-Brooks & Co, of Liverpool) appeared on behalf of the H appellants; Miss Linda Pearce (instructed by Ashby Cornforth & Co, of Liverpool) represented the respondent.

Giving judgment, LLOYD LJ said: This is an appeal from a decision of Her Honour Judge Downey sitting in the Liverpool County Court on August 11 1987. It concerns a house at 49 Cherry Avenue, Liverpool. The question is whether the defendant, Miss Florence Cairns, is entitled to protection under the Rent Acts, that is to say, whether she is a statutory tenant under section 2 of the Rent Act 1977. The answer depends on the meaning to be given to the word "family" in para 7 of Part I of Schedule 1 to that Act.

The facts are that the plaintiffs, Sefton Holdings Ltd, are the landlords of the premises in question. They let it to a Mr Richard J Gamble some time between 1939 and 1941 when the house was built. Mr Gamble died in 1965. His daughter, Ada, then succeeded to the tenancy. Miss Ada Gamble died in 1986. The defendant came to live with Mr and Mrs Gamble and their daughter Ada in 1941. She was then 23 and single. Both her parents had died. Her boyfriend had just been killed in the war. Miss Ada Gamble asked her parents if they would take the defendant in, which they did. They treated her as their own daughter. She called them "Mom and Pop". She has lived in the same house ever since. She is now some 70 years of age.

On June 6 1986, shortly after Miss Ada Gamble died, the plaintiffs served on the defendant a notice to quit. The defendant claims that she is entitled to remain on in the house as a statutory tenant under K para 7 of Part I of Schedule 1, which provides:

> Where paragraph 6 above does not apply but a person who was a member of the first successor's family was residing with him at the time of and for the period of 6 months immediately before his death then, after his death, that person or if there is more than one such person such one of them as may be decided by agreement, or in default of agreement by the county court, shall be the statutory tenant if and so long as he occupies the dwelling-house as his residence.

Miss Ada Gamble was the first successor within the meaning of that paragraph. So what we have to decide in this case is whether the defendant is a member of her family who was residing with her at the time of, and for the period of six months immediately before, her death. The defendant was clearly residing with Miss Ada Gamble at L the time of Miss Ada Gamble's death. But was she a member of Miss Ada Gamble's family? That is the question. The county court judge has decided that she was, and there is now an appeal to this court.

No court could help feeling sympathy for an elderly lady of 70 who is in danger of being turned out of the house in which she has lived for nearly 50 years. But it goes without saying that we have to put sympathy on one side and apply the law to the best of our ability. It has been held over and over again that, in deciding whether a person is a member of another person's family, we must give the word "family" its ordinary everyday meaning. We cannot extend that meaning in order to cover what might appear to be a hard case; we must not let affection press upon judgment.

M If the judge has given the word too wide a meaning, then she has erred in law and we are obliged in this court to correct her, however much we might like to agree.

We have been referred to a number of cases as to what is meant by saying that a person is a member of another person's family. The cases go back at least 60 years. Not all the cases are, at first sight at any rate, easy to reconcile. That may be because, as has been suggested, the meaning of the word "family" has broadened over the years. But we do not have to go into that question now.

The most useful passage from among the decided cases is to be found in the judgment of Russell LJ in the case of *Ross* v *Collins* [1964] 1WLR 425 at p 432. In that case Russell LJ said:

> A Granted that "family" is not limited to cases of a strict legal familial nexus, I cannot agree that it extends to a case such as this. It still requires, it seems to me, at least a broadly recognisable *de facto* familial nexus. This may be capable of being found and recognised as such by the ordinary man — where the link would be strictly familial had there been a marriage, or where the link is through adoption of a minor, *de jure* or *de facto*, or where the link is "step-", or where the link is "in-law" or by marriage. But two strangers cannot, it seems to me, ever establish artificially for the purposes of this section a familial nexus by acting as brothers or as sisters, even if they call each other such and consider their relationship to be tantamount to that. Nor, in my view can an adult man and woman who establish a platonic relationship establish a familial nexus by acting as a devoted brother and sister or father and daughter would act, even if they address each other as such and even if they refer to each other as such and regard their association as tantamount to such. Nor, in my
> B view, would they indeed by recognised as familial links by the ordinary man.

That passage was expressly approved and adopted by Lord Diplock giving the leading speech in the House of Lords in *Joram Developments Ltd* v *Sharratt* [1979] 1 WLR 928.

It seems to me that the facts of the present case are covered by the principle stated by Russell LJ. I have no doubt that the defendant and Miss Ada Gamble did regard each other as sisters and may well have called each other such. There is evidence, to which I have already referred, that the defendant called Mr and Mrs Gamble "Mom and Pop". But the fact remains that when the defendant was taken in nearly 50 years ago she was taken in, to use the language of Russell LJ, as a stranger; and however long she may have lived with the
C family and however kindly they may have treated her, and however close their friendship may have become, the defendant did not, and in my judgment could not have, become a member of Ada's family. As Miss Goodman put it in the course of the argument, length of residence cannot transform a resident into a member of the family.

Miss Pearce, in seeking to support the judge's judgment, argued that the defendant at any rate became a member of Mr and Mrs Gamble's family by adoption. If she had, then she would have been within the protection of the Act. For it has been held by this court in the case of *Brock* v *Wollams* [1949] 2 KB 388 that *de facto* adoption is good enough to make a child a member of the family in question.

But there are two difficulties with that argument. In the first place, the section requires us to ask whether the defendant was a member of
D Miss Ada Gamble's family, not the family of her parents. Even if we could surmount that difficulty, there is a second difficulty. There is no case which has been drawn to our attention in which the courts have held that the protection of the Rent Acts covers the adoption of an *adult* (if such a concept is possible). It will be noted that in the passage I have read from Russell LJ's judgment he refers to the adoption of a minor; and Lord Diplock (at p 929) refers to "adoption (*de jure* or *de facto*) during minority".

Miss Pearce sought to distinguish *Ross* v *Collins* and *Joram Developments* v *Sharratt* from the present case. She submits that in those cases a young person went to live with an older person and gradually took over the running of the house; here, by contrast, a
E young person, an orphan, was taken in by the family as a whole. It was not a case, says Miss Pearce, of one adult taking in another. The relationship therefore had, so she submits, the necessary quality to make it a familial relationship. Mr and Mrs Gamble should be treated as being *in loco parentis*.

I would, as I have said already, like to accept Miss Pearce's distinction if I could find a way of doing so. I agree that there are factual differences between the cases. There always are. But our approach should be the same as that laid down by Russell LJ and adopted by Lord Diplock. No case has gone as far as Miss Pearce would like us to go in the present case. The defendant was 23 when she came to live with Mr and Mrs Gamble and their daughter. It cannot be said that she was brought up as an adopted child of the family. It
F may be that things were different in 1941 from what they are now. But, if I ask myself the question whether the defendant became a member of Mr and Mrs Gamble's family (if that be the right question) by adoption, whether *de facto* or *de jure*, I am in all honesty compelled to answer that question "No".

Various attempts have been made by the courts from time to time to define the word "family", by identifying various categories within which a person would be a member of another person's family. But Lord Diplock did not embark on that task in *Joram Developments* v *Sharratt*, and I do not propose to embark on that task myself. All I would say is that, in approaching this case, I have found it useful to bear two matters in mind. First, there is the distinction drawn by Viscount Dilhorne in *Joram Developments* v *Sharratt* between being a member of the family and being a member of the household. G
Second, there is the distinction between *being* a member of the family and living *as* a member of the family. There is no doubt that the defendant lived as a member of the family and that may be why the judge decided this case in her favour. But the question we have to ask ourselves is not whether she lived as a member of the family but whether she was a member of the family. I am clear that she was not and that the man in the street would take the same view.

We should in this court be slow to reverse a county court judge on a matter of this kind. But, for the reasons which I have attempted to give, I for my part feel compelled to allow this appeal.

Agreeing, "with some feeling of regret", SIR ROUALEYN CUMMING-BRUCE said: I have been aided by the guidance given by H
Lord Dilhorne in the speech to which Lloyd LJ has referred. It seems to me clear beyond a peradventure that the plaintiff in this case became a member of the household of Mr and Mrs Gamble and, in spite of her age of 23, was offered and accepted a high degree of kindness and support which led the learned judge to make the finding that she reminded herself that in 1941 young women of 23 did not by and large live in the independent way that so many young women do today. Mr and Mrs Gamble were giving her a home and they took her in as a daughter and in every way treated her as such.

Like Lloyd LJ, it is with diffidence that I am moved to differ from the view of a learned county court judge familiar with the local circumstances of the Liverpool environment; but the question is a question of law, although, like all questions of law, it involves a nice J
appreciation of the relevant facts. Throughout the now mounting line of cases which deal with the meaning of the word "family" in the context of the relevant schedule it is to be observed that in no case has the court found it possible to identify the necessary ingredient or quality that distinguishes a familial nexus from a nexus less than familial. The approach of the courts has been to look at all the circumstances and then to seek to answer the question: what would an ordinary person characterise the relationship as? For the reasons explained by Lord Dilhorne, I do not myself think that, putting the test in that way (accepting that it has been so frequently), it really adds very much to the question: what is the ordinary meaning of someone being a member of somebody else's family? To be a member of a family is different from being treated as a member of the family. K

I would hold that the judge was wrong as a matter of law and that, like Lloyd LJ, the appeal should be allowed.

The appeal was allowed; legal aid taxation of respondent's costs ordered. Leave to appeal to the House of Lords refused but stay granted pending the petition for leave. Directions given in regard to possession.

Court of Appeal

February 12 1988

(Before Lord Justice SLADE, Lord Justice RUSSELL and Lord Justice TAYLOR) L

INVESTMENT & FREEHOLD ENGLISH ESTATES LTD v CASEMENT

Estates Gazette April 16 1988

[1988] 15 EG 99-105

Landlord and tenant — Rent Acts and Leasehold Reform Act 1967 — Long tenancy at a low rent — Rent payable of less than two-thirds of rateable value — Effect on the right to charge premiums — Parties under a misapprehension as to legal M
position when lease was entered into in 1972 — Lease, which was for a term of nearly 33 years from March 25 1968, provided for a service charge to meet the cost of lighting and cleaning common areas, outside painting and repairs to main structure and external parts — It contained a provision that the ground rent plus the lessee's proportion of the service charge would in no event exceed two-thirds of the rateable value of the premises and added a proviso that, if in any year it seemed likely to so exceed, the lessor would not be under an obligation to carry out any work or perform any duties which would impose on him a liability in excess of the amounts to be

collected from the lessees — There was also an agreement that "if the Leasehold Reform Act 1967 shall have been amended at any time in the future by legislative action or by a decision of the High Court" to the effect that service charges would not be taken into account so as to prevent the lessor or lessee from charging a premium on the grant or assignment of the tenancy, the lessor would then be at liberty to demand a full contribution to the service charges — These provisions appeared to indicate that the parties were unaware that the Housing Act 1969 had, by section 80(1), already required sums payable *inter alia* for services to be disregarded in determining whether a tenancy was a tenancy at a low rent for the purposes of the Rent Acts

The lessors sought by originating summons a declaration that, on the true construction of the relevant clauses in the lease, they were entitled to demand a full contribution from the lessee towards money spent in relation to services and repairs as set out in the lease — Judge Paul Baker QC, sitting as a judge of the High Court, decided in their favour and the lessee appealed

The lessee argued that in a strict literal sense the Leasehold Reform Act 1967 had not been "amended" by legislative action, but this argument was dismissed by the Court of Appeal as largely semantic and involving a narrowly literal interpretation — The lessee's more important submission was that the relevant legislative action could not be said to have occurred "at any time in the future" within the meaning of the lease because the 1969 Act was already law before the lease was executed in 1972 — The court held, however, that this argument ignored two important points — The first was the use of the future perfect tense in the words "shall have been amended" in contrast with the future tense "shall be amended" — The second point, a crucial one, was that the argument failed to do justice to the manifest intention of the parties — The parties at the time of execution of the lease wrongly believed that the then existing law would prevent the charging of premiums if the sum of the rent and the service charge exceeded two-thirds of the rateable value — Their intention was that, if the law as thus understood were to be altered, the lessors were to be free to demand a full contribution to service charges — The provisions in the lease were unhappily drafted, but the landlords' construction, without doing unacceptable violence to language, accorded more closely than the tenant's with the intentions which it was reasonable to impute to the parties — Appeal dismissed

The following case is referred to in this report.

Property Holding Co Ltd v *Clark* [1948] 1 KB 630; [1948] 1 All ER 165; (1948) 64 TLR 76, CA

This was an appeal by the lessee, Ann Casement, from the decision of Judge Paul Baker QC, sitting as a High Court judge, in favour of the lessors, Investment & Freehold English Estates Ltd, the present respondents (plaintiffs below). The flat which was the subject of the proceedings was Flat no 4, one of eight in a block known as 5 Kensington Park Gardens, London W11. The decision of Judge Paul Baker is reported at [1987] 2 EGLR 112, (1987) 283 EG 748.

Nicholas Patten (instructed by Freed Stone Goodman) appeared on behalf of the appellant; David Neuberger QC (instructed by Julian Holy) represented the respondents.

Giving judgment, SLADE LJ said: This is an appeal by Miss Ann Casement from an order of His Honour Judge Paul Baker QC, sitting as a High Court judge, made on February 17 1987. It raises a point of construction of a lease. The issue is whether the landlord is entitled to demand a full service charge from the tenant or is restricted to recovering a much smaller sum.

The premises demised by this lease (which I will call "the premises") are Flat no 4 in a building, 5 Kensington Park Gardens, London W11. The building is a block of eight flats. The reversion to the lease is now vested in Investment & Freehold English Estates Ltd, the respondents to this appeal. The lease itself is vested in the appellant tenant.

The lease was dated May 18 1972 and was made between Cityland Securities Ltd as lessor and Daniel Joseph as lessee. By clause 1 the lessor, in consideration of a premium of £9,000 and of the rent and the lessee's covenants therein contained, demised to the lessee the premises together with certain rights and subject to certain exceptions and regulations for a term from March 25 1968 to March 15 2001:

> . . . paying therefor during the said term the yearly rent of seventy five pounds (£75) and also a sum of thirty five pounds (£35) per annum on account of the proper proportion of expenses referred to in Clause 2(3) hereafter and so in proportion for any less time than a year the said rent to be paid without any deduction (except for Landlord's Property Tax) by equal quarterly payments on the usual quarter days the first of such payments after the date of these presents to be made on the 24th day of June One thousand nine hundred and seventy-two and the last payment to be made in advance one calendar month before the expiration of the said term or immediately on the determination thereof in the event of the same being determined by re-entry under the power in that behalf hereinafter contained and also paying by way of further or additional rent from time to time a sum or sums of money equal to the amount which the Lessor may expend in effecting or maintaining the insurance of the said premises or any part thereof against loss or damage by fire (including Architect's fees) and such other perils as are commonly included in similar policies such last mentioned rent to be paid without any deduction on the quarterly day for payment of rent next following each such expenditure by the Lessor as aforesaid provided always that the ground rent hereby reserved together with the proportion of expenses referred to in Clause 2(3) of these presents will in no event exceed two-thirds of the rateable value of the premises but Provided nevertheless that should the said expenses together with the ground rent be likely in any one year to exceed two-thirds of the rateable value the Lessor shall not be under any obligation to carry out any work or perform any duties (notwithstanding his covenant to that effect) which would impose upon him a liability in excess of the amounts to be collected from the Lessee and the lessees of other flats.

By clause 2(3), so far as material, the lessee covenanted with the lessor:

> To pay due proportion of all proper expenses of the Lessor in connection with the said property including lighting and cleaning of the common areas and the cost of painting in every third year of the said term and also in the last year of all the outside wood iron stucco cement and other like work in or upon the said premises or any addition thereto and of keeping in substantial repair the main structure of the building and all the external parts thereof including the common area in front and at the rear of the building the entrance passage and entrance hall the common staircase landings and passages and all the drains ducts conduits and water pipes such proportion to be determined by the Lessor's Surveyor and to be calculated by reference to the rateable value of the said premises . . .

Clause 12 of the lease took an unusual form:

> The Lessor and the Lessee hereby agree and covenant accordingly that if the Leasehold Reform Act 1967 shall have been amended at any time in the future by legislative action or by a decision of the High Court to the effect that service charges do not for the purpose of the Rent Acts as amended by the Leasehold Reform Act 1967 attract the operation of the Acts to estop the Lessor or the Lessee wishing to assign the residue of the term from charging a premium on the grant of the Lease or assignment (as the case may be) then and in such event the proviso hereinbefore contained shall be of no effect and the Lessor shall be at liberty to demand a full contribution from the Lessee as provided in Clause 2(3) of these presents notwithstanding that such contribution together with the ground rent hereby reserved exceeds two-thirds of the rateable value.

The wording of clause 12 of the lease indicates the belief and understanding of the draftsman and of the parties that, at the time of its execution, the effect of the Rent Acts, as amended by the Leasehold Reform Act 1967 ("the 1967 Act"), would have precluded the lessor from charging a premium on the grant of the lease and the lessee from charging a premium on the assignment of the lease if the lease had provided for the aggregate of the rent and service charges payable by the lessee to exceed two-thirds of the rateable value of the demised premises.

In order to explain the references to the Rent Act legislation, Mr Patten on behalf of the appellant has taken us on a helpful journey through the relevant statutory provisions relating to the charging of premiums. For present purposes, however, I think it will suffice to summarise very briefly the relevant legal position as it was immediately before the passing of the 1967 Act.

As at that date section 2(1) of the Landlord and Tenant (Rent Control) Act 1949, as amended by section 37(1) of and the 5th Schedule to the Rent Act 1965, precluded a person from requiring the payment of a premium as a condition of the grant of a tenancy to which the section applied. Section 2(2) of that Act, as amended, precluded a person from requiring the payment of a premium as a

A condition of the assignment of a tenancy to which the section applied. Section 2(3) in effect rendered the section applicable to any tenancy of a dwelling-house falling within the Rent Acts. However, the section did not apply if the tenancy was outside the Acts, that is to say if the rent was less than two-thirds of the rateable value (see section 12(7) of the Increase of Rent and Mortgage Interest (Restrictions) Act 1920, as amended by section 26(3) of and the 8th Schedule to the Rent Act 1957) or if the term of the tenancy was for more than 21 years (see sections 21(1) and 25(1) of the Rent Act 1957).

The term of the lease in the present case exceeded 21 years. Accordingly, if the lease had been executed before the 1967 Act was passed, the statutory provisions precluding the demand of a premium by the landlord on the grant of the lease or by the tenant on its B assignment would not have applied, even though the rent had exceeded two-thirds of the rateable value of the premises.

However, the 1967 Act brought tenancies for more than 21 years under Rent Act protection. Section 39(1)(a) of that Act provided:

> the Rent Act 1965 shall have effect as if at the end of section 1(2) (which lists the cases in which the Rent Acts are to apply as modified by that Act) there were added —
> "or (d) that the tenancy is a long tenancy within the meaning of Part I of the Landlord and Tenant Act 1954";

A "long tenancy" within that meaning was "a tenancy granted for a term of years certain exceeding 21 years".

Thus, for the time being, the effect of this amendment of section C 1(2) of the Rent Act 1965 was to preclude a landlord from charging a premium on the grant of a lease or a tenant from charging a premium on the assignment of a lease in any case where the aggregate of the rent payable by the tenant exceeded two-thirds of the rateable value of the demised premises, *even though the term exceeded 21 years.* Furthermore, it was well established by judicial authority that for the purpose of the Rent Acts the word "rent" included payments made by a tenant to his landlord by way of service charge (see, for example, *Property Holding Co Ltd* v *Clark* [1948] 1 KB 630).

On June 8 1968 the Rent Act 1968 ("the 1968 Act") came into force. It was a consolidating Act which replaced the previous statutory provisions of the 1967 Act to which I have referred but effected no change in the law material for present purposes. Part VII D contained provisions prohibiting the requirement of premiums on the grant and assignment of protected tenancies broadly corresponding with, though not identical to, those in force before the statute came into effect. It specifically repealed section 39(1)(a) of the 1967 Act.

The result of section 39(1)(a) of the 1967 Act and the reproduction of its substantive effect in the 1968 Act was that, if the rent payable under a lease was more than two-thirds of the rateable value of the demised premises, then, whatever the length of the term, the lease could not be granted or assigned for a premium and, furthermore, the decided cases showed that for this purpose service charges were to be reckoned as part of the rent.

E In so far as this state of the law applied to long leases, it was radically altered by section 80(1) of the Housing Act 1969 ("the 1969 Act") which, so far as material, provided:

> (1) In determining whether a long tenancy is, or at any time before the commencement of this Act was, —
> (a) a tenancy at a low rent within the meaning of the Rent Act 1968 . . .
> there shall be disregarded such part (if any) of the sums payable by the tenant as is expressed (in whatever terms) to be payable in respect of rates, services, repairs, maintenance, or insurance, unless it could not have been regarded by the parties as a part so payable.

Section 80(3), with an immaterial exception, defined "long tenancy" as meaning "a tenancy granted for a term certain exceeding 21 years".

F The tenancy granted by the lease in the present case is a "long tenancy". Accordingly, the effect of section 80 was that, in determining whether that tenancy is, or was at any time before the commencement of the 1969 Act, "a tenancy at a low rent" within the meaning of the 1968 Act, the service charges payable by the lessee fell to be disregarded. It is, therefore, clear that, contrary to the apparent belief and understanding of the parties as expressed in this lease, even at the time when it was executed, there would have been nothing to prevent the lessor from demanding a premium on the grant of the lease, nor the lessee from demanding a premium on its assignment, even though the lease had not included either of the two provisos to clause 1 or clause 12. It is clear that to this extent the parties entered into the lease under a complete misapprehension of the relevant legal position. It is in these circumstances that dispute arises as to the G construction and effect of clause 12 of the lease.

By the originating summons the respondent landlord sought a declaration that, upon the true construction of the proviso to clause 1 of the lease and pursuant to clause 12, it is entitled to demand a full contribution from the lessee in relation to moneys expended by it in performing its covenant contained in the lease in accordance with the provisions of clause 2(3). The learned judge granted a declaration to this effect and the tenant now appeals from his judgment.

The learned judge himself summarised the arguments submitted to him and his conclusions on them as follows in the course of his judgment:

> By the time this lease was granted in 1972, the danger which had obviously H been looked at and been guarded against of having the rent go over the two-thirds and thereby bring about the premium restrictions among other things, had already gone by legislative action. The parties nevertheless put a clause in on the basis that such alteration and change in the law had not taken place, and until it had, they would restrict the rent to within two-thirds of the rateable value. Mr Patten's argument on this is quite simple; that when you look at the earlier proviso it is clear and unambiguous, as I have noticed, and when you come to clause 12 — it was a provision that was out of date when it was put in. The Leasehold Reform Act could not be amended at any time in the future — it had already been amended in the past. Therefore, there is nothing for this clause to operate on. There is obviously some misapprehension here which may form the basis of some action for rectification or may not; but as a matter of construction it says you cannot apply it to these circumstances which have arisen because it had no operation. The Act of 1967 was not amended or judicially construed at any time after 1972 in any relevant sense, and therefore J one is left with the proviso through the whole term.
>
> There is considerable force in that argument, but I have come to the conclusion that as a matter of construction I must find that that is not the true meaning of the clause when one comes to attend to the wording more closely. First of all, one sees that it says, reading the terms of it overall — that it operates at the time, in my judgment, of the grant of the lease because what it says is:
>
> "The Lessor and the Lessee hereby covenant accordingly that if the Leasehold Reform Act shall have been amended at any time in the future . . ."
>
> "In the future" I would take to be simply an expression that means in the future following the passing of the Act. I would not have taken that view in the ordinary course, but the rest of the clause would seem to indicate it. Apart from the words "shall have been amended", which would seem to indicate that it had already been amended at the grant of the lease — those words by K themselves — one then goes on to describe what the effect of the amendment shall have been. It says:
>
> ". . . to the effect that service charges do not for the purpose of the Rent Acts as amended by the Leasehold Reform Act attract the operation of the Acts to estop . . ."
>
> or "prohibit", as I would read that word,
>
> ". . . the Lessor or the Lessee wishing to assign the residue of the term from charging a premium on the grant of the lease or assignment, as the case may be."
>
> It does not refer to the position of lessors and lessees generally; it refers to this particular lessor as defined. Yet the position of the lessor in this connection is only relevant at the grant of lease, not during its course. Then the clause goes on:
>
> "And in such event the proviso hereinbefore contained shall be of no effect." L
>
> It is not that it shall *become* of no effect, but it shall from all times, as I would read that, *be* of no effect.
>
> It is of course on the face of it a very singular way of providing for this because one would expect the parties at the time when the lease was granted to have made up their minds whether or not the provisions applied or the law was in such a state as to allow for the grant of a premium. That is obviously the sensible and efficient way to draft such a lease. There may be some excuse for it. As I have said, this is a block of eight flats and this formula was devised because there was going to be a series of grants, or there might be a series of grants, and the legislation might change during the course of the grants. It would not be known precisely when the legislation would have changed. It has this consequence; that if the lease was granted before any relevant amendment lifting the premium provisions, it would mean that that particular lease would be subject to the restrictions and continue to be so — or might be so. That is M not a case I have to consider before me. I read these two provisos together. At the grant of the lease one asks oneself, to my mind, whether the Leasehold Reform Act has already been amended and, therefore, with the consequence both that the Lessor can take a premium and the Lessee can subsequently assign it with the benefit of a premium.
>
> As I say, I have not found that at all an easy point of construction, but doing the best I can with the language which I am confronted and particularly relying on the words "shall have been" and the references to the Lessor in the clause, I am of the opinion that the provisos in clause 1 have no operation. Therefore I shall make the declaration that is asked for.

I now turn to the submissions advanced on behalf of the tenant by Mr Patten in support of this appeal. He pointed out that, as was common ground before Judge Baker, there has been no judicial decision to the effect that service charges are not to be treated as "rent" for the purpose of the Rent Acts. Indeed, it was the absence of any such decision which rendered the enactment of section 80 of the 1969 Act necessary. Therefore, as he pointed out, the only question for the court is whether the contingency referring to legislative action, mentioned in clause 12 of the lease, has in fact occurred, within the meaning of that clause.

One of the subsidiary grounds upon which Mr Patten criticised Judge Baker's decision was that, in answering this question, the judge suggested that the form of lease was prepared at a time before section 80 of the 1969 Act had taken effect. He suggested that, in referring to this block of eight flats, "this formula was devised because there was going to be a series of grants, or there might be a series of grants, and the legislation might change during the course of the grants". This hypothesis may or may not be well founded. Some slight support is perhaps lent to it by the chosen date of the commencement of the term, March 25 1968, which was a long time before the date on which the lease was actually executed. Nevertheless, with respect to the learned judge, his hypothesis is pure speculation and whether or not any such evidence would be admissible for the purposes of construing the lease, the speculation is not, I think, supported by any firm evidence. In those circumstances, I agree with Mr Patten that we must construe the lease without conjecture of this kind.

I now turn to the two principal propositions submitted by him in support of his appeal. As his first proposition he submitted that there has never been an amendment of the 1967 Act by "legislative action" of the nature referred to in clause 12. He put this point in two ways. First, he pointed out, the only relevant section of the 1967 Act, section 39(1)(a), was repealed by the 1968 Act. By the time that section 80 of the 1969 Act became law, section 39(1)(a) of the 1967 Act was no longer on the statute book. It cannot, in his submission, be properly said that a subsection has been "amended" by new legislation if it is no longer on the statute book when that new legislation is passed.

Further or in the alternative, he pointed out that section 39 of the 1967 Act, by adding a new subparagraph to section 1(2) of the 1965 Act, merely restored the pre-1957 position whereby the mere fact that the tenancy was a long tenancy did not disqualify it from Rent Act protection. This remains the position today. The effect of section 80 of the 1969 Act merely was to reverse the judicial interpretation of "rent" within the meaning of the Rent Acts. Thus, it was submitted, it could not be said to involve an amendment of the 1967 Act.

The first principal proposition of Mr Patten which I have attempted to summarise involves reading the reference in clause 12 of the lease to an "amendment" of the 1967 Act in the strictest and narrowest possible sense. As he accepted, it would mean that, even if the lease in the present case had been executed before the passing of the 1969 Act, the lessee would still have been able to assert that the lessor had no right to invoke clause 12 after the 1969 Act was passed, because it could do so only if the relevant change in the law had been brought about by an "amendment" of section 39(1)(a), no more and no less. I, for my part, decline to read the words of clause 12 in so narrow a sense. I think it would produce an absurd result. In this context the relevant words of clause 12 are:

> . . . if the Leasehold Reform Act 1967 shall have been amended at any time in the future by legislative action or by a decision of the High Court to the effect that . . .

If the coming into operation of a statute has the legal effect stated in clause 12, so as to negative the state of the law brought about by the enactment of section 39(1)(a) of the 1967 Act, the passing of such statute, in my judgment, amounts to "amendment" of the 1967 Act by legislative action within the meaning of clause 12. Accordingly, the passing of section 80 of the 1969 Act was, in my judgment, legislative action of the kind contemplated by the clause.

However, even if this be so, Mr Patten submitted that it cannot be said that the relevant legislative action occurred "at any time in the future" within the meaning of clause 12, because the 1969 Act had already become law prior to the execution of the lease on May 18 1972. This was his second main proposition and, to my mind, by far the most important point in the case. Judge Baker himself construed the words "shall have been amended at any time in the future" as meaning in the future following the passing of the 1967 Act. This, Mr Patten argued, was inconsistent with the actual date borne by the lease. By 1972 there was no need for the restrictions in the lease limiting service charge contributions to two-thirds of the rateable value, because by then section 80 of the 1969 Act had become law. As Mr Patten pointed out, the parties were obviously unaware of this and that was why they inserted the proviso to clause 1.

It follows, in his submission, that clause 12 can only have been inserted in the lease to cater for a change in the law taking place *after* the grant of the lease. In his submission, the words "shall have" in their context look simply to the future and mean simply "shall". This interpretation of the words as having merely prospective effect is, in his submission, consistent with the stated consequence that the proviso to clause 1 "then in such event . . . shall be of no effect".

There is no doubt that the original parties to this lease were labouring under a misapprehension as to the true state of the law when they executed the lease in 1972. Possibly the lease may be rectifiable but, in Mr Patten's submission, the language used is unambiguous and the court must not attempt to remedy the mistake by a process of construction, particularly as between parties now before the court who are not the original parties to the lease. The language of the clause may not have been appropriate, but the parties, he submits, are stuck with it and the court cannot rewrite its provisions.

On any footing, clause 12 is not happily drafted. Quite apart from its apparent misapprehension as to the state of the relevant law, the reference to the possibility of amendment to the 1967 Act "*by decision of the High Court*" reads very oddly. So, too, does the reference to the possibility of changes in the law enabling *the lessor* to charge a premium on the grant of the lease, bearing in mind that the lessor has already demanded and received a premium of £9,000 on the grant of this lease.

Mr Patten's argument, however, based as it is on the supposed literal and grammatical meaning of the words used, in my judgment, fails to give due weight to two important features of clause 12. The first, which is itself a grammatical feature, is the apparently deliberate use by the draftsman of the *future perfect* tense in the phrase "shall have been amended" in contrast with the *future* tense "shall be amended". The use of the future tense "shall be" is apt simply to describe an action or event yet to happen. The use of the future perfect tense "shall have been" likewise looks to a point of time in the future but denotes a completed event or action viewed in relation to that future point of time. The relevant future point of time designated by clause 12 is the time when the lessor wishes to exercise his right to demand a full contribution from the lessee, as provided by clause 2(3) of the lease. While the introductory condition of clause 12 is by no means elegantly or precisely worded, it is, in my judgment, grammatically capable of covering a relevant change of the law which shall have taken place at any time whatever before the lessor seeks in the future to demand a full contribution from the lessee.

Some slight grammatical support for this construction which treats the opening condition of clause 12 as capable of having some retrospective content is to be found in the use by the draftsman of the phrase "shall be of no effect" as opposed to the phrase "shall become of no effect".

However, the second crucially important feature of the clause to which Mr Patten's argument, in my judgment, gives quite inadequate weight is the obvious intention of the parties as manifested by clause 12. This clause, as I have already said, clearly demonstrates their belief, at the time of the execution of the lease, that the effect of the Rent Acts as amended by the 1967 Act would have precluded the lessor from charging a premium on the grant of the lease and would have precluded the lessee from charging a premium on the assignment of the lease, if the lease had provided for the aggregate of the rent and service charges payable by the tenant to exceed two-thirds of the rateable value of the demised premises.

Furthermore, no less clearly, clause 12 demonstrates the parties' intention that, if the law were to be changed so that service charges payable by the lessee no longer fell to be taken into account in deciding whether the rent payable under the lease exceeded two-thirds of the rateable value of the premises, for the purpose of applying the statutory provisions precluding the charge of premiums on the grant or assignment of a lease such as this, then the lessor was to be free to demand a full contribution of service charges from the lessee as provided by clause 2(3).

Let it be supposed that the parties had been asked at the time of the execution of the lease this question: "If, contrary to your understanding, there has already been amending legislation of the relevant nature, is the lessor to be free to demand a full contribution

A from the lessee in respect of service charges?" In my judgment, the only reasonable inference which one can draw from the language used in clause 12 is that they would have assuredly answered this question "Yes". Thus, I have no doubt that the construction which the respondent landlord attributes to clause 12 accords more closely with the intentions which it is reasonable to impute to the parties than does the tenant's construction. Nor, in my opinion, for the reasons which I have given, does the landlord's construction do violence to the language of that admittedly ill-drafted clause as a matter of grammatical construction.

Indeed, contrary to Mr Patten's submission, I do not find it fanciful to suppose that the draftsman could have deliberately used the future perfect tense to cover the eventuality that at some future B date it might be discovered that legislation or judicial decision (in particular, judicial decision) of which he was unaware had already effected the relevant change in the law before the lease was executed.

On the other hand, the construction which the appellant tenant attributes to the clause would render it totally incapable of having any effect at all.

I, for my part, feel unable to construe it in the way which the tenant propounds. In my judgment, the contingency referring to legislative action, mentioned in clause 12 of the lease, has occurred. It makes no difference that, unknown to the parties, the legislative action had already taken place before the lease was effected. The proviso referred to in clause 12 is, in my judgment, of no effect. The lessor is C entitled to demand a full contribution from the lessee in relation to moneys expended by it in performing its covenant contained in the lease in accordance with the provisions of clause 2(3).

By a respondent's notice the landlord submitted that, even if all the points argued by the appellant on this appeal were well founded, the learned judge's judgment should be affirmed on a further ground, that is to say that the 1967 Act has been amended by legislative action consisting of section 78 of the Housing Act 1980. As things were, we did not find it necessary to hear argument on this point raised by the respondent's notice.

For the reasons which I have given, the learned judge, in my judgment, reached the right conclusion. Mr Patten has argued this appeal with great ability, but I, for my part, would dismiss it.

D RUSSELL and TAYLOR LJJ agreed and did not add anything.

The appeal was dismissed with costs.

Court of Appeal

December 2 1987

(Before Sir John DONALDSON MR, Lord Justice GLIDEWELL and Sir Denys BUCKLEY)

GENT AND OTHERS v DE LA MARE

E Estates Gazette May 7 1988
[1988] 18 EG 63-67

Rent Act 1977 and Housing Act 1980 — Problem concerning a protected shorthold tenancy — Landlords agreed to let to appellant a property on a protected shorthold tenancy — The tenancy fulfilled all the conditions set out in section 52(1) of the 1980 Act — It was to be for a term of one year subject to a provision "that the landlords shall give to the tenant three months' notice in writing prior to the expiration date in the event of their not wishing to renew this agreement for a further term" — Landlords in fact gave no such notice and tenant F remained in occupation — Landlords brought proceedings for possession, alleging as a ground arrears of rent and also the mandatory ground for possession under Case 19 of Schedule 15 to the 1977 Act — An order for possession was made in the absence of a defence, but tenant applied to set it aside and, on the judge's refusal, the matter came before the Court of Appeal

So far as the claim on the ground of arrears of rent was concerned, the possession order was defective, as it purported to take effect forthwith, contrary to section 138(3) of the County Courts Act 1984 — That left for consideration somewhat more difficult questions under section 52(5) of the G 1980 Act and Case 19 in Schedule 5 to the 1977 Act — Section 52(5) provided, so far as relevant, that if a protected tenancy was granted for a term certain in accordance with section 52(1), and thereafter from year to year, it was a protected shorthold tenancy until the end of the term certain — The tenant argued that the tenancy in the present case did not fall within section 52(5) at all, from which he submitted that it was not a protected shorthold tenancy; alternatively, if the first year was for a protected shorthold tenancy, it continued after payment and acceptance of rent as a tenancy from year to year at common law — As such a tenancy could be terminated only by six months' notice to quit expiring at the end of a year of the H tenancy, the tenant was holding over as a tenant from year to year — The landlords' submission, which was broadly accepted by the Court of Appeal was that the tenancy for the first year was a shorthold tenancy within section 52(1) of the 1980 Act, but at the end of that year a tenancy arose under the tenancy agreement which was a protected tenancy but not a protected shorthold tenancy — The agreement was renewed for a further year, including the provision for three months' notice

Glidewell LJ doubted whether the original year's tenancy fell within section 52(5), but it was in any case clear that Case 19 in Schedule 15 to the 1977 Act applied — The landlords had J given the proper notices and there were no grounds for setting aside the judgment for possession which had been given by the judge below — Appeal dismissed

No cases are referred to in this report.

This was an appeal by the tenant, Richard St Clair de la Mare, from a decision of Judge Bloomfield at Newbury (Berks) County Court dismissing an application by the appellant to set aside an order for possession previously made against him. The landlords, respondents to the present appeal, were Robin Charles Frederick Lewis Gent, Elizabeth Mary Gent and John Charles Gent. The property concerned was South Stanmore Farmhouse, Beedon, near Newbury, K Berkshire.

The appellant appeared in person; the respondents were represented by Edward Hess (instructed by Charles Lucas & Marshall, of Newbury).

Giving the first judgment at the invitation of Sir John Donaldson MR, GLIDEWELL LJ said: By an agreement made on August 7 1983 the present plaintiffs, Mr Robin Charles Frederick Lewis Gent, his wife and another member of his family, whom I shall call "the landlords", agreed to let to the defendant, Mr de la Mare, the property known as South Stanmore Farmhouse, Beedon, near Newbury, Berkshire. The agreement specifically provided that it was the landlords' proposal to let the house and premises on a protected L shorthold tenancy as defined by the Housing Act 1980 and recited that "the appropriate notice has been served on the Tenant prior to the commencement of the said tenancy as the Tenant hereby acknowledges". It is right to say that Mr de la Mare has continued to acknowledge that the necessary preliminary notice was served upon him.

Clause 2 of the agreement provided:

> THE tenancy shall be for the term of One Year from the Seventh day of August One thousand nine hundred and eighty three subject to the provision that the Landlords shall give to the Tenant three months notice in writing prior to the expiration date in the event of their not wishing to renew this Agreement for a further term.

The rent reserved was £3,250, payable by equal instalments at M three-monthly intervals, and there were the usual provisions whereby the tenant agreed to pay the rent at the times agreed and for the landlords to re-enter upon the premises if at any time any part of the rent reserved was in arrears for 15 days, whether formally demanded or not.

After August 7 1984, the landlords not having given any notice in writing or at all that they wished to renew the agreement, the tenant, Mr de la Mare, remained in occupation. By a summons for possession issued out of the Newbury County Court on August 14 1985 the landlords claimed possession of the premises from Mr de la Mare on two separate grounds. The first was that he was in arrears with his rent and the second was that, since the tenancy was originally

A a shorthold tenancy, it fell within Case 19 of Schedule 15 to the Rent Act 1977, under which the landlords alleged that they were entitled, quite apart from the question of arrears of rent, to possession of the premises.

No defence was entered to the claim. At the date of hearing fixed, September 30 1985, Mr de la Mare did not appear. The learned judge then made an order for possession of the premises, which was expressed to be an order for possession to be given forthwith.

In October 1985 the defendant applied to set aside that order on the ground that he had not received the original summons and that he had a defence. The application to set aside, after what seems to be a considerable lapse of time, came before the judge on June 30 1986. He did not deliver any extensive judgment, but he did find as a fact,
B and so indicated, that Mr de la Mare had received the original summons. I should say that Mr de la Mare had given evidence to the effect that he had been out of the country at the material time when the summons was served, and no doubt there was a considerable amount of mail waiting for him upon his return. Be that as it may, as I say the judge found that he had received the summons. He made no other comment, but he dismissed the application to set aside.

Mr de la Mare then applied to the judge for leave to appeal to this court. That application was heard on August 7 1986, which, coincidentally, of course, was the third anniversary of the original tenancy agreement. On that hearing the judge, of whose judgment on that occasion we have a fairly full note, considered not merely
C whether he should alter his finding that Mr de la Mare had received the original summons but also the prospects of success of an appeal if he were to grant leave to appeal. He specifically said that he remained of the conclusion that, as a finding of fact, Mr de la Mare had received the original summons. He then went on to consider the arguments advanced about the two issues that were raised in the particulars of claim.

The first point which was made on behalf of Mr de la Mare was that the judge had no power to make an order for possession forthwith because section 138(3) of the County Courts Act 1984 provides:

> If . . . the court at the trial is satisfied that the lessor is entitled to enforce the right of re-entry . . . the court shall order possession of the land to be given to the lessor at the expiration of such period, not being less than 4 weeks from the
> D date of the order, as the court thinks fit, unless within that period the lessee pays into court all the rent in arrear and the costs of the action.

So the tenant in default is given a time within which he may make good his default if non-payment of rent is the only issue. It is clear that the order for possession in this case, if it depended wholly upon the arrears of rent, was not in accordance with section 138(3) of the Act, and therefore on that ground the judge concluded, as he said more or less, that an appeal must have a good chance of success. If the matter had stopped there, for my part I would certainly allow the appeal, despite the counterargument which Mr Hess bravely put forward.

However, there remains the more complex issues which arise under
E the provisions of the Rent Act 1977 and the Housing Act 1980. The point which eventually arises is one on which there appears to be no authority. It is not an easy point. It has been argued, if I may say so, with economy and clarity, both by Mr de la Mare himself (who had had the forethought, with I think a little assistance, to prepare what he had to say in writing beforehand, which was of the greatest help to this court and is an example which most litigants in person could follow with advantage) and by Mr Hess, counsel for the landlords, who has addressed us with equal brevity and clarity.

One starts by looking at section 1 of the Rent Act 1977, which provides:

> Subject to this Part of this Act, a tenancy under which a dwelling-house . . . is let as a separate dwelling is a protected tenancy for the purposes of this Act.

F So *prima facie* this house was let on a protected tenancy.

Section 98 of the Act provides:

> Subject to this Part of this Act, a court shall not make an order for possession of a dwelling-house which is for the time being let on a protected tenancy . . . unless the court considers it reasonable to make such an order and either . . .
> (b) the circumstances are as specified in any of the Cases in Part I of Schedule 15 to this Act.

I go to Schedule 15. One of the Cases is non-payment of rent, but I need no longer concern myself with that. The Case which is now relevant is Case 19, which was introduced by section 55 of the Housing Act 1980. That Case relates to a dwelling-house let under what is called a "protected shorthold tenancy". There is a short group of sections in the 1980 Act which describe and deal with this then new concept of a protected shorthold tenancy. It is, I emphasise, G
a species of protected tenancy — it is one of the types of protected tenancy within section 1 of the 1977 Act.

By section 52(1) of the 1980 Act it is provided:

> A protected shorthold tenancy is a protected tenancy granted after the commencement of this section which is granted for a term certain of not less than one year nor more than five years and satisfies the following conditions, that is to say, —
> (a) it cannot be brought to an end by the landlord before the expiry of the term, except in pursuance of a provision for re-entry or forfeiture for non-payment of rent or breach of any other obligation of the tenancy; and
> (b) before the grant the landlord has given the tenant a valid notice stating that the tenancy is to be a protected shorthold tenancy; and
> (c) either a rent for the dwelling-house is registered at the time the tenancy is H
> granted or —

certain other conditions are met.

It is accepted by Mr de la Mare that the requirements of section 52(1) were all fulfilled. This on its face was a tenancy granted for a term of one year. It could not be brought to an end by the landlord before the expiry of the one year except for forfeiture or re-entry for non-payment of rent and the other conditions were satisfied. However, Mr de la Mare referred us to section 52(5), which provides:

> If a protected tenancy is granted after the commencement of this section —
> (a) for such a term certain as is mentioned in subsection (1) above, to be followed, at the option of the tenant, by a further term; or
> (b) for such a term certain and thereafter from year to year or some other
> period; J
> and satisfies the conditions stated in that subsection, the tenancy is a protected shorthold tenancy until the end of the term certain.

Mr de la Mare submits that this tenancy did not fall within para (a) of that subsection, which is clearly right, and moreover did not fall within para (b) because it was not a tenancy for a term certain "and thereafter from year to year". Thus it is not within subsection (5) at all. He further submits that, if it is not within subsection (5), it is not a protected shorthold tenancy. Alternatively he submits that, even if the original year was for a protected shorthold tenancy, what happened after that was that he, the tenant, held over: he continued to pay, at least for a time (because he did fall into arrear), and the landlords continued to accept, rent at the contractual rate, and that
created at common law a tenancy from year to year which, being not K
governed by the agreement, could be determined only by a six months' notice to quit expiring at the end of the first or any subsequent year of the tenancy from year to year. He says that this notice was not a six months' notice — it was only a three months' notice — and therefore he is still holding over as tenant from year to year.

In answer to that Mr Hess submits that this was in effect a tenancy granted for a term certain and thereafter from year to year, but in any case he submits that it does not greatly matter — that is my phraseology, not his — whether that is correct or not, because what one has to do is to look at Case 19 to find out the circumstances under
which the order for possession may be made under that Case. That L
Case provides, so far as is material:

> Where the dwelling-house was let under a protected shorthold tenancy . . . and —
> (a) there either has been no grant of a further tenancy of the dwelling-house since the end of the protected shorthold tenancy or, if there was such a grant, it was to a person who immediately before the grant was in possession of the dwelling-house as a protected . . . tenant; and
> (b) the proceedings for possession were commenced after appropriate notice by the landlord to the tenant and not later than three months after the expiry of the notice.

the Case applies.

The notice is defined as being appropriate to this Case if M

> (i) it is in writing and states that proceedings for possession under this Case may be brought after its expiry; and
> (ii) it expires not earlier than three months after it is served . . .

Mr Hess submits that the tenancy under which this dwelling-house was let originally for the first year — the year ending on August 6 1984 — quite clearly was a shorthold tenancy within the meaning of section 52(1) of the 1980 Act. In that respect I agree with him. He then submits that at the end of the first year the tenant became tenant under a tenancy not arising by operation of law but arising by virtue of clause 2 of the tenancy agreement itself because in effect, he submits, the provision that "the Landlords shall give to the Tenant three months' notice in writing prior to the expiration date in the event of their not wishing to renew this Agreement for a further term'' means that if such notice be not given the agreement will be renewed for a further term, that is to say for a further year, and it will be this agreement which will be renewed, with the provision of three months' notice.

With that argument also I myself agree. In my view, in other words, for the year from August 7 1984 to August 6 1985 Mr de la Mare was in occupation of the house and premises as a tenant in all ways in accordance with the terms of the agreement of August 7 1983 save only that that tenancy would not itself be a protected shorthold tenancy within the meaning of section 52 of the Act. But it was a tenancy that was subject to three months' notice — not notice to quit, I emphasise, but to a notice from the landlords that they wished that the tenancy should not be renewed, in other words that they wished to determine the tenancy. In my view, whether or not such a tenancy, that is to say the original year's tenancy, fell within section 52(5) — I am inclined to think that it did not — it certainly did and does fall within the words of Case 19 of the Act to which I have referred.

Accordingly, in my view the judge was right in his conclusion that Case 19 did apply; he was entitled to make an order for possession under Case 19; the notice which had been given by the landlords, being a three months' notice, was valid to bring an end to this tenancy on August 6 1985; and accordingly that there was no valid ground for setting aside the judgment which he had entered on September 30 1985.

Accordingly I would dismiss this appeal.

Agreeing, SIR DENYS BUCKLEY said: The one point which has given me some cause for consideration in this case is the rather unusual language of the closing words of clause 2 of the tenancy agreement, which provides that the grant of the initial term of one year shall be ''subject to the provision that the Landlords shall give to the Tenant three months' notice in writing prior to the expiration date in the event of their not wishing to renew this Agreement for a further term''.

That provision, it seems to me, is clearly not contemplating what would normally be called a notice to quit. It is, I think, rather contemplating a notice to determine the agreement at the expiry of the current year of the tenancy. Such a notice was given in the present case at the end of the second year of the tenancy when, as I agree with Glidewell LJ, the tenancy had ceased to be a shorthold protected tenancy but was nevertheless a protected tenancy for the term of the second year of its subsistence. That notice having been given, the contractual tenancy, in my judgment, came to an end upon the expiry of that notice.

I agree with the reasons which have been put forward in the judgment which has been delivered by Glidewell LJ and reach the same conclusion as he for those reasons.

SIR JOHN DONALDSON MR also agreed and did not add anything.

The appeal was dismissed with costs.

Court of Appeal

October 26 1987

(Before Lord Justice NEILL and Lord Justice RALPH GIBSON)

BRICKFIELD PROPERTIES LTD v HUGHES AND OTHERS

Estates Gazette June 18 1988

[1988] 24 EG 95-102

Rent Act 1977 — Statutory tenancy — Whether tenant continued to occupy flat as his residence — Statutory tenant of a London flat began in 1978, after his retirement as a teacher, to stay with his wife in a cottage in Lancashire which his wife had inherited and which had previously been used for holidays only — Until he came to London for the trial of the present action in 1987 the husband had not been back to London at all and his wife had been back on three occasions only, to visit a sick child, to supervise rewiring and after a burglary — Since 1978 the flat had been occupied by the couple's four children and three of them, together with the husband of one of them, still lived there or had done until recently — The question was whether, in these circumstances, the defendants could resist an action for possession by the landlords on the ground that the husband's statutory tenancy continued to subsist — The county court judge decided in the defendants' favour — He found no difficulty about the establishment of a *corpus possessionis* in the shape of the furniture and the children left behind in the flat — As regards the *animus revertendi,* the judge took account of the husband's statement in evidence that he continued to regard the flat as his London home, that he would probably go back to the flat if his wife died first, and that in a few years' time he and his wife (both now in their seventies) might well have to move back to the flat, as it would be difficult to cope with the more primitive and uncomfortable conditions in the cottage, despite its scenic attractions — There was some evidence by a daughter that her parents would not be able to manage at the cottage and that she and a sister would like to help to look after them in the London flat — There was some other evidence of less weight, but altogether the judge was satisfied that the *animus revertendi* was established — The landlords appealed

In the course of his judgment Neill LJ reviewed a number of authorities, in particular *Gofor Investments Ltd* v *Roberts, Brown* v *Brash* and *Tickner* v *Hearn*, and formulated four guidelines as to the correct approach to be adopted by a county court judge — Held that, although this was a borderline case, and a troublesome one, it was impossible on the authorities to interfere with the decision reached by the judge after a careful consideration of the law and the facts — Ralph Gibson LJ expressed some sympathy with the view that some of the cases were going too far in discerning an intention to return in spite of a prolonged voluntary absence, but accepted that the judge's conclusion had, on the authorities, been open to him — Appeal dismissed

The following cases are referred to in this report.

Beck v *Scholz* [1953] 1 QB 570; [1953] 2 WLR 651; [1953] 1 All ER 814, CA
Brown v *Brash* [1948] 2 KB 247; [1948] 1 All ER 922, CA
Cove v *Flick* [1954] 2 QB 326n; [1954] 3 WLR 82n; [1954] 2 All ER 441, CA
Gofor Investments Ltd v *Roberts* (1975) 29 P&CR 366, CA
Hall v *King* [1987] 2 EGLR 121; (1987) 283 EG 1400, CA
Hampstead Way Investments Ltd v *Lewis-Weare* [1985] 1 WLR 164; [1985] 1 All ER 564; [1985] 1 EGLR 120; (1985) 274 EG 281, HL
Tickner v *Hearn* [1960] 1 WLR 1406; [1961] 1 All ER 65, CA

This was an appeal by the landlords, Brickfield Properties Ltd, from a decision of the late Judge John Edwards, at West London County Court, dismissing their action to recover possession of a first-floor flat at 20 Ashburnham Mansions, Ashburnham Road, London SW10. The respondents consisted of the first defendant below, John Glynne Hughes, the tenant to whom the flat had been let, the second and third defendants, a daughter and her husband, and the fourth defendant, a son.

David Neuberger QC (instructed by Memery Crystal & Co) appeared on behalf of the appellants; M R Dencer (instructed by Oliver O Fisher & Co) represented the respondents.

Giving judgment, NEILL LJ said: This is an appeal by Brickfield Properties Ltd, the landlords, from the decision of the late Judge John Edwards in the West London County Court on April 2 1987 whereby he dismissed the landlords' claim for possession of a flat on the first floor of 20 Ashburnham Mansions, London SW10.

Mr J G Hughes, the first defendant, who is now aged 74, was granted a lease of the flat in July 1965 by the predecessors in title of the landlords for a term of seven years from September 29 1961. The lease expired on September 29 1968, and it is common ground that thereafter, until the summer of 1978, Mr Hughes remained in the flat as a statutory tenant. It is to be noted that Mr Hughes had in fact lived at the flat with his wife since their marriage in 1954.

The question at issue before the judge was whether Mr Hughes continued to be a statutory tenant after the summer of 1978 and could therefore rely on the protection of the Rent Act 1977. The judge decided this issue in his favour.

The primary facts were not in dispute.

In or about 1970 Mrs Hughes inherited a cottage at Carnforth in

A Lancashire. For the next few years Mr and Mrs Hughes used the cottage for their holidays. In 1978, however, Mr Hughes retired from his work as a teacher. After spending the summer holiday in the cottage that year Mr and Mrs Hughes stayed on. They have been in the cottage at Carnforth ever since. Indeed, until he came to London at the end of March 1987 for the purpose of attending the trial, Mr Hughes had not been back to the flat at all. In those eight and a half years or so Mrs Hughes has been back to the flat on three occasions only, but on each occasion there was a special reason for the visit and all the visits were before 1982. On one occasion she came to visit a sick child; on another occasion she came so that there could be someone in the flat when rewiring was being carried out; and on the third occasion there had been a burglary.

B Since 1978 the flat has been occupied by the four children of Mr and Mrs Hughes. It was the home where they were brought up and some of them have stayed on since they became adults. Mrs Nicholas, the second defendant, a daughter now aged 32, lives there with her husband, the third defendant. Two other children — Jonathan (33), who is the fourth defendant, and Polly (29) — also live there or have done until recently. David Hughes, now 31, lives in Lancashire, not far from his parents.

What, then, is the law, because, as the judge put it in his judgment: "in ordinary terms and language" Mr and Mrs Hughes were living in the cottage between 1978 and the date of the trial?

Section 2 (1) of the Rent Act 1977 is in these terms:

C Subject to this Part of this Act —
(*a*) after the termination of a protected tenancy of a dwelling-house the person who, immediately before that termination, was the protected tenant of the dwelling-house shall, if and so long as he occupies the dwelling-house as his residence, be the statutory tenant of it.

It is also necessary to read section 2 (3). That subsection provides:

In subsection (1) (*a*) above and in Part I of Schedule 1, the phrase "if and so long as he occupies the dwelling-house as his residence" shall be construed as it was immediately before the commencement of this Act (that is to say, in accordance with section 3 (2) of the Rent Act 1968).

That takes one to the 1968 Act. Section 3 (2) of the 1968 Act is in these terms:

In paragraph (*a*) of subsection (1) above and in Schedule 1 to this Act, the D phrase "if and so long as he occupies the dwelling-house as his residence" shall be construed as requiring the fulfilment of the same, and only the same, qualifications (whether as to residence or otherwise) as had to be fulfilled before the commencement of this Act to entitle a tenant, within the meaning of the Increase of Rent and Mortgage Interest (Restrictions) Act 1920, to retain possession, by virtue of that Act and not by virtue of a tenancy, of a dwelling-house to which that Act applied.

The effect of section 3 (2) of the Rent Act 1968 was considered by Cairns LJ in *Gofor Investments Ltd* v *Roberts* (1975) 29 P&CR 366. At p 369 the learned lord justice said:

So the effect of that subsection was that in considering whether a person did retain a dwelling-house as his residence, one had to look at the authorities prior to 1968 to see what the position was.

E It was laid down in a number of cases from the 1920s onwards that the fact that the person was not physically in residence at any particular time did not mean that the house had ceased to be his residence. I take as a convenient summing-up for the purpose of this case one sentence from the judgment of Ormerod LJ in *Tickner* v *Hearn* [1960] 1WLR 1406

as expressed in the headnote:

In such cases

that is cases of the kind to which I have been referring

there must be a real hope of return coupled with the practical possibility of its fulfilment within a reasonable time.

Cairns LJ then went on to cite a passage from the judgment of the Court of Appeal delivered by Asquith LJ in *Brown* v *Brash and* F *Ambrose* [1948] 2 KB 247. *Brown* v *Brash* can be regarded as the leading authority in this branch of the law, and I therefore propose to cite the same passage. In *Brown* v *Brash* Asquith LJ, delivering the judgment of the court, said at pp 254–5:

We are of opinion that a "non-occupying" tenant prima facie forfeits his status as a statutory tenant. But what is meant by "non-occupying"? The term clearly cannot cover a tenant who, for however short a time, or however necessary a purpose, or with whatever intention as regards returning, absents himself from the demised premises. To retain possession or occupation for the purpose of retaining protection the tenant cannot be compelled to spend twenty-four hours in all weathers under his own roof for three hundred and sixty-five days in the year. Clearly, for instance, the tenant of a London house who spends his week-ends in the country or his long vacation in Scotland does not necessarily cease to be in occupation. Nevertheless, absence may be sufficiently prolonged or unintermittent to compel the inference, prima facie, G of a cesser of possession or occupation. The question is one of fact and of degree. Assume an absence sufficiently prolonged to have this effect: The legal result seems to us to be as follows: (1) The onus is then on the tenant to repel the presumption that his possession has ceased. (2) In order to repel it he must at all events establish a de facto intention on his part to return after his absence. (3) But we are of opinion that neither in principle nor on the authorities can this be enough. To suppose that he can absent himself for five or ten years or more and retain possession and his protected status simply by proving an inward intention to return after so protracted an absence would be to frustrate the spirit and policy of the Acts, as affirmed in *Keeves* v *Dean* [1924] 1 KB 685 and *Skinner* v *Geary* [1931] 2 KB 546. (4) Notwithstanding an absence so protracted the authorities suggest that its effect may be averted if he couples and clothes his inward intention with some formal, outward, and visible sign of it; that is, installs in the premises some caretaker or H representative, be it a relative or not, with the status of a licensee and with the function of preserving the premises for his own ultimate home-coming. There will then, at all events, be someone to profit by the housing accommodation involved, which will not stand empty. It may be that the same result can be secured by leaving on the premises, as a deliberate symbol of continued occupation, furniture; though we are not clear that this was necessary to the decision in *Brown* v *Draper* [1944] KB 309. Apart from authority, in principle, possession in fact (for it is with possession in fact and not with possession in law that we are here concerned) requires not merely an *animus possidendi* but a *corpus possessionis*, namely, some visible state of affairs in which the *animus possidendi* finds expression. (5) If the caretaker (to use that term for short) leaves or the furniture is removed from the premises, otherwise than quite temporarily, we are of opinion that the protection, artificially prolonged by their presence, ceases, whether the tenant wills or desires such J removal or not.

In addition, I propose to refer to the judgment of Upjohn LJ in *Tickner* v *Hearn* [1960] 1 WLR 1406. I can take the facts of that case conveniently from the headnote:

An elderly lady, the statutory tenant of a rent-controlled house in which she lived with her unmarried daughter, went on a temporary visit to another daughter in July 1954. While there, she had a breakdown and was admitted to a mental hospital under a summary reception order, suffering from paranoid schizophrenia. She remained in the hospital continuously thereafter, though by November 1959, when she was 73, she had so far improved as to be regraded to voluntary status and became entitled to discharge herself at short notice. Her unmarried daughter continued to maintain the house as their home. K

In December 1959, landlords, who had purchased the house in January of that year, brought proceedings against the tenant and her daughter, claiming that in view of her long absence the tenant was no longer entitled to the protection of the Rent Restriction Acts, and as the daughter's right to occupation depended on that of her mother, she was similarly disentitled from remaining in possession.

The county court judge found that, though it was unlikely in view of the tenant's age and condition that she would ever leave the hospital, she had always regarded the house as her home and frequently expressed the hope of going back there, and that in view of the advances in medical treatment and her own improvement the possibility of her return to her home one day could not be ruled out. He held that the evidence was just sufficient to discharge the onus on the tenant, and had satisfied him that there had been no real abandonment of possession. On the landlords' appeal: L

Held, dismissing the appeal, that there was evidence on which the judge could find an intention to return to live in the house.

Among the issues which the judge had to decide were whether, in view of her mental condition, Mrs Hearn was capable of forming an intention to return and whether there was a real possibility of her return within a reasonable time. At p 1416 Upjohn LJ dealt with the second of these issues as follows:

. . . I am of opinion that to establish an intention to return, by itself, is not necessarily sufficient to obtain the protection of the Acts. The ultimate fact to be established is the fact of occupation, for it is only those in occupation whom the Act protects. If a tenant is out of occupation for some time he must at least prove an intention to return; but if the circumstances of the case are such that by reason of mental or physical illness or for some other reason the M intention, even if clearly and bona fide held, seems most unlikely to be achieved within a reasonable time, it must surely be open to the judge in a proper case to find as a fact that the tenant is not in occupation.

A little later, having referred to a passage in the judge's judgment, Upjohn LJ continued at p 1417:

. . . As I read his judgment

that is the judgment of the judge

the conclusion which he reached was that, looking at all the facts during her mental illness, he regarded it as a borderline case, but he came to the conclusion on those facts that she had established an intention to return and, at least, a real possibility of her return within a reasonable time.

From these passages, and from other passages to which we were

A referred by counsel in the course of their very helpful submissions, I would extract the following guidelines as to the correct approach to be adopted by a county court judge:

(1) Where the tenant's absence is more prolonged than is to be explained by holiday or ordinary business reasons and is unintermittent, the onus lies on the tenant of establishing an intention to return if he seeks the protection of the Act.

(2) An inward intention, however, is not enough. It must be accompanied by some outward and visible sign of the tenant's intention. The continued occupation by a caretaker or relative or the continued presence of furniture may be sufficient, but in each case the question is whether or not the person or furniture can be regarded as a genuine symbol of his intention to return "home".

B (3) In addition the tenant must show that there is a "practical possibility" — to use the phrase used in *Tickner* and by Ormerod LJ in *Gofor* — or a "real possibility" — to use Upjohn LJ's expression in *Gofor* — of the fulfilment of that intention within a reasonable time. What is a reasonable time depends on the circumstances, but it is to be noted that in *Gofor* the Court of Appeal declined to interfere with the judge's finding that in the circumstances of that case 10 years was a reasonable time.

(4) The protection of the Act can be claimed even though the tenant has another home or residence — the two words appear to be synonymous in this context (see Lord Brandon in *Hampstead Way Investments Ltd* v *Lewis-Weare* [1985] 1 WLR 164, 169)* — but the
C court will look with particular care at two-home cases.

Before turning to the facts of the present case, I propose to remind myself finally of the words used by Lawton LJ in *Gofor* when considering the approach of an appellate court where the judge had correctly directed himself as to the law. Having decided that the judge had asked himself the correct question, Lawton LJ continued (at p 375):

> This court next has to inquire whether the evidence was sufficient to support his findings. Another way of putting this issue is to ask whether his findings on the evidence were so unreasonable as to be perverse. The judge gave most careful attention to the evidence in this case and he came to the conclusion, surprising, perhaps, to some, that there was a real intention to return and a sufficient *corpus possessionis*.

D I return to the case presently before the court. I can deal shortly with what is called the *corpus possessionis*. Mr Hughes relied on the fact that the flat was occupied by members of his family and on the fact that, though he had taken his clothes with him, he had left his furniture and books behind. It was argued on behalf of the landlords that, though these facts were not disputed, there was no evidence that either the family or the furniture were there for the purpose of the alleged homecoming, and the only reason why the furniture had not been taken was because the cottage was fully furnished.

The judge, however, disposed of this matter in a few words. According to the note of his judgment with which we have been provided he said:

> E I move to the *corpus possessionis;*, it *is* in dispute. However I feel I can deal with that matter pretty shortly on a simple finding. Listed in the further and better particulars are the items of furniture etc left by the first defendant in the flat. Mr Hughes spoke of them in evidence. That is the first element. Secondly, quite apart from that, it cannot be disputed that he left behind some of his children. There is a dispute as to their status in the flat. I believe, however, that I am entitled to find, and do find from the facts that the bulk of his possessions and his loved ones are left in the flat, that Mr Hughes does quite easily establish *corpus possessionis*.

It seems to me that there was evidence on which the learned judge could reach that finding, and it is not a finding with which I think this court can or should interfere.

The arguments concerning Mr Hughes' intention, however, are more difficult. Here it is necessary to cite a substantial passage from
F the judge's judgment. First the judge set out the test which he had to apply. He put it in this way:

> Does the defendant prove
> (i) an intention to return;
> (ii) plus a real possibility of returning;
> (iii) within a reasonable time.

Lower down on p 46 he continued as follows:

> That leaves *animus revertendi*. I have already said that I accept everything that Mr Hughes told me and think it fair to comment that he was a conspicuously honest witness. It would have been only too easy for him to say that he was intending to go back to the flat some time. However, he has not said that. He says that he intends to live and to continue to live indefinitely in G Carnforth in the bungalow. However, he has also said that he has continued to regard the flat as his London home, and says he did not positively say or do anything to assign it over to his children.
>
> It does count a little that he has continued to pay for the flat, although he has been taking contributions from his children. Relevant and marginally weighty are six factors linking him to London. They are not worth much, as both counsel are agreed.

The learned judge then set out a number of matters, such as the fact that his pension book showed his London address and his entry in the London telephone directory. However, as the judge regarded them as of only limited importance I do not think it is necessary for the purposes of this judgment to refer to them in detail. The judge continued: H

> These are some practical things. Now I turn again to what he said and what he meant.
>
> Though he lived in the bungalow as his current main home since 1978, he *said* that he has considered 20 Ashburnham Mansions as his London home throughout. That may be of light weight, but it is relevant.
>
> Secondly, he has said that he intends to go back to the flat one day to live. He said that most clearly, I regret to say, in answer to a question from me, and I am not quite sure what his exact answer was. The defendants' solicitors' note, which roughly accords with mine, is that "I did have a basic intention to go back". My own note reads: "Basic intention was to go back to the flat one day". I do not remember if "basic" was his word or mine. He went on to say that he would like to go back to the flat, but his wife was happier in Lancashire and he was too. Counsel for the plaintiffs had a slightly different recollection and note. Anyhow, I take from Mr Hughes a statement that he does intend to J go back to live in the flat.
>
> There are two clear circumstances which he indisputably said would involve him going back to the flat.
>
> The first, and especially notable, is if his wife should die first. He would probably have to go back to the flat then; he is 74 and she is 71, but, as he told me, not well. The point is that if he was left alone in the bungalow he could not cope on his own. He would have to go back to London to be looked after by his daughters.
>
> The second circumstance in which he might have to go back is in a few years' time — he mentioned a period of 5 years from now — when it might come about that he and his wife could not continue to cope at Carnforth. There is plenty of evidence about Carnforth. It is attractive scenically, but uncomfortable and primitive. What he means and what I accept is that the time cannot be far away when both he and his wife, ageing as they are, will K have to move back to the flat.
>
> A third element comes from the evidence of his daughter Jane. I find as a fact, if necessary, that the family is an affectionate and close one and they get on well together.
>
> Jane says: "I should like to take care of my father at the flat . . . I think that will have to happen pretty soon . . . I'd rather like my father and my mother to come here and live with us. The cottage looks pretty but I don't think they could manage much longer or last another winter there. Polly would help."
>
> I think I am entitled in my judgment to give weight to Jane's evidence at least on the second aspect of the observations I have cited from Upjohn LJ, namely the "real possibility" of the first defendant returning to 20 Ashburnham Mansions.
>
> It seems to me that my task is to weigh up these considerations, which I accept, against the potent effect of evidence that the first defendant has lived L elsewhere for nearly 9 years and has never come to London. It is a matter of balancing the evidence, and there is a conflict of inference. The proper basis on which to approach the matter is the balance of probabilities. The burden is on the first defendant, as I have said. All I can say is that the matter is not simple and is very borderline, but in all these circumstances, I have reached the clear conclusion that the first defendant has established what he has to, namely that he has an intention to return. Although he has been living in Lancashire for nearly 9 years, my decision is that he has a sufficient *animus revertendi*, that is an intention coupled with a real possibility of perfecting that intention, and that he is therefore still in occupation of 20 Ashburnham Mansions as his residence, and so his statutory tenancy still survives.

It was argued on behalf of the landlords that the two possibilities mentioned by Mr Hughes — that his wife might predecease him and the risk that, with increasing age, they might not be able to cope at M Carnforth, even if they both survived — showed that his intention was conditional and was too vague and uncertain to satisfy the test of intention laid down in the cases.

The question which the judge had to consider was whether, on the facts, there was a real possibility that Mr Hughes' intention to return would be fulfilled in a reasonable time. As to this the judge was clearly impressed by the daughter's evidence that she did not think that they could last another winter in the cottage. It seems to me that, though the prospect of Mr Hughes coming back because his wife had predeceased him might well have failed the test of a real possibility, the judge was entitled to infer and did infer that there was a real possibility that with advancing years they will both have to come back

* Editor's note: Also reported at [1985] 1EGLR 120.

to the flat quite soon. As the judge put it, "The time cannot be far away."

I confess that I have found this a troublesome case, but in the end I have come to the conclusion that it is impossible, in the light of the authorities, to interfere with the decision which the judge reached after what was clearly a very careful review both of the law and the facts. Accordingly, I for my part would dismiss the appeal.

I would add this. As the judge recognised, this is a borderline case. It seems to me that in a case such as the present the burden on the trial judge is an onerous one, because the scope for an appellate court to interfere is very limited. The court is clearly under a duty to extend the protection of the Rent Act to all those who need it. But it is to be remembered that the policy of the Act is to protect a personal right of occupation and that the words in section 2(1) of the Rent Act 1977 are "if and so long as he occupies the dwelling-house as his residence". It seems to me, therefore, that it is particularly important in cases where it is said that the tenant has two homes to scrutinise the evidence with care in order to distinguish between the case where the second "home" is indeed a present residence and the case where the second "home" is retained merely as a convenience for possible occasional visits or as premises which one day in the future may become a residence. I have in mind the words of Sir Raymond Evershed MR in *Beck* v *Scholz* [1953] 1 QB 570 where, at p 575, he said:

> . . . I cannot see that it would be in accordance with the main principle and purpose of the rent restriction legislation to hold that a man may have statutory protection for any premises to which he may occasionally find it convenient to resort, and in which he may keep furniture and install a caretaker, when in no true sense of the term are those premises his "home", for example, where they are merely used as a matter of convenience for occasional visits.

Nevertheless, on the facts of the present case, and in the light of the judge's findings, for the reasons which I have endeavoured to outline, I for my part would dismiss this appeal.

Agreeing, RALPH GIBSON LJ said: There is in my judgment only one real issue in the appeal, namely whether such intention to return to the flat as Mr Hughes has had over the years is sufficient for the purposes of the Rent Act 1977. On this matter I have found the case of some difficulty.

The learned judge reached the "clear conclusion" that Mr Hughes had established an intention to return. The matter was, in the judge's view, very borderline. The intention to return was sufficient because the intention was coupled with a real possibility of perfecting the intention.

The nature of Mr Hughes' intention which the judge found to be established is as follows. Since leaving the flat in 1978 Mr Hughes:

(1) Has had no wish to return to the flat. He and his wife prefer to live in Lancashire.
(2) After eight and three-quarter years without personally visiting the flat Mr Hughes still has no wish to return there: he and his wife still intend to live in the cottage in Lancashire indefinitely for so long as they are able to do so.
(3) The "basic intention" to return which the judge accepted Mr Hughes had always had would be carried out only, first, if his wife should die before him and, second, when he and his wife could no longer cope with living at the cottage. Mr Hughes thought that might happen in five years from April 1987, but the judge accepted that that time could not be far away.

It is clear that when a tenant claims to be the statutory tenant of a dwelling-house on the ground that, although he is not residing there, nevertheless he "occupies it as his residence" by reason of his intention to return to it, the question whether any intention he has is sufficient for the purposes of the Rent Act is one of fact and degree to be decided by the judge in all the circumstances of the case: see *Gofor Investments Ltd* v *Roberts per* Cairns LJ at p 373 and Lawton LJ at p 374.

It is open to this court to set aside the finding of the county court judge on that issue if it should appear to the court that, on the evidence, the intention which the tenant had was not such as to satisfy the requirements of the Rent Acts because, for example, it was too "contingent" or remote as to implementation (see *Cove* v *Flick* [1954] 2 QB 326 *per* Somervell LJ at p 328).

I think that the phrase "real intention to return" used by Lawton LJ in the *Gofor Investments* case at p 375 includes not only the reality of any purpose of returning which the defendant may assert but also the remoteness in time and contingency of any likely implementation of that purpose.

References have been made to remoteness in terms of the number of years: thus in *Brown* v *Brash*, already cited by Neill LJ, Asquith LJ said:

> . . . To suppose that

a non-occupying tenant

> can absent himself for 5 or 10 years or more and retain possession and his protected status simply by proving an inward intention to return after so protracted an absence would be to frustrate the spirit and policy of the Acts.

But Asquith LJ continued:

> . . . Notwithstanding an absence so protracted . . . its effect may be averted if he couples and clothes his inward intention with some formal, outward and visible sign of it.

In the *Gofor Investments* case the tenant had gone out of occupation of the dwelling-house, a flat, in 1970, and her intention was to return after a period of "not more than 10 years". It was submitted to the county court judge in that case that that was not a reasonable time, but he held that the reasonableness of the time over which the intended absence would extend was to be judged in the context of all the circumstances. Since the husband was "on a pension" it was not unreasonable to intend to be away for about 10 years. The county court judge thought it might be different "if they were merely going for pleasure or something of that sort", but there was nothing unreasonable "in saying that this flat does not cease to be their home for the time that the children are being educated in Malta".

Cairns LJ, after reciting those parts of the judgment of the county court judge, said that in his opinion the judge was applying the right test. Lawton LJ noted that counsel for the landlords had conceded that the judge had asked himself the right questions, but it is clear I think that Lawton LJ did not question the correctness of the concession. Accordingly, it seems to me to be impossible for this court to say that some limit can be set to the length of intended absence: any absence must be judged in the context of all the circumstances.

I said that I had found this part of this case to be difficult. My reason is that I find it hard to see by reference to what principles the county court judge is required to make his decision "in the context of all the circumstances". He is concerned with the fact of occupation as a residence and not with general reasonableness or greater hardship. Parliament has not enacted that a statutory tenant may retain a dwelling-house "as his residence" if he uses it in fact to provide a place to live for his children. The test to be applied is that the tenant will remain a statutory tenant under the Act "if and so long as he occupies the dwelling-house as his residence", but that phrase is to be construed and applied in accordance with the case law established immediately before the commencement of the Rent Act 1968: see *Hall* v *King*, June 22 1987,* decided in this court by Sir John Donaldson MR, Lloyd LJ and Balcombe LJ.

In the *Gofor Investments Ltd* case the county court judge thought that for the wife of a man on a pension to go with him to Morocco or Malta to live for 10 years while the children were educated in Malta was more "reasonable" than if the visit abroad was for "mere pleasure". That showed — since his view was approved by this court — that some prolonged absences may be seen as more meritorious than others for the purposes of the Rent Acts.

It seems to me to be clear that there is a relevant distinction between an absence caused by employment or government service or through illness, on the one hand, and an absence which is more truly voluntary, on the other. Examples of less voluntary absences are the sea captain mentioned in the early cases and, in *Tickner* v *Hearn* [1960] 1 WLR 1406, to which Neill LJ has referred, the tenant, a voluntary patient in a hospital, who wished to return to her home but remained in hospital at all times after a short period of certification because she was advised to stay and was too unwell to leave.

Mr Neuberger submitted that the distinction between the absence caused by work or illness and what he called a purely voluntary absence was of crucial importance in the judging of a long-intended absence. He argued that the intended absence of Mr Hughes in this case from 1978 until such time as he and his wife found themselves unable to cope with life in the cottage in Lancashire was voluntary and that, having regard to its probable duration in 1978 and the time over which it has in fact continued, this court should rule that it was not open to the county court judge to find that Mr Hughes had, for the purposes of the Rent Acts, a sufficient intention to return.

*Editor's note: Reported at [1987] 2 EGLR 121; (1987) 283 EG 1400.

A The learned judge did not in his judgment expressly consider the matter of voluntariness. Mr Neuberger, however — rightly, as I consider — disclaimed any contention that the learned judge had in that regard failed properly to direct himself. Mr Neuberger had submitted to the judge below the argument that he addressed to us, and the judge's use of the word "sufficient" indicated that he was finding not only that Mr Hughes had at all times had the intention described above but that, in the light of the authorities, it had the necessary quality. The learned judge had, in short, considered it in the light of all the circumstances in accordance with the guidance given by the *Gofor Investments* case and found it sufficient. Mr Neuberger submitted, nevertheless, that the judge was wrong. The cases had, he said, gone too far and this court should draw the line so
B as to exclude Mr Hughes from the protection of the Rent Acts.

I listened to this submission with some sympathy. On first reading the papers I found it surprising that this learned judge reached the conclusion which he did reach in this case. It is not easy to say of Mr Hughes that, by reason of an intention to return to it, he has throughout been occupying the flat (the dwelling-house) as his residence. Mr Hughes has in fact at no time had a present desire or purpose to return to the flat: he has had no more than what the judge called "the basic intention", that is to say, as it seems to me, the recognition and acceptance that the time will come when he will have to return to be cared for there by his children, and a purpose to return when that time arrives. The distinction between Mr Hughes and the C
tenant in *Tickner* v *Hearn* is obvious. The only dwelling-house of her own which that tenant had was the house of which she was tenant, and throughout it was her immediate wish to return to it; but Mr Hughes has throughout lived in the dwelling-house which he and his wife have preferred to occupy as a residence, namely the cottage in Lancashire.

I see force in this submission; and I think that a judge, in trying such cases, might on similar facts find such a submission decisive; but in my judgment it is impossible for this court to apply the point made by Mr Neuberger so as to rule that the judge's conclusion was not open to him as a matter of law. The absence of 10 years intended by Mrs Boyd, the tenant in the *Gofor Investments* case, was, so far as I
can see, in no relevant sense less voluntary than the absence of Mr D
Hughes. Mrs Boyd and her husband found it financially convenient, and, no doubt, agreeable, to live in Morocco and Malta. Mr and Mrs Hughes had found it agreeable, and it may be also cheaper, to live during their active retirement in a cottage in Lancashire. In the light of the decision of this court in the *Gofor Investments* case it was open to the learned judge to decide the case as he did, and I therefore agree that this appeal should be dismissed.

The appeal was dismissed with costs; leave to appeal to the House of Lords was refused.

LANDLORD AND TENANT

RENT REVIEW

Chancery Division

July 30 1987

(Before Mr E G NUGEE QC, sitting as a deputy judge of the division)

PANAVIA AIR CARGO LTD v SOUTHEND-ON-SEA BOROUGH COUNCIL*

Estates Gazette January 9 1988

[1988] 01 EG 60-64

Landlord and tenant — Rent review provisions in lease — Construction — Lease with "a number of glaring inconsistencies" — Rent reviewable, after an initial period, at five-year intervals — Following a somewhat complex statement of the review machinery it was provided that if a review was not completed within 12 months of the commencement of a relevant period the rent payable during that period would be the current rent increased by 25% — The main issue was as to the true construction of this provision — The tenants argued that, as the review in question in the present case had not in fact been completed in time, an automatic increase of 25% in the current rent took effect — The landlords contended that the 25% provision was merely an interim arrangement and that, if a review were subsequently completed, the reviewed rent would take effect if it produced a figure of more than 125% of the rent for the previous relevant period — It was pointed out that it was stated in the review provisions that time was not to be deemed of the essence — Held (rejecting the tenants' construction) that the 25% increase did not come into operation automatically if the review was not completed within 12 months of the completion of the relevant period — Time was not of the essence in that respect — It was open to the tenants to give notice making time of the essence, but unless that were done the rent remained payable at the current rate until the surveyor appointed under the lease should determine a new rent

The following cases are referred to in this report.

Lewis v *Barnett* (1981) 264 EG 1079

Smith's (Henry) Charity Trustees v *A WADA Trading & Promotion Services Ltd* (1983) 47 P&CR 607; [1984] EGD 103; 269 EG 729, CA

United Scientific Holdings Ltd v *Burnley Borough Council* [1978] AC 904; [1977] 2 WLR 806; [1977] 2 All ER 62; (1977) 33 P&CR 220; [1977] EGD 195; (1977) 243 EG 43 & 127, HL

This was an originating summons by which the plaintiff tenants, Panavia Air Cargo Ltd, sought a determination of the true construction of rent review provisions in a lease by the defendant landlords, Southend-on-Sea Borough Council, of about $2\frac{3}{4}$ acres of land forming part of Southend-on-Sea Municipal Airport, with a hangar and buildings thereon.

Jonathan Gaunt (instructed by Cameron Markby) appeared on behalf of the plaintiffs; David Neuberger QC (instructed by Rowe & Maw) represented the defendants.

Giving judgment, MR EDWARD NUGEE QC said: This summons raises a question on the construction of the rent review provisions contained in a lease granted by the defendant council, Southend-on-Sea Borough Council, to the plaintiff company, Panavia Air Cargo Ltd, formerly known as British Air Ferries Ltd. The lease is dated October 11 1978 and comprises about $2\frac{3}{4}$ acres of land forming part of Southend-on-Sea Municipal Airport, together with an aircraft hangar and certain other buildings erected thereon.

By clause 1 of the lease the council demised the property to the company

> to hold the same unto the company from the 1st day of January 1978 to the February 13 2035, yielding and paying therefor during the said term for the period January 1 1978 to February 13 1985 the annual rent of £2,700 plus any additional rent payable under the provisions for review contained in the schedule hereto without any deduction whatsoever by equal quarterly payments in arrear to be made on the usual Quarter Days.

The schedule contains provisions for reviewing the rent, of which the following are the most material:

> (1) During the seventh, twelfth, seventeenth, twenty second, twenty seventh, thirty second, thirty seventh, forty second, forty seventh and fifty second years of the said term for each successive five-year period of the said term and during the fifty second year of the said term for the remaining period of the said term, each of such periods being hereafter referred to as a relevant period, the Council and the company shall agree, or failing agreement shall determine by arbitration the sum total of the then current rack rent . . . and one quarter of the sum total so ascertained, or £2,700, whichever is the greater, shall be the rate of rent reserved by this lease in respect of the then next succeeding period . . . (3) If such agreement has not been made six months before the commencement of the relevant period the Council may require an independent surveyor (hereinafter called the surveyor) to be appointed to determine the new rent. The surveyor may be nominated by agreement between the Council and the company or appointed by the President for the time being of the Royal Institution of Chartered Surveyors on the application of the Council . . . (5) Notice in writing of his appointment by the said President shall be given by the surveyor to the Council and company, inviting each to submit within a specified period which shall not exceed four weeks a valuation accompanied if desired by a statement of reasons. (6) The surveyor shall act as an expert and not as an arbitrator. (7) The surveyor shall give notice in writing of his decision to the landlord and the tenant within two months of his appointment, or within such extended period as the Council may agree. (8) If the surveyor comes to the conclusion that the current market value of the demised premises is less than the rent operative for the period preceding the relevant period (hereinafter called the current rent) the new rent shall nevertheless be the same as the current rent, and the decision of the surveyor shall so state. (9) If the surveyor shall fail to determine the new rent and give notice thereof within the time and in the manner hereinbefore provided, or if he shall relinquish his appointment, or die, or if it shall become apparent for any reason he will be unable to complete his duties hereunder, the Council may apply to the said President for a substitute to be appointed in his place, which procedure may be repeated as many times as necessary . . . (11) Rent shall not be due at the rate of the new rent until after the tenant has been given such notice thereof as is hereby provided, and in the event of the relevant period starting before such notice has been given to him, the rent shall continue to be due at the rate of the current rent on each day appointed by this agreement for payment of rent until the said notice is given to him. On the first day after the notice is given to him, which is a day appointed by this agreement for payment of rent, there shall fall due for payment the appropriate instalment at the new rate together by way of additional rent a sum equal to the difference between the new rent and the rent actually paid for any part of the relevant period in respect of which a rent less than the new rent has been paid . . . (13) As respects the periods of time referred to in this schedule, time shall not be deemed to be of the essence of the contract. (14) If for any reason whatsoever no review of rent takes place in respect of any relevant period, or a review is not completed within twelve months of the commencement of any relevant period, then the rent payable during the relevant period under the provisions of this lease shall be increased by 25% of the rent then currently payable.

As can be seen, there are a number of glaring inconsistencies in the lease. Thus the reddendum in clause 1 states that the initial rent is to

*Editor's note: For Court of Appeal decision see p 124 *post*.

A be payable until February 13 1985. Para (1) of the schedule clearly indicates that each relevant period is to begin on the anniversary of the commencement of the term, that is to say, on January 1. Para (1) states that the rent shall be determined in default of agreement by arbitration, but para (6) clearly provides that the surveyor is to act as an expert and not as an arbitrator. Para (1) states that the rent during each relevant period shall be one quarter of the sum total ascertained or £2,700, whichever is the greater, and para (8) compares the current market value of the demised premises and the rent operative immediately preceding the relevant period, while para (14) provides that in certain circumstances the rent payable during the relevant period shall be the current rent increased by 25%.

B The question before me turns on the effect of para (14) construed in the context of the lease as a whole and particularly in the light of para (13).

The facts giving rise to the present dispute are as follows. The seventh year of the term was the calendar year 1984. On September 13 1984 the council's borough valuer and surveyor wrote to the company drawing attention to the fact that the first rent review under the lease was to operate from January 1 1985 and asking whether the company would be appointing a professional adviser to agree terms on its behalf. A rather desultory correspondence followed, until on January 17 1986 the borough valuer and surveyor wrote to the company's surveyors saying:

> This rental review is now considerably overdue and if the negotiations do not
> C progress in the very near future I shall have to report accordingly to the Council with a view to proceeding towards the appointment of an independent valuer in accordance with the schedule to the lease.

On February 17 1986, four days after the expiration of 12 months from the date mentioned in the reddendum, the company's surveyors wrote back saying:

> We have now taken our client's solicitor's advice, which is as follows. Under clause 14 of the schedule to the lease, which contains the relevant provisions, there will be an automatic 25% increase in the rent if, *inter alia*, a review has not been completed within twelve months of the commencement date. The automatic increase seems to happen whether or not a disputed increase has been decided by a surveyor at the time. It is arguable as a result of clause 1 of the lease the relevant rent review date is February 13 1985 rather than
> D December 31 1984, and we confirm that if the rent review has not been completed by twelve months from the relevant date the rent should merely be increased by 25%

and they ask the council to confirm the rent payable from the review date was to be increased to £3,375 pa, which is a 25% increase on the initial rent of £2,700. Since that letter the council has applied to the president of the Royal Institution of Chartered Surveyors for the appointment of a surveyor, but he has declined to appoint one in view of the dispute between the parties.

Rather more than a year after the last letter to which I have referred, this summons was issued by the company. It seeks a declaration that the rent payable by the plaintiff to the defendant under the above-mentioned lease with effect from February 13 1985
E is, in the events which have happened, to be assessed in accordance with para (14) of the schedule to the said lease.

Mr Gaunt, counsel for the company, points to the fact that all the initiatives under the schedule lie in the council's hands. Only the council can apply for the appointment of a surveyor under para (3); only the council can extend the time-limit for the surveyor's valuation under para (7); only the council can apply under para (9) for the appointment of a substitute surveyor. Para (14), he submits, was intended to give the company some measure of certainty. The schedule gives the council machinery which it can operate comparatively quickly and which should result in a new rent being determined some time before the commencement of the relevant
F period. The council is allowed to operate it more slowly, but it says if for any reason the rent is not determined by the surveyor within 12 months after the commencement of the relevant period, then the automatic increase of 25% takes effect. If para (14) does not mean what it says, he submits it does not mean anything at all.

Mr Neuberger, counsel for the borough council, submits first of all that para (14) is an interim provision which takes effect if no review has taken place or been completed within 12 months from the commencement of the relevant period, and that if a review is subsequently completed, the reviewed rent takes effect if it results in a rent greater than 125% of the rent payable for the previous relevant period, and the excess is payable on the next quarter day under para (11) for the whole of that part of the relevant period up to that quarter

G day. If it results in a rent less than 125% of the rent payable for the previous relevant period, his preferred solution is that the council is entitled to a 25% increase throughout the relevant period. Alternatively, he says that a term must be implied under which the council repays the excess that the company has paid under para (14). As a second alternative, he submits that the provision in para (14) that the 25% increase is to be payable if the review of the rent is not completed within 12 months of the commencement of any relevant period is a provision in respect of which time is not of the essence, and that in order that the 25% increase shall take effect the company can serve a notice on the council making time of the essence and only if the council fails to get the review completed within the period specified in the notice does the 25% increase take effect.

H The *locus classicus* on the question whether time is of the essence in relation to rent review provisions is the speech of Lord Diplock in *United Scientific Holdings Ltd* v *Burnley Borough Council* [1978] AC 904. At p 930 Lord Diplock said:

> So upon the question of principle which these two appeals are brought to settle I would hold that in the absence of any contra indications in the express words of the lease or in the interrelation of the rent review clause itself and other clauses or in the surrounding circumstances the presumption is that the time-table specified in the rent review clause for completion of various steps for determining the rent payable in respect of the period following a review date is not of the essence of the contract.

The approach which is adopted to the construction of rent review provisions was well set out by Griffiths LJ in *Trustees of Henry* J *Smith's Charity* v *AWADA Trading & Promotion Services Ltd* (1983) 47 P&CR 607 at p 616:*

> When they enter into a lease such as this the expectation of both landlords and tenants is that a fair market rent will be paid throughout the lease and unless driven to do so by the wording of the lease I am loathe to construe the machinery provided for arriving at a fair market rent as forcing either the tenant to pay an exorbitantly high rent, or the landlord to receive a ridiculously low rent, neither of which bears any relation to the fair market rent, because one or other of them was one day late in observing the timetable set out in the rent review provisions of the lease.

In the present case there is evidence to suggest that a 25% increase on the rent payable for the first seven years of the lease will result in the landlord's receiving what might be termed a ridiculously low rent. K

The lease is, as I have said, not well drafted, and Mr Neuberger's first submission that para (14) is intended to provide for an interim rent is one that is difficult to extract from the wording of para (14) itself. Moreover, the suggestion that if the period of 12 months referred to in para (14) expires without a review being completed the rent is automatically increased by 25% of the rent then currently payable, and the landlord is entitled to maintain that increase even if the review is subsequently completed and results in the surveyor's valuation at a figure lower than a 25% increase on the previous relevant period, is one that I would not readily accept. The alternative suggestion that it is a provision for interim payment which will be adjusted when the surveyor has completed his review, with some repayment possibly being due from the landlord if the surveyor L determines a rent smaller than 125% of the rent payable for the previous relevant period, is one which requires a good number of words to be written into the schedule which are simply not there. In my judgment, it is not possible to construe para (14) as providing for an interim increase of 25% with or without any adjustment if a review is subsequently completed. On the other hand, if Mr Gaunt's submissions are correct, one does find that the express words of para (13) which provide that "as respects the periods of time referred to in this schedule, time shall not be deemed to be of the essence of the contract" have to be qualified in some way so as not to apply to the 12-month period referred to in para (14). If the language is strong enough, that is perhaps not impossible, and Mr Gaunt referred me to M a decision of the Court of Appeal in *Lewis* v *Barnett* (1981) 264 EG 1079. In that case there was in fact no provision that time should not be of the essence. I need not state the full facts of the case, but there was a para 6 which provided that, if the landlord and the tenant should not have carried out steps set out in earlier paragraphs of the schedule to the lease there under consideration, the notice already given by the landlord to the tenant under those provisions should be void and of no effect. In that case the Court of Appeal held that the effect of para 6 could only be to make time of the essence in relation to the steps to which it referred. The earlier paragraphs in the schedule in that case had required the landlord to take certain steps within

*Editor's note: also reported at (1983) 269 EG 729.

A specified periods. Para 6 provided that if those steps were not taken within the specified periods the landlord's notice should be void, and there was no proper way of construing that, except by holding that it meant what it said, and therefore the consequence of not complying with the time-limits in the earlier paragraphs was to deprive the landlord of the benefit of the rent review clause. Mr Gaunt submitted that that case was, if not directly applicable, of relevance here, in that the machinery set out in paras (1) to (10) provided certain time-limits, to which I have already referred, and para (14) then provided what was to happen if those time-limits were not complied with. The distinction from *Lewis* v *Barnett*, in my judgment, is that the time referred to in para (14) of the present case is not linked with the time-limits in the earlier paragraphs of the schedule. It is a general B provision that if a review is not completed within 12 months of the commencement of the relevant period, then a different method of ascertaining the rent shall apply. In my judgment, that provision is one which is readily distinguishable from the one considered by the Court of Appeal in *Lewis* v *Barnett*.

Mr Gaunt submits that the purpose of para (14) is to give the tenant certainty, and that he should not be left without knowing what the rent is to be for the relevant period for longer than 12 months after the commencement of the period. That is already, as he points out, a considerable time after the landlord could have operated the machinery. But in the present case the machinery is not solely under the control of the landlord. The surveyor is required by the schedule C to take certain steps within specified times, but the landlord is not in a position to ensure that he does so, and, indeed, the schedule recognises in para (9) that unforeseen circumstances may arise which would prevent the surveyor completing his task.

In my judgment, the effect of para (14) is not to deprive the landlord of the benefit of the rent which para (1) *prima facie* entitles him to. I accept, and adopt gratefully, what Griffiths LJ says in the passage to which I have referred. To give para (14) a meaning which resulted in an automatic application of a 25% increase on the completion of 12 months from the commencement of the relevant period would be unlikely to result in the landlord's receiving or the tenant's paying one quarter of the fair market rent as provided in para (1), which is the principal direction in the schedule for a rent D review. Although the position of para (13) is perhaps a little strange, and it might more naturally have been found at the end of the schedule, to hold that in para (14) the period of time was of the essence of the contract would be to contradict para (13), which refers to periods of time referred to in the schedule and not to periods of time referred to in the preceding provisions of the schedule. Both because of the wording of para (13) and because of the general principles laid down by the House of Lords in *United Scientific Holdings Ltd* v *Burnley Borough Council*, I hold that the period of 12 months in para (14) is not one in respect of which time is of the essence. The consequence is that if a rent review does not take place, or is not completed within 12 months of the commencement of the relevant period, the provision that the rent shall be increased by 25% of the rent then currently payable does not automatically come into effect. It is open to the tenant, as Lord Diplock pointed out in *United Scientific Holdings Ltd* v *Burnley Borough Council*, to give a notice making time of the essence, with the result that if before the expiration of the notice the review had not been completed, para (14) would operate and the tenant's liability would be limited to an increase of 25% of the rent. But until such a notice has been given — and I appreciate, as both counsel recognise, that there may be difficulties in determining precisely what form any such notice should take — until such a notice has been given, in my judgment the rent remains payable at the rate currently payable, that is to say, the rent payable during the immediately preceding relevant period, until the surveyor has determined a new rent. When the surveyor has determined the new rent, that becomes the rent for the whole of the relevant period, and the tenant on the next quarter day will be obliged under para (11) to make good the difference between the rent he has paid up until then and the rent due at the rate fixed by the surveyor. He does in this lease have no obligation to pay interest on the balance of the rent which he is obliged to pay on that quarter day, and until that quarter day arrives he can continue paying at the old rate.

The result, in my judgment, accords with the justice of the case. It accords also with the express provision in para (13) that time is not to be of the essence of the contract, and so far as the other provisions in the schedule are concerned, it does not, in my judgment, conflict in any decisive way with any of those provisions.

There remains the question of the date at which the new rent is to G commence when it has been determined by the surveyor. Mr Neuberger submitted that the reddendum was capable of being construed so as to oblige the tenant to pay the additional rent from January 1 1985. Mr Gaunt submitted that the relevant period defined in para (1) of the schedule, in the light of clause 1 of the lease, should be taken as commencing on February 13 in each of the years there mentioned. This is not a case in which the draftsman has earned any accolades. More care, I think, has been put into the provisions of the schedule, although not sufficient to make it clear beyond argument. Para (1) of the schedule, in my judgment, does provide that each successive five-year period of the term should begin on January 1, on the anniversary of the commencement of the term, and in my judgment that just prevails over the inferences one might otherwise H draw from the reddendum. The reddendum itself is capable of meaning that from January 1 1985 to February 13 1985 the tenant should pay the annual rent of £2,700 plus any increase payable under the first schedule, and although I am certain that the draftsman had not directed his mind to the point, I think that that is more in accord with the scheme of the lease as a whole than to hold that the rent remains at £2,700 a year until February 13 1985. So I will declare that, in answer to the question in the summons, the rent payable by the plaintiff to the defendant under the lease with effect from January 1 1985 will, if and when the surveyor has determined the rent, be assessed in accordance with paras (1) to (10) of the schedule to the lease. That will be without prejudice to the right of the tenant to serve J a notice making time of the essence under para (14) if the council delays unreasonably in proceeding with the rent review. Until the new rent has been assessed, as I have said, the rent will remain payable as each quarter day arrives at the existing rate, with the difference, if any, between that rent and rent at the rate determined by the surveyor being payable as provided in para (11) of the schedule.

Queen's Bench Division

July 10 1987

(Before Mr Justice HUTCHISON) K

GLOFIELD PROPERTIES LTD v MORLEY AND ANOTHER

Estates Gazette January 16 1988

[1988] 02 EG 62-66

Landlord and tenant — Forfeiture — Application of rent review clause — Action by landlords of business premises claiming forfeiture for non-payment of rent — The question whether the tenants had paid the amount of rent due depended L on how a rent review clause had operated — The rent review clause was in a familiar form, with provisions for a trigger notice by the landlords specifying the amount of the review rent, for agreement between the parties, within three months of such notice, substituting a different sum, or at the election of the tenants, not later than three months after the landlords' notice, for determination of the review rent by an independent surveyor — Landlords gave a trigger notice in compliance with these provisions and there was no agreement as to the rent between the parties — The crucial question then was whether a letter from the tenants, sent within the time-limit, constituted a counternotice invoking the determination M of the rent by an independent surveyor — The material part of the tenants' letter said, after stating the landlords' proposed rent, "Please accept this letter as formal objection and counternotice. We would suggest an early meeting to discuss the matter in detail" — After considering a number of authorities, in particular the judgment of Templeman LJ (as he then was) in *Amalgamated Estates Ltd* v *Joystretch Manufacturing Ltd*, Hutchison J held that the letter did constitute a sufficiently clear election by the tenants in favour of the determination of the review rent by an independent surveyor — He expressed doubt as to whether there was a real

conflict between the decision of McNeill J in *Edlingham Ltd* v *MFI Furniture Centres Ltd* and that of Sir Nicolas Browne-Wilkinson V-C in *Nunes* v *Davies, Laing & Dick Ltd*, but if there was indeed a conflict he would follow the Vice-Chancellor's decision — Judgment accordingly for tenants

The following cases are referred to in this report.

Amalgamated Estates Ltd v *Joystretch Manufacturing Ltd* [1981] EGD 84; (1980) 257 EG 489, CA
Colchester Estates (Cardiff) v *Carlton Industries plc* [1986] Ch 80; [1984] 3 WLR 693; [1984] 2 All ER 601; [1984] EGD 461; (1984) 271 EG 778
Edlingham Ltd v *MFI Furniture Centres Ltd* (1981) 259 EG 421
Nunes v *Davies Laing & Dick Ltd* (1985) 51 P&CR 310; [1986] 1 EGLR 106; 277 EG 416

This was an action by the landlords, Glofield Properties Ltd, against the tenants, David Morley and Brian Stewart, for forfeiture of the lease, dated December 8 1977, for a term of 15 years of premises at 532 Kingsbury Road, London NW9.

Paul Dickens (instructed by Druces & Attlee) appeared on behalf of the plaintiffs; Kirk Reynolds (instructed by Fayers & Co) represented the defendants.

Giving judgment, HUTCHISON J said: This is in form an action by the landlords of premises, let for business purposes, for forfeiture for non-payment of rent. In reality, however, it is a dispute, to be decided without any oral evidence having been called, between the landlords and the tenants as to whether there has been compliance by the tenants with the provisions of a clause of the lease bearing on the question of the assessment of rent for the second five-year period of a 15-year lease. If the landlords are correct, then it is common ground that the tenants have not paid the full amount of the rent. Conversely, if the tenants are correct, it is, as I understand it, common ground that the tenants have paid or tendered the appropriate sum by way of rent, pending the determination of the new rent under the rent review clause.

The matter arises in this way. By a lease dated December 8 1977 the landlords (the plaintiffs) let premises at 532 Kingsbury Road, London NW9, to three persons, two of whom are surviving as tenants and defendants, for a term of 15 years at a rent of £3,400 per annum exclusive, subject to revision. The lease, as is common in these days of rapidly changing values, reserved the stated rent only for the first five years and contained provisions for the ascertainment of rent for the second and third five-year periods. Although it may seem strange, bearing in mind the dates I have given, and the date at which this dispute is being tried, we are in fact concerned with the break clause in relation to the second five-year period, that which began on December 7 1982.

The relevant provisions around which the dispute in this action centres are contained in clause 5 of the lease. That provides, in so far as it is material, as follows (and I interpolate the observation that it is a very common form of clause, identical in all practical respects to the clauses found in a number of reported decisions, and coming from a well-known book of precedents):

> The expression "the open market rental value" means the sum in relation to the review period, or the second review period, as the case may be, determined in manner hereinafter provided as being at the time of such determination the annual rental value of the demised premises on the open market.

The provision in the lease reserving the rent indicates that for the second period it shall be such rental as is determined in accordance with the provisions of the lease.

By subclause (ii) of clause 5 one finds that the open market rental value is to be determined as follows:

> (a) it shall be such sum as shall be specified in a notice in writing by the lessor to the lessee at any time not earlier than twelve months prior to the commencement of the review period, or the second review period, as the case may be, or (b) as shall within three months of such notice be agreed between the parties in writing in substitution for the said sum or (c)

and this is the crucial clause for present purposes

> it shall be determined at the election of the lessee by counternotice in writing to the lessor not later than three months after the lessor's said notice, time to be of the essence hereof, by an independent surveyor appointed for that purpose by the parties jointly in writing or, upon their failure to agree upon such appointment within one month after the date of the said counternotice, then an independent surveyor appointed in the manner specified.

So there is a provision which, as I understand it, enables the landlord to serve a notice specifying whatever rent he chooses. That is to be the rent for the next review period unless either some different figure is agreed (subclause (ii) (b)) or the tenant elects to have the matter determined by an independent surveyor, such election being made in accordance with the provisions of para (c). In the present case the dispute is as to whether the tenant did made a proper election and give a proper counternotice for such independent determination as they contend; or whether, as the landlord contends, there never was such a counternotice. The relevant documents are extremely short.

On August 20 1982 the agents for the landlords, a firm of surveyors, wrote to the tenants by recorded delivery service in the following terms:

> Dear sirs, re: 532 Kingsbury Road, London, NW9. We hereby give you notice in accordance with the provisions of the lease dated December 8 1977, under which you hold the above premises, that as from the day of December 8 1982 the rent payable for the five years of the term commencing on December 8 1982 will be £9,750 per annum exclusive,

and they sign the letter.

On September 23 1982 the tenants' surveyors and agents responded in a letter, on which the tenants rely, saying that it constitutes a counternotice within clause 5(ii)(c). It is addressed, quite properly, to the landlords' surveyors, and it reads as follows:

> Dear Sirs, re: 532 Kingsbury Road, London, NW9

and it gives the names of the tenants

> We act for the tenant of the above premises. We have been passed a copy of your letter dated August 20 in which you propose a rent of £9,750 per annum at the forthcoming rent review. Please accept this letter as formal objection and counternotice. We would suggest an early meeting to discuss the matter in detail.

That is signed by the writer. That is the document which is alleged to satisfy the provisions of clause 5(ii)(c) of the lease, and to constitute a counternotice invoking the determination of the rent by an independent surveyor.

It is common ground between the parties, not surprisingly, that the letter from the landlords was adequate compliance with clause 5(ii)(a). Accordingly, it follows that if there was neither an agreement nor, within the three months' period, a counternotice requiring independent determination, the rent specified by the landlord would have become the rent for the first review period.

That is all I need to say about the facts, and I turn to consider the law. I have had helpfully cited to me a number of authorities, three or four of which perhaps I ought to refer to. The first question is: what is the correct approach to determining whether a communication said to amount to a counternotice under a clause such as this does indeed do so? In that regard I adopt the language of Templeman LJ in a decision of the Court of Appeal in *Amalgamated Estates Ltd* v *Joystretch Manufacturing Ltd* (1980) 257 EG 489. I take a passage from his judgment beginning at the bottom of the left-hand column on p 493:

> Mr de la Piquerie's second point, namely that the tenant's letter dated December 12 1978 constitutes a counternotice, also seems to me to be untenable. By that letter the tenant refused to agree the landlord's figure of £10,000 and asked for an explanation. By the lease the tenant had three months from December 8 1978 in which to serve a counternotice electing for the rent to be determined by arbitration. I am unable to find in the tenant's letter of December 12 any words which enable me to construe it as a counternotice of the kind required by the lease. It is true that no magic formula is required but, in my judgment, the tenant must make it clear to the landlord that he proposes to have the rent decided by arbitration in accordance with the provisions of the lease. The landlord and tenant may reach a compromise before arbitration has been reached, but, nevertheless, on the plain words of the lease it seems to me that the tenant is directed to serve a counternotice making it clear to the landlord that he proposes to go to arbitration if that is his intention. I cannot spell that out of the tenant's letter, which simply said that he did not agree with the landlord's £10,000 and asked for an explanation.

In case it is material, I just mention the precise terms of the letter to which Templeman LJ was referring. That is to be found set out at the bottom of the right-hand column at p 489 of the report. The material part reads:

> We cannot agree with your rent increase, for, as you know, they are building a new centre which is going to push Regent Street right out of position. There are shops very close to our own that are being offered for around £4,000 per annum. Perhaps, therefore, you could explain how you arrive at £10,000.

Those words of Templeman LJ are the words which, as I say, I propose to follow in seeking to answer the question whether the letter in the present case can be said to constitute a compliance with the provisions of the relevant clause in the lease. I am encouraged to

A adopt that course because in a recent decision in *Nunes* v *Davies, Laing & Dick Ltd* [1986] 1 EGLR 106 the Vice-Chancellor, Sir Nicolas Browne-Wilkinson, took substantially the same view, as I read his judgment. It is necessary that I should digress for a moment, because of the course the argument has taken, just to deal with what it was suggested may have been a difference of opinion between certain of the judges who decided the cases that were referred to me on what was the correct approach. In that connection I read a passage on p 107 of the Vice-Chancellor's judgment in *Nunes*. He says this:

In a number of those decided cases the question for decision has been characterised as a question of construction of the letter from the tenants which is alleged to be an election under paragraph (c); but, as the argument in this case proceeded, it emerged that there are in reality two questions: (a) what
B in law is required to constitute a valid election under paragraph (c)? (b) does the letter satisfy those requirements?

I will in due course consider those separate points in turn but, before doing so, I should explain why, in my judgment, those are the two relevant questions. In an ordinary case of construction the court is bound either to give a document under construction some meaning or to hold that it is void for uncertainty, against which the court always leans. Often the true construction of a document in court is a marginal decision which the court reaches with some doubt. In other types of case — for example, notices to quit and, in the employment field, notices of dismissal — the law requires not simply that a document on its true construction has a particular meaning but that the document should clearly have that meaning so as to bring home to the mind of the recipient of the notice what is intended by it. So, in cases of the kind before
C me, the first question is whether the law does require of a notice under para (c) not merely that on its true construction it communicates an intention to have the rent fixed by an independent surveyor but that it should satisfy some higher standard, such as that it should be clear or unequivocal. What are the legal requirements of a valid notice under para (c)?

he asks, addressing the first of the two questions he has postulated.

In the reported cases this question has not been addressed separately. It has been treated as rolled up in the question of the construction of the alleged counternotice. But, in my judgment, the authorities do disclose that the law requires the alleged counternotice to have a certain degree of clarity in its phraseology. It is not merely a question of the judge construing the document; the question is whether the document brings home to the mind of the ordinary landlord the fact that the tenant is exercising his rights under para (c).

But the authorities disclose two possible tests. The first is that the alleged
D counternotice must constitute "a clear indication" to the landlord as to the tenant's intention, or must make it clear to the landlord that he proposes to go to arbitration (see per Lawton LJ and Templeman LJ in *Amalgamated Estates Ltd* v *Joystretch Manufacturing Ltd*). However, in two later cases the test has been expressed to be that the letter alleged to constitute the counternotice must be "unequivocal" (see *per* McNeill J in *Edlingham Ltd* v *MFI Furniture Centres Ltd* and *per* Nicholls J in *Sheridan* v *Blaircourt Investments Ltd).** It may be that in those two cases the learned judges were not meaning to apply any more stringent test than that of clarity laid down by the Court of Appeal in *Amalgamated Estates*.

This case discloses that there may be a difference between what is clear and what is unequivocal. If the expression "unequivocal" means "incapable of bearing any other meaning" then the way is open for arguments based on remote possibilities of construction, however unlikely they may be, in
E construing the counternotice.

In my judgment, the test is that applied by the Court of Appeal in the *Amalgamated Estates* case, namely that the counternotice should be in terms which are sufficiently clear to bring home to the ordinary landlord that the tenant is purporting to exercise his right under para (c), and that is the test that I propose to apply.

Pausing there, it is for that reason that I respectfully suggest that the Vice-Chancellor is adopting, as I have said I propose to adopt, the approach indicated as appropriate by Templeman LJ in the *Amalgamated Estates* case.

Later in his judgment, he does say this about McNeill J's decision in the case of *Edlingham*, which was also referred to me, where the relevant letter was in the following terms:

F I refer to your letter dated February 7 and return herewith one copy of the notice duly acknowledged. Will you please accept this letter as counternotice to the effect that we consider that the rent of £50,000 is excessive and will appreciate it if you will kindly forward to us comparables on which you have based this figure.

McNeill J in the case of *Edlingham* held that that was not a sufficient compliance with the provisions of para (c). As to that, in *Nunes* the Vice-Chancellor said this:

In the *Edlingham* case the learned judge does not explain what force he gave to the words "counternotice" in the letter there under consideration; there may well have been special features in that case which forced him to a contrary conclusion. If not, I must respectfully decline to follow that decision, which may have been influenced by his formulation of the test as a requirement that G
the letter should be unequivocal. It is certainly arguable in the present case that the letter of December 6 is, in one sense, equivocal.

But as appears from the rest of the judgment the Vice-Chancellor held that it was in fact an adequate compliance with the provisions of para (c).

At some length, I am afraid, I have cited those passages from the decisions in order that I may say this about them. Speaking for myself, I do not really find any inconsistency between the decision of the Vice-Chancellor and the decision of McNeill J. It is clear from what the Vice-Chancellor had to say that he took the view that McNeill J's decision and that of Nicholls J could be construed on a basis which made them consistent with his own approach. The reason H
I say that is that in a passage towards the end of McNeill J's decision in *Edlingham* he says this:

I do not think it is necessary for the purposes of this judgment to express a view as to whether or not a counternotice, to be effective, must expressly refer to the relevant clause of the lease which is in point. My inclination would be to say that that degree of formality is probably unnecessary. But the notice must identify the form of relief which the tenant is seeking and must do so unequivocally. If a notice may be read as an invitation to negotiate or as an election to arbitrate it seems to me to be a defective notice. It must, to adopt Goulding J's words, contain that which is specific and make it clear to the landlord that this is the tenant's election to choose his, in this case, 1(3)(c) course, and not his 1(3)(b) course. It is not a question, as Templeman LJ put it, of a "magic formula", but in his words "the tenant must make it clear to the landlord that he proposes to have the rent decided by arbitration in J
accordance with the provisions of the lease". This document does not do that and I do not think, on its proper construction, that it is, or is effective as, a counternotice under the subclause.

So McNeill J, too, was relying on the passage in Templeman LJ's judgment, and it seems to me that on a legitimate construction of his judgment he could well be taken to be deciding simply that the counternotice in that case was not a clear expression conveying to the mind of the landlord what it was intended to elect. If I am wrong about that, and there is a conflict between the decisions of the Vice-Chancellor and McNeill J, then following the guidance given by Nourse J in the case of *Colchester Estates (Cardiff)* v *Carlton Industries plc* [1986] Ch 80, I consider that I should follow the K
decision of the Vice-Chancellor, since I am certainly very far from being convinced that he was wrong in not following the decision of McNeill J. The Vice-Chancellor's decision, given subsequently, and in the light of a full argument and a careful consideration of McNeill J's decision is, in the circumstances, as Nourse J indicates, the one that I should follow, and I do. Having said all of that, I wish to reiterate that I do not think that it is productive to seek to resolve the question which this case gives rise to by taking the letters in each of the other cases which has been cited to me, and making some detailed grammatical and analytical comparison between the terms of those letters and the terms of the letter written by the defendants in the present case. Each case has to be considered on its own facts; each letter has to be considered in the light of the particular language of L
that letter. What I conceive my legitimate task to be is to approach the question with the principles enunciated by Templeman LJ, which I have said I will treat as guiding me, in mind and ask myself the simple question: does the letter in the present case clearly convey to the landlord that it is an election to proceed in accordance with clause 5(ii)(c)? In support of the argument that it is not, Mr Dickens, on behalf of the landlords, raises essentially two arguments, the second of which, as I understood it, he relied on more strongly than the first. He pointed to the fact that the letter makes no mention of arbitration, or of the relevant clause in the lease, or anything of that sort. It simply, in so far as it is material, contains the words, "Please accept this letter as formal objection and counternotice." Moreover, he points to the fact that the next sentence suggests a meeting which he M
argues is equally consistent, perhaps more consistent, with an intention to exercise a para (b) right to attempt to secure an agreement of the relevant figure than a para (c) right to attempt to have it determined. In connection with that last point, however, I would draw attention to the fact that para (c) in terms provides that the independent surveyor is to be appointed by the parties jointly in writing or, upon their failure to agree, by the president of the Royal Institution of Chartered Surveyors. Accordingly, there was every reason, as it seems to me, why the writers of the letter of September 23 should have been inviting a meeting to discuss matters, and that sentence could easily be taken to be applying, and I should have thought was intended to apply, to a meeting for that purpose.

* Editor's note: Reported respectively at (1981) 259 EG 421 and (1984) 270 EG 1290.

As to the failure to mention arbitration or the provisions of the lease, those are objections which seem to me to be without foundation. If, as Templeman LJ says, no magic formula is required and all that is necessary is that the matter should be brought clearly to the attention of the landlord in terms which leave him in no doubt as to what is intended, that can plainly be done by a letter which omits such details.

Mr Reynolds on behalf of the defendants made five points which appear to me to be validly taken in this connection. He pointed out, first, that the letter was in terms a response to the landlords' letter, which was a formal notice under the clause. Second, that the material sentence in the response contained the word "formal" — "Please accept this letter as formal objection" — and, as I read it, as formal counternotice. He suggests that that is a strange word to use if what is intended is anything other than the service of a formal document equivalent to a notice. Third, he points to the obvious fact that the very word "counternotice" is used in the letter — a word which echoes the words of the relevant clause, and, indeed, the lease refers to one counternotice only; there can be no other provision of the lease to which reference was being made. Fourth, Mr Reynolds argues that there is nothing in the letter which is inconsistent with its being a counternotice. There is nothing which clouds the issue and suggests that the intent is to invoke some other clause. I have already said, and I do not intend to depart from it, that I did not intend to make a detailed comparison between the letter in this case and the letters in the other cases that were cited to me. But I will simply say, without elaboration, that in every one of those, including, if I may respectfully say so, the letter in the case decided by the Vice-Chancellor, where he nevertheless found in favour of the tenants, the language tends to be much more equivocal, because there are phrases which tend to cloud the meaning of the simple invocation of a counternotice. Fifth, Mr Reynolds says that if you look at the letter as a whole, bearing in mind its heading, the reference to the premises, and the reference to tenants, it being a response to a formal notice, and its simple language, it is really impossible to give it any other meaning than that it is intended as a formal counternotice. All of those arguments appear to me, as I say, to be well founded and to dispose of Mr Dickens' first point to the effect that this letter is not sufficiently specific or clear to amount to a counternotice.

Mr Dickens' second point, and the one on which, as I say, I understand he places the greatest reliance, really hinges upon the fact that the relevant sentence begins with the words "Please accept this letter as formal objection and counternotice." He argues that any letter containing a sentence beginning with such a precatory invocation has to be construed entirely independently of that sentence. So, says Mr Dickens, you look at the rest of the letter to see whether the rest of the letter is capable of amounting to a counternotice. Plainly, and in this I agree with him, it does not; there is not a word about a counternotice in the first or third sentences of that letter. The landlord, Mr Dickens suggests, who receives that letter will be left in a state of complete doubt as to whether the tenant is intending to serve a counternotice, and if he looks at the language of the first and third sentences will conclude that he is not, because nothing is said which could justify the implication that he is. Alternatively, Mr Dickens says that if one treats the sentence beginning "Please accept" as part of the material at which one can look to see whether the letter conveys an adequate intimation of an intention to serve a counternotice, nevertheless, it does not do so because it is insufficiently clear that the tenant is electing; he is merely inviting the landlord and giving him a choice as to whether to accept the letter or not. I am sorry to say that I am wholly unable to accept that these arguments have any shadow of validity. The phrase "Please accept this letter as notice" of this, that or the other is, as I conceive, a very common one to find in letters from solicitors and agents and others. It is generally, in my understanding, used to excuse the sending of a more formal notice when the writer intends the letter itself to be treated as a notice. After all, the object of notices is to convey information; and if, as Templeman LJ indicates, no magic formula is required, and no formality is required, then it seems to me that if somebody says "Please accept this letter as formal objection and counternotice", that is in effect saying that the letter as a whole, including that sentence, is to be treated as though it were a formal notice. The difficulties that face Mr Dickens are perhaps exemplified by the fact that he made it clear that even had the sentence continued "and counternotice pursuant to clause 5(ii)(c) of the lease dated" such and such, it would still have been open to the same objection. I cannot accept that those arguments are valid.

Accordingly, in my judgment, the specific objections raised on behalf of the landlords to this letter as a counternotice are unsustainable. I come back to seek to answer what I conceive to be the relevant question for me, namely whether the letter is in terms which are sufficiently clear to bring home to the ordinary landlord that the tenant is purporting to exercise his right under para (c) — I quote the words of the Vice-Chancellor in *Nunes*. I answer that question without hesitation. In my judgment, the letter is certainly sufficiently clear to convey that. Indeed, looking at it in a sensible way I cannot conceive that it can be said realistically to bear any other meaning. As I have indicated, I accept, in addition to his first four, the fifth point advanced by Mr Reynolds. In the circumstances, there was an adequate counternotice within time, and accordingly there must be judgment for the defendants.

Chancery Division

May 14 1987

(Before Mr Justice MILLETT)

LEIGH AND OTHERS v CERTIBILT INVESTMENTS LTD

Estates Gazette January 30 1988

[1988] 04 EG 127-131

Landlord and tenant — Rent review clause — Construction — "Primitive form of rent review clause" contained in the lease of an industrial estate for 84 years from August 3 1965 at £75,000 per annum, with rent reviews every 21 years — Review rent to be equal to one half of "the estimated aggregate rack-rental value of the demised premises" — The estate consisted of 18 separate units, the majority of which were individually sublet, but some units were vacant — The purpose of this rent review clause was obviously not to maintain the real value of the original rent but to enable the landlord to share with the lessee the increase in the underlying rack-rental value — It had to be noted that the lease provided that the review rent should be the aggregate rack-rental value of the demised premises and not, as one might have expected, the aggregate of the rack-rents actually received from the sublet units plus the rack-rental values of the parts not sublet — A number of construction questions were raised and the main matters determined by the judge were the following — The valuer was not called upon to value the demised premises on the basis of a single hypothetical letting of the whole estate — He had to treat the estate as divided at the material date and substitute for the actual rents, where appropriate, the rental value of each separate unit — In so far as there were voids he must embark on the normal inquiries and ask how the landlord would best deal with each void — The valuer must assess the rack-rental value of each unit by assuming the unit to be available for subletting with vacant possession — As regards an option given to the lessee in certain circumstances to repair a building or demolish it, the valuer must treat the lessee as exercising the option in his best interests from the point of view of seeking to obtain the maximum rack-rental value of the property — Some questions raised by the summons did not require an answer as they were based on the false premise that the valuation was to be on the basis of a single hypothetical lease of the estate as a whole

No cases are referred to in this report.

This was an originating summons seeking the determination by the court of questions of construction in regard to the rent review clause of a lease of the Thames Side Industrial Estate at Silvertown, West Ham, London E16. The plaintiffs were Adelaide Kennedy Leigh, Leila Irene Foux, Lesley Doris Berman, Gerald Walter Leigh and

Geoffrey Goldkorn, the lessees, and the defendants were the lessors, Certibilt Investments Ltd.

David Neuberger QC (instructed by Baileys Shaw & Gillett) appeared on behalf of the plaintiffs; Kim Lewison (instructed by Helder Roberts & Co, of Epsom) represented the defendants.

Giving judgment, MILLETT J said: This summons seeks the determination by the court of certain questions of construction which arise on a rent review clause in a lease. The lease was granted on September 3 1965. It demised the Thames Side Industrial Estate at Silvertown, West Ham, London E16, which extends to some 11.3 acres and consists of 18 separate units and a vacant plot of 1.3 acres, together with its service roads. At present 15 units are individually sublet and three are vacant. Some of the units were sublet at the date of the lease, which was expressly granted subject to and with the benefit of one underlease and three agreements for underleases mentioned in the second schedule. The lease clearly contemplated that individual units would or might be separately sublet.

The lease was a full painting, repairing and insuring lease and was granted for a term of 84 years from August 3 1965 at a rent of £75,000 per annum, with rent reviews every 21 years. The first rent review thus fell due at the end of the first rental period on August 3 1986. The rent review clause is short, succinct and primitive. The crucial subclause is clause 2(2), which is in the following terms:

> During each Rental Period the amount of the rent shall be whichever one shall be the greater of the two following sums that is to say: (a) £75,000 per annum or (b) such yearly sum (hereinafter called "the Revised Sum") as shall be either agreed or assessed and determined in accordance with subclause (3) of this Clause as being a sum equal to one half of the estimated aggregate rack-rental value of the demised premises at the date of the commencement of such Rental Period.

The lessee's covenants in clause 3 include a covenant in para (xii) in the following terms:

> Not to underlet the whole or any part of the demised premises at any rent less than the prevailing market rack-rental value and any underlease for a term of and exceeding 21 years to include provision for the review of such rent by reference to the market rack-rental value current at the date of such review and not to take any premium or capital payment whatsoever upon the grant of any underlease.

The lessee had the right at any time to demolish any part of the demised premises so long as at least 250,000 sq ft of total floor area remained and it was entitled to demolish more with the prior consent of the lessor, such consent not to be unreasonably withheld.

The lessee's repairing covenant was subject to a proviso that if upon notice from the lessor requiring compliance with the covenant in respect of any building the lessee should give notice in writing of its intention to demolish such building, then the lessor should allow a period of 12 months and the lessee should, at the expiration of such period, either repair the building or demolish it.

The estate was already elderly in 1965. Internal evidence in the lease itself suggests that some of the buildings were at least 30 years old at the time the lease was granted. Judging by the respective attitudes adopted by the parties, some at least of the units appear to be let at rents higher than could now be achieved on new lettings if they were available with vacant possession.

The first question that has been argued before me is this.

Question 1:

> (a) Can the demised premises be valued on the basis of the aggregate rent which would be achieved of hypothetical lettings of individual parts of the demised premises or (b) must the demised premises be valued on the basis of a single hypothetical letting of the whole?

Of course, even on the latter hypothesis that the demised premises must be valued on the basis of a single hypothetical letting of the whole, it has to be remembered that they consist of an industrial estate, divided into separate units, with its existing service roads, adapted, therefore, for separate subletting in individual units; and any single potential lessee of the entirety would no doubt contemplate such subletting and would negotiate the rent he would be prepared to pay for the entirety on that footing.

The question turns on the meaning of the phrase "The estimated aggregate rack-rental value of the demised premises". The phrase "the demised premises" is defined as "the whole of the land and buildings comprised in the estate". The words "estimated" and "value" show that some process of estimating what could be achieved on a hypothetical letting or lettings is envisaged. The question is: what is the hypothesis upon which the valuer must work?

This question is determinative of most — though not quite all — of the other questions that have been argued, and it is not an easy one to answer. It is very much a matter of impression. I have come to the clear conclusion that in fact the answer is neither of those which have been argued before me.

The landlord, contending for the first of the two alternatives in the summons, fastened on the word "aggregate", which means "total", and posed the question: what are the component elements which go to make up the necessary total? If the hypothesis is of a single hypothetical letting of the whole, then the word "aggregate" would appear to be completely otiose.

The tenant was unable to give any satisfactory explanation for the presence of the word "aggregate", and Mr Lewison submitted that it really stressed the fact that a single overall rack-rental value was to be arrived at, and he pointed out, correctly, that it was to be the aggregate rack-rental value of the demised premises, that is to say, of the entirety of the demised premises and not of the individual parts.

In my judgment the difficulty with the tenant's explanation, so far as the actual wording goes, is that the word "aggregate" suggests not merely a single figure for the whole but also some process of collection or addition in order to arrive at that single figure.

In my judgment, however, the answer to this question is not to be found or to be found exclusively in the word "aggregate" but is to be found by reading the clause as a whole and by seeking to answer the question: what is the purpose of the clause? Both counsel urged on me certain general principles upon which they were agreed, and both submitted that the object of a rent review clause in general is to ensure that in real terms the rent does not fall below the rent originally reserved. They also agreed that there should be as little hypothesis as possible and that the terms and conditions and situation of the hypothetical negotiating parties should bear as close a resemblance to reality as possible, though they both conceded that it would not be possible to proceed by reference to reality either where the lease on its true construction contained an express provision to the contrary or where such a provision must necessarily be implied. Otherwise, they agreed, there should be as little departure from reality as possible.

Given the first of those general principles, I would agree that the rest would follow; but care must be taken not to allow general principles to override the actual language used in a particular case. Each case must depend upon its own language, the circumstances in which the lease was granted, and the purpose which it appears from the clause itself the clause was intended to achieve.

I bear in mind particularly these facts. This was an industrial estate, already elderly at the date of the original lease, already then adapted for subletting in parts, probably (although this cannot be certain: it is not fully set out in the evidence) not fully let at the date when the lease was originally granted, but certainly capable of being fully let or envisaged as being capable of being fully let, and no doubt expected to be fully let by the time of the first review date in 21 years' time. Next, there is no evidence that the original lessee was the developer or that it had spent substantial capital sums in improving the property prior to obtaining the lease. There is no recital to this effect in the agreement for lease: it does not appear to be a building lease or anything of that kind. Finally (and most importantly), when the estimated aggregate rack-rental value of the demised premises (whatever that may mean) has been agreed or assessed by the valuer in accordance with the provisions in that behalf contained in the lease, the landlord does not obtain the benefit of the whole (or anything like it) of the difference between that and the original rent, but only half.

In those circumstances, it seems to me that the purpose of this clause cannot possibly have been to maintain the real value of the original rent, for the landlord does not obtain, but shares with the lessee, the increase in the underlying rack-rental value. I have to ask myself why: and the only answer which suggests itself to me is that this was in order to compensate the lessee for the fact that it was taking a lease of an entire estate at a rent fixed originally or from time to time for the whole, and that it needed some compensation for potential voids, for the expense of management, including the cost of reletting units from time to time, and a profit rental.

If the lessee's arguments are correct, then the lessee gets a double benefit. First, the hypothetical single lessee in the market for a single lease of the estate as a whole will discount the total break-up rentals for these factors and will offer only a discounted figure for the lease of the estate as a whole. Having ascertained that discounted figure, then the valuer has done his work, but the lease makes a further deduction of 50% before comparing the result with the original rent of £75,000 per annum.

In my judgment, therefore, the purpose of this clause cannot have been to maintain the real value of the original rent but rather to give the landlord 50% of the underlying rack-rental value of the estate. The verbal difficulty in this approach that the lessee pointed out (that it is the rack-rental value of the demised premises and not of the units) is, I think, met by the argument — which goes back to the word "aggregate" — that what the the valuer is required to do is to take the rack-rental values of the individual parts of the estate, and that when he has aggregated them what he has got is a single total for the whole demised premises.

The lessee's argument that the demised premises are defined to include service roads as well as lettable units and that the aggregate would not therefore be an aggregate rack-rental value of the demised premises but only of the lettable parts of the demised premises is, in my view, wrong because the service roads could always be thrown in with any one or more of the units, but in any event if they have any value of their own or if they go towards increasing the value of any particular part of the premises, then the valuer would have to take that into account.

I ask myself why the lease provides that the figure should be the estimated aggregate rack-rental value of the demised premises rather than, as one might have expected, the aggregate rack-rents actually being received for such parts of the demised premises as are let and rack-rental values only of such parts as would, at the date of the rent review, be void or occupied by the lessee itself. The answer, in my judgment, is partly the fact that this is a very primitive form of rent review clause and partly by reference to clause 3(xii) of the lease, which I have read. It may well have been assumed that if any of the units were let below their rack-rental values that would be in breach of covenant and accordingly it would be right to take the rack-rental value rather than the actual rack-rents being received. It may also have been thought that the formula coped not only with that problem but also the problem of voids and with owner-occupied or lessee-occupied units. If so, it was a short, succinct and very effective way of dealing with the problem.

In fact, of course, this formula overlooked the fact that rents of the individual units when granted may have been in accordance with the rack-rental value at the date of the grant, and therefore in compliance with the covenant in that behalf, but may have fallen below the market rent at the date of the review; and it almost certainly overlooked the fact that, as appears to have happened, market rents generally have, in fact, fallen below the rents at which individual units were let, so that the substitution of rack-rental value for the actual rents may work to the landlord's disadvantage.

However that may be, I can give no sensible meaning to this clause, and in particular to the word "aggregate" and to the fact that only one half of the rack-rental value of the demised premises is to be taken in the primary calculation by adopting either construction that the landlord or lessee has put before me.

In my judgment, the valuer is not required to value the demised premises on the basis of a single hypothetical letting of the whole, nor is he entitled to do so. Nor is he entitled to embark upon a wide-ranging inquiry as to how the landlord would be advised to let the estate if he were free to let it in parts or as a whole. What the valuer must do is to treat the estate as divided at the rent review date; and he must then substitute, where appropriate, for the actual underlying rents, the rental value at the rent review date of each separate unit. In so far as there are voids, he must embark upon the normal inquiries: how the landlord would best deal with each void, what the rack-rental of that void is and, of course, he may take it to be subdivided or let as a whole. So far as the service roads are concerned, he is bound to add to the total anything which in his view the landlord could have obtained by letting them separately or throwing them in with one or more of the units, but only if, in the valuer's judgment, the service roads would add to the rack-rental value of the whole.

I come back to the fact that what the valuer has to determine is the aggregate rack-rental value of the entirety, that is to say, of all the units together with the service roads, but that it is an aggregate of the rack-rental values of the component parts.

Question 2 was whether the hypothetical lease was to contain rent review clauses, as the landlord contended, only once in every 21 years or, as the lessee contended, every five years. Question 3 raised the question whether the hypothetical lease was to be on (a) the same terms as the actual lease, as the landlord contended, or (b) on some other and, if so, what terms, as the lessee contended. Both questions, of course, are based on the false premise that there is only one lease to be considered, whereas now that I have answered question 1 as I have, the valuer's approach is essentially different and assumes a number of hypothetical underleases.

Accordingly, as was conceded by counsel in the course of argument, the valuer has a free hand to assume whatever term, duration and provisions would be appropriate in an underlease of the appropriate part.

Question 4 no longer arises. Leaving aside question 5 for the moment, question 6 is this: is the hypothetical lease assumed to be granted (a) subject to the actual underleases as they exist at the review date or (b) with vacant possession? Again, as in the case of questions 2 and 3, the premise upon which the question is asked turns out to be false and it is conceded by counsel that the result of the answer to question 1 is that the answer to question 6 must be in the sense of (b), that is to say, with vacant possession.

I now finally turn to question 5, which asks this: for the purpose of assessing the rent is it to be assumed (a) that the lessee has complied with his covenant so that the demised premises are fully repaired and decorated as at the review date (as the landlord contended) or (b) that the demised premises are in their actual condition as at the review date (as the lessee contended)?

That would normally be a very simple question to answer. It is well established that the lessee cannot rely, in diminution of the rent, upon his own breach of covenant. But the question is complicated in the present case by the proviso to the repairing covenant and the lessee's right to demolish. In the course of the argument before me, while neither counsel resiled from his first choice, both adopted as a second choice the following answer to the question: (c) that the demised premises are fully repaired and decorated to the extent of 250,000 sq ft of buildings, or such lower number of square feet as the lessee is lawfully entitled to leave undemolished; the balance of actual buildings being treated either as demolished or as fully repaired and decorated, depending upon whether a prudently advised lessee would have been advised as at the rent review date to demolish or not in all the circumstances, other than the circumstances of the rent review.

The lessee submitted that the landlord has no right in the present case to have more than 250,000 sq ft kept in repair in all and that therefore there was no right to make any false assumption about the state of repair of more than that area. But the landlord has a right to have the surplus either repaired or demolished and, for that reason alone, I see no ground for accepting the lessee's primary contention that the demised premises should be taken to be in their actual condition at the review date, that is to say, standing and possibly in disrepair. That must be, one way or another, in breach of covenant.

The real question is this: is the landlord right in saying that, because at the relevant review date the premises were not in fact demolished, therefore they ought to be assumed to be in good repair? One might say by way of contrast, because they were not in fact repaired, therefore they ought to be treated as demolished.

The landlord submitted that, on the true construction of the relevant covenants and provisions of the lease, the lessee has an obligation to repair with an option to demolish, and the lessee not having exercised that option, the obligation remains. But, of course, the reason why the lessee has not exercised that option, at least in respect of much of the property, is that it is sublet. If the hypothetical basis upon which the rack-rental value of individual units is to be taken as with vacant possession, then the only obstacle to possible demolition has gone as well.

In my judgment, the narrow approach that the landlord has taken, that it is an obligation to repair with an option to demolish, rather than an obligation to demolish with an option to repair, does not meet the problem. The lessee's option was still open on the hypothetical basis of vacant possession for the individual unit as at the review date, and therefore there is no reason to rely upon the fact that it had not, at that date, been exercised.

There is, of course, also the difficulty on this basis of appropriating the 250,000 sq ft to any particular part of the property.

In my judgment, the position is this. In regard to each unit the valuer must assess the rack-rental value of that unit first of all by assuming the unit to be available for subletting with vacant possession. Second, he must treat the lessee as exercising that option which is in his best interests to exercise from the point of view of seeking to obtain the maximum rack-rental value of the property, that is to say, ignoring his position as leaseholder under the lease, where his interest may well be in reducing the rack-rental value of the underlying units in order to reduce the rent payable by him to the landlord. It might well be right for the valuer to treat the lessee in effect as if he were the freeholder and with an option either to demolish the building — taking account, of course, of the expense of demolition — with a view to letting it with vacant possession or to repair the property at, again, the appropriate expense with a view to letting the repaired building with vacant possession.

In my judgment, the valuer must do that, in theory at least, with every unit, but of course on the footing that at the end of the day the lessee must repair at least 250,000 sq ft unless the landlord could not reasonably refuse his consent to the demolition of more than that. So the valuer will have to look at it overall again at the end of the day if his approach results in less than 250,000 sq ft being left standing.

If that is the case, he will then have to decide which units should be treated as demolished and which repaired, and he must approach that again on the basis that it is the lessee who has the option. Provided the lessee leaves not less than 250,000 sq ft standing, it is entirely up to the lessee to decide which part should be left standing.

There was no order, as to costs.

Chancery Division

November 3 1987

(Before Mr Justice MERVYN DAVIES)

SI PENSION TRUSTEES LTD v MINISTERIO DE MARINA DE LA REPUBLICA PERUANA

Estates Gazette April 2 1988

[1988] 13 EG 48-50

Landlord and tenant — Rent review clause in lease — Construction — Effect of user covenant — Effect of restrictions in licence to assign to defendants — The hypothetical letting was to be a letting in the open market for a term equal to the residue of the lease with vacant possession without a premium and subject to provisions similar to those contained in the lease — The lease contained a user covenant restricting the use of the demised premises to offices in connection with the lessee's business of mortgage finance and insurance consultants — The licence to assign to the present defendants restricted the user to offices occupied for the purposes of the assignee's diplomatic mission — It was argued on behalf of the defendants that the user should be deemed to be restricted to use as offices for the purpose of the diplomatic mission of [blank] with the blank to be filled in when the name of the hypothetical diplomatic lessee was known — The judge rejected this submission, holding that the "provisions of the lease" in the rent review clause meant the original provisions unaffected by the personal arrangements made in the licence respecting the current tenants — *The Law Land Co Ltd* v *Consumers Association Ltd* considered and in part distinguished — Held that, in applying the rent review clause, it should be assumed that there has been found a mortgage finance and insurance consultant intending to use the premises as offices; that the hypothetical tenant would be bound not to use the premises save in connection with his business as such a consultant; and that any assignee from the hypothetical tenant would be subject to the same user provision — The court refused to give directions to the arbitrator as to the kind of comparables that he might consider; this was a matter for his professional judgment

The following cases are referred to in this report.

Law Land Co Ltd v *Consumers' Association Ltd* (1980) 255 EG 617, CA

This was an originating notice of motion raising questions as to the construction of the rent review clause in a lease for 21 years and one day from June 18 1973 of the ground floor of 41 Lowndes Street, London SW1. The applicants were the landlords, SI Pension Trustees Ltd, and the respondents were the tenants, the Ministerio de Marina de la Republica Peruana.

Robert W Kirk (instructed by J E Kennedy & Co) appeared on behalf of the applicants; John R Cherryman QC (instructed by Amhurst Brown) represented the respondents.

Giving judgment, MERVYN DAVIES J said: This is an originating notice of motion. The applicant is SI Pension Trustees Ltd (the "landlord"). The respondent is the Ministerio de Marina de la Republica Peruana (the "tenant"). The questions arising involve construing some terms of a lease for the purpose of applying a rent review clause. By a lease dated October 1 1973 the ground floor of 41 Lowndes Street, London SW1, was demised for 21 years and one day from June 18 1973 at a rent of £12,000. The demise was by M D Johnson and C A Fry to Johnson Fry & Partners Ltd. Mr Johnson and Mr Fry are referred to in these terms:

(hereinafter together called "the Lessor" which expression shall wherever the context so admits be deemed to include the person or persons for the time being deriving title under the Lessor).

Johnson Fry & Partners Ltd are described in these terms:

(hereinafter called "the Lessee" which expression shall whenever the context so admits be deemed to include its successors in title and assigns).

The material provisions of the lease are as follows:

USER COVENANT

2. The Lessee hereby covenants with the Lessor as follows that is to say . . . .

(12) Not to use the demised premises otherwise than as offices in connection with the Lessee's business of Mortgage Finance and Insurance Consultants . . .

COVENANT CONCERNING ASSIGNMENT

14. . . .

(B) Subject to sub-clause (D) hereof not to assign the whole or charge underlet or part with possession of the whole (or any part) of the demised premises without the previous written consent of the Lessor which shall not be unreasonably withheld . . .

THE RENT REVIEW CLAUSE

5. (1) The Lessor may require the rent payable hereunder to be revised by serving upon the Lessee notice in writing to that effect not more than 12 nor less than 6 months before the expiration of the 5th 10th and 15th years of the said term (hereinafter referred to as "the review date(s)" and "the relevant review date" shall be construed accordingly) and as and from the review date the revised rent (calculated and determined in acccordance with the following provisions of this clause) shall become payable in all respects as if it were the rent hereby reserved.

(2)The revised rent shall be the greater of either

(i) the rent payable hereunder at the time of the service of the notice requiring the rent to be revised or

(ii) the market rent of the demised premises at the relevant review date.

(3)The expression "the market rent" shall be deemed in this clause to mean the yearly rental value of the demised premises (having regard to rental values current at the relevant time for similar property used for any purpose within the same use class under the Town and Country Planning (Use Classes) Order 1963 as that which includes the use of the demised premises permitted hereunder let with vacant possession without a premium and subject to provisions similar to those contained in this lease) taking no account of any goodwill attributable to the demised premises by reason of any trade or business carried on therein.

(4) [A provision that failing agreement the revised rent shall be determined by reference to the arbitration of an arbitrator nominated by the president of the RICS.]

The original parties to the lease are changed. The applicant is now the landlord. The tenant is the respondent. There is a licence dated July 22 1976 and made between C A Fry of the first part, Johnson Fry & Partners Ltd of the second part and "Ministerio de Marina de la Republica Peruana represented by the Peruvian Naval Attaché in Great Britain and having an office in the United Kingdom at 41 Lowndes Street in the City of Westminster (hereinafter together called the assignee)" of the third part. By the licence it was witnessed that the landlord granted licence to Johnson Fry & Partners Ltd to assign. As to the assignee — that is the minister — there was for him "after the lease has been assigned to him consent for him to use the premises demised by the lease as offices in one single occupation for the purpose of the assignee's (here meaning the from time to time accredited) lawful diplomatic mission in lieu of the user at present carried on by the Lessee [ie Johnson Fry]". Clause 2 of the licence was a covenant by the assignee

for himself and his successors in such office . . . (b) not to use or permit the premises demised by the lease or any part thereof to be used otherwise than for offices in one occupation for the purpose of the assignee's lawful diplomatic mission.

Clause 3 of the licence is as follows:

This licence is restricted to the particular assignment and change of user hereby authorised and save to the extent that the same are hereby expressly varied the covenants conditions provisions and agreements contained in the lease (as varied by this licence aforesaid) shall remain in full force and effect.

The rent review machinery was invoked by a letter dated December 10 1982. There was no agreement as to a revised rent. Pursuant to clause 5(4) of the lease Mr A G Salata [FRICS] ("the arbitrator") was appointed. He agreed to act as arbitrator to determine the revised rent. Negotiations ensuing between the parties then disclosed that questions of law respecting the construction of the lease arose. Arbitration was not regarded as suitable for the determination of those questions and so the questions have with the consent of the parties and Mr Salata been made the subject of the originating notice of motion.

The questions now in issue appear in subparas (ii), (iii) and (iv) of para 3 of the notice of motion. They are as follows:

(ii) Whether the "yearly rental value" of the demised premises has to be ascertained:
(a) on the basis of a hypothetical letting of the premises with vacant possession without a premium; or
(b) on the basis that the premises are already let to the tenant in occupation on all the terms of the lease as varied by the terms of a licence dated the 22nd July 1976 or on some other and, if so, what terms.
(iii) If the answer to question (ii)(a) is in the affirmative, whether the "yearly rental value" of the demised premises has to be ascertained:
(a) on the basis that the premises can be let to any lessee whatever his business but can only be used by that lessee as offices and for the purposes of his business and further on the basis that the lease would contain provisions similar to the present lease and in particular the same provisions as to change of use and assignment; or
(b) on the basis that the premises can be let to any lessee with diplomatic status but can only be used by that lessee as offices in one occupation for the purposes of his or its lawful diplomatic mission and further on the basis that the lease would (apart from the user clause) contain the same provisions as are contained in the present lease.
(iv) Whether the comparable rental values to which by subclause 5(3) of the lease the arbitrator is directed to have regard are:
(a) rental values current at the relevant time for similar property used for office purposes generally; or
(b) rental values current at the relevant time for similar property used for office purposes let with vacant possession without a premium subject to provisions similar to those contained in the lease as varied by the 1976 licence and in particular with the office use restricted to a named lessee.

Mr Kirk appeared for the landlord. Mr Cherryman appeared for the tenant. I turn to question 3(ii). The phrase "yearly rental value" is, of course, in clause 5(3) of the lease. Mr Cherryman submitted that 3(ii) should be answered as in 3(ii)(b). Despite the inclusion of the phrase "market rent" in clause 5(2)(ii) and (3) of the lease, it was said that the arbitrator was not required to act on the assumption that the rent was to be fixed by reference to any "market". It was said that nowhere in 5(3) is there necessity to assume any new letting, or any letting with vacant possession, or the existence of an open market. The reference to "vacant possession" in 5(3) was acknowledged but was said to enable the arbitrator to take account of letting values of similar properties then being let with vacant possession with the arbitrator then discounting those values in relation to the demised premises by reason of the fact that the present tenant is *in situ* in the demised premises. It was said further that it was significant that 5(3) does not expressly require tenants' occupation to be disregarded although there is to be an express disregard of goodwill. The position was contrasted with section 34 of the Landlord and Tenant Act 1954. I do not accept Mr Cherryman's submission. Clause 5 refers to a "market rent" which suggests that the revised rent is to be at a figure derived from the market; not fixed by reference to a single individual already in occupation. One goes further and reads 5(3). One sees that "the market rent" means the yearly rental value of the demised premises "having regard to rental values current . . . for [similar office property] let with vacant possession without a premium and subject to provisions similar to those contained in this lease . . ." To my mind that means that (1) the arbitrator is to ascertain the current rental values for similar office properties which have been let with vacant possession without premiums in terms akin to those in the lease; (2) the arbitrator is then to assess the revised rent with those rental values in mind; (3) it seems to me that if assessments are directed as in (2) it must be for the purpose of assessing a rent for the demised premises on a vacant possession basis: *a fortiori* in that 5(3) does not suggest any discounting of a vacant possession rent of the demised premises by reference to the fact that the tenant is in occupation. I answer 3(2) in the sense of (a) — that is to say the yearly rental value is to be ascertained on the basis of a hypothetical letting of the premises with vacant possession without a premium.

I go now to question 3(iii). This involves considering the difficulties that arise when applying the rent revision clause in the light of the existence of the user clause. Before me neither side, in the end, contended for the exact wording of either (a) or (b) in 3(iii). Mr Cherryman (on the assumption that he was unsuccessful on question 3(ii)) referred to the user covenant and to the licence dated July 22 1976. He said the arbitrator was to assume an hypothetical lease which restricted the use "to use as offices in one occupation for the purpose of the diplomatic mission of [blank] with the blank to be filled in when the name of the hypothetical diplomatic lessee is known". As will be evident from that formulation, Mr Cherryman regards 5(3) as now to be read as if "the provisions contained in the lease" were the provisions of the lease as varied by the licence. I do not think that is right. In the first place 5(3) refers to the provisions of the lease, not to the provisions of the lease as they may be varied from time to time. Second, when one reads the licence one sees that the variations there made are personal to the persons there contracting. If and when the minister ceases to be a tenant the terms of the original lease revive. See licence, clause 3. In these circumstances "the provisions of the lease" referred to in 5(3) seem to me to be the original provisions of the lease unaffected by the personal arrangements made respecting the current tenant.

It follows that one has to consider the impact of the user covenant 2(12) as originally expressed in the lease on the operation of the rent review covenant. It was common ground that the express provisions of a user clause are not impliedly revoked by reference to the provisions of any assignment clause. It was also common ground that where there is a user clause the arbitrator is not to have regard to the fact that the landlord may relax the user clause. Those common grounds are derived from *The Law Land Co Ltd* v *Consumers Association Ltd* (1980) 255 EG 617 at p 621, foot of the left column, and p 621 right column with its references to the words of Brandon LJ. Where, as here, one is faced with the task of assessing, on the basis of a hypothetical letting with vacant possession, a revised rent for premises subject to a user covenant, guidance is afforded by the *Law Land* case. As I understand, in the circumstances of that case, one supposes a lease about to be granted with vacant possession for a term equal to the residue of the term of the actual lease. In other respects the hypothetical lease reads as does the original lease save that the name of the lessee is blank and the user provision, if and in so far as it refers to a personal user by the original lessee, is to be regarded as modified so as to allow a similar personal user by the hypothetical lessee but not to allow such user by any assign of the hypothetical lessee. It is on that basis that the valuer proceeds in a *Law Land* situation. See, in connection with my reference to an assign, the words of Templeman LJ in the *Law Land* case in the paragraph beginning "The second declaration", right-hand column, p 621.

I must follow the *Law Land* guidance unless the circumstances of the situation before me differ significantly from the *Law Land* circumstances. There is, I think, a significant difference. In *Law Land* — see p 617 — the user covenant was that the lessee:

. . . would not, without the prior written consent of the landlord use or permit the demised premises or any part thereof to be used other than as offices of the Consumers Association and its associated organisations.

In the present case the user covenant — 2(12) — is as set out above.

One sees that the *Law Land* covenant permits office user only by an individual, ie the Consumers Association, but 2(12) names no individual. The reference is to "the lessee's business". It was contended that the phrasing of 2(12) was apt to be read as confining user to the original lessee — Johnson Fry: the context of 2(12) requiring the lease to be construed individually rather than in accordance with the definition in the lease of "lessee" which says that "lessee" means the lessee and its successors in title. I do not accept that view. To my mind, one reads 2(12) giving "lessee" its defined meaning; so that a successor in title of the original lessee may use the demised premises as offices in connection with his business if his business happens to be that of a mortgage finance and insurance ("MFI") consultant. It follows that in this case the hypothetical lessee is to be regarded as being granted a lease with 2(12) therein verbatim but on a basis that 2(12) is to be construed as meaning that the demised premises may be used as MFI offices not only by the original hypothetical lessee but also as MFI offices by assigns of the original hypothetical lessee.

I should say that Mr Kirk desired to construe 2(12) more liberally than I have done. As well as pointing to the definition of "lessee" in the lease, it was said that the words "of [MFI Consultants]" in 2(12) were surplusage or, at any rate, merely descriptive of or informative as to the original lessee. At first I was inclined towards that view, but when one sees from the *Law Land* case that the hypothetical lease is to reproduce the original lease save as to necessary blanks, I do not see myself at liberty to visualise the hypothetical lease without therein including 2(12) verbatim.

In the light of what I say above, it will be appreciated that I do not answer question (iii) either as (a) or (b) therein. But I cover the ground of the question by setting out guidelines for the arbitrator. In my view, clause 5(3) is to be applied as follows:

(a) there is to be assumed an hypothetical letting in the open market for a term equal to the residue of the lease "with vacant possession without a premium and subject to provisions similar to those contained in this lease taking no account of any goodwill attributable to the demised premises by reason of any trade or business carried on therein" — see 5(3).
(b) there is to be assumed there has been found a willing lessee who is an MFI consultant intending to use the premises as offices.
(c) the hypothetical lease will of course include 2(12). 2(12) is to be regarded as obliging the hypothetical tenant not to use the premises save in connection with his business as an MFI consultant.
(d) if the hypothetical tenant is allowed to assign pursuant to 2(14) the new tenant will be subject to the same user provision as the hypothetical lessee — that is to say the new tenant will not be able to use the premises save as offices in connection with his business as an MFI consultant.
(e) regard must be had — see 5(3) — to rental values of similar property used for office purposes.

As to (a) above, neither counsel contended for any other term than the residue of the lease.

Question (iv) in the notice of motion is answered by item (e) above. I do not think that any experienced arbitrator should be directed as to the kind of comparables he may consider. He will, I imagine, consider the level of office rents in the locality of Lowndes Square. Against that background he will fit in items (a)-(d) above, making such adjustments by reference to (a)-(d) as he in his professional judgment thinks appropriate.

Chancery Division

December 8 1987

(Before Mr Justice WARNER)

BRITISH HOME STORES PLC v RANBROOK PROPERTIES LTD

Estates Gazette April 23 1988

[1988] 16 EG 80-87

Landlord and tenant — Rent review clause in lease — Construction — Lease for a term of 99 years from August 11 1977 — Provision that the hypothetical tenancy was to be on the terms and conditions contained in the actual lease "save as to rent" — Issue as to whether the hypothetical tenancy should itself include the rent review provisions contained in the lease — Authorities considered — Tenants relied particularly on the guidelines laid down by Sir Nicolas Browne-Wilkinson V-C in *British Gas Corporation* v *Universities Superannuation Scheme Ltd* — In the absence of clear words compelling the construction that the rent review provisions should be disregarded, however wayward the results, effect should be given to the underlying commercial purpose of a rent review clause by requiring that future rent reviews should be taken into account — It was argued on behalf of the landlords that the construction of a document should be approached without any preconceived idea as to the purpose of any particular provision in it — It was only when there was a doubt or ambiguity that it was proper to look at the commercial purpose — The Vice-Chancellor's guidelines in *British Gas* were, on this view, "only a guide to the perplexed" — Applying this suggested principle, Warner J pointed out that its logic required the landlords to show that the provisions under consideration were so clear as to be susceptible of only one meaning, namely, that the rent review provisions were to be excluded from the terms of the hypothetical lease — Warner J held that the words "the terms and conditions (save as to rent) herein contained" were at least ambiguous, so that they did not come within the category mentioned by the Vice-Chancellor which "must be given effect to, however wayward the results" — It was therefore legitimate to look at the underlying commercial purpose and, incidentally, to avoid the "undesirable semantic distinctions" to which the Vice-Chancellor had referred — Declaration accordingly in favour of tenants' construction

The following cases are referred to in this report.

Basingstoke and Deane Borough Council v *The Host Group Ltd* [1987] 2 EGLR 147; (1987) 284 EG 1587, CA
British Gas Corporation v *Universities Superannuation Scheme Ltd* [1986] 1 WLR 398; [1986] 1 All ER 978; [1986] 1 EGLR 120; (1986) 277 EG 980
Datastream International Ltd v *Oakeep Ltd* [1986] 1 WLR 404; [1986] 1 All ER 966; (1985) 51 P&CR 218; [1986] 1 EGLR 98; 277 EG 66
Equity & Law Life Assurance Society plc v *Bodfield Ltd* [1987] 1 EGLR 124; (1987) 281 EG 1448, CA
MFI Properties Ltd v *BICC Group Pension Trust Ltd* [1986] 1 EGLR 115; (1986) 277 EG 862
Philpots (Woking) Ltd v *Surrey Conveyancers Ltd* [1986] 1 EGLR 97; (1985) 277 EG 61
Pugh v *Smiths Industries Ltd* (1982) 264 EG 823

This was an originating summons for the construction of rent review provisions in the lease of a building in Barnsley. The plaintiffs were the tenants, British Home Stores plc, under the lease, which was for 99 years from August 11 1977. The defendants, the landlords, were Ranbrook Properties Ltd.

Nigel Hague QC and N J Patten (instructed by Clifford Chance) appeared on behalf of the plaintiffs; David Neuberger QC (instructed by Grangewoods) represented the defendants.

Giving judgment, WARNER J said: This is an originating summons for the construction of a rent review clause in a lease. The tenant under the lease is the plaintiff, British Home Stores plc. The landlord is the defendant, Ranbrook Properties Ltd. It is a lease of a building in Barnsley which, I believe, is a shop. It is dated July 4 1978 and is for a term of 99 years from August 11 1977.

The rent review clause provides for rent reviews every five years. The first took place on August 11 1982. The second was due on August 11 of this year and it is with that review that these proceedings are concerned. It will be observed that the lease still has 89 years to run.

I need not go through many of the provisions of the lease. I will read, however, first, the reddendum, in clause 2 of the lease, which is in these terms:

paying therefore first until the rent commencement date

the precise meaning of that phrase does not matter

a peppercorn if demanded and then for the remainder of the first five years of the said term the yearly rent of £210,000 and thereafter such yearly rent as shall become payable under and in accordance with the provisions set out in the Schedule hereto to be paid by equal quarterly payments in advance on the usual quarter days in each year clear of all deductions whatsoever the first of such payments however to be made on the rent commencement date and to be for the period from the rent commencement date to the quarter day next following and second by way of further rent a yearly sum equal to the premium or premiums paid by the lessor

and then I can summarise in respect of certain insurance policies.

In clause 3 one finds the lessee's covenants and the first of those is a covenant in common form for the payment of the rents. In clause 3(16)(g) there is a covenant that:

. . . in any underlease of the demised premises or of any part thereof there shall be included provisions for review of the rent reserved by the underlease corresponding both as to terms and dates with those in the Schedule.

In clause 5(6) there is a provision for the suspension of the rent or an appropriate part of it in the event of the demised premises or any part of them being at any time during the term damaged or destroyed by any of the risks insured against.

Last, so far as the body of the lease is concerned, clause 6(i) contains a proviso for re-entry in common form.

Then one comes to the schedule, which is headed "Rent Review Provisions". Paras 1 to 5 of that are as follows:

1 In this Schedule the following expressions shall have the following meanings:

A (a) "Review date" shall mean the date of the expiration of each 5 years of the term hereby granted.

(b) "Open market rent" shall mean the yearly rent for which the demised premises could reasonably be expected to be let by a willing lessor to a willing lessee with vacant possession on the relevant review date in the open market for a term equal to the unexpired residue on the relevant rent review date of the term hereby granted (which unexpired residue shall be deemed to be a term of not less than 20 years) upon the terms and conditions (save as to rent) herein contained on the assumption (if not of fact) that all the lessees covenants shall have been complied with but there being disregarded

(i) any effect on rent of the fact that the lessee or any under-tenant may have been in occupation of the demised premises or any part thereof

(ii) any goodwill attached to the demised premises by reason of any trade or business carried on there by the lessee or any under-tenant and

B (iii) any effect on rent of any improvement carried out by the lessee otherwise than in pursuance of an obligation to the lessor under this lease and other than the improvements carried out by the lessee prior to or at the commencement of this lease.

2 From and after each review date the rent first hereinbefore reserved shall be whichever is the higher of

(i) the yearly rent operative immediately before that review date, and

(ii) the open market rent of the demised premises on the review date (hereinafter referred to as "the new rent").

3 If three months before any review date (hereinafter referred to as "the relevant review date") the lessor and lessee shall not have agreed on the new rent to operate from the relevant review date the lessor or the lessee may at any time thereafter but before the review date next following the relevant review date require an independent surveyor (hereinafter called "the surveyor") to C be appointed to determine the open market rent.

4 The surveyor shall be appointed by the President for the time being of the Royal Institution of Chartered Surveyors.

5 (a) Notice in writing of his appointment shall be given by the surveyor to the lessor and the lessee. He shall invite each to submit within a specified period which shall not exceed four weeks a valuation accompanied if desired by a statement of reasons.

(b) The surveyor shall act as an expert and not as arbitrator. He shall consider any valuation and reasons submitted to him within the said period but shall not be in any way limited or fettered thereby and shall determine the open market rent in accordance with his own judgment.

(c) The surveyor shall give notice in writing of his decision to the lessor and lessee within two months of his appointment.

(d) If the surveyor comes to the conclusion that the open market rent of the D demised premises on a relevant review date is or was less than the earlier rent operative immediately before the relevant review date (hereinafter called "the current rent") the current rent shall be the new rent and the decision of the surveyor shall state.

(e) The decision of the surveyor shall be final on matters hereby referred to him.

I need not read any of the subsequent paragraphs of the schedule.

Now the question that has arisen between the parties arises on the construction of the words "upon the terms and conditions (save as to rent) herein contained" in para 1(b) of the schedule. It is the familiar question whether the hypothetical lease on the basis of which the surveyor is to determine the new rent should or should not be treated as containing the rent review provisions themselves.

E What has happened here is that an independent surveyor was appointed on October 13 1987 and under the timetable contained in para 5 of the schedule as altered by consent, he is required to give notice of his decision by Monday next, December 14. The parties are anxious to have a decision on the question I have mentioned this week, so that the surveyor can give his decision in the light of it.

Mr Hague, who appears for the plaintiff, relies on a number of decisions including my own decision in *Datastream International Ltd* v *Oakeep Ltd* [1986] 1 WLR 404*, Hoffmann J's decision in *MFI Properties Ltd* v *BICC Group Pension Trust Ltd* [1986] 1 All ER 974 and, more especially, the judgment and decision of Sir Nicolas Browne-Wilkinson V-C in *British Gas Corporation* v *Universities Superannuation Scheme Ltd* [1986] 1 WLR 398, the guidelines laid F down in which were approved by the Court of Appeal in *Equity & Law Life Assurance Society plc* v *Bodfield Ltd* [1987] 1 EGLR 124. He also referred me to *Basingstoke and Deane Borough Council* v *Host Group Ltd*, in which the judgment of the Court of Appeal was delivered on November 3 1987 (it has not been reported yet),† which also approves the approach of the Vice-Chancellor in the *British Gas* case.

Putting it shortly, what Mr Hague says is that the construction put upon para 1(b) by the landlord, according to which the hypothetical lease is to be regarded as not containing the rent review provisions, is contrary to the commercial purpose of the clause as recognised in those cases, that purpose being that, under such a lease as this, the tenant gets security of tenure for a long period, but that, in order to protect the landlord from the possible consequences of that in an age of inflation and of rising property and rental values, the rent is reviewed periodically. It would not be consistent with that purpose, says Mr Hague, and in this he is supported by the authorities, that the rent review provisions should operate so as to require the tenant to pay for something that he is not going to get, namely, a lease for 89 years without rent review provisions. Mr Hague points to the guidelines in the Vice-Chancellor's judgment, which the Vice-Chancellor introduced with these observations (at p 401):

> Counsel agreed that there were three possible constructions of the rent exclusion provision in this case, viz, (1) that it requires the valuer to ignore *all* provisions relating to rent in the lease; (2) that it requires the valuer to ignore those provisions which relate to the quantification of rent, ie, the rent payable immediately before the relevant review date *and* the provisions for future rent reviews; (3) that it requires the valuer to ignore the rent actually payable before the review date only, ie, he must take into account the provisions for future reviews of the rent.
>
> The first of these constructions is the only one that the words literally bear.

The words that the Vice-Chancellor had to consider were "a lease containing the same provisions (other than as to the yearly rent) as are herein contained".

> But the result of such a literal construction

the Vice-Chancellor went on

> is so surprising as to offend commonsense. The covenant for payment of rent is a "provision . . . as to yearly rent". So is the right of re-entry reserved for non-payment of rent. Accordingly, if the literal construction were right, the valuer would be faced with the wholly unrealistic (and I would have thought impossible) task of fixing the open market rental for a large office block on terms which contained no covenant for payment of rent nor any right of re-entry in the event of non-payment of rent. Mr Barnes, for the defendant, to my mind rightly did not contend for this fully literal construction. Once one departs from the literal meaning of the words, one is necessarily implying some limitation into the words actually used in the lease. The only question is whether such implication should be that inherent in construction (2) or construction (3) above. At this stage, in my judgment one is necessarily forced to seek to discover the underlying purpose of the provisions for rent review so as to give effect to that purpose.
>
> There is really no dispute that the general purpose of a provision for rent review is to enable the landlord to obtain from time to time the market rental which the premises would command if let on the same terms on the open market at the review dates. The purpose is to reflect the changes in the value of money and real increases in the value of the property during a long term. Such being the purpose, in the absence of special circumstances it would in my judgment be wayward to impute to the parties an intention that the landlord should get a rent which was additionally inflated by a factor which has no reference either to changes in the value of money or in the value of the property but is referable to a factor which has no existence as between the actual landlord and the actual tenant, ie, the additional rent which could be obtained if there were no provisions for rent review. Of course, the lease may be expressed in words so clear that there is no room for giving effect to such underlying purpose. Again, there may be special surrounding circumstances which indicate that the parties did intend to reach such an unusual bargain. But in the absence of such clear words or surrounding circumstances, in my judgment the lease should be construed so as to give effect to the basic purpose of the rent review clause and not so as to confer on the landlord a windfall benefit which he could never obtain on the market if he were actually letting the premises at the review date, viz, a letting on terms which contain provisions for rent review at a rent appropriate to a letting which did not contain such a provision.
>
> Therefore, apart from authority I would hold that on a true construction of paragraph 3 of the second schedule the rack rental value has to be fixed on the basis that the hypothetical letting is on the terms of the actual lease excluding only the rent actually quantified and payable before the review date but including the provisions for five yearly rent review.

The Vice-Chancellor then examined the authorities that had been cited to him and concluded (at p 403), in a passage which was approved by the Court of Appeal in the *Equity & Law* case:

> In these circumstances, there are in my judgment conflicting decisions as to the correct approach to the construction of these clauses. I am accordingly free to adopt the approach I prefer. In my judgment the correct approach is as follows: (a) words in a rent exclusion provision which require *all* provisions as to rent to be disregarded produce a result so manifestly contrary to commercial commonsense that they cannot be given literal effect; (b) other clear words which require the rent review provision (as opposed to all provisions as to rent) to be disregarded (such as those in the *Pugh* case, (1982) 264 EG 823) must be given effect to, however wayward the results; (c) subject to (b), in the absence of special circumstances it is proper to give effect to the underlying commercial purpose of a rent review clause and to construe the

* Editor's note: The first three decisions mentioned are reported also at [1986] 1 EGLR 98, [1986] 1 EGLR 115 and [1986] 1 EGLR 120 respectively.

† Subsequently reported at [1987] 2 EGLR 147.

A words so as to give effect to that purpose by requiring future rent reviews to be taken into account in fixing the open market rental under the hypothetical letting.

I am conscious that such an approach is perilously close to seeking to lay down mechanistic rules of construction as opposed to principles of constuction. But there is an urgent need to produce certainty in this field. Every year thousands of rents are coming up for review on the basis of clauses such as the one before me: witness the growing tide of litigation raising the point. Landlords, tenants and their valuers need to know what is the right basis for their valuation without recourse to lawyers, let alone to the courts. The question cannot be left to turn on the terms of each lease without the basic approach being certain. It is in my judgment most desirable that this, or some other case, should at an early stage be taken to the Court of Appeal so as to resolve the conflicting judicial approaches that have emerged.

B Later, in the final paragraph of his judgment, the Vice-Chancellor referred to the undesirability, both legally and commercially, of narrow semantic distinctions in this field.

Mr Neuberger, who appears for the defendant landlord, submits that I should not approach the construction of any document with any preconceived idea as to the purpose of any particular provision in it. He submits that it is only when one has a doubt as to the meaning of such a provision that one may look at the commercial purpose of the document to resolve that doubt. He also submits that it is not right to base a decision on the interpretation of what the Vice-Chancellor helpfully called a "rent exclusion provision" on what has been said about other rent exclusion provisions in other cases, though he
C concedes, quite rightly of course, that I should have regard to general observations which have been made in other cases to the principles upon which the construction of such provisions should be approached.

He relied also on a decision which had not been cited by Mr Hague, namely, *Philpots (Woking) Ltd* v *Surrey Conveyancers Ltd*. That is a decision of the Court of Appeal, reported in [1986] 1 EGLR 97. Although this is not apparent from the report of the *MFI* case, I was told by counsel that it was cited to Hoffmann J. It does not appear to have been cited to the Vice-Chancellor in the *British Gas* case or to the Court of Appeal in the *Equity & Law* case.

It seems to me that that decision does not help very much, because in that case, which was not concerned with a rent exclusion provision
D but, so far as relevant here, simply with general principles of construction, the Court of Appeal was confronted with an ill-drafted clause which could only mean one thing. It was rather like the case before Goulding J of *Pugh* v *Smiths Industries Ltd** which is referred to in the judgment of the Vice-Chancellor and, indeed, also in my judgment in the earlier *Datastream* case. It supports proposition (b) in the paragraph of the Vice-Chancellor's judgment where he states the guidelines, viz that "other clear words which require the rent review provision (as opposed to all provisions as to rent) to be disregarded (such as those in the *Pugh* case) must be given effect to, however wayward the results". There cannot be any doubt that the result in the *Philpots* case was wayward, but neither can there be any doubt that the words that the parties had chosen put it beyond
E question that that was the result that was prescribed. The same is true of the actual decision in the *Equity & Law* case.

It seems to me that, in the sense illustrated by those decisions, Mr Neuberger's submission must be right, that it is only where there is some uncertainty as to the meaning of a particular provision in a lease, or indeed in any other document, that one can resolve the uncertainty by looking at its commercial purpose.

Turning to the authorities relied on by Mr Hague, Mr Neuberger submitted that my own decision in the *Datastream* case was distinguishable because there the ambiguous word was the word "amount", which we do not have in the present clause, and also because the clause in question there did not contain the words "as to"
F which we do have in the present clause. As regards the *MFI* case Mr Neuberger submitted that a passage in the judgment of Hoffmann J supported his argument. That was the passage at p 976, where Hoffmann J said:

> Given that the words then are reasonably capable of bearing both the wider and the narrower meaning I think that I am entitled in choosing between them to have regard to the fact that they appear in the context of a rent review clause and to have regard to the normal purpose of such a clause.

As regards the *British Gas* case, Mr Neuberger concedes that he cannot say the Vice-Chancellor's guidelines are wrong because he accepts that they are self-evidently right and, in any case, they have been approved by the Court of Appeal. But, he says, they are only a guide to the perplexed and the words in the lease that the Vice- G
Chancellor was considering were not identical with those in the present case.

In order to apply the principle on which he relies, and which seems to me right, Mr Neuberger has to demonstrate that the provision in the present lease is only susceptible of one meaning and that that meaning is such as to exclude the rent review provisions from the terms of the hypothetical lease. As to that, Mr Neuberger's starting point is of course that the rent review provisions are terms as to the rent so that they are excluded by the phrase "upon the terms and conditions (save as to rent) herein contained". In support of that he relies upon the words of Peter Gibson J in the *Equity & Law* case ([1985] 2 EGLR 144) where he said (at p 146):

> I turn to Mr Lewison's submission. He says that if one poses the question H
> "What are the terms of the lease as to rent?", the answer that would naturally be given would not be limited to the terms reserving the fixed quantified rents but would be bound to include a reference as to the terms as to the revised rent payable on a rent review. I agree.

In answer to Mr Hague's argument that the covenant for payment of the rent and the proviso for re-entry for non-payment of rent were terms and conditions as to rent which clearly could not be thought appropriately excluded from the hypothetical lease, Mr Neuberger submitted that the proviso for re-entry was not a term or condition as to rent. That I find difficult to accept because it is a term to the effect that if the rent is unpaid the lease is forfeit. If it is not a term as to the rent, it seems to me to be a condition as to the rent. Then Mr Neuberger submitted that the covenant for payment of the rent was J
not a term or condition as to the rent but a term or condition as to payment of the rent. With all respect to Mr Neuberger I do not think that that distinction will do. His submission was that "terms or conditions as to rent" meant what is in the reddendum and he pointed to the fact that the reddendum refers to the contents of the schedule. When I asked him about the further rent equal to the insurance premiums, which is also referred to in the reddendum, Mr Neuberger said that that was a further rent, not a rent in the narrow sense in which the word is used in the provision that I have to construe. He submitted that if one looked carefully at the words in that provision "the terms and conditions (save as to rent) herein contained" they were not apt to cover a mere figure and that it was not open to me to
"chop the reddendum in two" so as to include in the reference to rent K
in the provision in question the £210,000 mentioned in the reddendum but not "the provisions set out in the schedule" also mentioned there. Of course, it is not quite right that Mr Hague's argument would lead to such a chopping in two of the reddendum, because one has to bear in mind that there has already been a rent review which produced a figure in 1982 of £270,000. So the construction that Mr Hague advocates does not simply mean translating the words "the terms and conditions (save as to rent) herein contained" as "the terms and conditions (save as to the figure of £210,000) herein contained".

Mr Neuberger, with his usual fairness and candour, accepted that the construction for which he was pressing would be inequitable, L
because it would mean that the tenant would be paying for something he was not getting, and that it was against the weight of the authorities.

The more I listened to Mr Neuberger struggling with those provisions of the lease which mentioned rent but which he contended were not terms and conditions of the lease as to rent the more I became convinced that at the very least those words "the terms and conditions (save as to rent) herein contained" are ambiguous and that this is not a case which falls within limb (b) of the Vice-Chancellor's guidelines. That being so I must go — and I am glad to do so, because I think it is the justice of the case — to limb (c) in those guidelines, "in the absence of special circumstances it is proper to give effect to the underlying commercial purpose of the rent review M
clause and to construe the words so as to give effect to that purpose by requiring future rent reviews to be taken into account in fixing the open market rental under the hypothetical letting". I think it all the more right that that should be the result because it also avoids the undesirable semantic distinctions to which the Vice-Chancellor referred at the end of his judgment.

In the result I will make a declaration in the sense sought by Mr Hague. I will hear counsel as to what the precise terms of it should be.

Declaration made to the effect that the words "the terms and conditions (save as to rent) herein contained" included the rent review provisions in the schedule to the lease. Defendants ordered to pay plaintiffs' costs.

* Editor's note: Reported at (1982) 264 EG 823.

A # Court of Appeal

March 10 1988

(Before Lord DONALDSON of LYMINGTON MR, Lord Justice PARKER and Lord Justice TAYLOR)

PANAVIA AIR CARGO LTD v SOUTHEND-ON-SEA BOROUGH COUNCIL

Estates Gazette June 4 1988

[1988] 22 EG 82-86

B **Landlord and tenant — Rent review clause in lease — Construction — Whether time of the essence for a particular provision despite a general statement to the contrary — The lease in question was a long lease at a rent of £2,700 per annum plus any additions under review clause — Reviews at seventh year and subsequently at five-year intervals — In the schedule to the lease various time-limits were laid down for particular steps but para 13 provided that "as respects the periods of time referred to in this schedule time shall not be deemed to be of the essence of the contract" — However, para 14 provided that if for any reason no review took place in respect of any relevant period or a review was not completed within 12**
C **months of the commencement of any relevant period, then the rent for that period would be increased by 25% of the rent currently payable — In fact a review was not completed within 12 months of the commencement of the first review period — The tenants submitted that para 14 applied, that time was of the essence in relation to that para, and that the rent was due to be increased by 25% only, ie from £2,700 to £3,375 per annum (in contrast with a rent of probably over £20,000 if the review provisions in the schedule had taken effect) — The landlords argued that, although para 13 preceded para 14 in the schedule, it was not intended to apply only to time-limits in the**
D **earlier provisions; if it had been so intended it would have been easy to say so, but the plain words used clearly applied to the whole schedule — This construction had been adopted by Mr Edward Nugee QC, sitting as a deputy judge of the Chancery Division, and was accepted by the Court of Appeal on appeal by the tenants — The court rejected the tenants' argument that para 14 should be given its literal effect on the ground that its object was to give certainty and finality and to protect a tenant from inordinate delay in pursuing a rent review, perhaps resulting in a large amount of back-dated rent — It was pointed out that the tenants could have made time of the essence by serving the appropriate notice — The case of *Lewis***
E ***v Barnett*, cited on behalf of the tenants, was distinguishable — Held that on the wording of the schedule, and in the light of the general presumption in *United Scientific Holdings v Burnley Borough Council*, time was not of the essence for the purpose of para 14 — Hence the landlords were still entitled to have the rent determined by an independent surveyor in accordance with the provisions in the schedule to the lease — Appeal dismissed**

The following cases are referred to in this report.

Forbes v *Git* [1922] 1 AC 256
Lewis v *Barnett* (1981) 264 EG 1079
F *Smith's (Henry) Charity Trustees* v *AWADA Trading & Promotion Services Ltd* (1983) 47 P&CR 607; [1984] EGD 103; 269 EG 729, CA
United Scientific Holdings Ltd v *Burnley Borough Council* [1978] AC 904; [1977] 2 WLR 806; [1977] 2 All ER 62; (1977) 33 P&CR 220; [1977] EGD 195; (1977) 243 EG 43 & 127, HL

This was an appeal by the tenants, Panavia Air Cargo Ltd, from a decision of Mr Edward Nugee QC, sitting as a deputy Chancery judge, in favour of the construction put forward by the landlords, Southend-on-Sea Borough Council, of certain rent review provisions in the lease of some 2¾ acres of land at Southend Municipal Airport. The judgment of Mr Nugee was reported at [1988] 01 EG 60.*

Jonathan Gaunt (instructed by Cameron Markby) appeared on behalf of the appellants; David Neuberger QC (instructed by Rowe & Maw) represented the respondents.

Giving the first judgment at the invitation of the Master of the Rolls, TAYLOR LJ said: This is a tenant's appeal from a decision dated July 30 1987 of Mr Edward Nugee QC sitting as a deputy judge in the Chancery Division. The problem before him and before this court is the construction of a lease containing apparently contradictory clauses as to rent review. The lease was granted by the defendant council to the plaintiff company, Panavia Air Cargo Ltd, formerly British Air Ferries Ltd. It was dated October 11 1978 and comprised about 2¾ acres of land at Southend Municipal Airport together with ancillary buildings. The terms of the lease, so far as are relevant, were as follows. By clause 1 the council demised the property to the company:

> TO HOLD the same unto the Company from the First day of January [1978] to the thirteenth day of February [2035] YIELDING AND PAYING therefore during the said term for the period First day of January [1978] to the thirteenth day of February [1985] the annual rent of two thousand seven hundred pounds plus any additional rent payable under the provisions for review contained in the Schedule hereto without any deduction whatsoever by equal quarterly payments in arrear to be made on the usual quarter days.

The rent review provisions are contained in the schedule. It is necessary to set out some of the paragraphs and to summarise others. Para 1:

> During the seventh, twelfth, seventeenth, twenty-second, twenty-seventh, thirty-second, thirty-seventh, forty-second, forty-seventh and fifty-second years of the said term for each successive five-year period of the said term and during the fifty-second year of the said term for the remaining period of the said term (each of such periods being hereinafter referred to as a "relevant period") the Council and the Company shall agree or failing agreement shall determine by arbitration the sum total of the then current rack rent ([as defined] . . .) and one-quarter of the sum total so ascertained or Two thousand Seven hundred Pounds (whichever is the greater) shall be the rate of rent reserved by this Lease in respect of the then next succeeding relevant period.

Para 3:

> If such agreement has not been made six months before the commencement of the relevant period the Council may require an independent surveyor . . . to be appointed to determine the new rent. The Surveyor may be nominated by agreement between the Council and the Company or appointed by the President for the time being of the Royal Institution of Chartered Surveyors on the application of the Council.

Para 5:

> Notice in writing of his appointment by the said President shall be given by the Surveyor to the Council and the Company inviting each to submit within a specified period (which shall not exceed four weeks) a valuation accompanied if desired by a statement of reasons.

Para 7:

> The Surveyor shall give notice in writing of his decision to the Landlord and the tenant within two months of his appointment or within such extended period as the Council may agree.

Para 8:

> If the Surveyor comes to the conclusion that the current market value of the demised premises is less than the rent operative for the period preceding the relevant period (hereinafter called the current rent) the new rent shall nevertheless be the same as the current rent and the decision of the Surveyor shall so state.

Para 9:

> If the Surveyor shall fail to determine the new rent and give notice thereof within the time and in the manner hereinbefore provided or if he shall relinquish his appointment or die or if it shall become apparent that for any reason he will be unable to complete his duties hereunder the Council may apply to the said President for a substitute to be appointed in his place which procedure may be repeated as many times as necessary.

Para 11 deals with the date of commencement of the new rent and with payment retrospectively from the date of review. There then follow paras 13 and 14, which are the two paragraphs at the heart of this case. Para 13 reads:

> As respects the periods of time referred to in this Schedule time shall not be deemed to be of the essence of the contract.

Para 14:

> If for any reason whatsoever no review of rent takes place in respect of any relevant period or a review is not completed within 12 months of the commencement of any relevant period then the rent payable during the relevant period under the provisions of this lease shall be increased by 25% of the rent then currently payable.

*Editor's note: See also p 111 *ante*.

A The short point in the case is whether time is of the essence as to the 12-month period in para 14, despite the language of para 13.

I take the facts of the case from the judgment of the learned deputy judge, where they are succinctly stated. The seventh year of the term was the calendar year 1984. On September 13 1984 the council's borough valuer and surveyor wrote to the company pointing out that the first rent review under the lease was to operate from January 1 1985 and asking whether the company would be appointing a professional adviser to agree terms on its behalf. A rather desultory correspondence followed.

On January 17 1986 the borough valuer and surveyor wrote to the company surveyor saying that, as the rental review was now considerably overdue, if the negotiations did not progress in the very
B near future, he would have to report accordingly to the council with a view to the appointment of an independent valuer in accordance with the schedule.

On February 17 1986 the company surveyors wrote back saying that under clause 14 of the schedule there was to be an automatic 25% increase in the rent if *inter alia* a review had not been completed within 12 months of the commencement date. Since that was the situation, they therefore asked the council to confirm that the rent payable from the review date was to be increased to £3,375 per annum, a 25% increase on the initial rent of £2,700.

The council rejected that suggestion and thereafter applied to the president of the Royal Institution of Chartered Surveyors for the
C appointment of a surveyor, but he declined to appoint one in view of the dispute between the parties. (There was a subsidiary dispute as to whether the rent review date in 1985 was January 1 or February 13. The learned deputy judge held that it was January 1 and there is now no dispute about that.)

On March 4 1987 the originating summons was issued. The company sought a declaration that the rent payable to the council for the relevant period beginning in 1985, which was January 1 1985, is, in the events which have happened, to be assessed in accordance with para 14 of the schedule. The declaration sought would have required a holding that para 13 of the schedule does not apply in this context and therefore that the time-limits provided in para 14 should be strictly enforced.

D The council argued, first, that para 14 was purely an interim provision which did not deprive them of their right to review the rent. The learned deputy judge rejected this argument and there is no cross-notice challenging that finding. Second, the council argued that para 13 was clear in its terms and is expressed to apply to the periods of time referred to in the schedule without exception. Therefore the council can still have a rent review despite the expiry of 12 months referred to in para 14.

The company's prime argument was that, if para 14 did not mean what it said, it had no sensible function. The reply to that was that, although by reason of para 13 time was not of the essence under para 14, the company could make it so by serving a notice. Therefore,
E although the company argue that they could have served such a notice anyway, para 14 did have a function, namely, to provide a trigger date following which a tenant's notice by the company could nominate a reasonable time and make it of the essence. If even after that the council failed to have the rent reviewed, then para 14 would be effective to give them an automatic 25% rent increase.

The learned deputy judge found in the council's favour on these arguments. He started by considering the broad principles applicable to rent review provisions, as stated in two leading cases. The first was *United Scientific Holdings Ltd* v *Burnley Borough Council* [1978] AC 904. At p 930 Lord Diplock said:

> So upon the question of principle which these two appeals were brought to settle, I would hold that in the absence of any contra-indications in the express
> F words of the lease or in the interrelation of the rent review clause itself and other clauses or in the surrounding circumstances the presumption is that the timetable specified in a rent review clause for completion of the various steps for determining the rent payable in respect of the period following the review date is not of the essence of the contract.

In *The Trustees of Henry Smith's Charity* v *AWADA Trading & Promotion Services Ltd* (1983) 47 P&CR 607, Griffiths LJ (as he then was) said at p 616:

> When they enter into a lease such as this the expectation of both the landlords and tenants is that a fair market rent will be paid throughout the lease and unless driven to do so by the wording of the lease I am loth to construe the machinery provided for arriving at a fair market rent as forcing either the tenant to pay an exorbitantly high rent or the landlord to receive a ridiculously low rent, neither of which bears any relation to the fair market rent because one or other of them was one day late in observing the timetable set out in the G
> rent review provisions of the lease.

There was evidence in the present case that a review such as was contemplated by the schedule would have resulted in a rent of over £20,000, which, as the learned deputy judge remarked, may well have made the £3,350, if para 14 is applied, seem a ridiculously low rent.

Focusing on the two vital paragraphs in the case, the learned deputy judge concluded that para 13 governed the whole schedule, including para 14. He relied on the wording of para 13 itself and on the principles stated by Lord Diplock. He accepted the council's argument that the company could serve notice so as to make time of the essence. Accordingly, his order declared *inter alia* that the defendant council is entitled to have the rent payable from January 1 H
1985 determined by an independent surveyor nominated by agreement or appointed by the president of the Royal Institution of Chartered Surveyors pursuant to paras 1 to 10 of the schedule. He also declared that it was open to the company to make time of the essence for the purposes of para 14 by serving notice.

Before this court Mr Gaunt in his able argument for the company submits that para 14 should be given its literal effect and its time-limit should not be relaxed by applying para 13. Its object was, he says, to give certainty and finality. It was to protect the company from the possibility that the council might delay for years, then pursue a rent review, and require the company to pay a large amount of back-dated rent. This could happen, he says, because all the initiatives under the review procedure are with the council, which can therefore speed or J
delay a rent review at will. Therefore, Mr Gaunt argued that para 13 should be read as applying only to the time provisions in the paragraphs which precede it. Para 14 should be read as if it were a qualification or exception to the generality of para 13. As authority for reading the two paragraphs in that way he cited the case of *Forbes* v *Git* [1922] 1 AC 256. There a contract provided that, in consideration of 3,000 Canadian dollars payable by three instalments, the appellant agreed to furnish materials and perform certain services. A later clause, however, provided for the payment to the appellant of the value of the materials and the labour done at 12½% over cost "whether that value exceeded or was less than $3,000". The court held that the later clause prevailed, since the first clause did not necessarily mean the contract price was to be $3,000. K

But that case was totally different from the present, and the context of the two clauses was totally different from that of the two paragraphs under consideration. In the cited cases there was no repugnancy between the two clauses. Here, if para 13 applies to para 14 at all, its only effect could be that contended for by the council, unless words are added.

Mr Gaunt further relies on *Lewis* v *Barnett* (1981) 264 EG 1079. That was also a landlord and tenant case concerning rent review provisions. The lease provided in paras 2 and 3 for certain steps to be taken within certain time-limits. Para 6 provided that, if the parties had not carried out those steps within the specified periods, the landlord's notice already given under those provisions should be void L
and of no effect. The court held that the consequence of not complying with the time-limits laid down in paras 2 and 3 was to deprive the landlord of the benefit of the rent review provisions.

The learned deputy judge distinguished that case from this on the grounds that the time-limit in para 14 here is not linked with the time-limits in the earlier paragraphs. Mr Gaunt submits that that is not a valid or relevant distinction. But, quite simply, in that case there was no express provision that time was not to be of the essence: here there is. The question has been raised in argument as to whether, if a similar clause to para 13 had been included in the lease in the cited case, the outcome would have been the same. Mr Gaunt submits that it would. Without deciding what is a hypothetical case, I consider that it may be that the distinction made by the learned deputy judge in the present M
case would then have been of relevance, that the provisions of clause 6 in that lease did relate to provisions as to time in earlier clauses, which is not the case with para 14 here.

In my view, the wording of para 13 is crucial. Had it been intended to apply only to the foregoing or previous provisions of the schedule, that could have been said in terms. However, the plain words used clearly apply to the whole schedule.

In the *United Scientific Holdings* case both Lord Diplock and Lord Salmon in the context of rent review said that the parties should make it clear as to whether time was to be of the essence or not. At p 936G Lord Diplock said:

> The best way of eliminating all uncertainty in future rent review clauses is to

state expressly whether or not stipulations as to the time by which any step provided for by the clause is to be taken shall be treated as being of the essence.

At p 947E, Lord Salmon said:

> Such clauses could easily be drafted so that they state expressly whether time is or is not to be treated as of the essence. So drafted they would present no difficulty. Unfortunately they rarely are. They should be, for if they were, a great deal of expensive litigation would be avoided. If, eg, the parties to the present appeals had expressly stated whether or not they intended the provisos as to time in the rent revision clauses to be of the essence, there would have been no litigation between them let alone litigation fought up to your Lordships' House for the purpose of deciding what the rent revision clauses mean.

Their lordships' decision was in January 1977. This lease was made in January 1978. It contains in para 13 just the sort of indicative clause which their lordships recommended should be included. In my judgment the wording of that paragraph, para 13, is decisive of this appeal. I agree with the decision of the learned judge. I further agree with him that it accords with the justice of the case. There are provisions in the schedule which might well make the landlord's right to review the rent hazardous if the time-scale of para 14 were rigidly applied. For example, under para 9, should the appointed surveyor for any reason have to be replaced, perhaps more than once, the time-limits would render the landlord vulnerable in a situation beyond his control. Moreover, the flexible approach is more likely to achieve that fair market rent to which Griffiths LJ referred. I would therefore dismiss this appeal.

PARKER LJ agreed and did not add anything.

Also agreeing, LORD DONALDSON OF LYMINGTON MR said: *United Scientific Holdings Ltd* v *Burnley Borough Council* affirms that there is a presumption that parties to rent review clauses do not intend time to be of the essence. In para 13 of this lease the parties have expressly declared that it is their intention that time shall not be of the essence. They have not, as they could have done, made it clear in clause 13 that that applies only to the preceding clauses. Accordingly, I entirely agree that time is not of the essence until one of the parties declares it to be, and accordingly I, too, would dismiss the appeal.

The appeal was dismissed with costs.

Court of Appeal

February 12 1988

(Before Lord Justice SLADE, Lord Justice NEILL and Lord Justice MANN)

LADBROKE GROUP PLC AND ANOTHER v BRISTOL CITY COUNCIL

Estates Gazette June 11 1988

[1988] 23 EG 125-128

Landlord and tenant — Rent review clause — Construction — Provision for delay of first rent review date in the event of certain road works not being completed — Question as to effect of provision on first review date — Lease was for 125 years from July 5 1973 pursuant to a building agreement — There was to be a basic rent for the first seven years of the term, with subsequent reviews — Clause under dispute provided that if the road works had not been carried out by March 31 1973, then the first review date should be delayed "one year for every year or part of a year" until completion of the works or notification by the landlords that the works would not be carried out, whichever was the earlier — The lease when granted repeated this provision except that the tense was altered to the past ("as" for "if") because by March 31 1973 the works had in fact not been carried out — Eventually, in 1981, the landlords gave notice that the works would not be carried out — In their notification letter the landlords contended that the first review date, which would otherwise have been July 5 1980, would be postponed to July 5 1981 — The controversy before Judge Blackett-Ord and the Court of Appeal was as to the starting point for the period of delay — It was common ground that the period of time encompassed by the words "every year or part of a year", by reference to which the period of delay had to be measured, ended on February 10 1981, the date of the landlords' notification that the works would not be carried out — There were, however, two divergent views as to the starting point

The landlords' view, which was accepted by Judge Blackett-Ord, was that the starting point was July 5 1980, being the earliest review date provided for by the lease — Part of one year elapsed between July 5 1980 and the landlords' notice of February 10 1981 and the result was, according to the formula, that the first review date fell to be deferred by one year, ie from July 5 1980 to July 5 1981

The tenants' view, which was approved by the Court of Appeal, was that the first review date, in the events which had happened, was July 5 1988 — The weakness of the landlords' argument, the court pointed out, was that it gave no meaning to the reference in the formula to March 31 1973 — It was possible that the parties had expected that the lease would be executed before this date — It was permissible to look at the draft lease for the purpose of discovering the parties' intentions, including the relevance of March 31 1973 — At any rate, taking March 31 1973 as the starting point, the application of the formula resulted in July 5 1988 being the delayed date of the first rent review — Appeal allowed and declaration accordingly

No cases are referred to in this report.

This was an appeal from the decision of Judge Blackett-Ord, sitting as a judge of the High Court, reported at [1987] 2 EGLR 139; (1987) 283 EG 1071. The matter was raised by the plaintiffs, Ladbroke Group plc and English & Overseas Hotels Ltd, by originating summons seeking determinations as to provisions in a lease dated April 30 1974 following a building agreement for the erection on a site at Bristol of offices and a hotel. The defendants, now respondents to the appeal by the plaintiffs, were Bristol City Council, the site owners and lessors.

Nicholas Dowding (instructed by L A Goldsmith, solicitor's department, Ladbroke Group) appeared on behalf of the appellants; D H Fletcher (instructed by D W P Lewis, City Clerk, Bristol) represented the respondents.

Giving the judgment of the court, SLADE LJ said: This is an appeal by the plaintiffs in an action, Ladbroke Group plc ("Ladbroke") and English & Overseas Hotels Ltd ("the tenants"), from an order of His Honour Judge Blackett-Ord made on February 5 1987. It raises yet another short but not entirely easy question of construction concerning a rent review clause in a lease.

On February 8 1971 a building agreement ("the building agreement") was entered into between the defendants, the City Council of Bristol ("the council") of the first part, Ladbroke (as surety) of the second part and the tenants of the third part, though all three parties were therein, and in the subsequent lease, described by names different from those which they now bear.

Under the building agreement the council granted to the tenants a licence to enter certain land in Bristol for the purpose of enabling them to build a hotel and offices, those works of construction being together defined as "the works". The land in question had a frontage to Redcliffe Way. The licence was expressed (by clause 1) to endure "until a lease (in the form set forth in the Third Schedule hereto) shall be executed".

Clause 3(iii)(a) placed an obligation on the tenants at their own cost to commence by March 31 1971 and to execute on the site the works with due expedition and in accordance with a specified outline planning permission. Clause 3(iii)(b) obliged them to complete the works relating to the hotel before the expiry of the period of $2\frac{1}{4}$ years from the date of the building agreement, and the works relating to the offices before the expiry of the period of two years from the date upon which detailed planning permission should be granted for the offices. Clause 3(iii)(c) provided for the possible extension of these periods. By Clause 3(xxiii) the tenants covenanted with the council to comply with "the special conditions and stipulations" contained in the second schedule to the building agreement.

Clause 4 provided for the grant of a lease of the site and the works erected thereon by the council to the tenants for a term of 125 years as soon as the works had been completed, "such lease . . . to be in the form . . . contained in the Third Schedule hereto".

The second schedule, so far as material, provided as follows:

2 The lands which will be surplus to highway requirements consequent upon the re-alignment to Redcliffe Way by the Corporation and which will not be included in the proposed demise contained in the Lease contained in the Third Schedule hereto shall be suitably laid out by the Corporation in consultation with the Licensees as amenity areas and the appropriate areas shall thereafter during the term of the said Lease be maintained by the Licensees as an adjunct of the works and the Lease shall so provide.

The third schedule set out the form of lease referred to in clauses 1 and 4 (which we will call "the draft lease"). We need only refer to a few of its provisions.

Clause 1 provided:

In pursuance of an Agreement made between the parties hereto on the . . . day of February One thousand nine hundred and seventy one (hereinafter called "the said Agreement") and in consideration of the rent hereinafter reserved and of the Tenants' covenants hereinafter contained the Corporation hereby demise unto the Tenants ALL THAT the plot of land buildings and premises described in the First Schedule hereto (which said plot of land buildings premises with any alterations or additions thereto easements and rights are hereinafter referred to as "the demised premises") but EXCEPTING AND RESERVING unto the Corporation their lessees tenants and occupiers the easements and rights specified in Part I of the Second Schedule hereto and also for the benefit of the public in general the rights specified in Part II of the said Second Schedule and subject to the covenants affecting part of the demised premises referred to in Clause 3(xxx) hereof TO HOLD the same (except and reserved as aforesaid) unto the Tenants for the term of One hundred and twenty five years commencing on the . . . day . . . of . . . One thousand nine hundred and . . . yielding and paying during the said term the yearly rent following that is to say during the first seven years of the said term the yearly rent of Fifteen thousand five hundred pounds (hereinafter in Clause 2 referred to as "the basic ground rent") and during the next period of Eighteen years and the four successive periods of Twenty five years each a yearly rent to be assessed in accordance with Clause 2 aforesaid such rent to be paid in advance by equal quarterly payments on the Twenty fifth day of March the Twenty fourth day of June the Twenty ninth day of September and the Twenty first day of December in every year during the said term the first payment in respect of the period from and including the . . . day of . . . One thousand nine hundred and . . . to be payable on the execution hereof and for the purpose only of assessing the amount of such first payment the rent year shall be deemed to run from the quarter day immediately preceding the commencement of the term.

Clause 2, so far as material, provided:

(a) The yearly rent to be paid under this Lease during the period of Eighteen years immediately following the first seven years of the said term and during each of the four successive periods of twenty-five years of the said term shall be such sum as represents the same proportion of the rack rental value of the demised premises as defined in sub-clause (c) hereof at the time of assessment as the basic ground rent bears to the basic rack rental as defined in sub-clause (b) hereof and such yearly rent shall be agreed between the parties hereto during the last six months of the Seventh Twenty-fifth Fiftieth Seventy-fifth and One hundredth years of the said term respectively in respect of the appropriate period immediately ensuing such respective year PROVIDED nevertheless that the yearly rent shall at no time be less than the yearly rent payable during the period immediately before the year of assessment PROVIDED ALSO THAT if the realignment of Redcliffe Way and the landscaping of the lands south thereof to the site has not been carried out by the Thirty first day of March One thousand nine hundred and seventy three then the first rent review date hereinbefore stipulated shall be delayed one year for every year or part of a year until such realignment and landscaping shall have been completed or the Corporation declares by notice in writing to the Tenants that Redcliffe Way will not be realigned whichever is the earlier and accordingly the Eighth year referred to in sub-clause (c) hereof shall be delayed in like manner.

(b) The basic rack rental value of the demised premises shall be such sum as shall be agreed between the parties hereto as representing the total rack rental value of the demised premises either on the date when the demised premises are first fully occupied or three years from the date of this Lease whichever is the earlier AND when so agreed a Memorandum denoting the basic rack rental value shall be endorsed upon this Lease and the counterpart thereof and signed by or on behalf of the parties hereto as acceptance of such sum as the basic rack rental value.

(c) The rack rental value of the demised premises shall be such sum as represents the total rack rental value of the demised premises at the commencement of the Eighth Twenty-sixth Fifty-first Seventy-sixth and One hundred and first years of the said term and shall be agreed between the parties hereto within the six months immediately preceding the commencement of those respective years . . .

Clause 4(xxvii) contained a covenant by the tenant to comply with the "special covenants and conditions contained in the Third Schedule hereto".

The third schedule to the draft lease began as follows:

The lands which are surplus to highway requirements consequent upon the realignment of Redcliffe Way by the Corporation so far as they lie between the northern boundary of the demised premises and the realigned highway are not included in this demise but shall be suitably laid out by the Corporation in consultation with the Tenants as amenity areas and the Tenants hereby covenant with the Corporation that the Tenants will at the Tenants' expense after such amenity areas have been so laid out during the term granted by this Lease . . .

The third schedule then set out particulars of the tenants' obligations concerning the amenity areas.

We pause to make certain comments in regard to the building agreement. Beyond the references in that document and in the subsequent lease, there is no evidence before us to explain the nature of the contemplated realignment of Redcliffe Way and landscaping of neighbouring lands (which we will call collectively "the roadwork"), or the significance of these references. Nevertheless, it is plain from the references to the roadwork in the second schedule to the building agreement and in the draft lease set out in the third schedule that the parties were contemplating that (1) the roadwork would possibly or even probably be carried out in due course; (2) the doing of the work would be for the benefit of the tenants; (3) accordingly, in certain contingencies, the tenants should be entitled to rent concessions if the roadwork was not carried out.

There is likewise no evidence before us to explain the significance of the date March 31 1973 in the second proviso to clause 2(a) of the draft lease. Mr Fletcher, on behalf of the council, directed our attention to clause 3(iii)(a) and (b) of the building agreement and suggested that this date was inserted in the draft as representing the *earliest* date upon which the parties contemplated that the lease might in due course be executed. The purpose of the second proviso to clause 2(a) of the draft lease, he suggested, was simply to indicate the parties' intentions that the provision for postponement of the first rent review date (though it would still have required some amendment) should be inserted in the lease if, but only if, the roadwork had not been carried out when the lease came to be executed.

We are quite unable to accept this particular submission. It has to be borne in mind that the parties had contractually bound themselves to execute a lease in the particular form of the draft lease, subject of course to any subsequent mutual agreement for amendment. In our judgment, contrary to the council's submission, the only inference which we can draw from the form of the second proviso to clause 2(a) is that the parties had agreed this form on the footing and in the expectation (albeit not the certainty) that the lease would be executed *before* March 31 1973. On any other footing, the words "if [the roadwork] has not been carried out by 31st March 1973", looking as they do to the future, would have been manifestly inapposite.

We revert to the history of the matter. On September 6 1971 and June 28 1972 the parties executed two deeds of variation varying the provisions of the building agreement in a manner not material for present purposes.

On April 30 1974 a lease was executed by the same three parties and in the same order. The draftsman or draftsmen of the lease appear to have followed the form of the draft lease very closely, departing from it only so far as was necessary to give effect to certain agreed amendments and to take account of the present position in regard to the roadwork.

Clause 1 of the lease departed from the form of clause 1 of the draft lease in only a few respects which we need mention. It expressed the lease to be made in pursuance of the building agreement as varied by the two deeds of variation. It provided for the yearly rent payable during the first seven years of the term to be £8,000 (instead of £15,000) and defined this rent as "the basic rent" instead of "the basic ground rent". It stated the date of commencement of the term as July 5 1973.

Clauses 2(a), (b) and (c) of the lease departed from the agreed form of draft lease in only two respects which we need mention. First, references to "the basic rent" were substituted for references to "the basic ground rent". Second, and much more important, a small variation to the agreed form was made to take account of the fact that in the event the roadwork had not been carried out by March 31 1973. No doubt with this fact in mind, the draftsman substituted the word "as" for "if" when introducing the reference to the roadwork in Clause 2(a) of the lease.

A In the result, the second proviso to clause 2(a) of the lease, with which we are primarily concerned on this appeal, read as follows:

PROVIDED ALSO THAT as the re-alignment of Redcliffe Way and the landscaping of the lands south thereof to the site has not been carried out by the Thirty-first day of March One thousand nine hundred and seventy-three then the first rent review date hereinbefore stipulated shall be delayed one year for every year or part of a year until such re-alignment and landscaping shall have been completed or the Corporation declares by notice in writing to the Tenants that Redcliffe Way will not be re-aligned whichever is the earlier and accordingly the eighth year referred to in sub-clause (c) hereof shall be delayed in like manner.

We pause to say that Mr Dowding, in the course of his argument on behalf of the tenants which we will summarise later, attempted to B derive some support from the use of the word "then" in the second proviso. In our judgment, the presence of this word assists the case of neither party. It is simply a relic from the equivalent provision, clause 2(a), of the draft lease. In clause 2(a) of the draft, read in conjunction with "if", the word "then" was quite appropriate. In the lease itself, read in conjunction with "as", it is manifestly inappropriate. It cannot, in our judgment, be read as "therefore", as Mr Dowding suggested. All that can be said is that its presence appears to indicate the intention of the parties, in agreeing a revised form of clause 2(a) in 1974, to follow the form of the originally agreed draft with the minimum amendments required to make sense of the clause in the circumstances as then known, namely, that the roadwork had not been carried out by March 31 1973.

C In all other material respects, clause 2(a), (b) and (c) of the lease as executed followed the equivalent clauses of the draft. The lease included a third schedule in identical terms to the third schedule of the draft lease (quoted above). This third schedule was, at least in part, manifestly inappropriate in the events which had happened, but once again the parties (or their legal advisers) chose to follow the agreed draft, on this occasion without amendment.

It appears from an exhibit, though this is not otherwise referred to in evidence, that in or about 1979 an agreement was reached between the parties that the yearly rental of £8,000 should be increased to £9,038 as from June 3 1979, as a result of a change of use of part of the premises. Subject to confirmation that this increase duly took place, D references in the rest of this judgment to the basic rent are intended as references to this figure of £9,038.

Under the terms of the lease as executed, there was no question of any rent review taking effect before the expiration of the first seven years of the term, which expired on July 4 1980. By that date the council had still neither done the roadwork nor given notice of its intention not to do so. It is therefore common ground that, by virtue of the second proviso to clause 2(a), the first rent review date (which would otherwise have been July 5 1980) fell to be postponed by at least a year and that the basic rent continued to be payable at least up to and including July 4 1981.

Meantime, however, on February 10 1981 the council sent a letter E giving written notice to the tenants under clause 2(a) of the lease, declaring that Redcliffe Way would not be realigned. In this letter the council contended that, in the light of that notice, the first rent review date would be July 5 1981.

The plaintiffs, however, did not and do not accept this contention. On July 30 1986 they therefore issued an originating summons seeking declarations that on the true construction of the lease and in the events which have happened, the first rent review falls due on July 7 [sic] 1988 and the yearly rent payable thereunder until July 5 1988 is £8,000. (It appears that the references to July 7 1988 and £8,000 should be read as references to July 5 1988 and £9,038.)

The learned judge was asked to construe the lease against the background only of the building agreement. As he commented in his F judgment: "Properly, there is no evidence about what the point of it all was or what people meant".

It was common ground at the hearing that the effect of the second proviso to clause 2(a) of the lease, in the events which had happened, was that the first rent review date was to be deferred. It was also common ground that the period of time encompassed by the words "every year or part of a year", by reference to which the period of delay had to be measured, ended on February 10 1981. The dispute turned on the question when the period of delay was to be treated as beginning, for the purpose of applying the proviso.

The case for the tenants was and is essentially to the following effect: since the roadwork had not been carried out by March 31 1973, the first rent review date falls to be postponed by one year for every year or part of a year elapsing between *March 31 1973* and the G council's notice of February 10 1981. Seven years and a part of a further year elapsed between those two dates. Therefore, in the tenants' submission, by virtue of the second proviso to clause 2(a) of the lease, the starting point of the relevant period of delay is *March 31 1973* and the first rent review date falls to be postponed by eight years from July 5 1980 until *July 5 1988*.

The case for the council was and is essentially to the following effect: for the purpose of applying this second proviso, the starting point of the period of delay is simply to be treated as beginning on July 5 1980, being the earliest rent review date provided for by the lease. On this footing the proviso causes the first rent review date to be postponed by one year for every year or part of a year elapsing between *July 5 1980* and the council's notice of February 10 1981. H Part of one year, but no more, elapsed between the two dates. Therefore, in the council's submission, the first rent review date falls to be deferred by merely one year from July 5 1980 to *July 5 1981*.

The learned judge accepted the council's submission and made a declaration accordingly. The plaintiffs now appeal from his decision.

The learned judge, having summarised the contentions of the parties, expressed the ratio of his decision very concisely as follows:

But the proviso in the lease reflects the wording of the form of the lease in the schedule to the 1971 agreement as it was bound to do and remained, I think, bound to do even though the date March 31 1973 was passed. No one could have objected to the initial wording in the proviso, and the draftsman and, I J suppose, the solicitor on the other side to whom he submitted it did not think that there was anything to be gained in starting to cut the agreed draft about. The meaning of the proviso, in my judgment, is that, as it says, the first rent review date is to be delayed; that is to say, you look at that date, which was July 5 1980, and see that it is to be delayed by, roughly, the period until the roadworks are done or the notice is given. It is, of course, the latter which operates. The resulting period is one year.

We have had the benefit of excellent argument on both sides, as no doubt did the judge, before whom the same counsel appeared.

In the forefront of his argument on behalf of the council in support of the judge's decision, Mr Fletcher has relied on what he submits is the clear and inescapable meaning of the words of the second proviso beginning "the first rent review date . . ." down to the end. He K submits that these words, if read by themselves, with their use of the word "until", provide quite clearly for the *suspension* of the first rent review date previously stipulated (ie July 5 1980) for one year for every year or every part of a year until the roadwork has been completed or until the council serves the contemplated written notice. In his submission, these words of the second proviso do not permit the treatment of the period of delay as beginning on any date earlier than or other than July 5 1980. Furthermore, he contends the tenants' construction leads to absurd results. If it were correct then, since March 31 1973 had already passed before the lease was granted, the first rent review date could never in practice be July 5 1980, but would inevitably be deferred until at least July 5 1982, so that the designation of July 5 1980 as the first rent review date would be L meaningless. On the tenants' construction, Mr Fletcher submitted, a string of other provisions in the draft lease (for example, references to the seventh year in clause 2(a) and to the eighth year in clause 2(c)) would have required amendment in the final form of the lease, but were not amended. If one ignores the opening words of the second proviso to clause 2(a), in his submission this proviso gives rise to no difficulties of construction. These opening words, in his submission, are inserted merely to explain the presence of the provision for suspension of the first rent review date; they have no operative effect whatever.

We see much force in these submissions advanced on behalf of the council, but are ultimately not persuaded by them. The essential weakness in the council's construction of the lease is, in our M judgment, that it discounts — indeed, almost entirely disregards — the force of the reference to March 31 1973 in the lease. We cannot, in our judgment, disregard it.

If one reads the lease in isolation, this reference is quite inexplicable. However, we are, in our judgment, entitled and bound to look further to explain it. The lease was and is expressed to be granted in pursuance of the building agreement. Reference to the draft lease scheduled to the building agreement shows that it contained a clause 2(a) in more or less exactly the same form, with the presence of the word "if", instead of the word "as". The proper inference thus must be that, in executing the lease, the parties were content to leave the agreed draft of clause 2(a) alone in all substantial

A respects, merely replacing the word "if" with "as" so as to reflect the fact that in the event (contrary to the apparent contemplation of the draftsman of the original draft lease) March 31 1973 had already passed when the lease was executed. We agree with the judge's comment that no one could have objected to the initial words in the proviso, and his supposition that the legal advisers of the respective parties did not think that there was anything to be gained in starting to cut the agreed draft about.

In our judgment it is therefore permissible and relevant to look at the draft lease for the purpose of discovering the parties' intentions in including the reference to March 31 1973 in clause 2(a) of the lease as executed. In our judgment, a study of clause 2(a) of the draft lease in B its context makes reasonably clear the parties' intentions that the reference to this date was intended to designate the date upon which the period of delay was to be treated as beginning to run, for the purpose of applying the second proviso to clause 2(a). We do not know why that particular date was selected. As we have already indicated, we regard as mere speculation, and cannot accept, the suggestion that it was chosen as representing the earliest date upon which the parties contemplated that the lease might in due course be executed. On the contrary, in our judgment, the only inference which we can draw from the form of the second proviso to clause 2(a) of the draft lease is that the parties had agreed this form on the footing and in the expectation (albeit not the certainty) that the lease would be executed before March 31 1973. However, it was clearly chosen with C deliberation and inserted in the draft form of lease for a specific purpose.

If the council's construction of the wording were correct, the careful reference in the draft lease to March 31 1973 would have been utterly pointless. The date chosen might just as well have been July 4 1980, when the first seven years of the term would have expired. If such had been the intention of the parties, the obvious form of the opening words of the second proviso to clause 2(a) of the draft lease and of the equivalent proviso to clause 2(a) of the lease would have been some such form as this: "Provided also that if the re-alignment of Redcliffe Way and the landscaping of the lands south thereof to the site has not been carried out by the expiration of the seventh year of the said term . . .".

D If the proviso had taken this form, without any reference whatsoever to the date March 31 1973, their legal effect would have been precisely the same as that for which the council contends. Accordingly, the reference to this particular date in either document would have served no useful purpose at all.

While seeing the force of Mr Fletcher's contentions, we cannot accept such a construction either of clause 2(a) of the draft lease or of clause 2(a) of the lease which reflects its terms. In our judgment, as Mr Dowding submits, the only way to giving some force to the reference to March 31 1973 and to make consistent sense of these two subclauses is to treat the reference to March 31 1973 as designating the commencement of the relevant period of delay for the purpose of E applying the second proviso to clause 2(a). Contrary to Mr Fletcher's submissions, we do not regard this construction as doing violence to the language of the proviso or as involving the addition of words to that proviso by a process of construction. The proviso does not explicitly designate the *commencement* of the period of "every year or part of a year until . . .". On reading the second proviso to clause 2(a) of the lease as a whole, particularly when read in conjunction with the building agreement and draft lease scheduled to it, we do not find difficulty in accepting the construction submitted on behalf of F the tenants, which treats the reference to March 31 1973 as intended to specify the date of such commencement.

Mr Dowding accepted that if the tenants' construction of the lease is correct, the first rent review date would inevitably be delayed until at least July 5 1982. However, as the learned judge recognised, the parties' respective legal advisers, in finally settling the terms of the lease in 1974, appear to have decided not to amend the agreed draft more than was necessary. It might have been better if they had gone beyond substituting the word "as" for "if" and had incorporated a number of additional consequential amendments to take account of the fact that, in the event, the lease was to be executed after March 31 1973. In our judgment, however, the meaning of clause 2(a), when G read together with the equivalent subclause in the draft lease, is reasonably clear.

Though we do not base our conclusion on this consideration, the tenants' construction of the lease has this further merit which Mr Dowding drew to our attention. The council's construction of the lease could be said to lead to prolonged uncertainty, both as to the period for which the basic rent is to be payable and the date of the first review. Neither party to the lease would know until July 5 1980 actually arrived whether or not the first review would take place on that date. To take an extreme example, if the council were, on July 4 1980, to give notice of its intention not to do the roadwork, the rent review would, on the council's construction, be due the next day. In H contrast, the effect of the tenants' construction is that whatever happens the landlord and the tenant will have the benefit of knowing in advance at least seven years from the date of execution of the lease when the first rent review will be.

The learned judge rightly considered that the form of the draft lease must be looked at for the purpose of explaining the otherwise inexplicable reference to March 31 1973 in clause 2(a) of the lease. However, in our judgment, with all respect to him, having looked at the draft lease, he failed to appreciate the light which the reference to that specific date in that document throws on the parties' intentions.

It would be a brave judge who said he could be sure what the parties to this lease actually intended. There are many gaps in our J knowledge. In particular, the reason why the reference to the specific date March 31 1973 was included in the draft lease in the first place can be no more than a matter of conjecture. The lease itself could have been better drafted. However, we have to judge the parties' intentions from the words they used. On this footing, we have come to the clear conclusion that the tenants are right in submitting that, on the true construction of the lease and in the events which have happened, the first rent review date falls due on July 5 1988 and that the yearly rent payable thereunder until July 5 1988 is the basic rent.

Though we differ from this learned judge with hesitation in a matter of this kind, we therefore allow this appeal. We set aside his order and grant declarations in the terms of paras (1) and (2) of the originating summons, with the substitution of a reference to £9,038 K for the reference to £8,000, subject to confirmation from counsel on both sides that this is the correct figure, and with the further substitution of a reference to July 5 1988 for the reference to July 7 1988.

The appeal was allowed with costs in the Court of Appeal and below.

NEGLIGENCE

Queen's Bench Division

June 11 1987

(Before Mr Justice EWBANK)

NASH AND ANOTHER v EVENS & MATTA

Estates Gazette January 30 1988

[1988] 04 EG 131-137

Negligence — Action by owners of flat against surveyors alleging negligence in a valuation report to a building society in connection with plaintiffs' purchase of the flat — Plaintiffs purchased the flat in 1981 for £13,750, the value placed upon it by the defendant surveyors — The building society made a loan of £11,500 to the plaintiffs, but with a provision that the maximum amount advanced on the security of the flat, if no other security was taken, was £10,312 — The surveyors' report was not given to the plaintiffs and they decided against having an independent survey, relying on the fact that the building society was willing to lend money on the security of the property — In the report the defendants said that the external and internal condition of the property was good and that there was no evidence of movement, subsidence or landslip in the property — Some two years later trouble arose when the plaintiffs decided to sell the flat and the prospective purchasers obtained a report and valuation from another surveyor — He noticed some cracks (vertical, not horizontal) and advised that a structural survey should be obtained — A report, commissioned jointly by the plaintiffs and the prospective purchasers, from a consultant civil engineer revealed that the walls were cavity walls and that the wall ties had rusted — The rusting resulted in expansion, producing hairline cracks — At this point the prospective purchasers withdrew, leaving the plaintiffs with the flat — It appeared from evidence at the trial that the problem of rusting wall ties was not recognised until the 1970s and that surveyors who were primarily estate agents did not really become aware of it until later — When the plaintiffs began to find damp in the flat and some rotting floorboards they decided to sue the defendants — At the trial there was some conflict of evidence both as to what a reasonably competent surveyor should have seen in 1981 and as to the cost of the remedial works — The judge accepted that the defendants owed a duty of care to the plaintiffs — In the end the judge concluded on the evidence that there was nothing visible in 1981 to alert the defendants as to the potential problems which might arise from defects in the flat, nothing to indicate wall-tie rusting or structural movement — The defendants fulfilled their duty to take reasonable care to give a fair valuation on the basis of the limited inspection which they were expected to make — They had not been negligent — Action dismissed

No cases are referred to in this report.

This was an action by Ian Nash and his wife, Lynne Beverley Nash, against Evens & Matta, a firm of surveyors and estate agents, of Tonbridge and Tunbridge Wells, Kent, alleging a breach of duty of care to the plaintiffs in the valuation for mortgage carried out by the defendants on the instructions of the Eastbourne Mutual Building Society. The subject property was a flat at 211 Railway Cottages, Bexhill Road, St Leonards on Sea, East Sussex.

T Straker (instructed by Durnford Ford) appeared on behalf of the plaintiffs; D I Ridd (instructed by Barlow Lyde & Gilbert) represented the defendants.

Giving judgment, EWBANK J said: In 1981 Mr and Mrs Ian Nash decided to buy a home. Up to that point they had been living with a member of the family. They looked at several places and eventually decided to buy 211 Railway Cottages in Bexhill Road, St Leonards. The asking price was £13,750.

The estate agent who was trying to sell the house told them that it was a snip and that if they did not buy it someone else would snap it up. They agreed to buy it for the asking price. They asked for a mortgage and applied to the Eastbourne Mutual Building Society for a mortgage. The mortgage company instructed Evens & Matta to make a valuation inspection of the property.

It was a ground-floor flat and part of two semi-detached houses which had been converted to make four flats in all, nos 209 to 212 Railway Cottages. Mr R F Matta went along with his assistant and spent about 20 minutes there. He says this is the average time for this type of valuation of a small flat. He passed the flat and gave it a value of £13,750, saying to the building society that there was a great deal of property on the market but few sales and that a price war was likely. He thought it was prudent to suggest that the total loan should not exceed £11,500.

The Eastbourne Mutual Building Society made an offer to Mr and Mrs Nash of £11,750 and they accepted that offer. Among the documents that were sent to Mr and Mrs Nash by the building society was a notice that the maximum amount which the society would be prepared to advance on security of the property if no other security was taken, was £10,312.

The report by Mr Matta concerning the condition of the property was not sent to Mr and Mrs Nash. Mr Nash said that although there was a warning given by the building society that the valuer's duty was limited to making a valuation and that he might have difficulties with the property which would be revealed by a structural survey, he decided not to have any independent surveyor's report. He said that his solicitors did not advise him that he ought to get an independent surveyor's report. He says that he thought, as so many people do, that the fact that the building society were willing to lend money on the security of the property meant that the property was "OK" as he put it. He did not think that there was any need for a structural survey.

Although he did not know that it was Mr Matta who had done it, he thought that if Mr Matta had looked at the property and made sure that it was all right, then it would be all right. He drew no inference from the fact that — if there was any inference to be drawn — the maximum amount which the building society were prepared to advance on the security of the property was £10,312. So they bought this flat and moved in.

In due course, in 1983, they decided that they would like to move to a bigger place and they put the house on the market. Mr and Mrs Pilbeam agreed to buy the flat for £18,000 and they made application for a mortgage. Mr M W Rapley who, like Mr Matta, is a fellow of the Royal Institution of Chartered Surveyors, made a valuation at this time. He saw some cracks in the walls and he said that there should be a structural survey before the money was loaned. In due course an investigation took place into the condition of the house. The house had been built in 1895 and it had been built by the Southern Railway for their servants. I am told that the Southern Railway had a high reputation for high-quality work. Unknown to any of the surveyors who had so far looked at the house was the fact that it was a cavity-wall house because it was thought that, from the date of the house, it would be a solid-wall house. It was some way in advance of the general standards of the time. It was a cavity-wall house with wall ties made, I think, of iron.

What had happened, as it was subsequently discovered, was that these wall ties were rusting. This was causing cracks which would result in the walls falling down. The cost of repair was quite substantial.

The problem of rusting wall ties does not seem to have been recognised until the 1970s. It had not been recognised by the building people. There was an article in the *Sheffield Morning Telegraph* of May 30 1983. In Sheffield a special type of mortar had been used which was more deleterious to iron than other types of mortar, and houses in Sheffield were experiencing problems with rusting wall ties in advance of houses generally.

In this article it explained how the trouble had occurred and it said that at this date, 1983, statistics compiled by the Royal Institution of Chartered Surveyors suggested that failure in private houses was not yet a serious problem. The heading of the article stated that small pieces of metal hidden in cavity walls could be seriously weakening the brickwork of thousands of homes according to research by government building experts.

Surveyors dealing with building were aware of the problem, apparently, in the 1970s, but surveyors who were primarily estate agents, I am told, did not really become aware of the problem until later. Houses on the seafront of St Leonards had experienced this problem because of the effect of sea water carried on the wind, I suppose, so that the fact that wall ties could rust was known by then, but it was not the case further inland. The Railway Cottages had the railway between them and the sea and estate agents would not have expected, I am told, the rusting of wall ties at this time.

Mr and Mrs Nash also found damp in the house and there were also some rotting floorboards some of which were replaced by Mr Nash. When Mr Nash found out the extent of the problem which he faced he decided to sue Mr Matta.

Mr Matta owes a duty in preparing the report to exercise that degree of skill and care which is to be expected to be exercised by an ordinary, competent surveyor. In my judgment, he owes that duty not only to the building society but to those people who are likely to rely upon his report and that means, in the context of this case, Mr and Mrs Nash. The plaintiffs say that he is in breach of that duty because he ought to have realised in 1981 when he made his report not necessarily that the wall ties were rusting but that something was wrong and it needed to be looked at. He ought to have recognised, as Mr Rapley did in 1983, that a structural survey should be carried out. It was not enough merely for the building society to say in their valuation report that there may be defects in the property which would be revealed by a full structural survey and you can engage a certified surveyor who will tell you what his fee will be.

Mr Matta says that he saw no cracks when he examined the place in 1981 and there was nothing to alert him to the need for a structural survey. The plaintiff says that is not the case because he says himself that he saw some cracks at the time, and Mr D A Foster, who is an experienced building surveyor, says that by extrapolation, in the circumstances when he saw the flat later on, it is probable that there were cracks when Mr Matta saw the property.

The first report which I have seen is that of Mr Rapley, made in 1980. He was involved in the sale to the previous owners. His report states under general conditions that they were satisfactory, the internal decoration was good and the external decoration was good. Under general remarks he says that the building is in quite good condition, but that there was an area of rendering at the top which will require repairing. This item did not affect the valuation and presumably it will be dealt with during the normal course of maintenance. He said that the building was of 9in solid-brick walls rendered externally, as everyone so believed until the full investigation was made.

Mr Rapley saw the building from a footpath and other points of vantage. He saw two cracks in the rendering of the window of flat 210 at the far end from the Nashes' flat. Visible from the footpath were vertical cracks and they were over the window of no 210. He did not even mention them in his report because he thought they were of no significance. The vertical cracks were not significant, particularly in somebody else's flat. He saw no vertical or horizontal cracks in the Nashes' flat or in the flat above the Nashes'. When the Nashes decided to buy, Mr Matta, as I have said, made the valuation report and the fee for this was £29. In answer to the question whether there was any evidence of movement, subsidence or landslip in the property or its immediate vicinity, he said "no". As to the external condition of the property, it was good as was the internal condition.

Mr Matta said that the flat had been recently redecorated outside and it was in good condition. The redecoration, I am told, was in 1979. He did not see any significant cracking nor did he see anything to indicate damp. It is not normal in this type of valuation to lift fitted carpets near to night storage heaters nor to use damp meters in order to get a reading as to damp penetration.

He said that his inspection was not a cursory one and it would have been impossible for him to miss seeing cracks which were visible. He did not miss any cracks and he did not see any cracks. There was no smell of damp in the house and no sign of damp discoloration nor any indication of damp.

When the Pilbeams decided to buy the house in 1983 Mr Rapley was asked to make a report and valuation. I have indicated that he advised a structural engineer to make a survey because he saw, on this occasion, cracks. He said that they were all vertical cracks and not horizontal cracks and they were about 2mm to 3mm wide. He thought, as he had done previously and as Mr Matta had done, that it was a solid wall. He thought it indicated movement of the structure and that is why he thought there ought to be a structural survey. He said he did not see any horizontal cracks.

The report he wrote for the Eastbourne Mutual Building Society, who were asked to give a mortgage, was that there was evidence of structural movement resulting in the cracking of three external walls. He said that further investigation was essential and the inspection must be carried out by a structural engineer. He made a sketch of the position of the cracks that he saw in 1983 which alerted him. It is a little plan intended for his own use only. It indicates one crack on each of the external walls running forwards and upwards.

The Nashes decided to ask their upstairs neighbour to write a structural report as Mr Rapley had advised. It was suggested that the general idea was to allay fear rather than to provide a proper structural report. Mr Nash said that he agreed with this report. It describes the side elevation and a hairline crack below the sill of bedroom 1, a hairline crack by the windows of bedroom 2 and a hairline crack on each side of the window on the front elevation.

The report from the neighbour indicated cracks in the rendered surface only which did not extend to the building structure, caused by settlement over the years. There is no reason to suppose that the description of the cracks as hairline at the time he saw them is incorrect and it is consistent with Mr Rapley's description. Hairline apparently means up to about 2mm wide and Mr Rapley had found them to be 2mm to 3mm wide on average. Mr Rapley would not accept this report as a structural report. It merely told you what could be seen from the outside which Mr Rapley had seen for himself. So Mr Shepherd was appointed jointly by Mr Nash and Mr Pilbeam. He is a consultant civil engineer and he wrote a report. He made a survey of the cavity walls which involved boring a hole in the wall and looking through. He made the discovery, by looking through the holes he made in the wall, that the wall ties were rusting. He accepts that the wall ties had expanded, forming visible cracks in the same area on the outside. Mr Shepherd also described the cracks at this time as being hairline.

The Pilbeams, not unnaturally, dropped out of the sale and the plaintiffs remained in the house. Nothing has been done to repair the corrosion and they have been waiting, I suppose, for this case to be heard. It is in fact the landlord's obligation to repair and all four tenants are equally liable to pay for any repairs to the external walls. The landlord wished to make the repairs, but the Nashes have refused, on the advice of their solicitor, to allow the repairs to be done. There has been a dispute as to what should be done.

Mr Foster, who is a chartered building surveyor, says that the only proper way of dealing with the problem is to take down the outer wall, remove the rusted ties and rebuild the wall with new ties. The system generally adopted in 1981 was to take out the ties and replace them with new ones. It is suggested by Mr Foster that this method was a much cheaper method and longer lasting and I dare say he is right about that.

When one comes to damages the question is what would be the reasonable thing to do in order to sell the house, since the plaintiffs wished to sell it. So far as I can see, there is much to be said for doing

[A] limited repairs, as it has been suggested, and selling the problem on to somebody else. That would certainly appear to be considerably cheaper.

In a subsequent letter, Mr Shepherd says that Mr and Mrs Nash have a statement from their neighbour that the redecoration of the outside walls was carried out in about 1979. He says that presumably at that time the cracks were filled up. He says that it is possible at the time of the plaintiffs' purchase of the house the repairs may have been visible, but he says it is impossible for him to state in 1983 whether any open cracks would have been apparent.

Mr Foster takes a different view. He thinks that there probably would have been horizontal cracks visible by 1981 because of the nature of the defect. He agrees that the vertical cracks were of no [B] significance as far as wall-tie rusting was concerned. He says that he saw the place in the winter of 1983 and there were several cracks, both horizontal and vertical, to the external walls at high and low level. He said that the cracks would develop and become worse and the penetration of rising damp will affect the fabric. If it is allowed to continue the walls would become structurally unstable and the problem ought to be dealt with within six months. As I have mentioned, that has not been done.

He says that he cannot say from his own knowledge that there were cracks in the wall visible in 1981. In relation to the wall ties it is corrosion over the years and they were already badly rusted by 1981. He says that the cracks in the structure had been filled in probably in [C] 1979 and that as vertical cracks were seen by Mr Rapley on the other side of the place in 1980, he believes those were old cracks reopened. He agrees that it would not be apparent that there were cavity walls and he would not have expected the houses to have cavity walls.

He says that the cracking resulting from wall-tie rust is progressive and can increase rapidly. A single, small crack can develop into a bold and obvious crack within a year. It is suggested that, in certain circumstances, it could be a good deal less than a year. He thought in 1983 that there were signs of rising damp and decay. He puts the cost of repair at £8,000 in 1983. He says that now the cost would be around £15,000, if I understood him correctly. That cost would be divided among four people. In his report he says that taking into account the defects in the property the value which a competent surveyor ought to [D] have assessed was £9,000 as opposed to £13,750.

As far as repairs are concerned, I have estimates from three firms. The first is from Mr Lee and he describes himself as a replacement cavity wall tie firm. The way he would do it is not the way which Mr Foster would do it. He would charge £3,500 for repairs to the house. Mr Foster regarded it as a bodger's job. He said you have to regard all these estimates as fair. Mr Barker is another builder and he would charge £5,000 and there was Mr Boswell, who would charge something around £2,200. Mr Foster is particularly sceptical of Mr Boswell's estimate, which he believes does not go far enough and he believes it is wrong about some of the descriptions of the problem.

It is to be noted that these firms who are estimating to do the repair [E] job are estimating at a much lower figure than Mr Foster's solution. Mr Foster says that there are firms which have sprung up once the problem was known about and it is necessary to watch very carefully the firm which is employed. I am sure he is right.

Mr Rapley made a further report in September 1986. He says that at the time of his inspection in June 1980 the external decorative condition and the condition of the building was good. He said that when Mr Matta inspected the property it was also likely to have been good. He believes it was probable that the external condition was subtantially the same in July 1981 as it had been in June 1980.

He said that in 1980 there was no sign of rising damp and no sign of damage to the floor timbers although again he made only a limited visual inspection. In 1983 he noticed damp as Mr Foster had done in [F] 1986. He takes the view that in 1981 the trouble with damp and the floor defects would not have been readily apparent to a surveyor.

Mr Rapley wrote a letter on October 2 1986. He thought that £13,750 as a price was not unreasonable subject to this trouble being dealt with. He said that in 1981 when Mr Matta inspected the house, no wall-tie work would have been necessary. Mr Foster does not agree with that. He believes that you cannot deal with the problem in a piecemeal way. Mr Rapley says that it is only necessary to do the upper part and that may be right as far as the external sides are concerned. He costs the repairs that would have been necessary in 1981 at £3,000 and the liability, accordingly, to the Nashes of £750. He would assess the value of the purchase price which would have been appropriate, because of the defects, at £12,250. Mr Foster asserts that, having regard to the condition when he saw it, it is [G] probable that the cracks that should have alerted Mr Matta were visible in 1981.

Mr Nash says indeed that when they saw the house there was a crack from his window upwards and horizontal. I have to consider that evidence in the light of the other evidence I have heard. The evidence of Mr Foster on this aspect is, in my judgment, speculative. I have the evidence of Mr Matta that there were no cracks together with other evidence contained in various reports, to set against the evidence of Mr Nash that cracks actually existed, and horizontal cracks at that, in 1981.

I have to say that I am unable to accept Mr Nash's evidence that there were any horizontal cracks visible in 1981. It is quite contrary to all the other evidence in the case. I accordingly accept Mr Matta's [H] evidence that there were no cracks. That being the case, in my judgment there was nothing to alert Mr Matta as to the potential problem which might exist because of the defects of some sort or other in the house. Certainly there was nothing to indicate wall-tie rusting, in my judgment, and nothing even to indicate structural movement.

Mr Matta has described to me the nature of his visual inspection and valuation. As I have said, in my judgment I think he owes a duty to a potential purchaser to take reasonable care to give a fair valuation. The purchaser was only aware of the offer made by the building society and the figures that they provided. He did not and could not have relied on the valuation by Mr Matta in the figure of [J] £13,750 because the building society offer was at the figure of £11,750 and he had notice that the maximum amount they would advance on the security of the property was in the sum of £10,312. Even if Mr Matta had been in breach of his duty in making the valuation, in my judgment, on the evidence I have heard, the loss to the plaintiffs would not reduce the proper price of the house.

In those circumstances, sorry as I am for the unfortunate position Mr and Mrs Nash are in with this defective house, the valuation by Mr Matta of the flat in 1981 was not done negligently. It was within the range of valuations that would be expected from an ordinary, competent surveyor given the limited nature of the inspection he was asked to make. In those circumstances I am obliged to dismiss the action. [K]

The action was dismissed with costs; the order for costs not to be enforced without leave.

Court of Appeal

December 17 1987

(Before Lord Justice KERR, Lord Justice NOURSE and Mr Justice CAULFIELD)

HARRIS AND ANOTHER v WYRE FOREST DISTRICT [L] COUNCIL AND ANOTHER

Estates Gazette February 6 1988

[1988] 05 EG 57-68

Negligence — Duty of care — Disclaimer of liability — Mortgage valuation — Appeal from decision of Schiemann J in favour of houseowners — Review of authorities — Although Court of Appeal unanimous in allowing appeal, leave given to appeal to House of Lords in view of importance of issues — (As the House has itself given leave to appeal in the case of *Smith* v *Eric S Bush*, it seems likely that most aspects of [M] liability in respect of negligent mortgage valuations will be thoroughly explored) — The present decision of the Court of Appeal is of particular importance in relation to disclaimer — Among cases discussed was *Yianni* v *Edwin Evans & Sons*, but no final opinion was expressed upon it, although there was a reference to "its inherent jurisprudential weakness in any ordinary situation" — The present appeal was by the local authority and its staff valuer against the judgment of Schiemann J for £12,000, the judge having found the local authority both vicariously liable for the negligence of the valuer and primarily liable for offering a loan and mortgage

which under the relevant legislation they were not empowered to grant — The plaintiffs, present respondents, were a young couple buying their first house with the aid of a loan from the local authority — The house had suffered from extensive settlement and there was a tie bar round it — Valuer noticed these signs but concluded that the settlement was a thing of the past — He valued the house at £9,450 (the asking price) and recommended a 90% loan — In his report, under "essential repairs", he merely referred to the need for a report from the local electricity board on the electrical system and the carrying out of any recommendations and mentioned some minor works of repair — The respondents were charged a fee of £22 for the valuation — It was made clear to the respondents that the valuation was confidential and intended solely for the information of the authority and that "no responsibility whatsoever is implied or accepted by the council for the value or condition of the property by reason of such inspection and report" — The vendors agreed to a reduced price of £9,000 and the respondents purchased with the aid of a loan of £8,505 — Three years later the respondents wished to move but found, after inspections and two surveys, that the house was in fact unsaleable — The respondents then brought proceedings against the local authority and the staff valuer — In finding them both liable for negligence Schiemann J referred to the disclaimer mentioned above but decided that it did not relieve either the valuer or the authority from any liability for a negligent valuation

On the appeal, in the course of which a considerable amount of case law was reviewed, two questions were argued, namely, whether the appellants owed a duty of care to the respondents, and whether the appellants' disclaimer was effective or was invalidated by the Unfair Contract Terms Act 1977 — In fact the two questions were found to be intimately connected and the disclaimer, which had been dismissed by Schiemann J, was held to be of fundamental importance in deciding whether the appellants owed such a duty of care — By the terms of the mortgage application and acceptance no responsibility was assumed by the appellants for the value or condition of the property by reason of the inspection and report — Applying the principle of *Hedley Byrne & Co Ltd* v *Heller & Partners Ltd*, the result was that the appellants never came under a duty of care to the respondents — A disclaimer is not merely a possible answer to a duty of care which already exists; it may prevent a duty of care from arising — The court distinguished the case of *Yianni* v *Edwin Evans & Sons* by treating it as a case where there was no disclaimer

As regards the Unfair Contract Terms Act 1977, it can only apply where a duty of care has been established — If, as here, a notice of disclaimer has prevented the duty of care from coming into existence, there is nothing on which section 2 of the Act can bite — Appeal allowed

The following cases are referred to in this report.

Anns v *Merton London Borough Council* [1978] AC 728; [1977] 2 WLR 1024; [1977] 2 All ER 492; [1977] EGD 604; (1977) 243 EG 523 & 591, HL

Candler v *Crane, Christmas & Co* [1951] 2 KB 164; [1951] 1 All ER 426, CA

Curran v *Northern Ireland Co-ownership Housing Association Ltd* [1985] 8 NIJB 22, Carswell J; April 28 1986, CA, NI, unreported; [1987] 2 WLR 1043; [1987] 2 All ER 13, HL

Donoghue v *Stevenson* [1932] AC 562, HL

Hedley Byrne & Co Ltd v *Heller & Partners Ltd* [1964] AC 465; [1963] 3 WLR 101; [1963] 2 All ER 575; [1963] 1 Lloyd's Rep 485, HL

Ministry of Housing and Local Government v *Sharp* [1970] 2 QB 223; [1970] 2 WLR 802; [1970] 1 All ER 1009; (1970) 68 LGR 187; 21 P&CR 166; [1970] EGD 139; 213 EG 1145, CA

Odder v *Westbourne Park Building Society* (1955) 165 EG 261

Smith v *Eric S Bush (a firm)* [1987] 3 WLR 889; [1987] 3 All ER 179; [1987] 1 EGLR 157; (1987) 282 EG 326, CA

Stevenson v *Nationwide Building Society* [1984] EGD 934; (1984) 272 EG 663

Ward v *McMaster, Louth CC and Nicholas Hardy & Co Ltd* [1985] IR 29; [1986] ILRM 43

Westlake v *Bracknell District Council* [1987] 1 EGLR 161; (1987) 282 EG 868

Yianni v *Edwin Evans & Sons* [1982] QB 438; [1981] 3 WLR 843; [1981] 3 All ER 592; [1981] EGD 803; (1981) 259 EG 969

Yuen Kun Yeu v *Attorney-General of Hong Kong* [1987] 3 WLR 776; [1987] 2 All ER 705, PC

This was an appeal by Wyre Forest District Council and Trevor James Lee, a valuer on the staff of the council, against the decision of Schiemann J (reported at [1987] 1 EGLR 231), in favour of Adam Charles Harris and his wife, Kim Harris, plaintiffs in an action alleging negligence on the part of the council and Mr Lee in relation to a mortgage offered to the plaintiffs for the purchase of a house at 74 George Street, Kidderminster. Mr and Mrs Harris are the respondents to the present appeal.

Piers Ashworth QC and Nicholas Worsley (instructed by Lawrence Graham, agents for Rowleys & Blewitts, of Birmingham) appeared on behalf of the appellants; R L Johnson QC, Malcolm Stitcher and Bruce Speller (instructed by Thursfield & Adams, of Kidderminster) represented the respondents.

Giving the first judgment at the invitation of Kerr LJ, NOURSE LJ said: This is another in the line of recent cases in negligence where the question for decision has been whether a duty of care is owed to someone to whom it was formerly assumed that no duty was owed. Here it is a house purchaser who, in reliance on a valuation made without due care by an officer of a local authority advancing money on mortgage under the Housing Acts, proceeds with his purchase and suffers damage in consequence. In *Odder* v *Westbourne Park Building Society* (1955) 165 EG 261, a comparable action against a building society was described by Harman J as unheard of. But it is said that decisions subsequent to *Hedley Byrne & Co Ltd* v *Heller & Partners Ltd* [1964] AC 465, in particular *Yianni* v *Edwin Evans & Sons* [1982] QB 438, have now established that the action will lie. That was the view of Schiemann J, who decided this case in favour of the purchasers. The local authority and their surveyor, who was also made a party to the action, have now appealed to this court. It must be emphasised at the outset that this is a case where the local authority expressly disclaimed liability to the purchasers in respect of the valuation.

The material facts, most of which I take from the necessarily longer statement of the learned judge, are these. In 1978 the plaintiffs, Mr and Mrs Harris, then a young couple who were thinking of getting married, were looking in Kidderminster for a house costing less than £10,000. They were looking at Victorian terraced houses. They knew that they would have to obtain a mortgage for nearly all of the purchase price. They came to the conclusion that the most suitable source of finance was the first defendants, the Wyre Forest District Council, who at that time, like many other local authorities and with the encouragement of successive governments, had a policy of making loans to intending purchasers of older, small residential properties, on the security of which banks and building societies were in general unwilling to lend. The details of the council's scheme were set out in a document which was available to interested parties, but the judge found that the plaintiffs never saw it and that no steps were taken by the council to bring it to the attention of potential mortgagors.

Eventually the plaintiffs decided that they would like to purchase 74 George Street, Kidderminster, which was for sale at a price of £9,450. The selling agent, who evidently had a supply of the council's standard mortgage application forms, filled in one of them with the plaintiffs' help, indicating that they were prepared to contribute £500. That form was signed by the plaintiffs on August 23 1978. Immediately above their signatures appeared these words:

> TO BE READ CAREFULLY AND SIGNED BY ALL APPLICANTS.
> I/WE enclose herewith the Valuation Fee & Administration Fee £22.
> I/WE understand that this fee is not returnable even if the Council do not eventually make an advance and that the Valuation is confidential and is intended solely for the information of Wyre Forest District Council in determining what advance, if any, may be made on the security and that no responsibility whatsoever is implied or accepted by the Council for the value or condition of the property by reason of such inspection and report.
> (You are advised for your own protection to instruct your own Surveyor/Architect to inspect the property.)
> I/WE agree that the Valuation Report is the property of the Council and that I/WE cannot require its production.

The judge found that neither of the plaintiffs read those words, although they did enclose the fee of £22 with the application form. Nobody suggested to the plaintiffs that they should have their own survey done. They themselves did not think of it, although they knew

A that such a thing could be arranged at a price, which the judge said would have been somewhere between £75 and £100. The plaintiffs themselves did not make any inspection of the outside of the house.

In due course the council arranged for the second defendant, Mr Lee, to inspect and value the property. He was employed by the council as a valuation surveyor. He was not a building surveyor. He inspected the property on September 8 and subsequently prepared his report, which was dated September 13. He valued the house at the asking price of £9,450 and recommended a loan of 90% of the valuation for a maximum period of 25 years. Under the heading "Essential Repairs" he wrote:

> Obtain report for District Council from MEB regarding electrics and carry out any recommendations. Make good mortar fillets to extension.

B The judge held that it was implicit in Mr Lee's report that structural repairs were not in his view essential. He also found that Mr Lee knew that his report would not be forwarded to the plaintiffs but that they would be told the amount of any loan and of any repairs which he regarded as being essential. Mr Lee also knew that the typical applicant was a first-time buyer of modest means.

By September 25 the vendors had agreed to a reduced price of £9,000. Two days later the council wrote to the plaintiffs and told them that they were prepared to make an advance of £8,505 repayable over a period of 25 years, subject to a number of conditions, one of which was that the plaintiffs should sign an undertaking on completion "to carry out within 12 months C therefrom the work detailed overleaf". Overleaf appeared the following:

> Essential repairs (1) Obtain report for District Council from Midland Electricity Board regarding electrics and carry out any recommendations. (2) Make good mortar fillets to extension.

Completion duly took place on November 23 1978, when the plaintiffs signed an undertaking in the form required.

Three years later, in November 1981, the plaintiffs put 74 George Street on the market. In April 1982 some prospective purchasers applied to the council for a similar loan in respect of the property. Mr Lee made a further inspection and valued it at £13,750. But because of a problem with settlement at the next-door property which had been experienced in February 1981 he recommended that the whole D of the proposed loan should be retained until a structural survey report by an independent architect or structural engineer had been obtained and any recommended work carried out. A structural survey was duly obtained and it became clear that the cost of the recommended works would run into many thousands of pounds. That meant that the council were unable to lend money on the security of the property and that in consequence it was unsaleable. In June 1982 the plaintiffs instructed their own building surveyor, who reported that the building was unsaleable and required considerable underpinning and other works. His report was sent to the council who prepared a specification, which included, but went beyond, the plaintiffs' surveyor's suggestions. One contractor who was E approached was not prepared to tender, since he regarded the proposed work as impractical and unsafe. Another quoted over £13,000 to carry out the necessary works. Formal application was made for a repair grant, but the council had no surplus funds. The house was taken off the market in July 1982 and the plaintiffs have continued to live there ever since. The writ in the action was issued on September 16 1983.

The claim made in the reamended statement of claim can for present purposes be summarised as follows: (1) In carrying out his inspection and valuation in 1978 Mr Lee owed a duty of care to the plaintiffs; (2) the inspection and valuation were carried out carelessly by Mr Lee, for whose negligence the council is vicariously liable; (3) Mr Lee and the council, or one of them, carelessly misrepresented in F the valuation report that the property was worth more than the amount of the loan requested by the plaintiffs; (4) the mortgage offer made to the plaintiffs by the council was made in the light of and in reliance upon the valuation report and in the knowledge that the plaintiffs might also rely on it; and (5) the plaintiffs did rely on the report and thereby suffered damage. By their defence the defendants deny that they were negligent. They further deny that the plaintiffs placed any reliance upon the inspection or valuation. So far as necessary they rely on the disclaimer contained in the mortgage application form of August 23 1978. By their reply the plaintiffs contend that the disclaimer is ineffective to exclude liability for negligence and, so far as is necessary, they rely on section 2 of the Unfair Contract Terms Act 1977.

The council's statutory power to advance money on the security of G the property (see below) could be exercised only if they had satisfied themselves that it was, or would be made, in all respects fit for human habitation. Schiemann J found that the house was not fit for human habitation at the material time. I do not think that that finding can affect the outcome of this case. The judge also found that Mr Lee's valuation of the property was made without due care. The defendants' appeal against those two findings has not been pursued. Moreover, they accept that the plaintiffs did rely on the valuation and have suffered damage in consequence. The two questions which have been argued on this appeal are, first, whether the defendants or one of them owed a duty of care to the plaintiffs and, second, whether the disclaimer, if it is otherwise effective to exclude liability for negligence, is invalidated by section 2 of the 1977 Act. H

I turn first to the alleged duty of care. In 1978 the council's power to advance money on the security of the property was contained in section 43 of the Housing (Financial Provisions) Act 1958. It is the primary submission of Mr Ashworth, for the defendants, that the terms of that section, which provides, among other things, for a valuation to be obtained and acted upon by the local authority, show that the sole purpose of those provisions is to protect the local authority's funds and thus their ratepayers. It is not their purpose, or part of their purpose, to protect prospective mortgagors. But that submission, although undoubtedly correct, does not solve the problem, because it is still possible for a local authority to put themselves into a position where they come under a duty of care to a J prospective mortgagor. To take an extreme and improbable example, they might offer to make their valuation available to him for the express purpose of helping him to decide whether the price was right. Whether they have come under such a duty or not depends on the facts of the individual case. There are now at least two authorities where, on facts comparable with those of the present case, it has been held that a duty of care was owed to a prospective mortgagor.

The convenient course is to go straight to the decision of Park J in *Yianni* v *Edwin Evans & Sons*. There are a number of significant differences between that case and this. First, the mortgagee was a building society and not a local authority. Second, the valuation was not made by an employee of the building society but by an independent firm of valuers and surveyors instructed by them. Third K and most important, although the plaintiffs acknowledged receipt of (a) a notice excluding the building society's statutory warranty under section 30 of the Building Societies Act 1962 that the purchase price was reasonable and (b) an explanatory booklet containing a paragraph to much the same effect as the mortgage application form in the present case, no disclaimer of liability was made by or on behalf of the valuers. Fourth, no doubt as a consequence of that, the plaintiffs sued the valuers and not the building society. There was also unchallenged independent evidence from the chief surveyor to the Abbey National Building Society to the effect that for some six years or so over 90% of applicants for a building society mortgage had relied upon the building society's valuation, as represented by L the society's offer of an advance, as a statement that the house in question was worth at least that sum and, further, that applicants seeking to buy houses at the lower end of the property market did not read building society literature or, if they did, ignored the advice to have an independent survey and also the terms of the statutory notice; see [1982] QB 438 at pp 444H to 445H, 447D-F and 455F-H. It is important to note that counsel for the defendants accepted that the defendants did not make any disclaimer of responsibility; see [1982] QB at p 440E.

On those facts Park J held that the valuers owed a duty of care to the prospective mortgagors. The essence of his decision is expressed at [1982] QB at p 454C-D where, after reading at length from the dissenting judgment of Denning LJ in *Candler* v *Crane, Christmas & Co* [1951] 2 KB 164 and from the speeches in the House of Lords in M *Hedley Byrne & Co Ltd* v *Heller & Partners Ltd* and also a well-known passage from the speech of Lord Wilberforce in *Anns* v *Merton London Borough Council* [1978] AC 728 at p 751, he said:

> Accordingly guided by the passages in the judgment of Denning LJ in *Candlers* case and in the speeches in the House of Lords cases, I conclude that, in this case, the duty of care would arise if, on the evidence, I am satisfied that the defendants knew that their valuation of 1 Seymour Road, in so far as it stated that the property provided adequate security for an advance of £12,000 would be passed on to the plaintiffs, who, notwithstanding the building society's literature and the service of the notice under section 30 of the Building Societies Act 1962, in the defendants' reasonable contemplation

would place reliance upon its correctness in making their decision to buy the house and mortgage it to the building society.

Mr Ashworth submitted that *Yianni* v *Edwin Evans & Sons* was wrongly decided. Alternatively, he submitted that it is distinguishable, principally by reason of the disclaimer of liability contained in the mortgage application form of August 23 1978. The correctness or incorrectness of these submissions can be judged only after it is seen how the facts of the present case stand in relation to existing principles of law.

It is not, I think, suggested that the defendants owed a duty to the plaintiffs to make a valuation of 74 George Street. What is said is that, being under a duty to others to make a valuation which they knew would necessarily come to the knowledge of the plaintiffs, the defendants owed to them a duty to make it with due care. In negative form, the duty was not carelessly to misstate the value of the property to the plaintiffs. As I have said, it is now accepted that the valuation was carelessly made; in other words that there was a breach of the duty, if one was owed. It is also accepted that the plaintiffs relied on the valuation and have suffered damage in consequence. The only question therefore is whether a duty was owed.

This question must be answered by an application of established principles relating to negligent misstatement, which, being only one instance of the negligent acts and omissions of which the law takes account, depends on there being the necessary proximity between the maker and the recipient of the statement. It is here that difficulty is most often caused, sometimes because the statement lacks the necessary gravity of purpose or occasion and at others because it is available to a wider class of recipients than those towards whom a duty could with justice be constructed. In attempting to resolve one or other of these difficulties in the case before him, more than one eminent judge has based his decision on a more general statement of what is required to constitute the necessary proximity. Thus in *Hedley Byrne & Co Ltd* v *Heller & Partners Ltd* itself Lord Pearce said [1964] AC 465 at p 539:

> Was there such a special relationship in the present case as to impose on the defendants a duty of care to the plaintiffs as the undisclosed principals for whom the National Provincial Bank was making the inquiry? The answer to that question depends on the circumstances of the transaction. If, for instance, they disclosed a casual social approach to the inquiry, no such special relationship or duty of care would be assumed (see *Fish* v *Kelly*). To import such a duty the representation must normally, I think, concern a business or professional transaction whose nature makes clear the gravity of the inquiry and the importance and influence attached to the answer.

That, I respectfully think, is a valuable statement because, in dealing with the first and perhaps less familiar kind of difficulty, it highlights a requirement which is common to all those cases where the necessary proximity has been found. The circumstances must be such that the maker of the statement ought reasonably to recognise both the importance which will be attached to it by the recipient and his own answerability to the recipient in making it. In *Hedley Byrne & Co Ltd* v *Heller & Partners Ltd* itself and in other cases this requirement has been expressed as a voluntary assumption of responsibility or the like, but it has been held that in some circumstances the liability can extend to a case where the maker of the statement is under a duty to make it: see *Ministry of Housing and Local Government* v *Sharp* [1970] 2 QB 223.

I would therefore frame the essential question which we have to decide in these terms: were the circumstances such that the defendants ought reasonably to have recognised both the importance which would be attached to Mr Lee's valuation by the plaintiffs and their own answerability to them in making it? By now it will be obvious that that question cannot be answered without considering the terms and effect of the disclaimer contained in the mortgage application form of August 23 1978. So far as material, that form, which was signed by the plaintiffs, (1) expressed an understanding on the part of the plaintiffs that the valuation was confidential and intended solely for the information of the council and that no responsibility whatsoever was implied or accepted by the council for the value or condition of the property by reason of the inspection and report; (2) contained advice to the plaintiffs, for their own protection, to instruct their own surveyor or architect to inspect the property; and (3) expressed an agreement by the plaintiffs that the valuation report was the property of the council and that they could not require its production.

In the court below, Schiemann J held, first, that since the plaintiffs had signed the form immediately below the material words they must, notwithstanding his finding that neither of them had read them, be treated as if they had. Mr Johnson, for the plaintiffs, did not seriously seek to disturb that holding before us. Mr Ashworth submitted that it did not affect the outcome either way. Second, the judge held, apparently as a matter of construction, that the material words did not relieve either of the defendants from liability and he therefore ignored them in considering that question. He gave no reasons for that view and Mr Johnson has not seriously sought to support it. With all due respect to the judge, I do not think that it can be supported. In my view the material words in the mortgage application form are directly in point and are of fundamental importance in deciding whether the defendants owed a duty of care to the plaintiffs.

I am prepared to assume that, notwithstanding the terms of the disclaimer, the defendants ought reasonably to have recognised the importance which would be attached to Mr Lee's valuation by the plaintiffs. That assumption would be well justified by a commonsense assessment of the mental processes of prospective mortgagors in this category, as evidenced by the testimony of the chief surveyor to the Abbey National Building Society in *Yianni* v *Edwin Evans & Sons*. But, however that may be, in my judgment it is impossible to say that the defendants ought reasonably to have recognised their own answerability to the plaintiffs in making the valuation. By statute they owed no duty to the plaintiffs. By the terms of the mortgage application form it was clear that no responsibility whatsoever was implied or accepted by them for the value or condition of the property by reason of the inspection and report, even if they were made without due care. There is no rule, whether of public policy or otherwise, against the defendants' acting in that way. Indeed, it is of the essence of the decision in *Hedley Byrne & Co Ltd* v *Heller & Partners Ltd* that they are at liberty to do so.

For these reasons I am of the opinion that the defendants never came under a duty of care to the plaintiffs. While others might frame the essential question in different terms, I do not see that they could answer it in a different sense. Moreover, the disclaimer was as effective in the case of Mr Lee as an individual defendant as it was in the case of the council themselves. He at all times acted as a servant of the council, who could only reasonably have been expected to act through their servants. The references to the council therefore included Mr Lee as an individual.

It follows that *Yianni* v *Edwin Evans & Sons*, where there was no disclaimer of liability by or on behalf of the defendants, is in my view distinguishable from the present case. Mr Ashworth has nevertheless pressed us to consider the correctness of that decision. I for my part do not think that that is an inquiry on which we ought to embark. It is true that the well-known passage from the speech of Lord Wilberforce in *Anns* v *Merton London Borough Council*, on which it was partly based, has since undergone a process of reassessment in a series of decisions at the highest level, culminating in the recent decision of the Privy Council in *Yuen Kun Yeu* v *Attorney-General of Hong Kong* [1987] 3 WLR 776. Mr Ashworth also referred us to the judgment of the late Sir Maurice Gibson LJ in *Curran* v *Northern Ireland Co-Ownership Housing Association Ltd* (April 28 1986 — unreported), a contribution to the subject which repays careful study. That was a case where, on comparable facts but with no disclaimer of liability, the Court of Appeal in Northern Ireland held that the defendants owed no duty of care to the plaintiffs. But Gibson LJ treated *Yianni* v *Edwin Evans & Sons* as distinguishable and he certainly proceeded on the footing that it was correctly decided. While that can no doubt be explained partly by the conventions which are respected between judges sitting in different jurisdictions, I believe that it confirms that it is undesirable, even for appellate courts, to express a view as to the correctness of other decisions which have in large part depended on other facts. Moreover, *Yianni* v *Edwin Evans & Sons* has been followed in one reported decision at first instance and perhaps in other, unreported, cases as well. The reported decision is *Westlake* v *Bracknell District Council* (1987) 282 EG 868*. Finally, it is clear that *Yianni* v *Edwin Evans & Sons* has led to a wide change of practice among local authorities and building societies, who now usually, perhaps invariably, make a disclaimer of liability in cases of this kind. I think that we should be content to make a decision which may give some general guidance in other cases where there is a disclaimer. We should not seek also to deal with a different class of case whose numbers are likely to decrease.

---

* Editor's note: Also reported at [1987] 1 EGLR 161.

Before leaving this question, I would make three further points. First, in holding in favour of the plaintiffs, Schiemann J relied mainly on another part of the decision in *Curran* v *Northern Ireland Co-Ownership Housing Association Ltd*, which has since been reversed by the House of Lords: see [1987] 2 WLR 1043. (There was no cross-appeal by the plaintiffs in that case against the main part of the decision above referred to.) Second, the main thrust of Mr Johnson's argument on this question was that the disclaimer was, on the facts as a whole, overborne or reduced in significance by the council's duty, under the encouragement of successive governments, to provide young couples like the plaintiffs with decent homes. While I recognise the force of that argument and have very great sympathy with the plaintiffs, I have to say that it is not for this court to deny to the defendants a limitation of liability which the House of Lords have expressly allowed them. It is difficult to see what more the council could have done in order to limit their liability in the present case. Third, in *Stevenson* v *Nationwide Building Society* (1984) 272 EG 663, another decision on comparable facts, the late Mr John Wilmers QC, sitting as a deputy judge of the Queen's Bench Division, was confronted with a disclaimer in almost identical terms to those contained in the mortgage application form in the present case. Although it appears that it was conceded that, subject to the 1977 Act, it constituted an effective disclaimer, the learned deputy judge, who referred to passages in the speeches in *Hedley Byrne & Co Ltd* v *Heller & Partners Ltd*, clearly thought that that was the correct view: see (1984) 272 EG at p 671. A similar view was taken by this court in *Smith* v *Eric S Bush (a firm)* [1987] 3 WLR 889*, where, at p 898, Glidewell LJ said:

> Mr Seabrook's argument is that the disclaimer did not cover liability for negligence. If that is correct, the disclaimer did not cover anything. The plaintiff was never in a contractual relationship with the defendants. If they were or could be liable to her, that liability could only be in the tort of negligence — which is how the plaintiff's claim is framed.

I shall have to revert briefly to both those cases later.

I now turn to consider the application of the Unfair Contract Terms Act 1977, whose long title, so far as material, is:

> An Act to impose further limits on the extent to which . . . civil liability for breach of contract, or for negligence or other breach of duty, can be avoided by means of contract terms and otherwise . . .

Mr Johnson submits, first, that the Act applies to the disclaimer contained in the mortgage application form as a notice excluding the defendants' liability for negligence and, second, that it is not fair and reasonable to allow reliance on it. Mr Ashworth submits that the Act has no application to this case.

The provisions of the Act on which these submissions primarily depend are:

Section 1(1):

> For the purposes of this Part of this Act, "negligence" means the breach (a) of obligations in contract
> (b) of any common law duty to take reasonable care or exercise reasonable skill (but not any stricter duty); . . .

Section 2:

> (1) A person cannot by reference to any contract term or to a notice given to persons generally or to particular persons exclude or restrict his liability for death or personal injury resulting from negligence.
> (2) In the case of other loss or damage, a person cannot so exclude or restrict his liability for negligence except in so far as the term or notice satisfies the requirement of reasonableness.

Section 11(3):

> In relation to a notice (not being a notice having contractual effect), the requirement of reasonableness under this Act is that it should be fair and reasonable to allow reliance on it, having regard to all the circumstances obtaining when the liability arose or (but for the notice) would have arisen.

The governing provision in this case, being one where there is no liability for death or personal injury resulting from negligence, is section 2(2). When adapted to the present case, and with the definition of "negligence" in section 1(1)(b) read into it, the subsection provides that a person cannot, by reference to a notice given to particular persons, exclude his liability for the breach of any common law duty to take reasonable care, except in so far as the notice satisfies the requirement of reasonableness.

Mr Ashworth submits that when section 2(2) is spelled out in this way it is seen to apply only where a duty of care already exists. A notice which has prevented it from coming into existence is not one upon which the subsection bites. Mr Johnson submits that the wording applies as naturally to the latter case as it does to the former. He also relies on the words in section 11(3) "having regard to all the circumstances obtaining when the liability arose or (but for the notice) would have arisen" as showing that the Act expressly contemplates the inclusion of a case where a notice has prevented the duty from coming into existence.

This is a short question of statutory construction of wide general importance. Only two views are possible. Either the 1977 Act applies to all cases of negligent misstatement where a disclaimer has prevented a duty of care from coming into existence or it applies to none.

I have found this question to be both difficult and uncomfortable: difficult because either construction of section 2(2) appears to be tenable; uncomfortable because Mr Ashworth's argument could have been advanced, but evidently was not, by either of the experienced counsel who represented the defendants in *Stevenson* v *Nationwide Building Society* and *Smith* v *Eric S Bush (a firm)*. Giving it the best consideration I can, I have come to the conclusion that the argument of Mr Ashworth represents, on balance, the more natural reading of section 2(2). The contrary argument seems to involve reading the words "liability for the breach of any common law duty to take reasonable care" as including "liability for the breach of what without the notice would have been a common law duty to take reasonable care". Moreover, the words in section 11(3) on which Mr Johnson relies are found in a provision defining the requirement of reasonableness and, as such, are peripheral to the governing provisions of section 2(2).

I would therefore hold that the 1977 Act has no application to this case. My grave doubts about the matter are relieved by the belief that my lords may take a more sanguine view of it than I.

I must mention two further points. First, once Schiemann J had held that the material words in the mortgage application form did not relieve either of the defendants from liability, it became unnecessary for him to consider the application of the 1977 Act and he did not do so. Second, although in *Smith* v *Eric S Bush (a firm)* this court refused leave to appeal, the Appeal Committee of the House of Lords have now allowed a petition by the defendants for leave: see [1987] 1 WLR 1510. We have been informed that the only question which it is at present intended to raise on the appeal is whether the requirement of reasonableness under section 11(3) of the 1977 Act was satisfied in that case or not. It is not for us to express a view as to whether some other course ought now to be taken in that case.

I would allow this appeal and dismiss the plaintiffs' action.

Agreeing, KERR LJ said: This is a case in which one must have the greatest sympathy with the plaintiffs. In his judgment, to which I turn in a moment, Schiemann J said that Mrs Harris had "cried for hours" when she saw the report of the surveyor whom they had instructed in 1982 when they wished to sell the house. Their situation is doubly distressing as the result of our conclusion that this appeal must be allowed. Not only is their house virtually unsaleable, but in addition they remain liable to the first defendants under the mortgage. We were therefore glad to be told by counsel on behalf of the local authority, as I now put on record, that everything within their statutory powers and discretion will be done to assist Mr and Mrs Harris in so far as this is possible.

The judgment of Schiemann J has been reported in [1987] 1 EGLR 231. I do not propose to analyse it in detail, But if one considers the passages following the heading at p 235F "*Are either or both of the defendants in breach of any duty owed to the plaintiffs?*" one can see that the reasoning involves a composite consideration of what should in truth be two separate questions. The first is whether in all the circumstances either of the defendants owed any duty of care to the plaintiffs. Second, and only if the first question is answered affirmatively, one comes to the question whether or not any breach of any such duty has been established.

It seems to me that the judgment has largely elided the first of these questions. Thus, following this heading, the judge proceeded immediately to his conclusion that both Mr Lee and the council had been "negligent", and the subsequent passages show that he was concerned that in the light of his findings about the condition of the house (which have not been challenged) the council had acted *ultra vires*. He pointed out that if a local authority advances money for the purchase of a house without having satisfied itself that it is in all

*Editor's note: Also reported at [1987] 1 EGLR 157 and (1987) 282 EG 326.

respects fit for human habitation, the authority would be acting *ultra vires* section 43(1) of the Housing (Financial Provisions) Act 1958. And a passage on p 235J of the report shows that the judge was concerned whether liability could attach to a statutory body "on the basis that it had negligently acted *ultra vires* or had negligently impliedly represented that it was acting *intra vires*".

But how can the conclusion that the council had exceeded its statutory powers in making this advance be of any relevance to determine whether or not the council had owed an initial duty of care to the plaintiffs in valuing the property? As shown by *Curran* (*infra*) in the House of Lords, statutory provisions such as section 43(1) are designed for the protection of the public revenue or the general body of ratepayers. Whether or not the defendant authority acted *ultra vires* section 43(1) in the present case is therefore irrelevant to the question whether or not the council owed a duty of care to the plaintiffs. That is the first question which must be answered. Acts by the council which exceeded its statutory powers could at most provide some evidence of a breach of duty, but only if the council had been under a duty of care in the first place.

In fairness to the judge, however, his approach to this case was clearly conditioned by the state of the authorities as they then stood. Thus, the well-known passage in the speech of Lord Wilberforce in *Anns* v *Merton London Borough Council* [1978] AC 728 at p 751, to which the judge referred, had then not yet been subjected to the further restricting analysis which is to be found in the decision of the Privy Council in *Yuen Kun Yeu* v *Attorney-General of Hong Kong* [1987] 3 WLR 776. Above all, however, the judge was understandably influenced to a great extent by the second part of the decision of the Court of Appeal of Northern Ireland in *Curran* v *Northern Ireland Co-Ownership Housing Association Ltd*. He referred to this in considerable detail from p 236J to p 237A. But that part of the decision went to appeal and has since been reversed by the House of Lords in [1987] 2 WLR 1043. It must be highly doubtful whether the judge would have reached the same conclusion if he had had the benefit of that decision.

Turning to other more general matters, the present case shows how much has been built in this field of litigation upon the decision of Park J in *Yianni* v *Edwin Evans & Sons* [1982] QB 438. Schiemann J understandably relied on the authorities stemming from that decision. In particular he relied on the judgment of Costello J, who followed it in the High Court of Ireland in *Ward* v *McMaster and Others* [1985] IR 29 and on the Court of Appeal of Northern Ireland in *Curran* (*supra*). I agree with Nourse LJ that in view of the superstructure of decided cases and changes in practice which have followed *Yianni* it would not be appropriate for this court to consider, on the very different facts of this case, whether that decision can stand. But its inherent jurisprudential weakness in any ordinary situation is clear. Suppose that A approaches B with a request for a loan to be secured on a property or chattel — such as a painting — which A is proposing to acquire. A knows that for the purpose of considering whether or not to make the requested loan, and of its amount, B is bound to make some assessment of the value of the security which is offered, possibly on the basis of some expert inspection and formal valuation. Then assume that B knows that in all probability A will not have had any independent advice or valuation and is also unlikely to commission anything of the kind as a check on B's valuation. B also knows, of course, that any figure which he may then put forward to A by way of a proposed loan on the basis of the offered security will necessarily be seen to reflect B's estimate of the minimum value of the offered security. Suppose that A then accepts B's offer and acquires the property or chattel with the assistance of B's loan and in reliance — at least in part — on B's willingness to advance the amount of the loan as an indication of the value of the property or chattel. Given those facts and no more, I do not think that B can properly be regarded as having assumed, or as being subjected to, any duty of care towards A in his valuation of the security. Even in the absence of any disclaimer of responsibility I do not think that the principles stated in *Hedley Byrne & Co Ltd* v *Heller & Partners Ltd* [1964] AC 465 support the contrary conclusion. B has not been asked for advice or information but merely for a loan. His valuation was carried out for his own commercial purposes. If it was done carelessly, with the result that the valuation and loan were excessive, I do not think that A can have any ground for complaint. And if B made a small service charge for investigating A's request for a loan, I doubt whether the position would be different; certainly not if he were also to add a disclaimer of responsibility and a warning that A should carry out his own valuation.

It may be, but I agree that we should not decide this general question on the present appeal, that the particular circumstances of purchasers of houses with the assistance of loans from building societies or local authorities are capable of leading to a different analysis and conclusion. But be that as it may. A crucial distinction between *Yianni* and the present case is that in *Yianni* there were no express disclaimers of responsibility and warning to potential purchasers to carry out their own inspection, as there were here. In my view such disclaimer and warning must be of the utmost relevance in answering the primary question whether or not a duty of care can properly be held to exist. Or, more accurately in contexts such as the present, whether or not such a duty had been assumed.

The point where I part company with the judge's reasoning is at p 235L of the report of his judgment, where he dealt with what he referred to as the exemption clause. The terms of the disclaimer and warning have been set out in full in his judgment and again in the judgment of Nourse LJ on this appeal, and I therefore do not repeat them. But it must be remembered that it was stated in clear terms that (i) the valuation was solely for the benefit of the council, (ii) the council accepted no responsibility whatever for the value or condition of the property, and (iii) that the plaintiffs were advised to have the property professionally inspected for their own protection. The judge expressed the view that the terms of this provision were not effective to afford any protection to either of the defendants in negligence, and that it did not empower the council to make any loan which, but for this disclaimer, would be statutorily prohibited. The latter is obviously right, and since *Smith* v *Eric S Bush* [1987] 3 WLR 889 will evidently be considered by the House of Lords in the near future I express no view on the first part of this statement. However, in my view none of this matters. What matters, and where I part company with the judge's reasoning in any event, is in the following sentence:

> In those circumstances I propose to ignore the exemption clause when considering whether or no liability attaches to either defendant.

With respect, I think that this involves a confusion of thought. In order to decide the primary question whether or not the defendant owes any duty of care to the plaintiff it is not relevant to determine the precise legal effect of the terms of any disclaimer or other warning, on their true construction, let alone by the application of the Unfair Contract Terms Act 1977. Considerations of the legal effect of any such provisions arise only if and when the existence of a duty of care has been established. However, in determining whether or not the circumstances warrant the inference that any such duty of care was owed by the defendant to the plaintiff, any disclaimer of responsibility and warning addressed by the defendant to the plaintiff may be of the greatest importance. The correctness of this proposition is borne out by all the citations from *Hedley Byrne* which Caulfield J has set out in his judgment (*infra*) and which I therefore do not repeat.

Of course, disclaimers of responsibility and warnings will be ineffective to the point of irrelevance in situations where the existence of a duty of care is not open to doubt. Thus, to take an absurd illustration, it would obviously be irrelevant for a driver sued in negligence to say that he had displayed a sign on the top of his car "No liability accepted for any collision — keep out of the way". The reason is that anyone driving a car is *ipso facto* subjected to an inescapable duty of care towards other users of the highway which he necessarily assumes by driving. And there are obviously countless examples of activities affecting others in relation to which any purported disclaimers of liability would be equally irrelevant. But in regard to statements, ie words as opposed to deeds, the position is inevitably more complex. In such cases the entirety of the statements made must be considered in order to determine whether or not the maker should be held to have assumed, or for some reason to be subjected to, a duty of care to the person addressed. A disclaimer of responsibility for the accuracy of the statement, or a warning that its reliability should be checked independently, must then be of the greatest relevance.

In *Smith* v *Eric S Bush (supra)* this aspect appears to have been disregarded in the defendants' submissions, whatever may have been the reason. The existence of a duty of care was conceded or assumed, and the court was merely asked to decide the effect of the Unfair Contract Terms Act 1977. In the present case, on the other hand, the position was different, but Schiemann J nevertheless disregarded the disclaimer and advice altogether, as shown by the passage quoted above. In my view, this was an erroneous approach to the question whether or not the defendants were under a duty of care to the plaintiffs. For that purpose the council's disclaimer of responsibility and advice to have an independent inspection must be considered together with the council's statement that it was willing to advance £8,505 on the property subject to a number of essential repairs. When these statements are considered as a whole, as they should be, then it is in my view not open to conclude that the council assumed any responsibility for the reliability of the valuation, nor that it can for some other reason be held to have been subject to a duty of care to the plaintiffs.

The contemplation of an intervening inspection qualifies the principle of *Donoghue* v *Stevenson* [1932] AC 562, as Mr Ashworth QC rightly reminded us. And I would not accept that the charge of £22 paid by the plaintiffs can by itself create a duty of care in the face of the disclaimer and warning.

For these reasons I agree with the judgments of Nourse LJ and Caulfield J that the effect of the Unfair Contract Terms Act 1977 upon the disclaimer of responsibility and warning is of no relevance to the present case. One never reaches that issue, since it arises only if the existence of a duty of care and a breach of it have first been established. Mr Ashworth rightly conceded that if the effect of the 1977 Act were relevant, then the burden of proof would be upon him under section 11(1) and that there was no evidence which enabled him to contend that this had been discharged. His contention, however, was that the Act had no relevance to the primary question whether or not the defendants owed any duty to the plaintiffs. The speeches in *Hedley Byrne* make it clear that in order to answer that question the disclaimer of responsibility and warning are part of the circumstances which have to be taken into consideration. When this is done, it is clear that the defendants never came under any duty of care to the plaintiffs. Mr Lee owed no duty other than as an employee of the council. The council's advance to the plaintiffs was *ultra vires* section 43(1) of the Act of 1958, but only the ratepayers would have been entitled to complain about that. The council assumed no duty of care to the plaintiffs not to overvalue the property, as it made clear to the plaintiffs by the disclaimer and warning. Accordingly, no further question arises, and I agree — despite my sympathy for the plaintiffs — that this appeal must be allowed.

Agreeing, CAULFIELD J said: I agree with the judgment of Nourse LJ, but as he forecast I am more sanguine than he appears to be in his reasoning on the Unfair Contract Terms Act.

I have reached my conclusions by examining the position of the second defendant, who must be the *fons et origo* for an assessment of the question of whether the first defendant owed a duty of care to the plaintiff, disclaimers apart. If the first defendant is to be liable in negligence the second defendant must be liable as well. The notice in the valuation form to which Nourse LJ has referred made it abundantly clear that the valuation was confidential; was intended solely for the information of the council; that no responsibility whatsoever was implied or accepted for the value or condition of the property and that the plaintiffs were advised to consult their own surveyor for their own protection. On the ratio of *Hedley Byrne* I do not think in these circumstances any duty of care was assumed by the first defendant to the plaintiffs. In *Hedley Byrne*, although it was held that a duty of care could arise in respect of negligent statements giving rise to damage, the plaintiffs failed because the defendants had not on the facts assumed the duty of care. Lord Reid said at p 483:

> The most natural requirement would be that expressly or by implication from the circumstances the speaker or writer has undertaken some responsibility

and later, at p 493, Lord Reid said:

> . . . the respondents never undertook any duty to exercise care in giving their replies.

Lord Morris at p 504 said:

> . . . the bank in the present case, by the words they employed, effectively disclaimed any assumption of a duty of care. They stated they only responded to the inquiry on the basis that their reply was without responsibility. If the inquirers chose to receive and act upon the reply they cannot disregard the definite terms upon which it was given.

Lord Devlin at p 533 said:

> A man cannot be said voluntarily to be undertaking a responsibility if at the very moment when he is said to be accepting it he declared that he in fact is not.

Lord Pearce said at p 540:

> But in any event they clearly prevent a special relationship from arising. They are part of the material from which one deduces whether a duty of care and a liability for negligence was assumed.

In my judgment it is implicit from these quotations that a potential tortfeasor can avoid the incidence of a duty of care and in my judgment, as Nourse LJ has explained, no duty was assumed by the first defendant. So far I am only introducing my reason for concluding here that the second defendant would not on these facts have assumed a duty because I think it is implicit in *Hedley Byrne* that the speaker or a writer of a statement should have the opportunity to make clear whether he is assuming any duty as the first defendant did in this case. The second defendant, by virtue of his position as a servant of the first defendant, is not given any opportunity at all to decide whether or not he wishes to assume a duty of care towards the plaintiff, so in my judgment he is well remote from assuming any duty of care towards the plaintiff. The most natural requirement mentioned by Lord Reid cannot possibly exist as regards the second defendant. Accordingly, in my judgment the second defendant owed no duty because he never assumed a duty expressly or by implication towards the plaintiff.

In my judgment the 1977 Act is irrelevant. The reference in both subsections of section 2 is to liability for negligence. A person cannot be liable in the tort of negligence unless, being under a duty of care to the plaintiff, he is negligent in the discharge of that duty and his negligence caused damage to the plaintiff. If a person does not owe a duty he cannot have any liability and therefore he has no potential liability to be restricted; he does not need to reflect on the 1977 Act. That Act, in my judgment, can only be relevant where there is on the facts a potential liability. In my judgment the second defendant did not have any such potential liability. I would allow the appeal.

The appeal was allowed. No order was made as to costs. Leave to appeal to the House of Lords was granted.

Court of Appeal

October 27 1987

(Before Sir Nicolas BROWNE-WILKINSON V-C, Lord Justice NOURSE and Lord Justice NICHOLLS)

GP & P LTD v BULCRAIG & DAVIES AND ANOTHER

Estates Gazette March 26 1988

[1988] 12 EG 103-111

Negligence — Damages — Costs — Appeal by defendant solicitors against the computation of damages and against an order to pay the costs of the second defendants, a firm of surveyors, in an action by advertising agents for negligence — Appeal arose from a decision of Mr John Gorman QC, sitting as a deputy High Court judge, who found the solicitor defendants liable for negligence and ordered them to pay substantial damages to the plaintiffs; the claim against the surveyors was dismissed — Plaintiffs had instructed the surveyors to find suitable office premises to rent and had retained the solicitors to act in connection with the acquisition — The trouble arose when the plaintiffs' receiver (the company having gone into liquidation) discovered that the ground floor of the premises acquired was subject to a planning condition restricting part to use for offices for the printing trade only — The receiver found it impossible to dispose of the lease and it was eventually surrendered to the landlords for a payment — A crucial finding was that if the plaintiffs had known of the restriction they would never have taken a lease of the premises —There was no appeal by the solicitor defendants against the finding of negligence, but they submitted that the judge below had assessed the damages partly on the wrong basis and also that they ought not to have been ordered to pay the costs of the surveyors —It was accepted that in calculating the damages the plaintiffs were bound to give credit, *inter alia*, for the beneficial occupation of the premises which they had enjoyed for some 18 months —

The main dispute was as to the correct valuation of this occupation —The judge below had arrived at it on the basis of the difference between the improved rental value of £118,000 per annum and the unimproved rental value of £88,000, ie £30,000 per annum — Defendant solicitors contended that the plaintiffs should give credit for the difference between the improved rental value of £118,000 and the contractual rent actually payable to the landlords of £35,000, ie a difference of £83,000 — Held by the Court of Appeal that the defendants' submission was correct, rejecting an argument that this was in effect giving them the benefit of the favourable rent which the plaintiffs had negotiated with the landlords — The court dismissed the defendant solicitors' appeal against the order that they, not the plaintiffs, should pay the defendant surveyors' costs and refused a suggestion that these costs should be apportioned — Appeal allowed to the extent proposed in relation to damages, but dismissed in relation to costs.

No cases are referred to in this report.

This was an appeal by Bulcraig & Davies, a firm of solicitors, against the decision of Mr John Gorman QC, sitting as a deputy judge of the Queen's Bench Division, in so far as the decision concerned the computation of damages and the order that the appellants should pay the costs of the second defendants, John D Wood & Co, in the action brought by the plaintiffs, GP & P Ltd, for negligence. The plaintiffs' claim against John D Wood & Co had been dismissed.

The decision of the deputy judge was reported at [1986] 2 EGLR 148; (1986) 280 EG 356.

W A Blackburne QC and John V Martin (instructed by Barlow Lyde & Gilbert) appeared on behalf of the appellant; Elizabeth Appleby QC and J Chichester (instructed by William F Prior & Co) represented the respondent plaintiffs.

Giving judgment, SIR NICOLAS BROWNE-WILKINSON V-C said: This is an appeal from a decision of Mr John Gorman QC, sitting as a deputy judge of the Queen's Bench Division. The learned deputy judge held the first defendant, the solicitors, Bulcraig & Davies, liable in professional negligence and ordered them to pay damages in the sum of £195,621, together with interest of £137,572.49 and costs. There was also before the deputy judge a claim against the second defendant, John D Wood (the well-known surveyors), alleging that that firm had been professionally negligent. The deputy judge dismissed the plaintiff's claim against the second defendant and ordered the first defendant to pay the costs of the second defendant.

There is no appeal by the first defendant against the deputy judge's finding that it was liable in negligence. The only two grounds of appeal by the first defendant are: (1) that the deputy judge's computation of the amount of damages was in one respect on the wrong basis; and (2) that the deputy judge erred in ordering the first defendant to pay the costs of the second defendant.

The background facts are these. The plaintiff was a firm of advertising agents which was carrying on business from offices in Hans Crescent, Knightsbridge. The offices had an area of 6,259 sq ft and the plaintiff was paying a rent at the rate of £6.50 per sq ft. The lease was due to expire in June 1978, and the landlords had served the requisite notices under the Landlord and Tenant Act 1954. Accordingly, the plaintiff needed alternative premises. The judge found that its requirements were for premises approximately twice as large as its existing premises. In the circumstances the plaintiff retained the second defendant as surveyors to locate suitable office premises. The plaintiff also retained the first defendant to act on its behalf in the acquisition of new premises. The second defendant sent particulars of the premises which are the subject-matter of the present action to the plaintiff. Those premises are 127-130 Long Acre, Covent Garden, London. Those premises amounted to some 16,800 sq ft of office space on the ground, first and second floors, together with a basement of some 8,450 sq ft, which could be used for storage only. The rent being asked was remarkably moderate, namely £1.50 per sq ft. But the premises were in need of redecoration and refurbishment. The plaintiff negotiated, with the assistance of its advisers, to take a 15-year lease of those premises, and it was agreed it should take such a lease at the rent of £35,000 per annum. The first defendant acted in negotiating the terms of the lease and carrying out its ordinary duties as solicitors.

On May 17 1978 completion of the transaction took place on the basis of undertakings. The actual lease was not granted until August 14 1978. Between July 1978 and the end of January 1979 the plaintiff carried out very substantial works of refurbishment on the Long Acre premises at a cost in the region of £350,000. It started to pay rent to the landlords as from July 15 1978. The plaintiff went into occupation of part of the premises, the second floor, in October 1978 and went into full occupation of the whole of the premises on February 5 1979. From July 15 1978 onwards, right down to the date of the surrender of the lease (which I will shortly mention) the plaintiff paid rent for the premises at the rate of £35,000 per annum.

On April 8 1980 a receiver was appointed of the plaintiff company's business, and it went into liquidation on October 8 1980. It was then that the trouble, which has given rise to the present litigation, emerged for the first time. The receiver appointed a planning specialist to investigate the position. He discovered that as to the ground floor of the Long Acre premises the planning permission was subject to a condition restricting use of part of it to use for offices only in connection with the printing trade. Until that was discovered neither the plaintiff, nor indeed its landlord, nor the first and second defendants were aware of that fact.

The consequences of the restriction on the planning permission were that the receiver had considerable difficulty in disposing of the lease and indeed found it impossible to do so. He took the decision in December 1980 to surrender the lease to the landlords. That surrender took place and the plaintiff received £145,812 from the landlords as consideration for the surrender.

The plaintiff started proceedings against the first and second defendants, claiming that the first defendant had been negligent in failing to discover the existence of the restriction on the planning permission and making similar allegations against the second defendant, the surveyors. The deputy judge held, as I have said, the first defendant (the solicitors) liable in negligence for its failure to discover the condition, but dismissed the claim in negligence against the second defendant. Those are the background facts and I will now turn to the specific ground urged in the appeal relating to the deputy judge's assessment of the damages payable by the first defendant.

The cardinal fact, which in my judgment is of decisive importance in this case, is the finding by the deputy judge that if the plaintiff had been informed, as it should have been, by the first defendant of the existence of the condition restricting the planning consent, the plaintiff would not have taken the lease nor would it have refurbished the premises at the great cost that it did. That was the plaintiff's own case to the deputy judge, and he accepted it.

As to the basis of computation of damages, the parties were agreed both before the deputy judge and before this court on two matters: (1) that the damages should be computed on the basis of the out-of-pocket expenditure by the plaintiff which had been wasted by taking the lease and effecting the improvements when, if it had known the true circumstances, it would not have done either of those acts; (2) that in calculating the amount of recoverable damages the plaintiff was bound to give credit against that gross expenditure both for the sum received on the surrender of the lease to the landlords and, more pertinently for present purposes, for the benefit enjoyed by the plaintiff by reason of its occupation of the premises for a considerable period. The deputy judge found the plaintiff's expenditure out of pocket to be £386,433. From that figure he deducted the £145,812 received by the plaintiff on the surrender of the lease, thereby reducing the plaintiff's loss to £240,621.

The major dispute between the parties was as to the quantification of the second matter that had to be credited against the expenditure by the plaintiff, namely the benefit enjoyed by the plaintiff by reason of its occupation of the premises.

A number of points arose on that question. First, the deputy judge had to determine the period during which the plaintiff was enjoying beneficial occupation of the Long Acre premises. He determined that the period started on February 5 1979 and ended at the beginning of August 1980 being, as I understand it, the date on which the receiver would have been able to sell the lease if there had been no planning restriction as to the use of the ground floor. That is a perod of 18 months. I express no view as to whether that was or was not the right period. It has been accepted by the parties, and there is no appeal against the deputy judge's determination that that was the relevant period.

Next on this aspect of the case the judge had to determine whether the value of the occupation should be assessed by reference to the open market rent of the premises from time to time after February 5 1979 or at the rental value fixed on February 5 1979. He held that the latter was correct: there is no appeal against that.

He had next to determine whether the hypothetical rental had to be assessed on the basis that there was a restriction on use or not. He held that the beneficial occupation had to be fixed on the basis of a rent taking into account the existence of the planning restriction. I again express no view on the correctness of that; there is no appeal against his decision on that point.

Having decided those points on the basis of expert evidence the deputy judge held that the rental value of the premises on the open market on February 5 1979, taking into account the existence of the planning restriction and also taking into account the improvements done to the premises, was a rental at the rate of £118,000 per annum, which he called "the improved rental value". Again there was no appeal as to that figure.

On those figures it was the defendant's contention that in assessing damages there had to be brought into account against the gross expenditure incurred by the plaintiff on the lease and the premises the value of the plaintiff's beneficial enjoyment of the premises for the period of 18 months. The defendant said that the value of that beneficial enjoyment for that period was to be assessed at the improved rental value, that is to say a rental at the rate of £118,000 per annum, less the rent which the plaintiff actually had to pay to the landlords under the lease, namely a rent at the rate of £35,000 per annum.

The plaintiff, on the other hand, while accepting that the value of that beneficial enjoyment had to be brought into account, contended that the benefit should be assessed on a different basis. It relied on the fact that the contractual rent of £35,000 per annum for the premises was substantially below the open market letting value of the premises, even without the improvements, that is to say the unimproved rental value of the premises as at February 5 1979 was in itself substantially above £35,000 per annum. The deputy judge found that the unimproved rental value at the relevant date was £88,000 per annum. He held in favour of the plaintiff's contention and held that in computing the value to the plaintiff of its beneficial occupation of the premises (for which it had to give credit) the sum for which credit had to be given was the difference between the improved rental value, that is to say £118,000 per annum, and the unimproved rental value, £88,000 per annum, that is to say a figure of £30,000 per annum. On that basis he fixed the amount which had to be brought into account for the period of 18 months of beneficial occupation at the figure of £45,000, and reduced the damages otherwise payable by £45,000. After making certain other minor adjustments, he therefore entered judgment for the plaintiff for the figure that I have mentioned.

The defendant on this head of the appeal says that the plaintiff was bound to give credit for a substantially larger sum, namely the difference between the contractual rent payable to the landlords of £35,000 per annum and the improved rental value of £118,000 per annum.

It is difficult, especially in a case which has become as complicated in its concepts as this one, to do justice to the deputy judge's reasons in seeking to summarise them. Broadly, his reasoning for his conclusion was as follows. In his view the plaintiff had to give credit only for the value of its occupation of the premises so far as that value was attributable to the expenditure on the improvements. He held that the plaintiff should not be bound to give credit for the value of the occupation of the premises which was attributable to the fact that the open market rental value in its unimproved condition, namely £88,000 per annum, was substantially greater than the contractual rent payable under the lease of £35,000 per annum. That element of benefit, thought the deputy judge, flowed from the favourable bargain which the plaintiff had managed to achieve with its landlords and was in no way attributable to the expenditure on the premises, which expenditure was being recouped to the plaintiff by way of damages. The deputy judge accepted the plaintiff's submission that to make the plaintiff give credit for the difference between the rent payable under the lease and the improved rental value would be to allow the defendant to profit from its own wrong, which he held would be to assess damages on a wrong principle. He therefore took the view that the right approach was to require the plaintiff to give credit only for that part of the rental value of the premises occupied which was attributable to the making of the improvements, that is to say to make it give credit for the difference between the improved and the unimproved rental value.

Miss Appleby in her very able argument has supported the deputy judge's reasoning. I am unable to accept that reasoning. As I have said, to my mind the fundamental point in this case is the finding that if the defendant had performed its duty and informed the plaintiff of the true planning position, the plaintiff would not have gone on with any part of the transaction, either to carry out the improvements or indeed to take the lease at all. It was on that basis and that basis only that the damages fell to be quantified by reference to the expenditure thrown away. Therefore, to my mind, the damages must be directed to putting the plaintiff in the position in which it would have been if it had not taken the lease at all.

It there had been no lease at all, the plaintiff would not have had any bargain, let alone a favourable bargain, with the landlords as to the occupation of the premises. If there was no bargain and no lease it would not have been entitled to occupy these premises at £35,000 per annum or any other rent. It would have been faced with a position where it had to accommodate itself elsewhere at an open market rent. Therefore, in my judgment, the right approach in seeking to see what sum the plaintiff ought to bring into account as against its claim for expenditure wasted is to discover the rental payable on the open market for the premises in their improved condition, being the premises which it actually occupied. To do that would be to debit the plaintiff with the improved rental value of the premises, less the rent it actually paid out to the landlords. In my judgment that is the true reflection of the benefit it received, bearing in mind the finding that it would not have taken the lease had it known the true position.

In my judgment, the submissions of Miss Appleby and the deputy judge's reasoning are flawed in that they assume that, if the true facts had been made known to the plaintiff, the plaintiff would have continued to enjoy the favourable contract rent under its bargain with the landlords. But the finding that it would not have taken the lease had it known the true position is quite inconsistent with that approach. As to the deputy judge's view that to make the plaintiff give credit for the unimproved market rental would be to allow the defendants to profit from their own wrong, for myself I have never heard of that doctrine previously being applied in connection with the assessment of damages and we were not referred to any authority in support of its relevance. In the assessment of damages the only concern is to assess the loss suffered by the plaintiff. I find it hard to see how damages can be increased above the loss actually suffered by a plaintiff.

In my judgment the argument on the assessment of damages generally in this case has become unnecessarily complicated largely because of the way in which the case was pleaded. In the statement of claim the claim for gross expenditure included rent paid by the plaintiff down to February 5 1979, but did not claim, by way of out-of-pocket expenditure thrown away, the rent paid after February 5 1979. Miss Appleby explained that this was because in the plaintiff's view the rent payable after February 5 1979 was not wasted expenditure. But given that if the plaintiff had known the true facts it would never have taken this lease at all, to my mind the approach suggested by Miss Appleby is not a legitimate one. The claim being that the lease would not have been taken, it necessarily follows that all rent payable under the lease was out-of-pocket expenditure wasted. If, as in my judgment it should have been, the actual rent paid under the lease throughout its continuance had been claimed as damages, I think none of the difficulties which have been discussed in this court and before the learned judge would have arisen. There would be no doubt that such a claim for rent would have been unanswerable. The only question then would have been: what value was attributable to the plaintiff's occupation of the premises? If the matter had been pleaded in that way, the answer, I would have thought, would have been beyond doubt, namely, the open market value of its occupation of the premises, that is to say the improved market rental.

Miss Appleby has pressed us with a number of arguments based on the hardship of this conclusion to the plaintiff. They are indeed arguments *ad misericordiam*. She first urged that the lease of Long Acre being of a substantially larger area than that being enjoyed in Hans Crescent or being looked for by the plaintiff provided too large a space, and by reason of the first defendant's negligence the plaintiff has been led into taking too large an area which it only did because of the low rent available. Such a submission is to my mind inconsistent with the deputy judge's findings of fact that the plaintiff company

A was expanding rapidly, and indeed when it moved into the Long Acre premises it occupied the whole of the office space available.

The second plea *ad misericordiam* was that the plaintiff company took the lease and invested in improvements by way of a long-term capital investment, and that it had lost the value of that investment in the circumstances. But there were, I confess, times during the argument when I felt it was being suggested that it was the first defendant's duty not simply to draw attention to the flaw in the planning position but positively to provide premises which had no condition attached to the planning consent. That plainly, of course, was not its duty. Even if the first defendant had carried out its duty to the full, there was no way in which the plaintiff company could have made a basically sound investment in these premises, since there was
B the flaw in the planning position. I cannot, therefore, see how it could be said that the plaintiff had made a loss on the investment in the Long Acre premises. Such a sound investment was never available. The only way in which it can be put, and there may be force in this, is that if it had not tied up its money in investing, mistakenly, in the Long Acre premises, there would have been other ways in which the money could have been used profitably, and that opportunity had been lost. No such claim was referred to in the pleadings or considered before the deputy judge, and it would be too late, even if Miss Appleby had applied so to do, to consider it in this court.

The fact of the matter in my judgment is that in the way the matter was pleaded the finding that the plaintiff company would never have
C taken these premises if it had known the true position and would therefore have been forced to find accommodation elsewhere on the open market shows that the deputy judge's decision on the point is not right, and that the appellant's contention is correct. On this aspect of the case, therefore, I would allow the appeal and reduce the gross damages payable by a sum representing the difference between the contract rent of £35,000 per annum and the improved rental value of £118,000 per annum over a period of 18 months. Counsel will, no doubt, in due course tell us what that amount should be.

The second aspect of the appeal is as to costs. As I have said, the deputy judge ordered the first defendant to pay the second defendant's costs. The first defendant obtained leave to appeal from the deputy judge against that order.

D The plaintiff started its action against both defendants simultaneously. The defence of the second defendant, which was the last defence to be served, in para 6 pleaded as follows: that shortly after a particular date Mr Procter (the chairman of the plaintiff company) and Mr Burford (a partner in the first defendant):

> . . . agreed that the first defendant would deal with conveyancing matters only and that matters such as planning would be left to the second defendant and to Alexander, Reece & Thompson, a firm of surveyors engaged by the plaintiff in connection with the acquisition of the lease. The first defendant cannot give further particulars of the agreement or the occasion on which it was made.

The defence denied liability.

E The pleading in para 6 plainly indicated that the first defendant was contending that it was not liable, as solicitors would normally be, for failure to draw attention to a detectable flaw in the planning position because there was a special agreement under which the second defendant was to undertake those functions. That plea rested on an alleged conversation between Mr Burford (who has since died) and Mr Procter. No Civil Evidence Act notice was served until the first day of trial. On the third day of the trial, just before the cross-examination of Mr Procter, the first defendant abandoned the allegation in para 6 of the defence. However, the first defendant did not admit liability as such. After the deputy judge had delivered judgment the first defendant submitted either that it should not pay any of the costs of the second defendant or, at least, that there should
F be an apportionment of those costs, ie the first defendant should pay the costs of the second defendant down to the abandonment of the para 6 claim but the costs thereafter should be payable by the plaintiff and not by the first defendant. The deputy judge dismissed both those applications and ordered the first defendant to pay the whole of the second defendant's costs. His grounds for ordering the payment of the whole costs were contained in a separate judgment in which he said he accepted a submission that they were under a duty of care separate and independent from that of the solicitors, and indicated that if the matter had rested there he would have accepted the argument of the first defendant that these were not alternative claims and, as a result, that the plaintiff would have had to pay the costs of the unsuccessful action against the second defendant. However, he was very much influenced by the pleading in para 6 (to which he G
referred) and expressed his conclusions by saying:

> In those circumstances there clearly was a defence by the first defendants which justified the plaintiffs in maintaining the action against the second defendants in the light of that special defence. I do find that it was a reasonable course for the plaintiffs to take.

Mr Blackburne, with his usual engaging frankness, accepts that that is pure discretion and that we cannot interfere with it unless there is either a misdirection by the judge on the point or the result which he reached is plainly unreasonable. For myself, I think the judge was plainly entitled to reach that conclusion. So long as para 6 was in issue, there plainly was a real quandary facing the plaintiff as to whether or not the undoubted damage that it had suffered by the breach of somebody's duty was to be legally the responsibility of the H
solicitors or the surveyors. I can see no ground for interfering with the deputy judge's discretion to the extent that he declined to order the plaintiffs to pay all the costs of the second defendants.

Though it was not raised on the notice of appeal, we permitted Mr Blackburne also to appeal against the second of the deputy judge's decisions, namely the refusal to apportion the costs. The deputy judge was faced with the argument that once the special plea in para 6 had been abandoned on the third day of the trial, thereafter for the plaintiff to continue with the action as against the second defendant was unreasonable and that, accordingly, the first defendant should not be required to pay those later costs. The deputy judge on that issue said: J

> I am not impressed by that argument, especially in the context of the first defendant's track record in relation to this particular allegation. It seems unrealistic to me that, when all the costs of getting the parties before the court had been incurred, the plaintiffs, faced, as I have said, in the defence with a moving target, should be penalised by an advocacy decision halfway through their case in the light of the movements of the first defendants. It seems to me that the substantial question is whether the plaintiffs were justified in bringing the second defendants as defendants to this court in the light of what appeared to be the position on the documents and appeared to be the defence of the first defendants, and in my view they were.

Only the trial judge can appreciate the way in which a case has developed, the steps which have occurred and the attitudes adopted by the parties. It is a dangerous process (even if I felt tempted to do so) K
on appeal to try to disentangle the detail of the matter. In my judgment, the deputy judge's approach to that matter could be attacked only if it could be shown that he was wrong in taking the view that it was unrealistic to expect the plaintiff to discontinue at that stage as against the second defendant. I certainly cannot reach that conclusion. Indeed for myself, on the limited facts known to me, I would have thought that the deputy judge was quite right. True it is that once para 6 has gone, the special reason to which the deputy judge had previously referred had also gone. But in the course of litigation matters are not as clear as they sometimes seem with hindsight. Moreover, I think that if a plaintiff (as this plaintiff has) has reasonably brought the defendant in the first place, a tactical decision by the first defendant in the course of the trial would not L
normally mean that to continue against the second defendant would be unreasonable. I believe the deputy judge was fully entitled on the facts, which he knew better than me, to reach his conclusion that it was not an appropriate case for apportionment. I would dismiss the appeal on that aspect of the case.

Agreeing, NOURSE LJ said: The learned deputy judge gave a very careful judgment in which he decided many points. There has been an appeal on only one of substance, and since we differ from him on that point I will add some observations of my own.

In her primary argument Miss Appleby seeks to draw a distinction between the costs of the acquisition and refurbishment, which she says were wasted, and the rent payable under the lease, which she says M
was not wasted because the plaintiff got value for it through its occupation of the premises. While that was fundamental to the approach of the plaintiff's valuer, Mr John Trustram Eve, in making his second or open market valuation, it is clear to me that it is inconsistent with the case which has throughout been put forward by the plaintiff and which was accepted by the learned deputy judge. That case is that if it had known about the planning restriction on the ground floor the plaintiff would not have taken these premises. Not only would it not have gone ahead with the refurbishment it would never have taken a lease in the first place. The plaintiff's case has never been, as Miss Appleby's primary argument and Mr Trustram Eve's open market valuation would necessarily have it, that it would

A have taken a lease of the premises but not refurbished them. Nor, with respect to the alternative argument which Miss Appleby sought to put forward this morning, has the plaintiff's case ever been that, if it had known of the restriction, it would have sought smaller accommodation elsewhere.

The true view is that on the plaintiff's case, as it has throughout been put, it is not only the costs of the acquisition and refurbishment but also the rent which were wasted, in the sense that they were expenditure which would not have been incurred if there had been no negligence. I also accept Mr Blackburne's basic submission that if the rent had been regarded, as it ought to have been, as part of the plaintiff's gross expenditure, rather than as an element in the computation of benefit for which it had to give credit, the necessary
B distinction between expenditure and benefit would have been maintained. The gross expenditure would have included both the costs of the acquisition and refurbishment and the rent. Against that the plaintiff is required to give credit to the first defendant for a number of items, including the only one which is now in dispute, namely the value of its occupation of the premises between February 1979 and August 1980. That can only be the equivalent of the improved rental value of £118,000. The plaintiff cannot, however, be required to give the full credit because, having expended £35,000 a year in rent as the price of its occupation, it has not enjoyed the full benefit. Accordingly, the £35,000 is deducted from the £118,000, and you arrive at the basis for calculation of the credit for which the first
C defendant has successfully contended on this appeal.

If you look at it in that way, the *unimproved* rental value is seen to be a complete red herring. It is no part of the calculation at all. But the fact that it is a red herring has not prevented it from being the villain of the piece. It is not for us to reopen the figure of £88,000, although it does seem a little surprising when the evidence showed that the premises were put on the market at a rent of £75,000 about a year before May 1978, when the plaintiff finally agreed to take them, by which time it had been reduced to £35,000. That was May 1978. Even allowing for the fact that this is a part of London where rents rose very considerably during the material period, it is rather surprising to find that the valuation evidence supported an unimproved rental value of £88,000 in February 1979, a mere nine months later.

D I have called it the villain of the piece because it is the figure of £88,000 from which the plaintiff's complaint really stems. It is that figure which has caused it to say, erroneously in my view, that the first defendant has been seeking to take the theoretical profit resulting from the plaintiff's bargain with the landlords. It is true that it has taken a benefit, in the sense that the value of the plaintiff's occupation was inflated by this surprising rise in rental value over the material period. But once the plaintiff debits the first defendant with the costs of the improvements, it has no alternative but to credit it with the improved market value of the premises which was current during the period of its occupation.

In all the circumstances, it seems to me that there is no logical
E answer to the first defendant's case on this item of damage, and for these reasons, as well as for those given by the Vice-Chancellor, I also would decide that point in favour of the first defendant. I do not wish to add to his reasoning on the question of costs. Accordingly, I, too, would allow the appeal to the extent indicated by him.

Also agreeing, NICHOLLS LJ said: I add a few words on the principal point raised by this appeal only because we are differing from the deputy judge. He valued the benefit obtained by the plaintiff from its occupation of the property as its offices for the 18-month period in question at £45,000. That sum represented not the rental value of the property at that time (£118,000 per annum), less rent paid at the rate of £35,000 per annum, but only the enhancement (£30,000 per annum) in rental value attributable to the plaintiff's
F expenditure. The enhancement comprised the difference between £118,000, being the rental value of the premises as improved, and £88,000, being the rental value of the premises at the relevant time if they had not been improved.

The main thrust of Miss Appleby's argument was that it would be wrong to charge the plaintiff with having received as a benefit from its occupation a rental value over and above the enhancement in the rental value attributable to the plaintiff's expenditure. She submitted that it would be wrong to charge the plaintiff with any further benefit because that would be to give the first defendant the benefit of the beneficial bargain made by the plaintiff with its landlord: the further benefit put forward by the first defendant being the difference between the rental value of the premises being occupied by the plaintiff, if they had not been improved by the plaintiff (£88,000), G
and the rent actually paid under the lease (£35,000). It was submitted that it would be wrong to give the first defendant the benefit of that bargain, because that would enable the first defendant to profit from its own wrongdoing.

I, too, am unable to accept this submission. I can see no place in this case for any application of the principle or maxim that a defendant cannot be permitted to take advantage of his own wrong. Nor do I see how the first defendant's case is one which involves it doing so. Reduced to its essentials, the plaintiff's case is that by the first defendant's negligence it was led into an unprofitable transaction, a transaction it would not have embarked upon if the first defendant had properly advised it. The plaintiff claims damages on that footing. Thus, in calculating the plaintiff's financial loss, the H
basic comparison required is between (a) the plaintiff's financial position as it turned out in the events which happened and (b) what the plaintiff's financial position would have been had it not acquired the lease and spent money on improvements. That comparison in this case involves considering on one side all the expenditure incurred by the plaintiff which it would not have incurred if it had not entered into this transaction. That expenditure includes the cost of acquisition, the cost of improvements and the money paid under the lease by way of rent. On the other side the plaintiff did obtain some benefit from all this expenditure. In addition to the price received on the surrender of the lease the plaintiff had the use of the premises as its offices for 18 months. The benefit of that use is to be assessed, J
prima facie at all events, at the current market rental value of the premises in question. I do not see how, in seeking to uphold that approach, the first defendant is seeking in some way to take advantage of its own wrong. What it is doing is requiring the plaintiff to bring into the computation of its loss one of the benefits which the plaintiff obtained under the impugned transaction at its full current value.

I can see no reason why the first defendant should not do this. In principle it seems to me that this is neither unfair nor unreasonable. If, as it was entitled to do, the plaintiff chose, in effect, to repudiate the transaction, overall unprofitable, into which it was led by the first defendant's negligence, then the benefits enjoyed by the plaintiff under the lease are to be given their full value in carrying out the K
comparison exercise. That value cannot be equated, where the benefit consists of use of property, with the rent payable under the lease, when *ex hypothesi* that lease would not have been entered into on the footing on which the plaintiff is claiming damages.

I agree that, for these reasons and those given by my lords, this appeal should be allowed to the extent proposed, so far as it relates to damages, but that, for the reasons given by the Vice-Chancellor, this appeal should be dismissed so far as it relates to costs.

The appeal was allowed to the extent proposed by the Vice-Chancellor in relation to damages, but dismissed in relation to the payment of the second defendants' costs.

L

Official Referees' Business

January 13 1988

(Before His Honour Judge Lewis HAWSER QC)

R L POLK & CO (GREAT BRITAIN) LTD AND ANOTHER v EDWIN HILL & PARTNERS

Estates Gazette May 7 1988

[1988] 18 EG 71-76

Negligence — Preliminary issue as to liability — Whether a cause of action passed to a successor in a case where damage occurred during the period of ownership of predecessor — Judge's review of authorities and statement of the law — Judgment in preliminary issue based on assumptions as to nature of damage and other facts — There were in fact two actions, but the preliminary issue related to the second action in which the plaintiffs were an American corporation, the parent of the English company which were the first plaintiffs in the first action — The litigation arose out of a defective perimeter wall which eventually collapsed and the allegation of negligence was made against a firm of chartered surveyors M

and architects who had been retained for a development of land belonging to the English company on an industrial estate — The land on which the defective wall was situated was acquired by the American corporation from the English company on a date which was subsequent to the date on which the damage (in the sense used in the relevant case law) had occurred — The English company had therefore acquired a cause of action prior to the acquisition of the land by the American company — No assignment of any right of action was made to the American company at this time — (A deed of assignment made much later, years after the issue of the writ, is mentioned in the judgment, but disregarded) — The preliminary issue as formulated was whether, in view of the fact that damage occurred to the perimeter wall before the American corporation owned or occupied the premises, the corporation had any cause of action for negligence or breach of duty against the defendants — The American company, which must be assumed to have known of the facts giving rise to the damage, acquired the land and the development at full market value, no allowance being made for the defective condition of the wall — Immediately after the acquisition a liquidator was appointed for the English company — The judge's answer to the specific issue raised was that the American company did not acquire a cause of action — In the course of his judgment the judge considered a number of authorities and stated his understanding of the law to be as follows

1 A person in the position of the defendants owes a duty in tort not only to the original owner with whom he has contracted but also to the owner's successors in title — 2 If damage occurs a cause of action arises in favour of the person who is the owner at that time, but if he sells his property at full market value (ie without any allowance for the defect) he will not suffer any loss and will have no claim under this head — 3 The position of a successor is more difficult — On the most favourable view he could sue if he purchased at the full market value (as above) and did not know, and could not by the exercise of reasonable care have known, of the existence of the damage which had occurred during his predecessor's ownership — It may be, however, that in the absence of the assignment of a right of action he would have no title to sue — 4 If a right of action were assigned to the successor he would be able to sue — The objection against the assignment of a right of action in tort could be met by treating the assignment as relating to rights under the original contract with the predecessor — In the present case there was no effective assignment of a right of action

The judge emphasised that no evidence had been called before him and that his judgment had necessarily to be based on assumptions and was subject to evidence and additional argument if the matter proceeded further — Judgment for defendants

The following cases are referred to in this report.

Anns v *Merton London Borough Council* [1978] AC 728; [1977] 2 WLR 1024; [1977] 2 All ER 492; [1977] EGD 604; (1977) 243 EG 523 & 591, HL 1133

Bowen v *Paramount Builders (Hamilton) Ltd [1977] NZLR 394*

Donoghue v *Stevenson* [1932] AC 562, HL 1134

Dutton v *Bognor Regis Urban District Council* [1972] 1 QB 373; [1972] 2 WLR 299; [1972] 1 All ER 462; (1971) 70 LGR 57; [1972] 1 Lloyd's Rep 227, CA

Mount Albert Borough Council v *Johnson* [1979] 2 NZLR 234

Perry v *Tendring District Council* [1985] 1 EGLR 260; (1984) 30 BLR 118

Pirelli General Cable Works Ltd v *Oscar Faber & Partners* [1983] 2 AC 1; [1983] 2 WLR 6; [1983] 1 All ER 65; [1983] EGD 889; (1982) 265 EG 979, HL

Sparham-Souter v *Town & Country Developments (Essex) Ltd* [1976] QB 858; [1976] 2 WLR 493; [1976] 2 All ER 65, CA

The present proceedings dealt with only one preliminary issue arising in one of two actions against the defendants, Edwin Hill & Partners, the first action being brought by R L Polk & Co (GB) Ltd as first plaintiffs, the second plaintiffs being the company's American parent, R L Polk & Co, a corporation. The second action was by R L Polk & Co against the same defendants. The preliminary issue was raised in the second action. The land which was the subject-matter of the litigation was at the Chiltern Industrial Estate, Sudbury, Suffolk.

Jonathan Cole (instructed by Blakes) appeared on behalf of the plaintiffs; P Boulding (instructed by Ince & Co) represented the defendants.

Giving judgment, JUDGE LEWIS HAWSER QC said: There are two preliminary issues, but I am required to deal only with one, which reads as follows:

> . . . whether (i) in view of the fact that damage occurred to the perimeter wall in or about August 1981 before the plaintiff owned or occupied the premises, the plaintiff had any cause of action for negligence and/or breach of duty against the defendant in respect of the said damage as at July 26 1983, the date of the issue of the writ herein.

That is the preliminary issue raised in the action between R L Polk & Co (Corporation), plaintiff, and Edwin Hill & Partners (a firm), defendants.

R L Polk & Co (GB) Ltd, the first plaintiff in the first action (hereinafter called the "English company"), was an English company which owned and developed certain land at the Chiltern Industrial Estate, Sudbury, in the county of Suffolk. The defendants were retained as the chartered surveyors and architects for this development.

R L Polk & Co (a corporation), the second plaintiff, was and is the American parent of the English company (hereinafter called the "American company") and acquired the land and the development from the English company by a conveyance dated October 26 1981. The following day a liquidator was appointed to the English company. The defendants are alleged to have been negligent and in breach of their duties in respect of that development.

The works were practically completed on May 8 1979. In or about August 1981 the perimeter wall started to lean outwards, and by August 5 1982 the chief environmental officer of the local authority served a dangerous structure notice. By that time, sections of the wall had collapsed.

The American company acquired the land and the development at full market price, no allowance being made for the defective condition of the wall. The defendants submit that August 1981 was the date when the damage occurred, and that seems to be borne out by the correspondence and the decision in *Pirelli General Cable Works Ltd* v *Oscar Faber & Partners (a firm)* reported in [1983] 2 AC 1.

I should read a little of the correspondence relating to this matter. The first relevant letter is dated August 24 1981, written [by Edwin Hill & Partners] to the English company:

> Dear Bill,
> *re: Sudbury*
> I have just returned from holiday and have learned from my Assistant, Brian Read, that there is a problem with one of the boundary walls around the new development.
> I do not think that this is a matter to be referred direct to the builder in the first instance and I am therefore arranging to visit Sudbury on Friday of this week to see the problem for myself.

On September 2 1981 a further letter was written to the English company [by Edwin Hill & Partners], and the last paragraph of that reads as follows:

> Referring to the retaining walls around the car park and vehicle yard, I am very surprised to see the apparent movement in this large wall that does not appear to be related to any collapse of the adjoining surfaced areas. I have yet to locate the details of the construction of these walls but will pursue the problem until I establish the cause.

The English company wrote to Mr Julian J Bishopp of Edwin Hill & Partners:

> Dear Mr Bishop[p],
> I would like to confirm our telephone conversation of today's date, with respect to the perimeter wall.
> The retaining perimeter wall on the opposite side of the gate to the wall which you inspected, is now showing signs of movement. This is shown by an approximate $\frac{1}{2}''$ crack which has appeared in the tarmac surface, running parallel to the retaining wall.
> I trust that we shall be seeing some action on this in the very near future.
> At the time of writing I have not heard anything from Mr Ellmore.

That was on September 9 1981. The reply is dated September 11 1981, and the writer says:

> Thank you for your letter dated September 9, addressed to Mr Bishopp.
> We have passed a copy of the letter to the contractors concerned, requesting

A them to carry out the remedial work as quickly as possible, together with asking the consultant engineers for the contract, K A Lock & Partners, for their view as to the cause.

As soon as we have any further information we will be in contact with you again.

And on September 11 1981 the defendants wrote to the builders:

Further to our telephone conversation yesterday, I have received the enclosed letter from Polk's confirming further movement in the boundary wall.

As agreed, you will contact Ken Lock with a view to him inspecting the wall and giving his expert opinion, together with instructing a local builder to carry out the necessary remedial work to make the wall safe.

We can only assume that until the cause of the movement has been ascertained, R L Polk will be responsible for meeting any expenses incurred.

B On September 22 1981 the contractors, Deacon Contractors Ltd, wrote to Mr Bishopp:

I confirm our telephone conversation, wherein I advised you that myself and Ken from K A Lock & Partners would be visiting the above site as of September 24 1981, together with our local contractor to open part of the work up for a thorough inspection, with a view to advising yourself, and in turn, our Client as to the reason why the fault has occurred, and as to how it is best considered to solve the problem and budget costs for carrying out the necessary work.

On October 5 1981 there is a letter written by the defendants to Nigel Hamilton who afterwards became the liquidator of the first-named plaintiffs.

Thus the effect of the correspondence appears to be that damage in
C the strict sense in which it has been used in the cases had occurred by October 26 1981, the English company had acquired a cause of action in respect of the perimeter wall. A writ was issued on July 26 1983, claiming the cost of repairs to the wall, which were estimated to be about £30,000, a sum afterwards increased to £45,000.

On February 12 1986 a deed of assignment took place between Nigel James Hamilton (the liquidator), the English company and the American company, whereby the rights in the action were specifically assigned to the American company. No further reliance is placed on this and I need not consider it for the purpose of the judgment.

It seems to me to be clear that the damage for the purpose of the cause of action had occurred prior to October 26 1981 and that (if it be material) the American company knew, or ought to have known, of
D the fact of the matters giving rise to such damage when it acquired the property. However, no evidence has been called before me and accordingly I do not pronounce a definite decision on these two matters. It may be subject to subsequent evidence and argument, but my judgment proceeds on the assumption they are correct.

It is also common ground that no specific assignment of a right of action took place prior to October 26 1981.

The defendant's contention is that the person who owned the building at the time when the damage occurred which gave rise to the action on the contract or in law is the only person who can sue for such damage in the absence of any assignment of that right of action to a subsequent owner. Until *Pirelli* v *Faber* the law appeared to be that
E the date from which the period of limitation ran was the date at which the damage was discovered or could have been discovered by the exercise of reasonable care (discoverability). However, *Pirelli* v *Faber* decided that the date was when the damage actually occurred, irrespective of discoverability. It was not until the Latent Damage Act 1986 that the law was altered so that, in broad terms, discoverability became the test. This formulation of the effect of *Pirelli* v *Faber* appears to negative the necessity for the vendor or the purchaser to know of the damage, if in fact it had occurred before the purchaser obtained the property.

The defendants rely upon the judgment of His Honour Judge Newey in the case reported in (1985) 30 BLR 118 called *Perry* v *Tendring District Council**. The passage relied upon begins at p 141
F and the authorities are there in view. It should be noted that the passage is obiter. It is summarised in the headnote at p 120. P 141 reads as follows, starting at the bottom of the page:

Whether an owner or occupier may sue in respect of a cause of action relating to his property which accrued before his ownership or occupation without assignment to him of the right of action.

It is well established that everyone concerned with the design and erection of a building owes a duty of care, not merely to first owners or occupiers, but to their successors: *Dutton* v *Bognor Regis UDC* [1972] 1 QB 373, (CA) and *Anns Case* in the House of Lords. The cause of action does not arise at the date of the design or construction but when an owner or occupier has suffered damage.

* Editor's note: Also reported at [1985] 1 EGLR 260.

The Court of Appeal in *Sparham-Souter* v *Town & Country Developments (Essex) Ltd* [1976] 1 QB 858 decided that damage was suffered when it was detected or could have been with reasonable skill and diligence. Lord Denning MR at p 868 made it plain who was entitled to sue. G

And then he quotes:

"The only person who has a cause of action is the owner in whose time the damage appears. He alone can sue for it unless, of course, he sells the house with its defects and assigns the cause of action to his purchaser."

Counsel submitted that an existing cause of action might "come with the property" on its conveyance from one owner to another, but Roskill LJ thought that this was "an impossible argument". He said at p 873:

"There is no relevant estate contract here. There is no assignment of any pre-existing cause of action in tort in the plaintiffs' favour from their predecessors in title. Nor do I understand how, as the arguments presupposed, there can be H some inchoate or floating cause of action in tort existing *in vacuo* which can suddenly enure to the plaintiffs' benefit upon their acquisition of a legal or equitable title to the property in question . . . Furthermore, the present plaintiffs have clearly not acquired such a benefit by contract or by statute. I fail to see upon what principle they can be said to have acquired it by operation of law."

Roskill LJ at p 875 and Geoffrey Lane LJ, at p 880, each held that the earliest moment at which time would begin to run against each successive owner of defective property was when he bought or agreed to buy it.

In *Pirelli* v *Oscar Faber* in [1983] 2 AC 1; 21 BLR 99, the House of Lords disapproved the decisions of the Court of Appeal in *Sparham-Souter* so far as they concerned the date of accrual of a cause of action and date from which time could run against a new owner. The House held that the date of accrual was that on which the damage came into existence, regardless of whether it J was then known or could have been discovered. Lord Fraser of Tullybelton said, at p 18:

"I think the true view is that the duty of the builder and of the local authority is owed to the owners of the property as a class, and that if time runs against one owner, it also runs against his successors in title. No owner in the chain can have a better claim than his predecessor."

I think it follows from the words of Lord Denning and Roskill LJ in *Sparham-Souter* which I have quoted, and which were not disapproved in *Pirelli*, and from Lord Fraser's words in *Pirelli* that when damage to a building occurs its owner acquires a cause of action immediately, even though he is not aware of it, and that unless he assigns that right of action, when agreeing to convey or conveying the new building to a successor, or possibly transferring the occupation of it to him, the successor has no right to sue.

I do not know how an assignment could be effected. For assignment under K section 136 of the Law of Property Act 1925, notice to the "debtor" is required, but an assignor could not give notice to a tortfeasor if unaware of the existence of his cause of action. Perhaps there could be an equitable assignment: the assignee then suing in the name of the assignor. I am also uncertain as to what damages the assignee could recover, since the assignor would not have expended money on the remedying of undiscovered defects and would presumably have obtained market price for his property.

If this passage correctly states the law, then the plaintiff's claim must fail. (I should mention that section 3 of the Latent Damage Act 1986 deals with this situation, which appears to have been created by the *Perry* v *Tendring* decision.)

The defendants also rely upon *Sparham-Souter* v *Town & Country Developments (Essex) Ltd* and in particular the words of Lord L Denning at p 167G to 186G, and Lord Wilberforce's words in *Anns*, reported in [1978] AC 728 under the title of *Anns* v *Merton London Borough Council*. There Lord Wilberforce on p 758 at D in his opinion says:

To whom the duty is owed. There is, in my opinion, no difficulty about this. A reasonable man in the position of the inspector must realise that if the foundations are covered without adequate depth or strength as required by the byelaws, injury to safety or health may be suffered by owners or occupiers of the house. The duty is owed to them — not of course to a negligent building owner, the source of his own loss. I would leave open the case of users, who might themselves have a remedy against the occupier under the Occupiers' Liability Act 1957. A right of action can only be conferred upon an owner or occupier, who is such when the damage occurs (see below).

Again at p 759G, he says: M

Subject always to adequate proof of causation, these damages may include damages for personal injury and damage to property. In my opinion they may also include damage to the dwelling-house itself; for the whole purpose of the byelaws in requiring foundations to be of a certain standard is to prevent damage arising from weakness of the foundations which is certain to endanger the health or safety of occupants.

To allow recovery for such damage to the house follows, in my opinion, from normal principle. The classification is required, the relevant damage is in my opinion material, physical damage, and what is recoverable is the amount of expenditure necessary to restore the dwelling to a condition in which it is no longer a danger to the health or safety of persons occupying and possibly (depending on the circumstances) expenses arising from necessary displacement.

A The plaintiffs, the American company, argue that the passage in *Perry* v *Tendring* was obiter, and it is based on words used in previous cases which are only relevant to the situation in which the Public Health Act background, with its requirement that the damage must be such as to affect health or safety, is the governing factor. Mr Cole for the plaintiffs argued that the words do not apply to the ordinary duty of care, the breach of which gives rise to a claim in damages.

He also argued that the question of assignment was really a red herring, that assignment related to contract and it was applicable only in rare cases in tort which do not arise in the present circumstances. Mr Cole's submissions are helpfully set out on pp 12 and 13 of his typed argument. He says that you cannot assign a cause
B of action in tort. He cited a number of cases both before and after *Sparham-Souter* v *Town & Country Developments (Essex) Ltd* in which the owner of a property was held to be entitled to sue, even though the damage had occurred before he occupied the property. In particular, he relied upon *Dutton* v *Bognor Regis Urban District Council* [1972] 1 QB 373, a similar case. He contended that the duty is owed to the class of people who may be affected, that is to say, those owners who, subject to limitation, would include all the successors in title. He relied upon the words of Lord Wilberforce in *Anns* at p 753H. The sentence reads:

> And as the building is intended to last, the class of owners and occupiers likely to be affected cannot be limited to those who go in immediately after construction.

C Of course I entirely accept that statement, but it does not appear to me to be relevant in the present case.

Mr Cole summarised his submission in this way: a successor in title has a right to sue, even when damage occurs during the predecessor's ownership.

It seems to me that the relevant law on the case can be summarised as follows.

1 We start with *Donoghue* v *Stevenson* [1932] AC 562: that is the classic statement of the law and has been applied subsequently in cases which affect property. The case is reported under the heading of *M'Alister (or Donoghue) (Pauper) and Stevenson*. The relevant passage begins in the opinion of Lord Atkin about two-thirds of the way down the page on p 580. It is a passage which has been cited on
D many occasions and at the moment I will simply refer specifically to a particular passage in that section on p 582 about seven lines down, where Lord Atkin says:

> I draw particular attention to the fact that Lord Esher emphasises the necessity of goods having to be "used immediately" and "used at once before a reasonable opportunity of inspection". This is obviously to exclude the possibility of goods having their condition altered by a lapse of time, and to call attention to the proximate relationship, which may be too remote where inspection even of the person using, certainly of an intermediate person, may reasonably be interposed.

2 A person in the position of the defendants owes a duty in tort to a class of persons, not merely to the person with whom he contracts,
E the class being the original owner and his successors in title.

3 If and when damage occurs, a cause of action arises under which the person in whose ownership the property then lies can sue.

4 A successor to that person could sue only if he purchased at the full market price and did not know, and could not by the exercise of reasonable care have known, of the existence of the damage. That is putting it as high as it possibly can be put against the defendants who may well only be liable if the damage actually occurred during the successor's title, irrespective of discoverability.

5 Where damage occurs during a person's title he has a cause of action, though if he sells at the full market price he will not sustain any actual loss.

F 6 Where the successor knows of his damage, then under the principles laid down in *Donoghue* v *Stevenson* which I have referred to above, he will not be able to sue. In my judgment, even if he did not have that actual knowledge, but he knew or ought to have known of the damage, he would still be unable to sue. He can sue if the cause of action has been assigned to him by his predecessor. In my judgment, if there is a conveyance which includes or accompanies an assignment of the right to sue, then that right is not a right of the assigned cause of action in tort but is an integral part of the conveyance and, further, is an assignment of the rights under the original contract which are not open to the objection put forward by Mr Cole.

I would also mention that I have been assisted in the conclusions which I have come to by the observations in the case of *Bowen* v *Paramount Builders (Hamilton) Ltd* reported in [1977] NZLR 394, particularly at p 410, lines 35-45, p 413, lines 30-45, and p 414, lines G
12-25, and by the New Zealand decision in *Mount Albert Borough Council* v *Johnson* reported in [1979] 2 NZLR 234 and in particular the joint judgment of Messrs Cooke and Richardson at p 238, lines 35-40, and p 242, lines 5-25.

Therefore, as the matter stands, and on the basis of the assumptions as to the facts which I have made (relating to the nature of the damage and the American company's knowledge, actual or imputed), I would hold for the defendants. I would only add that I have been greatly indebted to counsel for the clarity of their submissions and their most useful written summaries.

The defendants were awarded the costs of the preliminary issue and certain other costs previously reserved. Leave to appeal was H
given.

[An appeal was not proceeded with — Ed]

Queen's Bench Division

December 18 1987

(Before Mr Justice NOLAN)

KITNEY v JONES LANG WOOTTON

Estates Gazette May 21 1988
[1988] 20 EG 88-92 J

Limitation of action — Preliminary issue — Whether for the purpose of section 2 of the Limitation Act 1980 the cause of action arose prior to the period of six years before the issue of a writ for negligence — The present proceedings were not concerned with the question of liability, which was not decided, but only with the date when the cause of action, if any, arose — The allegation against the defendants was that they were negligent in carrying out instructions given by the plaintiff to ensure that certain repair works required to property in which the plaintiff had a leasehold interest were K
properly carried out — The importance of this was that the plaintiff was entitled to extend the term of his lease, which was shortly coming to an end, provided that he had complied with all the covenants — One of these covenants was a repairing covenant to be performed during the last year of the term — Such performance was therefore of critical importance for securing to the plaintiff an extension of 21 years

The plaintiff met with difficulties when he endeavoured to exercise his option — A claim against the assignees of his original lessors was dismissed because the option had not been registered as an estate contract — A claim against the original L
landlords for damages for breach of covenant failed because it was found that the repairing covenant, performance of which was a condition precedent to the extension, had not been complied with — The plaintiff now sued the defendants, a firm of chartered surveyors, seeking under the head of negligence damages representing the loss of the extension of the lease — The preliminary issue was whether the cause of action was statute-barred — It was agreed that the cause of action accrued when the plaintiff suffered damage caused by the alleged negligence — Was this on December 25 1974 when the stipulated time for the plaintiff to comply with the repairing covenant expired or was it May 9 1984 when the M
judge in the previous proceedings held that the covenant had not been complied with, the former date being outside and the latter date within the limitation period?

In support of the latter date the plaintiff submitted that the damage only crystallised at the date of the judge's decision — Before that there was no certainty that the plaintiff would fail in his action against the original lessors — The court rejected this submission on the ground that it adopted a subjective assessment of the prospect of success as a criterion for postponing the date of accrual of damage — The judge's decision merely confirmed that the right to a new lease or

A **damages in lieu had been lost on December 25 1974 — Hence the cause of action was statute-barred**

The following cases are referred to in this report.

Dove v *Banhams Patent Locks Ltd* [1983] 1 WLR 1436
Moore (D W) & Co Ltd v *Ferrier* [1988] 1 WLR 267; [1988] 1 All ER 400, CA
Pirelli General Cable Works Ltd v *Oscar Faber & Partners* [1983] 2 AC 1; [1983] 2 WLR 6; [1983] 1 All ER 65; [1983] EGD 889; (1982) 265 EG 979, HL

In this action the plaintiff, Mr Alan William Kitney, sued the defendants, Jones Lang Wootton, a firm of chartered surveyors, alleging negligence in failing to ensure that certain repair work had
B been carried out to a property at 5 and 6 Granville Parade, College Road, Harrow, Middlesex, in which the plaintiff held a leasehold interest. The present proceedings were concerned with a preliminary issue only, namely the date when a cause of action arose for the purpose of the Limitation Act 1980.

Ian Romer (instructed by Dale & Newbery) appeared on behalf of the plaintiff; Mark Strachan (instructed by D J Freeman & Co) represented the defendants.

Giving judgment, NOLAN J said: This case comes before me for the determination of a preliminary issue. The question at issue is whether, for the purposes of section 2 of the Limitation Act 1980, the cause of action arose prior to the period of six years before the issue of
C the writ. Section 2 reads as follows:

> An action founded on tort shall not be brought after the expiration of six years from the date on which the cause of action accrued.

The writ in this case was issued on January 14 1985. The plaintiff claims that the defendants were negligent in failing to ensure that certain repair work was carried out to property in which the plaintiff held a leasehold interest. The work was required in order to satisfy the obligations of the plaintiff under a repairing covenant in the lease. It was of critical importance because of the facts which are set out in a statement prepared by Mr Romer for the plaintiff and accepted by Mr Strachan for the defendants.

Mr Romer's statement reads as follows:

D 1 By a lease dated April 20 1933 and made between Greater London Properties Ltd of the one part and Gee Bee (Outfitters) Ltd (therein called "the lessees" which expression included successors and assigns) of the other part, premises known as nos 5 and 6 Granville Parade, College Road, Harrow, Middlesex were demised to the lessees at a yearly rent of £700 for a term ("the original term") expiring on December 25, 1974.

2 By clause 5(2) of the said Lease Greater London Properties Ltd covenanted that if the lessees should be desirous of extending the lease for a further term of 21 years from the expiration of the original term and of such desire should give six months' previous written notice then provided the lessees should have paid the rents thereby reserved and observed and performed all the covenants and conditions on the lessees' part therein contained the said lease should be continued for such further term of 21 years at the same rents and subject to the same provisions except such covenant for renewal.

E 3 By clause 3(5) of the said lease the lessees entered into a covenant ("the repairing covenant") that during the last year of the original term they would (*inter alia*) in good and workmanlike manner paint all the parts of the said premises usually painted and that at the expiration of the said term they would yield up the said premises well and substantially repaired and painted as therein mentioned.

4 In 1966 the said term became vested in the plaintiff and remained so vested until its expiration on December 25 1974.

5 In May 1974 the plaintiff duly served notice of his desire to extend the same lease. By that time the reversion on the lease had become vested in MEPC.

6 The defendants are and have at all material times been a firm of chartered surveyors and on April 9 1974 they were instructed by Dale & Newbery (the plaintiff's solicitors) to prepare a schedule of works required to be undertaken at the said premises to comply with the repairing covenant. In May 1974 the
F defendants produced a schedule of works ("the May schedule") to be carried out at the said premises.

7 In October 1974 the defendants accepted instructions from the plaintiff to put in hand and to supervise the carrying out of the works required as aforesaid.

8 In or about October 1974 the defendants made certain amendments to the May schedule and the work specified in the said schedule as so amended was completed in early December 1974.

9 In March 1977 the plaintiff's claim against MEPC for a renewal of lease was dismissed by the Court of Appeal* because the option in the lease had not been registered as an estate contract.

10 On May 9 1984 it was held by Mr Michael Wheeler QC (sitting as an additional judge of the Chancery Division) in proceedings in the Chancery Division intituled: re the said lease: Kitney v Greater London Properties Limited† that "the repairing covenant had not been complied with and that by reason thereof the plaintiff was not entitled to have the said lease extended and that his claim against Greater London Properties for damages for breach of its said covenant to extend the said lease should therefore be dismissed." G

In other words, the plaintiff's assertion that the repairing covenant had been complied with could not be raised against MEPC with a view to getting a new lease because of the lack of registration. It could be raised only against the original lessor (whom I shall call "GLP") by way of a claim for damages for breach of the covenant to extend the lease.

The action against GLP having been dismissed by Mr Wheeler, the present action is proposed to be brought against the present defendants, again for damages representing the loss of the extension H to the lease but on the basis now that, as decided by Mr Wheeler, the covenant to repair had been broken by the plaintiff. The plaintiff says that the defendants were responsible for the breach by reason of their negligence.

It is common ground that the action can be brought in tort. It is common ground that the cause of action accrued for the purposes of section 2 of the Limitation Act 1980 when the plaintiff suffered damage caused by the negligence alleged. When was that? It was either (1) on December 25 1974, when the stipulated time for the plaintiff to comply with the repairing covenant expired, or (2) on May 9 1984, when Mr Wheeler held that the covenant had not been complied with. J

In support of the latter date, which is of course within the six-year period, Mr Romer submits that the damage crystallised only at the date of this decision. Until then, he submits, there was no certainty or even probability that the plaintiff would fail in his action against GLP and therefore it would not be commercially realistic to say that the plaintiff had suffered the damage which he now seeks to lay at the door of the defendants.

Mr Romer referred me to a number of authorities, notably the *Pirelli* and *Dove*‡ cases, in which a critical distinction had been drawn between defective advice or work by the defendant and the onset of actual damage resulting from that defective advice or work. The cases are helpfully summarised and analysed in *D W Moore & Co Ltd* v *Ferrier,* which was decided by the Court of Appeal on July 29 1987 K but is not yet reported save in *The Times* newspaper.§ If an analogy were drawn with the present case, it would be on the basis that the defendants' failure to carry out the necessary repair work before the lease expired was the defect and the decision of Mr Wheeler represented the accrual of the damage. But to my mind the analogy is far-fetched and unrealistic and I think in the end both counsel were minded to agree that the facts of the present case break new ground.

The principle of seeking out the date of accrual of the damage none the less still governs the matter. Mr Romer invokes a phrase used by the deputy judge in the *Moore* case, whose decision was upheld by the Court of Appeal, for the adoption of a commercial approach to the identification of the date of accrual of the damage. Following that approach, he submits as I have indicated that, until Mr Wheeler's L decision was given, its result was not certain or even probable. In factual support of that proposition, he cites Mr Wheeler's view that there was a narrow margin between the two sides and he relies on the facts that MEPC had been satisfied with the state of repair of the premises in 1974, that GLP had conceded that the state of repair was not bad, and that the present intended defendants still maintained that the work was carried out adequately. Thus the plaintiff might well have succeeded before Mr Wheeler and he could not definitely have been said to have suffered damage until he failed.

The difficulty with that approach to my mind is that it adopts a subjective assessment of the prospect of success as a criterion for postponing the date of accrual of the damage. I think that the difficulty is compounded by the fact that the assessment of the possibility or likelihood of success is bound to be coloured by the fact that, *ex hypothesi*, it is one which has been proved by events to have been wrong. That is the foundation of the present action.

With respect, I think that Mr Romer's approach could be sustained only on the broader ground that, irrespective of the prior expectation of the plaintiff, it was the decision of Mr Wheeler which precipitated the damage. But I prefer Mr Strachan's submission that Mr

* Editor's note: *Kitney* v *MEPC Ltd* [1978] 1 All ER 595.

† Editor's note: Reported at (1984) 272 EG 786.

‡ Editor's note: *Pirelli General Cable Works Ltd* v *Oscar Faber & Partners* [1983] 2 AC 1 and *Dove* v *Banhams Patent Locks Ltd* [1983] 1 WLR 1436.

§ Editor's note: Now reported at [1988] 1 WLR 267.

Wheeler's decision merely confirmed that the right to a new lease or damages in lieu had been lost on December 25 1974. The judgment established what the situation had been at the end of that day. It did not create the situation.

Mr Strachan also submitted by analogy with earlier authorities that the well-established doctrine that damage may accrue before it is discovered or discoverable is in point here. So he submits the fact that the plaintiff did not discover that he had lost the right until the date of Mr Wheeler's judgment did not prevent it from having been lost on the earlier date. I think that is a helpful analogy, but I base my decision on the simple former ground put forward by Mr Strachan.

For these reasons, the preliminary question for determination must be resolved in favour of the defendants.

Queen's Bench Division

March 29 1988

(Before Mr Justice McNEILL)

DAVIES AND ANOTHER v PARRY

Estates Gazette May 21 and 28 1988

[1988] 20 EG 92-100 and 21 EG 74-80

Negligence — Action against valuer in respect of mortgage valuation — Comprehensive review of authorities — Contrasting judicial approaches in *Smith* v *Eric S Bush* and *Harris* v *Wyre Forest District Council* (both of which are likely to be considered by the House of Lords) — "A plain difference of judicial opinion" — A fundamental question in regard to the effect of disclaimer and the application of the Unfair Contract Terms Act 1977

In the present case the defendant valuer, a chartered surveyor but not a building surveyor, was instructed by a building society to carry out a valuation in connection with the proposed purchase by the plaintiffs, a husband and wife, of a bungalow — The proposed purchase had been negotiated at a figure of £26,500 and the plaintiffs sought an advance from the society of £12,000 — The valuer's fee, of £45 plus VAT, was paid to the society by the plaintiffs and, according to the current practice, they received a copy of the report — Apart from the notice given under section 30 of the Building Societies Act 1962 disclaiming any warranty that the purchase price was reasonable, the form of application for the advance and the valuer's report itself emphasised that the valuation was for mortgage purposes only; that the report was not a structural survey; that the purpose was to decide whether the society should make an advance on the security of the property and on what terms; and that it should not be assumed that any defects mentioned in the report were the only defects in the property — The valuer's report valued the property for mortgage purposes at £26,500, recommended insurance cover of £36,000 and stated that "the purchase price was very reasonable" — In answer to a question, the valuer stated that in his opinion the property was not affected by settlement — Unfortunately, some cracks in various parts of the bungalow were evident from the start of the plaintiffs' occupation and a lump under the carpet in the lounge revealed, on investigation, an alarmingly large crack in the floor — It appeared that the concrete floor of the bungalow had been laid directly over burnt colliery shale which produced a sulphate attack on the concrete, leading to expansion, displacement of brickwork, and disturbance of internal walls

There was some conflict of expert evidence at the trial, but the judge preferred the evidence of the plaintiffs and their witnesses — He came to the conclusion that, if a duty of care was owed by the defendant, he had been unquestionably negligent — There had been sufficient visible indications of settlement or other structural damage to have caused a reasonable valuer to appreciate that there was a problem — To have "passed" the bungalow and to have described the price as "very reasonable" showed a lack of due care (if the duty of care existed) — The main question, therefore, was whether such a duty did exist in the circumstances of the case — In considering this question the judge discussed a large number of cases, both those dealing specifically with mortgage valuations, such as *Yianni* v *Edwin Evans & Sons* and the more recent cases mentioned below, and cases concerned with the general law of negligence, such as *Hedley Byrne & Co Ltd* v *Heller & Partners Ltd* and *Anns* v *Merton London Borough Council* — The judge was satisfied that in the present case the relationship between the defendant valuer and the plaintiffs was sufficiently "proximate" to create in the circumstances a duty of care, other things being equal — Recent cases, however, had raised difficult questions as to the effect of disclaimer and the application of the Unfair Contract Terms Act 1977

Analysis of *Smith* v *Eric S Bush* and *Harris* v *Wyre Forest District Council* appeared to disclose a clear difference of approach — In the *Bush* case the court, although accepting that the disclaimer prevented the defendants from owing any duty of care, nevertheless felt able to decide that the disclaimer did not satisfy the reasonableness test of the Unfair Contract Terms Act — In the *Wyre Forest* case the court held that the Act applied only if a duty of care was established; if a disclaimer prevented the duty from arising at all, the Act could have no application — McNeill J preferred the view, which he derived from the *Bush* case, that where a duty of care was independently established by the special relationship between a professional adviser and the recipient of advice the effect of the disclaimer was perhaps only to impose limitations on the extent to which the recipient can reasonably rely on the advice — McNeill J considered that this view was supported by section 13 of the 1977 Act and by the decision of the Court of Appeal in *Phillips Products Ltd* v *Hyland* — In giving the judgment of the court in that case, Slade LJ said that it would make nonsense of the 1977 Act if a condition disclaiming liability were itself to defeat the common law duty of reasonable care — It would mean that the very term which was pre-eminently suitable for review for reasonableness under the Act would be removed from its scope — The court, in applying the 1977 Act, should leave out of account at this stage the term which is relied on by the defence as defeating the plaintiff's claim for breach of the relevant obligation or duty

Held that in the present case the duty of care was not defeated by disclaimer and that the defendant had not discharged the onus upon him of satisfying the requirement of reasonableness under the 1977 Act — The judge rejected a submission that the plaintiffs should be held guilty of contributory negligence — Judgment accordingly for plaintiffs, with damages to be assessed

The following cases are referred to in this report.

Anns v *Merton London Borough Council* [1978] AC 728; [1977] 2 WLR 1024; [1977] 2 All ER 492; [1977] EGD 604; (1977) 243 EG 523 & 591, HL

Curran v *Northern Ireland Co-ownership Housing Association Ltd* [1985] 8 NIJB 22, Carswell J; April 28 1986, CA, NI, unreported; [1987] 2 WLR 1043; [1987] 2 All ER 13, HL

Harris v *Wyre Forest District Council* [1987] 1 EGLR 231

Hedley Byrne & Co Ltd v *Heller & Partners Ltd* [1964] AC 465; [1963] 3 WLR 101; [1963] 2 All ER 575; [1963] 1 Lloyd's Rep 485, HL

Nance v *British Columbia Electric Railway Co Ltd* [1951] AC 601; [1951] 2 All ER 448; [1951] 2 TLR 137, PC

Peabody Donation Fund Governors v *Grant* (1982) 264 EG 925, CA

Phillips Products Ltd v *Hyland* [1987] 1 WLR 659; [1987] 2 All ER 620, CA

Smith v *Eric S Bush (a firm)* [1987] 3 WLR 889; [1987] 3 All ER 179; [1987] 1 EGLR 157; (1987) 282 EG 326, CA

Ward v *McMaster, Louth CC and Nicholas Hardy & Co Ltd* [1985] IR 29; [1986] ILRM 43

Westlake v *Bracknell District Council* [1987] 1 EGLR 161; (1987) 282 EG 868

Yianni v *Edwin Evans & Sons* [1982] QB 438; [1981] 3 WLR 843; [1981] 3 All ER 592; [1981] EGD 803; (1981) 259 EG 969

Yuen Kun Yeu v *Attorney-General of Hong Kong* [1987] 3 WLR 776; [1987] 2 All ER 705, PC

This was an action by David Roland Davies and his wife, Joan Davies, against Robert Idris Parry FRICS, alleging negligence by the defendant in carrying out a valuation for mortgage, on the instructions of the Britannia Building Society, of a bungalow at 2 Berwyn Avenue, Chirk, Clwyd, purchased by the plaintiffs.

W R Wood (instructed by Stevens & Co, of Chirk) appeared on behalf of the plaintiffs; G W Little (instructed by John C Gittins & Co, of Oswestry) represented the defendant.

Giving judgment, McNeill J said: In 1983 the plaintiffs were minded to move house. They were already house owners. They saw and liked a bungalow, 2 Berwyn Avenue, Chirk, in the County of Clwyd, then the property of an elderly lady, Mrs Cater, who lived there on her own, and advertised for sale by Hall Wateridge & Owen, of Oswestry. A sale to the plaintiffs subject to contract was on August 30 1983 negotiated at a price of £26,500: the asking price had been £27,500. The plaintiffs sought mortgage facilities from the Britannia Building Society; they wanted an advance of £12,000. They proposed to find the balance from their own resources, presumably from the proceeds of sale of their existing house. On September 15 1983 the building society sent instructions to the defendant to value the property. The building society required the plaintiffs to pay to it the defendant's fee, £45 plus VAT, which they did. On September 19 or 20 the defendant visited the premises and inspected them: Mrs Cater was still in residence. At about this time her son was visiting her and assisting her with her move and the activities incidental to the sale of her house. The plaintiffs recall meeting him there after they had applied for the mortgage.

The defendant made his report (undated) to the building society. As had become the practice after the decision of Park J in *Yianni* v *Edwin Evans & Sons* [1982] QB 438, the defendant was required to provide three copies of the society's form of report, one of which (which omitted reference to the valuer's fee) was for the applicants. This copy was sent to the plaintiffs together with the society's "offer of advance".

Omitting matters not directly relevant to the issues between the parties, the society indicated its willingness to make the advance sought on various conditions. The document was endorsed with the notice required to be given by section 30 of the Building Societies Act 1962 (where liability was to be excluded) that making the advance implied no warranty on the part of the society that the purchase price of the property was reasonable. The defendant does not by his amended defence contend that this notice relieves him from any liability, if any, to the plaintiffs.

In the society's form of application for an advance the following notes appear:

1 Applicants are recommended to read the Society's mortgage leaflet and rules.

3 A copy of the Valuer's report will be supplied to the applicant where a property is being purchased. It is emphasised that the purpose of the Valuer's report is to enable the Society to decide whether or not to make an advance, and if so, on what terms. It is not a structural survey and there may be defects in the property which could only be revealed by a detailed survey. It cannot therefore be assumed that the repairs or defects shown in the report, if any, are the only defects which may be present in the property. Should the applicant require a comprehensive survey, he should make arrangements to have an independent inspection undertaken by a qualified surveyor.

There is then provision for the applicant to sign a declaration including the words: "I/we have read and agree with notes 1-8 above". I have not read notes 2 or 4 to 8 inclusive since they are not relevant to this dispute. The plaintiffs agree that they signed the declaration. It is not suggested on the defendant's behalf that anything here turns on anything in the mortgage leaflet or rules: it is on note 3 of this document that reliance is placed.

The first plaintiff's evidence, in cross-examination, was that the building society's local manager told him that the purpose of the valuer's report was to make sure that the house was worth £26,500. The plaintiff accepted that he was told of the content of note 3 and the purpose of the report there expressed.

The offer of advance bears on its face the following wording — apparently impressed and not part of the printed material on the form:

The applicant's attention is drawn to the fact that the inspection made by the Society's valuer was for the purpose of assessing the value only. There may be defects in the property which will only be revealed by a structural survey and the applicant is advised to have an independent inspection made.

The defendant's report, sent to the plaintiffs, contained details of the property, its area, accommodation and services. In a space numbered 8 the following appears: "In your opinion is the property affected by settlement? If so give details including the extent and effect on saleability." To this the defendant had answered "No". Then in a space headed "General Remarks" he recorded the following:

The property was generally in satisfactory condition but the attention of the applicant should be drawn to the following: 1. The facings of bricks below the front steps are damaged. 2. The front gutter was leaking over bedroom window and rain water was spraying against the brick work. 3. The brick work required replacing below damp course at the rear corner of the stores. 4. There was slight timber decay at the bottom of the door frames . . . All the above matters should be carried out as acts of husbandry and no retention was recommended. The property was situated on a corner site and the purchase price was very reasonable. The valuation on the documents for mortgage purposes was £26,500, the recommended insurance cover £36,000.

Finally on this document, under the heading "Notes for Applicants", the following appears:

1. This document is a report from the Society's valuer to the Society giving the valuation for mortgage purposes only. It is not a market valuation and should not be taken as such. It has been obtained from the valuer to comply with the Building Societies Act 1962. The purpose of the report is to enable the Society to decide if an advance should be made and upon what terms. 3. It is stressed that the report is not a structural survey. There may be defects in the property which could only be revealed by a detailed inspection. You should not assume that any repairs referred to in the report are the only defects in the property, or if no repairs are mentioned, that no defects exist. 4. If you are in any doubt as to the value of the property or the soundness of any part of the structure you should consult a qualified surveyor to advise. 5. If you are purchasing the property you will receive a notice that the Society does not warrant the purchase price as reasonable. 6. The contents of this report are for your information only and should not be disclosed to anyone other than your professional advisers.

With reference to no 4 of those notes, it was not suggested that the plaintiffs were in any doubt as to the value or soundness of the structure.

The plaintiffs accepted the building society's offer of an advance and the property was conveyed to them on December 12 1983. The mortgage was completed on the same day.

The evidence of the first plaintiff, subsequently confirmed in detail by the second plaintiff, as to what they saw when they visited the house was this:

When we first looked over the property we were just looking at it as a bungalow. We noticed some small cracks in the plaster work. We paid a second visit. Mrs Cater and her son were there. There was a crack horizontally across the chimney breast in the lounge but he said if he'd been selling the house he'd have decorated it and then it wouldn't have been seen. The block floor was raised under the radiator in the hall but he said that was probably the heat. There were cracks to be seen in the bedrooms in the positions to be seen in the photographs — A6, A7, A8, A9 and C8. What we saw put us off a bit but we'd paid for the survey and the surveyor would tell us if there was a problem or if it was not all right . . . When we got the report we were quite pleased: there was nothing about any problem inside so we went ahead and bought it. We were no longer bothered about the soundness of the bungalow. We were satisfied about it . . . We had no reason to believe that there was anything unsound structurally about the property.

The plaintiffs got the key of the house upon completion and went that very day to it. They had bought Mrs Cater's carpets as part of the bargain, but the premises were otherwise empty of furniture and furnishings. The second plaintiff's mother went with them. In the lounge she stumbled or tripped on what they thought at first was some rucking up of the underlay, a sort of lump under the carpet. However, when they investigated there was a crack in the floor. "We were astounded by the size of it," said the first plaintiff. "We were also very concerned and went to see the solicitors immediately."

On January 17 1984 the solicitors wrote to the building society in these terms, having identified themselves:

Our clients have discovered major problems with the floors in the property and we believe that this may have been caused by a chemically active material being used under the concrete flooring with the result that it has expanded and cracked the floor and walls. Our clients are in the process of obtaining detailed builder's estimates as to the total cost of repair, but it is clear even at this stage that the cost will be very considerable.

That, as I say, was on January 17 1984 and the letter was acknowledged and then a response was dated January 24. The writer of the letter said:

First of all we would like to mention the limitations of our own valuer's inspection. When your clients completed the application form on September 10 1983 it was emphasised that the purpose of our valuer's report was to enable the Society to decide whether or not to make an advance, and if so, on

> what terms. We went on to stress that our valuer was not asked to carry out a structural survey and that there may well be defects in the property which would only be revealed on a detailed inspection.

He went on to say that when the offer of advance was made these remarks were repeated and advice was given to have an independent inspection. Nevertheless, the matter was referred to the valuer who was identified as the defendant.

Solicitors for the plaintiffs then wrote to the defendant on February 2 making the claim against him and inviting his comments. He replied on February 9 denying negligence, enclosing a copy of his instructions. That letter was acknowledged on February 14. The solicitor said:

> The cracks to the walls and floors of the building are readily apparent and it should have been clear that the property was worth very considerably less than the figure at which it was valued.

That letter was acknowledged and noted by the defendant on February 16. He did not agree.

It was accepted on the plaintiffs' behalf that the defendant cannot be said to have been negligent in failing to take up and look under the carpet. Nevertheless, the fact is that the concrete floor-slab of the bungalow had been laid directly on to compact burnt colliery shale. That material is notorious for containing soluble substances which attack concrete. It is unnecessary to go into the chemistry. Laboratory tests on the shale here indicate its high potential for attacking concrete. The consequence of attack is that the concrete slab expands with a powerful outward push against the walls just below damp-proof course level. Brickwork is displaced and cracks appear until the resistance of the surrounding soil prevents further lateral displacement. The next stage is for the slab to burst upwards. Disturbance of the internal walls occurs as a consequence of horizontal, and perhaps also of vertical, pressures.

The dangers of sulphate attack on concrete have been well known for over 20 years and laying concrete floor-slabs directly on burnt shale is contrary to practice and the standard form of building byelaws. None of this was known to or appreciated by the plaintiffs.

When he was cross-examined, the first plaintiff said that he knew that the building society would require a valuer's report to assess the value of the security and the local manager told him that the report was to make sure that the house was worth £26,500. This plaintiff accepted that para 3 of the notes was drawn to his attention and he knew that the report was not a structural survey but was a report on the property. He knew it was not a "market valuation" but was to enable the society to decide whether or not to make an advance. He said:

> I could have had a structural survey but we weren't recommended to have one. If the valuer had seen what was there to be seen, he should at least have recommended us to have a structural survey. I expected him to pick up what we'd seen and I relied on him to advise on the property. I relied on what he did and we were satisfied from his report that he hadn't picked up any major defects. If he had, £26,500 would not have been a fair price. I suppose I took a risk not having a structural survey but I didn't think, having seen his report, that there was any risk. I relied on him. In particular I relied on his statement that the purchase price was very reasonable, that the recommended insurance cover was £36,000 and that he had specifically drawn attention to minor defects.

in the box to which the attention of the applicants should be drawn.

> Moreover, I did not think that the building society would lend money on a defective house.

The second plaintiff gave evidence broadly complementary to her husband's evidence. Both thought the cracks had got a little worse over the years. She said that if there had been anything seriously wrong she thought the surveyor would pick it up and if he had advised a full survey it would have been done. If he had told us that the cracks were not serious we would just have filled them in and redecorated. "I suppose," she said, "that if we'd been at all concerned about the property we would have had an independent surveyor, but we weren't." The property was (as she believed it) only 18 years old. In fact the defendant said he thought it was 12 years old.

A short statement by Mrs Cater was put in under the Civil Evidence Act. She verified the existence of a crack across the chimney breast in the lounge and that there was a crack in the wall above the bathroom door. What she said was this: "There were some cracks in the walls but they did not seem to get any worse while I was living in the house. The cracks did not worry me because I did not think that it mattered. I cannot remember if I had a survey done when I bought the house."

The plaintiffs had clearly had some professional advice before the solicitors' letter of January 17 1984, but Mr W D Bell FRICS was instructed on April 11 1984 and he gave evidence before me. Thereafter, in October 1985 the defendant, having seen Mr Bell's report reinspected the property. On October 5 1984 and again on April 24 1986 Mr J F W Baines FRICS inspected the property on the defendant's solicitors' instructions and on April 22 1986 Mr C V Knipe BSc C Eng MIME MIMM MIGeol FGS and chartered engineer inspected the property and dug exploratory pits on the instructions of the plaintiffs' solicitors. Both Mr Baines and Mr Knipe, in addition to Mr Bell, gave evidence before me. There are in the bundle reports by Mr R L Filde FSVA and Mr D W Hughes of the Industrial Research Laboratories of the City of Birmingham, but neither of these gentlemen was called to give evidence. The content of the report of Mr Hughes confirming the sulphate levels of the burnt shale and the concrete floor-slab was not in dispute.

Issues of fact arise as to what was to be seen when the defendant inspected the property on September 19 or 20 1983. He spent about half an hour to three quarters of an hour there. The defendant is a Fellow of the Royal Institution of Chartered Surveyors, a chartered valuation surveyor and additionally qualified as an agricultural valuer. He has been in practice for 27 years in Oswestry. About half his practice is in estate agency, the other half in building society and house purchase reports. He is not a qualified structural surveyor and does not do structural reports. He is on the panels of valuers of a number of substantial building societies and banks and in 1983 he did 173 building society valuations. He said:

> Normally building societies are limited to advancing 80% of the value of the property unless there was some additional security. My responsibility is to value the property and I am only addressing my comments to the society so that they can assess what mortgage advance can be offered. I also have to advise on insurance and the practice is to take reconstruction cost assessed at a scale figure of £30 per foot square. I look at the interior first, the accommodation, lay out and general condition, then I look at the exterior. Apart from what is in my general remarks in my report I saw nothing of significance. I saw no cracking inside except some hair line cracks of the sort you see in every property. I saw no cracks in the bedroom; if there had been any I would have seen them. I believe I saw a crack in the kitchen right above the flue to the central heating boiler but I took no particular notice of it. It might have been a reflection from the boiler. There was nothing in the kitchen that I should have seen and reported on but didn't. There was nothing wrong with the skirting board and I wouldn't take notice of the doors unless there was something very obvious. Externally there was no bulging of the brick work and no significant movement of the bricks. There were some minimal cracks in the outside walls but nothing to affect my opinion of the property. Sometimes you get mortar which is stronger than the bricks so you get cracked bricks rather than cracked mortar.

The defendant agreed that he knew now of sulphate action but he had never come across it in his career. He had inspected one or two other houses in the same estate as this house and sold one of them in 1984 for £28,000. He agreed that across the board only some 10% of purchasers have a structural survey: 50% of those not seeking mortgage facilities do and about 20% of those seeking only a half advance. Such cracks as he saw, he said, could have been filled in on redecoration. He did not regard a crack as significant unless it was wedge-shaped, due either to foundation failure or to pressure from above.

When he was cross-examined the defendant agreed that he kept himself up to date by reading his professional journals. He knew of *Yianni*'s case (*Yianni* v *Edwin Evans & Sons* [1982] QB 438) and was aware of it in 1983. He knew that that case was of great importance to his work.

> I knew one copy of my report was for the plaintiffs, but it didn't cross my mind to consider why they got a copy. But I know from my reading that I had a duty of care in law to the applicants. In my opinion this property was purchased at a reasonable price but I would know that there was only a one in four chance that the plaintiffs would get a structural survey. I was only inspecting and reporting for valuation purposes and I was expected to look for signs of settlement. I saw no sign of lipping which I agree might point to trouble, depending on the degree, but not here as this was a cavity wall. I thought there was some bad construction.

He looked at photograph C11 showing a brick cracked through vertically. He said:

> If I'd seen that I should not have noted it as significant. The crack in the bedroom (photograph A7) was not visible to me; it may have been behind the curtains or hidden by the bed.

This, I remind myself, was one of the cracks observed by both plaintiffs on their first visit to the house; so too, they said, the crack in A8 which the defendant said he saw on his second visit in October

A 1985. As to the crack in A6 which the plaintiffs had seen, the defendant said:

> If I'd seen that I'd have recommended a mortgage and I would not have recommended a structural survey. It is a 1mm crack which could be due to plaster drying out or in the skim finish to a plaster board backing, but you'd expect such a crack to follow the edges of the board. It is not more likely to be a settlement crack. I'd still pass it.

The defendant's attention was drawn to photograph A4 showing the crack behind the boiler flue. He said:

> I didn't look in the lounge on the other side of that wall. There is a crack on the other side but it is not opposite the crack in the photo. I should have looked on the other side. I walked through the lounge and I saw nothing. That crack is not connected with damage to the walls.

B Both plaintiffs said that they had seen the crack in the lounge and there is beyond question a crack there. Both Mr Baines and Mr Bell agree that that is so, although there is some difference between them as to its precise position and significance.

Mr Baines has been in practice doing building society valuations for nearly 40 years. He does about 300 a year. He also does structural surveys. The building society's valuer's inspection is cursory, he said, compared with a structural survey. He regarded the defendant's report as fair for the property, "as I saw it", he said, based on making a reasonable job of his inspection. He said:

> On my first inspection I saw signs of movement in the external brickwork over eight courses below the cill of the front window. There were hair cracks below C the lounge window and in the soldier arch but no cracks below the damp proof course. There was some lipping of the bricks below the damp proof course to a maximum extent of 3/8ths of an inch displacement. None of this was significant, nothing indicated settlement. They were minor matters; the building was not structurally unsound and there was no indication of collapse.

Internally, Mr Baines found a slight vertical crack below the point where the flue pipe entered the brickwork to which he attached no significance. It could have been caused by heat or by some work near to it. He did not agree that the crack went right through the chimney breast. The crack in the lounge, he said, does not correspond directly with the crack in the kitchen and there was no lateral movement of the chimney breast. He agreed that the crack in the bedroom (on D photographs A7 and C8) was not a hair crack but an expansion crack. Even if the valuer had seen the crack as in C8 he could have passed the property without being in dereliction of duty at all, but he accepted that the crack in C8 was caused by some nominal movement in that corner of the house. In C8 the crack seems considerably larger than in A7. That suggests continuous movement, he said. "Knowing what I know now, that doesn't surprise me," he said. Externally, he agreed that the crack in the brick in photographs B11, 12 and 13 would admit the blade of a penknife, though he still called it a hair crack. So far as the floor in the lounge was concerned, there was a ruptured area some 2 to 3 feet square, raised some 3/8th of an inch with powdered concrete below, then a series of 1/8th-inch-wide serious cracks E radiated from that. It was not the worst case of sulphate damage which he had seen.

Mr Baines accepted that there was a duty of care on the valuer and that if the valuer were negligent the prospective purchaser might suffer loss. In his view, no professional man would disclaim liability for negligent advice.

Mr Bell has been in practice as a chartered surveyor in the area since 1970. He prepares valuations for building societies as part of his general practice. He said that "settlement" included lateral structural movement and not only downward movement of a building. This was with reference to question 9 on the valuer's report: "In your opinion is the property affected by settlement?" His definition of settlement was not challenged. Here, said Mr Bell, he F found that the brickwork in the vicinity of the damp-proof course had been pushed laterally outwards. He said:

> At the time of my inspection this defect was readily apparent; it merited further attention. If I had been the building society valuer here I would have drawn attention to it myself or advised the building society either to decline the proposal or to recommend further reports. If the cracks which I found here were taken singly each on its own would not be serious but if, as here, there was a lot of cracks, as a surveyor I'd wonder why; so too with the distinct hump in the hall in the vicinity of the radiator. On visual examination alone I attributed that to hard core expansion. In the lounge I found severe cracking to the floor structure. As to the crack in the kitchen (photograph A4) if I'd been the building society valuer I'd have noted it. I'd have said, that looks interesting, is there a crack on the other side? and I'd go and look either immediately or later in my inspection. That isn't a heating crack or a drying out crack, it's bigger than that and it's gone through the wall paper. I found severe cracking on the chimney breast in the lounge. If that had been all I found I might have passed the house but only with a qualification, perhaps to answer question 9 "Yes" or to say something in the general remarks, perhaps suggesting that there might be need for a structural report, but by that stage I would have become very suspicious.

He went on to describe finding the cracking in photographs A7 and C8 which is, he said, the inside of the cracked brick in photograph C11. "Finding these would have reinforced my view," he said, "that there had been some settlement. In the valuer's shoes I would have told the building society that there was a problem or that they should decline the proposal. The property should not have been passed," were his words. He referred to additional factual information in his report which I need not set out in detail, and to his supplementary conclusions at p 15, para 18. Mr Bell would not have valued the bungalow at more than £9,000. There was no formal challenge to that figure, as the parties were agreed that I was to try liability only, but it was not in dispute that the bungalow was worth significantly less than £26,500 and it was not argued that £26,500 was a reasonable price to pay for it.

Cross-examined about his experience Mr Bell said that he did some 500 building society valuations and some 200 structural surveys a year and had done so since 1970 and before that he had acted as assistant on similar work to his late principal. He accepted that only in some 10% of cases did prospective purchasers get independent surveys; the remainder relied on the building society's valuation. He did not consider that the condition of the property between September 1983 (the defendant's inspection) and April 1984 when he inspected would have altered significantly. A building society valuer should look behind and over and under furniture; even a hair-line crack may be suspicious. He said:

> I don't agree that it was reasonable for the defendant to report as he did. These defects were such as would be revealed by a building society valuation. They were evident and not concealed and in my experience applicants normally rely on the valuer's report. The practice is to accept it. If he used the words "very reasonable" the applicant would think he was getting a bargain.

Mr Knipe, who did not inspect the bungalow until June 1986, broadly confirmed Mr Bell's findings of fact and described certain additional features, in particular cracking in the mortar joints below the lounge window and around the window head, with displacement of the brickwork below the damp-proof course there and on the side wall, and stepping on the damp-proof course along the rear half of the left side wall. I need not repeat the whole of the details in his report, particularly as he is dealing with a time about three years after the defendant's inspection. Mr Knipe was in the main confirming the sulphate attack and advising on remedial measures.

From his experience, Mr Knipe said the damage from sulphate attack in unprotected circumstances, as here, normally becomes visible in 5 to 10 years and it was so improbable as to be untenable that it was not there to be seen in September 1983 when the defendant made his inspection. The problems caused by colliery shale were well known before 1966 and you would expect that a surveyor familiar with his area would know that it was recognised as a suspect material. He identified the crack in the lounge chimney breast area as a "matching crack" to the crack in the kitchen. It indicated movement of fairly severe magnitude and it would indicate structural damage to any reasonable surveyor. A single crack might be passed but not a series of cracks, particularly in the brickwork — a reasonable surveyor could not disregard them. You would expect some note on the surveyor's report even if he did not identify sulphate attack; at least he should suggest a further report, but he should not pass it.

In cross-examination Mr Knipe said that there was clear movement of the brickwork below the damp-proof course; a valuer in half an hour should note the outward displacement of brickwork and would be alerted to something. The defendant indicated in his report that he had looked at brickwork but does not mention this displacement. "What was to be seen in 1983," he said, "applying my experience, was not consistent with minor shrinkage cracking."

Having had the opportunity of reviewing and reconsidering the whole of the evidence, I have concluded that I prefer and accept the evidence of Mr Knipe and Mr Bell and of the plaintiffs themselves to that of the defendant and Mr Baines. It follows that I find that there were in fact on September 19-20 1983 when the defendant inspected the premises cracks in the walls and brickwork and displacement of the brickwork visible to the naked eye of a size and to an extent that, at the very least, a reasonable surveyor carrying out a building society

A valuation would and should have seen them and recognised that, at the least, they indicated the real possibility of settlement or other structural damage: that, in Mr Bell's words, there was a problem. In those circumstances to "pass" the house for mortgage purposes or to describe the price payable as "very reasonable" was, on the assumption that the defendant owed a duty of care to the plaintiffs, unquestionably negligent. Mr Little, in his careful and comprehensive submissions on the defendant's behalf, while inviting me to consider, as I have done, the possibility that the plaintiffs' expert witnesses might have been applying a higher test than was required of the defendant as a building society valuer, did not dispute on the facts which I have found them to be (rejecting as I do Mr Little's submissions upon the facts which I have rejected) negligence B would be established.

The main question here, negligence being established, as Mr Little contended, was whether or not on the facts the defendant owed any duty of care in law to the plaintiffs.

That a duty of care can exist in similar circumstances is illustrated by the decision of Park J in *Yianni* v *Edwin Evans & Sons* [1982] QB 438. The question there in broad terms, as the learned judge described it, was whether surveyors who in a valuation report on a dwelling-house for a building society negligently misrepresented its value are liable to purchasers who in reliance upon the statement as to its value purchase the house and in consequence suffer damage. The defendants, the surveyors there, admitted negligence but contended C that they owed no duty of care to the plaintiff. Park J, relying on *Hedley Byrne & Co Ltd* v *Heller & Partners Ltd* [1964] AC 465 and *Anns* v *Merton London Borough Council* [1978] AC 728 (particularly the latter), concluded that the duty of care would arise if on the evidence he was satisfied, as he was, that the defendants knew that their valuation of the house in so far as it stated that the property provided adequate security for an advance of £12,000 would be passed on to the plaintiffs who, notwithstanding the society's literature and the service of the notice under section 30 of the Building Societies Act 1962, in the defendant's reasonable contemplation would place reliance on its correctness in making their decision to buy the house and mortgage it to the building society. He applied and adopted the words of Lord Wilberforce in *Anns* at p 751H:

D There was a sufficient relationship of proximity such that, in the reasonable contemplation of the defendants, carelessness on their part might be likely to cause damage to the plaintiffs.

Park J went on to consider what has been called "the second stage" of Lord Wilberforce's formulation of principle: whether or not there were considerations which ought to negative or to reduce or limit the scope of the duty of the class of persons to whom it is owed or the damages to which a breach of it may give rise and determined that there were not. I note at once that there was no suggestion in *Yianni*'s case of anything of the nature of a disclaimer of liability except to the extent that the building society's notice under section 30 of the 1962 Act excluded any warranty that the purchase price was reasonable. It E is to be remembered that in the *Hedley Byrne* case in fact there was an effective disclaimer of responsibility.

The speech of Lord Wilberforce in *Anns* has been subjected to more recent analysis: see eg Lord Keith of Kinkel in *Governors of the Peabody Donation Fund* v *Sir Lindsay Parkinson & Co Ltd* [1985] AC 210 at p 240:

There has been a tendency in some recent cases to treat these passages as being themselves of a definitive character. This is a temptation which should be resisted. The true question in each case is whether the particular defendant owed to the particular plaintiff a duty of care having the scope which is contended for, and whether he was in breach of that duty with consequent loss to the plaintiff. A relationship of proximity in Lord Atkin's sense must exist before any duty of care can arise but the scope of the duty must depend on all F the circumstances of the case.

Most recently, in *Yuen Kun Yeu* v *Attorney-General of Hong Kong* [1987] 3 WLR 776, Lord Keith of Kinkel, giving the judgment of the Judicial Committee of the Privy Council, reasserted that foreseeability does not of itself and automatically lead to a duty of care: "Foreseeability of harm is a necessary ingredient of such a relationship" ie one apt to give rise to a duty of care "but not the only one" — at p 783D.

In the same context, reference should be made to the speech of Lord Bridge of Harwich in *Curran* v *Northern Ireland Co-Ownership Housing Association Ltd* [1987] 2 WLR 1043 at p 1049 and to the speech of Lord Brandon in *Leigh & Sillavan Ltd* v *Aliakmon Shipping Co Ltd* [1986] AC 785 at p 815.

Some criticism of the decision in *Yianni*, based upon these most G recently cited cases, is to be detected in the decision of the Court of Appeal in *Harris* v *Wyre Forest District Council*, reported in the *New Law Journal* for January 22 1988 at p 15 and in *The Times* newspaper*. I have had sight of a transcript of the judgment. I should interpolate that since this judgment was written the case has been reported in [1988] 1 All ER 691. In that case the court, although minded to do so, was not prepared to overrule *Yianni,* which, as Nourse LJ put it, had been followed at first instance in *Westlake* v *Bracknell District Council* (1987) 282 EG 868 and in Ireland by Costello J in *Ward* v *McMaster* [1985] IR 29, had been treated as correctly decided but distinguishable on the facts in the Northern Irish Court of Appeal by the late Gibson LJ in *Curran* v *Northern Ireland Co-Ownership Housing Association Ltd* and had been H accepted as good law (though not mentioned by name) by and before the Court of Appeal in *Smith* v *Eric S Bush* [1987] 3 All ER 179†. The point does not seem to have been argued in the House of Lords in *Curran* [1987] 2 WLR 1043.

Mr Little recognised that the defendant here knew that his report would be disclosed to the prospective purchaser, who was an identifiable person. Since *Yianni* the evidence was that the practice has grown up, and was followed in this case, that building society valuation surveyors provide three copies of the report, one being expressly for the prospective purchaser, who actually pays for it and is known to pay for it. Nevertheless, the valuer is not contractually bound to the prospective purchaser; nor indeed, save for providing J the report, does he have any contact with the prospective purchaser. The relationship is not one of choice, he said; it results solely from the obligation imposed on every director of a building society to satisfy himself that satisfactory arrangements are made for assessing the value of the security on which the society is to lend money: section 25 of the Building Societies Act 1962. I cannot accept this: the surveyor is not obliged to accept instructions, but if he does he accepts the consequential responsibilities, whatever they may be. Mr Little did not contend that the mere existence of the statutory duty excluded liability: indeed, section 30 makes it clear that liability on the part of the society is not excluded unless the statutory notice is given.

There is no doubt in my mind but that a sufficiently proximate K relationship existed here in so far as the defendant well knew that his report would be made available to the plaintiffs. He said: He knew that his report would be seen by the prospective purchasers. He knew that he had a duty of care towards the applicants. He knew that there was only a 1 in 4 chance that the plaintiffs would get a structural survey. He didn't go to the property to check on its condition only on its value. He was expected to look for signs of settlement. He saw no signs of lipping of the brickwork: if there was lipping and depending on the degree it would point to trouble. He thought it was badly constructed. I repeat that passage from his evidence which I have already included in the recital of the facts.

Mr Baines told me that he thought the defendant made a reasonable job of his inspection and that his report was fair to the L property. A valuation survey is, and again I repeat, cursory compared with a structural survey. In cross-examination he agreed that a valuer in these circumstances has a duty of care to an applicant for a mortgage and that if the valuer is negligent the applicant may suffer loss.

The thrust of the defendant's case, in addition to the dispute as to the condition of the house with which I have dealt, was that, despite what I have just recorded, no duty of care arose in the circumstances. In Mr Little's words: "The circumstances in which the valuation report came to be made (and, I add, to be provided to the plaintiffs) were such taken as a whole as to negative the coming into existence of a duty of care." M

My attention was invited to two decisions of the Court of Appeal, one or both of which are, I understand, to be considered in the House of Lords.

In *Smith* v *Eric S Bush* [1987] 3 WLR 889† the plaintiff applied to a building society for a mortgage to assist her in buying a house. Certain aspects of the facts there, taken from the judgment of Dillon LJ, are material here, and I read from p 892E:

It has always been the case that applicants for mortgages are required by the building society to pay the fees for the valuation reports which the societies are by the statute required to get. It used for many years to be the practice that the

*Editor's note: Reported at p 132 *ante*.

†Editor's note: Also reported at [1987] 1 EGLR 157; (1987) 282 EG 326.

A building societies refused to disclose the reports to the applicants, on the footing that they were only obtained for the building societies' own purposes. But this practice gave rise to widespread discontent in that applicants could not see why they should not have copies of reports for which they had paid. Before October 1980, therefore, the building societies had adopted instead a new practice of making copies of the reports available to the applicants for mortgages, but with extensive disclaimers of liability.

In accordance with this new practice, there was included in a somewhat lengthy set of declarations just above the place where the plaintiff signed the mortgage application form, the following:

> "I/We have read the notes for the guidance of mortgage applicants and accept that if the loan is approved the mortgage guarantee policy premium will be added to the loan. I/We accept that the society will provide me/us with a copy of the report and mortgage valuation which the society will
> B obtain in relation to this application. I/We understand that the society is not the agent of the surveyor or firm of surveyors and that I am making no agreement with the surveyor or firm of surveyors. I/We understand that neither the society nor the surveyor or the firm of surveyors will warrant, represent or give any assurance to me/us that the statements, conclusions and opinions expressed or implied in the report and mortgage valuation will be accurate or valid and the surveyor(s) report will be supplied without any acceptance of responsibility on their part to me/us."

Just below the plaintiff's signature it is recorded that she had paid the inspection fee which was £36.89. There is a somewhat similarly worded disclaimer of liability in printed "Notes for the Guidance of Mortgage Applicants" which it was the practice of Abbey National to supply to all mortgage applicants, but I need not refer to those notes because the disclaimer there adds nothing to what appears in the mortgage application form, as
C above set out, and in the copy of the defendants' report which was supplied to the plaintiff as mentioned below.

The terms of what was described in that case as "the disclaimer" were as follows:

TO THE MORTGAGE APPLICANT(S): IMPORTANT

1. THIS DOCUMENT IS NOT A MARKET VALUATION, IT IS NOT AND SHOULD NOT BE TAKEN AS A STRUCTURAL SURVEY. It has been obtained by the society from a qualified surveyor or firm of surveyors to comply with section 25 of the Building Societies Act 1962.
2. If you are purchasing the property, you will receive a notice that the society does not warrant that the purchase price is reasonable.
3. This is a report to the society by its surveyor(s) and neither the society nor the surveyor(s) give any warranty, representation or assurance to you that the
D statements, conclusions and opinions expressed or implied in the documents are accurate or valid.
4. The surveyor(s) has/have made this report without any acceptance of responsibility on his/their part to you.

Dillon LJ recorded that the plaintiff admitted that she had read that and the word "disclaimer" was used, as I read that report, in the context of the notice of appeal. Again reading from the report, the matter came before the Court of Appeal as follows:

> By a notice of appeal dated May 14 1986 the defendants appealed on the grounds, *inter alia*, that (1) the recorder had erred in law and/or in fact in appearing to hold that the disclaimers contained in the mortgage application form and/or the valuation report were not to be construed as sufficient to exclude the defendants' liability to the plaintiff for negligence; (2) the
> E recorder had erred in law in appearing to hold that the disclaimers were subject to an implied condition that there should have been a complete visual inspection of all parts in the fabric, and in holding that the defendants had been in breach of that condition, and had thereby rendered the disclaimers ineffective in excluding the defendants' liability in negligence; (3) the recorder had erred in law in appearing to hold that the disclaimers did not prevent a duty of care to the plaintiff arising; (4) the recorder had erred in law in failing to find that the provisions of the Unfair Contract Terms Act 1977 were inapplicable to the disclaimers; and (5) the recorder had erred in law and/or in fact in appearing to find that it was not fair and reasonable for the defendants to rely on the disclaimers, and in particular had failed to give sufficient weight to certain evidence.

The conclusions of Dillon LJ on this aspect of the case are set out at p 895C and following:

> F Reference was made in a general way in the course of argument to the well known line of authorities, stemming from *Alderslade* v *Hendon Laundry Ltd* [1945] KB 189, in which the courts have considered the principles which are applicable to clauses which purport to exempt one party to a contract from liability to the other party for the consequences of the first party's own negligence. Assuming, however, that those are the principles which ought to be applied to the construction of these disclaimers, I find it impossible to conclude that the disclaimers do not cover tortious liability of the defendants for negligence. The final words in the notice on the copy of the report supplied to the plaintiff, "The surveyor(s) has/have made this report without any acceptance of responsibility on his/their part to you", and the corresponding words at the end of the declaration in the form of mortgage application are really meaningless if they do not extend to responsibility or liability in negligence. If authority is required, it is to be found in the speeches of Lord Upjohn and Lord Pearson in *Spalding* v *Tarmac Civil Engineering Ltd* [1967] 1 WLR 1508, 1526, 1529.
>
> At common law, as is recognised in *Hedley Byrne & Co Ltd* v *Heller & Partners Ltd* [1964] AC 465 where there is no contract the assumption of a duty of care can by appropriate words be avoided, eg where a reference is given "without responsibility", and where a report is supplied subject to stipulations disclaiming responsibility the recipient cannot accept the report and ignore the stipulations: see *per* Lord Reid at p 492, and *per* Lord Morris of Borth-y-Gest at p 504. Apart, therefore, from the Act of 1977, the disclaimers relied on would, in my judgment, have provided an effective defence for the defendants against the plaintiff's claim.
>
> I turn, therefore, to consider issue (2), the effect of the Unfair Contract Terms Act 1977. Section 2 of the Act provides:
>
> "(1) A person cannot by reference to any contract term or to a notice given to persons generally or to particular persons exclude or restrict his liability for death or personal injury resulting from negligence. (2) In the case of other loss or damage, a person cannot so exclude or restrict his liability for negligence except in so far as the term or notice satisfies the requirement of reasonableness . . ."
>
> Section 11(3) provides:
>
> "In relation to a notice (not being a notice having contractual effect) the requirement of reasonableness under this Act is that it should be fair and reasonable to allow reliance on it, having regard to all the circumstances obtaining when the liability arose or (but for the notice) would have arisen."
>
> Under section 11(5) the onus is on those claiming that a contract term or notice satisfies the requirement of reasonableness to show that it does.

He continued:

> Whether in any particular case it is fair and reasonable in all the circumstances to allow reliance on a notice excluding liability for negligence must, of course, depend on the circumstances of the particular case. In that sense no case is a precedent for any other case. But the circumstances of the present case are very ordinary, and we are told that there were many other cases waiting the outcome of this appeal. That is not surprising considering the number of building society surveys that are carried out. It appears that it is the practice of very many, if not all, building societies to include as a matter of course in their printed forms disclaimers, to protect the society itself and its surveyors from all liability, in substantially the same terms as the disclaimers in the present case.
>
> It is common ground that it was not part of [the surveyor's] duty to carry out a full structural survey of the house, and this was understood by the plaintiff. She knew that she was taking a chance on there being no hidden defects in the house which would not have been apparent on a reasonably careful visual inspection of the house. To hold that the surveyor in such a case as the present cannot rely on the disclaimers does not involve extending the surveyor's obligations so as to make it incumbent on him to detect such hidden defects. His obligation is still merely to carry out a reasonably careful visual inspection.
>
> But where he is dealing with a property at the lower end of the market and he knows that the purchaser is likely to rely on his report, and not instruct his or her own surveyor, I find it very difficult to see why it should be fair and reasonable to allow him to rely on an automatic blanket exclusion of all liability for negligence if his visual inspection of a property turns out not to have been reasonably careful.

Glidewell LJ summarised as the two issues: (1) On their face, did the words of the disclaimer suffice to prevent any duty of care being owed by the defendants to the plaintiff; (2) If so, did the disclaimer satisfy the requirement of reasonableness under sections 2(2) and 11(3) of the 1977 Act?

At p 898E he said:

> In my view, the wording of the disclaimer set out in the notice on the defendants' report and valuation did suffice to prevent the defendants owing any duty of care to the plaintiff. In other words, it was so worded as to relieve them of all liability in negligence.

He then went on to consider whether the notice satisfied the condition of reasonableness and held that it did not: accordingly it was ineffective as a disclaimer.

Sir Edward Eveleigh regarded the words in red on the notice as not merely a disclaimer of liability but also a warning that the report is of limited value as a survey report. He continued, at p 899H:

> If the plaintiff had relied on the report to an extent which would be unreasonable, bearing in mind the warning, there would be no liability on the defendants and no need to consider the efficacy of a notice excluding liability.

Here there seems to be an echo of Lord Wilberforce's "second stage".

At p 900 Sir Edward said this:

> The report did not reveal what a valuation report would have revealed if properly prepared, albeit for its limited purpose. On the contrary, it asserted that the property was readily saleable for the purpose of owner occupation. The plaintiff relied on it in this respect but the report did not merit even that degree of confidence. The warning, in my opinion, was not such as to make it unreasonable for the plaintiff to rely on the report as she did. Therefore, the

defendants will be liable to the plaintiff unless the notice is effective to exclude liability.

In this court counsel has not sought to argue that the defendants would not have been liable in the absence of a notice of disclaimer. None the less, I have analysed the position because I think it is important to bear in mind the basis of the defendants' liability when considering whether or not it would be fair and reasonable to allow reliance on the notice, having regard to all the circumstances obtaining when the liability arose.

A relevant part of the circumstances obtaining when liability arose was the fact that the report stated that the house was readily saleable for the purpose of owner occupation and was likely to remain so. There was no warning that this assertion was unreliable. It would have been perfectly simple to discover the dangerous state of the premises. No structural survey was needed for that. The least attention to the task that the plaintiff was entitled to expect from the defendants would have revealed the fault. The defendants' fault can hardly be regarded as a mere accidental error or omission.

I therefore am of the opinion that it would not be fair and reasonable to allow reliance on a disclaimer which might, I know not, be effective in other circumstances.

It has been necessary to set out extracts from those judgments at some length because of the observations upon the case in the Court of Appeal in *Harris* v *Wyre Forest District Council*.

Before turning to that case, however, I must first refer to the judgment of the late Mr John Wilmers QC, sitting as a deputy judge of this court, in *Stevenson* v *Nationwide Building Society* (1984) 272 EG 663. In that case, the prospective purchaser negotiated a loan on the basis of an application form which he signed, which included a clause stating, *inter alia*, that the inspection carried out by the society's valuer (who was in fact a staff valuer and an employee of the society) was not a structural survey and might not reveal defects which such a survey would reveal, inviting consideration of a structural survey, and concluding:

I/We understand that the report and valuation on the property by the society's valuer is confidential and is intended solely for the consideration of the society in determining what advance (if any) may be made on the security and that no responsibility is implied or accepted by the Society or its valuer for either the value or condition of the property by reason of such inspection and report.

The deputy judge decided that although the valuer was called on to do a building society valuation and not a structural survey he had not, on the facts, exercised reasonable skill or care and that if he were liable the society would be vicariously liable for his negligence. The question which remained, and leading counsel for the plaintiff expressly accepted it, was whether the words which I have set out constituted a valid disclaimer. If so, "unless something else operated to displace the effect of those words", said the deputy judge, "the defendants are entitled to rely on them". Counsel for the defendants invited attention expressly to the speeches in *Hedley Byrne* and the deputy judge accepted as apposite and adopted the words of Lord Morris of Borth-y-Gest at p 504:

The society in the present case by the words which they employed effectively disclaimed any assumption of a duty of care. They stated that they only supplied the report on the basis that it was without responsibility. If the applicant chose to receive it and act upon it he cannot disregard the definite terms upon which it was given.

The deputy judge found nothing to displace the effect of those words and so went on to consider whether or not section 2(2) of the 1977 Act applied. Counsel for the defendant contended that the disclaimer was a warning to the plaintiff that he should not rely on the words used in the report; a statement dissuading the recipient of a report from relying on it negatived any duty of care and so excluded or restricted liability. On that basis, counsel appears to have argued that section 2(2)(iii) would catch the disclaimer so that the effect would be the same. The deputy judge then turned to section 11(3) and concluded that the disclaimer passed the test of reasonableness. The plaintiff accordingly for that reason alone failed in his claim.

Some account of the facts in *Harris* is necessary. The plaintiffs were looking for a house at under £10,000. They sought the assistance of the defendant council, which operated a policy of making loans on older residential properties on the security of which banks and building societies were in general unwilling to lend. They provided standard mortgage application forms, one of which the plaintiffs obtained from the vendor's selling agent. The form which they signed read, so far as material:

. . . the valuation is confidential and is intended solely for the information of Wyre Forest District Council in determining what advance, if any, may be made on the security and that no responsibility whatsoever is implied or accepted by the Council for the value or condition of the property by reason of such inspection and report. (You are advised for your own protection to instruct your own Surveyor/Architect to inspect the property) I/We agree that the Valuation Report is the property of the Council and that I/We cannot require its production. G

Certain factual differences between *Harris* and the instant case are to be noted:

1 The claim there was brought against the lender as well as against the surveyor.
2 The surveyor was an employee of the lender.
3 The loan was (presumably) sought and made under some statutory power of the local authority, and
4 The surveyor's report was expressed to be and accepted as confidential to his employers.

Nourse LJ, giving the leading judgment, posed the question whether a duty was owed. He referred to the speech of Lord Pearce in *Hedley Byrne* at p 539 and went on to say: H

I would therefore frame the essential question which we have to decide in these terms: were the circumstances such that the defendants ought reasonably to have recognised both the importance which would be attached to the valuation by the plaintiffs and their own answerability to them in making it? By now it will be obvious that that question cannot be answered without considering the terms and effect of the disclaimer contained in the mortgage application form.

For reasons which I have already set out, I am satisfied that in the present case the defendant knew or, at best for him, would, if he had applied himself to the matter (assuming he did not), have known that the plaintiff would attach importance to his report in respect both of his valuation of the house and his statements as to its condition (including in this omissions of defects which in the context of the report should not have been omitted). Equally there can be no question but that he as a professional man would recognise his answerability for his report save only to the extent that he can shelter behind the disclaimer. J

Three points must be made clear.

1 This defendant was not the servant of the building society.
2 This defendant personally gave no disclaimer, nor was he, as I have said, in a contractual relationship with the plaintiff.
3 Mr Baines said in evidence, as I have already recorded, that no professional man would disclaim liability for professional advice. K One can see the force of that, for what is the advice worth if it is in effect said to be worthless.

The Court of Appeal in the *Harris* case concluded that the existence of the disclaimer prevented the creation of a duty of care. Nourse LJ put it like this:

For these reasons I am of the opinion that the defendants never came under a duty of care to the plaintiffs. While others might frame the essential question in different terms, I do not see that they could answer it in a different sense. Moreover, the disclaimer was as effective in the case of Mr Lee as an individual defendant as it was in the case of the council themselves. He at all times acted as a servant of the council, who could only reasonably have been expected to act through their servants. The references to the council therefore included Mr Lee as an individual. L

That Kerr LJ came to a similar conclusion, based on the "commercial purposes" of the lender (the surveyor being a servant and not an independent professional adviser) appears from this passage from his judgment:

I agree with Nourse LJ that in view of the superstructure of decided cases and changes in practice which have followed *Yianni* it would not be appropriate for this court to consider, on the very different facts of this case, whether that decision can stand. But its inherent jurisprudential weakness in any ordinary situation is clear. Suppose that A approaches B with a request for a loan to be secured on a property or chattel — such as a painting — which A is proposing to acquire. A knows that for the purpose of considering whether or not to make the requested loan, and of its amount, B is bound to make some assessment of the value of the security which is offered, possibly on the basis of some expert inspection and formal valuation. Then assume that B knows M that in all probability A will not have had any independent advice or valuation and is also unlikely to commission anything of the kind as a check on B's valuation. B also knows, of course, that any figure which he may then put forward to A by way of a proposed loan on the basis of the offered security will necessarily be seen to reflect B's estimate of the minimum value of the offered security. Suppose that A then accepts B's offer and acquires the property or chattel with the assistance of B's loan and in reliance — at least in part — on B's willingness to advance the amount of the loan as an indication of the value of the property or chattel. Given those facts and no more, I do not think that B can properly be regarded as having assumed, or as being subjected to, any duty of care towards A in his valuation of the security. Even in the absence of any disclaimer of responsibility I do not think that the principles stated in *Hedley Byrne & Co Ltd* v *Heller & Partners Ltd* [1964] AC 465 support the contrary conclusion. B has not been asked for advice or information but merely for a loan. His valuation was carried out for his own commercial purposes. If it was done carelessly, with the result that the valuation and loan were excessive, I do not think that A can have any ground for complaint. And if B made a small service charge for investigating A's request for a loan, I doubt whether the position would be different; certainly not if he were also to add a disclaimer of responsibility and a warning that A should carry out his own valuation.

It may be, but I agree that we should not decide this general question on the present appeal, that the particular circumstances of purchasers of houses with the assistance of loans from building societies or local authorities are capable of leading to a different analysis and conclusion. But be that as it may. A crucial distinction between *Yianni* and the present case is that in *Yianni* there were no express disclaimers of responsibility and warning to potential purchasers to carry out their own inspection as there were here. In my view such disclaimers and warnings must be of the utmost relevance in answering the primary question whether or not a duty of care can properly be held to exist. Or, more accurately in contexts such as the present, whether or not such a duty had been assumed.

The learned lord justice then went on to deal with the disclaimer which he referred to as the exemption clause. He set out the terms of the disclaimer in that case to which I have already referred and continued as follows:

The judge expressed the view that the terms of this provision were not effective to afford any protection to either of the defendants in negligence, and that it did not empower the council to make any loan which, but for this disclaimer, would be statutorily prohibited. The latter is obviously right, and since *Smith* v *Eric S Bush* [1987] 3 WLR 889 will evidently be considered by the House of Lords in the near future I express no view on the first part of this statement. What matters, and where I part company with the judge's reasoning in any event, is in the following sentence:

"In those circumstances I propose to ignore the exemption clause when considering whether or not liability attaches to either defendant."

With respect, I think that this involves a confusion of thought. In order to decide the primary question whether or not the defendant owes any duty of care to the plaintiff it is not relevant to determine the precise legal effect of the terms of any disclaimer or other warning, on their true construction, let alone by the application of the Unfair Contract Terms Act 1977. Consideration of the legal effect of any such provisions only arise if and when the existence of a duty of care has been established. However, in determining whether or not the circumstances warrant the inference that any such duty of care was owed by the defendant to the plaintiff, any disclaimer of responsibility and warning addressed by the defendant to the plaintiff may be of the greatest importance. The correctness of this proposition is borne out by all the citations from *Hedley Byrne* which Caulfield J has set out in his judgment and which I therefore do not repeat.

Of course, disclaimers of responsibility and warnings will be ineffective to the point of irrelevance in situations where the existence of a duty of care is not open to doubt.

Then Kerr LJ gave what he described as an "absurd illustration" — the driver of a motor car displaying a sign on the top of his car — "No liability accepted for any collision — keep out of the way."

Kerr LJ continued:

But in regard to statements, ie words as opposed to deeds, the position is inevitably more complex. In such cases the entirety of the statements made must be considered in order to determine whether or not the maker should be held to have assumed, or for some reason to be subjected to, a duty of care to the person addressed. A disclaimer of responsibility for the accuracy of the statement, or a warning that its reliability should be checked independently, must then be of the greatest relevance.

In *Smith* v *Eric S Bush* (*supra*) this aspect appears to have been disregarded in the defendant's submissions, whatever may have been the reason. The existence of a duty of care was conceded or assumed, and the court was merely asked to decide the effect of the Unfair Contract Terms Act 1977. In the present case, on the other hand, the position was different, but Schiemann J nevertheless disregarded the disclaimer and advice altogether, as shown by the passage quoted above. In my view, this was an erroneous approach to the question whether or not the defendants were under a duty of care to the plaintiffs. For that purpose the council's disclaimer of responsibility and advice to have an independent inspection must be considered together with the council's statement that it was willing to advance £8,505 on the property subject to a number of essential repairs. When these statements are considered as a whole, as they should be, then it is in my view not open to conclude that the council assumed any responsibility for the reliability of the valuation, nor that it can for some other reason be held to have been subject to a duty of care to the plaintiffs.

The contemplation of an intervening inspection qualifies the principle of *Donoghue* v *Stevenson* [1932] AC 562 as Mr Ashworth QC [counsel for the appellant defendants] rightly reminded us. And I would not accept that the charge of £22 paid by the plaintiffs can by itself create a duty of care in the face of the disclaimer and warning.

Kerr LJ went on to conclude that the Unfair Contract Terms Act 1977 had no relevance in that case. That issue never arose, since it arises only if the existence of a duty of care and a breach of it have first been established.

Caulfield J agreed and added this reason:

In my judgment it is implicit from these quotations

that is to say the quotations from *Hedley Byrne*,

that a potential tortfeasor can avoid the incidence of a duty of care and in my judgment, as my Lord has explained, no duty was assumed by the first defendant. So far I am only introducing my reason for concluding here that the second defendant would not on these facts have assumed a duty because I think it is implicit in *Hedley Byrne* that the speaker or writer of the statement should have the opportunity to make clear whether he is assuming any duty, as the first defendant did in this case. The second defendant by virtue of his position as a servant of the first defendant is not given any opportunity at all to decide whether or not he wishes to assume a duty of care towards the plaintiff; so in my judgment he is well remote from assuming a duty of care towards the plaintiff. The most natural requirement mentioned by Lord Reid cannot possibly exist as regards the second defendant. Accordingly, in my judgment, the second defendant owed no duty, because he never assumed the duty expressly or by implication towards the plaintiff.

Nourse LJ considered expressly the effect of the Unfair Contract Terms Act 1977 in this context. Having set out the words of sections 1(1), 2(1) and 11(3) he continued:

The governing provision in this case, being one where there is no liability for death or personal injury resulting from negligence, is section 2(2). When adapted to the present case, and with the definition of negligence in section 1(1)(b) read into it, the subsection provides that a person cannot by reference to a notice given to particular persons exclude his liability for the breach of any common law duty to take reasonable care, except in so far as the notice satisfies the requirement of reasonableness. Mr Ashworth submits that when section 2(2) is spelt out in this way it would seem to apply only where a duty of care already exists. A notice which has prevented it from coming into existence is not one upon which the subsection bites. Mr Johnson [counsel for the respondent plaintiffs] submits that the wording applies as naturally to the latter case as it does to the former. He also relies on the words of section 11(3) "having regard to all the circumstances obtaining when the liability arose or (but for the notice) would have arisen", as showing that the Act expressly contemplates the inclusion of a case where a notice has prevented the duty from coming into existence. This is a short question of statutory construction . . . . Only two views are possible. Either the 1977 Act applies to all cases of negligent misstatement where a disclaimer has prevented a duty of care from coming into existence or it applies to none. I find this question to be both difficult and uncomfortable: difficult because either construction of section 2(2) appears to be tenable; uncomfortable because Mr Ashworth's argument could have been advanced, but evidently was not, by either of the experienced counsel who represented the defendants in *Stevenson* v *Nationwide Building Society* and *Smith* v *Eric S Bush* (a firm). Giving it the best consideration I can, I have come to the conclusion that the argument of Mr Ashworth represents on balance the more natural reading of section 2(2). The contrary argument seems to require reading the words "liability for the breach of any common law duty to take reasonable care" as including "liability for the breach of what without the notice would have been a common law duty to take reasonable care". Moreover, the words in section 11(3) on which Mr Johnson relies are found in a provision defining the requirement of reasonableness and, as such, are peripheral to the governing provisions of section 2(2). I would therefore hold that the 1977 Act has no application to this case.

Respectfully, I am not sure that either Mr Moxon-Browne in *Stevenson* or Mr Hague QC in *Smith* v *Bush* did not advance the argument obviously advanced by Mr Ashworth in *Harris*, that the 1977 Act applied to none of the cases of negligent misstatement where a disclaimer prevented a duty of care coming into existence. In *Stevenson,* in passages of the report which I have already read, Mr Moxon-Browne appears to have done so. In *Smith* v *Bush*, apart from the reference in the grounds of appeal to which I have already referred, the argument for counsel for the defendant is summarised at p 894 in the judgment of Dillon LJ, firstly at p 894F:

On this appeal two points only are raised to which I shall come. The defendants do not dispute the recorder's findings that . . . in preparing his report Mr Cannell owed a duty of care to the plaintiff, apart from the disclaimers which I have mentioned

and at H

the defendants submit, however, that the disclaimers, ie the provisions set out above in the form of mortgage application signed by the plaintiff and in the notice in red on the copy of Mr Cannell's report which was supplied to the plaintiff, have the effect of absolving the defendants from all liability for negligence which would otherwise have fallen on them.

Glidewell LJ, in the passage which I have already read, described the first issue as this: "On their face, did the words of the disclaimer suffice to prevent any duty of care being owed by the defendants to the plaintiff?" Sir Edward Eveleigh, in the passage to which I have already referred, put the matter somewhat differently at p 899H.

> If the plaintiff had relied on the report to an extent which would be unreasonable bearing in mind the warning, there would be no liability on the defendants and no need to consider the efficacy of a notice excluding liability. However, in so far as the words constitute a warning, they do not, in my opinion, warn a person to place no reliance at all on the report. They do not say that it is dangerous to rely on the report for what it purports to be or for what it says. The report did not reveal what a valuation report would have revealed if properly prepared, albeit for its limited purpose. On the contrary, it asserted that the property was readily saleable for the purpose of owner occupation. The plaintiff relied on it in this respect but the report did not merit even that degree of confidence. The warning, in my opinion, was not such as to make it unreasonable for the plaintiff to rely on the report as she did. Therefore, the defendants would be liable to the plaintiff unless the notice is effective to exclude liability.

There is, as I see it, a plain difference of judicial opinion demonstrated in the two cases in the Court of Appeal, perhaps less eloquently expressed in this way: on the one hand (in *Harris*), the disclaimer being part of the material from which a duty of care is to be deduced prevents that duty ever being established: on the other hand (*Smith* v *Bush*), where a special relationship of professional adviser and recipient of that advice is established so as to impose on the former a duty of care, a disclaimer only operates either as a notice within section 2 of the 1977 Act or, perhaps, to limit the ambit of the duty by imposing on the latter limitations on the extent to which he can reasonably rely on the advice.

For a number of reasons, I have come to the conclusion for the purposes of this case that the second of these alternatives is that which I should follow.

First, there are substantial and significant differences between the facts of the present case and the facts of *Harris*: the facts here approximate more closely to those in *Smith* v *Bush*. Here there is an independent surveyor, not an "in house" employee; the report, far from being confidential, is intended to be seen by and is known to be likely to be relied on by the plaintiffs. The surveyor himself does not purport to limit his liability and makes no disclaimer. To deny the existence of settlement and to describe the price as very reasonable are two of the very things that the defendant knew or must have known would be relied on by the prospective purchaser and would make the prospect of his having a structural survey even more remote than usual, and that itself no more than a 1 in 4 chance.

Second, my attention was invited to section 13 of the 1977 Act; this does not seem to have been cited in any of the cases to which I have already referred, nor was the decision of the Court of Appeal in *Phillips Products Ltd* v *Hyland* [1987] 1 WLR 659 (though this case was decided on December 14 1984).

Section 13(1) of the 1977 Act reads:

> To the extent that this Part of this Act prevents the exclusion of any liability it also prevents — (*a*) making the liability or its enforcement subject to restrictive or onerous conditions; (*b*) excluding or restricting any right or remedy in respect of the liability, or subjecting a person to any prejudice in consequence of his pursuing any such right or remedy; (*c*) excluding or restricting rules of evidence or procedure; and (to that extent) sections 2 and 5 to 7 also prevent excluding or restricting liability by reference to terms and notices which exclude or restrict the relevant obligation or duty.

The facts in *Phillips*' case can be summarised. The plaintiffs hired a JCB excavator from the second defendants. It was a term of the contract of hire (condition 8) that the excavator driver (Mr Hyland, the first defendant) should be, for all purposes connected with the working of the plant, regarded as the servant or agent of the plaintiffs, who alone should be responsible for all claims arising in connection with the operation of the plant by the driver. The first defendant so negligently drove the excavator that it collided with and did considerable damage to the plaintiff's buildings.

The material issues were defined as follows by Slade LJ, giving the judgment of the court (the second defendant's name is Hamstead):

> These three issues are: (i) On the admitted facts of the present case, was there on the part of Hamstead "negligence" within the definition of that word contained in section 1(1) of the Act? (ii) If the answer to (i) is "Yes", is clause 8 a contract term which, apart from the effect of the Act, can properly be said to "exclude or restrict" Hamstead's liability for negligence within the meaning of these words in section 2(2) of the Act? In considering this issue, it is necessary to bear in mind the concluding words of section 13(1) which bring within the ambit of section 2(2) terms "which exclude or restrict the relevant obligation or duty". (iii) If the answers to (i) and (ii) are both "Yes", does clause 8 satisfy the requirement of reasonableness, within the meaning of that phrase as used in the Act?

Slade LJ continued:

> As to (i), the argument for Hamstead is simple, and runs on these lines. If a claim is based on contract, "negligence" within the definition of section (1)(*a*) can have occurred only if there has been a breach of "any obligation, arising from the express or implied terms of a contract, to take reasonable care or exercise reasonable skill in the performance of the contract". So, it is said, if in the case of such a claim the contract has by its express terms excluded liability for negligence, there can ex hypothesi have been no breach of any obligation of the nature referred to in section 1(1)(*a*). The claim in the present case, as it happens, is of the nature referred to in section 1(1)(*b*): the breach of a common law duty to take reasonable care is alleged. Here again a similar argument is advanced. It is suggested that there can be no breach of a common law duty to take reasonable care, within the meaning of section 1(1)(*b*), by a party to a contract which contains a condition which purports to absolve him from liability for negligence.
>
> These arguments, though superficially attractive, are in our judgment fallacious. If correct, they would make nonsense of the Act. They would mean that the very contractual term which pre-eminently is suitable to be subject to review for reasonableness under the Act would be taken out of its scope. The Act, however, is not nonsensical. Its purpose is not defeated by the wording of its first section. In our judgment, in considering whether there has been a breach of any obligation of the nature referred to in para (*a*) or of any duty of the nature referred to in para (*b*) or (*c*), the court has to leave out of account, at this stage, the contract term which is relied on by the defence as defeating the plaintiffs' claim for breach of such obligation or such duty, and section 1(1) should be construed accordingly.
>
> If any support were necessary for this construction of section 1(1), it is to be found in the concluding words of section 13(1) of the Act. For these words make it clear that section 2 is capable of negativing the effect of contract terms which purport to exclude or restrict "the relevant obligation or duty".

Even if the words "or of any duty of the nature referred to in para (*b*) or (*c*)" are obiter they are of strong persuasive authority.

Slade LJ continued:

> Accordingly, though the validity of clause 8 still remains to be considered, on the admitted facts of this case there was "negligence" on the part of Hamstead falling within section 1(1)(*b*) of the Act. This took the form of a breach (subject to the effect, if any, of clause 8) of Hamstead's common law duty to take reasonable care, by reason of the fact that Mr Hyland who, subject to clause 8, was Hamstead's servant, had caused the loss to Phillips by his negligence in the performance of his duties as such servant. Issue (i) therefore has to be answered "Yes".

That the words to which I have referred were not or may not have been obiter appears from Slade LJ's consideration of issue (ii). I read a short passage:

> On the particular facts of this case the effect of clause 8, if valid, is to negative a common law liability in tort which would otherwise admittedly fall on the plant owner. The effect of clause 8 making "the hirer alone responsible for all claims" necessarily connotes that by the clause the plant owner's responsibility is excluded. In applying section 2(2) it is not relevant to consider whether the form of a condition is such that it can aptly be given the label of an "exclusion" or "restriction" clause. There is no mystique about "exclusion" or "restriction" clauses. To decide whether a person "excludes" liability by reference to a contract term, you look at the effect of the term. You look at its substance. The effect here is beyond doubt. Hamstead does most certainly purport to exclude their liability for negligence by *reference* to clause 8. Furthermore, clause 8 purports to "exclude or restrict the relevant obligation or duty" within the provisions of section 13(1) of the Act. Issue (ii) must be answered "Yes".

Finally, the court went on to consider the third issue and concluded that the term was a fair and reasonable one.

Looking now at section 2, the words used are "contract term" and "notice"; notice must refer to some extra-contractual "disclaimer" or exclusion or restriction of liability. I cannot find, and the court in *Phillips* did not suggest, any different approach to the exclusion or restriction of liability where the "disclaimer" was a notice and not a term. Moreover, the use in section 13(1) with specific reference to section 2 of the words "prevent excluding or restricting liability by reference to terms and *notices* which exclude or restrict the relevant obligation or *duty*". I have underlined forensically the words "notices" and "duty". Excluding a duty postulates the existence of a duty in respect of which a defendant seeks by notice to exclude or restrict liability; in those circumstances the Act seems to me to say plainly that such a notice falls to be considered under the provisions of the Act.

For these reasons I have concluded that in the present case, on the facts, the disclaimer to be effective to relieve the defendant of liability must satisfy, if it can, the requirement of reasonableness: section 2(2). It is common ground that the onus of establishing reasonableness is on the defendant.

Mr Little relied on four facts:

(1) the number and prominence of the warnings;
(2) that the warnings were read by the prospective purchasers;
(3) the obligation to report arises from section 25 of the Building Societies Act 1962 and not from any personal relationship to the prospective purchaser: see also section 11(3);
(4) the report in any event could only be of limited coverage and was made at low cost.

He accepted that the onus was on him and contended that there was nothing wrong in principle (though contrary to practice) in a professional adviser limiting his responsibility for the contents of his report, which could still be of some assistance even if no reliance could be placed upon it.

However, Mr Little accepted that the 1977 Act could properly be regarded as being in the field of consumer protection so that "reasonableness" was to be considered in that context; further, that on very similar facts — on this point — Dillon LJ in *Smith* v *Bush* at p 896E regarded the circumstances of that case — as, in my view, this — as "very ordinary" and continued:

> But where he
>
> the surveyor
>
> is dealing with a property at the lower end of the market and he knows that the purchaser is likely to rely on his report, and not instruct his or her own surveyor, I find it very difficult to see why it should be fair and reasonable to allow him to rely on an automatic blanket exclusion of all liability for negligence if his visual inspection of a property turns out not to have been reasonably careful.

I respectfully agree and on the facts of this case am of the same opinion. The defendant has not discharged the onus upon him of satisfying the requirement of reasonableness.

Mr Little further contended that the plaintiffs should be held contributorily negligent. They were, he said, at fault in failing to take proper care to preserve the value of their property: see *Nance* v *British Columbia Electric Railway Co Ltd* [1951] AC 601. In a sense, as Mr Little conceded, the only thing the plaintiffs could and should have done was to commission a structural survey, but as one of the limbs on which the defendant's negligence was established was the defendant's knowledge that the plaintiffs would not, or were highly unlikely to, commission such a survey the point is not a valid one. Either the defendant was to blame or the plaintiffs were wholly responsible. There is, in my view, no middle way such as a division of fault would imply, and I can find no basis at all on the facts here for holding the plaintiffs at fault or guilty of contributory negligence and I reject this argument.

In the result, therefore, there will be judgment for the plaintiffs for damages to be assessed.

It was ordered that the damages, if not agreed, be assessed by the circuit judge taking official referee's business at Shrewsbury or such other venue as may be agreed between the parties. The plaintiffs were awarded their costs of the issue of liability against the defendant and there was an order for taxation of the plaintiffs' costs under the provisions of the legal aid scheme.

For further cases on this subject see p 231

RATING

House of Lords

February 11 1988

(Before Lord KEITH OF KINKEL, Lord ELWYN-JONES, Lord BRANDON OF OAKBROOK, Lord OLIVER OF AYLMERTON and Lord GOFF OF CHIEVELEY)

CLEMENT (VALUATION OFFICER) v ADDIS LTD

Estates Gazette March 12 1988

[1988] 10 EG 129-131

Rating — General Rate Act 1967, section 20(1)(a) and (b) — Valuation of factory situated outside, but in near neighbourhood of, enterprise zone — *Rebus sic stantibus* rule — Meaning of "state" in section 20(1) — Appeal from decision of Court of Appeal — Approach of local valuation court and Lands Tribunal vindicated by House of Lords — Subject hereditament situated within a mile of the Lower Swansea Valley Enterprise Zone — Common ground that one result of the zone was to depress rental values of commercial and industrial premises situated outside, but in the near neighbourhood of, the zone — This was because operation within the zone was attended by fiscal and numerous administrative and other advantages — The parties agreed that the appropriate rateable value of the subject premises would be £36,500 if the consequences of the zone were taken into account and £45,500 if they were not — The Court of Appeal had broadly accepted the argument of the valuation officer that the word "state" in section 20(1)(a) and (b) covered only physical factors or factors which affected the physical enjoyment of the hereditament itself or of other premises in the locality and did not extend to advantages or disadvantages resulting from primary or subordinate legislation — The appellant ratepayers submitted that the word covered the state of affairs created by the statutory designation of the enterprise zone, which produced advantages for those premises which were within the zone and corresponding disadvantages for those outside it, although in the same locality — Held, reversing the decision of the Court of Appeal, that the word "state" should be given a wide construction in both para (a) and para (b) of section 20(1), so as to include intangible as well as physical advantages and disadvantages — There was no good reason for disregarding any of the circumstances which would ordinarily be taken into account under the *rebus sic stantibus* rule — The advantages of the enterprise zone were an intangible factor which went to the state of the locality as regards other premises situated there — The value of hereditaments in the same locality which did not enjoy the advantages of the zone was reduced — Thus the effects of the zone on the value of the appellants' factory were properly to be taken into account in applying section 20 — Appeal allowed

The following cases are referred to in this report.

Dawkins v *Ash Brothers & Heaton Ltd* [1969] 2 AC 366; [1969] 2 WLR 1024; [1969] 2 All ER 246; (1969) 67 LGR 499, HL

Port of London Authority v *Orsett Union Assessment Committee* [1920] AC 273, HL

Seafield's Trustees v *Assessor for Inverness-shire* 1941 SC 271; 1941 SLT 270

This was an appeal by Addis Ltd, the occupiers of a factory outside, but in the near neighbourhood of, the Lower Swansea Valley Enterprise Zone, from the decision of the Court of Appeal (reported at [1987] 1 EGLR 168; (1987) 281 EG 683) reversing the decision of the Lands Tribunal (Mr J H Emlyn Jones) (reported at (1984) 271 EG 291) in favour of the appellants.

Guy Roots (instructed by McKenna & Co) appeared on behalf of the appellants; William Glover QC and David Mole (instructed by the Solicitor of Inland Revenue) represented the respondent valuation officer.

In his speech, LORD KEITH OF KINKEL said: The issue in this appeal concerns the proper interpretation of section 20 of the General Rate Act 1967. That section, so far as material, provides:

> 20(1) For the purposes of any alteration of a valuation list to be made under Part V of this Act in respect of a hereditament in pursuance of a proposal, the value or altered value to be ascribed to the hereditament under section 19 of this Act shall not exceed the value which would have been ascribed thereto in that list if the hereditament had been subsisting throughout the year before that in which the valuation list came into force, on the assumptions that at the time by reference to which that value would have been ascertained: *(a)* the hereditament was in the same state as at the time of valuation and any relevant factors (as defined by subsection (2) of this section) were those subsisting at the last-mentioned time; and *(b)* the locality in which the hereditament is situated was in the same state, so far as concerns the other premises situated in that locality and the occupation and use of those premises, the transport services and other facilities available in the locality, and other matters affecting the amenities of the locality, as at the time of valuation.
>
> (2) In this section, the expression "relevant factors" means any of the following, so far as material to the valuation of a hereditament, namely *(a)* the mode or category of occupation of the hereditament; *(b)* the quantity of minerals or other substances in or extracted from the hereditament; or *(c)* in the case of a public house, the volume of trade or business carried on at the hereditament; and in paragraph *(c)* of this subsection the expression "public house" means a hereditament which consists of or comprises premises licensed for the sale of intoxicating liquor for consumption on the premises where the sale of such liquor is, or is apart from any other trade or business ancillary or incidental to it, the only trade or business carried on at the hereditament.
>
> (3) References in this section to the time of valuation are references to the time by reference to which the valuation of a hereditament would have fallen to be ascertained if this section had not been enacted . . .

This section has been amended from time to time since its original enactment, but none of the amendments is relevant for present purposes.

The hereditament which is the subject of the appeal is a factory occupied by the appellants and situated near Swansea in South Wales. It lies just under a mile to the south of an area of land which by order dated May 18 1981 the Secretary of State for Wales, under powers contained in the Local Government, Planning and Land Act 1980, designated as the Lower Swansea Valley Enterprise Zone. The Designation Order 1981 (SI 1981 no 757) came into effect on June 11 1981. The nature of an enterprise zone is thus described in the decision of the Lands Tribunal [1984] RA 137, 139-140:

> The Government's purpose in setting up Enterprise Zones is to encourage industrial and commercial activity in run-down areas by the removal of certain tax burdens and by the relaxation or speeding up of the application of certain statutory and administrative controls. The benefits conferred on enterprises operating within an Enterprise Zone run for a period of ten years from the date when the zone comes into effect and cover both new and existing industrial enterprises within the zone. The available benefits are as follows: (1) Exemption from development land tax; (2) Exemption from rates on industrial and commercial property; (3) 100% allowances for corporation

and income tax purposes for capital expenditure on industrial and commercial buildings; (4) Applications for certain customs facilities from firms within the zones are processed as a matter of priority and certain criteria are relaxed; (5) Industrial development certificates are not needed; (6) Employers are exempt from industrial training levies and from the requirement to supply information to industrial training boards; (7) A greatly simplified planning regime; developments that conform with the published scheme in each zone will not require individual planning permission; (8) Those controls remaining in force are to be administered more speedily; (9) Government requests for statistical information are reduced.

It is agreed between the parties that one result of the setting up of the Lower Swansea Valley Enterprise Zone has been to depress the rental values of commercial and industrial premises situated outside, but in the near neighbourhood of, the designated area. This is because it is substantially more attractive to operate such concerns inside that area. The parties are also agreed that the appropriate rateable value of the appellant ratepayers' factory is £36,500 if the consequences of the enterprise zone are taken into account, and £45,500 if they are not. The question in dispute is whether or not, on a proper construction of section 20 of the Act of 1967, it is correct to take into account these consequences in determining, upon a proposal made by the ratepayers on March 18 1981, the rateable value of the factory at that time. It is common ground that although the proposal was made before the date of the order designating the enterprise zone, the prospect of it had already produced the result described above. The local valuation court and the Lands Tribunal decided that the consequences of the enterprise zone were properly to be taken into account, but the Court of Appeal (O'Connor and Woolf LJJ and Sir George Waller) took the contrary view. The ratepayers now appeal, with leave given here, to your Lordships' House.

The principal enactment prescribing the manner of valuation for rating of hereditaments in England and Wales is section 19 of the Act of 1967. There is no need to recite its relevant provisions, which are set out in the judgment of Woolf LJ in the Court of Appeal. One of the features of a valuation under section 19 is that the hereditament is to be valued in its actual state — *rebus sic stantibus* — at the date of the valuation, with all the advantages and disadvantages which it then possesses. It is clear that if the ratepayers' factory were to be valued under section 19 alone as at the date of the proposal in the present case the disadvantage which it suffers through being outside but close to the enterprise zone would be a factor properly to be taken into account. However, the proposal is one made under section 69 of the Act of 1967, for an alteration of the valuation list which came into force on April 1 1973, and in relation to such a proposal the effect of section 19 is modified by section 20. By virtue of that section the valuation is not to exceed that which would have been ascribed to the hereditament if it had been in existence throughout the year before April 1 1973. The principal purpose is to secure that where a hereditament comes to be valued for the first time or to be revalued as a result of alterations, in a year after that in which the valuation list came into force, the value is to be at the same general level as was applicable for purposes of the original list. As the sidenote to section 20 puts it, the valuation is to be according to the tone of the list. This avoids the injustice which would otherwise result in a period of inflation involving a general increase in rental values. If the valuation of a new hereditament were made during such a period on the basis of current rental values the occupier would have to pay higher rates than the occupier of a comparable hereditament subject to a valuation made in 1973. There is, of course, nothing to prevent a valuation officer making a proposal under section 69 of the Act for an increase in the valuation of a hereditament already in the list, but, since the general level of rental values to be applied would be that prevailing immediately before the list came into force, the proposal would have no actual effect unless there had been some material change of circumstances affecting the value to be arrived at upon that basis.

Although the valuation to be made under section 20 is not to exceed that which would have been ascribed to the hereditament if it had been in existence immediately before the valuation list came into force, subsection (1) provides that certain matters affecting the hereditament at the time when the valuation is actually made are to be assumed to have existed at the earlier time. These matters are specified in paras *(a)* and *(b)*. The question is whether the specification of these matters restricts the application of the *rebus sic stantibus* rule so as to exclude some factors which would properly have had to be taken into account if the valuation had fallen to be made under section 19 alone, in particular, in the present case, the effect on value of the enterprise zone.

The argument for the valuation officer, which was accepted by the Court of Appeal and renewed before this House, was that the word "state" in paras *(a)* and *(b)* of section 20(1) was intended to cover only physical factors or factors which affected the physical enjoyment of the hereditament itself or of other premises in the locality. The ratepayers, for their part, contended that it embraced also advantages and disadvantages resulting from primary or subordinate legislation. Reference was made to a passage in the speech of Lord Buckmaster in *Port of London Authority* v *Orsett Union Assessment Committee* [1920] AC 273, at p 305:

> The actual hereditament of which the hypothetical tenant is to be determined must be the particular hereditament as it stands, with all its privileges, opportunities and disabilities created or imposed either by its natural position or by the artificial conditions of an Act of Parliament.

The state of a hereditament, so it was maintained, included opportunities or disabilities created by Act of Parliament. The making of a demolition order under statutory powers came into this category, such an order having been held to be a factor properly to be taken into account in arriving at a valuation in *Dawkins* v *Ash Brothers & Heaton Ltd* [1969] 2 AC 366. Likewise, the state of a locality so far as concerns the other premises situated there included advantages to these premises resulting from the designation under statutory powers of an enterprise zone. It was these advantages which reduced the value of the subjects of appeal, which were admittedly in the same locality.

The broad purpose of section 20 was to secure that a hereditament which, for one reason or another, fell to be valued at a time after the coming into force of a valuation list should be valued on the basis of the general level of values prevailing when the list was made up. It is not apparent that there was any good reason for requiring the disregard of any of the circumstances which would ordinarily be taken into account under the *rebus sic stantibus* rule. If any of such circumstances were to be disregarded, the result would necessarily be to some extent artificial. The expression *rebus sic stantibus* raises the concept of a certain ascertainable state of affairs. It would involve no undue straining of language to conclude that when para *(a)* of section 20(1) refers to the hereditament being "in the same state as at the time of the valuation" the meaning is that the whole state of affairs affecting the hereditament is to be assumed to be the same, at the time of coming into force of the last valuation list, as it is at the time when the valuation is being made. The result would be that the *rebus sic stantibus* rule would fall to be applied to its full extent. But the reference to a defined category of specified relevant factors makes this difficult, and moreover it would make para *(b)* unnecessary. Since, however, there is no readily perceptible reason for limiting the application of the *rebus sic stantibus* rule, the word "state" should in my opinion be given a wide construction so as to include intangible as well as physical advantages and disadvantages. Thus I would think it appropriate to treat the making of a demolition order, such as was held to have a relevant effect on value in *Dawkins* v *Ash Brothers & Heaton Ltd* [1969] 2 AC 366, as going to the state of the hereditament. Another instance would be the prospect of requisition, which was held to be a relevant factor in the Scottish case of *Seafield's Trustees* v *Assessor for Inverness-shire* 1941 SC 271. Para *(a)* is not, of course, directly in point in the present case. It is para *(b)* which the ratepayers seek to apply, on the ground that the state of the locality in which the hereditament is situated includes the relative advantages enjoyed by some other premises there situated through the creation of the enterprise zone. "State" must have the same wide meaning in para *(a)* that it has in para *(b)*. I would therefore hold that the existence of these advantages is an intangible factor which goes to the state of the locality as regards other premises situated there. One effect of the circumstances affecting those other premises is to reduce the value of hereditaments in the same locality which do not enjoy the same advantages because they are not in the enterprise zone. I am therefore of opinion that on a proper construction of section 20 the effects of the enterprise zone on the value of the ratepayers' factory are properly to be taken into account, and would allow the appeal accordingly.

LORDS ELWYN-JONES, BRANDON OF OAKBROOK, OLIVER OF AYLMERTON and GOFF OF CHIEVELEY agreed with the speech of Lord Keith of Kinkel and would allow the appeal for the reasons given in it. They did not add anything of their own.

Appeal allowed.

House of Lords

December 10 1987

(Before Lord KEITH OF KINKEL, Lord BRANDON OF OAKBROOK, Lord GRIFFITHS, Lord OLIVER OF AYLMERTON and Lord GOFF OF CHIEVELEY)

WHITSBURY FARM & STUD LTD v HEMENS (VALUATION OFFICER)

Estates Gazette April 2 1988

[1988] 13 EG 43-48

Rating — General Rate Act 1967, section 26(3) and (4) — Rating Act 1971, sections 1 and 2 — Exemptions from rating — Whether premises occupied and used by the appellant ratepayers for the breeding and rearing of thoroughbred horses were entitled to the benefit of agricultural derating — The hereditaments in question were situated within, or attached to, land which was conceded to be agricultural land within the meaning of the rating legislation —The hereditaments comprised the usual kind of buildings found on stud farms, in particular stable blocks, loose boxes, covering yards and sheds, foaling boxes, veterinary rooms, hay stores and feed stores — The local valuation court had decided that the hereditaments in question were entitled to derating, but the Lands Tribunal (Mr V G Wellings QC) and the Court of Appeal decided that they were not — The issue turned on whether the buildings qualified either under the 1967 Act or the 1971 Act — Appellant ratepayers argued that, as the use of the land as pasture land admittedly made it agricultural land, the activity of pasturing the thoroughbred horses was necessarily an agricultural operation —That operation was part of the activity of raising and keeping thoroughbreds and, as the buildings were used for the same activity, the buildings were used (and solely used) in connection with agricultural operations on the land — In rejecting this line of argument the House of Lords (with the exception of Lord Goff of Chieveley, who reached the same ultimate conclusion in the appeal, but through a different course of reasoning: see below) held that "in connection with" in section 26(4)(b) of the 1967 Act meant "ancillary to" the agricultural operations on the pasture land — In the present case, however, the position was the reverse; the use of the pasture land served the purpose for which the buildings were used, namely, the breeding, rearing and keeping of thoroughbreds — Hence the buildings were not used "in connection with" agricultural operations on the pasture land —As regards the argument under the 1971 Act, the buildings would qualify as agricultural buildings if they were "used for the keeping or breeding of livestock" within the meaning of section 2(1)(a) of that Act — "Livestock" as defined in section 1(3) "includes any mammal or bird kept for the production of food or wool or for the purpose of its use in the farming of land" — The appellant ratepayers argued that this "inclusive" definition covered thoroughbred horses — The House (including Lord Goff) held on this point that it was unlikely that Parliament in this context would have intended to cover animals which were not normally regarded as ordinary farm creatures — The reasoning of Lord Parker CJ in *Belmont Farm Ltd* v *Minister of Housing and Local Government*, although in a planning context, was equally applicable here — Lord Goff of Chieveley, agreeing on the result under section 26 of the General Rate Act 1967, preferred to rest his conclusion, as did the Court of Appeal, on the view that the operations performed on the land were not agricultural operations. The operation of breeding and rearing thoroughbred horses for the purpose of recreation was not an agricultural operation. He questioned the view that the expression "in connection with" was equivalent to "ancillary to", as it may include cases where the uses of the land and buildings are complementary — Appeal dismissed

The following cases are referred to in this report.

Belmont Farm Ltd v *Minister of Housing and Local Government* (1962) 13 P&CR 417; 60 LGR 319, DC
Dilworth v *Commissioner of Stamps* [1899] AC 99
Eastwood (W & J B) Ltd v *Herrod* [1971] AC 160; [1970] 2 WLR 775; [1970] 1 All ER 774, HL
Forth Stud Ltd v *East Lothian Assessor* 1969 SC 1; 1969 SLT 75; [1969] RA 35
Gilmore v *Baker-Carr* [1962] 1 WLR 1165; [1962] 3 All ER 230; [1962] RVR 486; (1962) 9 RRC 240; [1962] RA 379, CA
Glanely (Lord) v *Wightman* [1933] AC 618
Hardie v *Assessor for West Lothian* 1940 SC 329

This was an appeal by Whitsbury Farm & Stud Ltd from the decision of the Court of Appeal (reported at [1987] QB 390; [1987] 1 EGLR 172) dismissing an appeal from the decision of the Lands Tribunal (reported at [1985] 1 EGLR 227) in regard to the rating of the appellants' hereditaments at Whitsbury, Hampshire.

William Glover QC and Alan Alesbury (instructed by Penningtons Ward Bowie) appeared on behalf of the appellants; Robert Carnwath QC and David Mole (instructed by the Solicitor of Inland Revenue) represented the respondent valuation officer.

In his speech, LORD KEITH OF KINKEL said: The question at issue in this appeal is whether or not premises occupied and used by the appellant ratepayers for the breeding and rearing of thoroughbred horses are entitled to the benefit of agricultural derating.

Four separate hereditaments, situated at Fordingbridge in Hampshire, are involved. Each of them is situated within or is attached to land which is conceded by the respondent valuation officer to be agricultural within the meaning of the relevant legislation, and each comprises buildings of the sort usually to be found on stud farms, in particular stable blocks, loose boxes, covering yards and sheds, foaling boxes, veterinary rooms, hay stores and feed stores.

The local valuation court held that the hereditaments were entitled to derating, but the Lands Tribunal and the Court of Appeal (Lawton, Lloyd and Balcombe LJJ) [1987] QB 390 decided that they were not.

The principal enactment concerned with the derating of agricultural premises in England and Wales is section 26 of the General Rate Act 1967, which provides:

(1) No agricultural land or agricultural buildings shall be liable to be rated or be included in any valuation list or in any rate.

(2) The gross value for the purposes of section 19(2) of this Act of a house occupied in connection with agricultural land and used as the dwelling of a person who (*a*) is primarily engaged in carrying on or directing agricultural operations on that land; or (*b*) is employed in agricultural operations on that land in the service of the occupier thereof and is entitled, whether as tenant or otherwise, so to use the house only while so employed, shall, so long as the house is so occupied and used, be estimated by reference to the rent at which the house might reasonably be expected to let from year to year if it could not be occupied and used otherwise than as aforesaid.

(3) In this section the expression "agricultural land" — (*a*) means any land used as arable meadow or pasture ground only, land used for a plantation or a wood or for the growth of saleable underwood, land exceeding 0.10 hectare used for the purposes of poultry farming, cottage gardens exceeding 0.10 hectare, market gardens, nursery grounds, orchards or allotments, including allotment gardens within the meaning of the Allotments Act 1922, but does not include land occupied together with a house as a park, gardens (other than as aforesaid), pleasure grounds, or land kept or preserved mainly or exclusively for purposes of sport or recreation, or land used as a race-course; and for the purposes of this paragraph the expression "cottage garden" means a garden attached to a house occupied as a dwelling by a person of the labouring classes; and (*b*) includes land occupied with, and used solely in connection with the use of, such a building as is mentioned in subsection (4)(*b*) of this section.

(4) In this section, the expression "agricultural buildings"[1] — (*a*) means buildings (other than dwellings) occupied together with agricultural land or being or forming part of a market garden, and in either case used solely in connection with agricultural operations thereon; and (*b*) includes a building which is used solely in connection with agricultural operations carried on on agricultural land and which is occupied either — (i) by the occupiers of all that land; or (ii) by individuals who are appointed by the said occupiers for the time being to manage the use of the building and of whom each is an occupier of some of the land or a member of the board of directors or other governing body of such an occupier who is a body corporate, where the number of occupiers of all the said land does not exceed 24 (two or more persons occupying jointly being counted as one, but as a separate person from any of them who are occupying any of the land severally).

A The scope of this enactment was extended by sections 1 to 4 of the Rating Act 1971. It is sufficient for present purposes to quote section 1(1)(*a*) and (3) and section 2(1) and (4):

> 1(1) In section 26 of the General Rate Act 1967 (in this Part of this Act referred to as "the principal section") — (*a*) the expression "agricultural buildings" shall include any building which is an agricultural building by virtue of section 2, 3 or 4 of this Act; . . .
>
> (3) In this Part of this Act "livestock" includes any mammal or bird kept for the production of food or wool or for the purpose of its use in the farming of land . . .
>
> 2(1) Subject to subsections (2) to (4) of this section, each of the following is an agricultural building by virtue of this section— (*a*) any building used for the keeping or breeding of livestock; and (*b*) any building (other than a dwelling) which is occupied together with one or more buildings falling within
> B paragraph (*a*) above and is used in connection with the operations carried on in that building or those buildings . . .
>
> (4) A building is not an agricultural building by virtue of this section unless it is surrounded by or contiguous to an area of agricultural land (as defined in the principal section) which amounts to not less than two hectares; . . .

The basis of the valuation officer's concession that the land occupied along with the stud farm buildings is agricultural is that the land is used for grazing by horses. He accepts that the land is used as "pasture ground only," within the meaning of that expression in section 26(3)(*a*) of the Act of 1967, notwithstanding that the horses put to pasture there are thoroughbreds which have nothing to do with farming in the ordinarily understood sense of the word. I am of
C opinion that the concession is rightly made. There is nothing in section 26(3)(*a*) or elsewhere which expressly limits use as pasture ground so as to exclude the pasturing of animals which are not reared as food or for use in the farming of the land, and I do not think that such an exclusion can be implied.

It follows that the buildings with which the appeal is concerned are occupied together with agricultural land, which is one of the relevant requirements of section 26(4)(*a*) if a building is to qualify as an agricultural building. The other requirement is that the building should be used solely in connection with agricultural operations on the land. Is that requirement satisfied? The expression "agricultural operations" is nowhere defined. In its ordinary meaning it indicates operations concerned with the raising of crops or the rearing of
D animals so as to produce or contribute to the production of "the means of human subsistence," to quote the words of Lord Robertson in *Hardie* v *Assessor for West Lothian* 1940 SC 329 at p334. In the present context, however, I think it must bear a meaning to some extent wider than that. Included in the meaning of agricultural land under subsection (3)(*a*) is land used for a plantation or a wood, and also certain cottage gardens and nursery grounds, where no doubt flowers and not any foodstuffs may be grown. No operations which are agricultural in the ordinary sense of the word would be carried out on such land, but one would not readily impute to Parliament an intention to exclude from the benefit of derating buildings used solely in connection with the carrying out of silvicultural or horticultural activities on these categories of "agricultural" land. Likewise, the
E special treatment for rating purposes of certain dwelling-houses, provided for by section 26(2), would, considering that a wood is to be treated as agricultural land, seem no less appropriate for a house occupied by a forestry worker than for one occupied by a farm worker.

The argument for the ratepayers proceeded on the lines that since the use of the land as pasture for thoroughbreds admittedly made it agricultural land, the activity of pasturing these horses was necessarily an agricultural operation; that operation was part of the activity of raising and keeping thoroughbreds; the same activity was what the buildings were used for; and hence the buildings were used, and solely used, in connection with an agricultural operation on the
F land.

It is necessary to give some consideration to the meaning of the words "in connection with". In *Gilmore* v *Baker-Carr* [1962] 1 WLR 1165 (a case under section 2(2) of the Rating and Valuation (Apportionment) Act 1928, the terms of which did not differ materially from those of section 26(3) and (4) of the Act of 1967) Donovan LJ said, at p 1175:

> But the clear impression which I receive from the statutory language is that the buildings exempted were to be ancillary or complementary to the agricultural purpose of the land, and not vice versa.

This passage was approved by Viscount Dilhorne in *W&JB Eastwood Ltd* v *Herrod* [1971] AC 160, at p 181, who said that to come within the statutory definition the buildings must be used as adjuncts to the agricultural operations on the land, and that it was G
wrong to consider whether the enterprise as a whole, the land and the buildings taken together, was "a combined agricultural operation". In the same case Lord Reid said, at p 168:

> Ordinary usage of the English language suggests that the buildings must be subsidiary or ancillary to the agricultural operations . . . I do not foresee serious difficulty if "used in connection with" is held to mean consequential on or ancillary to the agricultural operations on the land which is occupied together with the buildings.

Lord Morris of Borth-y-Gest expressed the view, at p 174, that the buildings must be such as were needed as an adjunct or necessary aid for agricultural operations taking place on agricultural land. The concept is thus that of principal and accessory. The use of the agricultural land is contemplated as being the principal use and that H
of the buildings as being an accessory use.

In the present case the use of the buildings is for all the most important operations concerned with the breeding and rearing and keeping of thoroughbreds. The use of the agricultural land is for the pasturing of the same horses. It is that use and nothing else which makes the land agricultural within the definition. The use of the pasture land serves the purposes for which the buildings are used, namely the breeding, rearing and keeping of thoroughbreds. The use of the buildings does not serve the purpose for which the pasture land is used. It is not ancillary to it. It is an end and purpose in itself. I would therefore hold that the buildings are not used in connection with agricultural operations on the pasture land and are certainly not J
used solely in connection with such operations.

Counsel for the valuation officer relied strongly on the Scottish case of *Forth Stud Ltd* v *East Lothian Assessor* 1969 SC 1, where under the Scottish provisions corresponding to section 26 of the Act of 1967 premises occupied and used as a commercial stud were held by the Lands Valuation Appeal Court not to be entitled to agricultural derating. These provisions, contained in section 7 of the Valuation and Rating (Scotland) Act 1956, differ in certain respects from those of section 26. No distinction is made between agricultural land and agricultural buildings. The whole hereditament, "land and heritage" in Scottish terminology, is treated as one. Derating is accorded to "agricultural lands and heritages," and the opening words of the definition of that expression, in section 7(2) of the Act of K
1956, are: "In this section — 'agricultural lands and heritages' means any lands and heritages used for agricultural or pastoral purposes only . . .". The subjects of appeal comprised 25 acres of grazing land, where the horses were grazed and exercised, and also steadings and an office. The court held that the subjects were not used for agricultural purposes only, in respect that racehorses were not associated with an ordinary farm and did not produce or contribute to the means of human subsistence. In relation to a subsidiary argument that the subjects were used for pastoral purposes only, Lord Hunter said, at p9:

> Let it be assumed in the appellants' favour that the use of the fields for grazing bullocks would, in appropriate circumstances and regarded by itself, be a use L
> for pastoral purposes. Let it also be assumed, though it is unnecessary for the purposes of the present case so to decide, that the use of the fields for the grazing of stallions, brood mares and the progeny of mares would also, in appropriate circumstances and regarded by itself, be a use for pastoral purposes. Even, however, upon these assumptions it is in my opinion clear that the subjects of appeal are not used for pastoral purposes only. They are used as a commercial stud in the manner described in the findings, and upon any view they are used in part, and indeed primarily, for that purpose.

Lord Fraser, at p 11, rejected the argument on the ground that the fields were used not only for grazing but also partly for exercising the horses and for growing hay which was cut and fed to the horses in the buildings. Lord Avonside, at p 13, took the view that "pastoral purposes" were limited to the grazing of animals used to produce the means of human subsistence. M

The differences in terminology between the relevant English and Scottish provisions have the result that the reasoning in *Forth Stud* is not capable of application to the present case, apart perhaps from the view of Lord Hunter that the subjects as a whole were used primarily for a purpose which was not pastoral. If the subjects as a whole were primarily used for a non-pastoral purpose, then the same was true of the buildings on their own.

The ratepayers, for their part, relied on *Lord Glanely* v *Wightman* [1933] AC 618, a decision of this House which appears to have resulted in English stud farms being derated for nearly 50 years. The only issue in that case, however, was whether or not fees received for the services of stallions kept on a stud farm in covering visiting mares sent there for that purpose were profits in respect of the occupation of the land, so as to be chargeable to income tax under Schedule B rather than Schedule D to the Income Tax Act 1918. It was held that they were. So the case is no authority on the proper construction of a statute concerned with agricultural derating.

Having reached the conclusion that the buildings here do not qualify as "agricultural buildings" under section 26(4) of the Act of 1967, it is necessary to consider whether or not they do so under section 2(1)(*a*) of the Act of 1971, as being buildings "used for the keeping or breeding of livestock". The answer to that question turns upon the meaning of "livestock". Does that word include thoroughbred horses? The word is defined by section 1(3) as including "any mammal or bird kept for the production of food or wool or for the purpose of its use in the farming of land". The definition is introduced by the word "includes". So it was argued for the ratepayers that the ordinary meaning of "livestock" is not thereby cut down, and that the ordinary meaning of the word is wide enough to embrace any animal which gains its sustenance from the land, in particular a thoroughbred horse.

It is true that what may be termed an inclusive definition is normally intended to widen the ordinary natural meaning of the word defined or at least to remove doubts as to the extent of that meaning. However, in *Dilworth* v *Commissioner of Stamps* [1899] AC 99 Lord Watson said, at p 106:

> But the word "include" is susceptible of another construction, which may become imperative, if the context of the Act is sufficient to show that it was not merely employed for the purpose of adding to the natural significance of the words or expressions defined. It may be equivalent to "mean and include," and in that case it may afford an exhaustive explanation of the meaning which, for the purposes of the Act, must invariably be attached to these words or expressions.

It is fair to say that it is hard to find any instance of such an exhaustive meaning being attributed to a definition which starts with the word "includes". Nevertheless, I think there can be no doubt that in some cases the language of an "inclusive" definition, considered with the general context, can have the effect that the ordinary natural meaning of a word or expression is to some extent cut down. The context here is agricultural, and the kinds of mammals and birds mentioned in section 1(3) of the Act of 1971 are those which one would expect to find on an ordinary farm. The Act was passed, as is well known, to overcome the effect of the decision in *W & J B Eastwood Ltd v Herrod* [1971] AC 160, where chicken broiler houses were held not to be agricultural buildings. Its provisions would also apply to buildings in which pigs or cattle beasts are permanently kept and fattened. An intention to cover animals not normally regarded as ordinary farm creatures is not readily to be inferred. On the other hand, the words "*any* mammal or bird" may well have been intended to cover the case where some unconventional type of mammal or bird came to be kept for food production, for example a deer. If "livestock" were to retain its full ordinary meaning, it would seem apt to include animals kept or bred as pets, for example dogs, cats and mice, so that, provided the building where that was done was surrounded by or contiguous to at least 2 hectares of agricultural land (see section 2(4)), that building would qualify for agricultural derating. That is unlikely to have been intended.

In *Belmont Farm Ltd* v *Minister of Housing and Local Government* (1962) 13 P&CR 417, a Divisional Court in construing the definition of "agriculture" contained in section 119(1) of the Town and Country Planning Act 1947 held that the words:

> the breeding and keeping of livestock (including any creature kept for the production of food, wool, skins or fur, or for the purpose of its use in the farming of the land)

did not cover the breeding and keeping of horses for show jumping. Lord Parker CJ said, at pp 421-422:

> It may be that this is rather a matter of first impression, but I confess that I approach it in this way: Of course, on one view "livestock" can be said to be used in contradiction to dead stock, and to include any animal whatsoever. In some contexts that might be so, but it seems to me that in the context of agriculture, as here, it has some less extensive meaning. What exact meaning should be given to it if it stood alone in this agricultural context, I do not propose to determine. I think that it is sufficient to say that there must be a limitation in that context on what I may call the wide dictionary meaning. I find it unnecessary to decide what it would mean if it stood alone because it does not stand alone, and the words in brackets that follow assist in determining what is meant by "livestock." The words are "including any creature kept for the production of food, wool, skins or fur." Pausing there, that is clearly an extension to cover, no doubt, an argument that, for instance, bees, possibly pheasants and fish are not livestock. It covers any creature kept for this purpose, and it then goes on to say "including any creature kept for the purpose of its use in the farming of land." Granting that the word "including" has been used in an extensive sense, it seems to me nonsense for the draftsman to use those words "any creature kept for the production of food, wool, skins or fur, or for the purpose of its use in the farming of land," if the word "livestock" was intended to cover the keeping of any creature whether for its use in farming land or not. It seems to me that those words show a clear intention that "livestock," however it is interpreted, does not extend to the breeding and keeping of horses unless it is for the purpose of their use in the farming of land.

In my opinion, that reasoning is equally applicable to section 1(3) of the Act of 1971, and I conclude that thoroughbred horses do not fall within the term "livestock."

My lords, for these reasons I would dismiss the appeal.

LORDS BRANDON OF OAKBROOK, GRIFFITHS and OLIVER OF AYLMERTON agreed that the appeal should be dismissed for the reasons given by Lord Keith of Kinkel and did not add anything of their own.

LORD GOFF OF CHIEVELEY in his speech said: I, too, agree that the appeal should be dismissed; but since I have reached that conclusion by a rather different route from that chosen by my noble and learned friend, Lord Keith of Kinkel, and (I understand) the remainder of your lordships, I propose very briefly to set out my own reasoning.

I agree that it was rightly conceded by the valuation officer that the land in question is "agricultural land" within the definition in section 26(3)(*a*) of the Act of 1967. This conclusion seems to me to be inescapable on the simple wording of the subsection. It leads, however, to what appears to me to be the main issue in the case, which is whether the buildings in question are "agricultural buildings" within the definition of section 26(4), and that in turn raises the question whether the buildings are used "solely in connection with agricultural operations" on agricultural land.

The argument before your lordships concentrated on the question whether the operations on the land are agricultural operations, the question whether (if they are agricultural operations) the buildings are used solely in connection with them being treated as a subsidiary question. I have come to the conclusion, in agreement with the Court of Appeal, that the operations in question are not agricultural operations.

Mr Glover, for the ratepayers, advanced the attractive argument that, since the land is agricultural land by virtue of a definition which characterises the land as agricultural by reference to its use, it must follow that the operations being performed on the land are agricultural operations. I find myself unable to accept this argument. I am not persuaded that the definition of agricultural land in section 26(3)(*a*) of the Act, which consists of a list of items not all of which are defined with reference to use, requires me to give to the word "agricultural" in the expression "agricultural operations" in section 26(4)(*a*) any other than its ordinary meaning in the broad sense of embracing both horticulture and forestry. On the contrary, it is the definition of "agricultural land" which seems to me to be artificial in the sense that it may require that land be classified as agricultural when one would not ordinarily describe its use as such —as in the case, posed in argument, of land used as pasture ground for exotic animals from a zoo; and if it was intended that this artificial meaning (for which there may be some good practical explanation) is to be imposed upon the words "agricultural" in the expression "agricultural operations," express provision could have been made to that effect.

If I turn to the operations being performed on the land in the present case, I accept Mr Carnwath's submission that those operations were the grazing of thoroughbred horses in the course of breeding and rearing. I do not think that, in describing these operations, it is sensible simply to focus upon the grazing, to the exclusion of the overall nature of the operation which was being performed on the land. If I then ask myself the question whether such operations are agricultural operations within the ordinary meaning of that expression, I would answer that question, in agreement with the Court of Appeal, in the negative. Like Balcombe LJ, I do not regard the operation of breeding and rearing of thoroughbred horses for the purposes of what can broadly be called recreation as an agricultural operation. Nobody would, I think, describe it as such in ordinary speech. On this part of the argument, I would for my part be content to adopt in its entirety the reasoning of Balcombe LJ.

On this approach the subsidiary question whether the buildings are

A used solely in connection with agricultural operations on the land does not arise, but I have to confess that I feel considerable doubt whether, if that question did arise, I would answer it in the negative. This is because I feel that it may not be right to restrict the expression "in connection with" to mean "ancillary to," for it may be appropriate to understand it as including cases where the use of the buildings and the land are complementary, both being used together in an agricultural operation, and neither being ancillary to the other. Mr Glover gave in argument the example of land and buildings being used together for the breeding and rearing of pedigree cattle. For such an operation, it may be that the land and the buildings are equally important: and I would have thought that, in such circumstances, it could properly be said that the buildings were used
B solely in connection with agricultural operations on the land. It is not to be forgotten that dairy farming has progressed considerably since the day of Tess of the d'Urbervilles, when the milkmaids went out to milk in the fields; it is difficult to imagine a modern dairy farm without a substantial building which acts not only as a mechanised milking parlour but also as a place in which the cattle may spend a substantial part of their time, especially in winter. Again, land and buildings may be regarded as equally important; and yet I would have thought that the buildings were agricultural buildings within the definition in section 26(4) of the Act. However, those cases are very different from the present case, for the simple reason, as I see it, that the operations here being performed on the land are not agricultural
C operations at all.

On the second limb of the argument, I agree, for the reasons given by Lord Keith of Kinkel, that the buildings in question do not qualify as agricultural buildings under section 2(1)(*a*) of the Act of 1971.

For these reasons, I, too, would dismiss the appeal.

Queen's Bench Division

January 28 1988

(Before Mr Justice KENNEDY)

CHANNEL SHIPPING (NEWPORT) LTD v NEWPORT
D BOROUGH COUNCIL

Estates Gazette April 23 1988

[1988] 16 EG 87-90

Rating — Rateable occupation — Whether occupation exclusive — Appeal by case stated from decision of magistrates holding that appellants' occupation of a warehouse and premises was exclusive — Four ingredients of rateable occupation — Authorities considered — Appellants in sole *de facto* occupation during relevant period — Complaint by appellants that magistrates failed to consider
E **anything other than the fact of sole occupation — It appeared from the case stated that the magistrates, because they believed that they had to consider actual occupation only, did not give any consideration to what estate or interest the appellants might have had in the premises — There was evidence that the owners of the premises had given the appellants a right to store goods there, for which a payment was made according to the tonnage stored, but that the owners had retained the right at their discretion to permit other persons also to store goods in the premises — The owners and the appellants shared control over the access to the premises —**
F **The magistrates had regarded the concession by the appellants that they had been in sole *de facto* occupation as conclusive — Held that it was clear from the authorities that the issue as to whether or not occupation is conclusive cannot be so easily resolved — The question involves consideration of rights and intentions as well as of the fact of sole occupation — The case stated showed that the magistrates had not gone on to consider such matters and had made practically no findings in relation to them — The question posed in the case was inappropriately expressed — The case must be remitted for further consideration in the light of the present judgment — Appeal allowed**

The following cases are referred to in this report. G

Cory v *Bristow* (1877) 2 App Cas 262, HL

Laing (John) & Son Ltd v *Kingswood Assessment Area Assessment, Committee* [1949] 1 KB 344; [1949] 1 All ER 224; (1949) 65 TLR 80; 47 LGR 64; 42 R&IT 15, CA

London County Council v *Wilkins (VO)* [1957] AC 362; [1956] 3 WLR 505; [1956] 3 All ER 38; (1956) 1 RRC 88; 49 R&IT 495, HL

Roads v *Trumpington Overseers* (1870) LR 6 QB 56

Rochdale Canal Co v *Brewster* [1894] 2 QB 852, CA

Westminster City v *Southern Railway Co Ltd, The Railway Assessment Authority and W H Smith & Son Ltd* [1936] AC 511; [1936] 2 All ER 322, HL

K Bush (instructed by Hopkins & Wood, agents for Edwards Geldard, of Cardiff) appeared on behalf of the appellants; Wyn H Williams (instructed by the Chief Executive, Newport, Gwent) represented the respondent council.

Giving judgment, KENNEDY J said: This is an appeal by way of case stated from a decision of the justices for the County of Gwent, sitting at Newport, who on November 21 1986 found that the rate duly made and demanded of the appellants in respect of a rating unit described as "Warehouse and premises" at East Dock Road, Newport, for the period from April 1 1984 to July 17 1985 was payable by that company.

One of the points taken before the justices was that the company could not be answerable for the entire period because it was not incorporated until July 23 1984 and did not commence trading until November 1984. That is now conceded by the rating authority, so if J this appeal succeeds the decision will relate only to the period after November 1 1984 and the sum at stake will have to be recalculated.

Most of the relevant facts are conveniently set out in para 2 of the case and have never been in dispute. Throughout the relevant period the premises were owned by W F Bartholomew & Sons, Property Investments Ltd and formed one unit for rating purposes with adjoining office premises. When the appellants began to trade they took over occupation from what may well have been an associated company and were in sole occupation of the premises until the end of the relevant period.

Para 2(e) of the case states:

> . . . Although others could have been let into occupation by the owner and the K appellant understood it could not have objected to this, no other person was actually let into occupation and any use of the premises by or on behalf of the owner was solely in connection with the upkeep and repair of the premises themselves.
>
> (f) Payment by the appellant to the owner in respect of its occupation of the premises was made on the basis of the tonnage of goods stored therein by the appellant and the period for which such goods were stored — a number of pence per ton per week. The tonnage stored, and hence the amount paid, varied from time to time depending upon the requirements of the appellant.
>
> (g) The appellant and the owner of the premises shared control over access to the premises. Each intended, throughout the period from April 1 1984 to July 17 1985, that the owner of the premises could terminate the appellant's occupation of the premises or alter the extent of its use of the premises upon reasonable notice. L

Section 16 of the General Rate Act 1967 provides that subject to the provisions of that Act "every occupier of property . . ." (such as that with which this case is concerned) ". . . shall be liable to be assessed to rates in respect of the hereditament or hereditaments comprising that property" according to the rateable value, and, as the justices found, a rate was duly made by the respondents, and for the period up to July 17 1985 it was duly demanded of the appellants.

The appellants then contended, and they contend before me, that they are not liable to pay the rates in respect of that period because they were not then the occupier of the premises for the purposes of section 16. The meaning in this context of the word "occupier" was said by the Court of Appeal in *John Laing & Son Ltd* v *Assessment Committee for Kingswood Assessment Area* [1949] 1 KB 344 to M contain four ingredients, namely: (1) there must be actual occupation; (2) it must be exclusive for the particular purposes of the occupier; (3) the occupation must be of some value or benefit to the occupier; and (4) the occupation must not be for too transient a period.

That analysis was approved by the House of Lords in *London County Council* v *Wilkins (Valuation Officer)* [1957] AC 362 and it is accepted by the parties in the present case. They are also agreed that as far as the appellants are concerned three of the four ingredients are satisfied and the dispute is confined to the question of whether or not the occupation by the company was exclusive for its particular purpose.

A For the rating authority Mr Wyn Williams submits that it was, because no one else was there. Even if the owners did have a right to require the company to share the premises with others it never exercised that right and the magistrates were entitled to disregard legal rights and to look at what happened in a rating period which was already past.

Certainly that is what they did. In para 6(b)(ii) of the case stated, after pointing out that in the circumstances the burden was upon the appellant company to show, on the balance of probabilities, that it had sufficient cause for not having paid the rate demanded, the magistrates say:

> As no other person was in fact in occupation of the premises, it was irrelevant B what intentions and beliefs the owner of the premises and the appellant held as to control over occupation, and irrelevant what method of payment for occupation was adopted. The appellant had exclusive occupation of the premises.

The question which I have to consider is then formulated by the magistrates as follows:

> Whether a reasonable bench of justices could have found that the company was in rateable occupation as aforesaid in the light of the evidence given on behalf of the respondent by William Bartholomew that: (i) he had not granted to the Company the exclusive right to store goods in the premises; (ii) he retained the right to permit other persons to store goods in the premises or any part thereof as he might in his discretion determine; (iii) the Company had no right to object to other persons storing goods in the premises under his C authority.

For the appellants Mr Bush submits that even in relation to a period which is already past, and even if there has been sole occupation, the question of whether or not there was during that period a right to exclusive occupation cannot be overlooked, and that on the facts of this case, as found by the justices, especially in para 2(e) of the case stated from which I have already quoted, there was in the relevant period no such right.

At para 6(b)(i) of the case stated the justices say in parenthesis:

> Because we were considering actual occupation, we did not go on to consider what, if any, interest or estate the appellant may have held in the premises.

That passage, submits Mr Bush, is a further illustration of the magistrates' mistaken approach, which, if correct, could mean that a D man allowed to park his wheelbarrow in the corner of an empty warehouse throughout a rating period could find himself liable for the whole of the rates.

Coming to the authorities, Mr Bush invited my attention first to *Roads* v *Trumpington Overseers* (1870) LR 6 QB 56 as support for the proposition that a person using land without an exclusive title to possession will be in rateable occupation if the character of his occupation is such that it does in fact exclude others from using the land in the same way. There the occupation of the land was for the purpose of raising coprolites. Here Mr Bush submits that the appellant company's occupation, on the facts as found, was not such as to exclude others from using the land in the same way — indeed the E company had no territorial interest in any particular part of the warehouse and paid for the facility of storing goods there by reference to the weight of goods and the period for which they were stored.

In *Cory* v *Bristow* (1877) 2 App Cas 262 the House of Lords was concerned with the occupancy of river moorings. At p 276 Lord Hatherley said:

> . . . The courts have not meant by the term "exclusively" that the interest may not be determined on certain terms and conditions, but merely that the person so occupying should have the right unattended by a simultaneous right of any other person in respect of the same subject matter.

A little later he said:

> F . . . a person sleeping at the hotel may have the exclusive use of his bedroom for the night, or the exclusive use of a sitting-room during the day, or a lodger the exclusive use of the chambers he occupies, still there is a concurrent right reserved by the landlord of the hotel or the person who lets the lodgings, of using the hotel or lodging-house for whatever purposes he may think fit for managing the establishment and all purposes connected with it. That is not such an occupation on the part of the lodger or the guest at the hotel as would make him liable to be rated.

In *Rochdale Canal Co* v *Brewster* [1894] 2 QB 852 the Court of Appeal was concerned with a berth and quay which had been set apart by the Mersey Docks and Harbour Board for the plaintiff canal company which had then been rated. At the beginning of his judgment at p 858 Lopes LJ stressed the importance of the intention of the parties, saying:

> The point in this case is whether such exclusive possession has been parted G with by the board as is necessary to make the respondents liable to pay rates. In determining this question, it is the intention of the parties which has to be looked at: it is not the words only that are to be regarded. The whole of the circumstances must be taken into consideration. It is the substance of the transaction, rather than the form that determines the question whether such an exclusive occupation exists as will make the property rateable.

In the present case Mr Bush submits that the magistrates closed their minds to everything other than the fact of sole occupation by the appellants throughout the relevant period. That was a relevant circumstance, but it was not the only relevant circumstance. It was also relevant to consider the terms of the contract under which the appellants came to be in occupation and the intentions of the parties, and those are matters which, it is clear from the case stated, the H magistrates deliberately decided to disregard.

In *Westminster City Council* v *Southern Railway Co Ltd* [1936] AC 511 the House of Lords was concerned *inter alia* with whether the occupation of railway property by a number of commercial undertakings such as booksellers, hairdressers and so forth, was exclusive. At p 567 Lord Wright MR emphasised the need to read the relevant contractual document as a whole.

Mr Wyn Williams recognises the weight of the authorities to which I have just referred, but submits that as the law has developed sole occupation during a particular period suffices to establish what he described as "*de facto* exclusivity".

In support of that submission he invited my attention to the decision of the Court of Appeal in *London County Council* v *Wilkins* J *(Valuation Officer)* [1955] 2 QB 653. There the issue was as to whether building contractors were in rateable occupation of their site huts, and Jenkins LJ considered separately each of those four ingredients of rateable occupation to which I have already referred.

At p 671 he said:

> Next, was the possession exclusive for the particular purposes of the contractors? In my judgment the answer to that question again must be yes. It is true that the architect had certain rights of access and it is also true according to the case, though we have not been referred to any express warrant for it in the contract, that the clerk of works could demand admittance to the huts at any time, but the huts were kept locked by the contractors and he could only enter by obtaining admittance from the foreman or some other employee who had the keys. In my view, the rights of access which were reserved to the K council or the architect or the clerk of works were certainly no more extensive than those which existed in the *Laing* case and were held by this court not to be sufficient to prevent the occupation from being exclusive.

The lord justice then went on to say that it was immaterial that the contractors, unlike a normal tenant taking possession, were given use of the site to erect buildings. He said:

> . . . the determining factor is actual *de facto* occupation and it matters not whether that occupation is the result of a licence or of a tenancy or other form of agreement.

At that point in his judgment Jenkins LJ had finished dealing with the question of exclusivity, and had turned to consider the relevance, if any, of the circumstances in which the particular occupiers came to L be in possession, and I can find nothing in that judgment to support the proposition for which Mr Wyn Williams contends.

The circumstances in the present case were such that, as the magistrates say at para 6(a) of the case stated, "it fell upon the appellant to show, on the balance of probabilities, that it had sufficient cause for not having paid the rate demanded". Therefore it was the appellants who undertook the burden of proving that their occupation was not exclusive for their particular purposes. They conceded that they were the sole occupiers and the magistrates, in my judgment wrongly, regarded that concession as decisive.

It must be clear from the authorities to which I have referred that the issue of whether or not occupation is exclusive cannot be so easily resolved. It involves consideration of rights and intentions as well as M consideration of the fact of sole occupation. Because the magistrates, as is clear from para 6(b)(i) and (ii) of the case stated, did not go on to consider those matters and made practically no findings in relation to them (save for the finding as to intentions set out in para 2(g)), all I can say at this stage is that the question which they pose seems to me to be inappropriately expressed. It recites some of the evidence given on behalf of the respondent but does not say whether or not that evidence was accepted. If it was, then it seems to me that the evidence of Mr Bartholomew as to the respective rights of the owner and the occupier, taken together with the magistrates' findings as to the intentions of those two parties, might well have led the magistrates to conclude that during the relevant period the appellants were not in

A rateable occupation because their occupation was not exclusive, but that is a matter which the magistrates must now decide.

The appeal was allowed with costs; case remitted to magistrates for further adjudication in the light of the judgment.

Queen's Bench Division

February 5 1988

(Before Mr Justice SCHIEMANN)

SHEAFBANK PROPERTY TRUST PLC v SHEFFIELD B METROPOLITAN DISTRICT COUNCIL

Estates Gazette June 25 1988

[1988] 25 EG 122-127

Rating — Rateable occupation — General Rate Act 1967, section 46A(1) — Whether sports ground purchased with a view to future resale but left unused was to be "treated as unoccupied" — Premises consisted of a clubhouse and other buildings and grounds of some 20 acres containing two football pitches and two bowling greens — Among equipment left behind by previous owners were furniture and sporting C equipment, freezer, dishwasher, refrigerator and television set — There was grass-cutting machinery in an outhouse and a part-time caretaker carried out maintenance of the pitches and greens

Section 46A(1) treats as unoccupied a hereditament which, apart from that section, would fall to be treated as occupied by reason only of there being kept on the hereditament plant, machinery or equipment which was used when the hereditament was last in use — This was the first occasion on which the High Court had been asked to construe the section — It was an appeal by case stated from the decision of magistrates who held that, despite section 46A(1), the D appellants, owners of the sports ground, were in rateable occupation — The magistrates had been persuaded that section 46A(1) did not apply — They considered that the maintenance carried out on the premises and the advantages derived from the continued storage of valuable effects were evidence of beneficial occupation — The rating authority took the view that section 46A(1) did not apply because the magistrates in reaching their decision took into account material other than the plant, machinery or equipment mentioned in the section — Held, after considering section 46A(1) and a number of authorities, that the magistrates had E fallen into error — They had taken the equipment and the maintenance activity into account — The proper approach should have been to ask whether, but for the presence of the plant, machinery or equipment, they would have found the hereditament unoccupied — Applying the law correctly, the only course which the magistrates could have taken was to find that there was no rateable occupation — Appeal allowed and magistrates' decision and distress warrant quashed

The following cases are referred to in this report.

Arbuckle Smith & Co Ltd v *Greenock Corporation* [1960] AC 813; [1960] 2 WLR 435; [1960] 1 All ER 568; (1960) 58 LGR 155; 6 RRC 95; 53 R&IT F 182, HL
Associated Cinema Properties Ltd v *Hampstead Borough Council* [1944] KB 412
1230
Harter v *Salford Overseers* (1865) 6 B&S 591; 34 LJMC 206; 29 JP 647
Townley Mill Co (1919) Ltd v *Oldham Assessment Committee* [1937] AC 419; [1937] 1 All ER 11; (1937) 156 LT 81; 53 TLR 205; 6 DRA 77, HL

This was an appeal by way of case stated by Sheafbank Property Trust plc from the decision of South Yorkshire magistrates holding that the appellants were in rateable occupation of a sports ground and buildings thereon at Oliver's Mount, Sheffield. The respondents were the rating authority, Sheffield Metropolitan District Council.

Stephen Sauvain (instructed by Addleshaw Sons & Latham, of Manchester) appeared on behalf of the appellants; Christopher G Lewsley (instructed by the solicitor to Sheffield Metropolitan District Council) represented the respondents.

Giving judgment, SCHIEMANN J said: This is an appeal by way of case stated against a decision of the magistrates sitting in South Yorkshire holding that the appellants were, in the rating period starting in April 1986, in rateable occupation of a sports ground including a clubhouse and a caretaker's flat all of which are described in the valuation list as "part of sports ground and premises . . . rateable value £3,350." The case is of some general importance because its facts are probably not all that unusual and partly because I am told that this is the first case in which section 46A of the General Rate Act 1967, which was inserted into that Act by the Rates Act H 1984, has fallen to be construed by the High Court.

The findings of fact in the case are as follows:

C The legal owner of the said premises from a time prior to the relevant rating period, and up to the determination of the Complaint, was Sheafbank Property plc of St Anne's Place, Manchester.

D General rates for the said period in respect of the said premises had been lawfully demanded of the Appellant and that at the date of making of the Complaint payment was still due from the Appellant more than seven days after such demand.

E At the date of determining the complaint, general rates of £19,080.22 remained outstanding in respect of the said premises. This figure comprised the said £11,452.31 for the year April 1 1986 to March 31 1987 together with £7,627.91 which was owed by the Appellant in respect of the same premises J for the period April 2 1985 to March 31 1986. This second figure is the net amount due for that period specified in a Distress Warrant granted to the respondent, which was unopposed by the Appellant, after the Appellant paid £2,000 for the previous period.

F The Appellant Company was formed in 1981 for the business of property investment and trading and had acquired the said premises in 1985 with a view to making a profit on resale.

G That on July 3 1987 Mr Russell Kelly, a Rating Inspector employed by the Respondent, visited Olivers Mount and found the premises comprised: a clubhouse, containing changing rooms; a lounge/bar; a kitchen; a snooker room; a caretaker's flat above; outbuildings for storage of equipment; grounds of approximately 20 acres containing two football pitches and two bowling greens.

H On July 3 1985 a caretaker employed by the Appellant was occupying the K aforementioned flat (which was furnished) and was present during Mr Kelly's visit. The snooker room contained a full size snooker table together with cues, rests and balls and chairs were set out around the room. The main lounge contained tables and chairs, music facilities, and a bar stocked with glasses and other bar equipment. There was no stock of alcohol. The grass on the sports pitches had been recently cut and goalposts were in position, but the pitches were not clearly marked out. There was grass cutting equipment in the outbuildings.

J On July 14 1986 Mr Kelly again visited the premises together with Mr Richard Sherwood, the Principal Rating Officer for the Respondent. They were accompanied by Mr M D Deakin, a Director of the Appellant Company. The grass on the two pitches was being maintained to quite a high standard and the goalposts were still in position but needed painting. The pitches were marked out but needed relining. The grass on the two bowling greens had been recently cut but one of the greens had quite a lot of moss on it. It was stated by L Mr Deakin that a part-time groundsman was being employed by the Appellant Company who worked mornings only. The changing rooms were locked: inside they were clean but empty.

The outbuildings contained grass cutting equipment. In the clubhouse the lounge contained furniture, a record deck with public address equipment, a bar and bar equipment. The kitchen contained a freezer, a dishwasher and other kitchen equipment. There was no evidence of the clubhouse actually being used. The caretaker's flat was not occupied: it contained a television, a fridge, chairs and a settee.

K The furniture and other equipment in the clubhouse, outbuildings and grounds had been included in the sale of the premises when purchased by the Appellant Company and had remained on the premises ever since.

L The Appellant had sought offers for the sale or lease of the premises as a whole. The equipment and other effects present on the premises could have M been included in the sale. Certain offers had been made to the Appellant but had not been accepted.

M The contents of the premises, particularly the snooker table, were of not inconsiderable value.

N No other person or body had occupied the premises during the relevant rating period.

O No caretaker now resided on the premises.

P Save for the grounds being maintained by use of the equipment described aforesaid and the caretaker's flat apparently being occupied at the first visit, there was no evidence of the premises being used during the Appellant's occupation.

These can, I think, be summarised without unfairness to either side by saying that the company bought the property and its contents in

1985 with a view to making a profit on resale, the contents have remained in the property ever since, the property and contents have not been used during the relevant period by anyone save for the purpose of maintenance and the appellant has sought offers for the sale or lease of the premises as a whole; the equipment and other effects present on the premises could have been included in the sale; certain offers have been made to the appellant which have not been accepted.

It is only necessary for me to refer to two cases which set out guidelines on how the question of rateable occupation is to be approached. The first of these cases is *Associated Cinema Properties Ltd* v *Hampstead Borough Council* [1944] 1 KB 412. That was a case where a company took a lease of some dwelling-houses during the war so that they might have accommodation available for use as offices in the event of their own premises being rendered unfit for use by enemy action or for use owing to the exigencies of their business. The company did not use their reserve premises for any purpose nor did they enter into physical possession of them, nor did they place any furniture or other chattels on the premises at any material time. It was held that in those circumstances they were not in rateable occupation of the premises.

Du Parcq LJ, delivering the reserved judgment of the court, said at p 414:

> It is well settled that, to constitute occupation under the statute of Elizabeth (43 Eliz C 2) there must be use and enjoyment of the hereditament. Thus in *Borwick* v *Southwark Corporation* [1909] 1KB 78 Bigham J defined occupation, for the purpose of rateability, as being "constituted of legal possession and of permanent (as distinguished from mere temporary) user". When the hereditament is a house, it is not rateable unless it is occupied as a house, unless, that is to say, it is used for one of the purposes for which a house may be used . . . It may be observed that every owner who has the legal possession of an empty house, may, if he pleases, occupy it or otherwise use it at any time, and it may ordinarily be assumed that he would occupy it if ever it suited his purpose to do so. If it were legitimate to regard him as an occupier for that reason, the rule stated by Lord Russell, which has been firmly established, would cease to have any meaning or effect.

That rule, which was cited earlier in the judgment, was that the owner of an empty house has the legal possession but he is not in rateable occupation.

Du Parcq LJ went on:

> It was contended, however, on behalf of the council that "the possibility of physical occupation coupled with the intention to occupy physically in case of need" constituted rateable occupation. In support of this submission much reliance was placed on the two cases which the justices had regarded as compelling them to hold that the company were occupiers. Of these the first was *R* v *Melladew*. The hereditament there in question was a warehouse. Now, it is the business of a warehouseman to let unoccupied space for the purpose of the temporary storage of goods. To be available for this purpose, the warehouse must be empty, in whole or in part. If the owner makes a public announcement that he is prepared to let space in an empty warehouse, he may reasonably be said to be using that warehouse for the purposes of his business . . . It is significant that no case could be cited in which occupation had been held to be established without proof of some overt act amounting to evidence of user. In *R* v *Melladew* the owner had advertised the premises as a warehouse. In most cases user has been proved by showing that the house has been furnished, or equipped for some business purpose. There is certainly no authority binding on this court for the proposition advanced on behalf of the council, and, in our judgment, a mere intention to occupy premises on the happening of a future uncertain event, cannot, without more, be regarded as evidence of occupation.

A little later he posed the case:

> A man has left furniture in a house, though he is not residing there. That fact shows an intention to return or to enter and so proves that he is enjoying the use of a furnished house. The tenant of a seaside boarding-house has shut it up for the winter. The fact that he intends to reopen it in the summer is proof that he is now using and occupying it as a boarding-house, although the seasonal nature of the trade causes it to be temporarily closed. It is never true to say that a mere intention to occupy in hypothetical circumstances which may never come into existence is equivalent to occupation.

That case cites some of the past authorities and the sort of principles that have been considered. The other case I propose to mention is a case called *Arbuckle Smith & Co Ltd* v *Greenock Corporation* [1960] AC 813. It is a decision of the House of Lords sitting on an appeal from Scotland, but decided in circumstances where their lordships proceeded on the footing that there was no difference between the laws of Scotland and England in the principles affecting rateable occupation.

That was a case where a company bought a warehouse with a view to using it as a bonded store. They bought it in March 1957. Approval of such use by the Customs and Excise was necessary and was applied for, but before it could be granted considerable alteration in the premises were required. From March 3 1958 to June 16 1958 that work was carried out and the company had not previously used the premises as a warehouse. The authority claimed that the company was in occupation and the company denied it.

Going through the judgments in the House of Lords, I stop first at some comments of Viscount Kilmuir LC at p 821, where he said this:

> . . . an owner of land must have made some actual use of the premises in the relevant year before he can be called on to pay rates. The crucial question, however, is what kind of use is contemplated or, in other words, whether it is sufficient to make alterations with the intention of carrying on a business when the alterations are completed.
>
> I do not think it is. I accept broadly the argument for the appellants which can be put in this way. The sole purpose of the appellants in acquiring the premises was to use them as a bonded store in connection with their business as warehousemen. The alterations were necessary in order that this purpose might, if the alterations were approved by the Customs and Excise, receive effect. Yet activity carried on in relation to premises, the sole object of which is to make the premises fit for the only use which is contemplated, does not amount to the kind of actual user as is essential to rateable occupation. So long as the activities were confined to making the premises fit for a contemplated purpose, the premises were not serving the appellants' purposes as warehousemen. The premises were not being applied to the purpose for which they existed but were in an antecedent stage. It must be remembered that under rating law it is open to the owner to sterilise a property — whether by leaving a house without furniture or otherwise — which is perfectly capable of being let for a valuable rent. If, therefore, there is no use of premises according to their nature, I find it difficult to see how there is occupation attracting liability for rates.

A little later on he says:

> I cannot, however, take their next step of saying that because it was an advantage to the appellants to alter the premises, therefore the appellants were making beneficial use of the premises. It is an undoubted advantage to the owner of an empty house to put on some slates and keep it weathertight, but the fact that he does so could not create rateable occupation where none existed. I cannot myself accept the view that when a person is repairing or altering something designed for a particular purpose, he is by that action making use of it.

Lord Reid, in his judgment at p 824, pointed out that:

> No one has contended that entry for inspection or cleaning or ordinary maintenance such as heating in cold weather or carrying out ordinary repairs to make the premises wind- and water-tight constitutes occupation, and it appears to me that occupation must involve something more than that.

At p 825, after referring to some cases involving statutory boards, he said:

> There statutory boards had made full use of their premises; they did all the things that a normal occupier would do, and the quality of their use was not in issue. Their case was that they did not have to pay rates because they were precluded from making profit. Their occupation was nevertheless held to be beneficial and rateable: it was beneficial because it was a property capable of yielding a net annual vaue. Since that time occupation of such property has been held rateable whether or not the occupier derives profit from it. But those cases throw no light on the question what kind of use amounts to occupation, and do not in the least support the fallacious argument: occupation to be rateable must be beneficial; these owners made beneficial use of their property; therefore they were in rateable occupation.

Lord Radcliffe echoes those observations and says, at p 828:

> It is evident, therefore, that there will not be occupation in the context of rating unless some use is made of the hereditament in the course of the relevant year. "Use" is not a word of precise meaning, but in general it conveys the idea of enjoyment derived by the user from the corpus of the object enjoyed. When rating law turns its attention from dwelling-houses to industrial or commercial premises the same conceptions are, in my opinion, applicable to determine whether or not occupancy ceases or begins, though they must be adapted to the particular circumstances of such premises, factory or warehouse, and the business for the purpose of which they are created. Nevertheless, I think that the words of Farwell LJ in *R* v *Melladew* provide a satisfactory indication of the correct test: "Rateable property has many varieties; of some the normal use is by personal occupation, eg, a dwelling-house, of others by occupation by live or dead stock, eg a linhay used as a shelter for cattle, or a barn; and the nature of the property and its mode of use must be considered in each case. The test, in a case like the present, of business premises, appears to me to be, Has the person to be rated such use of the tenement as the nature of the tenement and of the business connected with it renders it reasonable to infer was fairly within his contemplation in taking or retaining it."

Those cases set out the legal background against which I decide the present case.

It is in my judgment beyond argument that, leaving aside the presence of the furniture and grass-cutting equipment, none of the remaining facts found could give rise to rateable occupation. The groundsman's activities were limited to keeping the hereditament in good order or at best improving it.

The only matter which could result in the appellants being rateable is the presence of the furniture and grass-cutting equipment. There is an argument, apart from statute, that the keeping of those items on the premises resulted in the appellants being in rateable occupation — see the many cases dealing with furniture left in a house and *Harter* v *Salford Overseers* (1865) 6 B&S 591. That was a case where the rating authority rated a silk mill and some other buildings in circumstances where machinery had been left in the hereditament while the owner tried to sell it. The authority did not seek to rate the mill as a mill but rather as a warehouse — see the respondents' counsel's intervention at p 594 — and it was held that they were entitled so to do. I rather doubt whether, even on the law as it stood before the insertion of section 46A into the General Rate Act 1967, that case would have been similarly decided — see the remarks of Lord Russell of Killowen in *Townley Mill Co (1919)* v *Oldham Assessment Committee* [1937] AC 419 at p 431 — nor am I convinced that the present case is indistinguishable from it.

However, the matter is now governed by section 46A. This reads as follows:

> (1) For the purposes of this Act a hereditament to which this section applies shall be treated as unoccupied if, apart from this section, it would fall to be treated as occupied by reason only of there being kept in or on the hereditament plant, machinery or equipment, (*a*) which was used in or on the hereditament when it was last in use, or (*b*) which is intended for use in or on the hereditament.

It is common ground that the material left behind is plant, machinery or equipment which fell within (*a*) and that the section applies to this hereditament.

How, then, does it come about that the rating authority submitted that this section has no application? It argued that in the present case the magistrates took into account in reaching their decision material other than that relating to the plant, machinery and equipment and that if they were entitled so to do then the section has no application because the hereditament was not "treated as occupied by reason *only* of there being kept in or on the hereditament plant, machinery or equipment" but also by reason of other matters. I have already indicated my view that the magistrates were not entitled to take the groundsman's activities into account. However, it is clear that they *were* entitled to take into account the fact that the appellants owned the hereditament and no doubt the magistrates did take that into account and that therefore the hereditament was not treated by the magistrates as occupied by reason only of the presence of the furniture etc.

In my judgment, the authority is proceeding on a misconstruction of the section and persuaded the magistrates to do likewise. To adopt that construction would in nearly all, if not all, cases render the section inapplicable. There is practically always some other matter to be taken into account apart from the plant etc on the hereditament: in particular, as here, ownership. The proper approach to this section is for the rating authority or magistrates to ask themselves whether, but for the presence of the plant, machinery or equipment, they would have found occupation. In the present case the magistrates concluded, in para 6 of the case:

> We were of the opinion that: (a) By maintaining the premises in such condition that they were ready for occupation, the Appellant was in beneficial occupation of the rateable hereditament. (b) The Appellant derived a benefit from the continued storage of valuable effects on the premises. (c) Taken alone, the maintenance of the bowling greens and sports pitches was not sufficient evidence of occupation. Accordingly, we issued a Distress Warrant against the Appellant for the recovery of general rates of £11,452.31 plus £10 costs included in the Application and £30 legal costs against the Appellant for the Respondent's legal costs of the hearing.

It is clear, looking at the totality of that paragraph, that the magistrates must have taken the "equipment" into account and also taken into account the maintenance activity. However, in my judgment they fell into error in so doing. They had been bombarded by authority and my sympathies are with them in dealing with this branch of the law. I quite see how they fell into error. However, it is clear, and I think accepted by the respondents, that if the law be as I have held it to be then the only course which the magistrates could have taken on their findings of fact would have been to find that there was no rateable occupation. I therefore quash their decision to issue a distress warrant and their order as to costs. I am minded to give the appellants their costs of this appeal and also before the magistrates. I indicated to counsel arguing before me that costs would follow the event, and they agreed that seemed most probable. If there are no submissions on costs I make that order.

The magistrates' decision and distress warrant and costs order were quashed with costs.

Queen's Bench Division

February 19 1988

(Before Mr Justice KENNEDY)

CALMAIN PROPERTIES LTD v ROTHERHAM METROPOLITAN BOROUGH COUNCIL

Estates Gazette June 25 1988

[1988] 25 EG 127-134

Rating — Rateable occupation — Appeal by owners of warehouse from decision of magistrates holding that the owners were in rateable occupation of a particular bay of the warehouse — The warehouse consisted of four bays, three of which were occupied by a warehousing company — The fourth bay, the subject of the present appeal, had not been used for storage or any other purpose until after the end of the period in respect of which rates were demanded — The magistrates nevertheless decided that the appellants were in rateable occupation of this bay during that period — The magistrates held that the bay was open for business and that the appellants were in rateable occupation because of the terms in which they were advertising the property and because of the powers of the appellant company according to the magistrates' reading of its memorandum of association — The latter was couched in the very wide terms which are not unusual in memoranda of association and the magistrates took the view that it empowered the appellants to act as warehousemen — The appellants' submission was that the memorandum simply gave the appellants power as a property company to develop and manage premises in their capacity as owners

The magistrates also took the view that the estate agents' particulars advertising the bay indicated that the appellants were not merely offering to let the premises as an ordinary real estate transaction but were in effect saying that this bay was open for business, namely, the provision of warehousing services — The appellants submitted that this was an unnatural construction — The particulars contained a straightforward offer to let and this was supported, *inter alia*, by the absence of any reference to warehousing fees and by an indication of the user's responsibility for rent and rates

After consideration of a number of authorities, Kennedy J held that the magistrates were entitled on the evidence to come to the conclusion they did, namely, that the appellants had been in rateable occupation during the material period — The judge added that he was by no means convinced that in the place of the magistrates he would have reached the same conclusion, but the finding was a matter for them subject to the court's being satisfied that there was evidence on which they could so decide — Appeal dismissed

The following cases are referred to in this report.

Associated Cinema Properties Ltd v *Hampstead Borough Council* [1944] KB 412

Borwick v *Southwark Corporation* [1909] 1 KB 78; (1909) 99 LT 841; 7 LGR 10

Hewson, Chapman & Co Ltd v *Grimsby County Borough Council* (1953) 51 LGR 649; 46 R&IT 703, DC

Laing (John) & Son Ltd v *Kingswood Assessment Area Assessment, Committee* [1949] 1 KB 344; [1949] 1 All ER 224; (1949) 65 TLR 80; 47 LGR 64; 42 R&IT 15, CA

Liverpool Corporation v *Chorley Union Assessment Committee* [1912] 1 KB 270
London & North Eastern Railway Co's Appeal, Re [1946] KB 27; (1946) 173 LT 412; 109 JP 269; 43 LGR 191
R v *Melladew* [1907] 1 KB 192
R v *St Pancras Assessment Committee* (1877) 2 QBD 581

This was an appeal by Calmain Properties Ltd (previously known as NMT Properties Ltd) by case stated from a decision of Rotherham magistrates, who found the appellants liable to pay £90,224.02 in rates demanded by the respondents, Rotherham Metropolitan Borough Council, in respect of Bay 4 of the appellants' four-bay warehouse. The issue was as to whether the appellants had been in rateable occupation of the bay during the material time.

David Widdicombe QC and John Grove (instructed by Muscatt Walker & Co) appeared on behalf of the appellants; Patrick Hamlin (instructed by T C Mumford, director of legal and administrative services, Rotherham Metropolitan Borough Council) represented the respondent council.

Giving judgment, KENNEDY J said: This is an appeal by way of case stated from a decision of the justices sitting at Rotherham magistrates' court, who on April 11 1985 found the appellants liable to pay rates demanded by the respondents in the sum of £90,224.02 and therefore issued a distress warrant in that sum.

The premises in respect of which rates were claimed were Bay 4 of a four-bay warehouse completed in 1982 and owned by NMT Properties Ltd, which changed its name to Calmain Properties Ltd, the present appellants, on December 19 1984. Nothing turns on that change of name and for convenience I shall refer to the appellants hereafter as NMT.

From April 1 1982 onwards Bays 1, 2 and 3 were occupied by a warehousing company providing warehouse services for Carreras-Rothman Ltd. Until September 30 1983 the warehousing company which accepted responsibility for the rates was Industrial Storage & Forwarding Ltd, and thereafter it was Industrial Storage (Rotherham) Ltd.

Mr R N Khan was a director of each of those companies. He held 90% of the shares in Industrial Storage (Rotherham) Ltd and he was the sole director of NMT.

Bay 4 was not used for storage or for any other purpose until 1985, after the end of the period in respect of which rates were demanded. These respondents do not levy a rate on unoccupied property, having not availed themselves of the provisions of section 17 of the General Rate Act. The sole issue before the magistrates was whether NMT was in rateable occupation of Bay 4. The magistrates found that the company was in rateable occupation and the question for this court is whether that decision was right in law.

The law which the magistrates had to apply was not and is not in dispute. The dispute is as to whether, if the magistrates directed themselves correctly as to the law, there was any evidence entitling them to find as they did. But before I turn to the evidence I must say something about the law.

Of the four ingredients of rateable occupation identified in *Laing (John) & Son Ltd* v *Kingswood Assessment Area Assessment Committee* [1949] 1 KB 344, the one in issue in the present case is actual occupation or possession.

In *R* v *St Pancras Assessment Committee* (1877) 2 QB 581 Lush J said at p 588:

> Occupation includes possession as its primary element, but it also includes something more. Legal possession does not of itself constitute an occupation. The owner of a vacant house is in possession, and may maintain trespass against any one who invades it, but as long as he leaves it vacant he is not rateable for it as an occupier. If, however, he furnishes it and keeps it ready for habitation whenever he pleases to go to it, he is an occupier, though he may not reside in it one day in a year.
>
> On the other hand, a person who, without having any title, takes actual possession of a house or piece of land whether by leave of the owner or against his will, is the occupier of it.

That passage was cited with approval by Buckley LJ in *Liverpool Corporation* v *Chorley Union Assessment Committee* [1912] 1 KB 270. But before I come to the *Liverpool* case I must consider *R* v *Melladew* [1907] 1 KB 192, a decision of the Court of Appeal.

That case concerned an empty warehouse from which the owner had retreated into his adjoining premises. He was held to be still in occupation because of the extent to which he was willing and able to bring back into use the empty premises. At p 202 the Master of the Rolls, Sir Richard Henn Collins, said:

> I am aware that ownership is distinct from occupation, and that an owner does not make himself rateable by trying to let a house which he has ceased to inhabit. But the principle involved in that proposition does not apply to the case of the tenant of a vacant warehouse who retains control over it for the purpose of letting storage room. In the case of the owner seeking to find a tenant in the case put, no assertion of occupation by him is involved. If he gets a tenant, he substitutes the occupation of another person for his own. The warehouseman in getting a customer supplies visible evidence of his own occupation. In the case of the owner who tries to let the intention is not to occupy. In the case of the warehouseman it is exactly the reverse.

Here Mr Widdicombe submits that NMT was simply a property company seeking a tenant, not a warehouse company seeking customers, and if he is right about that the magistrates' decision cannot stand.

In *Borwick* v *Southwark Corporation* [1909] 1 KB 78 some manufacturers purchased a building to which they could transfer their factory in the event of a fire or other emergency and fitted it with some shafting and wooden benches. They were then held to be in rateable occupation. Bigham J issued a warning, which Mr Hamlin for the respondents invites me to heed. At p 84 the learned judge said:

> Cases of this kind depend much more on fact than on law. Whether a man "occupies" or not is in each case a question of intention to be ascertained with reference to the particular circumstances, and if there are facts which one way or the other can reasonably support the conclusion at which the justices arrive, I do not think this Court should interfere with that conclusion. It is a finding of fact.

The *Liverpool* case to which I have already referred was concerned with land adjacent to reservoirs and waterworks. In that case Buckley LJ in the Court of Appeal referred to the intention of the alleged occupier as a governing factor in determining whether rateable occupancy is established. Mr Hamlin also relies on that case to support the proposition that an owner in possession is prima facie the rateable occupier once beneficial occupation has begun. But of course the issue in this case, as Mr Widdicombe emphasised in reply, is whether beneficial occupation ever did begin.

In *Associated Cinema Properties Ltd* v *Hampstead Borough Council* [1944] 1 KB 412 the appellants took a tenancy of two dwelling-houses to use as offices if their premises were damaged by enemy action. They did not put anything into the houses and the appellants were held never to have gone into rateable occupation.

Referring to the *Melladew* case, Du Parcq LJ said at p 415:

> The hereditament there in question was a warehouse. Now, it is the business of a warehouseman to let unoccupied space for the purpose of the temporary storage of goods. To be available for this purpose, the warehouse must be empty, in whole or in part. If the owner makes a public announcement that he is prepared to let space in an empty warehouse, he may reasonably be said to be using that warehouse for the purposes of his business. This, briefly, was the ratio decidendi of *R* v *Melladew*.

In the present case it was certainly material for the magistrates to consider whether the owners made the sort of announcement to which Du Parcq LJ referred. Some doubt was cast on *Borwick's* case, at least in so far as it may suggest that a mere intention to occupy empty premises in a problematical contingency is equivalent to occupation. But no doubt was cast upon that passage from Bigham J's judgment in the *Borwick* case to which I have referred.

At p 416 in the *Associated Cinema* case Du Parcq LJ said:

> It is significant that no case could be cited in which occupation had been held to be established without proof of some overt act amounting to evidence of user.

Mr Widdicombe invites particular attention to that passage, but of course it has to be read in conjunction with the passage which I cited a moment ago.

My attention was also invited to the *LNER* valuation appeal [1946] 1 KB 27* and to *Hewson, Chapman & Co Ltd* v *Grimsby County Borough Council* (1953) 46 R&IT 703, but it seems to me that those decisions do not add significantly to the law as stated in the earlier cases.

How is the law to be applied to this case? This morning Mr Hamlin put before me four propositions: that in the light of the authorities, first, the nature of the occupation varies with the premises; second, that slight occupation will suffice; third, that for there to be beneficial occupation the benefit can be contingent; it is sufficient if it is part of the occupier's undertaking; and, fourth, the essential question is one of intention. In general terms that seems to me to be a reasonable summary of the authorities to which I have just referred.

*Editor's note: *Re London & North Eastern Railway Co's Appeal.*

A Undoubtedly Bay 4 was not in rateable occupation as soon as it was completed, although it was owned from the start by NMT. As it was a warehouse, the first signs of rateable occupancy would not necessarily be the arrival of fittings, but there would have to be something to indicate that it was open for business. That was the approach which the magistrates were invited to adopt and which they did adopt, and in my judgment it was the right approach.

For the appellants it was contended that Bay 4 never was open for business. It was simply being offered as a unit, possibly divided, to any tenant who cared to occupy it.

The magistrates found that the appellants were in rateable occupation because of the terms in which NMT was advertising this property of theirs, coupled with what the magistrates conceived to be B the powers of NMT as set out in its memorandum of association. Mr Widdicombe submits that they misinterpreted both the advertisement and the powers.

The memorandum of association is, as Mr Widdicombe says, a prolix and repetitive document. Para III sets out the objects of the company, and part of para III subpara (2) reads:

> . . . and to build, construct, alter, extend, equip, execute, carry out, improve, work, develop, remove, replace, maintain, manage or control all and any heritable or real estate, including but not limited to shops, stores, warehouses, depots, offices, works, plants, factories, roads and other buildings, structures or facilities of all kinds, whether for the purposes of the Company or for sale, letting or to hire or in return for any consideration from any company, firm or person, and to contribute to or assist in or carry out any part C of any such operation . . .

Mr Widdicombe submits that those words read in context do not, as the magistrates found, empower the appellants to act as warehousemen. Read in context and bearing in mind the title of this company, which was NMT Properties Ltd, they just give to NMT the power to develop and manage premises as an owner.

Simply by way of contrast, Mr Widdicombe invited my attention to the memorandum of association of Industrial Storage (Rotherham) Ltd, which does demonstrate how the memorandum might be worded if it were a memorandum being prepared in respect of what was intended to be a warehousing company. But of course it is relevant to note that the memorandum in respect of Industrial D Storage (Rotherham) Ltd is a very much shorter document, extending in all a little over three pages and, without intending to be in any way offensive, it was an English document and not a Scottish one.

Mr Hamlin submits that the memorandum of NMT does show not that the company would act as warehousemen but that it had the capacity to do so within the scope of its objects. He accepts that it is necessary for me to look at the document as a whole and that then, he submits, I must give the words of the document their ordinary meaning. Where one finds in the passage which I quoted words like "manage" or "control", that cannot, he submits, be restricted to managing or controlling, for example, a warehouse in the portfolio E sense. Similarly, where one finds verbs like "carry out" and "work", those verbs, if applied to the following types of enterprises, would be inappropriate if all that was intended was that this company should act as, what can conveniently be described as, a property company.

He invites me to look at the final words of the passage which I have quoted where there is a reference to the company acting in return for any consideration from any company, firm or person, and he invites me to consider what consideration there could be going beyond sale, letting or hire, which have already been referred to. He submits that the sort of consideration that the draftsman must have had in mind was the fees which would have been earned by storage, for example. But Mr Widdicombe answered that by saying that the consideration F might well have been, for example, shares in another company.

Mr Hamlin then invites my attention to the following clause or subclause in the memorandum which empowers this company as one of its objects to carry on any other business or activity whatsoever, and, submits Mr Hamlin, that must mean, even if it had not already been spelt out, that the company could within the scope of its principal objects have carried on the business of warehousemen.

Each of these objects was, he submits, a principal object. It comes, relatively speaking, at the beginning of this document, and further into the document one can find the powers which one would expect to find in the case of a company.

In clause III, para 2, there are, as is agreed on both sides, four parts. The first, second and fourth part are, Mr Hamlin submits, clearly objects of the company and principal objects of the company. G The first part deals with the acquiring and holding of real and personal property; the second part with the selling of such property; and the fourth part with carrying on business as builders and constructors. Mr Hamlin submits that it would be surprising indeed if sandwiched between parts of a clause which clearly set out objects of the company there were to be found a passage which is simply a power and not itself an object. Powers, he submits, are to be found much later on in clause V where there is a specific reference to powers.

In my judgment, the framer of the document in Scotland has deliberately chosen to cast his net very widely, no doubt to enable the company to operate in a flexible way. If this company did choose to run a warehouse, I do not see how it could be said to escape from its H obligations by contending that it was acting outside the scope of the memorandum of association. Accordingly, on this issue I conclude that the magistrates were entitled to find as they did.

I turn now to the estate agents' particulars which, of course, speak for themselves. There is no clear evidence before me as to when those particulars were published, but I do know, as I have already indicated, that the building with which I am concerned was finished in 1982 and it seems reasonable to infer — and both sides seemed content that I should infer — that these particulars were published and made available in late 1982 or in 1983.

Mr Widdicombe says that they constitute just an offer to let a building or part of a building and they are not in any way an indication that Bay 4 was a warehouse now open for business. He J points first of all to the agents who were being used by NMT — Merryweather Corbett & Dawson — who describe themselves as auctioneers, valuers, land and estate agents. And, submits Mr Widdicombe, that would be surprising if in fact what was intended was that a warehouse should be let in the sense that customers who require the use of warehouse space should be attracted by the advertisement.

Next Mr Widdicombe points to the lack of any reference to a charge being levied by reference to the amount of goods stored and the period during which the goods are to be stored, which he submits would be the normal way of raising revenue if what was being offered was warehousing facilities.

Then he points to the suggestion that the user of the warehouse may K have responsibility for rent and rates, which he submits would be unusual in the case of one normally using warehouse space.

He concedes that the particulars are poorly worded. Under the subheading "Lease" they read:

> Term by negotiation. Lessors are prepared to consider short, medium or long term period:
> Availability as a building lease or alternatively with full on site facilities provided by one of the leaders in the field of storage, materials handling and forwarding.

In this context a building lease clearly means a lease of the whole building, but it is the alternative offer upon which Mr Hamlin relies.

Mr Widdicombe submits that that alternative offer should be read L as if it went on to say "who would provide full on-site facilities for you".

Mr Hamlin asks by way of a rhetorical question: who is the leader in the field of materials, storage, handling and forwarding who is there being referred to? Was it not open to the magistrates to infer that it was indeed NMT who were placing the advertisement themselves or through some subsidiary whom they might nominate to carry out this work?

Then Mr Hamlin invites my attention, still in relation to these same particulars, to the way in which the particulars begin: 43,500 sq ft are being offered followed by the words "or may divide". He submits that there is an indication which comes within the precise scope of what was said in the case of *Melladew* to which I have already M referred.

At p 202 in that case Sir Richard Henn Collins MR said:

> If he holds himself out to let storage space not involving a demise of the whole warehouse, and, by securing exclusive control over the premises, has put himself in a position forthwith to give the accommodation required, is he to be deemed as not the occupier until some customer has been found to deposit goods for storage?

and he answered that question in the negative.

Mr Widdicombe in dealing with this point says that "or may divide" means no more than that the owners were prepared to consider separate leases. Mr Hamlin points to the words under the photograph "Available on short, medium or long term basis". He

A submits that a short-term basis would be a clear indication that the warehouse was open for business in the normal conventional sense and in the sense in which the magistrates were being invited to approach this question. Mr Widdicombe says that cannot be so. Here again all that is to be implied is that there could be, as the words suggest, a short lease.

Still on the same page, Mr Hamlin invites my attention to part of the description of the property which reads "Superb on site security, including controlled security boom". He says that of course the magistrates had the benefit of evidence in relation to what was being offered.

But Mr Widdicombe points out that the security boom can in fact be observed on the plan on the preceding page and is no more, he B submits, than a feature of the landscape. It was not an offer to provide some particular facility which NMT, as the owners of the building, were putting forward.

It seems to me that in the end it was open to the magistrates to read these estate agents' particulars as they did, namely, as an indication that the appellants were not just offering to let a building or a building in sections, and to find a warehouse company to run it if the prospective tenant so desired, but were actually indicating that the building as a warehouse was open for business.

As they said at p 12 in the case:

> At the stage in question, however, where no tenant had actually been found and the extent of the service requirements of a prospective tenant are not C known, we were of the opinion that there was nothing to prevent NMT Properties Ltd from providing certain service requirements themselves in respect of the warehouse.

Having regard to my finding in relation to the construction of the memorandum of association, that must be right.

The magistrates went on:

> Even if this did not eventually happen and that NMT Properties Ltd did not D directly provide the service facilities required, it was evident that they would indirectly provide those services by arranging for the property to be initially let to another company, such as Industrial Storage (Rotherham) Ltd, who would undertake to provide whatever services were required, and by introducing the prospective tenant to that company.

As to intention they then said this: "They were, in any event," — and "they" of course means NMT — "ready to make available warehouse space to take in the goods of a prospective tenant . . .".

That seems to me to be a finding of fact in relation to the intention of NMT which the magistrates, on the documentation which was before them and upon the oral evidence which they heard, were entitled to make and with which I should not interfere.

Mr Widdicombe has of course powerful points which he urged in E support of his clients' cause. He submits that these appellants had always operated as a property company. He points to the fact, which was given in evidence before the magistrates and apparently undisputed, that the appellants were not registered for value added tax and paid no corporation tax. He points to the assertion, apparently unchallenged before the magistrates and which I therefore, for present purposes, assume they found to be acceptable, that this appellant company, although it owns other properties, does not trade elsewhere as warehousemen.

But in all the circumstances it seems to me, as I have indicated, that, although I am by no means convinced that in the place of the magistrates I would have evaluated the matter as they did, they were entitled on the facts to make the findings that they did and F accordingly this appeal must fail.

The appeal was dismissed with costs. Leave to appeal, if necessary, was refused.

For further cases on this subject see p 219

REAL PROPERTY AND CONVEYANCING

Court of Appeal

December 9 1987

(Before Lord Justice SLADE, Lord Justice BALCOMBE and Lord Justice STOCKER)

ALLEN v SOUTH EASTERN ELECTRICITY BOARD

Estates Gazette February 20 1988

[1988] 07 EG 71-72

Deed of grant — Rights granted to Electricity Board on payment of capital sum — Further payment by way of compensation provided in the event of the landowner, having obtained planning permission for development, being unable to take full advantage of such permission as a result of the exercise by the board of the rights granted to it — The appellant purchased the farm which was the subject of the deed of grant and obtained a number of planning permissions for residential and office development — He then sold off parts of the farm for which planning permission had been obtained — He claimed that he had suffered a loss for which compensation was due because, owing to the presence of pylons and overhead cables, he was unable to obtain as good a price for the parts sold off as he otherwise would; he was thus unable to take full advantage of the planning permissions — Held by the Court of Appeal that Judge O'Donoghue (sitting as a judge of the High Court) was correct in rejecting the claim — The phrase "cannot take full advantage of the said planning permission" could not bear the wide meaning for which the appellant contended, equivalent to "cannot sell the property for as high a price as could have been obtained but for the rights granted to the board" — The compensation provision was directed to such consequences as a restriction on the density of housing development attributable to the existence of the pylons, towers and cables — It was common ground that the exercise of the board's rights in the present case did not in fact prevent the appellant from doing anything authorised by the planning permissions — The appellant had been compensated by the capital sum he received for any diminution of value due to the exercise of the rights granted — He was not entitled to further compensation, because he was in a position to "take full advantage" of the planning permissions — Appeal dismissed

No cases are referred to in this report.

This was an appeal by the plaintiff, Aubrey Paul Allen, from a decision of Judge O'Donoghue, sitting as a High Court judge, rejecting the construction which the plaintiff sought by originating summons to place on provisions for compensation in a deed of grant in favour of the South Eastern Electricity Board, the defendants and present respondents. The deed related to land at Southerham Farm, Lewes, East Sussex.

Michael King (instructed by Tuck & Mann, of Epsom) appeared on behalf of the appellant; Edward Faulks (instructed by the solicitor to the South Eastern Electricity Board) represented the respondents.

Giving the first judgment at the invitation of Slade LJ, BALCOMBE LJ said: This is an appeal from an order dated January 31 1986 of His Honour Judge O'Donoghue, sitting as a judge of the High Court, and it raises a short question of the construction of a deed of grant dated June 26 1978.

In 1978 certain trustees were the owners of freehold land known as Southerham Farm, Lewes, East Sussex. In 1978 the defendant, the South Eastern Electricity Board, erected and laid overhead and underground electric lines and towers on this land.

I now turn to the deed of grant and its terms. The parties were the trustee owners of the land, who were defined as the "Grantors", of the one part and the South Eastern Electricity Board, called "the Board" of the other part. It recited the seisin of the grantors of the property in question and also recited that they had already permitted the board to erect and lay the overhead and underground electric lines and towers thereinafter described. Clause 1 of the deed is, so far as material, in the following terms:

> In pursuance of the said agreement and in consideration of the sum of NINE THOUSAND SEVEN HUNDRED AND FIFTY POUNDS now paid by the Board to the Grantors (the receipt of which sum the Grantors hereby acknowledge) the Grantors as trustees hereby grant unto the Board full right and liberty for the Board and its successors in title to retain and maintain . . . inspect examine test and remove the overhead electric lines . . . on and across the said property and the underground electric lines . . . in and under the said property in the approximate position indicated . . . on the Plan . . . annexed hereto.

Then there were certain incidental rights such as the right to lop timber which might otherwise have interfered with the overhead lines and a right to pass and repass with vehicles in order to maintain the lines.

I need not refer to clause 2 of the deed. Clause 3 contains certain restrictive covenants by the grantors with the board to which I need refer only briefly:

> . . . the Grantors hereby covenant with the Board not knowingly without the previous written consent of the Board (such consent not to be unreasonably withheld) . . . (a) to erect any new or extend any existing dwellinghouse building or other erection or structure within Sixty feet on either side of the overhead electric lines . . . (b) to plant any timber or other trees within a distance of Sixty feet on either side of the overhead electric lines, (c) to alter the level of the ground within a distance of One hundred feet on either side of the overhead electric lines . . . (d) to allow any vehicle machinery or plant of any description to approach or pass under the overhead electric lines unless a clear space of not less than Twelve feet can be and is at all times maintained . . .

Clause 4 was a general restrictive covenant on the part of the grantors not to do anything on the cable route which would interfere with the underground cables. I need not refer to clause 5, but I will read in full clauses 6 and 7, which are the material clauses for the purposes of this case:

> 6 It is hereby agreed and declared that the said sum of Nine thousand seven hundred and fifty pounds paid to the Grantors by the Board as aforesaid shall include full satisfaction and compensation for all damage or injury sustained or to be sustained by the Grantors or other the owner for the time being of the said property (or any part or parts thereof) by reason of the injurious affection of the said property (or any part or parts thereof) by the grant of the rights and liberties hereby granted and the exercise of the same and the imposition observance and performance of the covenants on the part of the Grantors herein contained including (but entirely without prejudice to the generality of the foregoing) full satisfaction and compensation for all or any diminution in the value of the said property or any part or parts thereof by reason of such grant exercise imposition observance and performance.
>
> 7 It is further hereby agreed and declared that in the event of the Grantors obtaining planning permission for residential or industrial development of the said property or for mineral extraction thereon the Board shall pay to the Grantors (within six months of notification of the obtaining of such permission) compensation for the diminution in the value of the said property attributable to the exercise by the Board of the rights and liberties hereby granted with the consequence that the Grantors cannot take full advantage of the said planning permission: such compensation shall be in a sum agreed between the parties or settled by arbitration and shall take into account such

A part of the said sum of Nine thousand seven hundred and fifty pounds paid to the Grantors by the Board and of any compensation already paid under Clause 2 hereof or under this clause as shall be proper to be taken into account on the basis that the Grantors shall not be entitled to be compensated more than once for the same loss.

On May 30 1979 the appellant, the plaintiff in the original action, Mr Aubrey Paul Allen, purchased Southerham Farm for £290,000 subject to the deed of grant. He then obtained a series of planning permissions: two on September 9 1981, one in respect of a barn authorising a change from agricultural to residential use and the other in respect of another barn also authorising a change from agricultural to residential use; two on May 4 1983, a further planning permission in respect of one of the barns already mentioned for a
B change of use to two dwellings and a planning permission in respect of another building for a conversion of redundant agricultural buildings into a single dwelling with garages; and, finally, on May 2 1984, a planning permission in respect of a building on the land for change of use from a single dwelling to offices and parking for 17 cars. All those planning permissions were carried into effect from 1984 onwards.

The appellant has sold off parts of the land for which those planning permissions were granted. He estimates that, but for the existence of the pylons and the overhead cables, he could have secured a greater purchase price for the parts of the land which he sold off.

In those circumstances, he issued an originating summons, asking
C as the primary question whether, on the true construction of clause 7 of the deed of grant of June 26 1978 and in the events that have happened, compensation is now payable pursuant to clause 7 by the defendant, the South Eastern Electricity Board, to the plaintiff.

That originating summons came before Judge O'Donoghue on January 31 1986 and he gave judgment on the same day. He answered the question in the negative. The argument before him was, first, that the deed of grant should be read as a whole; second, that the words "full advantage of the planning permission" were not limited to physical implementation of the planning permission; and, third, that since any planning permission would inevitably take account of the existence of the pylons and the cables clause 7 would never have any
D effect if it were limited to the physical implementation of the planning permission.

The learned judge rejected these submissions and I read from the transcript of his judgment the substance of the reasons why he rejected them, starting at the top of p 6:

> In my judgment it is clear from the form of the Deed of Grant and from the fact that the Board was making a substantial payment (of £9,750.00) for the right to put a number of pylons on and to suspend cables across this land that it was envisaged that the land would thereafter be less valuable than it had been thitherto.
>
> Approaching the provisions of Clause 7 from this standpoint I ask myself how should the words which the parties chose to use in that Clause be construed. Prima facie it seems to me that the words "cannot take full advantage of the
> E said planning permission" should be read as a whole and those words suggest to me strongly that the parties envisaged a situation in which planning permission might thereafter be granted for a particular purpose but could not be fully enjoyed or given effect to (or implemented) in an actual or physical sense by reason of the rights and privileges granted to the Board which the Board insisted upon enforcing.
>
> Thus it might well frequently occur that a wayleave or similar grant is made to an Electricity Board or other person or body by an owner of land who later obtains planning permission for certain development, but who then finds that he is hampered in some way from taking full advantage of such permission because the Board chooses to require (for example) a pillar of support or a particular means of access to one or more of its pylons, cables, substations or similar works.
>
> Thus I am not convinced by the submission made by Mr King . . .

F I pause there to interpolate that Mr King appeared for the plaintiff below as he did for the appellant plaintiff before us.

> . . . that in no circumstances would a local authority grant planning permission in a manner which might interfere with the rights of other individuals or bodies. It seems to me that the prime concern of a planning authority is to decide whether or not a certain type of development should be allowed on a piece of land and not whether any person other than the owner would be entitled to stop it whether by restrictive covenant or by any other agreement. It is for this reason that the provision of section 84 of the Law of Property Act 1925 as amended and similar provisions are frequently used. Accordingly, when one considers that planning permission could be granted for residential or industrial development or mineral extraction on this land which development could be physically curtailed by the Board in exercise of its rights and privileges under the Deed then it seems to me that the most obvious and natural meaning of the words ". . . cannot take full advantage of such G
> planning permission . . ." would be the prevention by the Board of the owner enjoying the full physical benefit of such planning permission in the sense of being able fully to implement it or carry out the permitted development in full (or at all).
>
> In such circumstances is it right for the Court to give the words the much wider connotation sought by the plaintiff so as to give him the right to extra compensation for loss of a potentially higher profit which he might have obtained on resale with planning permission (even in the case where the development has been fully carried out in accordance with the planning permission)?
>
> In my judgment it would not be right to allow a grantor (or his successor in title) years after the Grant had been negotiated and made to seek a wholly new and additional payment by way of compensation years after the original Deed of Grant unless the parties had agreed to such a provision in clear terms in the H
> original Grant itself.
>
> In my judgment the relevant words ". . . cannot take full advantage of the said planning permission . . ." primarily and in their ordinary meaning envisage a situation in which the owner who later obtains planning permission is prevented in whole or in part from enjoying or implementing that planning permission *as such* in full.
>
> In those circumstances I can see no proper reason for construing those words in the wider "economic" sense urged by Mr King on behalf of the plaintiff.

I have read so much of the judgment in full because I wholly agree with what the learned judge there says and it seems to me he has put it in words which it is difficult to improve upon. Clause 6 of the deed of grant in this case was clearly intended to cover a diminution in the value of the land by reason of the grant of the rights to the board and J
the board exercising those rights by maintaining pylons, towers and underground cables. There is nothing in that clause to suggest to me that the compensation was intended to be limited to a diminution in the "existing use" value of the land. Clause 7 of the deed fits in quite naturally with this interpretation of clause 6, and the judge's words at the foot of page 6 and top of page 7 of the transcript seem to me to be completely correct. I repeat that paragraph of his judgment:

> It seems to me that the prime concern of a planning authority is to decide whether or not a certain type of development should be allowed on a piece of land and not whether any person other than the owner would be entitled to stop it whether by restrictive covenant or by any other agreement.

In the course of the argument before us various examples were K
given to show the manner in which clause 7 could operate. For example, outline permission for the development of land for residential purposes might be granted. It might be either for a particular density, so many houses per acre, or for a maximum number of houses upon the piece of land in question. Then, to apply that in practice, there might be an application for detailed planning permission, and the existence of the pylons, towers and cables would mean that the land could not be developed to the maximum density permitted by the outline permission. Another example was a case where detailed planning permission might be given with a condition as to landscaping. That condition might require the levelling of the ground within a distance of 100 ft of the overhead lines. The board might refuse to consent to the grant of such levelling under the L
provisions of the deed which I have read. Therefore, in that circumstance also, the owner of the land could not take full advantage of the planning permission granted thus again completely within clause 7.

It seems to me that there is no need to give any further examples. I am satisfied that the judge was right in his conclusion and for the right reasons. I would dismiss this appeal.

STOCKER LJ agreed and did not add anything.

SLADE LJ said: I will briefly express my reasons for dismissing this appeal in my own words, in deference to the argument of Mr King, who has said everything that could be said in support of this appeal.

The question asked by the originating summons was whether, on a M
true construction of clause 7 of the deed of grant and in the events that have happened, compensation is now payable pursuant to clause 7 by the defendant to the plaintiff. The relevant events, as set out in the plaintiff's affidavit, are that, since the deed of grant was executed, planning permissions have been obtained for four of the properties referred to in that affidavit. One of them has been sold for £109,000 and he has deposed to an estimate that, but for the presence of the pylons and overhead cables erected by the South Eastern Electricity Board in exercise of the rights given by the deed of grant, that property would have fetched £125,000. He claims that compensation is payable under clause 7 on the grounds that, in the events which have happened, there has been a "diminution in the value of the said

A property attributable to the exercise by the Board of the rights and liberties hereby granted with the consequence that the Grantors cannot take full advantage of the said planning permission", within the meaning of that clause.

The crucially important phrase in clause 7, as has appeared from Balcombe LJ's judgment, is the phrase "cannot take full advantage of the said planning permission". The plaintiff's claim in substance involves reading this phrase as meaning in its context "are unable to sell the said property for as high a price as the grantors would have been able to obtain but for the exercise by the board of the rights and liberties hereby granted".

In my judgment, this phrase simply will not bear this meaning. The natural and ordinary meaning of the words is, in my judgment, B "cannot do all those things which are authorised by the planning permission". I can see no sufficiently clear context to give them any other meaning in clause 7.

The closing words of clause 6 of the deed of grant beginning with the word "including" make plain the general intention of the parties that the sum of £9,750 payable by the board to the grantors on the execution of the deed is intended to be received by them in full satisfaction and compensation for any diminution in the value of the property by reason of the grant or exercise of the rights and liberties granted by that deed. Thus, if, on a future sale of the property, they or their successors receive a smaller purchase price than they would have received if the deed had not been entered into and the rights had C not been exercised, they are to receive no additional compensation. Compensation for such diminution in value has been paid already.

The general intention of clause 7, in my judgment, is merely to entitle the grantors or their successors in title to additional compensation for diminution in value if, but only if, the exercise by the board of their rights under the deed of grant prevents the grantors or their successors in title from doing all the things which they would otherwise be authorised to do under any future planning permission which they may obtain. In the present case it is common ground that neither the presence of the pylons and overhead cables referred to by the plaintiff in his affidavit nor any other exercise by the board of their rights under the deed of grant prevents the plaintiff from doing anything which is authorised by the planning permission which he D has obtained. In these circumstances, in my judgment, no diminution in value of the property has occurred in respect of which he is entitled to compensation under clause 7. He *is* in a position to "take full advantage of the said planning permission" in the relevant sense.

For these reasons and the further reasons given by Balcombe LJ and the learned judge, with which I find myself in full agreement, I would dismiss this appeal.

The appeal was dismissed with costs, not to be enforced without the leave of the court. Legal aid taxation of the appellant's costs was ordered.

E

Chancery Division

December 2 1987

(Before Sir Nicolas BROWNE-WILKINSON V-C)

STROVER v HARRINGTON AND OTHERS

Estates Gazette March 5 1988

[1988] 09 EG 61-72

Vendor and purchaser — Action by purchasers of house against vendors alleging misrepresentation and negligent F misstatement and against purchasers' surveyors alleging negligence — House was represented as having main drainage but in fact it drained into a cesspool through a joint system with a neighbouring property — Purchasers bought the house for £215,000 and sold it about a year later after the present litigation had started for £252,000, but claimed damages despite the profit on resale — The sequence of events was curious — In the sale particulars there had been a misrepresentation by the vendors' agents that the house had main drainage, but this error had been corrected by the agents in a letter to the purchasers' solicitors — Unfortunately this crucial letter had disappeared, no copy of it had ever been found on the file, and the purchasers had not been told of the G correction — The vendors, however, repeated the misrepresentation in a conversation between one of them and the purchasers' surveyor, who reported to his clients as a result that the house had main drainage — It was in fact repeated once more, this time by the vendors' solicitors in replying to a pre-contract inquiry — In due course the purchasers, not knowing that their own solicitors had been informed of the absence of main drainage, decided (in consultation with those same solicitors) to proceed with the purchase, relying on the answer to the pre-contract inquiry and on their surveyors' report — On the questions of liability, the Vice-Chancellor H decided first that the purchasers' surveyors had not been negligent — They were justified in accepting the vendors' assurance that there was main drainage — It would have been impossible without a special test (which the purchasers expressly declined) to ascertain the ultimate destination of the effluent and in any case it was normally a solicitor's function to check the type of drainage by a preliminary inquiry — There was no evidence that the surveyors had acted contrary to accepted good professional practice — As to the liability of the vendors under the Misrepresentation Act 1967, the difficulty in the way of the purchasers was the fact of their solicitors' knowledge that the drainage was not main drainage — Such J knowledge on the part of the purchasers themselves would clearly have been fatal to their case and the Vice-Chancellor held, after considering a number of authorities, that the knowledge of the solicitors must be imputed to the purchasers — He also held that the position was not altered by the later representation on the part of the vendors — Although the purchasers relied on it they would not have done so if they had known all the facts of which the knowledge was imputed to them — Thus any loss did not flow from the representations but from the failure of the purchasers' solicitors to convey the relevant information to their clients — The purchasers' K remedy, if any, was against their solicitors — The Vice-Chancellor concluded by expressing his view on damages in case the matter went further — He found on the evidence that the absence of main drainage did not affect the house's value; thus even if he had held the vendors liable no damage had been proved — Purchasers' claim dismissed against all defendants

The following cases are referred to in this report.

Arta, The (Markapper Inc v *N W Spratt & Son Ltd)* [1985] 1 Lloyd's Rep 534, CA

Ayrey v *British Legal & United Provident Assurance Co Ltd* [1918] 1 KB 136

Bawden v *London, Edinburgh & Glasgow Assurance Co* [1892] 2 QB 534 L

Blackley v *National Mutual Life Association (Australasia) Ltd* [1972] NZLR 1038

Newsholme Bros v *Road Transport & General Assurance Co Ltd* [1929] 2 KB 356

Thake v *Maurice* [1986] QB 644; [1986] 2 WLR 337; [1986] 1 All ER 497, CA

Wells v *Smith* [1914] 3 KB 722

This was an action by Mr and Mrs Strover, the purchasers of a house called "Greenbank" at Ashurst, East Sussex, against the vendors, Mr and Mrs Harrington, and against surveyors instructed by the purchasers, Cobbs Property Services Ltd.

M J Douglas (instructed by Farrer & Co) appeared on behalf of the plaintiff purchasers; P Goodenday (instructed by Thornton Lynne & Lawson) represented the defendant vendors; and D I Ridd M (instructed by Barlow Lyde & Gilbert) represented the third defendants, Cobbs Property Services Ltd.

Giving judgment, SIR NICOLAS BROWNE-WILKINSON V-C said: On April 23 1986, Mr and Mrs Strover entered into a contract with Mr and Mrs Harrington to buy a house, Greenbank, Ashurst, East Sussex, and certain fixtures and fittings, all at a price of £215,000. The contract was completed on May 21 1986. The particulars of sale, as a result of which they were introduced to the property and bought it, stated that the property had main drainage.

Certain answers given by the vendors' solicitors to inquiries before contract also stated that the property had main drainage. The Strovers engaged a firm of surveyors, Cobbs Property Services Ltd

— which I will refer to as "Cobbs" — to survey the property. The Cobbs survey report stated that the property had main drainage. Soon after completion the Strovers discovered that the property did not enjoy main drainage but drained into a cesspool. In this action, Mr and Mrs Strover as plaintiffs claimed damages for misrepresentation and negligent misstatement against the vendors, Mr and Mrs Harrington, and damages for negligence against their surveyors, Cobbs.

The proceedings started by a writ issued on November 1 1986. In December 1986, without changing the drainage system in any way, the Strovers put Greenbank on the market again. They effected a resale in June 1987 at a price of £252,000 — that is to say, a profit in the year of some £37,000.

Notwithstanding that profit, the plaintiffs, Mr and Mrs Strover, bring this action saying that if the property had had main drainage it would have been worth some £15,000 to £20,000 more than that.

The conveyancing background of the case is this. The house now known as Greenbank was formerly an old coach-house, and together with an adjoining property, Sheraton House, was formerly church property. In 1962 the church sold the two properties, Sheraton House and Greenbank, to two separate purchasers. At that time there was a joint drainage system used jointly by Sheraton House and Greenbank. Shortly stated, the system was this: the foul-water drainage from Greenbank drained into the grounds of Sheraton House, passed under or through Sheraton House itself out to the far side, together with of course the foul water from Sheraton House, into a field some distance the other side of Sheraton House, where there was a cesspool. The cesspool had a capacity of only some 1,000 gallons.

The conveyance of Greenbank to the predecessors in title of Mr and Mrs Strover contains the grant of an easement to drain Greenbank through the drainage system under Sheraton House to the cesspool.

It is common ground in this case that the actual cesspool with its capacity of 1,000 gallons is wholly inadequate to receive the sewage from the two houses under modern conditions, save in the way in which it is in fact operated. The cesspool was designed to contain all solid and liquids within the cesspool itself. Periodically the cesspool was pumped out. At some stage it has been "adequate" by permitting the fluids entering the cesspool to flow out of the cesspool into a trench. Once the foul liquids are in the trench, they then percolate through into the subsoil. There is no evidence that the percolation system is other than effective; the foul water does not emerge elsewhere but disappears from the trench directly into the subsoil, which has a very good porosity and absorbs all the liquids necessary. The cesspool has been pumped out for at least the last 20 years annually by the local authority, latterly at a moderate cost.

Mr and Mrs Harrington, the vendors, acquired the property in 1976. A Mr and Mrs Hutton acquired Sheraton House, the adjoining property, in 1980. In the pleadings and in certain parts of the evidence it was alleged that prior to the purchase by Mr and Mrs Strover there had been troubles relating to the joint sewage system as between the occupiers of Sheraton House on the one hand and the occupiers of Greenbank on the other. In the event, no direct evidence was led of there having been such troubles, nor was any direct evidence led that the drainage system had ever backed up into Sheraton House. For present purposes, those allegations must be competely ignored. There is no relevant evidence to the effect that the drainage system had previously given trouble. The most the evidence shows is that the system whereby the piping was carried under Sheraton House had been found defective and adapted in 1980.

Coming now to the particular circumstances of the present case, in early 1986 the vendors, Mr and Mrs Harrington, instructed agents, Parris & Quirk, to prepare particulars of sale. In those particulars of sale it was prominently said that the amenities included main drainage. An agreement subject to contract having been reached in February 1986, Parris & Quirk wrote to the purchasers' solicitors, Farrer & Co, confirming the sale subject to contract at a price of £215,000 to include the carpets, curtains, kitchen equipment and a garden tractor. A certain amount of time was taken up in the course of the hearing as to the value of those fixtures and fittings. It is I think for present purposes enough to say that they were taken by the purchasers as having a value of some £5,000.

That information having been sent to Farrer & Co on February 13, five days later Parris & Quirk wrote a letter of cardinal importance in this case, which I must read. The letter is addressed to Mr Hamnett of Farrer & Co and is addressed to him personally. It is headed: "In the matter of Greenbank", and reads as follows:

> Further to our telephone conversation today, I confirm that our sale particulars are incorrect inasmuch as that the above property is not connected to mains drainage but is served by septic tank.

He also enclosed three copies of a site plan. The statement that the property was not connected to the mains drainage was of course correct. The statement that it was served by septic tank was not correct; the drainage was not to a septic tank but to a cesspool.

That letter was acknowledged by letter from Farrer & Co dated February 20 1986, which thanked Parris & Quirk for the letter of February 18 with the enclosures. The letter is signed with an illegible signature.

The position therefore was that at that stage the purchasers' solicitors were clearly informed that the particulars were in error and that there was no main drainage attached to the property. Thereafter, that piece of information was completely lost sight of. Indeed, when the case started before me, the position being adopted by the plaintiffs was that they were still requiring the defendants to prove that that letter had ever been sent and received by Farrer & Co. In due course, I am glad to say that position was modified, and Farrer & Co accepted that that letter had been received and that the acknowledgement of February 20 had come from Farrer & Co. The fact remains that at no stage was the information contained in that letter ever communicated by the solicitors to their clients. The letters were not apparently found on the file and at no stage was there any reference apparently made in Farrer & Co's advice to the information contained in that crucial letter. That is the first of the unusual facts of this case.

Thereafter, Mr and Mrs Strover instructed Cobbs to conduct the survey. The letter accepting instructions expressly pointed out that Cobbs could, if the clients wished, arranged for specialist tests to be carried out on *inter alia* the drainage. In response, on February 27, Mr Strover wrote back to Cobbs saying that at that stage they did not require any specialist tests.

On March 3 1986 Mr N R R Butler [ASVA], the surveyor conducting the survey on behalf of Cobbs, went to Greenbank and surveyed the property. It is Mr Butler's evidence that he there met Mrs Harrington — which she agrees — and that he was told by Mrs Harrington that the property had main drainage and drained through Sheraton House. Mrs Harrington in her evidence denies that she ever said that to Mr Butler.

On the same day, March 3, Mr Butler met and had a long discussion with the purchasers, Mr and Mrs Strover, and discussed a number of things. Mrs Strover says that among the matters discussed was the question of drainage. Mr Butler does not recall that but is not prepared to say that it did not occur.

On March 6 Cobbs sent their survey report to Mr and Mrs Strover. It contained a number of limitations and provisos which at this stage at least I need not refer to. What is crucial is that at para 23 the following occurs:

> The foul drainage system. The property is drained to the main sewer by a drain line through the adjacent property. The drains are in good order where visible.

The matter then proceeded between solicitors; draft contracts and entries on the register were sent. The entries on the register disclosed the existence of the easement permitting drainage through the joint drainage system expressly to a cesspool on the adjoining land belonging to the church. On April 2 Farrer & Co submitted inquiries before contract. There were a number of inquiries relating to drainage. The most important was general inquiry no 6, the inquiry being this:

> 6A. Drainage. 1. Does the property drain, (a) directly into a local authority sewer, and (b) without any part of the drains other than the flat sewer passing through any neighbouring property? 2. If not, how is the property drained? Please supply details of any easement, licence or agreement relating to its drainage, and of the charges, if any, payable in respect thereof.

To that the vendors' solicitors gave this reply:

> The property has mains drainage. Please see the office copy entries for details of easements granted and reserved.

There is therefore, in answer to that inquiry no 6, a further repetition of the representation that there was mains drainage.

The pre-contract inquiries raised by Farrer & Co included certain other inquiries directed to the nature of the drainage. In particular, additional inquiry no 3 asked:

> Please specify on a plan the whereabouts of the cesspool referred in the

property register of title K133400, and confirm that the rights in respect of this have been exercised by the vendors without any dispute with the adjoining owners.

The response to that was, "Please inspect".

Farrer & Co were not satisfied with that answer to the inquiry and pursued it, but it is unnecessary for me to trace the matter through, since what was being pursued was the question of the exercise by the vendors without dispute of the rights over the cesspool. Since the whole question of there having been such a dispute has dropped out of this case, it is unnecessary for me to trace the further pursuit of that inquiry.

The conveyancing pre-contract inquiries having proceeded and the survey having been received, shortly before April 23 there was a meeting between the purchasers, Mr and Mrs Strover, and their solicitors. The state of the evidence as to that meeting is rather strange. I have certain information as to what took place. Mr Goodenday, for reasons which I did not quite understand, objected to various other pieces of evidence relating to that meeting being produced, since he said it was not evidence against him. On the limited information that I have as to that meeting, it appears that there was a report on the position prepared by Farrer & Co, which I have not seen. Among other matters that report raised the question of drainage. There was a discussion as to the position on drainage, the question being this: why was there a reference on the title to an easement of drainage to a cesspool while the vendors were representing in the answer to the inquiries that there was main drainage? I was told, and I accept, that the decision was taken by Mr and Mrs Strover, in the light of the advice then tendered, that they would rely on the unambiguous answer given to inquiry no 6, and on the statement contained in Cobbs' report that there was main drainage; the view being taken that presumably the drainage had been altered since the grant of the original easement.

On that footing, and plainly in my view in reliance on the misrepresentation, Mr and Mrs Strover exchanged contracts on April 23.

The National Conditions of Sale applied to the contract. National Condition 17 provides:

1. Without prejudice to any express right of either party, or to any right of the purchaser in reliance on the Law of Property Act 1969, Section 24, to rescind the contract before completion, and subject to the provisions of Paragraph 2 of this Condition, no error, misstatement or omission in any preliminary answer concerning the property, or in the sale plan, or the special conditions, shall annul the sale; nor, save where the error, misstatement or omission relates to a matter materially affecting the description or value of the property, shall any damages be payable or compensation allowed by either party in respect thereof.

Completion of the sale took place on May 21. Mr and Mrs Strover did not move in immediately but, as I understand it, at some stage late in June. Quite shortly thereafter, they had discussions with Mr and Mrs Hutton at Sheraton House, from which they discovered the true position as to the drainage. They were upset. They instructed Fox & Manwaring to prepare a report on the drains, which was submitted to them on July 18 1986. Nothing turns on the detail of that report; broadly it suggested the installation of a septic tank in the grounds of Greenbank separately from the joint drainage system under Sheraton House.

On July 28 Farrer & Co, on behalf of the plaintiffs, wrote to the vendors, to Parris & Quirk and to Cobbs, complaining of the misrepresentations as to the drainage of the property. The reply from Parris & Quirk is of some importance. On July 29 they replied, saying:

We wrote to Mr Hamnett of your company on February 18 advising you that our sale particulars were incorrect and that the property was not connected to main drainage.

It was apparently as a result of that that the long-forgotten letter of February 18 for the first time resurfaced; but, as I have said, Farrer & Co had apparently no copy of it on their file.

All parties denied liability and, as a result, on November 21 1986 the writ in this action was issued against the vendors, Mr and Mrs Harrington, as first and second defendants, and Cobbs as third defendants. In December 1986 the Strovers decided to put Greenbank on the market again, it not having proved satisfactory as a house to them — not because it was anything other than a very nice house but, so they told me, partly because of the drainage, partly because of difficulties of adaptation, and partly because country life had not appealed unduly to Mrs Strover.

It was put on the market in December 1986 at a price of £252,000. The true nature of the drainage (which remained the overflowing cesspool) was disclosed, as was the existence of this litigation. The property was resold in May or June of 1987 at a price of £252,000, plus £750 for the tractor.

One therefore has a change in the market price, or a change in the price, between what the Strovers paid for it — £215,000 — and the £252,000 obtained approximately a year later.

A strange feature of the evidence before me is that there is no explanation of a number of the cardinal questions in this case. First, I have been given no explanation as to how Parris & Quirk in the first instance came to include a statement that there was mains drainage. Mr R M Brocklebank [ARICS] from Parris & Quirk gave evidence. He was not the gentleman who took the details and particulars. He indicated by name the gentleman who did take the details, and said he was available to give evidence; but he was not called. Mr Brocklebank said that under the firm's usual practice statements of this kind were certainly not made up out of the draftsman's head and were normally obtained from the vendors. There is still no explanation as to how that original misrepresentation was made. No cause of action is based by the plaintiffs on the misrepresentation in Parris & Quirk's particulars.

The second matter that remains unexplained is how Farrer & Co completely lost sight of the letter from Parris & Quirk informing them of the true position that there was no mains drainage.

The third matter that is unexplained is how, given the true state of affairs, and given the fact that Parris & Quirk knew as a result of information communicated to them by the Harringtons that there was no mains drainage, it came about that Mr and Mrs Harrington's own solicitors made exactly the same misrepresentation thereafter in answer to the inquiries before contract.

The facts of the case are unusual to say the least.

I propose to deal first with the question of liability and then thereafter with the question of damages. First, are the surveyors, Cobbs, liable for professional negligence? The prime question on that issue is whether or not Mr Butler was told by Mrs Harrington that there was mains drainage. His evidence was that he had a conversation with her, either in the kitchen or the hall of Greenbank; that he put a question to her asking whether the property was connected to the mains drainage, to which she answered, "Yes; the drains run through our neighbour's property into the road", indicating the general direction of the run of the drain. In cross-examination, he accepted that she might have misheard his question, but had no doubt whatsoever that the answer was given that I have recounted.

Mrs Harrington on the other hand absolutely denied talking about drains to Mr Butler at all. She denied that she could have told him that there was mains drainage because she knew very well that there was not. She denied telling him that the drains ran through Sheraton House. There is therefore here a complete conflict of evidence.

Mr Butler impressed me as a careful and impressive witness. I have no doubt that he was telling me the truth and that he was in fact told by Mrs Harrington that there was mains drainage. I am prepared to accept that she may have misheard the question, but I find that he certainly thought he had been told there was mains drainage.

In case that is not sufficient and this case goes further, I must make a finding as to what the position is if there was no mishearing. On that I have to prefer Mr Butler's evidence. As I have said, he was an impressive witness. The immediate response of Cobbs when the matter was taken up with them was a response that Mr Butler had been told by Mrs Harrington that there was mains drainage. The survey report, as I have mentioned, refers to the fact that there is a joint drainage system through the adjoining property. Nobody has put forward any suggestion as to where Mr Butler could have got that information from, save through what he was told by Mrs Harrington; yet Mrs Harrington denies discussing drainage at all. Finally, there is the total background mystery as to how the question of mains drainage ever came into this case at all. The Harringtons' evidence is that they knew it was not mains drainage and knew it was cesspool drainage, and yet in some way they have twice managed through their agents to make the representation that it was mains drainage. I am not satisfied that I know the full story from the Harrington point of view. Therefore, if I have to make a finding, I find that I accept Mr Butler's evidence that he specifically asked the question and Mrs Harrington specifically gave the answer that there was mains drainage.

That being the position, the question is whether Mr Butler, and therefore Cobbs, can be held to be guilty of negligence in reporting as they did in categorical terms that there was mains drainage. Mr Butler in evidence said that it was the normal practice with questions of drainage to ask the vendor what the position was. As is manifest, it is impossible, without special tests, to discover where a drain goes. You can look at the manhole and see the direction in which it is running, but the ultimate destination of the sewage can be ascertained only if you know where to look for the ultimate destination. It was not his practice — and indeed it was unascertainable on his evidence — if told there was mains drainage, to seek to find the junction with the main sewer, because there was often not a visible junction in the road at all. There was no other evidence led before me as to what is sound, careful surveying practice. It has not been proved that the way in which Mr Butler proceeded in relying on what he was told by the vendors is contrary to good surveying practice. Therefore, in my judgment, there can be no question of his being negligent in so relying. The only question is whether, in the unambiguous way in which he asserted that there was mains drainage, he was negligent.

Mr Douglas, in his powerful submissions on behalf of the plaintiffs, submitted that since Mr Butler was in fact relying only on the word of the vendor and not on any objective observation or verification of the nature of the drainage, the report which he put in should not have contained a categorical statement as to the nature of the drainage but should have been qualified by such words as "it is thought that", or "it is said that" the property has mains drainage, thereby alerting Mr and Mrs Strover to the fact that he was going on secondhand information on the subject.

Mr Douglas points out many passages in the surveyors' report where such qualification is inserted in cases where it was impossible for Mr Butler to verify specifically whether or not a state of affairs existed. In cross-examination, Mr Butler accepted that with hindsight it might have been better to qualify the categorical statement in the way in which Mr Douglas suggests he should have done.

If this was a matter for me to decide, in the absence of evidence as to the practice of skilful surveyors, I would not hold Mr Butler liable in negligence for failing to qualify his report. Of its nature, the ultimate destination of the sewage is impossible to detect without sophisticated testing. The surveyor is therefore bound as a matter of commonsense to rely on what he is told. In doing an ordinary survey for a purchaser thinking of buying, the surveyor is bound to rely primarily on what he is told. The position is not the same as where he is unable to inspect, for example, to see whether there is a damp course behind the plaster. That is a matter which is capable of verification by objective viewing. It is not done because to do so means taking the plaster off. But without some limitation on the words used, it could be said that the surveyor had held himself out as having looked to see or represented that he had looked to see. In my judgment, the nature of drainage is different for the reasons that I have given.

Moreover, in the context of house purchase, it is obviously known to the surveyor, and one would have thought to the client, that it will be the solicitor's function in making preliminary inquiries to verify exactly this point. There is a common-form printed inquiry which relates to the nature of drainage which is made on every house purchase conducted by a solicitor. The surveyor's report is not the principal means for verifying the nature of the drainage.

The likelihood of a vendor's misrepresenting the nature of the drainage must be very remote indeed. Nobody, for example, has referred me to any other case where this has previously happened. In those circumstances, to hold a surveyor liable for failing simply to qualify the way in which he gives his answer, when he is, in accordance with good practice, relying on information communicated to him which in the ordinary run of events — indeed the invariable run of events so far as I am aware — has proved to be accurate, would be to be imposing too high a duty.

That would be my view of the matter if it were for me to decide; but in my judgment it is not for me to apply my yardstick of what is or is not negligent. As a professional man, Mr Butler can only be held liable for negligence in the conduct of his profession if he acted in a way which no surveyor of ordinary skill would be guilty of, if acting with ordinary care. No expert evidence was led as to the practice of careful surveyors, and the burden is on the plaintiffs to prove negligence.

Therefore, in my judgment, there is no evidence on which I could hold Mr Butler to have been negligent, even if, contrary to my own inclinations, I thought he had been.

Mr Douglas submitted that that was the wrong approach and that in the circumstances of this case I could decide the matter on the basis of Mr Butler's own evidence. He relied on the decision of the Court of Appeal in *Thake* v *Maurice* [1986] QB 644. That was a case in which a doctor was alleged to have been negligent in failing to give warning to the patient of a risk that the position might change thereafter. The defendant himself gave evidence that in cases of that kind it was his practice to give such warning, and that such warning was necessary. It was held that there was no need for expert evidence from other doctors to be called to prove what the defendant himself had accepted was the proper practice. That seems to me wholly different from the present case. Mr Butler did not give evidence that it was his or any other surveyor's practice to qualify their statements as to the form of drainage when relying on what they were told by vendors. The highest it can be put is that with hindsight — and I stress the word "hindsight" — he thought it might be better to qualify the statement. But the question is not what Mr Butler will think is desirable in the future as a result of the perhaps bitter experience of this case, but whether on March 4 1986 he fell below the then standards of a skilful and careful surveyor. To my mind there is no evidence that he did so. Therefore, in my judgment, the claim against Cobbs fails.

I turn then to the claim against the first and second defendants. The case is put in two ways. First, damages are claimed under the Misrepresentation Act 1967; and second, it is alleged that Mr and Mrs Harrington were guilty of negligent misstatement.

I need take no further time on the claim based on negligent misstatement. Mr Douglas accepts that if he cannot succeed on the claim under the Misrepresentation Act, he will not get home on the claim in negligence.

The misstatements relied upon are, first, the answer given by the Harringtons' solicitors in answer to the general inquiry and, second, the statement made by Mrs Harrington to the purchasers' agent Mr Butler, and by him further transmitted to Mr and Mrs Strover in the surveyors' report. The chapter of accidents in this case is so extraordinary that it is not perhaps surprising that the law does not clearly provide what the result should be. I will, at the risk of extending this judgment unduly, state again the relevant facts. First, the vendors' selling agents make a misrepresentation in the sale particulars that there is mains drainage. Second, the vendors' selling agents specifically inform the purchasers' solicitors that that statement was an error and that there is not mains drainage. Third, the purchasers' solicitors do not inform the purchasers of that fact. They do not inform them either immediately or at any time thereafter. Fourth, the vendors through Mrs Harrington repeat the misrepresentation by telling Mr Butler that there is mains drainage. Fifth, the vendors through their solicitors again repeat the misrepresentation that there is mains drainage, in answer to the pre-contract inquiries. Sixth, neither of those two latter misrepresentations — the second and third misrepresentations — is ever corrected. Seventh, the purchasers, in ignorance of the fact that their solicitors have been told that there is not mains drainage, then decide in consultation with those very solicitors to go ahead with the purchase in reliance on the misrepresentations given in answer to the pre-contract inquiries and in the surveyors' report that there are mains drains.

It is not really disputed — and I certainly find as a fact — that so far as the actual intention of Mr and Mrs Strover is concerned, they did rely on the misrepresentations contained in the answers to the pre-contract inquiries and in Cobbs' report, and that that was a material factor in their decision to go ahead with the contract notwithstanding the entries on the register referring to easements to a cesspool.

The Misrepresentation Act 1967, section 2(1), provides as follows:

> Where a person has entered into a contract after a misrepresentation has been made to him by another party thereto and as a result thereof he has suffered loss, then, if the person making the misrepresentation would be liable to damages in respect thereof had the misrepresentation been made fraudulently, that person shall be so liable notwithstanding that the misrepresentation was not made fraudulently, unless —

and there then follows a proviso which has been subsequently amended which enables the maker of the representation to escape liability if, shortly stated, he can show that he had good grounds and did in fact believe in its truth. That proviso is irrelevant in the present case, since no attempt has been made to show that Mr and Mrs Harrington had any grounds for the belief that the property did not have mains drainage.

The plaintiffs must therefore prove four things: first, that they entered into the contract; and plainly they did. Second, after a misrepresentation had been made to them by the vendors; plainly it was. Third, that as a result of that misrepresentation they suffered loss; that is a matter which I think may be of considerable importance and will consider later. Fourth, that the misrepresentations made by the Harringtons, albeit made innocently, would have given rise to damages if they had been made fraudulently. That last requirement requires one to go back to the old law of misrepresentation.

It is common ground that the person to whom a representation is made (the representee) cannot complain if he knows the misrepresentation to be false and knows the true position. The law is concisely stated in *Spencer Bower & Turner on Actionable Misrepresentation*, 3rd ed at p 214. Therefore, if Mr and Mrs Strover had known that there was not mains drainage, they could have no claim in misrepresentation. The first question is whether the knowledge of Farrer & Co, their solicitors, that there was not mains drainage, or the information which Farrer & Co had that the selling agents were saying that it did not have mains drainage, can be imputed to Mr and Mrs Strover for the purpose of considering their right of action in misrepresentation. Though Mr and Mrs Strover did not in fact know the contents of the letter, their agent who was responsible for communicating relevant information to them, namely, their solicitors, did have that information. In those circumstances, can the Strovers be heard to say they did not know?

If such knowledge can be imputed to Mr and Mrs Strover, the second question which arises is whether they had sufficient knowledge to be debarred from relying on the misrepresentations subsequently made by the Harringtons. Neither of those questions is easy and neither appears to be covered by any very satisfactory authority.

Spencer Bower at p 216 states quite unequivocally that for the purpose of deciding whether a representee knows the representation to be false, and knows the truth of the matter, the knowledge of the agent cannot be imputed to the principal. *Spencer Bower* says:

> Moreover, actual and personal knowledge must be proved. Constructive or imputed notice is entirely out of the question.

No authority is cited which supports that proposition nor have I been referred to any authority which directly covers it. As a matter of impression, in my judgment it ought to be sufficient if in a transaction for the sale of land a vendor who has made an innocent misrepresentation corrects that misrepresentation by informing the purchaser's solicitors of the true facts. To take not this case but one which must be of common occurrence, say the vendor's solicitor has given an erroneous answer to a preliminary inquiry. The vendor's solicitor then discovers his error and corrects it by a communication to his opposite number, namely, the purchaser's solicitor. It could not be right, so it seems to me, that in those circumstances the purchaser could be heard to rely on the original misrepresentation, which has been corrected by communication to his own solicitor, just because his own solicitor had failed to inform him of the correction. If that were to be the law, in such a case the vendor for his own protection would have to communicate not through the usual conveyancing channels with the other side's solicitor but with the other side personally. Otherwise, he would stand the risk that notwithstanding his express correction, he would still be held liable for the misrepresentation he had in fact corrected to the solicitors.

My first impression, therefore, is that the passage in *Spencer Bower* cannot be right in its full width.

The authorities cited to me certainly do not compel me to hold otherwise. No case has been cited which deals with the events in this case, namely, the correction by a vendor of a simple misrepresentation by communication of the true facts to the agent of the purchaser.

The case of *Wells* v *Smith* [1914] 3 KB 722 was one in which the defendant made a fraudulent statement to the plaintiff's agent, who communicated it to the plaintiff. The crucial fact was that the plaintiff's agent himself knew the fact to be untrue, and was himself a party to the defendant's fraud. It is not surprising that the court held that the knowledge of the true facts by the fraudulent agent of the plaintiff could not be imputed to the plaintiff. That case seems to me a hundred miles away from the case we have here, since it would be impossible to impute a fraudulent agent's knowledge to his principal.

The other cases cited are all concerned with insurance. If to the knowledge of the insurers' agent, but not of the insurers themselves, a proposal for insurance contains a misrepresentation of fact, can the insurers escape liability, or is the knowledge of the agent to be G
imputed to the insurers? In *Bawden* v *London, Edinburgh & Glasgow Assurance Co* [1892] 2 QB 534 and *Ayrey* v *British Legal & United Provident Assurance Co Ltd* [1918] 1 KB 136, in such circumstances the knowledge of the agent was imputed to the insurers, who were therefore unable to disclaim liability under the policy. The opposite conclusion was reached in *Newsholme Bros* v *Road Transport & General Assurance Co Ltd* [1929] 2 KB 356, in which the earlier cases were commented on unfavourably.

In my judgment the insurance cases turn on a different point to that before me. In making the insurance proposal, the insured accepts that the facts in the proposal itself are the basis of the contract and warrants their truth. The true analysis of those cases is that they turn on the breach or otherwise of a contractual warranty, not on the law H
of misrepresentation. In the case I have to deal with, I am dealing purely with a collateral representation which is in no sense a term of the contract. I do not think I get any help from those cases.

Next is the decision of the Court of Appeal in *The Arta* [1985] 1 Lloyd's Rep 534. In that case the defendant shipbrokers represented that a particular client was good for the financial obligations under a charter being negotiated with the plaintiffs. In a completely different transaction, the plaintiffs' brokers had acquired knowledge which cast doubt on the financial integrity of the proposed charterer; but the plaintiffs' agent did nothing more to find out fuller facts of the matter, nor did he inform his principal of what he had learned. The Court of Appeal held that the defendants were J
liable for the misrepresentation, and notwithstanding the state of knowledge of the plaintiffs' agent.

The main ground of the decision was that the plaintiffs' agent did not become aware of all the relevant circumstances until after the contract had been made and therefore his state of knowledge could not be relevant to the question of misrepresentation being actionable or not. However, there are *dicta* in the case to the effect that even if the plaintiffs' agent had known the full position before the contract was entered into, his failure to inform his principal would not have exempted the defendants from liability for their misrepresentation. Goff LJ says at p 538:

> The simple fact is that the assurance was given by the defendants: it was negligently given: it was obvious that it would be relied on by the plaintiffs: it K
> was relied on by them, and they had suffered damage in consequence. In these circumstances, there can be no question of any *novus actus interveniens* breaking the chain of causation.

I cannot regard that and other *dicta* in the case as throwing much light on the present case. There the Court of Appeal was considering the case where the representee's agent had received information not in the course of the particular transaction but in a different transaction; such information not being communicated to him by the other party to the transaction as a specific correction of an earlier misrepresentation. I read the *dicta* in that case as applicable to the facts of that case and not directed to a case such as the present where there has been a specific correction of a previous error made to the L
representee's agent, who was acting for the representee in the same transaction.

Last, I was referred to a New Zealand case, *Blackley* v *National Mutual Life Association (Australasia) Ltd* [1972] NZ Law Reports 1038. Although the decision in that case is not directly in point, it does to my mind provide some guidance as to the proper approach. In that case the insured had made an accurate and truthful proposal for insurance, but he was under an obligation to disclose any material change of circumstances which occurred after the date of the proposal but before the policy was issued. A material change of circumstances did occur. The full facts of that change of circumstances were made known to the insurers' agent but not communicated by the agent to the insurers themselves. The New M
Zealand Court of Appeal held that the disclosure to the agent was sufficient. They held that he had ostensible authority to receive such information. Therefore the insurers were estopped from denying that they had received it. Therefore, the knowledge of the agent was imputed to the insurers.

In my judgment, similar principles apply in a case such as the present. In this, as in all other normal conveyancing transactions, after there has been a subject-to-contract agreement the parties hand the matter over to their solicitors, who become the normal channel for communication between vendor and purchaser in all matters relating to that transaction. In so doing, in my judgment the parties impliedly give actual authority to those solicitors to receive on their

A behalf all relevant information from the other party relating to that transaction. The solicitors are under an obligation to communicate that relevant information to their own clients. At the very least, the solicitors are held out as having ostensible authority to receive such information. Whether there be express or ostensible authority, the purchaser is, in my judgment, estopped from denying that he received the information relating to the transaction which has been communicated to his solicitors acting in the same transaction. In my judgment, such knowledge should be imputed to the principal. If that were not to be so, the consequences to which I have previously referred would follow.

So in this case, in my judgment, Mr and Mrs Strover must be treated, contrary to the true facts, as though they in fact knew the B contents of the letter of February 18 from Parris & Quirk, namely, that the representation in the particulars of sale was erroneous in that the property did not have mains drainage.

However, the matter does not stop there. After the disaster of the failure to communicate the correction on the one side, we have the mirror disaster on the other, namely, the repetition of exactly the same misrepresentation by the vendors. What is the effect of that? It seems to me there are two possible views. Mr Douglas contends that all that the Strovers are to be taken as knowing is that the vendors were making conflicting representations, at one time saying there was not mains drainage, at another time saying that there was. That, says Mr Douglas, does not amount to knowledge of the full truth which is C necessary if you are to defend a claim for misrepresentation. He says — and to my mind rightly — that if it is once shown that a misrepresentation has been made, it is no answer for the representor to say that the representee has been negligent and could have found out the true facts if he had acted otherwise The representee is under no duty of care to the representor to check on the accuracy of the representation. The representor is bound by his representations, however careless the representee may have been. That is one view which has considerable force.

The other view is that if Mr and Mrs Strover had known, as they must be taken to have known, of the contents of Parris & Quirk's letter, they would never have entered into this contract at all. Therefore, the cause of the loss, if any, suffered by reason of their D entering into the contract must be taken to be the failure of Farrer & Co to inform them of the information they had received from Parris & Quirk, not the later misrepresentation made by the vendors. Although in fact Mr and Mrs Strover did rely on the later misrepresentation, if they had known all the facts knowledge of which is to be imputed, they would not have so relied. Thus the loss flows not from the later misrepresentation made by the vendors but from the failure of Farrer & Co to communicate the relevant information to their client.

I found this much the most difficult point in the case, and I have reached a conclusion with considerable hesitation. I find as a fact that if at the pre-contract meeting Mr and Mrs Strover had known of the E contents of the letter from Parris & Quirk, and Farrer & Co in advising them had recollected its contents, Mr and Mrs Strover would never have entered into the contract in the way they did. I am satisfied that the existence of mains drainage as an amenity at the house was important at least to Mrs Strover. The matter of drainage was specifically considered at the pre-contract meeting, and the decision was taken on balance to go ahead in reliance on the representation made in answer to the preliminary inquiries and on Cobbs' report. If it had been known to Mr and Mrs Strover that there was exactly the opposite information in their hands — or should have been in their hands — to the effect that there was no mains drainage, it defies all probability that they would have gone ahead and contracted, at least without fully investigating the position; and Mr F and Mrs Strover impressed me as careful people to whom this drainage was a matter of considerable importance. On the information and evidence before me, I reach the view that they would not have entered into this contract had they known the full facts.

On that footing, I have with some hesitation come to the view that the causation approach is the correct one. To succeed in an action for misrepresentation, the plaintiff must show that the loss he has suffered is caused by, and is the result of, the misrepresentation. In my judgment, the loss, if any, suffered by the Strovers in this case did flowed not from the later misrepresentation but from the failure of their own solicitors to inform them of the true position. On the hypothetical state of facts which I have to assume, namely, that Mr and Mrs Strover knew that the statement in the particulars that there was mains drainage had been corrected, they would not have relied G on the later misrepresentation so as to enter into the contract in reliance on it. In my judgment, the Strovers' remedy is not against the vendors but, if any, against their own solicitors. Of course, I have not heard the solicitors' case on the matter; I have not had their explanation; and so I cannot reach any view as to whether there is an answer to such a claim.

I therefore hold that the vendors are not liable for the misrepresentation contained either in the answer to preliminary inquiries or in the indirect representation via Mr Butler.

I should shortly notice two points. Mr Goodenday for the vendors relied on National Condition 17 as a defence, on the basis that the statement that the property had mains drainage was not a material misdescription of the property. H It is not necessary for me to decide that point, but I have great doubts whether such an argument could succeed. The type of drainage enjoyed by a property seems to me a part of the description of the property, and indeed a material one.

Second, Mr Goodenday expressly conceded that he was not putting forward any argument based on contributory negligence as against the claim under the Misrepresentation Act. Whether that concession was right or wrong it is unnecessary for me to express a view, and I do not do so.

Finally, therefore, I must come to the question of damages. I must express a view on this in case the matter goes further and my view on the law and on liability is held to be wrong. I must therefore shortly make the necessary findings as to whether there was any damage J suffered.

It is common ground, subject to one minor point, that the measure of damages is to be found by establishing the difference between the contract price, namely £215,000, and the actual value which the property had in May 1986 with the cesspool drainage that it had. Mr Ridd contended that because Mr Butler in his report valued the property at £210,000, the right figure to take as against Cobbs was that figure. I do not think that is right, but I do not believe that anything turns on it, because the only relevant evidence has been by way of deduction from or addition to whatever was the open market price.

Given the nature of the drainage in this case, namely, to a cesspool which operates as effective drainage only by permitting copious K overflow of foul water, I would have expected there to be a difference between the value of a property with mains drainage and the value with such adapted cesspool drainage. It is a system which depends on overflow and is operating not in accordance with the design. There was some evidence, which I am perfectly prepared to accept, that in the summer months the system smelt. I would have expected the price to reflect the fact that not everybody would find that system very satisfactory, and would at the lowest wish to install a proper septic tank system. My views on drainage obviously coincide fairly closely with Mr and Mrs Strover's in that we do not regard this as satisfactory drainage.

Unfortunately, that is not the question I have to decide. I have to L decide on the evidence before me what is the open market value of the property with the cesspool drainage that it has. That does not necessarily coincide with what I would buy or with what Mr and Mrs Strover would buy. It depends on whether purchasers can be found to buy at a particular price, notwithstanding the nature of the drainage.

All the experts agreed that the open market value, with mains drainage, in May 1986 was the contract price paid by Mr and Mrs Strover, namely £215,000. It was the evidence of Mr J S K Calcutt [FRICS CAAV] the plaintiffs' expert, that its value with the drainage it actually had, namely, cesspool drainage, was some £15,000 to £20,000 less than if it had mains drainage. The defendants' experts on the other hand say that the nature of the drainage, and indeed the form of the particular installation here, makes no difference to the M open market value of the property. They say that cesspool drainage in this area, together with the overflow system, is a very common form of drainage; that a number of purchasers would indeed expect to find something of this kind; that in early 1986 the market was extremely buoyant, and there were many purchasers around, and that some purchaser would be found who would be prepared to pay the same price, whether it be mains drainage, septic tank drainage, or cesspool drainage, provided the system was actually working in the sense of disposing of the sewage, as this system was.

I have to choose between those two views, and although it is contrary to my own instincts, I have to find that the evidence of the defendants' witnesses was most impressive and persuasive,

A particularly that of Mr C D Kinloch [FRICS] [for the third defendants].

Mr Calcutt's evidence seemed to me less satisfactory. First, his report was criticised as expressing a view that at the time of the report, namely, October 1987, the difference in value between Greenbank House with the drainage problems it has and with the drains properly connected to the main sewer is £15,000 to £20,000. That is plainly on the face of it a difference of £15,000 to £20,000 on the then October 1987 figure. I find it difficult to accept his evidence in the witness box that that differential remains the same whatever the basic value of the house. His evidence was that the difference was not only £15,000 to £20,000 now but was £15,000 to £20,000 in May 1986 at a time when B the market on his evidence was 30% lower. I find that difficult to accept. The discount one would have thought must vary according to the overall value of the commodity.

Second, taking Mr Calcutt's evidence as saying that the property was worth £20,000 less than £215,000 in May 1986, that gave it a current market value in May 1986 of £195,000. The one truly objective fact in this case is that it was sold, with its existing cesspool drainage, one year later for £252,000. On Mr Calcutt's valuation, the result would be that apart from the drainage factor, the value of the house had risen 30% over the year. Indeed, he gave evidence that that was the average increase in value of housing in that area. I am not satisfied with his answer on that point, which was given without research. Mr M P Brooker [FRICS] [for the defendant vendors] gave C evidence, which I accept, that 30% is way outside the increase. There was a very substantial increase, but it was at most 20% in the year in question.

If one takes an average — and I appreciate it is only an average — increase of 20%, 20% of £195,000, which is Mr Calcutt's 1986 value, would be £39,000 and would give an open market value in May 1987 of £234,000; yet it was actually sold for £252,000. All that seems to me to indicate that a deduction of £15,000 to £20,000 is out of line.

Finally, I found Mr Calcutt's evidence unconvincing as to why, even if a proper efficient septic tank were to be installed, there would still be a difference in the price of some £10,000. For the life of me, I cannot see why anybody should require to, or would, drop his offering price by £10,000 if there was a truly smell-free and efficient D drainage system by septic tank working. The evidence is that a septic tank could have been installed here for in the region of £5,500.

For those reasons, I am forced to prefer the evidence of the defendants' experts and to hold that, given the market conditions in May 1986, the fact that the property did not enjoy mains drainage but only the use of an overflowing cesspool did not affect its value, and whatever its form of drainage it was worth what Mr and Mrs Strover paid for it, namely, £215,000. Therefore, even if I were to have held that the defendants were liable, I would hold that no damage has been proved.

Accordingly, I dismiss the plaintiffs' claim against all three defendants.

E

Chancery Division

October 16 1987

(Before Mr Justice MERVYN-DAVIES)

WOOLF PROJECT MANAGEMENT LTD v WOODTREK LTD

Estates Gazette March 26 1988

[1988] 12 EG 111-115

F **Vendor and purchaser — Land Registration Act 1925 — Motion by purchasers for removal of caution — Whether certain sums payable by purchasers were part of the purchase price — Caution entered by vendors to protect vendors' lien — Justification for caution disputed by purchasers — Part of the present judgment deals with the problem of a caution relating to an earlier transaction in which a third party was concerned and the effect of a mistake by the Land Registry — The main issue, however, was as to the effect of special conditions in the contract of sale of the subject premises, a property in Hampstead, London NW3 — The vendors contended that additional sums payable by the purchasers in certain events under the special conditions were in fact part of the purchase** G **price and so proper subjects for protection by a caution — Mervyn Davies J analysed each of the relevant conditions in detail — The first merely provided for the vendors to act as the agents of the purchasers in connection with buying out subsisting leasehold interests in the property and was no more than an agency agreement — The second condition provided for the payment of an additional sum by the purchasers if they decided to carry out a development consisting of two garden flats, which they were under no obligation to carry out —The third condition provided for a payment by the purchasers if they were to carry out a 12-flat development, which, again,** H **they were under no obligation to carry out — None of these conditions was concerned with an addition to the purchase price of the property — The case of *Uziell-Hamilton* v *Keen*, cited on behalf of the vendor, was distinguishable — Held, applying the robust approach recommended by Megarry J in *Rawlplug Co Ltd* v *Kamvale Properties Ltd*, that an order should be made, in the terms of the notice of motion, for the vacation of the caution against the purchasers' registered title**

The following cases are referred to in this report.

Rawlplug Co Ltd v *Kamvale Properties Ltd* (1968) 20 P & CR 32; [1968] EGD 554; 208 EG 147 J

Uziell-Hamilton v *Keen* (1971) 22 P & CR 655; [1971] EGD 791; 219 EG 1603

This was a motion by the plaintiff purchasers, Woolf Project Management Ltd, seeking the removal of a caution registered by the defendant vendors, Woodtrek Ltd, at the Land Registry against the plaintiffs' registered title to property known as 12 Lindfield Gardens, London NW3.

J C Harper (instructed by Masons) appeared on behalf of the plaintiffs; E Cohen (instructed by Oliver O Fisher & Co) represented the defendants.

Giving judgment, MERVYN-DAVIES J said: This is a plaintiff's motion dated June 23 1987 for the removal of a caution registered on June 6 1985 by the defendant at HM Land Registry against the K plaintiff's registered title no 163257 to property known as 12 Lindfield Gardens, London NW3.

The plaintiff is Woolf Project Management Ltd. The defendant is Woodtrek Ltd. The writ was issued on June 16 1987 with a statement of claim endorsed. As well as removal of the caution it seeks damages. The motion, of course, is concerned only with the removal of the caution. An amended defence dated October 1 1987 has been served.

By a contract of sale dated December 20 1984 (Exhibit DHW1) the defendant ("Woodtrek") agreed to sell to the plaintiff ("Woolf") the freehold property 12 Lindfield Gardens, NW3, held under registered title 163257 for £180,000. A deposit of £28,000 was paid. The contract incorporates the National Conditions of Sale 20th L edition together with a number of special conditions. There will be occasion to refer to the following special conditions:

J. If the purchaser proceeds with the development in accordance with the existing planning permission for two garden flats at the rear of the property the purchaser shall pay to the vendor an additional sum equal to 50% of the amount by which the consideration for the sale of the two garden flats on 99 year leases exceeds either (a) £400,000 ("the sum") or (b) in the event that the purchaser's construction costs exceed £120,000 the sum plus the amount of the excess over £120,000.

L. (i) For six months after the date of completion hereof the vendors will be appointed to act as agents for the purchasers to negotiate the purchase of leasehold interest of the flats at the property.

(ii) The vendors confirm that they will at all times act in the best interests of M the purchasers and will not do any act or thing detrimental to the purchasers.

(iii) The vendors undertake not to harass any leaseholder at the property and at all times to act in a proper and professional manner.

(iv) The agents fee payable to the vendors in respect of any leasehold interest acquired by the purchaser shall be the greater of (a) one half of the open market value of the relevant leasehold interest at the property less the total cost of acquiring such leasehold interest or (b) 3% of the purchase price of the relevant leasehold interest.

(v) In the event that the open market value of the relevant leasehold interest cannot be agreed between the parties hereto the matter shall be referred to an Independent Chartered Surveyor agreed upon between the parties or failing agreement at the request of either party by the President of the Royal Institute (*sic*) of Chartered Surveyors. Any such surveyor shall act as an arbitrator under the Arbitration Act 1950 as amended and the decision of any such

A surveyor shall be final and binding upon the parties hereto.

(vi) The vendors covenant not to negotiate for the purchase of any of the leasehold premises at the property either on their own behalf or on behalf of any party other than the purchaser.

N. Should the vendors secure vacant possession of the property within six months from the date of completion and should planning consent for the construction of 12 flats at the property without any unduly onerous conditions attached thereto be obtained by the purchaser then the purchaser shall pay to the vendors 15% of the development profit. Within 28 days of obtaining vacant possession and the said planning consent the purchaser shall pay to the vendors 80% of the 15% of the development profit referred to above. The final 20% balance to be paid on completion of the sale of the final unit. In the event of any dispute pursuant to this clause the dispute shall be referred to a surveyor appointed in like manner as referred to in Condition
B L(v) hereof.

The contract of sale is, as I have said, dated December 20 1984. On the same day the parties executed a second agreement. The recitals in the second agreement indicate that the second agreement is supplemental to the contract of sale (called "the first agreement"). The operative parts of the second agreement are as follows:

It is hereby agreed that notwithstanding the terms of the first agreement,
(1) the vendors covenant to procure the discharge of the caution referred to in entry number 9 of the Proprietorship Register of the said Title Number on or before the contractual date for the completion of the first agreement;
(2) in the event that the said caution is not discharged on or prior to the contractual date of completion referred to above the purchaser is hereby authorised and entitled to pay to Cobstar Limited the necessary money out of
C the balance of the sale proceeds to satisfy the agreement in respect of which the caution is registered (being a sum not in excess of £29,000);
(3) in the event the sum of properly (*sic*) necessary to satisfy the agreement referred to above is in excess of £29,000 the purchaser shall be entitled to recover the difference from the vendors on demand by the purchaser and recover any proper and reasonable costs incurred by the purchaser in securing the discharge of the said caution.

The contract of sale was completed on January 8 1985 in the sense that on that day an executed transfer was delivered to Woolf, and Woodtrek received the balance of the purchase money (over the deposit) but less the sum of £29,000 referred to in the second agreement. Thus on January 8 1985 the money received by the
D vendors on account of the sale of the property was £151,000 with another £29,000 owing or the subject of the second agreement. In fact, Woodtrek had not by January 8 secured the removal of the caution referred to in clause (1) of the second agreement. This was a caution entered by a company called Cobstar Ltd in respect of some earlier transactions said to have affected title no 163257. In these circumstances, Woolf and Woodtrek were agreed that the £29,000 should for the time being remain with Woolf to await events. In an affidavit sworn on October 12 1987 Mr M G Jordan, a partner in the firm of Masons, acting for Woolf, states — see para 2 — that Woolf were prepared to give Woodtrek "a reasonable opportunity of reaching an agreement direct with Cobstar Ltd for the removal of the caution". The correspondence exhibited as MGJ1 to the affidavit
E confirms this. So the position was that Woodtrek wished itself to dispose of the Cobstar caution so that there would be no need for Woolf, using clause (2) of the second agreement, to pay off Cobstar; with the consequence that Woolf would pay the outstanding £29,000 to Woodtrek once Woodtrek had got rid of the Cobstar caution.

On April 2 1985 Woodtrek wrote to Woolf asking for the Woolf solicitors to be ready to pay the £29,000 to Woodtrek in the event of the Cobstar caution being removed on April 9 1985. Woolf replied on April 4 stating that they would make immediate arrangements to pay the £29,000 "on our return from the Easter holidays on April 15". There followed a difficulty arising from a mistake made at HM Land Registry. A Land Registry letter dated April 9 1985 written to Woodtrek states that the Cobstar caution had been cancelled. The
F Land Registry did effect a cancellation, but it seems that in doing so they acted in error. A letter dated April 12 written by an Assistant Land Registrar to Masons (for Woolf) reads as follows:

Dear Sirs. Title No 163257. 12 Lindfield Gardens, London NW3. I refer to my telephone conversation with your Mr Jordan on April 10.
I am writing to confirm what I said over the telephone. That is, that a letter was received from the cautioner's solicitors on April 4, within the time limit set. The letter was from Messrs Hammond, Fathers & Co of 314 High Holborn, London WC1V 7DT (Ref 10/X666) which firm the cautioner has now instructed in place of Messrs Barlow Liam. Unfortunately, because of the change of solicitors, the letter was not associated with the case until April 10. The caution was cancelled on April 9 because it was not then realised that the letter had been received.
In the circumstances, the caution was cancelled in error. Accordingly, I must G
inform you that an application has now been taken out for the rectification of the title by way of the restoration of the caution to the register. I have attempted to contact Mr Moschi by telephone to inform him of the position but have been unsuccessful. I have however sent a telex to Woodtrek Ltd and followed that with a letter.
I am now investigating the merits of the caution. I look forward to hearing from you as soon as possible as to the position regarding the payment of moneys by your clients.
In the meantime, may I apologise to yourselves and your clients for the error which has occurred. Yours faithfully.

In this situation Woodtrek contended that, since the Cobstar caution had in fact been vacated, the outstanding £29,000 was payable by Woolf to Woodtrek. But of course Woolf were concerned
by the fact that there was pending in the Land Registry an application H
to reinstate the Cobstar caution; and Woolf had always held the £29,000 for the purpose of obtaining a title cleared of the Cobstar caution.

As to the outcome of this situation it is sufficient to say that Woodtrek engaged themselves in negotiations with Cobstar. At first their negotiations appeared to be about to result in Woolf paying the £29,000 as to £14,000 to Cobstar and as to £15,000 to Woodtrek. But no such final result emerged. Then on or about June 3 1985, according to Mr Jordan, he (Mr Jordan) was telephoned by Mr Sachs, the solicitor acting for Woodtrek. Mr Jordan was told that Cobstar would agree to the withdrawal of their caution and the
cancellation of the application to rectify the register (by a J
reinstatement of the caution) in consideration of Cobstar being paid £20,000. The balance of the £29,000 was to go to Woodtrek. A meeting was arranged at Mr Jordan's office for June 3. The meeting took place and in attendance were Mr Jordan (for Woolf), Mr Sachs and Mr Moschi (for Woodtrek) and Mr Fathers of the firm of Hammond Fathers, acting for Cobstar. Mr Jordan had with him a banker's draft for £20,000 in favour of Cobstar and another in the sum of £9,000 for Woodtrek. Mr Fathers left the meeting with the Cobstar draft and the Woodtrek representatives took away the other draft. It is plain that the Woodtrek representatives knew that Mr Fathers received the Cobstar draft for £20,000. In his affidavit Mr
Jordan includes these words, at the end: K

. . . as the agreement for the removal of the caution in consideration of the payment of £20,000 to Cobstar was arranged by Mr Moschi and Woodtrek's solicitors, I considered that Woodtrek accepted that my clients were in those circumstances entitled to pay the £20,000 direct to Cobstar Ltd. Indeed, they requested us so to do.

Mr Jordan obtained of course authority from Hammond Fathers to secure the removal from the Land Register of the Cobstar caution and an assurance that there was no wish that the caution be reinstated. This authority was in the form of a letter dated June 3 1985 addressed to the Land Registry signed by Hammond Fathers & Co and given to Mr Jordan with authority to produce the same to the Land Registry. In consequence the long-delayed registration of
Woolf as registered proprietors of 12 Lindfield Gardens proceeded in L
the Land Registry without there appearing thereon the Cobstar caution.

With the Cobstar caution thus disposed of, Woodtrek were then minded to register a caution of their own. Mr Sachs left for registration a caution against 12 Lindfield Gardens on behalf of Woodtrek, the statutory declaration in support of the caution stating that Woodtrek was interested as "having a vendors lien by virtue of a contract dated December 20 1984 and made between Woodtrek Ltd and Woolf Project Management Ltd". I note that the statutory declaration in support of the caution is dated February 12 1985, so that Woodtrek must have had in mind the entry of a caution against
Woolf for some months and did in fact enter the caution as soon as M
Woolf were registered as proprietors. This Woodtrek caution is said by Woodtrek to have been entered to protect their vendor's lien, not only in respect of the unpaid £29,000 but also in respect of moneys payable or to be payable by Woolf to Woodtrek pursuant to the special conditions of the first agreement dated December 20 1984 already mentioned, such special conditions money being said to be part of the purchase price payable for the property. I will deal later with the issues arising out of these special conditions. For the present I will deal with the caution as said to be a protection of the vendor's lien for an unpaid £20,000.

In short, the Woodtrek contention is that despite the payment of £20,000 made by Woolf to Cobstar on June 3 there remains owing from Woolf to Woodtrek the sum of £20,000. The price of the property was £180,000 and, says Woodtrek, only £160,000 has been paid. See the Woodtrek letter to the Land Registry dated July 4 1985.

Mr Cohen appeared for Woodtrek. He took me carefully through the correspondence already referred to (in exhibit MDJ1) and drew attention to the fact that there was no express agreement between Woolf and Woodtrek to the effect that if Woolf paid £20,000 to Cobstar then Woodtrek, on receiving the balance of £9,000, would regard the £180,000 purchase price as fully paid. So, said Mr Cohen, one gets back to the contract of sale; and there we find that £180,000 (not £160,000) is payable. Furthermore, we see from the Woolf letter dated April 4 1985 that Woolf accepted that if Woodtrek succeeded in getting the Cobstar caution removed then Woolf would pay to Woodtrek all the outstanding £29,000. Woodtrek did so succeed, said Mr Cohen, because the Cobstar caution was removed on April 9 1985: see the Land Registry letter dated April 9.

I find this approach artificial in the extreme. The true position emerges, I think, as a matter of irresistible inference from the affidavit of Mr Jordan already referred to together with the correspondence exhibited thereto. I am satisfied that the £20,000 paid by Woolf to Cobstar was paid with the consent of Woodtrek, with Woodtrek well realising that Woolf was thereby paying £20,000 of the £180,000 purchase price to Cobstar. I see nothing in the affidavit evidence of Mr Moschi or Mr Sachs or Mr Fathers that in any way convincingly goes against that view. It is idle to contend that Woolf, having paid £20,000 to Cobstar with Woodtrek standing by, should now have to pay another £20,000 to Woodtrek simply because by an error of the Land Registry the Cobstar caution was taken off the register. Since an application to reinstate the caution was made immediately I cannot regard the Cobstar caution as ever having been effectively removed until after June 3, pursuant to the application sent to the Land Registry by Mr Jordan under cover of his letter dated June 3 1985. It follows that the defendants are not entitled to maintain a caution on the grounds that any part of the £180,000 purchase price is unpaid.

As I have mentioned, Woodtrek have sought to maintain their caution, not only by reference to the £20,000 dealt with above but also by reference to what they say are further sums of unpaid purchase money. As will be seen from the contract of sale p 1, the purchase price is said to be £180,000. However, in the Woodtrek amended defence that sum is alleged to be "merely the basic minimum purchase price". The purchase price, so it is said, "further consisted of the additional sums payable by the plaintiff to the defendant under the provisions of special conditions J, L and N of the contract of sale": see the amended defence para 1. It follows that one must examine the conditions referred to for the purpose of ascertaining whether or not all or any one of them have the effect of causing any money payable thereunder to be regarded as part of the purchase price paid or to be paid for the property sold. In other words, one must construe the special conditions against the background of the contract of sale as a whole.

Counsel before me considered at some length, and by reference to many documents, what had or ought to have been done by Woolf at the property since completion. It is not, in my view, necessary or appropriate to consider those matters. As I understand the position, one simply reads the contract of sale for the purpose of finding out whether or not the parties intended that any payments to fall due under the special conditions were to be regarded as part of the purchase price payable in the future.

Of the three special conditions relied upon I take first special condition L as set out above. This condition provides for Woodtrek acting as agents of Woolf in a matter of seeking to buy out the various leasehold interests that were subsisting in 12 Lindfield Gardens at the time of completion. For this work Woodtrek were to receive a fee: see L(iv). I do not read condition L as affecting any addition to the purchase price. It merely constitutes a supplemental agreement for agency.

As to conditions J and N, it will be seen that condition J contemplates the possibility of two garden flats being built at the rear of 12 Lindfield Gardens, whereas condition N contemplates the possibility of 12 Lindfield Gardens itself being altered so as to comprise 12 flats; there being originally therein (and now) eight flats.

Reading special condition J, as at the time when the contract of sale was made, one sees that Woolf comes under no obligation to build two garden flats; but if they do so, they will be obliged to pay "an additional sum". An additional sum, it is said, means an addition to the purchase price; so a sum payable under J is part of the purchase price and protectable by a caution. I do not read J as having that effect. It seems to me that at the time when the purchase was completed, Woolf was free to seek or not to seek to proceed with a two-flat garden development. If they were to do so, then some further money might become payable to Woodtrek. If they were not to do so, nothing would be payable to Woodtrek. In other words, J provides for a payment to Woodtrek emerging as a possibility only if Woolf, having acquired title to the property, decides to carry out a garden development, and does so with a certain margin of profit. That being the situation, I cannot regard any condition J payment as being part of the purchase price paid for the property. A condition J payment, if emerging, is a sum "additional" to the purchase price in the sense that, as well as the £180,000 purchase price, there is to be paid in addition a separate sum derived from any successful operation of condition J.

Going on to special condition N one sees again no obligation is imposed on Woolf to seek to carry out a 12-flat development. It follows that any payment that might arise under N is not to be regarded as payment of part of the purchase price; *a fortiori* since an N payment is not said (compare J) to be "an additional sum". Looking at the contract as a whole I see it as (a) providing for Woolf becoming owner of 12 Lindfield Gardens in return for a payment of £180,000, and (b) embracing three further separate agreements (in J, L and N) giving rise to the possibility of Woolf being contractually indebted (irrespective of the purchase price) to Woodtrek in the event of Woolf engaging, if so minded, in certain defined development activities (J and N) or in the event of Woodtrek (being obliged to do so) successfully acting as Woolf's agents under Condition L.

I was referred to *Uziell-Hamilton* v *Keen* (1971) 22 P&CR 655. That was a case concerning the sale of a solicitor's practice together with some office leases, furniture etc for the sum of £5,000. There was a further condition —condition 8 — in the sale agreement whereby the purchaser was obliged to get in and account to the vendor for money owing to the practice at the time of the sale. The agreement was construed as showing the intention of the parties to be that the sale transaction "should be subject, not only to the payment of £5,000, but to the observance by the purchaser of his obligations under, *inter alia*, clause 8 of the agreement . . .". In those circumstances, the lien there alleged prevailed. There the purchaser was obliged to get in the outstanding moneys and pay them over to the vendor. Here the purchaser is placed under no obligation under either J or N. The case is accordingly distinguishable on that, if not on other grounds. It does not bear upon condition L as to which I have already expressed my view.

I was reminded of the well-known words of Megarry J, as he then was, in *Rawlplug Co Ltd* v *Kamvale Properties Ltd* (1968) 20 P&CR 32:

> Accordingly, it seems to me that the speedy form of remedy by way of a motion ought to be available to a landowner in all cases where there are no substantial grounds for supporting the registration. I would thus favour a certain robustness of approach in these motions, of the type to be found in the administration of Ord 14. If there is a fair, arguable case in support of the registration, then the matter must stand over until trial. But if, though not cloudless, the sky has in it no more than a cloud the size of a man's hand, I would clear the register and leave the purchaser to seek such remedy by way of specific performance or damages as he may be advised.

By reference to those words, Mr Harper for Woolf urged me to be "robust", while Mr Cohen urged that there was "a fair arguable case in support of the registration". I say in answer to Mr Cohen that despite his persuasive submissions I do not see here a fair arguable case, in respect of either the £20,000 point or the points arising from the special conditions.

Mr Cohen also submitted that removal at this stage was a discretionary remedy and suggested certain considerations as bearing on a refusal of discretion in Woolf's favour. In this connection, in particular, he said that the plaintiffs' financial situation was unsatisfactory and that the plaintiffs were in breach of the sale agreement and so not fitted to be afforded discretionary relief. I do not find the sum of Mr Cohen's considerations in this regard as justifying the withholding of the relief which the plaintiffs claim.

I should add that the Woodtrek caution has been the subject of previous proceedings between the parties. See the action 1986 W 2339. That action was disposed of by a Tomlin Order dated May 19 1986. I raised the point that perhaps those proceedings precluded the

A plaintiffs from again litigating the removal of the Woodtrek caution. However, having read the statement of claim in the earlier case and the statement of claim in this action, it is apparent that the complaints made and damages claimed differ in the two actions. Taking that fact into account, and also that the matter was not fully argued before me with any authorities, I do not feel justified in withholding relief from the plaintiffs by further consideration of this point.

There will be an order in the terms of the notice of motion.

Court of Appeal

B March 21 1988

(Before Lord Justice DILLON, Lord Justice NEILL and Lord Justice WOOLF)

PRODUCTION TECHNOLOGY CONSULTANTS LTD v BARTLETT

Estates Gazette June 25 1988

[1988] 25 EG 121-122

Misrepresentation Act 1967 — Innocent misrepresentation — Purchase of freehold house at auction sale in reliance on C innocent misrepresentation in the particulars — Particulars stated that the first floor was let at a weekly inclusive rent of £2.55 and that application had been made for the registration of a fair rent — In fact the first floor was also the subject of a concurrent lease for a term of 99 years from September 25 1979 at a ground rent of £40 per annum — It would have been possible for the representative of the plaintiff purchasers who attended the auction to become aware of this lease of the reversion before the sale if he had inspected a copy of the charges register available in the auctioneers' office, but he did not take advantage of this facility — After the auction sale and D before completion the plaintiff company's solicitors were supplied with the relevant documents of title, including a copy of the lease of the first floor, but it appears that neither the vendor's nor the purchasers' solicitors fully appreciated the effect of that lease at the time

At the trial before Mr Robert Wright QC, sitting as a deputy judge of the Chancery Division, the plaintiff purchasers claimed damages for misrepresentation, relying on section 2(1) of the Misrepresentation Act 1967 — The judge held, however, that the claim for damages failed, on the ground that the purchasers had elected to complete: "by making the election to complete without protest [they] exhausted the E effect of the misrepresentation and cannot be said to have completed relying upon it."

The issue in the appeal was whether it is right that where there has been a misrepresentation before contract which has induced a party to enter into a contract, but the facts have become known to the party, or to his agent acting for him in the transaction, before completion, the purchaser is then bound either to rescind or to complete on the terms of the contract without any further remedy after completion by way of a claim for damages or compensation — The defendant vendor accepted that if there had been a fraudulent F misrepresentation inducing a purchaser to enter into a contract for the purchase of land and the fraud was discovered only after completion, the purchaser would have a claim for damages notwithstanding completion — It was also accepted that if the fraud became apparent after contract but before completion, the purchaser could reserve his cause of action for damages by electing to complete without prejudice to his claim for damages — The defendant, however, submitted that there had to be an express reservation of such a cause of action in damages, otherwise it would be lost — Held, rejecting this last submission, that it was contrary to the 1967 Act and to dicta in *Arnison* v *Smith* and *Gordon* v *Selico Co Ltd* — Accordingly the plaintiffs in the present case were not precluded from G claiming damages by the fact that the purchase had been completed — Appeal allowed

The following cases are referred to in this report.

Arnison v *Smith* (1889) 41 ChD 348
Gordon v *Selico Co Ltd* [1986] 1 EGLR 71; (1986) 278 EG 53, CA

This was an appeal by the plaintiffs, Production Technology Consultants Ltd, from a decision of Mr Robert Wright QC, sitting as a deputy judge of the Chancery Division, rejecting their claim as plaintiffs to damages against Derek Joseph Bartlett, the defendant and present respondent, for his misrepresentation in the particulars H at an auction sale of a house at 42 Culverden Road, Balham, London SW12.

C P F Rimer (instructed by A R Wilkin & Co) appeared on behalf of the appellants; J Littman (instructed by Talfourd & Co, of Hornchurch, Essex) represented the respondent.

Giving judgment, DILLON LJ said: This is an appeal by the plaintiff in the action, a company called Production Technology Consultants Ltd, whose moving spirit is a Mr Behcet, against a decision of Mr Robert Wright QC, sitting as a deputy judge of the High Court in the Chancery Division, as long ago as July 30 1986.

The judgment of the learned deputy judge was a reserved judgment; he appears to have been somewhat belaboured by the citation of numerous authorities which do not appear to have any J relevance to the very short point, which is the only point taken on this appeal.

The plaintiff company bought at an auction sale held on December 10 1980 lot 46, which is a freehold house known as 42 Culverden Road, Balham, London SW12. It was bought as an investment. It was divided into four flats on different floors. According to the particulars the top floor was leased for a term of 99 years from September 25 1979 at a ground rent of £40 a year. The first floor was let to a Mr Buttle at £2.55 per week inclusive and a registered rent had been applied for. The ground floor was leased for a term of 99 years from March 25 1976 at a ground rent of £30 per annum and the semi-basement was let at a weekly rent of £5 inclusive. On those figures a K calculation of total income was given and there was a reference to the landlord's paying the rates for the first and semi-basement floors.

Those particulars were misleading and inaccurate, in that the first floor was also subject to a separate lease for a term of 99 years from September 25 1979 at a like ground rent of £40 per annum.

Mr Behcet, having formed the view that Mr Buttle was unlikely to remain the tenant of the first floor for long, attended the auction and bid for the property. His was the highest bid. The conditions of sale for lot 46 provided that the property was sold subject to the entries in the charges register of the title; it provided also that a copy of the register might be inspected at the offices of the auctioneers during usual office hours prior to the auction, but Mr Behcet did not take advantage of that facility. The office copy entries would have L disclosed the existence of the 99-year lease of the first floor.

The deputy judge found as a fact that Mr Behcet bought relying on the particulars; he accepted Mr Behcet's evidence that he would not have bid, or at any rate the price at which he would have bid would have been much less, if he had known of the 99-year lease.

After the auction had been held Mr Behcet instructed solicitors to act for the plaintiff company. They were supplied by the defendant vendor's solicitors with copies of the three 99-year leases, including the lease of the first floor, but it appears that neither firm of solicitors fully appreciated the effect of the grant of the 99-year lease in reversion on Mr Buttle's weekly tenancy, because the correspondence between the vendor's and the purchaser's solicitors leading up to completion was concerned with the apportionment of M the rent payable by Mr Buttle. Of course, in view of the grant of the 99-year lease, which was a lease to the vendor, the defendant Mr Bartlett, and his son, the only rent to be apportioned was the rent under the 99-year lease, and the lessees under that lease were entitled to Mr Buttle's rent.

The purchase was completed by post on January 19 and 21 1981. On January 19 the solicitors for the plaintiff company sent a bank draft for the money on the basis that in return they were sent the land certificate, the three leases, authority to inspect and a deed of assignment in respect of arrears of rent. The reply on January 21 from the vendor's solicitors thanked them for the document, enclosed the land certificate and so forth, and then went on:

So far as the Deed of Assignment is concerned, unfortunately we have only at this late stage realised that the same is slightly wrong in the sense that there is a 99 Year Lease applicable to the first floor, ie, 42C. This means that the question of arrears or otherwise in that case are not applicable, since they are the responsibility of the Ground Rent Tenant of that flat . . .

The response to that was that on January 29 the plaintiff company's solicitors objected very strongly that the plaintiff company had a very substantial claim for damages on the ground of misrepresentation, which they were instructed to pursue; they suggested alternatively that the 99-year lease at a ground rent might be surrendered and extinguished, but that did not happen.

The action came on for trial. The learned deputy judge stated at the end of his judgment that he was satisfied that Mr Behcet relied upon the representation in the particulars in entering into the contract and thereby acted to his detriment. He also accepted that a certain general condition of the contract excluding the provisions of the Misrepresentation Act 1967 was avoided by the Unfair Contract Terms Act 1977, but nothing turns on that in this appeal. But he held that the claim for damages for misrepresentation must fail because the contract was completed with express notice of the long lease in issue. He says that the plaintiff had an election whether to rescind or go ahead,

and by making the election to complete without protest he exhausted the effect of the misrepresentation and cannot be said to have completed relying upon it.

The question, of course, is whether he entered into the contract relying upon it.

There is no doubt that the misrepresentation in this case was entirely innocent and that it was in no way fraudulent, but the Misrepresentation Act 1967 effected a very important change in the law. Previously, damages for an innocent misrepresentation could not be recovered; damages for misrepresentation could be recovered only if the misrepresentation was fraudulent.

Section 2(1) of the 1967 Act provides as follows:

Where a person has entered into a contract after a misrepresentation has been made to him by another party thereto and as a result thereof he has suffered loss, then, if the person making the misrepresentation would be liable to damages in respect thereof had the misrepresentation been made fraudulently, that person shall be so liable notwithstanding that the misrepresentation was not made fraudulently, unless he proves that he had reasonable ground to believe and did believe up to the time the contract was made that the facts represented were true.

The final condition is not relevant in this case and it is not suggested that it can be invoked. The position is, therefore, that under section 2(1) if the person making the misrepresentation would be liable for damages in respect thereof had the misrepresentation been made fraudulently, that person will be so liable notwithstanding that the misrepresentation was not made fraudulently.

I should, however, digress at this moment to deal with a point on which Mr Littman for the respondent/defendant sought leave to amend his respondent's notice. We refused that application; it concerns the words "and as a result thereof had suffered loss". Mr Littman submitted that the plaintiff company had failed to show that it had suffered loss. By agreement between the parties the trial before the deputy judge was conducted as a split trial on liability only, on the footing that if the plaintiff succeeded the court would direct an inquiry as to damages. The parties came to court, each armed with a surveyor's report as to the value of the property at the relevant times, but agreed that it was inappropriate to take up the time of the court with the evidence of value and loss until the issue as to liability had been decided and the basis, if any, on which there was liability had been determined by the judgment of the court.

Mr Rimer, who was engaged in the case in the court below, told us that Mr Eyre, who then appeared for the defendant, accepted that there was no question that they had not proved loss sufficient to warrant an inquiry as to damages if they were otherwise right in their submissions on law. As I have already mentioned, the learned judge held that in entering into the contract Mr Behcet had acted to his detriment. Prima facie there is considerable difference in value between a flat let at a weekly rent of £2.55 to an elderly tenant who may well quit, and a flat let for a term of 99 years at a ground rent of £40 a year only.

In those circumstances, when Mr Littman applied to amend his respondent's notice to include a paragraph as follows:

Alternatively, that the Plaintiff failed to prove that any loss (measured in the manner of damages awarded in the tort of deceit) had been incurred by it, alternatively had resulted from a misrepresentation made to it by the Defendant or from its entering into a contract after the making of such a misrepresentation.

we ruled that having regard to the way in which the case had been conducted in the court below, that was not open to him.

The only issue in this appeal, therefore, is whether it is right that where there has been a misrepresentation before contract which has induced a party to enter into a contract, but the facts have become known to the party, or to his agent acting for him in the transaction, before completion, the purchaser is then bound either to rescind or to complete on the terms of the contract without any further remedy after completion by way of a claim for damages or compensation.

Mr Littman accepts on behalf of the defendant that if there has been a fraudulent misrepresentation which has induced a purchaser to enter into a contract for the purchase of land and the contract is completed, and the fraud is discovered only after completion, the purchaser will have a claim for damages as against the vendor after completion and notwithstanding completion. He accepts also that if the fraud becomes apparent after contract but before completion, the purchaser could reserve his cause of action for damages by electing to complete without prejudice to his claim for damages. But he says that there has to be an express reservation of such a cause of action in damages, otherwise it will be lost.

I find it difficult to see the reasoning that is supposed to lie behind that. We have been referred to cases which are concerned with the waiver of objections to title and to purchasers' being bound to accept title to property if they fail to deliver, or follow up, requisitions on title according to the time-limits prescribed by the contract. But that seems to me to be a completely different territory. In so far as matters of title are concerned, the purchasers' rights depend particularly on the terms of the contract, but the 1967 Act gives a right which overrides any provision in the contract. In fact, section 3 of the 1967 Act, as amended by the Unfair Contract Terms Act 1977, expressly states that. It provides as follows:

If a contract contains a term which would exclude or restrict

(*a*) any liability to which a party may be subject by reason of any misrepresentation made by him before the contract was made; or

(*b*) any remedy available to another party to the contract by reason of such a misrepresentation

that term shall be of no effect except in so far as it satisfies the requirement of reasonableness as stated in section 11(1) of the Unfair Contract Terms Act 1977; and it is for those claiming that the term satisfies that requirement to show that it does.

As long ago as 1889 Cotton LJ, in his judgment in *Arnison* v *Smith* (1889) 41 Ch D 348, at the foot of p 371, said:

It was contended that going on after receiving the circular was an affirmance of the contract, with knowledge of the real facts, and that such affirmance is inconsistent with this action. But a man may affirm a contract and yet sue the person who by fraud induced him to enter into it.

We were referred also to the decision of this court in *Gordon* v *Selico Co Ltd* [1986] 1 EGLR 71. In giving the judgment of the court in that case Slade LJ said at p 77J:

Even if at the date of completion the plaintiffs had been fully aware of the fraudulent misrepresentation which had been made to them, this would not have precluded them from affirming the contract and seeking appropriate relief, though it would have prevented them from seeking to rescind it.

The statement of Slade LJ was also obiter, but in my judgment it was, if I may respectfully say so, a correct statement of the law.

Accordingly, the plaintiff company is not precluded from claiming damages by the fact that the purchase was completed in the condition of mutual muddle between the solicitors as to the effect of the 99-year lease which I have mentioned.

In these circumstances, with every respect to the learned deputy judge, who seems to have been somewhat led astray by the welter of old authority which was put before him, for my part I would allow this appeal, set aside the order of the deputy judge and direct an inquiry as to damages.

NEILL and WOOLF LJJ agreed and did not add anything.

The appeal was allowed; order of court below set aside; inquiry as to damages directed; plaintiff/respondent's costs of appeal, to be taxed and paid forthwith; costs in court below reserved to hearing of inquiry.

For further cases on this subject see p 251

TOWN AND COUNTRY PLANNING

Court of Appeal

September 23 1987

(Before Lord Justice GLIDEWELL)

FULLER v SECRETARY OF STATE FOR THE ENVIRONMENT AND ANOTHER

Estates Gazette January 9 1988

[1988] 01 EG 55-56

Town and Country Planning Act 1971, section 246(4) — Ex parte application for leave to appeal from a decision of Stuart-Smith J, who had dismissed an appeal from a decision of the Secretary of State upholding two enforcement notices served on the applicant — The enforcement notices alleged a material change in the use of agricultural land by the use of grain silos for the storage of grain not grown on the particular agricultural unit to which the notice related — The applicant occupied agricultural land (some as owner and some as tenant) in different areas which were near to each other, but not completely contiguous, eg in one case about two miles from another of his farms — The present dispute had narrowed down to the position created by the storage of grain produced on the applicant's land and stored on two of his farms but not intended to be used for the feeding of livestock at those farms — The difficulty was that the planning authority and the Secretary of State had treated the applicant's holding as divided into different planning units and had decided that "the bringing in of grain from other parts of the holding to the planning unit under consideration and its storage there for purposes unconnected with agriculture on the planning unit is a use for storage in itself and not a use for agriculture" — Stuart-Smith J had upheld this ruling — In applying for leave to appeal (in accordance with section 246(4) of the 1971 Act) it was urged that if, as in this case, there were areas sufficiently close to one another, albeit not contiguous, to be conducted as one agricultural holding, then they are one agricultural unit and whatever is carried on there is use for the purpose of agriculture and does not constitute development — On this view the question whether there is one planning unit is irrelevant — It was also submitted that the Secretary of State erred in law in holding that he was entitled to regard the planning unit as being something smaller than the totality of the agricultural holding — In refusing leave to appeal Glidewell LJ said that it was a question of fact whether land forms part of one agricultural unit or not and it was also a question of fact and degree for the planning authority and the Secretary of State to decide what constitutes an appropriate planning unit — Observations on "planning unit" as "a useful piece of shorthand" in *South Staffordshire District Council* v *Secretary of State for the Environment* referred to — Stuart-Smith J's decision correct — Leave to appeal refused

The following cases are referred to in this report.

Hancock v *Secretary of State for the Environment and Torridge District Council* [1987] JPL 360

South Staffordshire District Council v *Secretary of State for the Environment and Bickford* [1987] JPL 635, DC

This was an *ex parte* application by Richard Mackney Fuller for leave (required by section 246(4) of the Town and Country Planning Act 1971) to appeal from a decision of Stuart-Smith J dismissing an appeal from a decision of the Secretary of State for the Environment upholding two enforcement notices served on the applicant by Dover District Council in relation to grain silos installed by the applicant on farms which formed part of the 2,000 acres which he farmed in Kent near Dover. The decision of Stuart-Smith J is reported at (1987) 283 EG 847.

Barry Payton (instructed by Marsland & Barber, of Margate) appeared on behalf of the applicant.

Giving judgment, GLIDEWELL LJ said: The applicant, Mr Fuller, is a farmer. Quoting from the judgment of Stuart-Smith J:

> He farms over 2,000 acres in Kent near Dover. Some of the land is owned by him, the rest he holds as a tenant. There are two main areas of land; the largest is at Sutton Court Farm (which includes Langdon Abbey and Langdon Court Park) and the surrounding area;

and I interpolate the whole of that is more or less contiguous, though not completely contiguous

> the next largest is at Walmer Court Farm, separated from Sutton Court by about two miles as the crow flies. In addition to this there are a considerable number of other areas, varying from a few acres to some 300 acres (the latter being acquired after service of the enforcement notices).

The applicant has built, I believe, nine grain silos with drying facilities at Sutton Court Farm and a smaller similar installation at Walmer Court. In 1984 the respondent district council served two enforcement notices relating to the use of these storage installations for the storage of grain. Notice A, as it was called, was served on May 15 1984 and it related to a restricted area of land at Sutton Court Farm on which the silos actually stand. The development alleged in that notice was the "making of a material change in the use of the land to use for the purpose of agriculture and of the commercial storage of grain not grown on the agricultural unit comprising Sutton Court, Langdon Abbey, Langdon Court and Walmer Court Farms", and the notice required that use to be discontinued.

Notice B, served on July 3 1984, related to the whole area of Sutton Court, Langdon Abbey, Langdon Court and Walmer Court Farms. The alleged development was "the making of a material change in the use of the land and premises to use for the purpose of agriculture and of the commercial storage of grain not grown on the agricultural unit comprising Sutton Court, Langdon Abbey, Langdon Court and Walmer Court Farms", and the notice required that use to be discontinued.

The applicant appealed against both enforcement notices on most of the grounds in section 88(2) of the Town and Country Planning Act 1971. After receiving the report of an inquiry conducted by an inspector on his behalf, the Secretary of State amended both enforcement notices to allow more time for compliance, but otherwise upheld them and refused planning permission for what he found to be the development. The applicant appealed against that decision under section 246 of the 1971 Act. On May 22 of this year Stuart-Smith J dismissed that appeal. The applicant now applies for leave to appeal against that decision, being required to do so by subsection (4) of section 246.

In his judgment the learned judge categorised the purposes for which grain stored at Sutton Court and Walmer Court was so stored. First, he referred to grain produced on the appellant's own land and he said: "This may be either (a) grain produced on Sutton Court or

A Walmer Court land and stored on the farm where it is produced," missing out some words, "(b) grain produced elsewhere than at Sutton Court or Walmer Court but stored respectively at those farms and used for the purposes of agriculture at those two farms, for example in the feeding of stock there."

I break off to say that the judge found — and, in my respectful view, found entirely correctly — that the storage of grain for either of those purposes was within the meaning of "agriculture" within the Town and Country Planning Act 1971 because, if not part of the principal use, it was certainly ancillary to the use of the land for agriculture and, therefore, there could be no question of the use of the land for such storage constituting development.

B The third category of grain produced on the appellant's own land is "(c) grain produced elsewhere than at Sutton Court and Walmer Court Farm but stored respectively at those farms".

Category 2 is drying and storage of grain produced by neighbouring farmers; 3 is possible storage incidental and ancillary to the business of a corn merchant, at one time carried on by the appellant or his predecessors at Northbourne Farm, which is some distance from both Sutton Court and Walmer Court; 4 is storage of grain under Grade and Feed Trade Association agreement, and 5 is what is called intervention storage.

The effect of the Secretary of State's decision (which was upheld by the judge) was that only storage for purposes 1(a) and 1(b) did not constitute development and that storage for all the other purposes I C have enumerated did constitute development as being not use of the land for the purposes of agriculture and thus was caught by the enforcement notice. So far as the storage of grain under the GAFTA agreement and for intervention purposes is concerned, that decision is accepted by the applicant and so the area now in dispute is narrowed, and particularly relates to the storage of grain produced on the appellant's land at Sutton Court and Walmer Court Farms, stored at those farms but not intended to be used for the feeding of livestock at those farms. In other words, grain within category 1(c) in Stuart-Smith J's categorisation.

The parties at the hearing of the inquiry were apparently in agreement (according to the inspector) that it was right to treat all the land occupied as an agricultural holding by Mr Fuller as if it were one D agricultural unit and possibly one planning unit. The Secretary of State disagreed with that, and it is his disagreement with that which is really the subject of challenge both before Stuart-Smith J and as the basis of this present application. The Secretary of State said:

> The question which will first need to be decided in these appeals is whether or not any material change of use of land has occurred in breach of planning control. That is a question of fact and degree which cannot be considered in a vacuum: it can only be considered in the context of the relevant planning unit or units.

and he then went on to decide, upon the evidence as a matter of fact, that realistically Sutton Court, including Langdon Abbey and Langdon Court, could properly have been considered to be one E planning unit and Walmer Court, some two miles distant, a separate planning unit. But he said, in effect, "I cannot follow the logic of that view through because to do so would make the enforcement notice even more stringent than it is already, and I am not allowed to do that. The planning authority, in their Notice B, had treated as one planning unit Sutton Court with its adjacent farms and Walmer Court together and so I must do so too. But they do constitute one planning unit (this is the effect of his decision) and anything outside those areas, albeit forming part of the total holding farmed by Mr Fuller, is not within the planning unit and therefore the bringing in of grain from other parts of the holding to the planning unit under consideration and its storage there for purposes unconnected with agriculture on the planning unit is a use for storage in itself not a use F for agriculture and, therefore, constitutes development."

That is the logic of the Secretary of State's decision, and it is that logic which Mr Payton seeks to challenge. It is the logic which the learned judge has upheld.

Mr Payton has, if I understand him correctly (as I hope I do), two main arguments. The first is that an agricultural unit is, broadly speaking, the area of land occupied and farmed by one farmer at any one time, provided that its parts are not separated so distantly that he cannot realistically farm the whole as one operation. He accepts — to use my absurd example — that if a farmer occupies land in Kent, Cornwall and Northumberland he would be hard put to it to describe them all as forming part of one agricultural unit, because his management of them could not be conducted day to day or even week to week as one activity. But, he says, if you have areas sufficiently G close to each other, albeit not contiguous, to be conducted as one agricultural holding, then they are one agricultural unit. Second, he says, if that be so, then whatever is carried on there is use for the purposes of agriculture and does not constitute development by reason of section 22(2)(e) of the 1971 Act, and the question whether there is one planning unit is nothing to the point. Or, to put it another way, the unit to which the enforcement notice should relate must be the whole of the agricultural unit, that is to say the whole of the agricultural holding, and the Secretary of State is wrong to have restricted the phrase "planning unit" to something smaller than the agricultural holding.

There is no general definition of "agricultural unit" in the Town H and Country Planning Act. There is a definition in section 207, but that is only for the purposes of that part of the Act which relates to purchase notices, which is not in issue here. The most helpful definition to which Mr Payton referred me is in section 109 of the Agriculture Act 1947, which in subsection (2) describes "agricultural unit" as follows:

> means land which is occupied as a unit for agricultural purposes, including — (a) any dwelling house or other building occupied by the same person for the purpose of farming the land, and (b) any other land falling within the definition in this Act of the expression "agricultural land" which is in the occupation of the same person, being land as to which the Minister is satisfied that having regard to the character and situation thereof and other relevant circumstances it ought in the interests of full and efficient production to be farmed in conjunction with the agricultural unit, and directs accordingly. J

So it becomes a question of fact for the minister, even under that definition, to decide whether land forms part of one agricultural unit or not.

But the major attack which Mr Payton launches is on the second item of the Secretary of State's decision, that he can decide on what is a planning unit and choose something which may be more restricted than the whole of an agricultural holding. Mr Payton says, as a matter of law, that that is wrong.

In the very recently published decision of the Divisional Court — not, of course, binding on the Court of Appeal — in *South Staffordshire District Council* v *Secretary of State for the Environment and Bickford* K (which is reported in the *Journal of Planning and Environment Law* for September 1987 at p 635) I, giving the judgment of the court, said that the phrase "planning unit"

> is a useful piece of shorthand to describe the area of land or part of a building or group of buildings to which a local planning authority should properly have regard when considering whether there has been a "material change of use in the use of any buildings or other land".

I adhere to that; that is what the phrase — which, of course, is not to be found in the Act or regulations — means. I do not think Mr Payton would dissent from that. The question is whether there is an argument at law that in the context of a use for the purposes of agriculture the Secretary of State is in effect bound to regard one L agricultural holding, which consists of a number of pieces of land separate geographically but farmed as one business, as one planning unit. If he is, then all the activities going on on that unit for agricultural purposes are uses for the purpose of agriculture, and so storage of grain produced in any part of that holding or any other part would be within the meaning of the word "agriculture".

I can find no warrant for Mr Payton's submission that as a matter of law the Secretary of State was wrong to say that he was entitled to regard the planning unit as being something smaller than the totality of the agricultural holding. It is a question of fact and degree, first for the planning authority and later for the Secretary of State, to decide what constitutes an appropriate planning unit, and I do not accept that the concept does not apply in a situation such as this. M

Mr Payton also argues that his client is going to be disadvantaged because in a recent decision of November 28 1986 in *Hancock* v *Secretary of State for the Environment** Hodgson J decided that in a case in which a farmhouse was separated by some little distance from the land farmed by the farmer occupying the farmhouse, nevertheless the whole could properly be treated as both one agricultural unit and one planning unit. Mr Payton argues that since that decision is under appeal by the Secretary of State, his client should be allowed to appeal also so as to add weight to the defence by the respondent, Mr Hancock, to his appeal and if that defence succeeds Mr Payton

*Editor's note: see [1987] JPL 360.

A suggests that his case on appeal will be strengthened. It is with that last proposition that I respectfully disagree. Even if Mr Hancock succeeds (as he may well, even without the support of Mr Payton and his instructing solicitors), the proposition that it is proper to treat as one unit an area consisting of a piece of land separated by a short distance — because it was a short distance in that case — from the farmhouse and its immediately adjacent paddock seems to me to be not helpful in relation to the question whether a holding consisting in a number of disparate pieces of land, some of them separated from each other by several miles, can properly be considered to be one unit.

For those reasons I am quite satisfied that the decision to which Stuart-Smith J came was correct and I refuse leave to appeal.

B *Application for leave to appeal was refused.*

Court of Appeal

December 14 1987

(Before Lord Justice PARKER, Lord Justice Ralph GIBSON and Sir George WALLER)

WEALDEN DISTRICT COUNCIL v SECRETARY OF STATE FOR THE ENVIRONMENT AND ANOTHER

Estates Gazette February 27 1988

C **[1988] 08 EG 112-124**

Town and Country Planning Act 1971 — Stationing of caravan on agricultural land — Whether in the circumstances a material change of use — Caravan placed in order to store and mix calf feed in it and to provide shelter for farmer and his wife — The farm consisted of some six hectares and was used for raising beef cattle — There were some farm buildings on the land, not in weatherproof condition, but no farm house — The farm formed part of the High Weald area of outstanding natural beauty in East Sussex — The planning authority issued an enforcement notice alleging a material change of use in the D placing of the caravan without planning permission — The farmer appealed to the Secretary of State, invoking section 88(1)(a) of the 1971 Act, that is he claimed that planning permission ought to be granted and did not at this point claim that permission was not necessary — The inspector quashed the enforcement notice on the ground that there was no material change of use and his decision was upheld by Kennedy J when the planning authority appealed to the High Court — The authority now appealed to the Court of Appeal — It was submitted on behalf of the authority that the stationing of the caravan was itself a use of the land apart from E any activity carried on by the farmer in it; that it was a new primary use; that the use of the site had thus changed from the single primary use for agriculture to a mixed use for agriculture and for the stationing of the caravan; and that this change was a material change of use so as to constitute development and that the caravan was visually objectionable as an alien and incongruous feature in a rural area of outstanding natural beauty — Held, upholding the decision of Kennedy J, that the submission put forward by the planning authority was an artificial approach to the facts — The correct approach in principle was that of Forbes J in *Restormel Borough Council* v *Secretary of State for the Environment* — F It was not disputed that the farmer stationed the caravan on the land for the purposes of agriculture — Section 22(2)(e) of the 1971 Act applied where there was the "use of any land for the purposes of agriculture", which covered activities ancillary or incidental to the activities listed in section 290(1) — As the use of the caravan was use for the purpose of agriculture, section 22(2)(e) ensured that it did not constitute development — The question whether it would otherwise have been development by virtue of a material change of use did not therefore arise — Appeal dismissed

The following cases are referred to in this report.

Britt v *Buckinghamshire County Council* (1963) 14 P&CR 332; 60 LGR 430 G

Cheshire County Council v *Woodward* [1962] 2 QB 126; [1962] 2 WLR 636; [1962] 1 All ER 517; (1962) 13 P&CR 157; 60 LGR 180; [1962] RVR 148, DC

Guildford Rural District Council v *Fortescue* [1959] 2 QB 112; [1959] 2 WLR 643; [1959] 2 All ER 111; (1959) 10 P&CR 232; 57 LGR 169, CA

Restormel Borough Council v *Secretary of State for the Environment and Rabey* [1982] JPL 785

Trentham (G Percy) Ltd v *Gloucestershire County Council* [1966] 1 WLR 506; [1966] 1 All ER 701; (1966) 64 LGR 134; 18 P&CR 225, CA

This was an appeal by Wealden District Council from the decision of Kennedy J dismissing the council's appeal from the decision of the inspector appointed by the Secretary of State quashing an enforcement notice issued by the council. The enforcement notice required the removal of the caravan stationed by Mr Colin Day on his H land at Shepherd's Gate Farm, Coleman's Hatch, Hartfield, East Sussex, on the ground that the stationing of the caravan was a breach of planning control.

Michael Burrell (instructed by Cripps Harries Hall, of Crowborough) appeared on behalf of the appellants; Miss A M Williams (instructed by the Treasury Solicitor) represented the respondent Secretary of State.

Giving the first judgment at the invitation of Parker LJ, RALPH GIBSON LJ said: This appeal is concerned with a caravan which Mr Colin Day brought on to his land at Shepherd's Gate Farm, Coleman's Hatch at Hartfield in East Sussex. The Wealden District Council, the appellants in this court, contend that by stationing the caravan on his land Mr Day was in breach of planning control, in that J he thereby made a material change in the use of the land for which he had not obtained a grant of planning permission.

Mr Day's land is some six hectares in extent, or about 15 acres. He bought the land in 1983 with a view to raising beef cattle. The site plan shows the land as including two fields, each of a little over five-and-a-half acres, and two smaller fields. The land is in a rural area of deeply undulating countryside, forming part of the High Weald area of outstanding natural beauty. At the south-eastern end of the land there are some farm buildings, but there is no farmhouse.

In January 1984 it was observed on a site inspection by the council's enforcement officer that a caravan measuring 6.4 metres long and 2.8 metres wide had been placed by the farm buildings. Mr K Day informed the council's officer that because the existing farm buildings were in poor condition and not weatherproof the caravan had been placed on the land in order therein to store and mix calf feed and to provide shelter for himself and his wife.

On March 8 1984 the council resolved that enforcement action be authorised pursuant to section 87 of the Town and Country Planning Act 1971 ("the 1971 Act") to secure the removal of the caravan, which was considered to be unsightly, alien in character and appearance to traditional farm buildings, and detrimental to the visual amenities of the area. The enforcement notice was issued on April 30 1984. The breach of planning control alleged in the notice was: L

> The making of a material change in the use of the land namely to use for the purpose of stationing of a caravan thereon for the purpose of storage shelter and feed preparation.

Mr Day appealed to the Secretary of State on the ground that planning permission ought to be granted for the development to which the enforcement notice related — section 88(2)(*a*). His case on the facts was that the caravan was being used as a farm office and was an agricultural necessity. He used it to store milk powder and to prepare feeds. He regretted any blot caused upon the beauty of the area by the presence of the caravan and was willing to consider changing its colour.

The appeal was by consent conducted by written representations. M A number of residents near the site sent in objections to the presence of the caravan because of its obtrusive appearance. Mr Day had thus appealed on the ground that permission ought to be granted to him to retain the caravan and not on the ground that he did not in law need permission and had therefore not committed any breach of planning control. The written submissions of the council were therefore directed to the contention that there was no agricultural need for the caravan as a specialised and visually inappropriate structure to be on that piece of land. They said in para 5.2:

> On the County Development Plan the appeal site is shown to be within an Area of Great Landscape Value where, according to the Written Statement, it is the intention that the principal land uses shall continue to be agriculture and forestry and that development "not directly related to these pursuits shall in

A general only be permitted if a strong case for such development can be proved and subject in any event to appropriate siting and a high standard of design''.

The inspector appointed by the Secretary of State, after inspecting the site, allowed Mr Day's appeal on January 21 1985 and quashed the enforcement notice. The grounds of decision were that: (i) the caravan was used by Mr Day for storage, for preparation of animal feed and for shelter; (ii) those uses were ancillary to the agricultural use of the land and there had therefore been no material change of use; (iii) the caravan itself was not a structure and did not of itself constitute a material change of use because the land uses carried on in it were not materially different from the lawful and permitted use of the land. The inspector made no reference to section 22(2)(e) of the 1971 Act.

B The council appealed to the High Court against that decision under section 246 of the 1971 Act on the ground that the inspector had gone wrong in law. The council did not claim to rely on the fact that the inspector had decided the appeal on a ground different from that on which the appeal of Mr Day had been based and without notice to the council. Kennedy J dismissed the council's appeal and held that the inspector had directed himself correctly in law.

The council now appeals to this court. Before describing the submissions made on behalf of the council, some reference is necessary to the outline of the statutory provisions by which planning control of development is imposed by and may be enforced under the 1971 Act. Subject to certain exceptions not relevant in this case,
C ''development'' means the carrying out of building, engineering, mining or other operations in, on, over or under land or the making of any material change in the use of any buildings or other land: see section 22(1) which, in this regard, contains provisions substantially identical to those in section 12(2) of the original Act of 1947. Subject again to certain exceptions not relevant to this case, planning permission is required for the carrying out of any development of land: section 23(1). Where it appears to the local planning authority that there has been a breach of planning control, then, if it considers it expedient so to do, it may issue a notice requiring the breach to be remedied: section 87(1).

Returning to section 22, the definition of ''development'' in subsection (1), to which I have referred, describes two sorts of
D development: first, operational development by the carrying out of building or other operations in, on, over or under land, of which the most obvious example is construction of a building; and, second, the making of a material change in the use of any buildings or other land. There is then provided in subsection (2) a list of ''operations or uses of land'' which are not to be taken to involve development of the land and, of them, three must be noted:

(d) the use of any buildings or other land within the curtilage of a dwelling-house for any purpose incidental to the enjoyment of the dwelling-house as such;
(e) the use of any land for the purposes of agriculture or forestry (including afforestation) and the use for any of those purposes of any building occupied together with land so used;
E (f) in the case of buildings or other land which are used for a purpose of any class specified in an order made by the Secretary of State under this section, the use thereof* for any other purpose of the same class.

The contention is made in this court, as it was before the learned judge, for the Secretary of State that use made by Mr Day of the caravan was within (e) above, being use of land for the purposes of agriculture and therefore expressly provided not to be development by material change of use.

Next, reference must be made to a principle which has been developed as to the correct approach to analysis of the factual situation for the purposes of considering whether there has occurred a material change of the use of land, namely, identification of the
F appropriate planning unit: see *G Percy Trentham Ltd* v *Gloucestershire County Council* [1966] 1 WLR 506. It is common ground in this case that the question whether any material change was made in the use of the land by Mr Day is to be determined by reference to the whole occupation unit held by Mr Day, namely, the six hectares of agricultural land and not the specific piece of land on which the caravan stands. To that extent the enforcement notice was, as Mr Burrell has acknowledged, defective, but the notice was, of course, capable of correction: see section 88A(2) of the 1971 Act. Mr Burrell has submitted that, if the appeal succeeds and the matter is remitted for rehearing, the notice should be amended so as to read: ''The G
making of a material change in the use of the land to a mixed use for the purpose of agriculture and of stationing a caravan thereon for agricultural purposes, which stationing is not ordinarily incidental to the agricultural use.''

Mr Burrell's submission before the learned judge was in essence as follows:

(i) that the stationing of the caravan upon the land was itself a use of the land apart from any activity which Mr Day carried on inside it;
(ii) that use of the land for stationing a caravan upon it was not ordinarily incidental to the existing use of the site for agriculture because the caravan was designed for human habitation; and use of the land for stationing the caravan was therefore a new H
primary use and not the use of that piece of land for the purposes of agriculture within section 22(2)(e);
(iii) the use of the site was therefore changed from the single primary use for agriculture to a mixed use for agriculture and for the stationing of a caravan;
(iv) the change of use to that mixed use was a material change of use, so as to constitute development, of the site as the planning unit because the caravan was visually objectionable as an alien and incongruous feature in a rural area of outstanding natural beauty.

The learned judge rejected those submissions. First, he held that the caravan, for whatever purpose it was originally designed, was J
being used for an agricultural purpose, ie the mixing of feed and the provision of shelter for agricultural workers, which was ordinarily incidental to the agricultural use of the land. In that approach the learned judge followed the inspector in not separately considering the use of the land for the stationing of an allegedly unsightly caravan as distinct from the activities of Mr Day inside the caravan. Second, he held that the reasoning on which Mr Burrell's submissions were based was flawed by a misunderstanding of the provisions of the 1971 Act in that section 22(2)(e), to which reference has been made, ''is not concerned with aesthetics''. Kennedy J continued:

Perhaps it ought to be, but it is not. The fact that an item which is brought on to land is aesthetically objectionable does not of itself cast any light upon the question of whether the land is being used for the purposes of agriculture and K
whether the item complained of is contributing to that purpose.

Kennedy J referred to a decision of Forbes J in the case of *Restormel Borough Council* v *Secretary of State for the Environment* [1982] JPL 785, where a similar question was raised by an enforcement notice in relation to a caravan stationed on land comprising a hotel and used to house hotel waitresses during the holiday season. He cited a passage from the judgment of Forbes J as follows:

It did seem that whichever way one decided this matter difficulties were going to arise, but it appeared that the proper way of looking at this was to ask: what was, in planning terms, the effect on the planning unit? It was not, looked at in that way, sufficient to stop at the stationing of the caravan. You had to look further and say: for what purpose was the caravan to be stationed? If the stationing of the caravan for the purpose envisaged was not a material change L
of use of the land consisting of that particular planning unit, then it seemed that no breach of planning control existed.

Kennedy J accepted that that was a correct statement of the law and concluded that, on the face of it, nothing remained of the case for the Wealden District Council. He held, in accordance with the submission of Mr Laws for the Secretary of State, that, in order to determine whether there has been a change of use of land, it is first necessary to ascertain the existing lawful use of the planning unit: in this case it was agricultural. The next question to be considered is whether that use has been departed from by the event complained of in the enforcement notice. The arrival of the caravan on this land did not of itself signify a change in the use of the planning unit and since the caravan was used for agricultural purposes it actually contributed M
to that existing use.

On behalf of the Wealden District Council Mr Burrell has mounted an attack upon the process of analysis and reasoning by which the inspector reached his conclusion and by which the learned judge upheld it. The council maintains that the issues of control of the use of caravans in the countryside is important and that, if the learned judge has correctly stated the law, that law is deficient and should be amended by legislation. For my part, I agree with the council that the issue is of importance for all who care for the preservation of the traditional appearance of the countryside. Whether or not the law requires to be amended and, if yes, in what terms are matters of social policy for others to consider.

*Editor's note: The words ''the use thereof'' were replaced by ''the use of the buildings or other land or, subject to the provisions of the order, of any part thereof'' by the Housing and Planning Act 1986, s49 and Sched 11, para 1, from January 7 1987.

A Control of the placing of caravans upon the land has given rise to problems over the years. It was long ago established that a caravan with wheels was not a structure and the placing of it on land did not constitute operational development within section 12(2) of the 1947 Act, now section 22(1) of the 1971 Act: see *Guildford RDC* v *Fortescue* [1959] 2 QB 112 CA; *Britt* v *Buckinghamshire CC* [1963] 14 P&CR 332, Widgery J; *Cheshire CC* v *Woodward* [1962] 2 QB 126. Operational development, of course, requires planning permission (save in certain excepted cases) regardless of the use intended to be made of the land or of the building intended to be constructed. It followed that planning control of the placing of a caravan upon land depended solely upon the concept of the material change of use. A separate system of control with reference to caravans was enacted by
B the Caravan Sites and Control of Development Act 1960 to apply in addition to the control of development under the 1971 Act; but that system is concerned with caravans stationed for human habitation; and it has not been contended that its provisions are relevant to this case.

Since planning control of the placing of a caravan upon land depended solely upon the concept of the "material change of use", the extent and effectiveness of control by a planning authority depend in cases covered by paras (d), (e) and (f) of section 22(2) upon the proper construction of those special provisions, and in cases not covered by those paragraphs upon the general principles established in planning law for the construction and application of the phrase
C "material change of use".

Mr Burrell, in order to emphasise the importance of this appeal, explored the extent of the deficiency in the law which the Wealden District Council contends is apparent if the learned judge has correctly stated the law. The judge's view means, he said, that, by reference to the general principles, there can be no "material change of use" constituted by the stationing of a caravan upon the land, if the use made of the caravan, ie the activities within it, is incidental or ancillary to the existing primary use of the planning unit. Further, by reference to the special provisions of para (e) of section 22(2), there can be no development by "material change of use" constituted by the stationing of a caravan upon any land if the use of the caravan, ie, again, the activities within it, is "for the purposes of agriculture". In
D either case the planning authority is deprived of any control over the stationing of the caravan upon the land, however damaging to the appearance and visual amenities of the immediate area the presence of the caravan may be. The court's attention was drawn to what are said to be the unwelcome consequences of that view of the law: see, for example, the discussion at [1986] JPL 248 in the article by Mr L A Rutherford and Mr J D Peart.

It must be noted that such a lack of power of aesthetic surveillance and control by the planning authority over the consequences of the activities of occupiers upon their land is present at other points in planning law, probably, I think, because complete control over the visual appearance of such activities was regarded as both
E objectionable and as administratively impossible of achievement. An example concerned with operational development, as opposed to material change of use, which is immediately relevant to agricultural land, is provided by the provisions of the Town and Country Planning General Development Order 1977 (SI 1977 No 289), which consolidated with amendments earlier orders. The order was made under section 24 of the 1971 Act. By article 3 development of any class specified in Schedule 1 to the order is permitted by the order itself and may be undertaken without the permission of the local planning authority or of the Secretary of State, subject to the provision there stated.

In Schedule 1, Class VI is headed "Agricultural buildings, works and uses": the development permitted by the order under this class
F includes:

> the carrying out on agricultural land having an area of more than one acre and comprised in an agricultural unit of building . . . operations . . . *requisite* for the use of that land for the purposes of agriculture (other than the placing on land of structures *not designed* for those purposes or the provision and alteration of dwellings), so long as — (a)

deals with limitation on ground area covered by a building

> (b) the height of any buildings . . . does not exceed three metres in the case of a building . . . within three kilometres of the perimeter of an aerodrome, nor 12 metres in any other case.

Mr Burrell drew attention to the fact that the building operations must be both *requisite* for the use of that land for the purposes of agriculture and that any structure placed on the land within this permission must be *designed* for those purposes. It would, he said, be G
surprising if this consent for buildings extended only to buildings designed for the purposes of agriculture, so that planning consent would be required for any building not so designed, even if the use to be made of the building would be agricultural, but that the occupier of agricultural land were free without consent to bring on to the land a caravan which is not designed for the purposes of agriculture.

At this stage attention must be drawn to two matters. First, as Mr Burrell acknowledged, no requirement is imposed that any building erected under this permission must be well or appropriately designed, or that it must not be as to shape, materials or colour "unsightly, alien in character and appearance to traditional farm buildings or detrimental to the visual amenities of the area" to use the words H
applied by these appellants to Mr Day's caravan. These aspects of a new building permitted under the General Development Order are left to the good sense and judgment of the developer. But, second, as to the structure of control of the appearance of such buildings to be assumed, article 4 (omitting words irrelevant for this purpose) provides that if either the Secretary of State or the appropriate planning authority is satisfied that it is expedient that development of any of the classes specified in Schedule 1 should not be carried out in any particular area, or that any particular development of any of those classes should not be carried out, unless permission is granted on an application in that behalf, the Secretary of State or the planning authority (where necessary with the approval of the Secretary of State) may direct that the permission granted by article 3 shall not J
apply to all or any development of all or any of those classes in any particular area. By the Town and Country Planning (National Parks, Areas of Outstanding Natural Beauty and Conservation Areas, etc) Special Development Order 1985 (SI 1985 No 1012) (which replaced SI 1981No 246), Schedule 1 to the General Development Order of 1977 was modified so far as concerned land within such areas by restriction upon certain classes of development including enlargement etc of a dwelling-house, but the provisions of Class VI, Agricultural buildings, works and uses, are not thereby affected. Finally, special provision was made generally with reference to land within national parks and certain other areas listed by the Town and Country Planning (Agricultural and Forestry Development in National Parks etc) Order 1986 (SI 1986 No 1176). Thereby the K
carrying out of any development under Class VI of the General Development Order (Agricultural buildings, works and uses) is subjected to the condition that a 28-day notice of the proposed development must be given and, upon notice given by the planning authority within that period, the development may not be carried out without the prior approval of the planning authority to the siting, design and external appearance of any building which is to be erected, altered or extended.

In short, it seems that it has not yet been thought necessary to give general control over the appearance of new agricultural buildings, constructed within the permission given by the General Development Order of 1977, in areas designated as of outstanding natural beauty L
as contrasted with the national parks and other areas listed in Order No 1176 of 1986.

I have sympathy with the contention of the Wealden District Council that it is both surprising and a reasonable ground of concern if the occupier of agricultural land is free under planning law to station at any point upon his land one or more caravans, intended to serve the same purpose as farm buildings, regardless of the harm which the planning authority reasonably considers will be caused by the presence and appearance of the caravans in the place where they are stationed. Nevertheless, I agree with Kennedy J that such reflections upon apparent gaps in the extent of planning control cannot affect the proper construction of section 22(2)(e), because, in my view, the meaning of the words there used in the context of the Act M
as a whole is clear, nor are such reflections of any real assistance in determining whether, by stationing the caravan on the land, Mr Day may be held to have initiated a new primary use of land within the planning unit.

It is convenient to start with the proper construction of section 22(2)(e) and its application to the facts of this case. It is not in issue that Mr Day stationed the caravan on the land for the purposes of agriculture in the sense that he intended to use the caravan for those purposes and no others.

The effect of section 22(2)(e) operates where there is "use of any land for the purposes of agriculture". The word "agriculture" is defined by section 290(1) as including a list of agricultural activities among which are, for example, fruit growing and the breeding and keeping of livestock. No reliance was placed by Mr Burrell upon any argument to the effect that section 22(2)(e) can apply only to use of land for the purposes of one of the listed agricultural activities and not to use for the purposes of activities ancillary or incidental to those listed agricultural activities. I think he was right not to rely upon such argument. The definition is an inclusionary definition. Construed in its context, in my judgment, there is "use of land for the purposes of agriculture" where the land is used for activities in direct furtherance of an agricultural activity.

The only ground upon which Mr Burrell has submitted that section 22(2)(e) was not applicable to this case is, as submitted to the learned judge, that because this use of the land for stationing of the caravan upon the land by Mr Day was not ordinarily incidental to the agricultural primary use of the planning unit, therefore the use of the land effected by the stationing of the caravan could not be "use of land for agricultural purposes" within the meaning of section 22(2)(e).

The concepts used in that submission, namely, "ordinarily incidental", "primary use" and "planning unit" have become part of planning law by decisions of the courts and of the Secretary of State. I have referred above to "planning unit". Mr Burrell referred to the *Encyclopedia of Planning Law and Practice*, edited by Sir Desmond Heap, Mr Harold Brown and Dr Malcolm Grant, where it is pointed out (para 2-815) that, apart from the exclusionary definition of "use" in section 290(1), there is no further definition contained in the Act of the formula "material change of use" and that, in general, the application of it in individual cases contains often a significant element of subjective judgment and is regarded by the courts as being primarily a matter of fact and degree for the Secretary of State in the first instance. The passage continues:

> But certain limits to the formula, and certain tools of analysis, have emerged from decisions of the courts and of the Secretary of State. They have, for example, insisted that a distinction be drawn between the primary use of land and the ancillary uses which accompany it and are thus protected through the link with the primary use. Secondly, they have indicated that a change need not be a change to a different kind of use altogether (such as from residential to industrial) before it may be regarded as material, and that change in the "character" of the use may suffice.

There follow references to the concepts of intensification and of the planning unit.

The words "ordinarily incidental" do not occur in the 1971 Act but were used in the Town and Country Planning (Use Classes) Order 1972 (SI 1972 No 1385), which re-enacted with amendments the Use Classes Order of 1963 and which was made under, and for the purposes of, section 22(2)(f), which is set out above. If land is used for a purpose of any class specified in the order, section 22(2) provides that use of that land for any other purpose of the same class shall not be taken to involve development of the land, ie by the making of a material change of use. By article 3(3) of the order "A use which is ordinarily incidental to and included in any use specified in the schedule to this order is not excluded from that use as an incident thereto merely by reason of its specification in the said schedule as a separate use". Mr Burrell has contended that this phrase appears in article 3(3) because it is a concept generally applicable for the purposes of deciding what ancillary uses are to be regarded as included within a primary use. He relied upon the decision of this court in *G Percy Trentham Ltd* v *Gloucestershire County Council* [1966] 1 WLR 506, where Lord Denning MR said (at p 512D):

> In applying the Use Classes Order, you must consider the whole of the unit which is being used. I think that Diplock LJ indicated the right test towards the end of the argument. You should look at the whole area on which a particular activity is carried on, including uses which are ordinarily incidental to or included in the activity.

Diplock LJ (at p 513H) said:

> What is the unit which the local authority are entitled to look at and deal with in an enforcement notice for the purpose of determining whether or not there has been a "material change in the use of any buildings or other land"? As I suggested in the course of the argument, I think for that purpose what the local authority are entitled to look at is the whole of the area which was used for a particular purpose, including any part of that area whose use was incidental to or ancillary to the achievement of that purpose.

Mr Burrell cited a number of other decisions and submitted that they bear out his submission and show that the test is objective by reference to the facts and not to the intentions or purposes of the particular occupier. For my part, I would accept this part of the submission to the extent that the question whether there has been a material change of use is to be approached in accordance with the guidance given by this court in the *G Percy Trentham Ltd* case. I would accept also that the question whether a particular additional use is ancillary to or included within the primary use of the planning unit is to be determined objectively in the sense that the purposes for which the land is used must be determined according to all the evidence, in addition to evidence of what the occupier says is his purpose, including evidence of what is being done on the land and the assessment of the relationship of that activity to the primary use.

Mr Burrell's argument was further developed as follows. He contended that the stationing of this caravan upon the land was a use of land and was a new use. There was, he said, no justification for ignoring that particular use and for inquiring only what use would be made of the caravan or what would be done in it, as the inspector had done. So to do was to ignore a necessary step in the required analysis. No doubt in many cases the stationing upon the land of some vehicle, or equipment, is so obviously itself a use without separate significance with reference to planning considerations (for example an ordinary vehicle) that it will be safe to pass straight to the question as to the use for the purposes of which the vehicle has been placed there and to consider that use with reference to the primary use. But where, as here, the use by stationing this caravan on the land does have separate planning significance because of the effect upon the visual amenities of the area, then, he argued, it is error to ignore that step in the analysis.

There seemed to me to be force in this submission. It is not clear why the impact upon the land and its immediate area caused by the appearance of the vehicle or equipment and by the use of the land constituted by stationing the vehicle or equipment upon the land should be required by some rule to be treated as irrelevant even in the most extreme case provided that the use made of the vehicle or equipment is for the purposes of the existing primary use. One idea of an extreme case would be the stationing upon the land such as that at Hartfield in East Sussex of two double-decker buses as substitutes for agricultural buildings.

Miss Williams, for the Secretary of State, submitted that the stationing of the caravan by Mr Day was what she called a nil use — irrelevant provided that the use to be made of it is for the purposes of the existing primary use. Nothing in the 1971 Act expressly requires this rule to be applied. The concepts developed by the courts as stated above were devised, as I understand it, in part at least for the purposes of working out when a new or additional use of the land is to be treated as constituting a material change of use: if by the reasoning so developed the new use is to be regarded as ancillary, or ordinarily incidental, then the new use has been shown, having regard to the relevant planning factors, as of no materiality. It is not clear to me why such reasoning and concepts can be said to require in every case, provided that the new use of the land by stationing a caravan or caravans upon it is for the purposes of the existing primary use, that there be disregarded the damage caused by the new or additional use to the visual amenities of the area which in other situations is a relevant and material planning factor.

There was cited to the learned judge the decision of the late Forbes J in *Restormel Borough Council* v *Secretary of State for the Environment* [1982] JPL 785. Forbes J had long experience of planning law and I approach his decision with the very greatest of respect. I have set out above the passage from his judgment which Kennedy J cited. I agree with Kennedy J that, if that is the correct statement of the law, nothing remained of the case for the appellants. It was necessary, said Forbes J, to look further than the stationing of the caravan and to ask for what purpose it was stationed:

> If the stationing of the caravan for the purposes envisaged was not a material change of use of the land consisting of that particular planning unit, then it seemed that no breach of planning control existed.

The report in [1982] JPL 785 shows that the inspector had held that the use of the caravan at the hotel was not a separate self-contained unit of living accommodation, but solely as a bedroom for two waitresses, and that use was to be regarded as ancillary to the main use of the premises as a hotel. The argument for the planning authority was that as a matter of law the mere stationing of a caravan on land could amount to a material change of use of that land irrespective of any use to which the caravan was put. Caravans, it was said, had a visual impact on amenity and that that was a planning matter: it was the stationing itself which was the important issue. The argument for the Secretary of State was that it was necessary to ask what was the use of the caravan, because it was only by deciding to what use the caravan had been put that it was possible to decide the main question, namely whether there was a material change of use of land. The stationing of the caravan itself was an equivocal act. Forbes J acknowledged that the case raised difficult policy questions whichever way the case was decided. He concluded that the contentions for the Secretary of State were right. In addition to the passage cited above, he said:

> It was inappropriate, when a caravan was stationed on land for a particular purpose, to look at the stationing of the caravan separately and say that that was something which was development requiring planning permission because it made a material change of use.

Mr Burrell submitted that, assuming the decision of Forbes J to be right, it was distinguishable on the facts. He submitted that the stationing of a caravan was a use ordinarily incidental to the primary use of the premises as a hotel because the caravan was being used for the purpose for which it was designed, whereas in this case it was not. He submitted further that Forbes J was not right in that he asked the wrong question, namely, for what purpose was the caravan stationed? And that he should have asked: Was the stationing of the caravan for the purpose of the primary use ordinarily incidental to the primary use? The answer would have been yes, so the result of the case was right.

In the five years since Forbes J decided the *Restormel Borough* case there has been, as I understand it, no departure from it or criticism of of it. For the reasons I have given, I agree with Forbes J that the questions of policy raised by the decision were difficult, but I have reached the conclusion that his decision and his reasoning were right. It seems that the point raised in that case had not been raised for decision before 1982 in any reported case. A large number of caravans, vehicles and pieces of equipment must, I think, have been stationed upon land and permitted to remain without challenge in reliance upon the state of the law as it was submitted in that case on behalf of the Secretary of State to be. I find it impossible to hold that the decision or reasoning have been shown to be wrong, and I can see no reason to depart now from a principle which appears to have been long accepted. To subject the stationing on land for the purposes of a primary use of a caravan or other mobile equipment to possible control by reference to the concept of material change of use of land, depending upon the impact on visual amenities caused by the presence of the caravan or other equipment in the particular place chosen for stationing it, as assessed by the local planning authority, could have substantial effects upon the freedom of use of land for agricultural purposes, and I am not persuaded that such control was intended to be imposed by the words of the statute.

If, as I think, the decision and reasoning of Forbes J in the *Restormel Borough* case should be approved, then the argument that the principle does not apply in this case, by reason of distinctions of fact, cannot, in my judgment, succeed. Since the stationing of the caravan on the land was without doubt for the purposes of agriculture, there is no occasion to assess separately the use of the land by the stationing of the caravan upon it unless there is some aspect of the facts of this case which could require or justify application of a different principle. Mr Burrell's submission is that the stationing of the caravan on this piece of land takes the case outside the principle established in the decision of Forbes J. A typical caravan, which we can assume Mr Day's caravan to be, equipped with cooking and washing and sleeping facilities, as well as with space to sit, is, said Mr Burrell, designed for human habitation as a residence, and therefore the stationing of it on land cannot be ordinarily incidental to a primary agricultural use.

It is assumed in that submission that the degree of connection, between the land use in question and the primary agricultural use, is accurately expressed by the phrase "ordinarily incidental" if the land use is to be held to be "for the purposes of agriculture" within section 22(2)(e). I shall assume that this is so, but I do not find it necessary to decide in this case whether the connection expressed by the meaning contained in that phrase would in every case be necessary for this purpose. Upon that assumption I would hold on the facts of this case that these appellants have made out no ground for holding that this case is outside the principle decided by Forbes J. There is, as I think, nothing in the nature of the typical residential caravan, of which we have all seen large numbers both on the road and parked in caravan sites, which renders the use of such a caravan incapable of being properly regarded as ordinarily incidental to the agricultural use of land, that is to say, as an ordinary piece of equipment for stationing upon land and for use when so stationed for the purposes of agriculture. I accept that the typical caravan is primarily designed and equipped so as to be used for human habitation, usually over short periods of time; but the primary use of a planning unit for agriculture, whether, for example, it be called a farm or a smallholding, may in my judgment include many different incidental or ancillary uses of different characters for some of which the use of a typical caravan would seem to me to be an ordinary piece of equipment. I have in mind such ancillary uses as the provision of shelter and accommodation for the person in charge of parts of a farm or orchard where members of the public are allowed to pick fruit or vegetables for themselves and, to that end, are provided with containers and with facilities for weighing and for paying for what they have picked. I also see no reason why, for the activities carried on by Mr Day in his caravan, namely, storing and mixing of feed, some related office work and the taking of shelter, the typical caravan should not be regarded as an ordinary piece of equipment. The purpose of the caravan for such uses is to provide protection, for the things to be placed within and for the people using it, against the weather and against interference whether by people or by animals.

If this question were raised with reference to a four-wheeled farm cart with a canvas or wooden cover, it would, I think, in accordance with Mr Burrell's argument, admit of only one answer. Such a thing would, I think, be recognised as an ordinary piece of equipment for the purpose, although now in most parts of the country it might be impossible to find, whereas the typical caravan is readily available and reasonably cheap. There is, I think, in planning law, even with reference to the most beautiful parts of our countryside, no basis for excluding from the notion of ordinary equipment a useful and suitable article such as a caravan on the ground only that it is not traditional in construction or appearance for the particular purposes for which Mr Day has applied it.

At one stage of the argument Mr Burrell was pressed as to the status in his submission of a caravan of typical shape and size, but constructed with animal feed boxes within it, and a small desk for keeping stock and feed records, instead of the kitchen and washing facilities and beds usually supplied. He was, I think, constrained to concede such a piece of equipment on land could not be classified as incapable of being ordinarily incidental to a primary agricultural use on the ground that it was designed for human habitation; but he suggested that, if the caravan was a typical caravan when stationed on the land, the later removal of kitchen and washing facilities and of beds could not make it ordinarily incidental. In my judgment, this argument cannot be right. The typical caravan must, I think, be regarded as being as useful and convenient a means of providing protection and storage for things and shelter for people as would be a suitable building. I reject the contention that because of its nature and ordinary use and associations the stationing of a typical caravan on a planning unit used for agriculture must be regarded as incapable of being incidental to the primary use for agriculture. The risk of abuse of the limitations upon planning control over the placing of caravans upon agricultural land, as revealed in this case, is obvious. Forbes J referred to it in his judgment in the *Restormel Borough* case. If it should be thought necessary to impose further control, it seems to me that legislation will be necessary.

For these reasons, in my judgment, the appellant has failed to show that Kennedy J was wrong in dismissing its appeal from the decision of the inspector. I would dismiss the appeal.

Agreeing, PARKER LJ said: We are here concerned with a planning unit of some 15 acres in area, which, prior to the advent of Mr Day's caravan, was, it is common ground, being wholly used for the purposes of agriculture.

Mr Day brought on to a part of that land measuring some 21 ft × 9 ft a caravan for the purpose of providing a weatherproof place for the storage and mixing of food for his cattle and providing shelter for himself and his wife. He thereafter used it for such purposes.

The question for determination is whether such actions on Mr Day's part constituted a "development" within the definition of that word set out in section 22(1) of the Town and Country Planning Act 1971. They can only have done so if Mr Day thereby made a material change in the use of the land within the meaning of the subsection, because it is rightly accepted on behalf of the local planning authority that Mr Day's actions did not constitute an operational development within the subsection.

A The definition set out in the subsection is, however, expressly made subject to the following provisions of the section, and subsection (2)(e) provides:

The following operations or uses of land shall not be taken for the purposes of this Act to involve development of the land, that is to say . . .

(e) the use of any land for the purposes of agriculture or forestry (including afforestation) and the use for any of those purposes of any building occupied together with land so used.

If, therefore, Mr Day's actions constituted the use of the small area of land on which the caravan stood for the purposes of agriculture, they do not constitute a development and the question whether they would otherwise have constituted a development by material change
B of use does not arise.

That Mr Day's land was, before the advent of the caravan, being used for the purposes of agriculture is common ground. "Agriculture" is defined by section 290 of the Act to include the breeding and keeping of livestock and this was the activity being carried on by Mr Day.

The question is, therefore, whether Mr Day's actions were such that the plot of land upon which the caravan stood was, on its arrival, still being used for the purposes of agriculture or whether, as the planning authority contend, it was then being used for mixed uses, namely use for the purposes of agriculture and use for stationing a caravan.

This contention is in my view artificial in the extreme. If it were
C right, it would seem to follow that any land upon which a farmer placed a mixing trough must be regarded as being subject to mixed uses, namely, the stationing of a mixing trough and the mixing of food, the latter being a use for the purposes of agriculture but the former not being for such purpose.

To escape from this obvious absurdity, it was submitted that that case was different because the mixing trough would be ordinary agricultural equipment and therefore the stationing would itself be for an agricultural purpose, whereas a caravan is not such equipment.

I cannot accept this. If it were right, different results would follow if the farmer used for his mixing a trough sold for the purpose and if, to save expense, he used instead a container of the same size and
D appearance which had originally been sold for some other purpose. Such a result cannot, in my view, be extracted from the language used or be taken to have been intended by Parliament. In both cases the land is, in my view, being used for the agricultural purpose of keeping livestock. That purpose may be effected by mixing the food in a trough and it can make no difference whether the trough was itself purpose-designed or not.

The position would no doubt be different if a farmer merely parked his holiday caravan on the land. It could then be said that the land on which it stood was being used not for agricultural purposes but for parking of a residential caravan. It would, however, be unreal to say that the land was being used for parking a caravan if a farmer,
E his mixing shed having been destroyed by fire or storm, decided to forfeit his holiday caravan previously parked in his garage, stripped it of internal fittings, equipped it with storage bins and mixing troughs and then placed it in the farmyard, was using the land on which it stood for anything other than agricultural purposes.

On this simple ground I, too, would dismiss the appeal.

SIR GEORGE WALLER agreed that the appeal should be dismissed for the reasons given by Ralph Gibson LJ and Parker LJ and he did not add anything of his own.

The appeal was dismissed with costs. Application for leave to appeal to the House of Lords was granted.

Court of Appeal F

January 28 1988

(Before Lord Justice PARKER, Lord Justice RALPH GIBSON and Mr Justice CAULFIELD)

THRASYVOULOU v SECRETARY OF STATE FOR THE ENVIRONMENT AND ANOTHER

Estates Gazette March 12 and 19 1988

[1988] 10 EG 131-142 and [1988] 11 EG 83-88

Town and Country Planning Act 1971, sections 36 and 88 — Application of principle of issue estoppel — Appeal from decision of McCowan J — Important question of planning law — The question arose in regard to a number of enforcement notices served by the London Borough of Hackney — The notices alleged a breach of planning control consisting of the making of a material change in the use of the property to use as a hostel for homeless families — The notices required the termination of such use — The authority considered that such a use was inappropriate in an area which was primarily residential, with already a large number of hotels and boarding houses and where a proliferation of hostels for homeless families would put a strain upon local educational, social and medical services — McCowan J dismissed appeals against decisions of the Secretary of State, who had upheld the enforcement notices — McCowan J had rejected a submission that the decisions of a previous inspector in 1982 which went the other way rendered the matter *res judicata* — In the present appeal to the Court of Appeal, however, the effect of the 1982 decisions was presented in a different way and elaborated in relation to the doctrine of issue estoppel — The determination of the inspector in 1982 in regard to certain of the properties affected by the enforcement notices was that the use was that of an hotel — The circumstances in these cases in 1985 did not differ from those prevailing in 1982 — It was therefore contended that the principle of issue estoppel applied and the inspector in 1985 should have quashed the enforcement notices — It was submitted, on the other hand, that issue estoppel had no place in planning law — It was the duty of the Secretary of State or his inspector to decide in the light of all the evidence without such constraint — The Court of Appeal held, after considering a number of authorities, that the principle of issue estoppel was applicable to appeal proceedings under section 88 of the 1971 Act — It was also applicable to a section 36 appeal if the party seeking to raise it proved that the earlier decision was given in his favour against the planning authority upon an issue as to his existing rights of use of the property and was a determination of the same issue as that raised in the new proceedings — The court accordingly allowed the appeal in relation to four properties to which the issue estoppel point arose and remitted these cases for rehearing and determination in the light of the opinion of the court that the inspector was obliged to allow the appeals to the Secretary of State and quash the enforcement notices — In view of the importance of the matter, however, and despite the unanimity of the court, leave was given to appeal to the House of Lords

G H J K L

The following cases are referred to in this report.

Fidelitas Shipping Co Ltd v *V/O Exportchleb* [1966] 1 QB 630; [1965] 2 WLR 1059; [1965] 2 All ER 4; [1965] 1 Lloyd's Rep 223, CA
O'Reilly v *Mackman* [1983] 2 AC 237; [1982] 3 WLR 1096; [1982] 3 All ER 1124, HL
R v *Smith (Thomas George)* [1984] Crim LR 630; (1984) 48 P&CR 392, CA
Rockhold Ltd v *Secretary of State for the Environment and South Oxfordshire District Council* [1986] JPL 130
Tebbutt v *Haynes* [1981] 2 All ER 238, CA
Thoday v *Thoday* [1964] P 181; [1964] 2 WLR 371; [1964] 1 All ER 341, CA
Wandsworth London Borough Council v *Winder* [1985] AC 461; [1984] 3 WLR 1254; [1984] 3 All ER 976; (1984) 83 LGR 143, HL
Western Fish Products Ltd v *Penwith District Council* [1981] 2 All ER 204; (1978) 38 P&CR 7; 77 LGR 185; [1978] JPL 623, CA

This was an appeal by Aristophanes Thrasyvoulou from a decision of McCowan J affirming the decision of an inspector appointed by the Secretary of State for the Environment who had upheld enforcement notices served by Hackney London Borough Council in respect of properties in Wilberforce Road, London N4. The decision of McCowan J was reported at [1987] 2 EGLR 195; (1987) 284 EG 511.

Robert Carnwath QC and Joseph Harper (instructed by Frank E C Forney & Partners) appeared on behalf of the appellant; Duncan Ouseley (instructed by the Treasury Solicitor) represented the Secretary of State; Miss Anne Williams (instructed by the borough solicitor, Hackney) represented the borough council.

Giving the first judgment at the invitation of Parker LJ, RALPH GIBSON LJ said: This appeal is brought by Aristophanes Thrasyvoulou, the appellant, from the decision of McCowan J given on November 25 1986. Leave to appeal was granted by this court at the outset of the hearing. The respondents are the Secretary of State and the London Borough of Hackney, the local planning authority ("the council"). The appeal raises the important question whether the principle of issue estoppel has any place in the law regulating planning appeals.

The appeal concerns the rights of the appellant under planning law to use as accommodation for homeless families a number of properties which he owns in Wilberforce Road, London N4. By the decisions of the inspector, which were upheld by McCowan J, it was held that such use amounted to use of the properties as hostels for homeless families, that the appellant has no rights to use any of the properties for such purpose, and that planning permission to continue such use should not be granted.

Enforcement notices dated February 2 1985 were served by the council in respect of five separate properties and groups of properties. The effective notices and the properties to which they related were as follows: Notice B: 3, 5 and 7 Wilberforce Road; Notice C: 11, 13 and 15; Notice D: 18; Notice E: 20; Notice F: 25.

The notices were in the same terms. In each case the breach of planning control was alleged to be:

> . . . the making of a material change in the use (of the property) to use as a hostel for homeless families.

The steps required by the notices to be taken were stated to be: "To terminate the use of the said land as a hostel for homeless families." In the annex to each notice the council explained briefly why, pursuant to section 87 of the Town and Country Planning Act 1971, they considered both that there had been with reference to each property a breach of planning control after the end of 1963 and that it was expedient to take enforcement action. They said:

> The Council considers that the use of these premises for the almost exclusive accommodation of homeless families is materially different from their former use; and is a use which it is inappropriate to allow to continue in an area which is primarily residential, but where there are already a large number of hotel, guesthouse, boarding house, hostel, lodging and similar establishments and where a concentration or proliferation of hostels for homeless families produces disproportionate strain on local educational, social and medical services.

It is to be noted that nothing was said as to what the "former use" was or as to when the change was alleged to have been made. It is not suggested that the notices were thereby defective.

The case made for the council on the merits with reference to the planning issues was eventually successful. After an inquiry extending over five days in January and March 1986, and a detailed inspection of the properties carried out on February 3 1986, Mr C H Johnson, the inspector appointed by the Secretary of State to determine the appeals, held that all the properties were being used as hostels for homeless families; that the lawful or established uses of the various properties were either as hotels or as houses in multiple occupation; and that there had been in each case a material change of use. In addition he held that planning permission should not be granted for the continuance in any of the properties of the use as hostels for homeless families. It is right to emphasise, as was clear from the statement made in the annex to each notice and from its case at the inquiry, that the council was fully aware of the need for proper accommodation for homeless families and that accommodation for those families should be provided in part and for short periods in hostels, but the council was objecting on planning grounds to the excessive concentration of properties used for hostels in this area. It is common ground in this court that the inspector, so far as concerned the planning issues, had directed himself correctly in law and had ample evidence on which to find as he did.

The attack on the inspector's decision in this court is founded upon a novel ground. It is said for the appellant that, irrespective of the merits of his case on the planning issues, the council was estopped by issue estoppel from contending before the inspector that the use made by the appellant of his properties was as hostels for homeless families, or that the use actually being made constituted a material change of use, or that it was other than a lawful use. The issue estoppel is claimed to have arisen as a result of earlier enforcement proceedings taken by the council against this appellant in 1982 with reference to some but not all of the properties covered by the present five enforcement notices.

The earlier proceedings

The proceedings in 1982 were concerned with 11, 13, 15, 18, 20 and 25 Wilberforce Road, but the issue estoppel is said to result only from the contest and decisions with reference to nos 11, 13, 15 and 25. In 1982 the distinction between use of land as an hotel from use as a hostel and the possible materiality in planning terms of a change from one to the other were less clear than later they became.

Before dealing with the issues in the earlier proceedings it is necessary to state some facts which were not in dispute. All these properties had, since 1978, been used predominantly for providing temporary accommodation for homeless families referred by local authorities under their statutory duties. Prior to 1981 there had been grants of planning permission for the use of nos 3, 5, 7 and 20 for hotel use. In respect of no 11 there was an established use certificate for hotel use.

The issues in 1982 with reference to the relevant properties were as follows:

(i) By an enforcement notice dated October 12 1981 the council alleged that there had been development of no 13 without permission by the making of a material change of use to use as an hotel and required that use to be terminated. As an alternative, a second enforcement notice made the same allegations and requirement on the basis that the use of no 13 had been changed to that of hostel.

(ii) Four separate enforcement notices of October 12 1981 made identical alternative allegations and requirements with reference to nos 15 and 25.

(iii) With reference to all those enforcement notices the appellant appealed under section 88 of the 1971 Act, as amended, on various grounds which (so far as now relevant) can be summarised briefly. His main contention was that he was using the properties as budget hotels on the basis that almost all his customers were homeless people for whom bookings were made by local authorities. Any breach of planning control by change of use to use as an hotel had occurred before the beginning of 1964: section 88(1)(e). So far as concerned the allegation that there had been a change to use as hostels, the appellant contended that such a change did not constitute a breach of planning control (section 88(1)(b)) and/or that the alleged breach of planning control had not taken place (section 88(1)(c)) because that use was covered by the allegedly lawful use as hotels. Finally, in each case he claimed that planning permission ought to be granted so as to make lawful the use which he was making of the properties.

(iv) As to no 11, in respect of which there was an established use certificate for hotel use, an application for planning permission had been made in May 1979 by the appellant for the retention of a single-storey side extension and for its use as a reception area for that property as a bed-and-breakfast hotel. The permission granted on March 26 1981 had been for "retention of the extension and its use as a reception area in connection with the use of the premises as a *hostel* for homeless families". There was also a condition for removal of the extension by February 1986. The appellant appealed under section 36 of the 1971 Act against the wording of the planning permission and the condition. Again his main contention was that his use of the property was as an hotel.

It is important to note that there was no enforcement notice in respect of no 11. The attitude of the council apparently was that it was not seeking to stop the use of no 11 for use as a hostel for homeless families: indeed, the council has granted planning permission for the retention of the side extension in connection with that use. The alleged distinction between hostel use and hotel use was seen as important by both the council and the appellant. If the permission remained in the terms in which it had been granted, the appellant, while he could go on using that property as he then was using it, presumably feared that he might have difficulty in reverting to what the council would regard as normal hotel use if, for any reason, the supply of homeless families sponsored by local authorities should disappear.

The contest with reference to no 11 was therefore, in substance, that the appellant was asserting that the then use for homeless families was within the established hotel use and the council was contending that it was not. If the issues raised by the council had been decided in its favour, therefore, the appellant could have continued to use no 11 as he was then using it under the specifically permitted hostel use, but he would have had to cease using nos 13 and 15 for the same use.

An inquiry was held over four days in March 1982 by Mr C Butler-Stoney, who was appointed to determine the appeals. His findings were as follows:

(i) At the time of service of the enforcement notices, nos 11, 13, 15 and 25 were being used almost exclusively as accommodation for homeless people for whom bookings were made by local authorities.

(ii) The circumstances of occupancy were as follows: those staying in the properties were served breakfast, and the furnished rooms were cleaned by staff daily. There were no cooking facilities in rooms but communal kitchen space was provided to cater for the needs of small children. There was no supervision of occupants, who were bound by normal hotel rules.

(iii) The inspector concluded that the booking arrangements and the facilities provided were those of hotels and not of hostels.

(iv) Therefore the appeals in respect of nos 13, 15 and 25 succeeded so far as concerned any breach of planning control under those notices which alleged change to use as hostels, since that change of use had not taken place, and the enforcement notices alleging development by change to that use were quashed.

(v) Further, as to nos 13 and 15, which had been purchased by the appellant in 1977 for use in connection with no 11 as an hotel, the evidence showed that they had been in use as a guesthouse since 1960 and therefore the appeals in respect of those properties succeeded also on ground (e) under section 88(2), so that the alternative notices alleging change to use as hotels were also quashed.

(vi) As to no 25, which had been purchased by the appellant in 1978, there had been use of the property as a guesthouse in 1955, but that use had been abandoned. Therefore, there had been a change in 1978 to use (as the inspector held it to be) as an hotel, and the appeal on ground (e) under section 88(2) failed. The alternative enforcement notice alleging change to use as an hotel was therefore upheld.

(vii) Having held that no 11 was being used as an hotel and not as a hostel, the inspector, exercising his power under section 36(3) of the 1971 Act, varied the planning permission dated March 26 1981 by deleting the reference to use of the premises as an hotel for homeless families.

There was no appeal by the council under section 246(1) of the 1971 Act alleging error of law on the part of the inspector.

The submission for the appellant in this court by Mr Carnwath, in brief summary, was that the general principle of law known as issue estoppel applies to a decision made by the Secretary of State or by the inspector in an appeal against an enforcement notice under section 36 and section 88 of the 1971 Act. The decisions made in the 1982 proceedings estopped the council from contending in the 1985 proceedings that there had been a material change of use in respect of nos 11, 13, 15 and 25, since the circumstances in 1985 did not differ in any relevant respect from those proved in 1982; and there was no other ground upon which the council should be permitted to reopen the issues which had been then finally decided. The estoppel so arising in respect of those properties, which the inspector failed to recognise or apply, should be held to have affected also his decisions with reference to the other properties and that the decisions in respect of all the properties should be remitted for rehearing: RSC Ord 94, r 12(5).

For the Secretary of State it was submitted by Mr Ouseley that the principle of issue estoppel has no place in planning law and in particular no place in the decision of appeals under section 36 or 88 of the 1971 Act. If an enforcement notice was a valid notice, so as to give rise to appeal proceedings under section 88, then, as under section 36, it was the statutory duty of the Secretary of State, or of an inspector to whom determination of the appeal is delegated, to decide the appeal having regard to all the evidence submitted. Due regard would be paid to any earlier decision and such weight should be allowed to it as might appear to be appropriate in all the circumstances. Such proceedings were not private contests between the council and the appellant but concerned also the public interest and the rights of any person having an interest in the land: see *Western Fish Products Ltd* v *Penwith District Council* [1981] 2 All ER 204. Provision is made for notice to other parties interested in the outcome of the appeal and for the making of submissions by them. Objection might be taken to the issuing under section 87 of an enforcement notice by a local planning authority on the ground that it constituted an abuse of power, but in such case the only remedy is, said Mr Ouseley, by application for judicial review to quash the enforcement notice, and not by claiming to exclude evidence or particular submissions at the inquiry. Finally, he said that the principle of issue estoppel, if applicable in any form in these proceedings, applied only to nos 13 and 15 and could not be applied to any other property.

The exclusion of issue estoppel from planning appeals

It is necessary to decide first whether issue estoppel is applicable at all in the proceedings under consideration in this case before deciding whether it is shown that the inspector went wrong in law with reference to any of the appeals.

Mr Carnwath invited the court to apply the principles stated by Diplock LJ (as he then was) in *Fidelitas Shipping Co Ltd* v *V/O Exportchleb* [1966] 1 QB 630. At p 642E Diplock LJ said:

> This is but an example of a specific application of the general rule of public policy, nemo debet bis vexari pro una et eadem causa. The determination of the issue between the parties gives rise to what I ventured to call in *Thoday* v *Thoday* [1964] P 181, 198, an "issue estoppel". It operates in subsequent suits between the same parties in which the same issue arises. A fortiori it operates in any subsequent proceedings in the same suit in which the issue has been determined. The principle was expressed as long ago as 1843 in the words of Wigram VC in *Henderson* v *Henderson* which were expressly approved by the Judicial Committee of the Privy Council in *Hoystead* v *Commissioner of Taxation* [1926] AC 155, 170. I would not seek to better them: I believe I state the rule of the court correctly when I say that where a given matter becomes the subject of litigation in, and of adjudication by, a court of competent jurisdiction, the court requires the parties to that litigation to bring forward their whole case, and will not (except under special circumstances) permit the same parties to open the same subject of litigation in respect of matter which might have been brought forward as part of the subject in contest, but which was not brought forward, only because they have, from negligence, inadvertence, or even accident, omitted part of their case. The plea of res judicata applies, except in special cases, not only to points upon which the court was actually required by the parties to form an opinion and pronounce a judgment, but to every point which properly belonged to the subject of litigation, and which the parties, exercising reasonable diligence, might have brought forward at the time.

Issue estoppel applies to arbitration as it does to litigation.

In denying the applicability in any form of the principle there stated to appeals against enforcement notices, Mr Ouseley was forced to argue that if, without error of law, the Secretary of State or an inspector had determined in such an appeal, as a matter of fact and degree, that there had been no material change of use of a property as alleged by a local planning authority in an enforcement notice, that authority could issue a new enforcement notice in the same terms on the day following that determination and, in the absence of any change of circumstances or other justification, persuade the Secretary of State or a different inspector to conclude that there had been a material change of use. Mr Ouseley argued that, without importing into planning law the principle of issue estoppel, the occupiers of land are sufficiently protected by the discretion and sense of fairness of planning authorities, of the Secretary of State and of inspectors, and by the power to order costs against any planning authority which might take new proceedings to reverse an earlier decision without good cause.

I will first consider these submissions with reference to the section 88 appeals, which concern nos 13, 15 and 25, and then consider whether the conclusion reached with reference to section 88 appeals applies also to the section 36 appeal in respect of no 11.

McCowan J accepted the submissions made on behalf of the Secretary of State. It was argued by counsel then appearing for the appellant that the activities on the properties had not changed since 1982. The decisions of 1982 were made on the same facts as in 1986. The appeals should therefore be considered as *res judicata* and that the council "should not be permitted to keep issuing enforcement notices until they can find an inspector who holds in their favour". In the view of the judge, the doctrine of *res judicata* had no application. He accepted the argument of Mr Laws for the Secretary of State that, appeal rights having been exercised under section 88, the Secretary of State had the responsibility of determining the appeal, or his

A inspector if he delegated it. If the inspector were fettered by the decision of a previous inspector, so that he could not decide the issue for himself, he would not be discharging the statutory task placed upon his shoulders. McCowan J considered that that view of the law was confirmed by passages from the judgment of Forbes J in *Rockhold Ltd* v *Secretary of State for the Environment and South Oxfordshire District Council* [1986] JPL 130.

It must be made clear that the way in which the issue estoppel point has been taken in these proceedings has changed and developed over the progress of them to this court. The point has been presented to this court differently from the way in which it was presented to the inspector and thereafter to McCowan J. With responsible
B generosity, as it seems to me, the Secretary of State has not sought to argue that the point was not effectively taken below so as to preclude reliance upon it in this court. The council, which is the second respondent on this appeal, has taken the same position. That means that there is no need for this court to examine closely in what terms the point was raised, and it is sufficient to say that, if it had been argued below as it has been argued in this court, a different result might have emerged as to part of the case.

I have found the main point in this appeal to be difficult. It seems that it has not previously arisen for decision by the court over the years since 1947, during which time the successive planning legislation has been in force, save for the case of *Rockhold Ltd* mentioned above. That fact seemed surprising to me and it caused me
C to feel some doubt over the correctness of the answer which otherwise on general principle appeared to me to be clear. In the end, and for the reasons which follow, it seems inescapable to me that, upon the proper construction of the relevant provisions contained in the 1971 Act, Parliament is not shown to have intended entirely to exclude, from the working of the appeal procedure against enforcement notices, the application of the principle of law and public policy contained in the concept of issue estoppel. The principle is necessary for the protection, as a matter of right, of the occupier of land against the unfair, although well-intentioned, repetition of proceedings which, in my judgment, Parliament cannot be taken to have intended to authorise.

Mr Ouseley relied upon the reasoning and statements of this court
D in *Western Fish Products Ltd* v *Penwith DC* in support of his argument that issue estoppel cannot properly be extended to section 88 proceedings or to any appeal proceedings under the 1971 Act. In that case this court considered the application of proprietary estoppel and estoppel by representation with reference to the actions and statements of a planning authority and its officials. In the judgment of the court (Megaw, Lawton and Browne LJJ), delivered by Megaw LJ, the following passages appear (p 219C):

> An estoppel cannot be raised to prevent the exercise of a statutory discretion or to prevent or excuse the performance of a statutory duty (see *Spencer Bower and Turner on Estoppel by Representation* 3rd ed, 1977, p 141 and the cases there cited).

E Then later in the judgment, at p 219H:

> Counsel for the plaintiffs submitted that, notwithstanding the general principle that a statutory body could not be estopped from performing its statutory duties, there are exceptions recognised by this court.
>
> There seem to be two kinds of exceptions. If a planning authority, acting as such, delegates to its officers powers to determine specific questions, such as applications under sections 53 and 94 of the 1971 Act, any decisions they make cannot be revoked. This kind of estoppel, if it be estoppel at all, is akin to res judicata.

Then at p 221B:

> We can deal with the second exception shortly. If a planning authority waives a procedural requirement relating to any application made to it for the exercise of its statutory powers, it may be estopped from relying on lack of
> F formality . . .
>
> The extension of the concept of estoppel beyond these two exceptions, in our judgment, would not be justified. A further extension would erode the general principle as set out in a long line of cases of which the decision of the Privy Council in *Maritime Electric Co Ltd* v *General Dairies Ltd* . . . [1937] AC 610 and the judgment of the Divisional Court in *Southend-on-Sea Corporation* v *Hodgson (Wickford) Ltd* [1961] 2 All ER 46, [1962] 1 QB 416 are notable examples. Parliament has given those who are aggrieved by refusals of planning permission or the serving of enforcement notices a right of appeal to the Secretary of State: see sections 36 and 88 of the 1971 Act. He can hear evidence as to the merits and take into account policy considerations. The courts can do neither. The application of the concept of estoppel because of what a planning officer has represented could result in a court adjudging that a planning authority was bound to allow a development which flouted its planning policy, with which the courts are not concerned.
>
> There is another objection to any extension of the concept of estoppel G
> which is illustrated by the facts of the *Lever (Finance) Ltd* case ([1971] 1 QB 222). If the modifications which were permitted by the planning officer in that case had been properly to be regarded as immaterial, no problem of general principle would arise. But the court regarded itself as competent to decide as to the materiality and, despite the submission to the contrary by the successful plaintiffs, held that the modifications were material. On what basis of evidence or judicial notice the court reached that conclusion, we need not stay to consider. We assume both that the court had jurisdiction to decide that question, and that, on the facts of that case, their decision as to materiality was right. But then comes the difficulty, and the real danger of injustice. To permit the estoppel no doubt avoided an injustice to the plaintiffs. But it also may fairly be regarded as having caused an injustice to one or more members of the public, the owners of adjacent houses who would be adversely affected by this wrong and careless decision of the planning officer that the H
> modifications were not material. Yet they were not, and it would seem could not, be heard. How, in their absence, could the court balance the respective injustices according as the court did or did not hold that there was an estoppel in favour of the plaintiffs? What "equity" is there in holding, if such be the effect of the decision, that the potential injustice to a third party, as a result of the granting of the estoppel is irrelevant? At least it can be said that the less frequently this situation arises the better for justice.

It is clear that the court in that case was not considering issue estoppel arising from the decisions by the Secretary of State, or by an inspector, of issues relating to existing rights in an appeal under section 88. Such decisions are made in the procedure provided by the Act after due notice to all parties concerned, after provision of opportunity to interested parties to make representations, and with J
proper regard to all relevant circumstances. To attach to such decisions the capacity to give rise to issue estoppel, within the limits required by the context in which they are made, is to give to those decisions, as I see it, the force and effect which the statutory provisions require them to have and, for my part, I find nothing in the reasoning of this court in *Western Fish Products* to require any other view to be taken.

Mr Ouseley submitted that decisions on appeals under section 36 against refusal of planning permission have never been regarded as giving rise to issue estoppel against the appellant when the appeals are rejected and such appellants have always been regarded as free to keep making fresh applications. No doubt they seek to make a new application sufficiently different to improve, at least marginally, its K
prospects of success, but even on separate issues within successive applications no issue estoppel arises. Therefore, it is said, issue estoppel cannot properly be held to be applicable to section 88 appeals. It is in this context that the decision of Forbes J in the *Rockhold Ltd* case is relevant. In that case there had been three applications, made in succession, for planning permission to develop a site for housing purposes. Each application went to appeal, and the three inspectors who in turn rejected the appeals did so for seemingly different reasons.

It was submitted for the appellant that, in so far as the first inspector had decided that housing development on this site would not be prominent, consistency of approach demanded that all L
inspectors should follow that lead. Forbes J rejected that submission. He held that each inspector had to be free to exercise his own judgment on what were matters of opinion and planning expertise. For my part, I agree with the reasoning of Forbes J in that case, but it is not decisive of the point raised in this case. It seems right to me that, in general, issue estoppel has no application to the sort of section 36 proceedings with which Forbes J was concerned. There is a clear distinction, as it seems to me, between the rejection of an appeal against the refusal, in administrative discretion, of an application for planning consent to new development and the holding on an appeal against an enforcement notice that a use being made of land is a lawful use within the established rights of the occupier. It is an essential requirement for application of the principle of issue M
estoppel that the decision relied upon as giving rise to the estoppel be a final decision between the parties. A decision on an appeal under section 88 that a particular use of land by the occupier is a lawful use within the established rights of the occupier is, in my judgment, in that sense to be regarded as a final decision. A decision on appeal under section 36 rejecting an application for planning consent is not final in the same sense.

Next, the interest of individual members of the public, or of the public at large, in the result of an appeal under section 88 does not, as I think, exclude the application of the principle. In *Thoday* v *Thoday* [1964] P 181, referred to by Diplock LJ in the passage cited above from *Fidelitas Shipping Co* case, the assertion of issue estoppel said to arise from the decision by a judge of a divorce suit based upon cruelty was made with reference to new divorce proceedings based upon desertion.

In its divorce jurisdiction the court was required by section 4 of the Matrimonial Causes Act 1950, to exercise what Diplock LJ [in *Thoday* v *Thoday*] referred to as a "quasi-inquisitorial function". The statutory provisions were as follows:

> 4(1) On a petition for divorce it shall be the duty of the court to inquire, so far as it reasonably can, into the facts alleged and whether there has been any connivance or condonation on the part of the petitioner and whether any collusion exists between the parties, and also to inquire into any counter charge which is made against the petitioner.
>
> (2) If the court is satisfied on the evidence that — (a) the case for the petitioner has been proved . . . the court shall pronounce a decree of divorce . . .

At p 196 Diplock LJ said:

> To such a function it is not easy to adapt the concept of estoppel inter partes, which was developed under and is consistent only with the adversary system of legal procedure which it is the function of this court to apply except in the exercise of its jurisdiction in divorce.
>
> "Estoppel" merely means that, under the rules of the adversary system of procedure upon which the common law of England is based, a party is not allowed, in certain circumstances, to prove in litigation particular facts or matters which, if proved, would assist him to succeed as plaintiff or defendant in an action. If the court is required to exercise an inquisitorial function and may inquire into facts which the parties do not choose to prove, or would under the rules of the adversary system be prevented from proving, this is a function to which the common law concept of estoppel is alien. It may well be a rational rule to apply in the exercise of such an inquisitorial function to say that if a court having jurisdiction to do so has once inquired into the truth of a particular allegation of fact and reached a decision thereon, another court of co-ordinate jurisdiction in the exercise of its own discretion should not re-embark upon the same inquiry, but should accept the decision of the first court. But this is a different concept from estoppel as hitherto known in English law. It will be interesting to watch its development in future cases, but fortunately it is not, in my view, necessary to develop it in the present appeal.

Later in his judgment, having examined the relationship of the findings in the earlier decision to the issues sought to be raised in the current proceedings, Diplock LJ held, notwithstanding the inquisitorial nature of the proceedings, that (p 199):

> Had Collingwood J in the earlier litigation made a specific finding that the husband's conduct during the period before the wife left him did not amount to serious ill-treatment of the wife, irrespective of its actual or apprehended effect on her health, this would, in my view, have given rise to an "issue estoppel" and would have prevented her from relying upon the same conduct as constituting the necessary serious ill-treatment for the purposes of her defence of just cause for separating from her husband or her allegation against him of the matrimonial offence of constructive desertion.

Because no such finding had been made, there was no estoppel on which the husband could rely.

In my judgment, if there is shown to be sufficient identity of issue, the principle of issue estoppel is not prevented from being applied in and as arising from appeal proceedings under section 88 of the 1971 Act by reason of the statutory duty of the Secretary of State or of an inspector to have regard to matters outside the private or separate interest and submissions of the appellant and of the planning authority. The interest of the public and of other persons particularly affected by the decision should be properly protected according to law in the first proceedings, from which the estoppel may be held to arise. The interest of the public and of other persons affected by the issue raised in the later proceedings will receive proper protection by the application of the principle of issue estoppel in the later proceedings. The fact that such proceedings are not merely a private contest between the local planning authority and the appellant may, in a given case, properly affect the decision of the Secretary of State, or of the inspector, as to whether the issues in the two sets of proceedings are to be regarded as sufficiently identical and as to whether, on the facts of the particular case, any sufficient justification for permitting the issue to be reopened has been shown. There is no ground for holding that the principle has no application at all to such proceedings.

I see no force in the other grounds put forward for denying application of the principle. I accept that, in a proper case, the issuing of an enforcement notice might be quashed on judicial review on the ground of abuse of power. Such proceedings would not be excluded by the provisions of section 243(1) of the Act. But the availability of judicial review as a means of questioning an enforcement notice on the ground of abuse of power does not demonstrate that the principle of issue estoppel should not be applied in and as arising from appeal proceedings under section 88. In some cases, perhaps in most, it would not be clear whether any relevant issue estoppel arose from the earlier proceedings, or should in justice be applied in the new proceedings, until the allegations made in the new proceedings and the evidence proffered to prove them have been closely examined against the issues decided in the earlier proceedings. That is an investigation which, as it seems to me, can properly and in most cases will more conveniently be carried out in the new section 88 appeal than in separate and preliminary judicial review proceedings.

When asked why the remedy should be limited to judicial review, Mr Ouseley answered that he relied upon the principles established in *O'Reilly* v *Mackman* [1983] 2 AC 237. I take the main principle to be as stated by Lord Diplock at p 285D as follows:

> Now that . . . all remedies for infringements of rights protected by public law can be obtained upon an application for judicial review, as can also remedies for infringements of rights under private law if such infringements should also be involved, it would in my view as a general rule be contrary to public policy, and as such an abuse of the process of the court, to permit a person seeking to establish that a decision of a public authority infringed rights to which he was entitled to protection under public law to proceed by way of an ordinary action and by this means to evade the provisions of Order 53 for the protection of such authorities.

The protective provisions referred to are the requirement for leave to apply granted on an application made promptly and within a limited time, the requirement that the application be supported by sworn affidavit evidence complying with the *uberrimae fidei* rule and the provision of a speedy means of deciding the matter. In my judgment, the principle stated by Lord Diplock does not exclude the right of the person against whom enforcement proceedings are taken by a planning authority to raise, as a matter of defence in the appeal proceedings under section 88, a claim that the Secretary of State, or the inspector, should apply the principle of issue estoppel against the planning authority: *cp Wandsworth London Borough Council* v *Winder* [1985] AC 461.

A recent statement of the principle by this court provides, in my judgment, a secure basis for the decision of this case. *Tebbutt* v *Haynes* [1981] 2 All ER 238 was a case concerned with the applicability, in Chancery proceedings over ownership of a house, of an issue estoppel against a wife and in favour of a husband's mother, said to arise in earlier proceedings in the Family Division between wife and husband, in which the mother had been permitted to intervene.

Lord Denning MR, at p 242G, said:

> We considered the question of issue estoppel recently in *McIlkenny* v *Chief Constable of West Midlands Police Force* [1980] 2 All ER 227, [1980] QB 283. I ventured to suggest this principle: if there has been an issue raised and decided *against* a party in circumstances in which he has had a full and fair opportunity of dealing with the whole case, then that issue must be taken as being finally and conclusively decided *against* him. He is not at liberty to reopen it unless the circumstances are such as to make it fair and just that it should be reopened.

Brightman LJ at p 244B, used words to the same effect:

> The principle involved is that of issue estoppel. A person should not have to fight all over again the selfsame issue that has been decided before, provided that the party *against* whom the estoppel is raised has had a fair and full opportunity to contest the issue, so that it would not be just to allow him to reopen the matter in subsequent proceedings.

Griffiths LJ agreed with the reasons given by Lord Denning MR and Brightman LJ. I would hold the principle so expressed to be applicable in these proceedings with reference to section 88 appeals.

The principles by which effect should be given to an issue estoppel seem to be clear. The estoppel in this case is sought to be raised by one party to both sets of proceedings, that is to say the appellant, against another party to both sets of proceedings, the council. The burden of proving that the earlier decision relied upon as giving rise to an estoppel was given upon grounds and in circumstances which show it to have been a final decision upon the issue in the second proceedings, with reference to which the estoppel is raised, rests upon the party who claims to be protected by the estoppel: see *Halsbury's Laws of England*, 4th ed, vol 16, para 1551. It is for him to show that it is the selfsame issue.

An allegation made in 1985 that the current use of a property was use as a hostel for homeless families and constituted a material change of use of that property is capable of being shown to be the same issue as an allegation differing in no material respect with reference to the use of the same property current in 1982 which was

A decided in earlier proceedings. It will, in my judgment, be shown to be the same issue if there is no relevant difference in the use being made of the property at the time relevant to the later proceedings as compared with the use being made of the property at the time relevant to the earlier proceedings. If that be proved, and the issue were decided in favour of the occupier of the property in the earlier proceedings, and if there be nothing to justify permitting the issue to be reopened, an issue estoppel will prevent the planning authority from contending that such a change of use has taken place. It is not open to the planning authority to relitigate the same issue: that means that the estoppel cannot, for example, be displaced by a more copious citation of authority, or by demonstration of the fact that the current
B exercise of discretion or of judgment on matters of fact and degree has been operating differently from the exercise of those powers at an earlier time, or by more detailed evidence of the nature of the activities if there has been no material change in fact in their nature between the time of the first decision and the later proceedings.

That follows from the clear principles cited by Diplock LJ in *Fidelitas Shipping Co Ltd* from Wigram VC at p 642G:

> . . . the court requires the parties . . . to bring forward their whole case, and will not (except under special circumstances) permit the same parties to open the same subject of litigation in respect of matter which might have been brought forward as part of the subject in contest, but which was not brought forward, only because they have, from negligence, inadvertence, or even accident, omitted part of their case.

C The occupier of the property would, however, fail to prove sufficient identity of the issue, and would thereby have no protection from estoppel, if it is shown that the use of the property at the time relevant to the later proceedings is materially different from the use being made at the time relevant to the earlier proceedings. It is not necessary to consider the different matters which may help to make it fair and just to permit an issue, otherwise closed by issue estoppel, to be reopened. An obvious example is proof of false or misleading evidence as to the use being made of the property given at the time of the earlier decision.

It is necessary now to consider whether issue estoppel can arise out of the decision by the Secretary of State or an inspector upon a section 36 appeal. In my judgment, issue estoppel can so arise if the party
D seeking to raise it proves that the earlier decision in the section 36 appeal was given in his favour against the planning authority upon an issue as to his existing rights of use of the property and was a decision of the same issue as that raised in the new proceedings.

Upon the findings of the inspector in the present proceedings, the issue decided in the 1982 proceedings on the section 36 appeal with reference to no 11, namely, that the then use of the property was, and had been at the time of the grant of planning permission which was varied, within the established use of no 11 as an hotel, was capable of giving rise to issue estoppel in these proceedings in favour of the appellant.

The relevance of issue estoppel in this case

E It has been conceded that the point of issue estoppel was sufficiently taken by the appellant before the inspector. It is necessary to consider how he dealt with the point on the facts in order to decide whether he has been shown to have gone wrong in law and , if so, with reference to which of the various appeals before him.

The inspector set out his decisions with great care and clarity. Having regard to the way in which the case was submitted to him I would emphasise that, in my view, he is in no way to be criticised for his handling of the issue estoppel point. He noted at para 35 that the contentions made for the appellant included the following:

> The use of these properties had not changed since the two earlier appeal decisions in 1978 and 1982. Specifically in 1982 the inspector found the same
> F present-day use of all the properties, except no 20, and concluded that these were hotels. It would be unjust after these two earlier decisions to say now that the properties were not hotels . . . . It could not be in the public interest for the council to continue to pursue enforcement action on the basis of different legal arguments.

It must be noted that before the learned judge and in this court the appellant has placed no reliance upon the earlier decision in 1978.

The inspector also noted that the contentions made for the council included the following:

> 41 The first question

ie whether or not the present use was correctly described as a hostel

> could only be answered by reference to the facts and features pertaining to the actual use of the property. There was little dispute on this point, the properties were used predominantly for providing accommodation for homeless G
> families . . . . These were placed by local authorities and the position had not changed since the 1970s. There was some evidence that the occasional tourist also stayed at the properties.

Then:

> 45 A hostel was a place where people lived or stayed depending on the type of occupant. Students and nurses might be said to "live" in their hostels whilst people "stayed" in youth hostels. They had two essential features, namely communal facilities such as cooking and washing facilities which were shared and the standard of accommodation was modest. The properties contained kitchens for residents' use. Whilst these might be described as baby bottling sterilising rooms they were in practice used by adults for the preparation of food. At the time the notices were served no main meals were provided by the
> management. Food was provided by the homeless families and cooked for H
> themselves.

Finally, the inspector recorded as part of the submissions for the council the following:

> 48 The finding of the inspector in 1982 that a number of the properties were hotels could not be relied upon by the appellants because the question of whether the premises were a hotel or a hostel was one of fact or opinion or of mixed fact and opinion. Each inspector must reach his own decision and was not bound by the decisions of previous inspectors. *Rockhold Ltd* v *Secretary of State per* Forbes J. Furthermore, the 1982 decision did not in this respect appear to be based on a full citation of authority or of principle. Thus, the premises were all correctly described as hostels for homeless people.

The inspector's reasons for his conclusions, so far as they touch
upon the issue estoppel point, included the following: J

> 55 Although at my inspection I did not see every room in each property, I saw the majority and I was careful to note the mode of occupation and the standard of that accommodation. It was apparent that the larger families are allocated more than one room, most of which contain a bed or beds, wash hand basin, a cupboard or wardrobe, table and one or more chairs. These family units are not self-contained however. In many cases the accommodation was crowded, partly because of the number of persons present and partly because of the quantity of personal effects in the rooms. It was evident from the amount of stored food, household articles and cleansing materials, pots and pans, kettles, televisions and, in at least one instance, the presence of a washing line, that most of the families are occupying their rooms as their homes, at least for the time being. That some of them appeared to be
> well settled may be explained by the length of time they had been in residence. K
> 56 . . . . The baby bottle steriliser rooms had the appearance of conventional kitchens, each being provided with the basic facilities for preparing food, although I saw no one using them. In some cases the cookers are unusually large, having six burners . . . . It is clear from the evidence that although it is against the management's rules, these kitchens are used by residents for the cooking of meals.
>
> 59 Having regard to the case law and all the facts it seems to me that all the appeal properties are being used as hostels rather than hotels. My reasons for this conclusion are that . . . the majority of residents are homeless families referred to the premises and paid for on a daily basis by their local authorities' sponsors . . . the homeless have no homes elsewhere and are therefore living in the accommodation as opposed to staying away for holiday or business purposes. Their mode of occupation . . . suggests that they have made their
> homes there for the time being . . . Communal facilities are provided in the L
> form of washing facilities and, even if not the intention, kitchens are available and are used for the preparation of food, this last factor being unknown in my experience in a conventional hotel . . .
>
> 65 I consider that the intensity of occupation of the premises as hotels for homeless families and the effect of the demand for local services by this type of occupant amounts to a significant difference in character to a hotel or multiple-occupied residential use such as flats and, as a matter of fact and degree, I conclude that this constitutes a material change of use . . .
>
> 66 Whilst I appreciate that my findings in the above respects are at variance with those of the 1982 appeal decision, in respect of the majority of the properties, they have been reached on the basis of very considerable volume of evidence and submissions as to law put before me, much of which may not have been available to my predecessor.

I have already referred to the fact that McCowan J held that issue M
estoppel, in the form in which the matter was submitted to him, had no application in law. He found also another ground of decision, namely that the decision of the inspector in these proceedings did not appear to have been made on the same facts as that of the inspector in 1982. He said that he had in mind in particular the aspect of cooking. He compared the description of the communal kitchen in the circumstances of occupancy as found by the inspector in the 1982 decision with the description of the use of the communal kitchens in para 56 of the inspector's decision letter in the present proceedings, of which the substance is set out above. McCowan J took the view, as I read his judgment, that even if issue estoppel had been available as a matter of law in the form in which it was raised before him, there was

A a material difference in the user of the property in 1985 which would have rendered the estoppel inapplicable.

In my judgment, the appellant has demonstrated that, so far as concerns nos 13 and 15, the inspector misdirected himself in law by failing to consider, and properly to apply, the principle of issue estoppel. The allegations made in the 1982 proceedings by the council were that the then use of these properties by the appellant constituted a material change of use and was unlawful because not a permitted or established use. The council's case was that the then use was unlawful whether it was properly to be categorised as hotel use or as hostel use. The main issue that the then use was unlawful was decided against the council; the specific decision was that the use being made of nos 13 and 15 was properly to be described in law as hotel use, and as such B was a permitted or established use in respect of these properties. The allegations made in the 1985 enforcement notices were again that the use being made of nos 13 and 15 at the date of the notices constituted a material change of use.

It is common ground that the use being made of the properties in 1985 and that considered by the inspector in these proceedings is, in all material respects — subject to the matter of the kitchens to which I refer below — identical to the use being made of the properties in 1982, which was held not to have constituted a material change of use. Therefore the council, by the 1985 notices, was seeking to litigate again against the appellant the selfsame issue as had been decided against the council in the 1982 proceedings. The council had had, in C 1982, full and fair opportunity of dealing with the whole case, as had any other person then interested in the result. The decision on the issue given in 1982 must be taken to be final and conclusive against the council if there are no circumstances raised by the council or by any other person interested, such as would make it fair or just to permit the issue to be reopened. It was not suggested that there were any such circumstances.

As to the kitchens, although McCowan J (as I have said) regarded the difference in the evidence as to the use of the kitchens as between 1982 and 1985 as a possible answer to issue estoppel, if, contrary to his view, it was available to the appellant, it was not argued for the Secretary of State or for Hackney Council that there had been any material change of use since 1982 or that it was open to the inspector D to hold that there had been. I have seen in advance the judgment of Parker LJ, and I agree with him that on the material before him the inspector was bound to quash the notices relating to nos 13 and 15 and that any additional evidence as to the use of kitchens could not prevent application of the estoppel.

As to no 11, I would reach the same conclusion. For the reasons I have given, the decision on the section 36 appeal in the 1982 proceedings with reference to this property gave rise to an issue estoppel in favour of the appellant, namely that the use being made in 1985 had been previously held to be within the established use of that property. The inspector failed, as with reference to nos 13 and 15, to give effect to the issue estoppel claimed by the appellant.

E Last, as to no 25, I have found the case to be less clear. In the 1982 proceedings the purpose of the council was to demonstrate that the then use constituted a material change of use and was not lawful under any permitted or established use of the property. The issue of mixed fact and law raised by the allegation that the then use constituted use of the property as a hostel and not as an hotel was decided against the council after full and fair opportunity on its part to contest that issue. The alternative issue that the then use constituted use as an hotel and a material change of use because begun after 1964 without consent was decided in favour of the council. Can the decision on the first issue give rise to issue estoppel in the 1985 proceedings?

In *Thoday*'s case [1964] P 181, cited above, Diplock LJ considered F the nature of an issue capable of giving rise to issue estoppel under the common law principle which he regarded as an extension of the same principle which underlies cause of action estoppel. At p 198 he said:

> There are many causes of action which can only be established by proving that two or more different conditions are fulfilled. Such causes of action involve as many separate issues between the parties as there are conditions to be fulfilled by the plaintiff in order to establish his cause of action; and there may be cases where the fulfilment of an identical condition is a requirement common to two or more different causes of action. If in litigation upon one such cause of action any of such separate issues as to whether a particular condition has been fulfilled is determined by a court of competent jurisdiction, either upon evidence or upon admission by a party to the litigation, neither party can, in subsequent litigation between one another upon any cause of action which depends upon the fulfilment of the identical condition, assert that the condition was fulfilled if the court has in the first litigation determined that it G was not, or deny that it was fulfilled if the court in the first litigation determined that it was.

Diplock LJ later in his judgment applied that analysis to what he called "matrimonial offence estoppel". In short, an issue estoppel could arise, in his judgment, with reference to one of the two conditions necessary to proof of the matrimonial offence of cruelty, namely (i) conduct amounting to serious ill-treatment of the spouse, and (ii) adverse effect or reasonable apprehension of such effect upon the health of the spouse.

In my judgment, that analysis can properly be applied to issue estoppel arising in section 88 appeal proceedings. The issue of material change of use, as a matter of fact and degree, to a use properly to be categorised as hostel use, and materially different H from hotel use, was a condition necessary to be satisfied by the council in its enforcement proceedings under one of the then enforcement notices. That issue was decided against the council. The same issue was raised in the 1985 proceedings. I would hold that the 1982 decision gave rise to an issue estoppel. We do not know, and it is not relevant to explore, why the council did not take proceedings to force the appellant to cease the use which he was making of no 25 in breach of the enforcement notice which the inspector in 1982 had upheld. With reference to no 25 also I would hold that the inspector went wrong in law in failing to give effect to the issue estoppel claimed by the appellant.

As to the other properties, I see no ground for questioning the J validity of the conclusions of the inspector, which were upheld by McCowan J after consideration by him of the various criticisms directed at the inspector's findings on the merits of the planning issues. On those matters no appeal has been brought to this court. The failure by the inspector to give effect to the defences of issue estoppel with reference to nos 11, 13, 15 and 25 has no impact, in my judgment, on the other parts of the case. I would therefore allow the appeal to the extent indicated and no further, and I would remit the appeals relating to nos 11, 13, 15 and 25 to the Secretary of State for rehearing and determination by him in the light of the opinion of this court that the inspector was obliged to allow the appeals to the Secretary of State and quash the enforcement notices as proposed by Parker LJ in his judgment. K

Agreeing, PARKER LJ said: This appeal raises an important question of planning law. It arises most directly with regard to two of the several properties to which the appeal relates, all of which are situate in Wilberforce Road, London N4. I shall therefore, initially, consider the position in relation to such two properties. They are nos 13 and 15. Moreover, since the relevant facts are the same in relation to both of them, I shall set out such facts only in relation to one of them, namely no 13.

In October 1981 the Hackney Borough Council, which is the local planning authority for the area, served on the appellant two enforcement notices in relation to the property under section 87 of the Town and Country Planning Act 1971 as amended (the Act). The L power so to do arises where it appears to the authority that there has been a breach of planning control after the end of 1963. A breach of planning control consists, so far as immediately relevant, in the carrying out of development without the planning permission required by Part III of the earlier (1962) Planning Act or Part III of the Act. For present purposes the only kind of development which needs to be mentioned is "the making of any material change in the use of any buildings or other land": see section 22(1) of the Act.

Both of the notices alleged a material change in the use of the land. One specified the change of use as being "to use as an hotel", and required that use to be terminated. The other specified the change of use as being "to use as a hostel", and required that use to be terminated. I shall refer to the two notices respectively as "the hotel M notice" and "the hostel notice".

The appellant, as he was entitled to do under section 88 of the Act, appealed against both notices.

In the case of the hotel notice he appealed on grounds (a), (e) and (h) of the grounds specified in section 88(2) as the grounds upon which an appeal may be brought. Stated shortly, such grounds are respectively: that planning permission ought to be granted; that the breach of planning control alleged occurred before the beginning of 1964, and that the period specified for discontinuance of the use was too short.

It will thus be seen that the appellant did not seek to challenge the alleged use as an hotel but was contending that such use had commenced before the beginning of 1964. Only if this ground failed would ground (a) or (h) arise.

In the case of the hostel notice, the appellant appealed additionally on grounds (b) and (c), namely, that the use alleged did not constitute a breach of planning control and that the use alleged had not taken place. If ground (c) succeeded, the other grounds would not arise.

The appeal was duly heard and determined by an inspector, pursuant to Schedule 9 to the Act, in 1982. As to the hostel notice, he held that the property was being used as an hotel not as a hostel. He therefore quashed the hostel notice. As to the hotel notice, he held that change of use to an hotel had occurred before the beginning of 1964. He accordingly quashed the hotel notice also.

The appeal against both notices having succeeded, and the notices having been quashed, the council was entitled to appeal on a point of law under section 246 of the Act, but it did not seek to do so.

At this point it is convenient to refer to certain further provisions of the Act. Section 243 provides, subject to other provisions not presently material, that the validity of an enforcement notice shall not, except by way of an appeal under Part V of the Act, be questioned in any proceedings whatsoever on any of the grounds on which an appeal may be brought. The person on whom an enforcement notice is served must therefore, if he wishes to challenge it on any of the specified grounds, do so by way of appeal under section 88, albeit on other grounds he could challenge by way of judicial review. This restriction applies of course only to those seeking to challenge the notice, but once the inspector has made his decision, both sides are subject to a like restriction.

Para 2(3) of Schedule 9 provides:

> Where an appeal to which this Schedule applies has been determined by a person appointed under this Schedule, his decision shall be treated as that of the Secretary of State and
> (a) except as provided in Part XII of this Act the validity of his decision shall not be questioned in any proceedings whatsoever;

The relevant provision of Part XII in the present case is section 246, which gives both sides a limited right of appeal.

These provisions appear to me to show a very clear intention on the part of Parliament that a decision on an appeal under section 88 should be truly final and binding on both parties, subject only to a section 246 appeal.

In relation to the restrictive provision in section 243(1), this court has had to consider whether it prevented a person prosecuted under section 89 for failure to comply with an enforcement notice from challenging its validity by way of defence. The question arose in *R* v *Smith (Thomas George)* (1984) 48 P&CR 392. It was contended that the restriction prevented challenge only by way of judicial review and did not prevent challenge by way of defence in criminal proceedings. This contention was rejected.

At the conclusion of the 1982 proceedings, therefore, the parties were both prohibited from questioning the validity of the inspector's decisions, save under section 246 and the council, not having exercised its right of appeal under that section, the prohibition became unqualified.

I now return to the factual history. In February 1985 the council served an enforcement notice relating to the property. It again set up a material change of use as the breach of planning control relied on and specified the change as being "to use as a hostel for homeless families".

The last three words did not appear in the 1981 notice, but it has, in my view rightly, not been suggested that the use set up by the council in the 1985 notice was in any way different from the use which had been negatived by the inspector by his decision in 1982.

The notice itself is — whatever may be the effect of the 1982 decision — not open to objection because it must on any view be open to the council to complain that, although used as an hotel from before 1964 to 1982, there had since then been a change from use as an hotel to use as a hostel.

The appellant appealed, again relying on ground (c) and contending that the use was as an hotel and not a hostel, but on this occasion he failed.

At para 59 of his written decision the inspector said:

> Having regard to the case law and all the facts it seems to me that all the appeal properties are being used as hostels rather than hotels.

Then in para 62:

> Accordingly I am satisfied that the breaches of control alleged in notices. . .C . . .have taken place and the appeals on ground (c) fail.

Then in para 66:

> Whilst I appreciate that my findings in the above respects are at variance with those of the 1982 appeal decision, in respect of the majority of the properties, they have been reached on the basis of a very considerable body of evidence and submissions as to law put before me, much of which may not have been available to my predecessor.

Since it was and is common ground that there had been no change of use since 1982, it is thus clear that the inspector in 1985 came to the conclusion that the 1982 decision was wrong.

Before him the appellant relied *inter alia* on the 1982 decision, but not as a bar to any finding that the property was used as a hostel. The council's contention is recorded in para 48 of the decision, which reads:

> The finding of the inspector in 1982 that a number of the properties were hotels could not be relied upon by the appellants because the question of whether the premises were a hotel or a hostel was one of fact or opinion or of mixed fact and opinion. Each inspector must reach his own decision and was not bound by the decisions of previous inspectors. *Rockhold Ltd* v *Secretary of State*, Forbes J, July 26 1985 (document 60). Furthermore, the 1982 decision did not in this respect appear to be based on a full citation of authority or of principle. Thus, the premises were all correctly described as hostels for homeless people.

On his appeal under section 246 to the High Court it was argued for the appellant, although it was at best doubtful whether the point was fairly included in the grounds stated in the notice of appeal, that the use up to and including 1982 could not be challenged but was *res judicata* by reason of the 1982 decision.

McCowan J dealt with the matter shortly. He accepted submissions on behalf of the Secretary of State that once an appeal was instituted under section 88 it was the duty of the Secretary of State, or in this case the inspector, to determine it. If, it was said, the inspector was fettered by a previous decision so that he could not decide the issue for himself, he would not be discharging the statutory task placed upon his shoulders.

In this court the matter has developed differently, the appellant setting up issue estoppel. I agree with Gibson LJ that the principle is available on a section 88 appeal, and I agree with his reasons. If it is so available, then it is, in my judgment, clear that the inspector erred in law. There was no question of any change of use since 1982. The council was seeking to reopen the selfsame issue as had been decided against it in 1982, and seeking by more voluminous evidence and more fully developed legal argument to reverse that decision. On the facts of this case the inspector was, in my view, bound in law to quash the notice relating to no 13. The learned judge mentioned the question of cooking space and its use. There may well have been more evidence of the way in which the kitchen space was used than had been available in 1982 but, since it was not, and is not, contended that there had been a change of use since 1982, that cannot be relevant. It can only amount at most to this, that had the additional evidence been given in 1982, the inspector might then have reached a different decision.

Apart altogether from issue estoppel properly so called however, it appears to me that the council was directly questioning the validity of the decision in 1982, and this it is not entitled to do if the 1985 appeal consists in proceedings within the meaning of para 2(3) of Schedule 9 to the Act. If, as has been held, the proceedings referred to in section 243 of the Act include criminal proceedings in which a person seeks to defend himself from non compliance with a notice, I can see no reason why "proceedings" in para 2(3) of Schedule 9 should not include proceedings in which he is in effect seeking to defend himself from an attempt by the council to render him potentially criminally liable for continuing a use of premises, the continuance of which had previously been held incapable of creating such criminal liability.

It is true that a section 88 appeal is not a proceeding *in court*, but an appeal under section 246 is a proceeding in court and specifically recognised to be proceeding within the meaning of proceedings in section 243 for, were it not, there would be no need to exempt a section 246 appeal from the otherwise total ban. It appears to me somewhat unreal to suggest, in such circumstances, that a section 88 appeal is not itself a proceeding and thus to permit it to be used, via a fresh enforcement notice, to challenge the validity of an earlier decision which was plainly intended to be final subject only to the section 246 appeal on law.

Since no reliance was placed by the appellant on the restrictive provisions of section 243 and para 2(3) of the 9th Schedule, since we therefore heard no argument on it and since I agree that the appellant

A succeeds on the basis of the issue estoppel point, I express, however, no concluded view on the matter.

Even, however, if a section 88 appeal is not a "proceeding", the provisions of para 2(3) appear to me to afford a strong supporting reason for the applicability of issue estoppel to a section 88 appeal.

It is certainly clear that if the council wished to challenge the validity of the 1982 decision, in point of law the method of so doing is, and is only, by way of appeal to the High Court under section 246. When there is no appeal under that section, I cannot think that Parliament intended that the council should be entitled to serve a fresh notice and to contend, as it did, that the earlier decision was bad in law or because it had better evidence.

B For precisely the same reasons the inspector was, in my judgment, bound in law to quash the notice with regard to no 15.

The position with respect to no 25 is different. In that case also there were both a hostel notice and an hotel notice. In that case also the hostel notice was quashed. The hotel notice was, however, upheld because the contention that the use began before the beginning of 1964 was rejected. It is not clear to me why, having succeeded on the hotel notice, the council should now seek to reject what it asserted and the inspector upheld in 1982 and to reassert its alternative assertion which was rejected in 1982. Whatever the reason — hostel use having been rejected and hotel use having been confirmed in 1982 (and it being common ground that there has been no subsequent change of use) — it cannot again assert hostel use.

C The inspector was, in this case also, obliged to quash the notice.

I turn now to the fourth of the properties alleged to be covered by issue estoppel, namely no 11, as to which the factual and procedural history are different. At all material times there was an established use certificate in respect of this property for hotel use. In 1979 the appellant applied for planning permission for the retention of a single-storey side extension and for its use as a reception area for that property as a bed-and-breakfast hotel. On March 26 1981 permission was granted for "the retention of the extension and its use as a reception area in connection with the use of the premises as a hostel for homeless families". It was, however, a condition that the extension be removed by February 1986.

The appellant appealed under section 36 of the Act, seeking the D removal of the condition and variation of the wording of the permission by deletion of the words "in connection with the use of the premises as a hostel for homeless families". Again the issue between parties was whether the premises were being used as a hostel or as an hotel, and again the inspector held that the premises were being used as an hotel. Having so found, he allowed the appeal under both heads. He discharged the condition and deleted the words to which the appellant objected. He thus effectively granted permission unrestricted in point of time for the retention of the extension and its use as a reception area for the hotel.

In that case there was no enforcement notice, for the council was apparently content that the property should be used as a hostel, at all E events until 1986.

In 1985, however, the council had apparently changed its position. The enforcement notice alleged that there had been a breach of planning control by change of use to use as a hostel for homeless families. This too, in my view, gives rise to issue estoppel and questions the validity of the 1982 decision. The appellant had an established use certificate for use as an hotel and planning permission to retain the extension as a reception area for that hotel on the single ground that the property was then being used as an hotel. It being common ground that there had been no change of use since 1982, the inspector was bound in law to quash the enforcement notice.

With regard to the remaining properties, no issue estoppel is alleged to arise. However, since it was not suggested that there was F any difference between the uses of any of the properties, it was submitted that all the decisions should be referred back. There is, in my view, some merit in those submissions for, had the inspector realised that he was bound to accept hotel use and quash the notices in respect of nos 11, 13, 15 and 25, it is, as it seems to me, possible that in the interests of consistency, he might have adopted the same course with regard to the other properties.

It is, however, impossible to hold that in regard to the other properties his decisions — I use the plural because although embodied in one document, there were a series of decisions on a series of appeals — were erroneous in point of law.

I would therefore allow the appeal against the decisions with regard to nos 11, 13, 15 and 25, and remit those cases for rehearing and determination in the light of the opinion of this court that the G inspector was, in the circumstances, obliged to allow the appeals to the Secretary of State and quash the enforcement notices.

In conclusion I should perhaps mention the case of *Rockhold Ltd* v *Secretary of State for the Environment* [1986] JPL 130. In that case three successive planning applications in respect of the same development were rejected by three different inspectors. Two concluded that the development would not be prominent. The third considered that it would be damaging to the visual beauty of the area. That case does not appear to me to afford any assistance at all to the respondents in this case. All three inspectors dismissed the appeals, albeit they differed in one of their reasons.

In the present case the second inspector differed from the first on H the very point that formed the only ground upon which the earlier appeals were decided. If, at the opening of the 1985 appeals, the inspector had asked the council's representative the question: "Do you contend there has been any change of use since 1982?" he would have received the answer "No". If he had then asked the question: "Do you then contend that the inspector was wrong in concluding in 1982 that the properties were being used as hotels and not as hostels?" he would have received the answer "Yes". If he had asked finally: "On what grounds?" he would have received the answer: "Because we now have better legal arguments and better evidence." That is very different from the *Rockhold* case. Not only is it different, it demonstrates that the council was seeking to question the validity of the decision otherwise than in the only way in which Parliament J has permitted, namely, by challenge on a point of law under section 246.

Agreeing, CAULFIELD J said: I, too, would allow the appeal against the decision with regard to nos 11, 13, 15 and 25, and would remit these cases for rehearing and determination.

My Lords have rightly emphasised that it was, and is, common ground that there had been no change of use since 1982, and I think it is clear that the 1985 inspector ruled as he did because he thought the 1982 decision was wrong.

I agree, for the reasons advanced by Ralph Gibson LJ and Parker LJ that issue estoppel is available on a section 88 appeal, and on the concession that there had been no change of use since the 1982 K decision, the 1985 inspector erred in law.

The appeals were allowed in part, with costs in the Court of Appeal against both parties; no order for costs in the court below. Leave to appeal to the House of Lords was granted.

Queen's Bench Division

December 11 1987

(Before Mr Justice SIMON BROWN)

BELLCROSS CO LTD v MID BEDFORDSHIRE DISTRICT COUNCIL L

Estates Gazette April 16 1988

[1988] 15 EG 106-108

Town and Country Planning Act 1971, section 245 — Application by development company seeking to quash a tree preservation order in respect of a single mature yew tree — Applicants relied on the principles in *Associated Provincial Picture Houses Ltd* v *Wednesbury Corporation*, in particular that the planning authority had taken into account a matter which it ought not to have taken into account — The M applicants had obtained outline planning permission for residential development and subsequently applied for approval in regard to the reserved matters — At that time the plan for development involved the removal of the yew tree — The respondent authority, after considering reports from the district planning officer and the county forestry officer, made a tree preservation order in respect of the tree in question, based on the contribution made by the tree to the visual amenities of the area — The developers had in the meantime amended their application for approval of the reserved matters in order to avoid the removal of the tree — They

A **objected strongly to the making of the tree preservation order, observing that they had amended their application specifically to prevent the tree preservation order from being used as a basis for further delay of the development — They sent with their protest a copy of a tree surgeon's report, suggesting that the retention of the tree would be dangerous — They explained that they had not intended to lose the plot of land on which the tree was located — In challenging the order the applicants submitted that an immaterial consideration had been allowed to influence the authority's decision — They referred to comments by the district planning officer and the county** B **forestry officer about the applicants' amendment of the plan in order to exclude the tree and the suggestion that there was then no reason against the retention of the tree — Held, rejecting the developers' challenge, that the court had not been satisfied that the authority had been influenced by an immaterial consideration; in particular it was wholly unsatisfied that the authority had improperly taken account of the fact "that the tree was not directly to be damaged by the then proposed development" — Application dismissed**

The following case is referred to in this report.

C *Associated Provincial Picture Houses Ltd* v *Wednesbury Corporation* [1948] 1 KB 223; [1947] 2 All ER 680, (1947) 45 LGR 635, CA

This was an application by Bellcross Co Ltd, a development company, to quash under section 245 of the Town and Country Planning Act 1971 a tree preservation order made by the respondent authority, the Mid Bedfordshire District Council, in respect of a single mature yew tree at Denel End Farm, Flitwick, in Bedfordshire. The tree stood on the site of a proposed residential development for some 230 houses.

S Pickles (instructed by Benning, Hoare & Drew, of Shefford, Beds) appeared on behalf of the applicants; Christopher Lewsley (instructed by the district solicitor, Mid Bedfordshire District Council) represented the respondent council.

D Giving judgment, SIMON BROWN J said: This is a section 245 application by a development company seeking to quash a confirmed tree preservation order in respect of a single mature yew tree at Denel End Farm, Flitwick, Bedfordshire, on the site of an intended residential development for some 230 houses.

The challenge is founded upon *Wednesbury* principles. In particular, for reasons to which I shall come later, it is asserted that the respondent authority took account of an immaterial consideration when they confirmed the order.

The factual circumstances within which this challenge arises can best be related chronologically. On December 19 1985 the Secretary of State on the developers' appeal granted an outline planning E permission for this residential development. On March 27 1986 the developers applied for the approval of reserved matters, the plan at that point involving the removal of the tree.

On June 10 the respondents first considered making this TPO. The district planning officer's report to the relevant committee recommended making an order with immediate effect. It said:

An application has been submitted for approval of reserved matters for housing development at Denel End Farm and involves removal of the existing farmhouse and buildings and redevelopment of the site. A yew tree stands in the garden of the farmhouse and is under threat from the development proposals.

The county forestry officer has inspected the tree and advises that it is in very good condition and is probably one of the oldest trees in Flitwick. The F tree has been inspected by a planning officer and it is considered to make an important contribution to the visual amenities of the area.

On June 16 there was served upon the developers a notice of a tree preservation order in regard to the yew tree, together with a direction under section 61 of the Town and Country Planning Act 1971 indicating that it came into immediate effect. The grounds for making the order were said to be:

The tree makes an important contribution to the visual amenities of the area.

It is convenient at this stage briefly to allude to the relevant statutory provisions. By section 60(1) of the 1971 Act it is, so far as relevant, provided:

If it appears to a local planning authority that it is expedient in the interests of amenity to make provision for the preservation of trees or woodlands in their area, they may for that purpose make an order . . . (a) for prohibiting . . . the cutting down, topping, lopping . . . of trees. G

Section 60(4) provides that such an order shall not take effect until it is confirmed by the local planning authority.

However, by section 61 it is provided:

(1) If it appears to a local planning authority that a tree preservation order proposed to be made by that authority should take effect immediately without previous confirmation, they may include in the order as made by them a direction that this section shall apply to the order.

(2) Notwithstanding section 60(4) of this Act, an order which contains such a direction shall take effect provisionally on such date as may be specified therein and shall continue in force by virtue of this section until — (a) the expiration of a period of six months beginning with the date on which the order was made; or (b) the date on which the order is confirmed whichever H first occurs.

I need not refer to the TPO itself beyond saying that the second schedule, the terms of which are governed by regulations, includes the following:

This order shall not apply so as to require the consent of the authority to . . . (3) the cutting down, uprooting, topping or lopping of a tree . . . (c) where immediately required for the purpose of carrying out development authorised by the planning permission granted on an application made under Part III of the Act . . .

On June 20 the developers by letter submitted a robust objection to the making of the TPO, contending that it was too late in the day for such an order. The company having overcome the planning authority's earlier objections to development on the appeal to the J Secretary of State, the council's attitude was trenchantly described as, among other things, devious, undemocratic, mischievous and involving a misuse of power.

The respondent authority replied on July 4 1986. It is unnecessary to relate the terms of that letter beyond observing that it noted that the developers had amended their outstanding application for the approval of reserved matters in order to avoid the removal of the tree.

That amendment had in fact been made by redrawing the development plan on May 19 1986, the amended plan being received by the authority on June 23 1986.

On July 9 the developers wrote again in terms as emphatic as those of their earlier letter. They observed that they had amended the current application specifically so that the tree preservation order K could not be used as a basis for further delay in the development but added:

This does not mean in any way to be intended to lose that plot as the infrastructure of this development is costing a fortune and this plot with roads and services would cost this Company £25,000 to £30,000.

They further enclosed a copy of a tree surgeon's report and suggested that it would be dangerous to retain that tree. On that basis they were objecting to the order.

The tree surgeon's report, itself dated July 8, described problems affecting the health of the tree and said that although it could be safely left in its present environment: L

However, as there are to be houses built on this land and obviously children living in the immediate vicinity, we can only, with reluctance, recommend that this tree be felled, for the safety of all concerned.

On July 11 the developers' solicitors wrote essentially to confirm the gist of the developers' objection based on the expert's report regarding the danger the tree would constitute within the proposed development.

On July 15 the respondent authority approved the reserved matters pursuant to the amended application. The only relevant condition is no 16, which states:

The permission shall extend only to the application as amended by plans received 23.6.86.

These, as I have indicated, left the yew tree standing. M

It was some four months later that the question of the confirmation or otherwise of the TPO came before the respondent council's planning and development committee. The committee had before them a report of the council officers. That said that the district planning officer and the county forestry officer had been consulted upon the objections to the confirmation of the relevant tree preservation order and their comments were as follows:

District Planning Officer: The main point made by Bellcross is that we should have acted at the outline stage. In fact, the outline proposal suggested in no way that the part of the farm where the tree is situated would be affected by the development.

I observe parenthetically that it was of course always a possibility at the outline stage that the tree might be affected by the development, but nothing turns on that in respect of the motion presently before the court. The quotation continues:

When the details were submitted the original farm buildings were included in the redevelopment proposals and it was only at this stage that any threat to the tree was recognised. I see no reason why we should not proceed with the confirmation of the Order. Incidentally the application has been amended to avoid the tree.

County Forestry Officer: The tree is an old specimen with some decay in the main trunk. The crown of the tree was removed many years ago and extensive regrowth has occurred to form a new crown. If the site has been redrawn to retain the tree and there are no proposals for any development closer than 10 metres, then I see no reason why the tree should not be retained. It is probable that regular pruning work will be required to maintain the tree's safety but this would only be good management.

In the light of those comments it was recommended to the committee that the TPO be confirmed without modification.

The committee considered that report on November 18 1986 and, having duly considered the objections to the order in the light of those comments, they recommended to the council that the TPO be confirmed without modification.

There then occurred what Mr Pickles for the applicant company describes as an unfortunate episode, which consisted of the respondent authority treating that committee recommendation as itself the confirmation of the order and on November 28 1986 serving notice of confirmation upon the company.

On December 16 1986 the provisional TPO expired by effluxion of time pursuant to section 61(2). That, however, was not appreciated by the company, who believed that the notice had duly been confirmed. There is an affidavit before the court which deposes to the fact that but for their misapprehension in this regard they would in fact have felled the tree upon the expiry of the order.

On January 6 1987 the council itself duly accepted the committee's recommendation and adopted it. Thus, finally, the order came to be confirmed. Service of the notice of the confirmed TPO was then effected on January 9 1987. In an accompanying letter the authority apologised for their error in serving a purported notice in November. These proceedings were issued on February 17 1987.

I return to the basis of this challenge. In an admirably clear and succinct argument, Mr Pickles contends that an immaterial consideration was introduced into the decision to confirm this TPO. He refers particularly to passages in the respective comments made by the district planning officer and county forestry officer, notably where the former said: "Incidentally the application has been amended to avoid the tree", and where the latter said: "If the site has been redrawn to retain the tree and there are no proposals for any development closer than 10 metres, then I see no reason why the tree should not be retained."

He relies also upon the following passage in an affidavit sworn by Mr Adamson, the respondent's district solicitor, on March 27 1987:

The order does not apply so as to require the consent of the Council to the cutting down, uprooting, topping or lopping of a tree where immediately required for the purpose of carrying out development authorised by a planning permission (Sch 2 para 3(c) of the Order). Accordingly there would be no point in confirming the order in those circumstances. The fact that the proposed development avoided direct damage to the tree was therefore relevant and supported confirmation of the Order. Further, the report to committee . . . makes it clear that this was no more than an incidental matter.

The essence of Mr Pickles' submission is that the point that the proposed development would avoid direct damage to the tree was accordingly being invoked and deployed as a positive point in favour and support of the confirmation of the order, whereas it ought more properly to have been regarded either as a point militating against confirmation or at worst a neutral point. The fact that the tree was not to be felled pursuant to the amended plan did no more, contends Mr Pickles, than leave wholly open for the council's consideration the expedience or otherwise (in the language of section 60(1)) "in the interests of amenity" of making this preservation order.

Attractively though the argument has been put, in my judgment it fails. It seems to me critically important to place the officers' report submitted to the committee on November 18 1986 in its context. That context involves these considerations. First, the original recommendation for the making of the order had already specified the vital factors in the officers' minds for the making of such an order. That was in the report of June 10 1986 which I have already recited in full. Second, by November 1986 it was necessary to acquaint the committee with the changed situation. The committee, and in turn the council, would need to know whether or not the detailed approval by then already given would require the removal of the tree. If it did, then that would clearly be a factor against confirmation, having regard to the provisions of Schedule 2, para 3(c), to the order. As it was, of course, and as the district planning officer noted, the application had in fact been amended to avoid the tree. This comment, as it seems to me, is thus to be regarded as no more than an anticipatory response by officers to the committee's likely question whether or not it was expedient to confirm the order having regard to the actual development proposals for the site.

The real ground of objection that had been canvassed by the company was that, if there were to be development in the immediate vicinity of this tree, safety required its felling. That was the essential threat posed by the developer in the light of that objection. The passage in the county forestry officer's comments upon which Mr Pickles lays great stress is to my mind to be read merely as an answer to that point. He is there saying that, given that development will not come closer than 10 metres (and I interpolate that, although the scaling off from the plans suggests, according to the information before me, that in fact such development comes to within 7.5 metres at one point, that cannot assist the challenge on the grounds in fact advanced), there was no reason in the interests of safety for the tree to be lost. The introductory words to that comment, namely: "If the site has been redrawn to retain the tree", again to my mind do no more than pre-empt any suggestion such as might have been canvassed on behalf of the developers or raised by the council that this was in fact a case where, because of para 3(c) of Schedule 2 to the order, it was inappropriate to confirm the order. That of course is the essence of what Mr Adamson was saying in the paragraph I have quoted from his affidavit.

I recognise the force of Mr Pickles' comment — in the light of the amended legislation governing the exercise of control over tree preservation whereby there is no appeal to the Secretary of State at the initial stage but only upon a refusal of the authority to consent to work being done to protected trees —that it is particularly important that the council give proper consideration to the making of the order in the first place. I am satisfied, however, that in this instance they did. I reject the contention that they took account of immaterial considerations. In particular, I am wholly unsatisfied that the committee or council here improperly put into the balance as a positive factor in favour of confirmation the mere fact that the tree was not directly to be damaged by the then proposed development.

In these circumstances the application fails and must be dismissed.

The application was dismissed with costs.

Queen's Bench Divisional Court

March 30 1988

(Before Lord Justice BINGHAM and Mr Justice HUTCHISON)

PORTER v HONEY

Estates Gazette April 30 1988

[1988] 17 EG 129-135

Town and country planning — Control of advertisements — Estate agents' sale boards — Offences — Whether an estate agent commits an offence if, after he has lawfully erected a sale board, another agent, with or without the former's knowledge, erects a second board — Conflict of opinion among local authorities — A "hard rule" imposed by Parliament — The appellant estate agent appealed against his conviction by magistrates under the Town and Country Planning (Control of Advertisements) Regulations 1984 for displaying a sale board without consent

The appellant was in fact charged under two informations relating to two different houses — In each case after the appellant had placed a sale board on the property in question other agents (different agents in each case) placed a second board on it — The offence alleged against the appellant was that, in contravention of regulation 6 of the 1984 regulations, he had displayed an advertisement without the express or deemed consent of the local planning authority — There was

A **no question of any express consent and the deemed consent mentioned in the regulations was restricted to one advertisement consisting of a board of the prescribed dimensions — The magistrates concluded that deemed consent under the regulations was for the first and only board, not for one board per advertiser — Once a second board was added no deemed consent existed; it did not continue to be attached to the first board — The divisional court affirmed the magistrates' decision — The court pointed out that the 1984 regulations plainly limited the owner of the property to one board and in the court's view the same limitation applied to**
B **the estate agent — It could not be argued that each agent had displayed only one board in respect of his sale, because he was not selling at all but merely advertising the owner's willingness to sell**

The court rejected a submission that no offence was committed by the appellant in the absence of knowledge by him of the facts giving rise to the offence, namely, knowledge that he was one of two persons each displaying a separate advertisement in respect of a single sale — The answer to this was that the offence was one of absolute liability; there was no requirement of the existence of *mens rea* — Held accordingly that the appellant was properly convicted even though he
C **lawfully displayed the first advertisement with deemed consent and was in ignorance of the arrival of the second advertisement which destroyed that consent — It might seem strange that an act, lawful when done, could become unlawful as a result of something done by a stranger — However, it was noteworthy that new regulations, to come into operation in October 1988 (Town and Country Planning (Control of Advertisements) (Amendment No 2) Regulations 1987) (SI 1987 no 2227), showed in clearer language that Parliament was willing to impose a hard rule — Appeal dismissed (public interest certificate granted but leave not given to appeal to**
D **House of Lords)**

***Per* Hutchison J: It may be thought that if, despite penalties to which estate agents are subjected, offences are widespread, they would be more effectively curtailed if house-owners were also prosecuted**

The following cases are referred to in this report.

Maidstone Borough Council v *Mortimer* [1980] 3 All ER 552; (1980) 43 P&CR 67; [1980] EGD 941; 256 EG 1013, DC

R v *Wells Street Metropolitan Stipendiary Magistrate, ex parte Westminster City Council* [1986] 1 WLR 1046; [1986] 3 All ER 4, DC

This was an appeal by A H Porter, of Whitman Porter, estate
E agents, against his conviction by Richmond magistrates for two offences of displaying estate agents' sale boards without the consent of the local planning authority, contrary to regulation 6 of the Town and Country Planning (Control of Advertisements) Regulations 1984 (SI 1984 no 421). The prosecutions were brought by Michael John Honey, chief executive and town clerk of the London Borough of Richmond upon Thames.

T Bergin (instructed by Axelrods, Richmond, Surrey) appeared on behalf of the appellant; David Lamming (instructed by G S McGowan, solicitor to the London Borough of Richmond upon Thames) represented the respondents.

Giving judgment, BINGHAM LJ said: This appeal raises a vexed
F problem on the planning control of estate agents' boards. If an estate agent, instructed to sell or let a residential property, displays a board outside the property, no other board being at that time displayed, does he commit an offence if another agent (to his knowledge or without it) thereafter displays a second board?

I described this problem as "vexed" because different local authorities have favoured different solutions. Amended regulations, to take effect in October 1988, will impose a statutory solution. But I understand that a number of cases await the outcome of this one, so the case is of some general importance.

I would like at the outset to acknowledge our indebtedness to counsel for their clear, well-prepared and economical arguments. Both arguments were models of appellate advocacy. If we have reached the wrong answer, it is through no fault of counsel.

The appellant is an estate agent. The respondent is the chief G
executive and town clerk of the London Borough of Richmond upon Thames. In about November 1986 the appellant displayed a "For Sale" board outside 18 The Vineyard, Richmond. In about January 1987 he displayed a "For Sale" board outside 25-28 The Hermitage, Richmond. He erected his boards before any other was erected at either house. A second board was later erected at each house by another agent, so that there were two boards displayed at 18 The Vineyard on November 18 1986 and at 25-28 The Hermitage on January 25 1987. The appellant may have been aware of the existence of a second board at 18 The Vineyard before February 23 1987 when summonses were issued. He did not know if he was aware of the second board at 25-28 The Hermitage before that date.

Two informations were laid against the appellant by the local H
authority, one relating to each house. Each charged that the appellant:

> did . . . display an advertisement, namely an Estate Agents "Sale" Board being one of two such boards displayed at the said premises without consent having been granted in that behalf in accordance with Regulation 6 of the Town and Country Planning (Control of Advertisements) Regulations 1984 contrary to Regulation 8 of the said Regulations and section 109(2) of the Town and Country Planning Act 1971.

The informations were heard by Richmond justices on May 13 1987. They found the cases proved. The appellant accordingly appeals against those findings by way of case stated.

I shall refer to the Town and Country Planning Act 1971 as "the J
Act" and to the Town and Country Planning (Control of Advertisements) Regulations 1984 (SI 1984 no 421) as "the 1984 Regulations".

Section 63 of the Act provides for regulations to be made "for restricting or regulating the display of advertisements so far as appears to the Secretary of State to be expedient in the interests of amenity or public safety". Such regulations may provide "for requiring the consent of the local planning authority to be obtained for the display of advertisements" (section 63(2)(b)).

The offence of which the appellant was convicted was created by section 109(2) of the Act:

> Without prejudice to any provisions included in regulations made under section 63 of this Act by virtue of subsection (1) of this section, if any person K
> displays an advertisement in contravention of the provisions of the regulations he shall be guilty of an offence and liable on summary conviction to a fine of such amount as may be prescribed by the regulations, not exceeding £100 and, in the case of a continuing offence, £5 for each day during which the offence continues after conviction.

Regulation 8 of the 1984 regulations increased the maximum penalty on contravention.

Section 109(3) of the Act is also important:

> For the purposes of subsection (2) of this section, and without prejudice to the generality thereof, a person shall be deemed to display an advertisement if — (*a*) the advertisement is displayed on land of which he is the owner or occupier; or (*b*) the advertisement gives publicity to his goods, trade, business or other concerns: L
> Provided that a person shall not be guilty of an offence under that subsection by reason only that an advertisement is displayed on land of which he is the owner or occupier, or that his goods, trade, business or other concerns are given publicity by the advertisement, if he proves that it was displayed without his knowledge or consent.

The heart of the problem before the court lies in the 1984 Regulations. Regulation 6(1)(2) and (3), so far as relevant, provide:

> (1) No advertisement may be displayed without consent granted by the local planning authority or by the Secretary of State on an application in that behalf (referred to in these Regulations as "express consent"), or deemed to be granted under paragraph (2) below.
> (2) Consent shall be deemed to be granted for the display of any advertisement displayed in accordance with any provision of these M
> Regulations whereby advertisements of that description may be displayed without express consent; . . .
> (3) In so far as the nature of the consent permits, consent for the display of advertisements shall enure for the benefit of the site to which the consent relates and of all persons for the time being interested in that site; . . .

No express consent was in this case sought or given, so one must inquire whether under the 1984 regulations consent is to be deemed to have been granted. Regulation 14 specifies the classes of advertisements which may be displayed without express consent. Only one class is relevant. This is Class III headed "Certain advertisements of a temporary nature". Para (a) reads:

> Advertisements relating to the sale or letting of the land on which they are displayed; limited, in respect of each such sale or letting, to one advertisement

A consisting of a board (whether or not attached to a building) not exceeding 2 square metres in area, or of two conjoined boards, together not exceeding 2.3 square metres in area; . . .

Paras (b) and (c), dealing with sales of goods or livestock and building works, contain a similar limitation to one advertisement. Paras (d) and (e), dealing with certain local events, specify a maximum area of advertisement only.

Our attention was drawn to the Town and Country Planning (Control of Advertisements) (Amendment No 2) Regulations 1987 (SI 1987 no 2227). By virtue of regulation 3 of and Schedule 2 to those regulations, para (a) of Class III of regulation 14(1) of the 1984 regulations is, with effect from October 28 1988, to be replaced by a

B paragraph in these terms:

(a)(i) An advertisement relating to the sale or letting, for agricultural, industrial or commercial use or for development for such use of the land or premises on which it is displayed consisting of a single board not exceeding 2 square metres in area, or of two joined boards, together not exceeding 2.3 square metres in area, which, if displayed on a building, does not project from the face of the building by more than one metre.

(ii) An advertisement relating to the sale or letting for residential use or for development for such use of the land or premises on which it is displayed consisting of a single board not exceeding 0.5 square metre in area, or of two joined boards together not exceeding 0.6 square metre in area, which, if displayed on a building, does not project from the face of the building by more than one metre:

Provided that nothing in paragraph (i) or (ii) — (i) shall permit more than one

C advertisement to be displayed at any one time on the land or premises concerned: (ii) shall authorise the display of an advertisement indicating that land or premises have been sold or let other than by the addition to an existing advertisement of a statement that a sale or letting has been agreed or that the land or premises have been sold or let, subject to contract.

In this paragraph — "area" means the area of a single surface, whether or not both surfaces of the relevant board are used for advertising; and "joined boards" means boards joined at an angle so that only one surface of each is usable for advertising.

This amendment was relied on by the appellant as showing that under the 1984 regulations the limit is one board per advertiser, not one board per house. Otherwise there would have been no need for a change. I think this submission should be treated with caution. The amendment may well represent a departmental view on the effect of

D the 1984 regulations. But that view may not be correct, and the varying interpretations put on the 1984 regulations by different local authorities are good enough reason for putting the matter beyond doubt. So I think we must concentrate our attention on the proper construction of the Act and the 1984 regulations.

The appellant contended that the offence created by section 109(2) of the Act was one of displaying an advertisement in contravention of the 1984 regulations. Regulation 14 of the 1984 regulations defined the classes of display which were not to contravene the prohibition on display without consent. Class III(a) contained a limitation on display, to one advertisement per sale. The appellant was accordingly not in breach of the 1984 regulations because he displayed an

E advertisement of the type described in Class III(a) and restricted himself to display of one such advertisement in respect of his sale.

The respondent contended that Class III(a) limited deemed consent to one sale board per house. Even if several estate agents were instructed by the owner to find a purchaser, it was still one and the same sale. That construction accorded with the purpose of the 1984 regulations, which was (among other things) to prevent streets becoming disfigured by a proliferation of estate agents' boards.

The justices concluded that the consent deemed to be granted under Class III(a) of regulation 14 could not be for one board per advertiser. There would be deemed consent for the first and only board, but no deemed consent once a second board was displayed. I agree with that conclusion.

F The offence created by section 109(2) of the Act is to display an advertisement in contravention of the provisions of the 1984 regulations. There is a contravention of the 1984 regulations if an advertisement is displayed without express or deemed consent. There being no express consent, the appellant had to show deemed consent. Consent for the display of his board would be deemed to be granted only if his display was in accordance with any provision of the 1984 regulations permitting display of his board without express consent. Class III(a) of regulation 14 permits display without express consent of advertisements relating to the sale or letting of the land on which they are displayed limited to one advertisement in respect of each such sale or letting. The vendor or lessor is the building owner. He can only sell or let once. The 1984 regulations plainly limit the building

G owner to one board. In my view they limit estate agents similarly. If a second board is displayed the building owner and both agents commit an offence because the limitation to one advertisement in respect of each such sale or letting is exceeded. It cannot be said that each agent has only displayed one board in respect of his sale or letting, because he is not selling or letting at all but merely advertising the building's owner's willingness to sell or let. It is, initially, surprising that an act, lawful when done, can become unlawful as the result of something done by a stranger. But that is, in my view, the effect of the 1984 regulations, and the 1987 amendment, in my view, shows that Parliament is willing to impose a hard rule to achieve a result regarded as desirable.

H It was not argued (I think rightly) that a first board erected with deemed consent continues to enjoy that deemed consent even after the erection of a second board if the second board was erected without deemed consent.

The appellant advanced an alternative argument. If the correct interpretation of regulation 14 Class III(a) is that only one advertisement in total is permissible per sale or letting, then (it was said) no offence can be committed in the absence of knowledge by the defendant of the facts giving rise to the offence, namely that the defendant is one of two persons each displaying a separate advertisement in respect of a single sale or letting.

The respondent submitted that section 109(2) created an absolute offence. Despite the general presumption that a guilty mind must be

J proved to establish criminal liability, it was argued that other sections of the Act have been construed as not requiring proof of a guilty mind: *Maidstone Borough Council* v *Mortimer* [1980] 3 All ER 552; *R* v *Wells Street Metropolitan Stipendiary Magistrate, ex parte Westminster City Council* [1986] 1 WLR 1046. The proviso to section 109(3), it was argued, showed that an offence under section 109(2) was not to require proof of *mens rea*.

The justices held that section 109(2) created an absolute offence. I agree with them. The offence is committed "if any person displays" not "if any person knowingly displays". The very limited defence provided by the proviso to section 109(3), if the defendant can establish it, is inconsistent with any general requirement of *mens rea*. I cannot think Parliament intended to burden local authorities with

K the task of showing what one estate agent knew about the activities of another.

In agreement with the justices, I conclude: (1) that deemed consent is limited to one advertisement in total per sale or letting; (2) that a person alleged to be displaying in contravention of the 1984 regulations may be properly convicted even though he displayed his own advertisement with deemed consent and is ignorant of the existence of a second board.

I would, accordingly, dismiss this appeal.

Agreeing, HUTCHISON J said: It seems at first sight surprising that an estate agent who lawfully erects a board can find himself guilty of an offence when, subsequently and without his knowledge, another

L agent erects a further board. But I am persuaded, particularly on a consideration of the proviso to section 109(3), that Parliament must have intended that proof of *mens rea* was not required in the case of offences under section 109(2).

It would seem from what we have been told in this case that in general local authorities prosecute estate agents rather than prospective vendors. The latter, of course, since a board can only be erected by an agent with the consent of his principal (and, anyway, such boards are obvious) will invariably have knowledge of the commission of the offence. It may be thought that if, despite penalties to which estate agents are subjected, offences are widespread, they would be more effectively curtailed if houseowners were also prosecuted.

M Estate agents prepared to take the risk of erecting a second board, if they alone faced prosecution, might well be deterred if they knew that their principals might suffer the same indignity.

The appeal was dismissed with costs. Certificate was granted as to point of general public importance, but leave to appeal refused.

Court of Appeal

March 10, 1988

(Before Lord Justice SLADE, Lord Justice RUSSELL and Lord Justice TAYLOR)

BELL v CANTERBURY CITY COUNCIL

Estates Gazette June 4 1988

[1988] 22 EG 86-94

Town and Country Planning Act 1971 — Tree preservation order — Compensation claimed by landowner for refusal of permission to fell trees on part of woodland subject to a tree preservation order, the owner's intention being to grub out and clear scrub and coppice with a view to conversion into grazing land — Dispute as to correct basis of compensation — Appeal by local authority from decision of Lands Tribunal (Mr C R Mallett FRICS) — It had been agreed that if the approach of the tribunal were correct the compensation (being a variation of the tribunal's award) should be £38,851.27 — If, however, the appellant authority succeeded in their claim that compensation was limited to the commercial value of the timber which the landowner could not realise, the actual amount would be nil after appropriate deductions for acquisition and sale

The facts of the case were that after the landowner had begun to clear an area of woodland which he had recently purchased the local authority served a tree preservation order on some 39 acres of it in the form prescribed by the Town and Country Planning (Tree Preservation Order) Regulations 1969 — The order prohibited felling without consent — Article 6 provided that (with certain exceptions) the authority were required to give a direction to replant when consent was granted — Article 9 provided for compensation in respect of loss or damage in consequence of refusal of consent — The landowner's application for permission to grub out the 39 acres having been refused, he referred his claim for compensation to the Lands Tribunal — Before the hearing by the tribunal the authority passed a resolution to the effect that if they had given consent it would have been accompanied by a direction to replant under article 6 — The authority thus hoped to defeat any claim to compensation based on diminution of land value

The Lands Tribunal upheld the landowner's contention that compensation should be based on the diminution in the value of the land and was not restricted to loss caused by inability to fell timber — The tribunal found that the value of the land in its present state was £450 per acre, whereas if reclamation for grazing had been permitted it would have been £1,900 per acre

The Court of Appeal agreed with the tribunal's approach — They held as follows: (1) Article 6 derived its authority from section 60 (1) (*b*) of the 1971 Act and was confined to cases where the felling was "in the course of forestry operations", which did not apply here; the appellant authority were under a misconception if they thought that a direction to replant followed automatically upon giving consent; (2) Even if it had been open to the appellants to give such a direction its exercise contemporaneously with the granting of consent would have been unreasonable on the principles of *Associated Provincial Picture Houses Ltd* v *Wednesbury Corporation*; (3) Compensation for refusal of consent to felling was derived from article 9 of the 1969 regulations, which were made under the enabling powers of section 174 of the 1971 Act; "loss or damage in consequence of a refusal of consent" covered the diminution in the value of the claimant's land — There was nothing in an argument by the appellants that the claimant's loss was a consequence of the order itself as distinct from the refusal of consent — The loss was a natural and probable consequence of the refusal and was not too remote — Appeal dismissed

The following cases are referred to in this report.

Amalgamated Investment & Property Co Ltd v *John Walker & Sons Ltd* [1977] 1 WLR 164; [1976] 3 All ER 509; (1976) 32 P&CR 278; [1976] EGD 166; 239 EG 277, CA

Associated Provincial Picture Houses Ltd v *Wednesbury Corporation* [1948] 1 KB 223; [1947] 2 All ER 680, (1947) 45 LGR 635, CA

Hoveringham Gravels Ltd v *Secretary of State for the Environment* [1975] QB 754; [1975] 2 WLR 897; [1975] 2 All ER 931; (1975) 73 LGR 238; 30 P&CR 151; [1975] EGD 496; 235 EG 217 & 295, CA

This was an appeal by the compensating authority, Canterbury City Council, by way of case stated, from a decision of the Lands Tribunal awarding compensation to the claimant, Peter Joseph Bell, the present respondent, on the basis of the diminution in the value of his land following a refusal of permission to fell trees on a woodland subject to a tree preservation order. The woodland was part of the respondent's farm, Great Palmstead Farm, Upper Hardres, Kent. The decision of the Lands Tribunal was reported at [1986] 2 EGLR 289; (1986) 279 EG 767.

Robert Carnwath QC and Edward Cohen (instructed by Mr M G P Young, solicitors' department, Canterbury City Council) appeared on behalf of the appellants; Jonathan Milner (instructed by Kingsford Flower & Pain, of Ashford, Kent) represented the respondent.

Giving the first judgment at the invitation of SLADE LJ, RUSSELL LJ said: This case is concerned with the level of compensation to be awarded to a landowner who is refused permission by a local authority to fell trees on his land, when such land is subject to a tree preservation order. The appellants, the Canterbury City Council, contend that compensation is basically limited to the commercial value of the timber which the landowner cannot realise because of the refusal of permission. The respondent, a farmer, contends that if the value of the land itself is diminished as a result of the refusal of permission, such diminution in value of his interest in the land is a proper head of claim. The matter comes before this court on appeal by way of case stated from the Lands Tribunal, the member of the tribunal, C R Mallett Esq FRICS, having found in favour of the respondent's submissions as to the basis of compensation and having rejected the narrower approach of the local authority. Other supplementary matters were originally in dispute between the parties, but it is now agreed that if the basic approach of the Lands Tribunal was appropriate, the award of compensation should be £38,851.27. If the appeal succeeds, then the value of the timber involved is such that, after making the appropriate deductions for its acquisition and sale, no compensation is due to the claimant.

The facts fall within a comparatively small compass and are, for the most part, contained in the written decision of the tribunal. Mr P J Bell, the claimant, purchased Great Palmstead Farm, Upper Hardres, in 1970. It comprised a stock and arable farm of some 114 acres including a dilapidated farmhouse and derelict farm buildings. Mr Bell sought to improve the holding and grant approval was sought and obtained from the Ministry of Agriculture, Fisheries and Food to a six-year development plan dated July 31 1980, under the Agriculture and Horticulture Development Scheme.

In 1981 Mr Bell purchased some 88 acres of adjoining woodland. After consulting with his professional advisers, he applied to the Ministry for a variation in the development plan involving clearing scrub and coppice from 50 acres of the woodland and converting it to grazing land. He purchased a bulldozer and proceeded to clear two acres of woodland, whereupon the Canterbury City Council served on him the Canterbury City Council (Quilters Wood, Middle Wood, Featherley Wood, part of Mill Bank Wood and part of Atchester Wood, Kingston/Upper Hardres) Tree Preservation Order No 3 1982. Service of the order was effected on May 27 1982 before it was confirmed by the city council, but there is a provision in the Town and Country Planning Act 1971 (as amended) for this to be done in cases of urgency. The order was in the form prescribed by the Town and Country Planning (Tree Preservation Order) Regulations 1969 and, in general terms, it provided that, except with the consent of the authority and in accordance with conditions (if any) imposed on such consent, no person could cut down any tree specified in the first schedule to the order, or comprised in a woodland therein specified. The first schedule included what ultimately became the area involved

A in this appeal, namely Quilters Wood and Featherley Wood. Together they amounted to 39.1 acres of woodland.

In 1982 the claimant was successfully prosecuted for grubbing out 2.9 acres of the land covered by the order without permission.

On March 30 1983 Mr Bell submitted an application to the authority to grub out the 39.1 acres and, in accordance with the requirement of article 3 of the order, he indicated in writing to the authority his reasons for making the application, namely, to provide grazing for "a beef/sheep enterprise". In the application form itself, the proposed operation was described as "to grub out the trees so that the land may be used for agriculture".

On June 15 1983 permission was refused by the local authority on amenity grounds and also because the proposed works were said to be
B contrary to the Kent Structure Plan. The claimant appealed, but after a local public inquiry in February 1984 the Secretary of State dismissed the appeal.

On May 29 1985 the claimant referred his claim for compensation to the Lands Tribunal. Before the hearing, on November 27 1985, the Canterbury City Council passed a resolution in the following terms:

> That, as a matter of urgency because of the time limit involved, it be confirmed that had the Council granted consent to fell or grub out trees on part of the woodland the subject of the above Tree Preservation Order, a direction would have been made to replant the land in accordance with article 6 of the Tree Preservation Order.

C It will be necessary to return to this resolution later in this judgment. The decision of the Lands Tribunal is dated June 5 1986, after a hearing which occupied some four days. The fundamental basis of Mr Mallett's award was that the current capital value of the land in its present state was £450 per acre, whereas its value, if reclamation for grazing land had been permitted, would have been £1,900 per acre.

I turn to the statutory provisions affecting the appeal. Section 60 of the Town and Country Planning Act 1971 provides, so far as it is material:

> (1) If it appears to a local planning authority that it is expedient in the interests of amenity to make provision for the preservation of trees or woodlands in their area, they may for that purpose make an order (in this Act
> D referred to as a "tree preservation order") with respect to such trees, groups of trees or woodlands as may be specified in the order; and, in particular, provision may be made by any such order —
> (*a*) for prohibiting (subject to any exemptions for which provision may be made by the order) the cutting down, topping, lopping or wilful destruction of trees except with the consent of the local planning authority, and for enabling that authority to give their consent subject to conditions;
> (*b*) for securing the replanting, in such manner as may be prescribed by or under the order, of any part of a woodland area which is felled in the course of forestry operations permitted by or under the order; . . ..

The tree preservation order with which this case is concerned contained the following material articles to which hitherto I have made no reference. Article 4 is in these terms:

> E (1) Where an application for consent is made to the authority under this order, the authority may grant such consent either unconditionally, or subject to such conditions (including conditions requiring the replacement of any tree by one or more trees on the site or in the immediate vicinity thereof), as the authority may think fit, or may refuse consent.
> Provided that where the application relates to any woodland specified in the First Schedule to this Order the authority shall grant consent so far as accords with the principles of good forestry, except where, in the opinion of the authority, it is necessary in the interests of amenity to maintain the special character of the woodland or the woodland character of the area, and shall not impose conditions on such consent requiring replacement or replanting.

This article, therefore, expressly precluded the appellants from granting any consent conditional upon the replacement or replanting of trees in the woodlands with which this case is concerned. So much
F was acknowledged by counsel for the local authority.

Reliance, however, was placed upon article 6, which reads as follows:

> (1) Where consent is granted under this order to fell any part of a woodland other than consent for silvicultural thinning then unless —
> (*a*) such consent is granted for the purpose of enabling development to be carried out in accordance with a permission to develop land under Part III of the Act, or
> (*b*) the authority with the approval of the Secretary of State dispense with replanting,
> the authority shall give to the owner of the land on which that part of the woodland is situated a direction in writing specifying the manner in which and the time within which he shall replant such land and where such a direction is given and the part is felled the owner shall, subject to the provision of this order and section 175 of the Act, replant the said land in accordance with the direction. G

There follows an immaterial subparagraph dealing with the requirements of such a direction.

As to (*a*), it was not submitted to the Lands Tribunal that the operation contemplated by Mr Bell did amount to development under Part III of the Act (see section 22(2)). As to (*b*), there had, of course, been no dispensation approved by the Secretary of State.

In my judgment article 6 plainly derives its authority from section 60(1)(*b*) of the Act, and by virtue of its provisions is confined to cases where any part of a woodland area "is felled in the course of forestry operations". I strongly suspect that in the instant case the local authority felt that they were obliged, by virtue of article 6, to give the direction to replant in the event of giving consent. If that was the H motivation behind the passing of the resolution of November 27 1985 it was misconceived; if it was not, it is difficult to understand how any local authority could reasonably give consent to the felling of trees for the purposes of reclaiming the land for grazing while, at the same time, requiring replanting of those trees, thereby defeating the whole purpose of the operation and the object of the consent. In my view the position would be quite different in what I understand is meant by the phrase "forestry operations". No statutory definition of the phrase is to be found in the statute or regulations made under it, but I am firmly of the opinion that what was contemplated by the claimant did not involve forestry operations and that consequently article 6 did not require the local authority to serve a direction. That the article J was intended to relate exclusively to forestry operations is, in my judgment, clear not only from the provisions of section 60 itself but from the wording of article 6. "Silvicultural thinning" referred to in article 6 is an activity confined to what one would normally regard as forestry operations.

The local authority plainly took the view that by passing the resolution to which I have referred they were effectively preventing the claimant from mounting any claim based upon differences in land values, because no diminution in land value could arise if, when giving consent, there was a contemporaneous direction to replant. I repeat that, in my opinion, the proposals of the claimant did not constitute forestry operations which, I venture to suggest, involve the commercial exploitation of timber. K

That is enough to dispose of the submission made on behalf of the appellants that the effect of the resolution was to defeat Mr Bell's claim for compensation but, even if the giving of a direction to replant had been open to the appellants, its exercise to take effect contemporaneously with the consent would, in my judgment, have been *Wednesbury* unreasonable — as the Lands Tribunal concluded.

Before leaving this part of the case, it is worth observing that compensation is not payable in every case where the local authority refuses consent. Article 5 provides:

> Where the authority refuse consent under this order or grant such consent subject to conditions they may when refusing or granting consent certify in respect of any trees for which they are so refusing or granting consent that they are satisfied — L
> (*a*) that the refusal or condition is in the interests of good forestry; or
> (*b*) in the case of trees other than trees comprised in woodlands, that the trees have an outstanding or special amenity value.

The proviso to article 9 reads:

> . . . no compensation shall be payable in respect of loss or damage suffered by reason of such refusal or grant of consent in the case of any trees the subject of a certificate in accordance with Article 5 of this order.

I move on to consider other submissions of counsel. Section 174 of the 1971 Act provides:

> The matters for which provision may under section 60 of this Act be made by a tree preservation order include the payment of the local planning authority, subject to such exceptions and conditions as may be specified in the order, of M compensation in respect of loss or damage caused or incurred in consequence of the refusal of any consent required under the order, or of the grant of any such consent subject to conditions.

This, then, is the enabling provision. It is not the statute itself which sets out the terms upon which compensation is to be payable but the tree preservation order.

Article 9, the proviso to which I have referred, reads:

> Subject to the provisions of this order, any person who has suffered loss or damage in consequence of any refusal (including revocation or modification) of consent under this order or of any grant of any such consent subject to conditions, shall, if he makes a claim on the authority within the time and in the manner prescribed by this order, be entitled to recover from the authority compensation in respect of such loss or damage:

As to the measure of compensation, the operative words are "loss or damage in consequence of any refusal of consent".

Counsel for the appellants drew our attention to various provisions in the 1971 Act dealing with the measure of compensation consequent upon other refusals of consent by local authorities exercising planning functions. It is true that in some cases there is express reference to the depreciation of value of an interest in land as a basis for compensation. The absence of any such provision in section 174, submits counsel, is significant. I disagree. Section 174 does not seek to define the measure of compensation; that is the function of article 9, the interpretation of which should, in my judgment, be read without reference to other provisions in the Act of Parliament.

Counsel for the respondent submits that the words of article 9 should be given their ordinary meaning and that the proved diminution in the value of the claimant's interest in his land amounts to "loss or damage in consequence of a refusal of consent". Nothing is to be found in the order which could justify limiting compensation in the way suggested by the local authority. Nor was any such limitation to be found in the enabling statute. I agree with these submissions and, for my part, I would reject the argument advanced on behalf of the local authority by their counsel to the effect that the loss suffered by the claimant as a result of his inability to reclaim his land for agricultural grazing was a consequence of the order itself, as opposed to the refusal which, it was contended, simply prevented the cutting down of trees.

In my view the diminution in the value of the land was a natural and probable consequence of the refusal of consent, within the contemplation of the local authority refusing consent. It was not too remote and was, therefore, recoverable. Its proper assessment is to be derived from the words of article 9, not section 174 — hence the exclusion of section 174 from the provisions of section 178, to which our attention was directed.

Counsel for the respondent did seek to argue that the grubbing-out operation was an engineering operation within Class VI of the Town and Country Planning General Development Order 1977 and that consequently it did amount to development under Part III of the Act so as to exclude the operation of article 6 in any event. For my part I have my reservations about this submission, not least because we have insufficient evidence to rule definitively upon it. The point was not taken before the Lands Tribunal, and in the light of the views I have earlier expressed as to article 6 and its applicability, it is unnecessary to prolong this judgment by any further discussion of the true nature of the claimant's operations in a planning context.

I hope that in this judgment I have answered the relevant questions posed at the end of the case. Because the parties have come to terms on matters of detail, some of the questions are no longer appropriate. I would dismiss the appeal.

Agreeing, TAYLOR LJ said: I would add only two further reasons in support of what Russell LJ has said.

The first concerns the scope of article 6 of the order. I agree that it is included in pursuance of section 60(1)(*b*) of the Act and is therefore concerned only with forestry operations. That view is supported by the text of the article itself. After referring to the direction to replant, it continues:

> . . . and where such a direction is given and the part is felled the owner shall, subject to the provision of this order and section 175 of the Act, replant the said land in accordance with the direction.

Section 175 of the Act provides for compensation where a direction to replant has been given. Subsection (1) reads:

> The provisions of this section shall have effect where, in pursuance of provision made by a tree preservation order, a direction is given, by the local planning authority or the Secretary of State, for securing the replanting of all or any part of a woodland area which is felled in the course of forestry operations permitted by or under the order.

Thus, article 6, by invoking section 175, clearly proclaims itself to be related to forestry operations.

My second point concerns Mr Carnwath's submission that it was the making of the tree preservation order itself and not the refusal of consent which precluded Mr Bell from converting his land for grazing. Thus the claim for compensation fails, it is said, on grounds of causation.

If this argument be right, then it would apply equally to compensation for the commercial value of the timber. The inability of Mr Bell to cut down his trees must result from the same causative factor whether his complaint is that he cannot sell his timber or that he cannot graze his livestock. Yet Mr Carnwath concedes that Mr Bell is entitled to compensation for loss of the sale value of his timber.

In my judgment, that involves a distinction which cannot logically be justified. I consider both heads of loss resulted from the refusal of consent, so that the causation argument fails.

Also agreeing, SLADE LJ said: I will add some observations of my own only out of deference to Mr Carnwath's argument.

The member of the Lands Tribunal in his decision described the issue before him as being "determination of compensation payable under section 174 of the Town and Country Planning Act 1971". Strictly, section 174 does not itself afford any entitlement to compensation. All it does is to authorise the inclusion of provisions for compensation of a specified nature in a tree preservation order made under section 60 of that Act.

Article 9 of the Tree Preservation Order No 3,1982 states:

> Subject to the provisions of this order, any person who has suffered loss or damage in consequence of any refusal (including revocation or modification) of consent under this order or of any grant of any such consent subject to conditions, shall, if he makes a claim on the authority within the time and in the manner prescribed by this order, be entitled to recover from the authority compensation in respect of such loss or damage:

Article 9 ends with a proviso which does not affect the claimant's rights to compensation in the present case.

On June 15 1983 the appellant council refused its consent under the order to Mr Bell's application to grub out about 39.1 acres of woodland so that the land might be used for agriculture. The Lands Tribunal found that the value of the land in its current state is £450 per acre, but the value of the land as reinstated pasture land would be £1,900 per acre.

Prima facie, therefore, the loss or damage suffered by the claimant in consequence of the refusal of consent under the order was the difference between the value of the land in its present state and its value as reinstated pasture land and was recoverable by him under article 9 accordingly. This was in substance the basis upon which the member made his award.

If I correctly followed Mr Carnwath's argument, there were three principal reasons why he submitted that the member erred in law in reaching this conclusion, though I do not summarise these reasons either in the same order or in the same terminology as he employed.

First, he referred us to a number of sections in the 1971 Act (for example, sections 164, 169, 170 and 171) which explicitly provide for a person to receive compensation for a depreciation in the value of his interest in land in the contingencies there specified. He also referred to section 178 which, with certain stated exceptions of which section 174 is one, renders the provisions of section 5 of the Land Compensation Act 1961 applicable for the purpose of assessing any compensation which, under the preceding provisions of Part VIII of the 1971 Act, is payable in respect of depreciation of the value of an interest in land. He submitted that these sections of the 1971 Act show that where the value of an interest in the relevant land falls to be considered as a potential element in compensation, the 1971 Act specifically refers to this element.

I do not think the submission advances the council's case. Section 178(2) provides:

> This section applies to any compensation which, under the preceding provisions of this Part of this Act other than section 174, 175 or 177 is payable in respect of depreciation of the value of an interest in land.

Section 174(2) thus shows that the legislature contemplated that a tree preservation order made under the authority conferred by sections 60 and 174 might contain provisions rendering compensation payable in respect of depreciation of the value of an interest in land. It is not suggested that article 9 of the tree preservation order in the present case was *ultra vires*. The only question, therefore, is whether, on its true construction, article 9 does render compensation payable in respect of such depreciation. The language of the article is very broad and, subject to what is said below, I can see no sufficient grounds for restricting its meaning so as to exclude compensation in respect of this particular category of loss.

Second, Mr Carnwath correctly pointed out that no compensation is available for the imposition of the tree preservation order itself. He submitted that the claimant in the present case had purchased the land in question subject to the inherent risk that such an order might be made, just as the plaintiffs in *Amalgamated Investment & Property Co Ltd* v *John Walker & Sons Ltd* [1977] 1 WLR 164 had bought their land subject to the inherent risk that the building could

A be listed as of architectural or historic interest. In his submission, it was the making of the tree preservation order itself, not the refusal of the consent to felling, which precluded the claimant from converting his land to agriculture. In these circumstances, it was submitted, the refusal of consent caused no depreciation of the claimant's interest in the land. By way of analogy, some reliance was placed on the decision of this Court in *Hoveringham Gravels Ltd* v *Secretary of State for the Environment* [1975] QB 754.

In my judgment, the short answer to this point is that the tree preservation order carried with it (under article 9) the right for the claimant to receive the compensation properly due to him under the article, if and when an application was thereafter made under the B order to fell trees and that application was refused. Only when his subsequent application was refused on June 15 1983 did his loss occur — and his corresponding right to compensation arise.

It is now common ground that the amount of the compensation (if any) falls to be assessed as at the date of the refusal of consent, namely, June 15 1983.

Third, and as his principal point, Mr Carnwath submitted that even if his other submissions were ill-founded, the claimant, as at June 15 1983, had suffered no loss and accordingly would be entitled to no compensation because the council, if it had given consent, would have been entitled and bound to give him a direction for replanting of the relevant land under article 6 of the tree preservation C order. The council's resolution of November 27 1985, in his submission, accurately reflected what would have happened if consent had been given. I agree with Russell LJ and Taylor LJ that article 6, while not expressly so stating, must be confined to cases where consent is granted under the order to the felling of a part of a woodland in the course of forestry operations. Though Mr Carnwath drew our attention to the general words at the beginning of section 60(1) of the 1971 Act, I think it clear that article 6 was made pursuant to the authority given by the particular words of section 60(1)(*b*), and with those particular words in mind. The exception of "silvicultural D thinning" in article 6 lends support to the view that the article is dealing with consent given for the purpose of forestry operations. So does the reference in article 6 to section 175 of the 1971 Act, as Taylor LJ has explained.

I can see no compelling reason why the authority should need a power to impose conditions or directions requiring the replacement of trees comprised in a woodland where it gives its consent to the felling of such trees for the purpose of *non-forestry* operations, and I do not think that the prescribed form of order gives it any such power.

In cases where consent is given to the felling of trees not comprised in woodlands, the position in regard to replacement is sufficiently covered by article 4, which empowers the authority, in giving E consent, to impose conditions requiring the replacement of trees. In general terms, article 5 of the order (quoted in Russell LJ's judgment), read together with article 9, seems to me to indicate the clear intention of the order that a person who has suffered loss or damage in consequence of any refusal of consent under the order, or of any grant of consent subject to conditions, *in respect of trees comprised in woodlands*, shall be entitled to compensation unless the refusal or condition is in the interests of good forestry, and the authority has so certified under article 5(*a*). (The certificate procedure under article 5(*b*) is not available in respect of trees comprised in woodlands.)

For these reasons I am unable to accept what I have understood to F be the three principal submissions advanced by Mr Carnwath. As regards the other points arising on this appeal, I do not wish to add anything to the judgment of Russell LJ.

I would vary the decision of the Lands Tribunal by substituting a figure of £38,851.27 for the sum awarded to the claimant by way of compensation but, subject to this variation, would dismiss this appeal.

The appeal was dismissed with costs. Leave to appeal to the House of Lords was refused.

LANDS TRIBUNAL

COMPENSATION

Compensation claims

ADP & E FARMERS v DEPARTMENT OF TRANSPORT (REF/201/1984)

February 3 1988

(C R MALLETT FRICS)

Estates Gazette May 7, 14 and 21 1988

[1988] 18 EG 80-84, 19 EG 147-150 and 20 EG 104-108

Severance damage — Strip of land (red land) fronting on to Exeter bypass acquired in 1973 for M5 spur road — Retained land (blue land) ripe, in general terms, for housing development and as such valued at agreed £42,500 per acre — Claim for £750,079 on assumption of no prospect of development, alternatively for £726,647 if *Bwllfa* principle not applicable — Acquiring authority's figure of £98,010 based on assumption that alternative means of access available — Planning permission for residential development of subject land refused in 1962, 1967 and 1976 — Certificate of appropriate alternative development issued in 1983 by Secretary of State for the Environment under s18 of Land Compensation Act 1961

Major issues concerning application of s6 of Land Compensation Act 1961 and *Pointe Gourde* rule and what account should be taken of s18 certificate relating to retained land — Claimants' valuer argued that on severance of subject land there remained no practicable access to permit residential development — Only remote possibility that other developments or road improvements would eventually enable retained land to be developed — Therefore, in adopting *Bwllfa* principle he excluded "hope" value — Acquiring authority's valuer assumed that bypass would never have been free of bypass traffic in no-scheme world and would never have been detrunked — S18 certificate certified that conditional planning permission would have been granted for acquired land not sooner than 1976 — Conditions referred *inter alia* to means of access — S18 certificate for adjoining acquired land similarly certified for development not before 1981 — Reasonable to assume Secretary of State contemplated development of red and blue land would have been permitted when bypass detrunked and that development of adjoining land would have followed some years later — Claimants contended s18 certificates on both areas of acquired land would not have been issued if scheme for compulsory purchase had been thought to include any part of M5, since in absence of scheme there would have been no relief of bypass traffic and therefore no detrunking — Tribunal observes that certificate has no statutory force in relation to retained land but matters of fact or judgment which lead to its grant for acquired land may well be strong or decisive evidence in relation to retained land — *Abbey Homestead* principles followed — Tribunal not empowered to vary or go behind certificate — Therefore, tribunal concludes that values in before-acquisition world of red and blue land properly assessed on basis that planning permission for residential development would have been obtained not earlier than 1976 — Reasonable and proper assumption that access would have been obtained from bypass after detrunking — Tribunal would have reached similar conclusion apart from binding effect of s18 certificate

In real-world situation after acquisition blue land virtually incapable of being developed unless access arranged over red land — Some 12 years after acquisition a highway engineer asked by acquiring authority to find solution — What finally emerged was less-than-ideal "massive" roundabout probably well beyond capability of residential developer, costing over 10 times more than normal junction for residential development and with various other difficulties — Wholly incorrect, in tribunal's view, to calculate compensation for damages on alternative and imaginary scheme which in effect alters and extends scheme which gave rise to acquisition — As a matter of fact the massive roundabout a wholly improbable solution and unlikely to affect value of blue land — Prospect of gaining access by certain other routes too remote

Tribunal prefers claimants' valuer's approach of valuing interests separately "as statutes require", ie acquisition under rule (2) and damage under s7 of Compulsory Purchase Act 1965, rather than acquiring authority's valuer's approach of in effect valuing red and blue land together and apportioning result pro rata — *Stokes* v *Cambridge* payment at approximately one-third of full residential value of blue land — No deduction made for buffer strip on bypass frontage — No deduction for deferment — Proper to add agreed amounts for value of agricultural use — As matter of law *Bwllfa* principle capable of being applied in this case, but as matter of fact it should not — Compensation of £726,647 awarded

Costs — Introduction of "massive" roundabout scheme at 11th hour and not in final form until case part heard represented fundamental change in acquiring authority's approach to compensation assessment — Claimants put to "exceptional" cost — Costs awarded to claimants on "standard" basis with additional cost incurred because of late notice of scheme compensated on indemnity basis under RSc Ord 62 r 3(4) proviso

The following cases are referred to in this report.

Abbey Homesteads Group Ltd v *Secretary of State for Transport* [1982] EGD 795; (1982) 263 EG 983, 1095 & 264 EG 59, 151, LT

Bartlett v *Barclays Bank Trust Co Ltd (No 2)* [1980] Ch 515; [1980] 2 WLR 430; [1980] 2 All ER 92

Bolton Metropolitan Borough Council v *Waterworth* [1981] 2 Lloyd's Rep 625; [1981] EGD 586; (1981) 259 EG 625; [1982] JPL 33, CA

Bowen-Jones v *Bowen-Jones* [1986] 3 All ER 163

Bwllfa & Merthyr Dare Steam Collieries (1891) Ltd v *Pontypridd Waterworks Co* [1903] AC 426

Carslogie Steamship Co Ltd v *Royal Norwegian Government* [1952] AC 292; [1952] 1 All ER 20, HL

East End Dwellings Co v *Finsbury Borough Council* [1952] AC 109; [1951] 2 All ER 587; (1951) 2 P&CR 135; 49 LGR 669, HL

Margate Corporation v *Devotwill Investments Ltd* [1970] 3 All ER 864; (1970) 22 P&CR 328; 69 LGR 271; [1971] EGD 313; 218 EG 559, HL

Myers v *Milton Keynes Development Corporation* [1974] 1 WLR 696; [1974] 2 All ER 1096; (1974) 72 LGR 420; 27 P&CR 518; [1974] EGD 405; 230 EG 1275, CA

Pointe Gourde Quarrying & Transport Co Ltd v *Sub-Intendent of Crown Lands* [1947] AC 565, PC

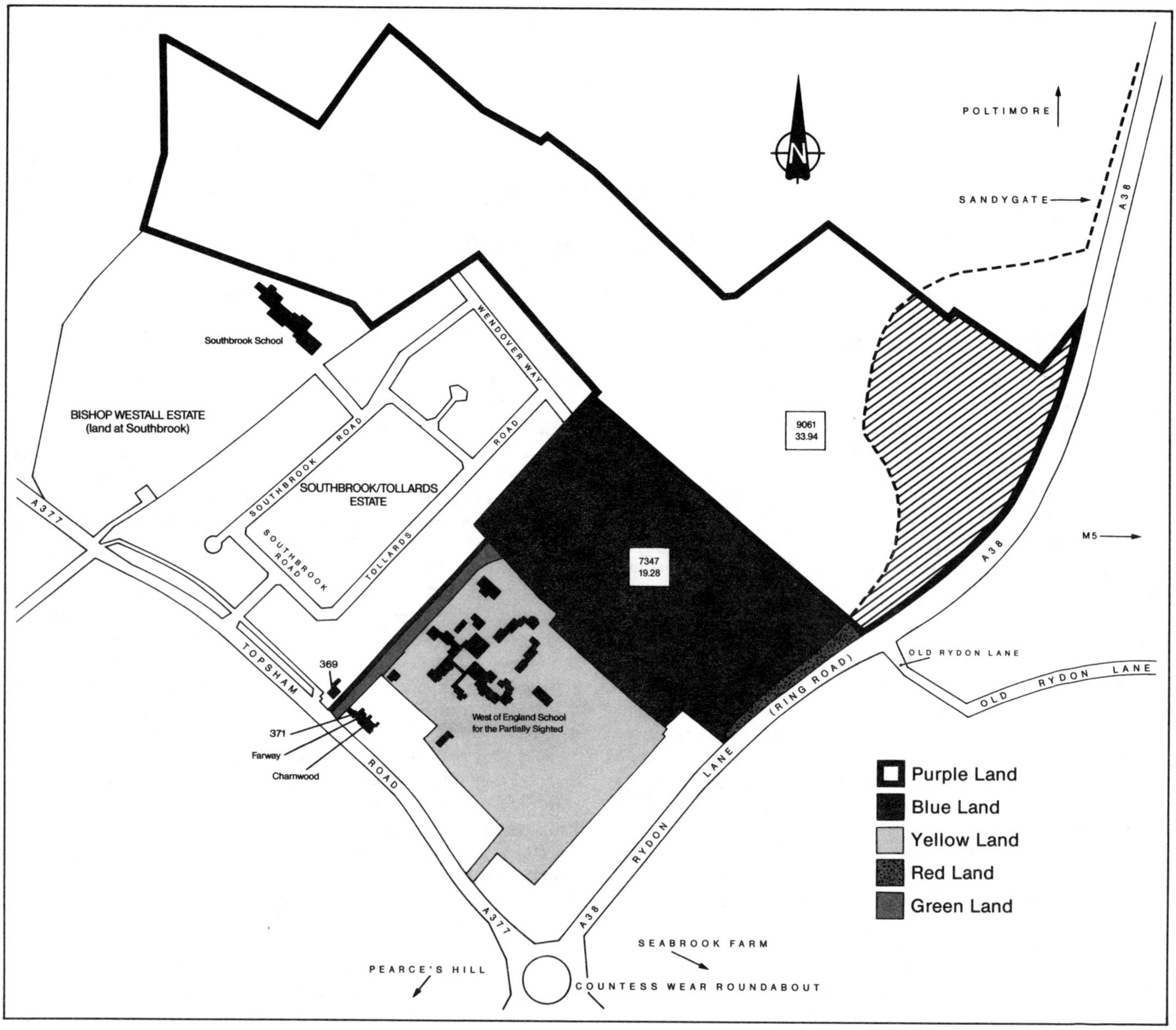

A simplified version of the map attached to the tribunal's decision as Appendix I

Sprinz v *Kingston upon Hull City Council* (1975) 30 P&CR 273; [1975] EGD
E 517; 235 EG 225, LT
Stokes v *Cambridge Corporation* (1961) 13 P&CR 77; [1961] EGD 207; 180 EG 839, LT

Michael Barnes QC and John Male (instructed by Asshetons) appeared for the claimants; Charles Fay (instructed by the Treasury Solicitor) for the acquiring authority.

Giving his decision, MR MALLETT said: This is a reference to determine the amount of compensation payable on the acquisition in 1973 of 0.555 acre of land, fronting on to what was then known as the Exeter Bypass, for the purpose of forming a spur road to the proposed M5 motorway.

F The acquired land formed part of some 19.28 acres of land which may be described in general terms as ripe for housing development, and the principal dispute arises as to the damage caused to the owner of the retained land due to severance.

The value of land ripe for housing development, without any abnormal costs, traffic problems or deferment, is agreed at £42,500 per acre as at the date of entry. The value of the land limited to agricultural use at the same date is agreed at £1,300 per acre over the area capable of cultivation. The claimants' figure of compensation is £750,079 on the assumption that after the acquisition of the 0.555 acre of frontage land there was no prospect of the remaining land being developed and its value did not exceed agricultural value. In the alternative, that figure is £726,647 if what they term the *Bwllfa* principle is not applicable in relation to certain events which occurred subsequent to the date of severance.

The acquiring authority's figure is £98,010, which is based upon the assumption that alternative means of access were available on the retained land so that it could have been developed. They also produce a number of additional valuations based upon different hypotheses as to the mode of access on the retained land.

Mr Michael Barnes QC and Mr John Male, counsel, appeared for the claimants and called Mr R J Challinor FRICS, a principal in the firm of Richard Challinor & Co, chartered surveyors, Mr E E Taylor FRICS FRTPI FRSA, a town planning consultant, Mr W Sommerville, a partner in the firm of Mander Raikes & Marshall, highway engineers, and Mr M M Dixon, one-time principal engineer employed by George Wimpey & Co Ltd but now retired.

Mr Charles Fay, counsel, appeared for the acquiring authority and called Mr N J Abernethy ARICS, the district valuer in the Exeter office of the Inland Revenue Valuation Office, Mr A Cartwright, a highway engineer in the firm of Freeman Fox Ltd, consulting engineers, Mr D N Beardmore MSc FRICS FRTPI, a town planning consultant, and Mr K A Burgess, a chartered engineer. I also received a written report from Mr E H Richmond, a chartered engineer in the employ of the Devon County Council Highways Division, who was unable to attend owing to illness.

Subject land

The subject land at the date of entry comprised some 19.28 acres of grade 2 agricultural land, shown as enclosure OS 7347 on the attached reproduction from the Ordnance Survey Map and coloured red and blue (Appendix I). At that time the claimant owned the red (acquired) and blue (retained) lands and had an option to purchase the land coloured green, owned by the West of England School for the Partially Sighted (the blind institution).

Access to the land was by two field gates on the Rydon Lane frontage. Rydon Lane formed part of the trunk road generally known as the Exeter Bypass.

Topsham Road is a radial road leading from Exeter and crosses the Exeter Bypass at the Countess Weir roundabout close to the subject land.

Leading from a frontage on to Topsham Road and forming part of OS 7347 is an access strip about 805 ft long, varying in width from 47 ft at the Topsham Road frontage to 90 ft at the north-east end. Only part of the access strip was in the ownership of the claimant, that part coloured blue on the attached plan. This comprised a strip of land approximately 16 ft wide over part of which, adjoining the western boundary, is a right of way for agricultural purposes and made, at the date of entry, in favour of the occupiers of land to the north-east of OS 7347 comprising OS 9061 and other land (the Langdon land).

That part of the access strip coloured green was (and still is) owned by the blind institution, who also own the land coloured yellow on the attached plan. Access to the blind institution land coloured yellow is along a metalled roadway on the green land from Topsham Road over a length of 570 ft.

The red land was acquired compulsorily under the M5 Motorway (Poltimore-Sandy Gate Section and Exeter Spur) Compulsory Purchase Order (No CSW8) 1972. Notice to treat was served on March 5 1973 and the date of entry was June 13 1973. This latter date therefore becomes the date of valuation. The various areas of land are set out separately in Appendix II.

APPENDIX II

Areas (OS 7347)		
Gross		19.28 acres
Owned by ADP & E Farmers		18.58 acres
Owned by blind institution		0.70 acre
Land acquired		0.555 acre
Land retained by ADP & E Farmers		18.025 acres
Area capable of agricultural cultivation		
(a) before acquisition		18.28 acres
(b) after acquisition		17.725 acres
Access strip:		
(a) owned by ADP & E Farmers		0.30 acre
(b) owned by blind institution		0.70 acre
	Total	1.00 acre
(c) Capable of residential development (if not required for access)		
(i) owned by ADP & E Farmers		0.08 acre
(ii) owned by blind institution		0.22 acre
	Total	0.30 acre

Adjoining land To the north of OS 7347 was an area of 56.5 acres of agricultural land owned by a Mr G E Langdon (the purple land).

To the south was an area of land owned and occupied by the blind institution (the yellow land). Beyond that was a row of houses facing on to Topsham Road.

To the west was a housing estate (the Southbrook/Tollards Estate) facing on to Topsham Road with access on to that road via a "T" junction at Tollards Road.

To the west of that estate was an area of land fronting on to Topsham Road and owned by the local authority. Well to the east of the junction of Topsham Road with Rydon Lane (Countess Weir roundabout) was an area of agricultural land facing on to Topsham Road (Seabrook Farm).

History It is convenient to set out in chronological order a summary of some of the major events which affected the land at OS 7347.

(a) In the 1950s a Mr J B Martin bought the subject land.

(b) In 1952 the City of Exeter Development Plan was approved.

(c) In 1962 a planning application for residential development of OS 7347 was dismissed.

(d) In 1963 the City of Exeter Development Plan was revised.

(e) On September 13 1966 Mr J B Martin entered into an arrangement with George Wimpey & Co Ltd, a national firm of housebuilders and developers, who sought to purchase the whole of OS 7347 (including that part of the access strip owned by the blind institution) conditional on obtaining planning permission to develop the land. At the same time George Wimpey & Co Ltd entered into two conditional contracts with the Countess Weir House Co Ltd (whose directors were the Martin family) for the purchase of parts of the front gardens of nos 369 and 371 Topsham Road, which adjoined the access strip facing on to Topsham Road.
At the same time there was a conditional contract between J B Martin and George Wimpey & Co Ltd for that company to purchase from Mr Martin that part of the front garden of Farway, 371a Topsham Road.
George Wimpey & Co Ltd were empowered to rescind all the conditional contracts on failing to obtain planning permission for the development of OS 7347.

(f) On November 14 1966 Mr J B Martin completed an agreement with the blind institution giving him an option to purchase all the land coloured green on the attached plan for the purpose of constructing an access to serve residential development of OS 7347.

(g) In 1967 a further residential planning application in respect of OS 7347 was refused.

(h) In 1969 OS 7347 was sold by Mr J B Martin to the claimants.

(i) On October 18 1971 the Development Committee of the Exeter City Council, reporting on the road proposal 10-year programme, considered that the earliest date for construction starting on the proposed Southern Exe Valley Primary Distributor Road was early 1975.

(j) On April 19 1973 a section 17 [Land Compensation Act 1961] certificate was issued by the Exeter City Council in respect of 10¾ acres of the Langdon land adjoining the Exeter Bypass to the effect that there was no immediate prospect of any development on the grounds that the land was shown on the development plan as "white" land used for agriculture and in view of its high agricultural value and the fact that other land was available for development.

(k) On March 5 1973 notice to treat was served on the claimants.

(l) On June 13 1973 the acquiring authority entered on the red land and works started almost immediately.

(m) On December 10 1974, regarding land known as Seabrook Farm, the Secretary of State for the Environment dismissed an appeal against the inspector's report following refusal of the former Exeter City Council to grant planning permission for residential development on that land.

(n) On December 16 1974 the Policy Committee of the Exeter City Council granted deemed planning permission for housing development of land belonging to the council on the Bishop Westall Estate adjoining the Southbrook/Tollards Estate. They suggested that the possibility of an alternative access be explored as the Topsham Road access was opposed by the county engineer. On March 18 1975 the Development Committee resolved to adhere to the original access to Topsham Road via the Southbrook Estate.

(o) On December 18 1974 there was a local inquiry into an appeal against the section 17 certificate on the Langdon land.

(p) In October 1975 the Exeter Spur Road to Sandy Gate was opened.

(q) On December 23 1975 the owner of the adjoining Langdon land obtained a section 18 certificate from the Secretary of State for the Environment (following an appeal against the section 17 certificate and a local inquiry) to the effect that in relation to the 10¾ acres of land to the west of the Exeter Bypass, planning permission for residential development might have been expected not sooner than January 1 1981.

(r) On June 21 1976 the Exeter City Council issued a notice of refusal of planning permission to develop OS 7347. The application was made by George Wimpey & Co Ltd and an appeal against the refusal was not pursued to a conclusion. In a letter dated June 1980 the Department of the

A Environment advised the Exeter City Council that the company had withdrawn from the appeal. One of the reasons for the refusal was "The use of access to Topsham Road . . . would cause conditions of danger and inconvenience to the free flow of traffic on that road . . .".

(s) In May 1977 the M5 Exminster section was opened.
(t) In January 1978 the Exeter Bypass was detrunked.
(u) On February 10 1983 a section 18 certificate on the subject land was issued by the Secretary of State for the Environment.
(v) In 1981 the County of Devon Structural Plan was approved.
B (w) In 1984 the Exeter Local Plan was adopted.

Principal valuations

It is common ground that there are two stages in the valuation:

(a) The value of the land acquired under rule (2) of section 5 of the Land Compensation Act 1961 and
(b) The valuation of the damage to the owner of the retained land under section 7 of the Compulsory Purchase Act 1965.

It is convenient at this stage to set out the principal valuations on which the parties rely in support of their cases and these are shown in Appendix III.

C

APPENDIX III

Valuations of R J Challinor FRICS for the claimants

Valuation A

Compensation for red (acquired) land under r(2) of s5 of 1961 Act		
Residential value:		
0.555 acre @ £41,770 per acre	£23,182	
Agricultural use:		
Value for 2½ years	45	
Value for access purposes to blue (retained) land at one-third of development value of blue land (see valuation of blue land) based upon principle of *Stokes* v *Cambridge Corporation* (1961)	241,787	£265,014
D *Compensation for blue (retained) land under s 7 of 1965 Act*		
Before value:		
Full residential value:		
18.025 acres @ £41,770 per acre	752,904	
Agricultural use:		
Value for 2½ years	1,430	
	754,334	
Less (i) Cost of exercising option over land owned by blind institution	4,500	
	749,834	
(ii) Allowance for purchase of access land (access being obtained over the red land) ie ⅓ of (£752,904, *less* £4,500, *less* agricultural value		
E £23,042)	241,787	
	508,047	
After value:		
Full agricultural value of area capable of cultivation:		
17.725 acres @ £1,300 per acre	23,042	485,005
		750,019
Disturbance		60
Compensation		£750,079

Valuation B

(Check valuation based upon actual loss to claimants)

F 1 *Value of both acquired and retained land prior to acquisition:*

(a) 18.58 acres @ £41,770 per acre	£776,086
(b) Agricultural use for 2½ years	1,475
	777,561
(c) *Less* Cost of exercising option over land owned by blind institution	4,500
	773,061
2 *Value of retained land after acquisition:*	
17.725 acres @ £1,300 per acre	23,042
	750,019
3 *Disturbance*	60
Compensation	£750,079

Valuation C G

This valuation is submitted should the Lands Tribunal consider that the *Bwllfa* principle (*Bwllfa & Merthyr Dare Steam Collieries (1891) Ltd* v *Pontypridd Waterworks Co* [1903] AC 426) is not applicable in relation to events concerning planning matters which occurred subsequent to the valuation date.

1 *Compensation for the acquired (red) land*			
(as for valuation A)			£265,014
2 *Compensation for the retained (blue) land*			
Before value			
(as for valuation A)		£508,047	
After value			
Agricultural value:			
(as for valuation A)	£23,042		
Hope value for future residential			H
development 18.025 acres @ £1,300 per acre	23,432	46,474	461,573
			726,587
3 *Disturbance*			60
Compensation			£726,647

Preliminary valuation 1

No acquisition world

Value of whole (Red land, together with blue land)

Developed with access to A38 Rydon Lane per s17 certificate

(a) Total area		19.28 acres	
Less Access to Topsham Road	1.00 acre		
Less For developable part access	0.30 acre		J
	0.70 acre		
Buffer strip/tree screen to A38	0.25	0.95 acre	
		18.33 acres	
(b) 18.33 acres @ £42,500 per acre			£779,025
(c) *Less* Cost of exercising option for Topsham Road access strip			4,500
			774,525
(d) *Less* For disability of access to and from A38 at peak traffic periods			
Say minimum 5% reduction in house prices,			K
ie 190 houses (mainly semi-detached) @ £11,750 av		£2,232,500	
or 125 houses (all detached) @ £17,500 av		2,187,500	
Adopt 5% of, say, £2,200,000			110,000
			664,525
(e) *Less* Abnormal costs			
(i) Tree screen — would be absorbed by developer as a nominal cost	Nil		
(ii) Roundabout at A38	£68,900		
Say shared 50% adj owner		34,450	
Less Costs of a normal junction		20,900	
			13,550 L
			650,975
(f) Deferment			
As planning permission not available until Jan 1 1976			
PV of £1 in 2½ years @ 10%			0.789
			£513,619
(g) *Add* Agricultural income (if any), say			Nil

(a) Value of whole, say £513,600, ie £28,019.64 per acre (net)

Preliminary valuation 2

Acquisition world

Value of blue land retained after severance M

Developed with access to Rydon Lane

(a) Total area		19.28 acres	
Less Red land taken	0.555 acre		
Access to Topsham Road	0.70 acre (net)		
Buffer strip/tree screen	0.25 acre		
New access to Rydon Lane	2.9 acres	4.405 acres	
		14.875 acres	
(b) 14.875 acres @ £42,500 per acre			£632,188
(c) *Less* Cost of exercising option over Topsham Road access strip			4,500
			627,688

A (d) *Less* Abnormal costs			
(i) Tree screen	Nil		
(ii) Roundabout to Rydon Lane/slip road (excluding earthworks)	£116,500		
say 50% shared adjoining owner		£58,250	
Less Costs of normal junction		20,900	
			37,350
			590,338
(e) Deferment			
To Jan 1 1976			
PV of £1 in 2½ years @ 10%			0.789
			465,777
(f) *Less* Earthworks cost	115,500		
B say 50% shared adjoining owner	57,750		
PV of £1 in 1½ years @ 10%	0.868		50,127
			415,650
(g) *Add* Agricultural income (if any), say			Nil
Value of blue land retained, say			£415,650

Valuation of N J Abernethy ARICS for the acquiring authority
Valuation and compensation payable

(a) Red land taken		
0.555 acre @ £28,019.64 per acre	£15,551	
(See preliminary valuation 1)		say £15,600
(b) Severance and injurious affection to blue		
C land retained		
Before acquisition		
Net developable area of whole	18.33 acres	
Less Red land	0.555 acre	
	17.775 acres	
17.775 acres @ £28,019.64 per acre	£498,049	
say	498,000	
After acquisition		
As revised preliminary valuation 2	415,650	82,350
(c) Disturbance to growing crop (agreed)		60
Compensation		£98,010

Check

D Valuation of whole before acquisition	£513,600	
Valuation of land retained after acquisition	415,650	
	97,950	
Disturbance to growing crop	60	
		£98,010

By way of introduction to the issues which arise in this case, it is necessary to state broadly the assumptions that have been made by the two valuers concerned. Mr Challinor, acting for the claimant, has held, and consistently held, the view that at the date of acquisition it has to be assumed that planning permission for residential development on the red (acquired) land would have been granted not E earlier than January 1 1976 in accordance with the section 18 certificate issued by the Secretary of State for the Environment. As the reasoning which gave rise to the certificate assumed that the red land would be developed in conjunction with the blue (retained) land, it was reasonable to assume that this land also would be the subject of planning permission for residential development granted on the same date. Development would be achieved by access from Rydon Lane by means of a simple roundabout (the Mander Raikes & Marshall roundabout) after the detrunking of Rydon Lane. It is also an essential part of this assumption that the scheme, which has to be ignored for the purpose of assessing compensation, included the spur road but not the M5 motorway, as per the Secretary of State for the F Environment in his letter of February 10 1983.

Mr Challinor formed the view that on the severance of the blue land there remained no practicable access to enable residential development of that land to take place. At the date of the acquisition there may have remained the hope that other developments or road improvements would eventually enable the blue land to be developed, but that was a remote possibility. Subsequent and relevant events showed that this hope was not justified. Therefore, in adopting what is generally termed the *Bwllfa* principle* he has excluded any hope value from his primary valuation.

Mr Abernethy for the acquiring authority took the view that the scheme underlying the compulsory purchase embraced the whole of G the M5 motorway or at least those sections between Poltimore and Sandy Gate and Sandy Gate to Pearce's Hill and the link with the A38 to the south. On this assumption Rydon Lane would never have been free of bypass traffic in the no-scheme world and would have never been detrunked.

He is of the view that in the acquisition or real world access to the blue land could have been obtained over the red land by means of a massive roundabout (the Cartwright roundabout), which could, he says, have been incorporated in the spur road connecting Rydon Lane with Sandy Gate.

In the alternative he is of the view that there were four separate possible means of access to the blue land from Topsham Road although all were subject to possible delays in obtaining planning H permission. Therefore he made a deduction in each case for "risk" which he said was almost entirely due to the likely delay in development.

Thus the major issues in this case are the application of section 6 of the Land Compensation Act 1961 and the common law principle, generally referred to as the *Pointe Gourde*† rule, to the facts of this case and what account should be taken of the section 18 certificate relating to the retained land.

A secondary issue concerns the application of what is known as the *Bwllfa* principle in valuing the damage caused by the severance of the blue land.

There are also several other comparatively minor issues on details J of valuation which I shall come to later. First I propose to refer to the issues concerning section 6 and the *Pointe Gourde* rule.

In the present case the claimant made an application under section 17 in respect of the red land, and on appeal the Secretary of State issued a certificate under section 18 certifying that planning permission for residential development in conjunction with the adjoining land to the west forming the remainder of OS 7347 (the blue retained land), subject to certain conditions, would have been granted in respect of the land acquired not sooner than January 1 1976 if it were not proposed to be acquired by any authority possessing compulsory powers.

The conditions referred to were:

> Details of the siting, design and external appearance of the buildings, the K means of access thereto. The landscaping of the site shall be as may be agreed with the local planning authority:

The decision of the Secretary of State was contained in a letter dated February 10 1983.

At the acquisition date a Mr G E Langdon owned an area of agricultural land including OS 9061 extending to approximately 56.5 acres and situated to the north of OS 7347 and shown edged purple on the attached plan.

Part of that land extending to approximately 10.75 acres, shown [purple land] hatched black on the plan, was acquired by the Department of Transport for the same purpose as that which led to the acquisition of part of OS 7347. The balance of Mr Langdon's land L remained in his ownership until his death in 1980. The date of entry on his land was June 7 1973, a few days before the acquisition date on the subject land, and the section 18 certificate issued by the Secretary of State in respect of 10.75 acres of acquired land certified that planning permission for residential development, subject to conditions, might reasonably have been expected to be granted at a future time not sooner than January 1 1981 if that land were not proposed to be acquired by an authority possessing compulsory purchase powers. The conditions referred to were similar to those contained in the section 18 certificate on the subject land.

Thus it is reasonable to assume that the Secretary of State contemplated that the development of the red and blue lands would M have been permitted when Rydon Lane was expected to be detrunked and that the development of the Langdon land would have followed some years later. The certificate on the part of the Langdon land that was acquired ([purple land] hatched black) did not state that the acquired land would have been developed only in conjunction with other land, but it is agreed between the parties that it would have been reasonable to expect that at least some of the retained land would have been developed in conjunction with the acquired land.

For the claimant it is contended that the certificates on both the red (acquired land) and the purple hatched black (acquired land) would

*Editor's note: From *Bwllfa & Merthyr Dare Steam Collieries (1891) Ltd* v *Pontypridd Waterworks Co* [1903] AC 426.

†Editor's note: From *Pointe Gourde Quarrying & Transport Co Ltd* v *Sub-Intendent of Crown Lands* [1947] AC 565.

not have been issued without assuming that the access was from Rydon Lane after detrunking and that that was the sole access. In other words, the certificates would not have been issued if the scheme for compulsory purchase had been thought to include any part of the M5 motorway. Because, if it had, in the absence of the scheme, there would have been no relief of traffic on Rydon Lane and therefore no detrunking. Furthermore, the reference to the date was "not before". So much is clear from the Secretary of State's letter dated February 10 1983, the relevant part of which is as follows:

Referring to the decision of the Exeter Council:

> The council agreed that the proposal to construct the spur road would have to be disregarded but took the view that in this case the Secretary of State must also disregard the intention to build the M5 itself. This argument has been considered but is not accepted. The Secretary of State would take the view that the principle that he must generally ignore the purposes for which appeal land is being acquired, would in this case require him to leave the spur road out of account. However, it is considered that this principle could not reasonably be extended to a road with which the spur road linked, in this case the M5.
>
> It is considered that by itself the site, because of its shape, its small size and its proximity to the busy Rydon Lane, could not have been satisfactorily developed. It is noted, however, that the parties agree that the site would have been developed, for residential purposes, in conjunction with adjoining land to the west which is in the same ownership. This view is accepted. However, your clients dispute the timing placed on such a development, namely not earlier than 1st January 1981, and the condition preventing vehicular access to Rydon Lane. In considering these issues it is noted from the Department of Transport's letter of 19th November 1982, that no alternative routes had been considered for the Exeter Spur Link. In these circumstances it is thought that northward bound traffic from Exeter and south bound traffic to Exeter would have used junction 29 of the M5 since this would have been the nearest northern junction to Exeter, and there is no alternative route for the spur road that the Secretary of State would be required to take into account. Rydon Lane would have been therefore one of a number of routes to the city centre used by traffic entering or leaving Exeter on its northern side. Traffic by-passing Exeter would have used the motorway instead of Rydon Lane and consequently it is thought that there would have been a reduction in traffic in both Rydon Lane and Topsham Road. Rydon Lane, however would have continued to carry a considerable volume of traffic until the completion of this part of the M5 bypassing Exeter. Therefore until that time no vehicular access would have been allowed to have been formed to Rydon Lane. The opening of this part of the M5 was due to take place in 1976 and this could be said to have been reasonably foreseeable at the relevant date . . . the conclusion reached therefore is that, in the absence of a proposal to acquire the appeal site by an authority possessing powers of compulsory purchase, residential development of the appeal site in conjunction with the adjoining land, in the same ownership, would have been allowed but not before January 1976. All the other arguments advanced by the parties have been taken into account, but are not thought to outweigh the considerations mentioned above.

In their submissions to the Secretary of State the Exeter City Council said:

> A copy of the county engineer's advice in respect of the application which is the subject of this appeal is attached. This indicates that the county engineer would object to an access to Rydon Lane or to Topsham Road for the reasons stated. The principal traffic route referred to is, in fact, Topsham Road. The city council are in full agreement with the county engineer's advice, but it seems that in 1973 the former city council may have allowed access onto Topsham Road by means of the private access road referred to by the county engineer, or possibly via Wendover Way and the Southbrook Estate Roads, subject in either case to satisfactory highway improvement being carried out or secured.

Therefore, it is clear that the possibility of an alternative access was before the Secretary of State but was rejected.

Furthermore, it is not disputed that Mr Abernethy's predecessor, Mr Ruffett, agreed compensation on the Langdon land on the basis that residential development would have been permitted not sooner than January 1 1981.

For the acquiring authority it is contended that the scheme contemplated in the section 18 certificate is not necessarily the same as the scheme to be assumed for the purposes of assessing compensation. The entire scheme was obviously the completion of the M5 motorway and link road and it was a matter of administrative convenience that the works and contract were subdivided into various sections. Furthermore, it says, any alternative route connecting Poltimore with Sandy Gate based upon the existing road networks could only connect with the Sandy Gate end of the motorway by passing over land forming part of that land which was already contained in the compulsory purchase order. This, it says, is fatal to the claimants' case.

The claimants maintained that although a section of the M5 motorway and the slip and spur road happened to have been built together under the terms of the same contract and under the same compulsory purchase order, in fact they served distinct and different purposes. The motorway, they say, was intended to cater predominantly for through traffic as part of a national network and the slip and spur road predominantly for traffic having a local function. The Secretary of State has already on two occasions, in December 1975 on the Langdon land and in February 1983 on the subject land, found that the spur road must be ignored for the purposes of the section 18 certificate but that it would be right to take account of the existence of the M5 motorway. Compensation in respect of the Langdon land has already been agreed on that basis.

I shall deal with each of these issues in turn.

Before acquisition

The section 18 certificate

The section 18 certificate relates only to the acquired land. In relation to the retained land, the certificate has no statutory force but the matters of fact or judgment which lead to the granting of a certificate on the acquired land may well be strong or decisive evidence in relation to the blue land. These principles were considered by this tribunal (Mr V G Wellings QC) in *Abbey Homesteads Group Ltd* v *Secretary of State for Transport* (1982) 263 EG 983, 1095 and 264 EG 59, 151 and I follow them.

The object of the certificate is to establish the assumptions to be made as to planning permission in the no-acquisition world in calculating compensation for compulsory purchase. This tribunal is not empowered to vary or go behind a certificate issued by the Secretary of State. Therefore I come to the conclusion that in the before-acquisition world the value of the red (acquired) land and the value of the blue (retained) land are properly assessed on the basis that planning permission for residential development of those lands would have been obtained not earlier than January 1 1976.

On the evidence before me, I think it is a reasonable and proper assumption that access for the development would have been obtained from Rydon Lane after that road was detrunked.

If I were not bound by the section 18 certificate, on the evidence before me I would have reached a similar conclusion on the grounds:

(a) The scheme under the common law *Pointe Gourde* principle should be considered separately from that under section 6.
(b) On that basis the scheme under the common law principle includes the slip and spur road but does not include any part of the M5 motorway.
(c) The proposed alternative route between Poltimore and Sandy Gate was feasible.
(d) The alternative route was substantially over different land than that contained in the compulsory purchase order and would have involved compulsory purchase of substantially different land for substantially different purposes. The fact that some of the land was common to the compulsory purchase order and the alternative scheme is not detrimental.

In reaching these conclusions, I have followed this tribunal's decision in *Sprinz* v *Kingston upon Hull City Council* (1975) 30 P&CR 273, where in a decision given by the president, Sir Douglas Frank QC, Mr Victor Wellings QC and Mr E C Strathon FRICS, all seated together, said at p 282:

> It is understandable that as a matter of convenience in judgments or decisions, where section 6 of the Act of 1961 and the *Pointe Gourde* principle cover the same ground, express reference should be made only to section 6 and that this may make it appear that the principle is limited to covering ground which is not within the section. However, it is our opinion that as a matter of strict law both the section and the principle must be applied, where on the facts they are capable of applying, independently of the other.

I have also followed the authority of the House of Lords decision in *Margate Corporation* v *Devotwill Investments Ltd* [1970] 3 All ER 864 and in particular Lord Morris' words at p 869 citing Lord Asquith's general observations in *East End Dwellings Co Ltd* v *Finsbury Borough Council* [1952] AC 109:

> If one is bidden to treat an imaginary state of affairs as real, one must surely, unless prohibited from doing so, also imagine as real the consequences and incidents which, if the putative state of affairs had in fact existed, must inevitably have flowed from or accompanied it.

And later (on p 109) in his own words dealing with the issues of the case:

> An assumption had to be made that the respondents' land was not going to be acquired so that a by-pass should there be constructed — but it was in no way

A an inevitable corollary that there would be a by-pass on some other line and in some different position. If there was not to be a by-pass on the respondents' land it by no means followed that there would inevitably be a by-pass somewhere else. There might be or there might not be. It might have been possible to have another route for a by-pass; it might have been quite impossible. It would be a question depending on topographical and various and many other factors whether there could be a by-pass somewhere else. It would be for consideration whether any alternative by-pass was or was not possible or probable and further whether its construction was or was not likely. These matters could not rest on any assumptions but rather on an examination of all the evidence.

Also the words of Lord Denning MR in *Myers* v *Milton Keynes Development Corporation* [1974] 1 WLR 696 where at p 704 he said:

B It is apparent, therefore, that the valuation has to be done in an imaginary state of affairs in which there is no scheme. The valuer must cast aside his knowledge of what has in fact happened in the past eight years due to the scheme. He must ignore the developments which will in all probability take place in the future 10 years owing to the scheme. Instead, he must let his imagination take flight to the clouds. He must conjure up a land of make-believe, where there has not been, nor will be, a brave new town: but where there is to be supposed the old order of things continuing — a county planning authority which will grant planning permission of various kinds at such times and in such parcels as it thinks best, but with an assurance that in March 1980 planning permission would be available for the residential development of the Walton Manor Estate.

After acquisition

C I now turn to the real world situation following the acquisition of the red land.

An alternative scheme

Without any access on to Topsham Road the blue land is virtually incapable of being developed after the compulsory acquisition of the red land unless some form of access could be arranged over the red land on to the slip and spur road that were about to be built. The acquiring authority say that this could readily have been achieved by the construction of a massive roundabout (the Cartwright roundabout), which would accommodate Rydon Lane, the junction with the spur road, the local traffic on Old Rydon Lane, the service road to adjacent houses on the south side of Rydon Lane and an D access to the blue and purple Langdon land to serve residential development on that land.

The possibility of such a means of access does not appear to have been considered by anyone until some 12 years after the date of acquisition, when Mr Cartwright was asked to find a solution. The roundabout did not emerge in its final form until the case was part heard. What finally emerged was a roundabout that was feasible as an engineering project but less than ideal from the point of view of a highway engineer and probably well beyond the ordinary capability of a residential developer. The total cost was estimated to be at least £232,000 or over 10 times the cost of a normal junction for a residential development of this nature. The cost would have been considerably greater if the earthworks could not have been carried E out at the same time that the spur road was being built.

There would have been considerable difficulty in accommodating such a massive roundabout on a sloping site confined by existing houses, the junctions with Old Rydon Lane and the spur road connecting with the M5 motorway. It would have involved the acquisition of some small amount of additional land belonging to some of the houses on the south side of Rydon Lane. It was not disputed that the cost of building such a roundabout could be justified only if it were shared "fifty-fifty" with the adjoining owner of the Langdon land. The roundabout would have been constructed after the spur road was built and it would have involved realignment of that road.

The proposed roundabout is an alteration and extension of the F approved scheme which is the subject of the compulsory purchase order. It is in effect a different scheme which was not considered by anybody at the time of the acquisition and did not materialise thereafter. In my view it would be wholly incorrect to calculate compensation for damages on an alternative and imaginary scheme which in effect alters and extends the scheme which gave rise to the acquisition.

If I am wrong, it is necessary to consider whether, as a matter of fact, the massive roundabout was a reasonable possibility.

For the cost of the roundabout to be economically viable the earthworks would have had to have been carried out at a time when the spur road was under construction. This would have left a matter of months for the owner of the blue land to have sought the co-operation and financial backing from the owner of the Langdon land G when planning permission for residential development of the Langdon land had been refused a few months previously on account of the agricultural quality of that land.

It would also have been necessary to have the co-operation of the owners of the existing houses on the southern side of the proposed roundabout and to seek and obtain approval or consent from the Ministry of Transport, the county council highway authority, the planning authority, the local authority and the Ministry of Agriculture, Fisheries and Food (in respect of the purple land).

It appears to me unrealistic to assume that all these matters could have been dealt with satisfactorily in the very short time available. Therefore I find as a matter of fact that the massive roundabout was a wholly improbable solution to the problem and unlikely to affect H the value of the blue land.

Access on to Topsham Road

I turn to consider the four alternative means of access on to Topsham Road put forward by the acquiring authority, three of which entail the improvement of one or other of the two existing accesses on to Topsham Road, at the Tollards Road junction and the blind institution junction, and the fourth assumes a limited development with access on to Tollards Estate at Wendover Way.

It is evident that the possibility of some form of access on to Topsham Road was before the Secretary of State when dealing with the section 18 certificate and it cannot be said that the possibility of access on to Topsham Road at a date later than January 1 1976 was J rejected. Therefore it is necessary to examine the evidence.

Wimpey's experience in 1976 was that the county council highway authority would not countenance the possibility of considering any form of access on to Topsham Road which would increase the amount of traffic. The acquiring authority's case is that the Exeter City Council, who were the planning authority at the date of acquisition, would have taken a more lenient view, but no witness from the city council was called in support. Mr Abernethy prepared separate valuations in respect of each of the four cases and in each case deducted 25% for risk, which he explained to be almost wholly if not entirely the risk of probable delay in obtaining planning permission rather than the risk of failing to obtain planning permission at all. He expressed no preference for any one scheme, but K when pressed in cross-examination and after considering the matter overnight, he finally came to the conclusion that the preferred scheme was that which involved exercising the option to purchase the green land, closing the blind institution's access on to Topsham Road, breaking through the gardens of houses in Tollards Road and providing an access via Tollards Road into Topsham Road.

There are several likely objections to this scheme, particularly from houseowners in Tollards Road, but the principle one is that the scheme could not have been achieved within the terms of the option agreement with the blind institution in that it would have closed up their right for a direct access on to Topsham Road which the agreement specifically reserved. L

Three of the four schemes involved taking parts of gardens, or in one case a whole house, in order to provide an access. They all involved negotiations with a number of local residents as to the land to be taken or the damage to retained land.

The highway and planning experts could not agree their preference for one scheme or another. All the schemes fall short of the ideal and whether or not any scheme would have been viable was to some extent dependent on the pressure placed on the planning authority at the relevant date to provide building land.

There is some evidence that at the date of acquisition the stock of building land in Exeter was adequate. In the inspector's report of October 1974 in respect of the inquiry concerning Seabrook Farm, having listed the various planning objections including the M unsatisfactory nature of a single access point on to Topsham Road for an estate with 400 to 500 dwellings, he concluded:

There remains the question of whether there is need for housing land sufficient to override what I consider to be exceptionally compelling planning objections to the release of this land. Choice of location and type of house site is presently somewhat restricted by the larger areas of available housing land being on the north and west fringes of the city on generally steep slopes, but there are smaller sites available all over the city, and bearing in mind that in total there is an estimated 8 years supply of housing land which has planning permission, and that the strategy towards housing requirements should be apparent as soon as 1975 with the publication of housing land feasibility studies, to be followed by the draft structure plan in mid 1976, I see no overriding need for the release of the appeal site.

A Any prospective developer purchasing the retained land at the acquisition date would have to weigh the chances of a delay in obtaining planning permission and the cost of purchasing the land and negotiating settlements with adjoining owners against the risk that, if planning permission could not be obtained, only the agricultural value of the land could be recovered. In addition to the purchase price, there would be the cost of ensuring that a proper access was available. It is almost certain that there would have been delays and Mr Abernethy has properly allowed for this; but there was a very real risk that planning permission might not be granted. It seems to me that the point had been reached where any further reduction in the purchase price could not adequately compensate a prospective developer for the risk of losing most of his capital, and as B there was other land in Exeter available for development he would be prepared to pay little more than agricultural value for the blue land.

For these reasons I find the prospect of gaining access on to Topsham Road, by any of the four proposed routes, to be too remote to have any measurable influence on the value of the blue land.

Valuation problems

I now turn to various problems that arise in the application of my findings to the valuation of the red land on acquisition and the assessment of damages to the owner of the blue land.

Method of valuation

C Mr Challinor has valued the two interests separately as statutes require: the acquisition of the land under rule (2) of section 5 of the Land Compensation Act 1961 and the damage under section 7 of the Compulsory Purchase Act 1965.

Mr Abernethy in effect values the red and blue land together and apportions the result pro rata to the respective areas. In the result the method adopted was of little consequence, but I prefer Mr Challinor's approach because it is that required by statutes. His method involves grappling with the problem of what is termed a *Stokes* payment (see *Stokes* v *Cambridge Corporation* (1961) 180 EG 839 or, in other words, the ransom payment that the owner of the red land would be able to exact from the owner of the blue land in order to enable the latter to gain an access for the residential development of D his land. Payment is added to the value of the red land and the same figure is deducted from the value of the blue land so that in the end result the total payment of compensation is not altered and the precise amount of the *Stokes* payment is not a critical factor.

I see no reason for differing from Mr Challinor's view that the payment would represent approximately one-third of the full residential value of the blue land.

Before acquisition

It is agreed that the open market value of OS 7347 for residential purposes as at the date of entry is £42,500 per acre exclusive of:

(i) abnormal costs;
(ii) disabilities (if any) relating to high traffic volume (if applicable) E at access point;
(iii) deferment (if any) of the ripe development value.

Mr Challinor therefore values 0.555 of an acre of the red land at £41,770 per acre and this figure takes account of the cost of exercising the option of obtaining the Topsham Road access strip which would release a small section of land for residential development and the abnormal cost of building the Mander Raikes & Marshall roundabout to gain access on to Rydon Lane. Both these cost figures have been agreed. By this means he arrives at a figure of £23,182.

Mr Abernethy arrives at a figure of £15,600 by taking the same acreage at £28,019.64 per acre and he arrives at this figure by making the same deductions as Mr Challinor, but in addition he excludes a buffer strip of some 0.25 acre which he contends would be required F on the Rydon Lane frontage and he makes a further deduction of £110,000 to allow for what he calls a disability of access to and from Rydon Lane at peak traffic periods. In addition he defers the value of the red land for two and a half years to take account of the difference between the acquisition date and the earliest date when planning permission could be obtained.

I shall deal with each of these points of difference in turn.

Buffer strip

Mr Abernethy was of the view that the strip was essential whether Rydon Lane was relieved of its by-pass traffic or not. Mr Challinor was of the view that such a strip was desirable but could easily be accommodated in the normal landscaping costs of development and would not decrease the number of houses that could be built on the G land. Mr Beardmore thought that careful siting of the new dwellings together with sensitive planting would achieve the same result and at the most would be likely to involve using a 6-m wide strip. In his view this would be regarded by a developer as normal landscaping cost.

Having heard the evidence, I am of the view that there is no justification for deducting 0.25 of an acre from the total area of land.

Disability of access

Mr Abernethy's deduction in respect of this item is on the assumption that Rydon Lane would continue to carry the by-pass traffic in a world where there was no M5 motorway. In view of my findings to the contrary I disallow this item. H

Deferment

Mr Challinor agrees with the principle of making a deferment and with Mr Abernethy's method of calculating the allowance. He maintains that, in the very special circumstances of this case, the certainty of planning permission in $2\frac{1}{2}$ years' time was a benefit rather than a disadvantage. In the preceding year the index of land values had increased from 147 to 290. The cost of borrowing money over a period until the land could be developed and sold was possibly somewhere between 12% and 14%. Land banks were a common feature and there were proposals published by the Chancellor of the Exchequer to introduce a land bank charge for failure to complete a development within a specified time (four years) of the granting of outline planning permission. Under these circumstances Mr J Challinor is of the view that there was a positive advantage in having the certainty of planning permission deferred for $2\frac{1}{2}$ years in a development of this size where certain preparatory works could be done in the interim period.

In the special circumstances of this case I agree that there should be no deduction for deferment.

Agricultural use

Mr Challinor makes an addition of £45 to the value of the red land (and £1,430 to the value of the blue land) to allow for the value of the agricultural use of the land over two and a half years. The amounts are agreed but the additions are disputed. Mr Abernethy contends that no one would trouble to let the land during such a short period K and run the risk of creating a protected agricultural tenancy. The claimants contend that such a tenancy is simply avoided by letting in a series of terms for one year and 300 days.

I accept that it is proper to add the agreed amounts for the value of the agricultural use of the land.

After acquisition

The Bwllfa *principle*

The final issue involves consideration of the true meaning of what is generally known as the *Bwllfa* principle. The *Bwllfa* case in 1903 has given its name to a principle relating to the assessment of damages in English law in existence well before the turn of the century. Put in L the simplest of terms, the principle is that in assessing damages it is unnecessary to speculate on what might be the measure of the loss if at the time of the trial the true loss has become a fact.

Mr Challinor freely admits that at the date of acquisition there remained with the retained land some faint hope of a future event which would allow residential development of that land. He discounted the immediate possibility of access on to Topsham Road and hope was centred on the remote possibility of a major road development to the south of Exeter (the Southern Exe Valley Primary Distributor Road) which would relieve Topsham Road of some of its traffic and thereby allow for additional access to that road, or a series of estate roads being developed to the north of the retained land which would eventually create an alternative means of access from M that direction. The hope was remote and he valued it at no more than double the agricultural value of the land. His estimate was based upon his experience and previous transactions. Mr Abernethy criticised that approach and was of the view that hope value should be calculated as a percentage of the possible development value. As the possibility of development is very remote and in the absence of evidence in support of Mr Abernethy's view, I prefer Mr Challinor's approach to the valuation of hope value.

By the time of the hearing a number of events had taken place, so that clearly there was no longer any hope of the blue land being developed for residential purposes; therefore, the claimants say, the latest situation must be taken into account.

There is no dispute between the parties that it is an obvious principle of law, founded on common sense, that in valuing land at a given date one has to have regard to all the expectations that could reasonably be supposed to be in the minds of persons at that date. Events which occur after that date cannot be taken into account save in so far as they are relevant to the anticipation. Many of the cases referred to me were concerned with the enunciation of that principle and there was no dispute in their application to the value of the acquired land in this case.

The claimants submit that a distinction must be drawn between the valuation of an interest in the land and the assessment of damages flowing from a given event. In the latter case, they say, it is proper to take account of events taking place between the date of the occurrence which gave rise to the claim and the actual calculation made by a court or tribunal of the damage or loss in so far as they show that it is a true loss which flows from the event. They say that damage which occurs under section 7 of the Compulsory Purchase Act 1965, be it for severance or injurious affection, falls within this rule.

I was referred to the well-known words of Lord Macnaghten in the decision of the House of Lords in *Bwllfa* referring to the arbitrator's duty:

> In order to enable him to come to a just and true conclusion it is his duty, I think, to avail himself of all information at hand at the time of making his award which may be laid before him. Why should he listen to conjecture on a matter which has become an accomplished fact? Why should he guess when he can calculate? With the light before him, why should he shut his eyes and grope in the dark?

The claimants maintained that this principle was well illustrated in the House of Lords decision in *Carslogie Steamship Co Ltd* v *Royal Norwegian Government* [1952] AC 292. A ship damaged in a collision through the fault of another ship received temporary repairs which rendered her seaworthy, but on the way to a port where permanent repairs were to be effected she encountered heavy weather and thereby suffered damage rendering her unseaworthy and requiring immediate repair. At her destination both sets of repairs were effected concurrently; the work occupied 30 days. Ten days would have been required for the collision damage if executed separately. It was held that the owners of the ship at fault in the collision were not liable for damages in respect of the 10 days' retention, since the heavy weather damage was not a consequence of the collision and the owners of the damaged ship sustained no loss of profitable time by reason of the fact that for 10 out of the 30 days occupied in repairing that damage she was also undergoing repairs.

The acquiring authority maintain that damages must be assessed as at the date of severance; therefore it would be wrong to look at future events which might themselves be the cause of damage.

In *Halsbury's Laws of England* vol 12 para 1134 the principle is stated thus:

> The damages that result from one and the same cause of action must be assessed and recovered once and for all, and the plaintiff must sue in one action for all his loss, past, present and future, certain and contingent. The sum awarded will be assessed on the basis of an immediate award even where, as in the case of a minor, there are possibilities of increase by investment and accumulation before the beneficiary will acquire the right freely to dispose of it. The difficulty of estimating future and contingent loss may in some cases be reduced by the facts which have actually happened down to the date of trial, for, while assessment should normally be as at the date of the breach of duty, notice can be taken at the trial of subsequent realities which reduce the limits of speculation. A second action cannot be brought in respect of the same cause of action, however great the loss that may be suffered in the future, and however unexpected the loss may be, since the cause of action has become res judicata.

That is a principle which applies to the assessment of damages generally; it is necessary to examine whether it can in law be applied to the assessment of damages due to the severance of land and if so how it applies to the facts of this case.

Bolton Metropolitan Borough Council v *Waterworth* [1981] EGD 586; (1981) 259 EG 625 in the Court of Appeal was concerned with the damages resulting from severance of land whereby the access to the retained land was removed. The facts were not dissimilar to those in the present case. The member of the Lands Tribunal, Mr E C Strathon FRICS, acting as arbitrator (1978) 251 EG 963, 1071, found that as a matter of law the *Bwllfa* principle could be applied but as a matter of fact was not relevant to the circumstances of that case. The Court of Appeal did not disagree, although they did not find the arbitrator's reasons for his finding of fact very fully or clearly explained.

On that authority I find that as a matter of law the *Bwllfa* principle is capable of being applied to this case and I turn to consider whether or not it should as a matter of fact.

Section 7 of the Compulsory Purchase Act 1965 reads as follows:

> In assessing the compensation to be paid by the acquiring authority under this Act regard shall be had not only to the value of the land to be purchased by the acquiring authority, but also to the damage, if any, to be sustained by the owner of the land by reason of the severing of the land purchased from the other land of the owner, or otherwise injuriously affecting that other land by the exercise of the powers conferred by this or the special Act.

Thus what has to be ascertained is the damage to be sustained by the owner of the land by reason of the severance. It is common ground that damage is to be measured by the depreciation in market value of the retained land as at the date of severance.

As the retained land was not actually bought or sold in the open market either immediately before or after that date, an estimate has to be made of the value of the land on consideration of all the factors "past, present and future, certain and contingent", which would be in the minds of those in the market for that land at that time.

Thus the damage in this case is the measure of the owner's loss as at a certain date. This can be distinguished from damage which results from an occurrence at a certain date but where the loss is experienced, or can be better calculated, subsequently. It is easy to imagine cases of injurious affection under section 7 of the 1965 Act coming within this second category. It seems to me that the damages in *Bwllfa* and *Carslogie Steamship Co Ltd* also come within the second category.

The damage due to severance in this case is not ongoing; both valuers approach the problem on a "before and after" basis of valuation as at the date of severance. The events which occurred after that date, the refusal of the Wimpey application, the abandonment of the Southern Exeter Valley Primary Distributor Road and the estate road to the north, the approval of the County of Devon structure plan and the adoption of the Exeter local plan and other matters are all factors which subsequently affected the hope value of the retained land, but it is not suggested that they had any bearing on the hope value estimated as at the date of severance.

In my view it would be wrong in valuing damages due to severance to take account of subsequent events which had no bearing on the valuation as at the date of severance. I would distinguish this case from cases where subsequent transactions in land might be called in aid to establish a valuation at a previous date or where damages for injurious affection can be better identified or valued some time after the date of severance.

For the reasons that I have set out above I accept Mr Challinor's alternative valuation C and award compensation in the sum of £726,647.

Summary

Section 18 certificate

In the before-acquisition world the value of the red (acquired) land and the value of the blue (retained) land are properly assessed on the basis that planning permission for residential development of those lands would have been obtained not earlier than January 1 1976 in accordance with the section 18 certificate.

Access on to Rydon Lane

It would be wholly incorrect to calculate compensation for damages based upon an alternative and imaginary scheme (the Cartwright Roundabout) which in effect incorporated and extended the scheme which gave rise to the acquisition.

In any event, I find as a matter of fact that the massive roundabout was a wholly improbable solution to the problem and unlikely to affect the value of the blue (retained) land.

Alternative access on to Topsham Road

I find the prospect of gaining access on to Topsham Road by any of the four proposed routes to be too remote to have any measurable influence on the value of the blue land.

Valuation problems

Method of valuation

I prefer Mr Challinor's approach of valuing separately the acquisition of the red land under rule (2) of section 5 of the Land Compensation Act 1961 and the damages due to severance of the blue land under section 7 of the Compulsory Purchase Act 1965.

Stokes *payment*

I accept Mr Challinor's view that the payment would represent approximately one-third of the full residential value of the blue retained land.

Before acquisition

Buffer strip

I find no justification for deducting 0.25 of an acre from the total area of the land.

Disability of access

In view of my finding on the section 18 certificate, the scheme does not include the M5 motorway and it cannot be assumed that Rydon Lane would continue to carry the bypass traffic after January 1 1976. Therefore there should be no allowance for disability of access.

Deferment

In the special circumstances of this case, I find that there should be no deduction for deferment.

Agricultural use

I find that it is proper to add the agreed amount for the value of the agricultural use of the land.

Bwllfa *principle*

I find as a matter of law that the *Bwllfa* principle is capable of being applied in this case but as a matter of fact it should not.

For these reasons I accept Mr Challinor's alternative valuation C and award compensation in the sum of £726,647.

Costs

Having read this part of my decision, I had opened in court the sealed offer made by the acquiring authority, which is less than my award.

I heard submissions as to costs. There is no dispute that costs should follow the event and, with one exception, that this should be on the High Court scale of costs on the "standard basis".

Counsel for the claimants submitted that costs should be recovered over and above the standard basis due to the Cartwright roundabout concept being introduced late into the case by the acquiring authority. That is to say that the proviso to rule 3(4) of Ord 62 should apply and that that part of the costs should be taxed on the indemnity basis. He recognised that the proviso is to be applied only in very special cases and referred me to *Bowen-Jones* v *Bowen-Jones* [1986] 3 All ER 163 and a passage from the decision of Knox J at p 165 as follows:

> On the basis of this change in the rules counsel for the plaintiff submitted to me that this presented an opportunity for a review of the basis on which orders for taxation should be made in favour of a successful litigant. He submitted that I should take this opportunity to redress what he described as having been commented adversely on by a succession of judges as something in the nature of an injustice, namely that a successful litigant who, under the old dispensation, was awarded costs on a party and party basis only recovered roughly two-thirds in practice of the costs to which he or she had been put. It was not disputed that, broadly speaking, two-thirds, as a matter of practical fact (although it might vary from town to country), was a fair measure of the difference between what a successful litigant would actually spend and what he might recover on a party and party taxation. It was submitted to me that a fairer result would be to limit that basis of taxation, which is now provided for by the standard basis under r12(1) of Ord 62, to those cases where there was what was described as a genuine difference of opinion or of recollection and there was no question of a deliberate attempt to keep the other party out of their money or a deliberate attempt to mislead the court.
>
> In my judgment this is not the occasion for reappraising the cases which should be dealt with on the standard basis, which in my judgment is the successor of party and party costs and common fund costs and those cases where costs should be awarded on the indemnity basis. It seems to me that there has been a rationalisation of the different bases of taxation rather than a complete restatement of the rules in such a way as to place the matter entirely at large as to how costs should, in general, be dealt with between a successful and an unsuccessful litigant.

It follows that, in general, I see no reason to depart from the practice that was stated by Brightman LJ in *Bartlett* v *Barclays Bank Trust Co Ltd (No 2)* [1980] Ch 515 at p 547; [1980] 2 All ER 92 at p 98, where he said:

> . . . the usual rule (subject to well-recognised exceptions) in the case of fiduciary, contractual or tortious wrongdoing is that the defendant pays to the plaintiff only party and party costs. It is not, I think, the policy of the courts in hostile litigation to give the successful party an indemnity against the expense to which he has been put and, therefore, to compensate him for the loss which he has inevitably suffered, save in very special cases. Why this should be, I do not know, but the practice is well-established and I do not think that there is any sufficient reason to depart from that practice in the case before me.

Of course, there has to be a departure now, in that party and party costs are no longer provided for by Ord 62, and so this passage needs to be read in the light of the new rule and one has to substitute "standard" for "party and party" costs. But it would seem to me that something much more radical in the way of recasting of the rules would be required before that very well-established practice was, as a matter of principle, departed from. There is nothing that I have been shown in the new Ord 62 that directly suggests that any new principle is to be adopted by the court as a matter of general policy. I therefore do not see that as a matter of general policy it would be right for me to depart from that principle so stated by Brightman LJ.

Counsel for the claimant contended that there were four matters relating to the Cartwright roundabout which could be regarded as very special circumstances as follows:

1 That both the acquiring authorities and the district valuer were in a special position in that it was in the public interest that they had a duty to ensure that the right compensation was paid.

2 That the concept of the Cartwright roundabout was introduced at the eleventh hour.

3 That as a result the proceedings were considerably prolonged.

4 That in my decision the Cartwright roundabout scheme was rejected.

Counsel for the acquiring authorities submitted that the true test was whether the matters referred to were "exceptional" and not just "special". There could be no suggestion that the introduction of the Cartwright roundabout was in any way an attempt to mislead and because as a matter of fact and law the scheme was rejected this did not qualify the matter to be treated as exceptional.

I was not referred to any examples of the "well-recognised exceptions" referred to by Brightman LJ in *Bartlett*. Therefore I am left to determine the matter on my own views as to what can be regarded as exceptional.

I have no doubt that the action by the acquiring authority to introduce a fundamental change in their case was made in good faith, but one hopes that to do so at the eleventh hour was an exception to customary practice. I think it follows that the claimants have been put to exceptional cost in order to counter the acquiring authority's submissions, but to the extent that costs follow the event they are in part compensated for the expenses that they have incurred.

Rule 3(4) of Ord 62 reads as follows:

> (4) The amount of his costs which any party shall be entitled to recover is the amount allowed after taxation on the standard basis where:
> (a) an order is made that the costs of one party to proceedings be paid by another party to those proceedings, or
> (b) an order is made for the payment of costs out of any fund (including the legal aid fund), or
> (c) no order for costs is required,
> unless it appears to the Court to be appropriate to order costs to be taxed on the indemnity basis.

The dispute which gave rise to this reference has been outstanding for many years. The Cartwright roundabout was not introduced until the eleventh hour and did not take its final form until the case was part heard. This represented a fundamental change in the acquiring authority's approach to the assessment of compensation. I think that it is right that the claimants should be compensated on the indemnity basis for the additional cost that they incurred because of the late notice of the scheme and for that only, and I so award.

LANDS TRIBUNAL

RATING

ALUWIHARE (VO) v MFI PROPERTIES LTD
(LVC/121/1986)
August 18 1987
(W H REES FRICS)

Estates Gazette January 9 1988
[1988] 01 EG 67-72

Validity of proposal and notice of appeal — Point not previously before Lands Tribunal or courts — Property in Old Kent Road, SE London, comprising 2-storey showroom building, separate warehouse building at rear and car-parking yard in between — Respondents (MFI), a firm selling large pre-packed items, occupied all parts except ground-floor showroom, which was in different occupation — VO's proposal was to insert three separately rateable hereditaments in valuation list in place of existing single hereditament (which was under reconstruction and had nil assessment) — The hereditaments proposed were the ground-floor showroom (with GV £29,000), the showroom occupied by MFI (GV £24,375) and the warehouse (GV £18,400) — (Present appeal not concerned with assessment on ground-floor showroom) — During LVC hearing VO became aware that under MFI's lease there were rights of car parking in, and provisions as to maintenance of, the yard — By date of present appeal hearing he had accepted that MFI's occupation was of one hereditament, but he contended for assessment of GV £34,000 against MFI's figure of not exceeding £24,375 — LVC had found functional dependency between parts of MFI's occupancy, since customers ordered and paid for goods in showroom, then went across car park to collect them from warehouse — Proposal held invalid by LVC and VO appealed — MFI contended his notice of appeal also invalid — Tribunal found proposal incorrect: it had a fundamental flaw in that it sought to assess as two hereditaments a property which comprised only one — *British Railways Board* v *Hopkins* distinguished — VO in his notice of appeal referred only to ground-floor showroom and omitted any reference to first-floor showroom and warehouse occupied by MFI — Tribunal held notice of appeal bad, since it did not describe accurately the hereditament or hereditaments referred to in proposal — As in *White* v *Dales*, tribunal had no jurisdiction to entertain present appeal — Costs awarded against VO

The following cases are referred to in this report.

Behrmann v *Seymour (VO)* [1972] RA 313; (1972) 17 RRC 333, LT
Brighton Marine Palace & Pier Co v *Rees (VO)* (1961) 9 RRC 77; [1961] RVR 614
British American Typewriters v *Hill (VO)* (1962) 9 RRC 353; [1962] RVR 374
British Railways Board v *Hopkins (VO) and Birmingham District Council* [1981] RA 328, LT
Eagle Construction Ltd v *Casey (VO) and Croydon London Borough Council* [1981] RA 347; (1981) 259 EG 871, LT
Ellerby v *March* [1954] 2 QB 357; [1954] 3 WLR 53; [1954] 2 All ER 375; (1954) 47 R & IT 335; 52 LGR 397, CA
Evans v *Lamb (VO)* [1973] RA 377, LT
Firkins v *Dyer (VO)* [1972] RA 209; (1972) 17 RRC 363, LT
Gilbert (VO) v *S Hickinbottom & Sons Ltd* [1956] 2 QB 40; [1956] 2 WLR 952; [1956] 2 All ER 101; (1956) 1 RRC 46; 49 R&IT 231; 54 LGR 215, CA
Jollys (Oldham) Ltd v *Almond (VO)* (1951) 14 R & IT 37
Morecambe & Heysham Corporation v *Robinson (VO)* [1961] 1 WLR 373; [1961] 1 All ER 721; (1961) 59 LGR 160; [1961] RVR 137; 8 RRC 33, CA
R v *Winchester Area Assessment Committee, ex parte Wright* [1948] 2 KB 455; [1948] 2 All ER 552, CA
Stirk & Sons Ltd v *Halifax Assessment Committee* [1922] 1 KB 264; [1921] All ER Rep 355; (1921) 126 LT 338; 38 TLR 83; 20 LGR 43
White v *Dales (VO)* Lands Tribunal (LVC/368/1984), unreported

The appellant valuation officer, T B Aluwihare FRICS, appeared in person; Nigel Hague QC (instructed by Cripps & Shone, of High Wycombe) for the respondent ratepayer company.

Giving his decision, MR REES said: This is an appeal by a valuation officer against a decision of a local valuation court which determined that a proposal to insert entries in a valuation list was invalid. The respondent sought to uphold the decision of the court below on that point but also contended that the notice of appeal to this tribunal, too, was invalid. The premises concerned are a showroom on a first floor (above another — ground-floor — showroom not in the respondent's occupation) and a warehouse at the rear with a yard used for parking vehicles in between. No witnesses were called: values were agreed.

A copy of the proposal referred to above and a copy of the local valuation court's decision are annexed hereto.* The authorities and decisions to which I was referred are:

Gilbert (VO) v *S Hickinbottom & Sons Ltd* [1956] 49 R & IT 231; *British Railways Board* v *Hopkins (VO)* [1981] RA 328; *R* v *Winchester Area Assessment Committee, ex parte Wright* [1948] 41 R & IT 348; *Firkins* v *Dyer (VO)* [1972] RA 209; *Behrmann* v *Seymour (VO)* [1972] RA 313; *Evans* v *Lamb (VO)* [1973] RA 377; *Eagle Construction Ltd* v *Casey (VO) and Croydon London Borough Council* [1981] RA 347; *Jollys (Oldham) Ltd* v *Almond (VO)* [1951] 14 R & IT 374; *British American Typewriters* v *Hill (VO)* (1962) 9 RRC 353; [1962] RVR 374; [1962] RA 298; 182 EG 761; *Ellerby* v *March* [1954] 2 QB 357; [1954] 2 All ER 375; *Brighton Marine Palace & Pier Co* v *Rees (VO)* (1961) 9 RRC 77; [1961] RVR 614; *Morecambe & Heysham Corporation* v *Robinson (VO)* [1961] 1 WLR 373; *Ryde on Rating* 13th ed p 136.

The grounds of appeal to this tribunal are that "the decision of the local valuation court is incorrect and bad in law".

By the date of the hearing before me, the appellant accepted that the occupation by MFI Ltd was of one hereditament (not two as in the proposal) and it was agreed that if an entry should be made in respect of that hereditament comprising the first-floor showroom, the warehouse and premises, the gross value should be £34,000. The respondent contended, however, that if the proposal is not invalid and there should be an entry in respect of that single hereditament, the gross value should not exceed £24,375. The appeal is not concerned with the assessment on the ground-floor showroom occupied by Carpetland Ltd.

In the notice of appeal to this tribunal dated March 10 1986 (there was an earlier notice) the address and description of the hereditament is given as: "Showroom and Premises 651/657 Old Kent Road, SE15 (Ground Floor)". The respondent contended that that is incorrect.

I consider first:

The proposal

The appellant's submission

The appellant submitted that the proposal satisfied all the requirements of section 69(5) of the General Rate Act 1967: it was in

*Editor's note: The copy of the proposal is not reproduced here. All the material information is given in the two opening paragraphs of the local valuation court's decision: see p 221 *post*.

A writing, it specified the grounds on which the proposed alteration was supported and complied with the relevant regulations. In that connection he referred to *R* v *Winchester* at p 189. It was clear from the proposal that the entirety of the MFI occupation had been accurately and adequately described with correct addresses and was identifiable although in two parts. There was no doubt as to the extent of the occupation and the purposes thereof. The fact that the valuation officer decided on two hereditaments and not one did not invalidate it. He had exercised his judgment and had decided upon two entries and it was wrong to say that because he had chosen that course the proposal was invalid. He relied particularly upon the following passage from *British Railways* at p 344:

B I have come to the conclusion that it was open to the valuation officer in the light of circumstances existing at the date of the proposals to make proposals in respect of the separate parts. By virtue of s69(2) a valuation officer may at any time make a proposal for any alteration of the valuation list and clearly he must exercise his judgment in so doing. In the instant case I think it would also have been open to him to make a single proposal in respect of the whole of the unoccupied parts and subsequently to carve portions out of that hereditament to accord with the lettings which subsequently occurred. In the same way it would have been open to the valuation officer to make two proposals in respect of the occupied part, that is to say one in respect of the lower ground, ground and part first floor and another in respect of the fifth to tenth floors inclusive. The fact that he chose one way rather than the other does not seem to me to lead inevitably to the conclusion that on one approach effect is to be given to a proposal but on the other no effect on the grounds that the proposal
C is invalid. It seems to me that there are many instances where two parts of a single hereditament may appear as two separate entries in the valuation list even though they could also and more appropriately be described as one. So much is implicit in the passages which I have quoted from the judgments of Denning LJ in *Gilbert* v *Hickinbottom* and of Waller LJ and Griffiths LJ in *Brent* v *Ladbroke Rentals.*

The position here was analogous to those circumstances: it was only later that it had been agreed that there should be one hereditament here. He referred to *Firkins* at p 218:

I think the proposal of March 27 1969 was bad because the grounds of the proposal were insufficient and incorrect and did not satisfy the requirements of s69(5) of the 1967 Act. Also I think the true facts and arguments were so far away from those indicated in the proposal that they cannot be regarded as fairly falling within the contentions of the valuation officer or as within the
D powers of the local valuation court or of the Lands Tribunal.

Here the facts and arguments were not far away from, and within the contentions of, the valuation officer. There was no change in the address or description of the hereditament. The agreed gross value of the single hereditament was below the proposed total of the two parts. The same comments could be made in respect of *Behrmann.* He referred to the headnote to *Evans*:

The simple increase of the single assessment to the £200 gross value determined by the local valuation court in place of the division of the assessment proposed could not be made on the valuation officer's proposal because:
(1) The grounds of proposal were not eked out by anything else in it which
E could be taken to support the alterations now sought by the valuation officer.
(2) As it was no longer contended that the property constituted two hereditaments the grounds stated were incorrect and the true facts and arguments were so far away from those indicated in the proposal that they could not be regarded as falling within the valuation officer's contentions.
(3) The proposal was therefore invalid.

He was not now contending for something completely different and there was one assessment in the valuation list. The proposal made here showed clearly what the valuation officer's intention was. He sought to distinguish *Eagle* on the facts: the proposal there was made after the new building was completed and then a further proposal was lodged. In *Eagle* there were several occupations, but here the proposal identified the subject hereditament and described the
F occupations even though MFI was shown as two. He referred to *Jollys* and to *British American* as examples of circumstances where proposals were found to be valid. With regard to the respondent's contention, the proposal as regards Carpetland was held to be valid: the proposal could not be good in part and bad in part at the same time. The respondent's submission that the ceiling was gross value £24,375 was not tenable, since it related to only part of the occupation: the ceiling should be the total of £24,375 and £18,400 (£42,775).

The respondent's submissions

As a preliminary point counsel pointed out that the appellant valuation officer's argument (above) was not put to the court below. He agreed that the valuation officer's decision to make MFI's
G occupation the subject of two assessments was reasonable at the time in the light of the information which the valuation officer then had, that it was *bona fide* but it was wrong. In effect the appellant was now seeking to amend the proposal: there was no power in the General Rate Act 1967 to do so. The warehouse cannot be treated as being included in the proposal under the description "Showroom & Premises, 651/7 Old Kent Road — MFI Ltd": that it could be was not suggested by the appellant. To support that submission he relied on *Evans.* He referred to *Ellerby* and to *Morecambe* with reference to his submission that the maximum gross value which could be inserted in the list was £24,375. *Winchester* and *Firkins* (where the circumstances differed) were concerned with grounds of proposals, not with the question "What is the hereditament?": they were of no
H assistance here. Sections 69(1), 76(5) and 79 of the General Rate Act 1967 referred to "hereditament", which was defined in section 115(1) of that Act.

He referred to the following passage in *Ellerby* at p 363 [*per* Somervell LJ]:

On the facts stated in the present case, the local valuation court could have done anything from directing that the figures should remain as they were, £27 gross value and £20 rateable value, up to the alteration proposed by the valuation officer of £140 gross value and £113 rateable value. Therefore, it is submitted that, however limited the ambit of the appeal might be in this or other cases, the final words of section 49(1)

of the Local Government Act 1948

J confer a statutory jurisdiction on the appellate tribunal, which must not be restricted by any procedural condition precedent, to hear and determine all the issues which were raised and raisable before the original local valuation court notwithstanding, as I have said, the ambit of the appeal.
I have come to a clear conclusion that these words do not support that contention. In my opinion their true meaning is to be found by referring back to section 48(4) and seeing what are the directions which the local valuation court might have given. They are: "such directions . . . as appear to them to be necessary to give effect to the contention of the appellant if and so far as that contention appears to the court to be well founded". In my judgment, one has to read that limitation into section 49(1), substituting, of course, for the county court the Lands Tribunal, which is now the appellate court. In my view, that is the natural meaning of those words, but if there is any ambiguity I think that it falls to be resolved by a consideration of the procedure in relation to rating appeals in the past, the nature of legal proceedings in
K general, and considerations of convenience.
The general principle on which all courts and tribunals found their jurisdiction is that they deal with matters which parties bring before them, and it would be quite exceptional to find them dealing with matters which parties have not brought before them. Matters are brought before them in some formal document such as a writ or summons or, in the case of an appellate tribunal, a notice of appeal. It would have needed very clear words to satisfy me that it could have been intended that an appellate court should deal with issues other than those which an appellant brought before it.

So the local valuation court and this tribunal could consider only the proposal which has been made. He read (4) of the headnote in *Brighton Marine*:

L (4) The appellants could not contend that the kiosks which had been let should have been split off and assessed separately, because that issue was not raised by the proposal; *R* v *Winchester Area Assessment Committee. Ellerby* v *March* and *R* v *Surrey (Mid-Eastern Area) Assessment Committee* applied.

and referred to the following at p 85:

Both valuations begin with gross receipts. They have been taken from the accounts for 1956 and are agreed at £148,704. Mr Dixon, but not Mr Stiles, proceeds at once to deduct £1,450. His ground for so doing is that certain properties have been wrongly included in the hereditament. I have mentioned that at the landward end of the pier is a forecourt; it is owned and repaired by the Pier Company. After crossing the forecourt from the promenade one comes to the gates of the pier and the turnstiles; beyond them is the pier proper. The appellants contend that some kiosks on the forecourt, which are let by the Pier Company, should be separately assessed.
M I am satisfied that the appellants' contention is bad in law; which being so, I need not deal with its merits. Neither in the appellants' notice of objection, nor during the hearing in the local valuation court, nor in the pleadings, is there any suggestion that part of the hereditament should be split off and assessed separately. It is well established in law that the only issue in the proceedings is that raised by the proposal. See *R* v *Winchester Area Assessment Committee*; *Ellerby* v *March*; and *R* v *Surrey (Mid-Eastern Area) Assessment Committee.* If the appellants wanted the hereditament divided they should have made a proposal to that effect: it is too late to raise that issue now.

That was the reverse of the instant appeal. He also relied on the following in *Evans* at p 381:

I think that, as in *Firkins'* case, the true facts and arguments are so far away from those indicated in the proposal that they cannot be regarded as fairly

A falling within the contentions of the valuation officer. Once the contention that the hereditament constitutes two separately rateable hereditaments is withdrawn, there are no grounds remaining on which the proposed alteration in the valuation list can be supported, as required by the Act. In my opinion, therefore, no effect can be given to the valuation officer's proposal of May 7 1970.

Eagle differed as to the facts but, so far as it applies here, it supported his argument. In *British Railways* part of the building was vacant: the valuation officer there did not know how it would be occupied, whether there would be one or more occupiers, so he had choices open to him. There was no choice open to the valuation officer once there was an occupier. Neither *Jollys* nor *British American* was of assistance here: they were only examples.

B Before coming to my decision, it is right that I set out the circumstances as put to me by the appellant in which the proposal was made. At that date only some of the facts as to occupation were known to him. He made inquiries and as a result he considered that MFI Ltd occupied two hereditaments. He was unaware until later that under a lease there were rights of car parking in, and provisions as to maintenance of, the yard between the showroom building and the warehouse. He only learned of this at the local valuation court: subsequently he obtained a copy of the lease and agreed that there should be only one hereditament, the value of which has now been agreed. He has since made another proposal, but this appeal is not concerned with that.

C

Decision on proposal

It is remarkable that this point has not come before this tribunal or before the courts before 1950. Counsel rightly stressed the reference to the word hereditament in the sections of the General Rate Act 1967 which relate to "alterations of current valuation list" (section 69 onwards). It seems to me that the correct identification of the hereditament or hereditaments is fundamental to the question. As I said in *Eagle* at p 352: ". . . it seems to me that a ratepayer is entitled to have his property correctly treated by a valuation officer when it is brought into assessment" and at p 353 (after a reference to *Stirk & Sons* v *Halifax Assessment Committee* [1922] 1 KB 264):

D There, the amount of the assessment was being considered but it seems to me that the principle is even more important when related to the question "What is the hereditament?" If so, it is not open to a valuation officer seeking to bring a property into assessment to "apply an arbitrary rule of thumb", to choose whether he does so separately or as part of another hereditament: faced with the task he must first consider whether it is a separate hereditament or whether it is part of another or whether it comprises several.

So here, the valuation officer did not correctly identify the hereditament. The proposal is incorrect: it seeks to assess as two hereditaments a property which comprises only one. That is a fundamental flaw. In *British Railways* the Member said that the valuation officer had a choice and that was not disputed by counsel as far as the vacant space there is concerned. As to the occupied portion

E of the building in that case, the circumstances were not the same as here, where I see no choice for the valuation officer. As to what should be the upper limit of the gross value is only one of the uncertainties which would follow from the proposal which was in fact made, assuming that it was valid. Moreover, the proposal did not with certainty tell the ratepayer the property in respect of which he was to be rated.

Notice of appeal

In a notice of appeal dated March 5 1986 the "address and description of hereditament" was given as:

Showroom & Premises 651/657 Old Kent Road, SE15
F Showroom & Premises 651/657 Old Kent Road, SE15 (Ground Floor)
Warehouse & Premises r/o 651/657 Old Kent Road, SE15
(Plots 4, 5 & 6 Ruby Street)

There followed communications between the registrar of this tribunal, the local valuation court and the appellant and, as a result, another notice dated March 10 1986 was given. The "address and description of hereditament" was stated to be "Showroom and Premises 651/657 Old Kent Road, SE15 (Ground Floor)". There is no dispute that MFI Ltd do not occupy a ground-floor showroom. The notice dated March 10 1986 was accompanied by a letter addressed to the registrar of this tribunal which reads:

Thank you for your letter of March 7 1986 and enclosures.
I enclose as requested two copies of my Notice of Appeal omitting any reference to the First Floor Showroom and the Warehouse occupied by MFI Ltd. G
From my telephone conversation today with Mrs Richards I understand that the fact that the Notice does not contain a reference to them will not preclude me from contending for the inclusion in the Valuation List of entries in respect of the MFI Ltd hereditaments.

Counsel informed me that the respondent was not told of the notice of March 5 1986 until after the respondent's reply to the appellant's statement of case had been made: he submitted that the notice of March 10 1986 (which was the notice which initiated this appeal) was not valid. MFI did not occupy any ground floor or showroom. He frankly admitted that he had not been deceived but the notice was incorrect.

I need only decide the matter concerning the notice of appeal to this tribunal should the quesion of the validity of the proposal go further and my decision on that were found to be wrong. Having considered the submissions, it seems to me that the notice of appeal is bad: it does not describe accurately the hereditament or hereditaments referred to in the proposal. That being so, there is no appeal to this tribunal against the decision of the local valuation court. H

The position of this tribunal and of the local valuation court in this matter is somewhat similar to that determined in *White* v *Dales (VO)* (LVC/368/1984 — VG Wellings Esq QC). There the learned Member decided that the appellant at the local valuation court "had no standing to make her proposal. . . . In effect the court declined jurisdiction in the matter." The learned Member went on to consider the relevant authorities and then concluded: J

For these reasons the Lands Tribunal has no jurisdiction to entertain the present appeal and it must be and is hereby struck out.

So here the tribunal has no jurisdiction to entertain the present appeal.

Having received written submissions I order that the appellant shall pay the respondent's costs, such costs if not agreed to be taxed by the Registrar of the Lands Tribunal on the High Court Scale.

APPENDIX

IN THE MATTER OF AN APPEAL BEFORE A GREATER LONDON (SOUTH-EAST) VALUATION COURT ON FEBRUARY 11 1986 K

BETWEEN

P CREASE VO (APPELLANT)

AND

MFI LTD (OBJECTOR)

AND

HARRIS QUEENSWAY LTD
T/A CARPETLAND LTD (OBJECTOR)

re: 651/657 OLD KENT ROAD, SE15

Reasons for the decision

The appeal arises out of a valuation officer proposal made on March 28 1985 seeking to alter the valuation list by the deletion of the existing entry or entries and the insertion of the proposed entry or entries shown below: L

	Description	*Address*	*Gross value* £	*Rateable value* £
Existing Entry or Entries	Showroom Workshops & Premises (under reconstruction)	651/657 Old Kent Road	NIL	NIL
Proposed Entry or Entries	Showroom & Premises	651/657 Old Kent Road, MFI Ltd	24,375	20,284
	Showroom & Premises	651/657 Old Kent Road, Ground Floor, Carpetland Ltd	29,000	24,138
	Warehouse & Premises	r/o 651/657 Old Kent Road (Plots 4, 5, 6 Ruby Street) MFI Ltd	18,400	15,305

M

The grounds on which the proposed alteration is supported are: "That structural alterations have been made and that the hereditament as at present shown in the Valuation List now constitutes 3 separately rateable hereditaments which should be assessed accordingly".

An objection was made to the proposed assessments on behalf of MFI Ltd, who were represented at the hearing by Mr Beaumont of King & Co.

A further objection was made in respect of the proposed assessment for the Showroom & Premises occupied by Carpetland Ltd, which was verbally agreed between the agents, Smith Melzack, and the valuation officer at gross

A value £24,875, rateable value £20,701. The objecting company, Carpetland Ltd, was not represented at the hearing.

Miss Andrew, representing the appellant valuation officer, submitted schedules of valuations of the appeal properties and of comparable properties to support revised valuations of:

	Gross value
Showroom & Premises, First Floor, 651/657 Old Kent Road	£17,850
Showroom & Premises, Ground Floor 651/657 Old Kent Road	£24,875
Warehouse & Premises, R/O 651/657 Old Kent Road	£18,600

She considered that the Showroom & Premises on the first floor were not contiguous with the Warehouse & Premises and not within the same curtilage
B because the car park in between the two premises was in shared occupation. She distinguished the functional dependency of the premises by degree from that in the Court of Appeal decision in *Gilbert (VO)* v *Hickinbottom* (1956), and she argued that, while it was convenient, it was not essential for MFI Ltd to have the Showroom and Warehouse close together. She therefore considered that the Showroom & Premises (MFI Ltd) and the Warehouse & Premises should be shown as two separate entries in the valuation list. She also cited the decisions in *Barker (JW) Ltd* v *Westbury (VO)* (1951) 44 R & IT 458, *Butterley Co Ltd* v *Tasker (VO)* (1961) 7 RRC 213 and *Scaife (VO)* v *Bird's Eye Foods Ltd* (1962) 9 RRC 219 in support of her contention.

Mr Beaumont submitted photographs, a plan, and a comparable assessment in support of his valuation at gross value £28,700, rateable value £23,888 for the whole of the premises occupied by MFI Ltd. He cited decisions of the Lands Tribunal in *Underwood & Co Ltd* v *Beer (VO)* (1981) 261 EG 577,
C *Lewis' Ltd* v *Grice (VO)* (1982) 263 EG 775 and *City of Westminster* v *Burton Group plc and Saul (VO)* (1986) 277 EG 757 to support the level of value adopted in his valuation of the first-floor Showroom & Premises.

He argued that the first-floor Showroom & Premises were contiguous with, and within the same curtilage as the Warehouse & Premises because MFI Ltd had paramount control of the car park, and he read from the lease granted to Carpetland Ltd in support of this. He added that the car park was signposted by "MFI" and gave the appearance of being in the sole occupation of MFI Ltd. He explained that MFI Ltd specialised in selling large prepacked items, and that a car park is essential to their method of trading. He considered the Showroom & Premises and Warehouse & Premises to be functionally dependent because customers order and pay for the goods in the showroom, then go across the car park to collect the goods at the "pick-up" point in the warehouse.

D He distinguished the facts in the appeal case from the facts in the decisions cited by the valuation officer and cited the Lands Tribunal decision in *Page* v *Whitfield (VO)* (1985) to support his argument that the premises occupied by MFI Ltd should be regarded as one hereditament. He therefore felt that there could be a case for the court to consider the whole proposal invalid and, E
because of this, confirm the figures in the list at gross value NIL, rateable value NIL, but he suggested the court should confirm the verbally agreed figures in respect of the entry proposed for the Showroom & Premises Ground Floor (Carpetland Ltd), and direct a figure of up to gross value £18,400 in respect of the premises occupied by MFI Ltd.

Having regard to section 76 (5) of the General Rate Act 1967, the court "shall give such directions with respect to the manner in which the hereditament in question is to be treated in the valuation list as appear to them to be necessary to give effect to the contention of the appellant *if and so far as* the contention appears to the court to be well founded". The court are clearly bound by the contention of the appellant, which in this case is the deletion of the existing entry and the insertion of the entries proposed.

There is no dispute between the parties that the existing entry should be deleted and the court believe this to be well founded. The valuation list should F
therefore be amended accordingly.

With regard to the proposed insertion of the entry for the Showroom & Premises, Ground Floor, the court accept that the hereditament is properly described and direct the figures verbally agreed between the valuation officer and the agents for Carpetland Ltd which are fair and reasonable at gross value £24,875, rateable value £20,701.

In respect of the proposed insertions for the Showroom & Premises (MFI Ltd) and Warehouse & Premises, the court distinguish the circumstances in the appeal case from those in *Barker* v *Westbury* because there is contiguity, the car park being in the control of MFI Ltd and the two buildings being additionally connected by their shared "sprinkler" system. In *Butterley* v *Tasker* functionally unconnected agricultural land separated the premises concerned and public rights of way appeared to have been established on the private road. G

Referring to *Scaife* v *Bird's Eye Foods Ltd*, the court accept that the premises could be separately let, but it is not axiomatic that therefore they should be assessed separately. The functional unity of the Showroom, Warehouse and customer car park is such that commonsense demands a single hereditament in this case as indeed it did in *Gilbert* v *Hickinbottom*.

The court find that in the appeal case there is functional dependency, the ordering and paying for the goods in the Showroom, and the collection of these goods from the Warehouse being part of the same process, ie the purchase of goods.

The court decide therefore that the premises occupied by MFI Ltd should be regarded as one hereditament for rating purposes. The proposed insertions in respect of the Showroom & Premises (MFI Ltd) and Warehouse & Premises are considered by the court to be invalid because they do not adequately or H
accurately describe the hereditament presently occupied by MFI Ltd. The court are therefore unable to make an appropriate insertion in the valuation list.

For further cases on this subject see p 157

LEASEHOLD VALUATION TRIBUNALS

LYNCH AND ANOTHER v CASTLEBEG INVESTMENTS (JERSEY) LTD

RENT ASSESSMENT PANEL FOR WALES (LVT/876/1/85)

June 3 1987

(MALCOLM HUSBAND DFC (chairman), RHYS DAVIES ARICS ARVA and E A DEPP)

Estates Gazette January 23 1988

[1988] 03 EG 101-103

Freehold price of house in Cardiff outer suburb with 85 years of lease unexpired — Agency commission on single insurance premium — Landlord's valuer's reliance on Lands Tribunal's decision in *Re Castlebeg Investments (Jersey) Ltd's Appeal* (1985) in which he gave evidence to effect that purchaser could obtain commission on single insurance premium — Lands Tribunal did not actually decide whether it was right to consider sale of one freehold reversion in isolation but did not think it was — On evidence adduced, however, that tribunal accepted YP which reflected insurance commission — Tenants, appearing in person, argued that once freehold was sold the element in agency commission to cover administration and collection costs would disappear and it was therefore unnecessary to pay landlord compensation for its loss — Present tribunal considered themselves bound by Lands Tribunal's decision, provided they had sufficient evidence that purchaser of freehold reversion could obtain commission on insurance premium — Under lease terms landlord could compel tenants to insure comprehensively in nominated office through landlord's agency — Tribunal must, in their view, be satisfied that purchaser of single freehold could obtain such commission — Although matter not entirely free of doubt, they concluded, on evidence before them (including evidence from landlord's valuer additional to his *Re Castlebeg* submissions), that they should properly admit that element of claim — Landlord's 20 YP multiplier not accepted — Contrary to landlord's valuer's view, tribunal considered negotiations in all but one of his comparables probably influenced by "Delaforce" effect — 7% applied in determination of £680 against tenants' figure of £650 and landlord's £800

The following cases are referred to in this report.

Castlebeg Investments (Jersey) Ltd's Appeal, Re [1985] 2 EGLR 209; (1985) 275 EG 469, LT

The applicant tenants appeared in person; Mr W K R Ricketts FRICS represented the respondent landlord.

Giving their decision, THE TRIBUNAL said: This is a decision on an application made by the tenants, Mr James Lynch and Mrs Jane Bernadette Lynch, for the determination of the price payable under section 9 (as amended) of the Leasehold Reform Act 1967 for the freehold of the house and premises, 4 Fairbrook Close, Rhiwbina, Cardiff.

We inspected the subject property prior to the hearing and viewed 2 Fairbrook Close from the Close.

The subject property is situated in an outer suburb of Cardiff comprising good-quality dwellings and according to Mr W K R Ricketts FRICS, who appeared for the landlord, Castlebeg Investments (Jersey) Ltd, the property has an RCA of 1,813 sq ft plus garage and porch. The subject lease is for a term of 99 years from March 25 1972 at a yearly rent of £40 and the relevant time for the purposes of section 37(1)(d) of the Act is May 28 1986.

At the hearing Mr and Mrs Lynch appeared in person. Mr Lynch capitalised the ground rent by using 7%, producing a multiple of 14.3 and giving £572 and also added a figure for the reversionary interest of £62. Mr Ricketts agreed to exclude such interest from the calculations. The total figure from Mr Lynch was £634, rounded off at £650. Although Mr Lynch referred to two comparables, he was unable to give details of them and Mr Ricketts invited us to exclude them. We do not think that they were particularly relevant, but Mr Lynch also referred to the price that he understood had been paid for the freehold of 2 Fairbrook Close in 1984, namely £600. While it seems probable that the term of years and ground rent were identical to the subject property, Mr Lynch produced no evidence in support of this, although Mrs Lynch thought that the owner of no 2 had paid £20 a half year ground rent.

As to the comparables introduced by Mr Ricketts, Mr Lynch did not accept them, since he felt that tenants were minded to compromise rather than have legal argument.

In the matter of the landlord's right to receive insurance commission Mr Lynch was not aware of any provision in the Act to cover this and stated that once the freehold was sold the element in agency commission to cover administration and collection costs would disappear and that therefore it was unnecessary to pay the landlord compensation for its loss. He could not see how an agency commission could be as legitimate a part of the landlord's income as the ground rent.

Mr Ricketts was relying on the authority of *Re Castlebeg Investments (Jersey) Ltd's Appeal* heard by the Lands Tribunal in June 1985 [1985] 2 EGLR 209.

Mr Lynch suggested that because the tenant was not represented at such hearing this tribunal should not be influenced by such decision, but he was unable to suggest any submissions that he would have made had he been present at such hearing.

The tenant introduced a cutting from a Sunday newspaper suggesting a price for a freehold of 5-8 years' ground rent. We are not influenced by this document. Although the tenant referred to advice from two chartered surveyors he did not call any witnesses.

Mr Ricketts' valuation adopted a 20 YP, giving a price of £800. He supported this figure by two groups of comparables. In the one group 20 YP had been used in the negotiations and in the other the price reflected 22 YP. Mr Ricketts could not say whether in such matters the tenants had received professional advice save that in one case in the former group he himself had advised the tenant. Mr Ricketts acknowledged the existence of what is known as the "Delaforce" effect and contended that this was represented by the difference of two years' ground rent between the two groups. In reply to an inquiry by the tribunal Mr Ricketts maintained that there was no "Delaforce" effect in the negotiations resulting in a 20 YP formula. We do not accept this and consider that in both groups "Delaforce" would probably have influenced the negotiations save in the instance in which Mr Ricketts participated.

Mr Ricketts relied on the decision in *Re Castelbeg Investments (Jersey) Ltd's Appeal* in which a multiple of 16 had been adopted. The Lands Tribunal in that decision had emphasised that it was reached on the evidence adduced and that there being no respondents the tribunal did not have evidence from any other expert or any argument on behalf of the tenants, but the tribunal added that it had no reason to reject any of the evidence in that instance. Mr Ricketts' evidence before the Lands Tribunal was to the effect that a purchaser could obtain a commission in respect of a single insurance premium and would take it into account and it seemed to the tribunal that it would be correct to take account of any advantage which a lease may have to the freeholder, and the commission on a fire insurance premium was such an advantage.

The Lands Tribunal did not actually decide whether it was right to consider the sale of one freehold reversion in isolation, although the tribunal did not think it was.

The question of whether the value of the right to receive an insurance commission should be taken into account in calculating the price of a freehold under the Act has been considered on several occasions by a leasehold valuation tribunal and has been rejected in the absence of satisfactory evidence. Indeed, the said decision of the Lands Tribunal followed a leasehold valuation tribunal decision rejecting such a claim.

However, we are now bound by such Lands Tribunal decision provided that we have sufficient evidence that a purchaser of a freehold reversion could obtain a commission in respect of the insurance premium. Certainly there is no doubt that under the terms of the lease the landlord can compel the tenant to insure the subject property comprehensively in a nominated office through the landlord's agency. Our own view, notwithstanding the reservation expressed by the Lands Tribunal, is that we must be satisfied that the purchaser of a single freehold could obtain such a commission. In addition to referring to the acceptance by the Lands Tribunal of such a submission, Mr Ricketts introduced further evidence, and although the matter is not, we think, entirely free of doubt we come to the conclusion that on the evidence before us we should properly admit this element of the landlord's claim.

Mr Ricketts had produced evidence of the cost of rebuilding the property at £70,000 and Mr Lynch assented to this figure. Mr Ricketts deduced that the commission at the relevant time would have been £16.80 pa and emphasised that this figure would increase as building costs (and possibly agency commission rates) increased.

Mr Ricketts was not persuaded to apportion the price between the two elements of ground rent and insurance but maintained that the multiple of 20 was the correct figure and was supported by the comparables.

We do not accept the YP of 20 and we allot a multiplier of 14.286 to the ground rent, and since the insurance commission is not as well secured as the ground rent we allot a YP of 6.5 for the insurance commission.

Accordingly our calculations are:

Ground rent	£40.00pa	
YP at 7%	14.286	£571.44
Insurance commission	£16.80pa	
YP	6.50	
		£109.20
		£680.64
		Say £680.00

We therefore determine the price under the subject application at £680.

For a further case on this subject see p 85

Note: The next case appears on p 227 of the green section

SUPPLEMENT
CASES NOT REPORTED IN "ESTATES GAZETTE"

LANDLORD AND TENANT
GENERAL

Court of Appeal

October 7 1986

(Before Lord Justice KERR and Lord Justice PURCHAS)

NEW PINEHURST RESIDENTS ASSOCIATION (CAMBRIDGE) LTD v SILOW

Landlord and tenant — Appointment of managing agents — Landlords a tenants' association with tenants who were also shareholders in the association — Lease, which had originally been granted by a development company, provided for the appointment and duties of managing agents, who were to act as experts and not as arbitrators — Issue as to whether association acted correctly in appointing as managing agents a committee of tenant shareholders — Appeal by a tenant from decision of county court judge — The leases provided for a service charge and for contributions from the tenants towards the expenses of repairs and services to be carried out by the landlords — Accounts were to be kept by the managing agents and audited, and the agents were responsible for making adjustments according to whether tenants had paid too much or too little by way of contributions — The main question was whether the appointment of the management committee of tenant shareholders was a proper appointment from the point of view of independence and expertise — The county court judge, after considering and distinguishing the case of *Finchbourne* v *Rodrigues*, held that the appointment was proper in both respects — Although not professional surveyors or estate agents, the members of the committee had experience of worldly affairs and in some cases of actual property management, including the bursar of a Cambridge college — They were people who might be expected to have some independence of mind and had sufficient expertise to discharge their duties — The Court of Appeal agreed and rejected a suggestion that their independence was open to challenge on the ground that the managers were themselves tenants — They were separate legal persons from the landlords and the kind of abuse which existed in the *Finchbourne* case was not present — As regards expertise, no formal professional qualification was required and the judge was justified in holding that they possessed sufficient expertise to carry out their duties — There was a subsidiary point in the case — The appellant tenant objected to pay as part of the service charge the cost of thermostatic radiator valves installed in some of the flats on the estate — The appellant did not want to have them fitted in her flat and they never were fitted — The county court judge held that as the landlords were required by covenant "to use their best endeavours to keep the radiators in the flats efficiently and adquately heated", they were entitled to decide that the most efficient and cheapest way of doing so was not merely to install new boilers but also to install the thermostatic heating valves — The Court of Appeal agreed with the judge on this issue also — Appeal dismissed

The following case is referred to in this report.

Finchbourne Ltd v *Rodrigues* [1976] 3 All ER 581; (1976) 238 EG 717, CA

This was an appeal by Mrs Bertha St Alban Silow from the decision of Judge McDonnell at Westminster County Court in favour of her landlords, the New Pinehurst Residents Association, a tenants' association of which Mrs Silow was herself a tenant and shareholder. The premises in question were in Westbury Court, Grange Road, Cambridge, one of three blocks owned and administered by the respondents.

Philip Mott (instructed by Coles & Stevenson) appeared on behalf of the appellant; Robert Lamb (instructed by Clarke Willmott & Clarke) represented the respondents.

Giving judgment, KERR LJ said: This is an appeal from part of a judgment given by His Honour Judge McDonnell OBE in the Westminster County Court on March 5 1986. He had before him three separate actions, but we are concerned with only two issues, which are common to all of them, though it appears that there were also numerous other matters of dispute.

The appeal is by the defendant, Mrs Bertha Silow, against her present landlords, the New Pinehurst Residents Association (Cambridge) Ltd, who are, as their name indicates, a tenants' association, she herself being one of the tenants and a shareholder in the association.

The premises in question, owned and administered by the plaintiffs, comprise three blocks of flats in Cambridge — Westbury Court, Marlborough Court and Amhurst Court. Mrs Silow is the tenant of two flats in Westbury Court in Grange Road. Nothing turns on the fact that she is the tenant of two flats; although two leases and two shares are involved, the issues are the same.

The original lease — I can refer to it in the singular — was by a developer, Contemporary Homes Ltd, in 1978. Then, in 1981, the freehold became vested in the respondents/plaintiffs and, as I have said, each of the tenants has one share per flat, which he or she is bound to assign to a successor if the lease is sold.

The first issue can be summarised as follows. Upon becoming the freeholders, the plaintiff residents' association appointed as managing agents, who were to act as experts and not as arbitrators in the management of the three blocks of flats, a committee of tenants/shareholders, who were no doubt appointed in general meeting after the plaintiff company had been established. The issue in that regard is whether that appointment, having regard to the composition of the committee and the qualifications of the persons involved, falls within a term of the lease dealing with managing agents, to which I shall come in a moment.

The second, and less important, issue concerns one particular item of expense which formed part of the total service charge which the defendant was obliged to pay under the lease, the cost of thermostatic radiator valves which were installed in many of the flats on the estate.

I turn first to the relevant provisions of the original lease. It was concluded on January 1 1968 for a term of 99 years; for present purposes I can go straight to clause 4(2)(a):

THE Lessee hereby covenants with the Lessor and with the owners and the Lessees for the time being of the other flats carports and garages comprised in the property that the Lessee will at all times hereafter . . . (2) (a) contribute and pay to the Lessor a due proportion (as hereinafter defined) of the costs expenses outgoings and matters mentioned in the Fourth Schedule hereto. The annual amount of such contribution (hereafter called "the service charge") shall be estimated and certified by the Lessor's managing agents acting as experts and not as arbitrators for each year commencing from the

A 29th September 1967 and such certificates shall contain full details and figures relating to all the component elements comprised therein . . . .

Then there are provisions for the payment by the lessee of the service charge so computed in certain instalments annually, to which it is unnecessary to refer.

Subclause (d) provides that:

the Lessor will procure that the Lessor's Managing Agents will keep proper accounts of the amount of the Service Charge paid by each Lessee,

and then it goes on, as well as subclause (e), to provide that the accounts so kept by the managing agents are to be audited, and it also provides for the subsequent adjustment of the accounts, according to whether each tenant has paid too much or too little or the correct B amount.

Although relating to the second issue, I then come to the provisions of the lease which deal with the components of the service charge. Clause 6 provides as follows:

THE Lessor Hereby Covenants with the Lessee that subject to the Lessee paying the contribution referred to in Clause 4(2) hereof the Lessor will

— and then one can go to subclause (5) —

maintain repair redecorate and renew as and when necessary:
(a) the main structure and in particular the main walls roofs gutters and rainwater pipes ventilation plant and machinery central heating apparatus and boilers in the property.

The subsequent subclauses deal with other common parts and C services in each of the blocks of flats. Then subclause (9), read with the beginning of clause 6, provides as follows:

THE Lessor Hereby Covenants with the Lessee that subject to the Lessee paying the contribution referred to in Clause 4(2) hereof the Lessor will . . . use its best endeavours to keep the radiators in the flat sufficiently and adequately heated between the 1st October and the 30th April in each year.

Finally in that connection I turn to the fourth schedule. I need only read 1(a) of the first provision, which is in the following terms:

The expenses of maintaining repairing redecorating and renewing:

and it then repeats verbatim the wording of clause 6(5)(a), which I have read and which refers to the central heating apparatus and boilers in the property, but it adds the words "and lifts". That had escaped attention until now, but I am not attaching any importance D to it, though it is a little mysterious why it should come in there and not elsewhere.

Two other provisions of the fourth schedule to which I should refer are clauses 4 and 5; these are also among the items comprised within the service charge referred to in clause 4(2)(a). Para 4 of the fourth schedule reads:

The cost to the Lessor of performing its covenants contained in clause 6 . . . (5) . . . and (9) of this Lease . . . ;

and para 5:

The cost of employing managing agents and accountants and auditors in connection with the Lessor's obligations under this Lease including the claims accounts and certificates referred to in clause 4(2)(a)

E and I need not read the rest. Those were the relevant provisions of the original lease, to which the defendant became a party when her landlords were the developers, Contemporary Homes Ltd.

In 1981, however, the position changed. By a supplemental deed of February 5 1981 the new arrangements were brought into force as between the present plaintiffs and the various tenants including the defendant. It is unnecessary to refer to any of the provisions of this short supplemental deed in detail. Its effect was that in consideration of an additional payment of £304 Mrs Silow became a tenant of this new company, the residents' association, and a shareholder on the basis to which I have already referred.

What then happened — and the documents show that these matters had been discussed quite openly and agreed upon well before F February 1981 — is that the new landlords, the present plaintiffs, had to deal with the question whom they should appoint as managing agents pursuant to clause 4(2)(a). We know nothing about the identity or nature of whoever acted as managing agents before February 1981. They decided to appoint six of the tenants/shareholders as managing agents, two from Westbury Court, two from Marlborough Court and the chairman and the sixth member from Amhurst Court. The formalities of the appointment are not in issue; only the identity of the appointees. The judge referred to the six persons who appear to have been in office when these disputes arose, although one has to bear in mind that presumably from time to time there were changes in the composition of what I shall call "the management committee". He said:

Now the first question which arises is whether the certificates given annually G by the managing agents of the Plaintiffs are void through lack of independence or expertise. The way in which that question arises is this: that when the Plaintiff Company came into existence it was in a sense a co-operative organisation with an incorporated persona and the shareholders were the tenants and the tenants were the shareholders, and what the Plaintiffs did was to appoint as managing agents some half dozen or so people whom they thought, rightly or wrongly, wisely or foolishly, would perform that function satisfactorily. It appears that none of the managing agents were professional estate agents or surveyors but at all material times I am satisfied that the managing agents so appointed, each of whom was a tenant of a flat on the property, included persons who had experience of worldly affairs and in some cases experience of actual property management. I need only mention that at the material times with which I am concerned the secretary of the managing agents was a Mr Abbot who had been for a number of years, Bursar H of a Cambridge College and I should have thought had the knowledge and experience ideally suited for this task. There was also a lady who is now the chairman who, it appears, took some part in the management of the property of another college. That was Lady Platt, a person who had had some experience in the academic field.

Lady Platt appears as the chairman, who signed one of the reports of this committee, which is included in the bundle before us. It is extremely detailed and shows a very considerable interest, as one would expect, in the property and considerable care in the consideration which was clearly given to its management. It concludes with an expression of thanks to the committee, and in particular to Mr Abbot with a reference to the sense of security which could be placed upon his knowledge and experience. J

What is said in relation to the first issue is that, having regard to the fact that its members were themselves tenants and shareholders, and in some cases directors, of the residents' association company, this management committee was not qualified to act as managing agents within the terms of clause 4(2)(a) of the lease. It will be remembered that the relevant words are that the annual amount of the service charge "shall be estimated and certified by the Lessor's managing agents acting as experts and not as arbitrators". What is said in regard to the management committee is, first, that it did not have the necessary degree of independence and, second, the necessary degree of expertise which is required by those words.

Before dealing with what the judge said in rejecting that submission I should refer to the only authority to which he and we K were referred, a decision of this court in *Finchbourne Ltd* v *Rodrigues*, reported in [1976] 3 All ER 581. The judge took the view that the facts of that case were quite different from the present one, and I entirely agree. I therefore deal with it before I come to those parts of the judgment in which he dealt with the facts of the present case. In that case Finchbourne Ltd were the owners of a block of flats and the leases in question comprised precisely the same provision concerning the appointment of managing agents for the purposes of ascertaining the service charge as in the present case. It was found as a fact that the plaintiff company were merely a "front" for a Mr Pinto, who carried on business as a developer, or estate agent, under the firm name of Pinto & Co. The plaintiff company purported to L appoint Pinto & Co as the managing agents. It was held that that appointment could not stand, because the managing agents necessarily had to be agents of the landlord in the position of a principal, and therefore separate legal personalities from the landlord. The county court judge found as a fact that Mr Pinto was a principal and that the company was his mere nominee. In those circumstances it is not surprising that it was held that the appointment of Pinto & Co as managing agents could not stand.

I should read one or two passages from the judgments. Cairns LJ said at p 586:

Undoubtedly under this lease the managing agents were to act as experts and not as arbitrators.

I can then omit a passage before Cairns LJ went on as follows: M

I express no opinion whether the function of the managing agents here went no further than to ascertain whether the sum said to have been spent had actually been spent. Even assuming that that is so, the managing agents were not a mere calculating machine. I note that they were to "certify as experts." Though the certificate, in contrast to the report of the surveyors referred to elsewhere, was not declared to be binding, the tenant is entitled to assume that an expert mind has been applied to the figures, and that unless there is something glaringly wrong with them he can rely on the certificate. It is said that if the lessor were a company the managing agents might be a subsidiary of that company. However that may be, I am quite satisfied that the managing agents are intended to be somebody different from the lessor.

Once it is found that the real lessor here was Mr Pinto, it follows that the managing agent must, according to the terms of the lease, be somebody

A different from Mr Pinto. Indeed, I would go further and say that it was a misdescription to describe Pinto & Co as agents at all.

Orr LJ agreed and said:

> Where, as here, the managing agent is an estate agent, two matters very much within his expertise are what work is reasonably required to be done in order to comply with the terms of the lessor's obligations under the lease, and what is the reasonable cost of such work, but merely to add up the figures of expenses incurred on the work done does not call for expertise.
> Counsel for the plaintiffs' answer to this point, as I understand it, was that the wording of the eighth schedule, "acting as experts and not as arbitrators", which I agree with him is a phrase commonly used in legal documents, is designed to do no more than to make it clear that the agent is not to act as an arbitrator, but this ignores the words "acting as experts" and I decline to
> B construe the passage as if it read "acting neither as experts nor as arbitrators".

Browne LJ also agreed and said:

> The intention clearly was that the tenant should be entitled to rely on the expertise of such a third person.
> On the judge's findings of fact here, which are not challenged, the managing agents and the lessor were in fact the same person.

I think that two principles are to be derived from that case. First, it is quite clear that the managing agents must be legally distinct from the lessors. Second, since they are to act as experts and not as arbitrators, I must accept from the judgments that a measure of expertise is to be required from them; but I would not accept that they have to be professional persons with professional qualifications.

C I think I can best deal with these points in relation to the present case by resuming quotations from the judgment. It contains a number of findings of fact, and Mr Mott, who argued this appeal with great lucidity on behalf of the appellant, rightly conceded that he could not challenge any of them. The judge continued as follows:

> The first issue that I have to decide is whether these managing agents, one or two of whom were directors of the Plaintiff Company, (indeed at one time —

and I correct a small error here; it should read "three out of six of them and the spouse of one of them" —

> were directors of the Plaintiff Company and there were only six directors) could properly fulfil the functions of managing agents for the purpose of the leases. Mr King who has argued this matter very fully on behalf of the Defendant, has submitted that those persons could not properly act and could
> D not therefore properly give certificates of expenditure to be incurred because they were too closely identified with the Plaintiff Company.

The judge then referred to *Finchbourne*, with which I have already dealt, and concluded that it did not govern the present case. He said, in regard to that decision:

> Now the reasoning is simple and that is because it is clearly outside the contemplation of the parties to that lease that the managing agent should be in effect the lessor himself because he would then be in a position to expend what money he liked as lavishly as he liked upon the property and to certify that it was reasonable and necessary.

He then turned to the facts of the present case, and he said:

> Now I accept entirely the reasoning in that case and I have to consider what is
> E the application to this case. There seem to me to be very special circumstances in this case. At the time of the lease of course, the reversioners were a Company —

meaning Contemporary Homes Ltd.

> I suppose I do not know whether they were trustees for anyone or who they were or what they were, but at the time of the supplemental deed the situation was that the reversion was held and to be held by the Plaintiff Company, that the Plaintiff Company was I suppose a private Company and all tenants were members and there were no outside members and the persons appointed by the Company, I suppose by the Board of Directors who would have the management of the affairs of the Company were all themselves tenants and in my judgment they were all people who I think might be expected to have some independence of mind. But they were all people who had a direct interest in keeping the amount of expenditure within reasonable bounds whilst at the
> F same time not neglecting items of necessary expenditure and I also bear in mind that it appears from the accounts that were shown to me that they made no charge for their services although there is a provision in schedule 4, I think it is paragraph 5, that the costs of employing the managing agents could be included in the service charge. And so I hold that the managing agents actually appointed in this case were not disqualified because they were to be identified with the Plaintiffs themselves.

That deals with the issues of independence.

Then he went on as follows:

> But it is submitted that the lease contemplates that they will act as experts and that involves their having some expertise —

and he then cited the short passage from the end of the judgment of Orr LJ which I have already read.

He went on as follows: G

> Bearing in mind what I said about the persons who were actually empowered by the Plaintiff Company to act as managing agents, in my judgment they had ample expertise to discharge their duties and I have no reason to suppose that any tenant has suffered anything from lack of expertise on their part. In these circumstances, I have come to the conclusion that the first issue must be decided in favour of the Plaintiffs.

In my judgment that decision is correct, on the aspects of both independence and expertise. Dealing first with independence, the word "independent" has naturally crept into the argument because it is settled law, both as the result of *Finchbourne* and indeed by necessary implication from the concept of principal and agent, that the managing agent must be somebody who is distinct from the lessor himself and must therefore be somebody independent in that sense. H Second, he must possess some degree of expertise. There again the concept of some independence of mind is implicit in this provision, because an expert is somebody who will exercise his judgment in an expert manner. He will do so in accordance with the merits of what he has to judge, and to that extent independently from whoever appoints him and also independently from what may be the wishes or interests of those who are affected by his decision. However, in my view the decision in *Finchbourne* in no way precluded the appointment of this managing committee. Its six individual members were clearly separate legal persons from the lessors, the company. Second, on the judge's finding they were persons selected on the basis that they could and would exercise an independent judgment in J relation to the management of these three blocks of flats.

In this court it was also suggested that their independence could be challenged on the ground that they themselves were tenants. That does not appear to have been argued below. I would only say that I do not see that this provides a valid objection. Leaving aside the question of expertise, what they have to be is agents appointed by the landlords. Therefore they must be separate legal persons from the landlords. I can see no other limitation upon the identity of the persons who can be appointed as managing agents, provided that they are, in fact and in law, distinct from the landlords. It may well be that relationships can be called into question as providing grounds for disqualification where there is a sufficiently close connection between the landlord's interests and those whom he had appointed as K managing agents, even without the facts being as stark as in the *Finchbourne* case. But I do not see why that has any relevance to a situation such as the present, which the judge recognised to be sensible, likely to operate for the benefit of the tenants themselves, and in no way liable to cause any mischief.

I therefore hold that so far as the words "managing agents" are concerned, and the degree of "independence" which they required, this management committee was not disqualified from being appointed.

That leaves the question of the level of expertise required by the words "acting as experts". I have read the relevant passages from the judgment in that regard. There is a finding that this management L committee had, to use the judge's words, "ample expertise to discharge their duties". Since no professional qualification is required where the words are quite general, as they are here — there is no reference, for instance, to accountants, surveyors, estate agents etc — it follows that on the judge's findings this management committee could also comply with the requirement that they were to act as experts. Accordingly I agree with the judge in his conclusion on this issue.

The second issue is quite separate. It was the fourth issue before the judge, and it is convenient to begin by reading what he says about it and the facts which he finds in that regard. He says:

> Now the fourth issue relates to the fitting of thermostatic radiator valves.
> What happened was this: that in, I think, about 1981, the boilers which M provided the water for the central heating were renewed. Thereafter, it was found, possibly because they were of larger capacity or more efficient, that some of the flats became rather too hot and the managing agents thought that it would be sensible and an efficient use of gas to fit thermostatic radiator valves. They were not fitted because the existing radiator valves were worn out or unserviceable, they were fitted as a matter of deliberate policy and were, in my judgment, an improvement rather than a repair. The question I have to consider is whether the cost of fitting them can be included in the service charge. I shall deal in a moment with the fact that the Plaintiff decided that she did not want those valves fitted in her flat and they never were fitted. That is a separate issue.

Pausing there, that is not an issue which is raised on this appeal.

The judge went on as follows:

> A Mr King based his argument that the cost of fitting these valves was outside the scope of the service charge upon clause 6 of the lease, paragraph 5 of which was a lessor's covenant to maintain, repair, redecorate and renew as and when necessary the main structure etc, and central heating apparatus in the property. His submission was the word "renew" should be construed *ejusdem generis* with the words maintain, repair and redecorate and that renewal in the landlord's repairing covenant only covered such renewal as was necessary to replace that which had either worn out or had become unserviceable and I think that submission is well-founded.

Pausing again, we were referred to a passage in *Woodfall: Landlord and Tenant*, vol 1 at p 599, under the heading "Repair to be distinguished from renewal", which bears that out, as well as to the paragraph on p 598, headed "Repair distinguished from
B improvement". No doubt the judge had those distinctions in mind.

He went on to say:

> One also has to look at clause 6(9) in which the landlord contracted to use its best endeavours to keep the radiators in the flats sufficiently and adequately heated between October 1 and April 30. Now in my judgment it is not right simply to look at what the lessor's obligations were, one has to look at what was the expenditure which was set out in the fourth schedule because it was a share of that expenditure which could be recovered from the lessee and Schedule 4 refers to the expenses of maintaining, repairing, redecorating and renewing the main structure etc and central heating apparatus and it is not without significance that the words "as necessary" do not appear there. Furthermore, the service charge includes by paragraph 4 the costs to the lessor of performing its covenants as contained in clause 6(9) *inter alia* and I have come to the conclusion that there is a deliberate contrast between these
> C provisions, that the landlord was only obliged to maintain, repair, redecorate or renew as necessary and therefore the lessor would not be liable for breach of covenant if he failed to do so. But if he chose to do some renewal which was not necessary, then in my judgment they — that is, the landlords — were entitled to include the cost of so doing in the service charge, provided, and provided only, that it was something which was reasonable and here, bearing in mind their obligation to provide central heating — that is a reference to clause 6(9) — I am satisfied that the most efficient and cheapest way of doing so was not merely to install the new boilers but to install the thermostatic radiator valves. I have come to the conclusion, after giving the matter considerable thought, that the Plaintiffs were entitled to recover the cost of the installation of these valves by way of including it in the service charge.

Mr Mott criticises those passages on two grounds. First, he says
D that in the earlier paragraph the judge found as a fact that these thermostatic valves were "an improvement rather than a repair". I am not convinced that the judge was there considering the supply of thermostatic valves to the blocks of flats served by the newly installed boilers as a whole. I think he was referring to the radiator valves in the defendant's flat, which he mentions in the next two sentences. E
Viewed on their own, it would of course be right to say that their replacement would be an improvement rather than a repair. However that may be, one also has to consider what follows. In regard to that, Mr Mott's second objection to the judge's reasoning is that he attached undue importance to the absence of the words "as and when necessary" in the fourth schedule, when they were present in clause 6(5).

I agree with that submission. I am not convinced that the draftsman of the lease, by the omission of those words, intended to produce a distinction between the landlord's obligations under clause 6(5) and his rights to charge the tenant with the cost of providing renewals, or possibly improvements, under the fourth schedule. It may be that that was the intention, but I would be very F
doubtful about placing any reliance on the absence of those words in the fourth schedule. It seems to me much more likely that it was merely intended to itemise the components which could fall within the service charge as derived from the body of the lease as a whole. But the judge based only part of his reasoning upon the absence of those words. His main point was that one has to bear in mind the landlord's covenant "to use its best endeavours to keep the radiators in the flats efficiently and adequately heated". He refers to that three times in the course of that paragraph. I agree with the sentence in which he says:

> . . . bearing in mind their obligation to provide central heating, I am satisfied that the most efficient and the cheapest way of doing so was not merely to install the new boilers but to install the thermostatic radiator valves. G

I think that that finding is quite sufficient to bring the expenditure on the thermostatic radiator valves within the provisions of clause 6 of the lease as well as within the fourth schedule. Looked at in that way, I cannot regard the installation of these valves, in the context of the perfectly proper renewal of the boilers, as an independent improvement which fell outside the words "repair and renewal". That, I think, was the crucial element in the judge's reasoning, and on that basis I would agree with it.

Accordingly, I agree with the learned judge's conclusion on the second issue as well, and would dismiss this appeal.

PURCHAS LJ agreed and did not add anything. H

The appeal was dismissed with costs, such costs to include costs of application for leave to appeal; application by appellant for leave to appeal to House of Lords refused.

For further cases on this subject see p 33

NEGLIGENCE

Queen's Bench Division

April 9 1987

(Before Mr E A MACHIN QC, sitting as a deputy judge of the division)

McINTYRE v HERRING SON & DAW
JOHNSON v HERRING SON & DAW

Rating valuation — Negligence — Two actions with same issues tried together — By agreement present judgment deals only with issue of liability, questions of causation and quantum not being considered — Plaintiffs, lessees of houses in Chelsea claimed that they had suffered loss through the breach of a duty of care on the part of a partner in the defendant firm in advising the plaintiffs and carrying out negotiations with the valuation officer in relation to their rating assessments — The losses claimed concerned overpayment of rates and, more important, the loss of a right of enfranchisement under the Leasehold Reform Act 1967 by reason of the rateable value being too high — The judge held that the plaintiffs had established that the defendants had fallen below the standard reasonably to be expected of them as competent surveyors in accordance with the accepted practice of the profession — The judge gave his reasons in a lengthy judgment in which, in addition to analysing the evidence in detail, he discussed the case law on the professional duty of care and examined methods of rating valuation — He held that the defendants were in breach of duty in advising the plaintiffs to agree to a rateable value of £1,638 in negotiations with the valuation officer, although this figure represented a considerable reduction from previous values — The advice given to the plaintiffs was uncompromising, as the partner dealing with the matter expressed himself as "certain that the change in assessment could not be improved upon if we went to the valuation court hearing" — In evidence to the court he admitted, however, that as a conscientious witness he could have advanced a case in the local valuation court which, if accepted, would have brought the rateable values below the enfranchisement limit, although he did not believe that he would have succeeded — The judge held that the plaintiffs should have been given all the information necessary to enable them to decide whether to take the matter to the local valuation court rather than to have been told in very definite terms that it would in effect be useless — The plaintiffs should also have been reminded in very clear terms of the vital importance of the £1,500 limit for enfranchisement — Judgment accordingly for plaintiffs on the issue of liability

The following cases are referred to in this report.

Barratt v *Gravesend Assessment Committee* [1941] 2 KB 107
Bolam v *Friern Hospital Management Committee* [1957] 1 WLR 582; [1957] 2 All ER 118
Maynard v *West Midlands Regional Health Authority* [1984] 1 WLR 634; [1985] 1 All ER 635, HL
Sidaway v *Board of Governors of the Bethlem Royal Hospital and the Maudsley Hospital* [1985] AC 871; [1985] 2 WLR 480; [1985] 1 All ER 643, HL
Singer & Friedlander Ltd v *John D Wood & Co* [1977] EGD 569; (1977) 243 EG 212 & 295
Whitehouse v *Jordan* [1981] 1 WLR 246; [1981] 1 All ER 267, HL

These two actions were brought by Mr and Mrs McIntyre in respect of 4 Petyt Place and by Mrs Johnson in respect of 3 Petyt Place, Chelsea, London SW3, against the defendants, Herring Son & Daw. The plaintiffs were long leaseholders of their properties and (subject to a residence question in the case of Mrs Johnson) could have acquired the freeholds in accordance with section 1(1A) of the Leasehold Reform Act 1967 if the rateable value had been reduced to £1,500 or less.

David Neuberger (instructed by Harbottle & Lewis) appeared on behalf of the plaintiffs; Nigel Hague QC and Miss Judith Jackson (instructed by Stephenson Harwood) represented the defendants.

Giving judgment, MR E A MACHIN QC said: These two actions, in which the issues raised are identical, have been tried together with the consent of the parties. During the course of the trial counsel asked me, and I agreed, to try the issue of liability only. Any questions of causation and quantum are therefore questions upon which I express no view.

Mr and Mrs McIntyre were at all material times the lessees of 4 Petyt Place, Chelsea. Mrs Johnson was at all material times the lessee of 3 Petyt Place. The defendants are and were chartered surveyors acting throughout by their partner, Mr G E Grieves [FRICS]. In May 1973 all the plaintiffs, and also the lessees of nos 1, 2 and 5 Petyt Place, retained the defendants to advise, negotiate and act for them in appeals against the 1973 rating valuation assessments of their respective premises. The nub of the plaintiffs' case is that Mr Grieves broke the duty which he owed to them as a careful and competent surveyor; it is said that he conducted his negotiations with the district valuer's representative, Mr J S Taylor [ARICS], in a manner which fell short of that required by professional prudence and that he wrongly advised the plaintiffs to accept a proposed settlement achieved by him with Mr Taylor, whereas he should have advised them that better terms could be obtained on appeal to the local valuation court.

The damages claimed fall under two heads. First, there is a claim for overpayment of rates since 1973, based upon the difference between the rateable value achieved by Mr Grieves' negotiation, accepted as it was by the plaintiffs and what the plaintiffs say it should have been. Second, there is a much larger claim for loss of the right at that time to acquire the freehold of the premises pursuant to the provisions of the Leasehold Reform Act 1967 as amended. Under the second head the plaintiffs contend that had the defendants negotiated and advised in accordance with their obligations the rateable value of the premises would have been reduced to £1,500 or less with effect from April 1 1973; this would have enabled the plaintiffs to serve a notice to enfranchise.

Whereas the plaintiffs' claim under the first head must be measured in hundreds of pounds only, their claim under the second head is pleaded at sums of between about £38,800 and £49,400. It is, therefore, the second head of damage which has nourished this litigation and which has generated a trial lasting seven days.

This judgment must necessarily be lengthy and detailed and I do not think that it is fair to keep the parties in suspense. I therefore say at the outset that I find that the plaintiffs have established a case of breach of duty on the part of the defendants. In what follows I give my reasons.

Before I embark upon this task, I wish to record how greatly indebted I am to Mr Neuberger and Mr Hague, counsel for the plaintiffs and the defendants respectively, for the manner in which they have conducted their clients' cases. Not only have they steered

A me courteously but firmly, and I trust successfully, through pathways beset with mystery and danger to one bred in the generous traditions of the common law, but they have consistently refused to take any point which did not seem to them to be a good point. In consequence, this case has been a pleasure to try. I am also very grateful to the solicitors for the parties for having neatly bundled and paginated the relevant documents, so that reference may be readily made to them.

The documents in this case are to be found in ring bundles nos 1 to 11, and I shall refer to the individual documents by quoting first the bundle number and then, after an oblique stroke, the number of the page.

B LEGAL PRINCIPLES

Under this heading I set out, first, such principles of rating and enfranchisement law as it is necessary to make use of for the purpose of the case; and, second, the duty of a competent surveyor, as I understand it to be. There was no difference in approach between counsel upon either of these topics.

Rating and enfranchisement principles

By the General Rate Act 1967, section 16, every house occupier is liable to be assessed for rates in respect of his hereditament, the rateable value whereof must be taken to be its net annual value ascertained in accordance with section 19(2) to (4) (see section 19(1)).

By section 19(2), in the case of houses without land other than
C gardens and other specified appurtenances, the net annual value is the gross value less a statutory deduction. By section 19(6) gross value means the rent at which the hereditament might reasonably be expected to let from year to year if the tenant undertook to pay all usual tenant's rates and taxes and the landlord undertook to bear the cost of repairs, insurance and other expenses necessary to maintain the hereditament in a state to command that rent.

By section 19(2A), added by the General Rate Act 1970, section 1(1), when the gross value of a hereditament which is a dwelling-house fell to be ascertained for the purpose of the new valuation list of April 1973, it was provided that the rents at which other houses had been let and the relationship between such rents and the gross values of the hereditaments shown in the expiring valuation lists should be
D regarded as relevant for the purpose of ascertaining the new values.

The state of affairs supposed by section 19(6) of the 1967 Act is referred to as the "statutory hypothesis".

In the course of the evidence, the "gross value" defined by section 19(6) has been referred to as the "gross rateable value" and the "net annual value" referred to in section 19(1) has been referred to as the "net rateable value" or the "rateable value" simpliciter.

To a householder it is the net rateable value which matters, for it is this figure which, by the application of poundage, produces the sum which by law he must pay to his local rating authority. Furthermore, power to enfranchise depends upon the net rateable value of premises and not upon their gross rateable value. Valuation of hereditaments,
E however, is the process of assessing their *gross* rateable value, and for the purpose of this case it has been the gross rateable values of the houses in Petyt Place, and of certain other houses said to be comparable, which have come under scrutiny.

Hereafter in this judgment I propose to refer to the gross rateable value of a house simply as its "rateable value" and if it is necessary to refer specifically to the net rateable value I shall use that expression.

Section 69 of the General Rate Act 1967 contains provisions which enable a person to make a proposal for the alteration of a valuation list if he is aggrieved by any value ascribed in the list to a hereditament and which enable the valuation officer himself to make a proposal for the alteration of a list. A valuation of a hereditament pursuant to a proposal must be assessed as at the date of the proposal (*Barratt* v
F *Gravesend Assessment Committee* [1941] 2 KB 107 at p 115). This principle is subject to the limitation imposed by section 20. By this section, putting matters shortly, for the purpose of any alteration to a valuation list pursuant to a proposal, the value is not to exceed the value which would have appeared in the list if the hereditament had been subsisting throughout the year before that in which the list came into force. For this purpose the hereditament is to be assumed to be in the same state as at the time of valuation (that is of the proposal) and the relevant features of the locality set out in section 20(1)(*b*) are to be assumed to be in the same state as at the time of valuation.

The proposals in respect of 3 and 4 Petyt Place were each dated September 13 1973 and it follows that any valuation pursuant to the proposal should be based upon the condition of the properties and their environment as at that date. Subsequent changes must be put on G
one side.

Finally, by section 79(1) of the Act, where the valuation list is altered following a proposal, the alteration is deemed to have had effect as from the commencement of the period in respect of which the rate was made. The practical effect of this subsection in the present case is that any alteration to the net rateable value of the subject properties took effect, or (as the case might be) would have taken effect, from April 1 1973, the date of the current valuation list.

I turn to enfranchisement. Part I of the Leasehold Reform Act 1967 granted to certain tenants of residential houses holding under a long lease the right to acquire on fair terms the freehold or an extended lease of their house. The combined effect of section 1 of that Act and amendments made by section 118 of the Housing Act 1974 H
may, so far as is necessary for the present purpose, be simply stated. If the net rateable value of the house (being in Greater London) was not on April 1 1973 more than £1,500, then each of the tenants of 1-5 Petyt Place had the right, after the passing of the 1974 Act, to acquire their freeholds upon the terms set out in section 1A of the 1967 Act. A tenant desiring to acquire the freehold must serve a notice in the form prescribed by Part II of Schedule 3 to the 1967 Act; at the time he gives that notice he must have been a tenant under a long tenancy at a low rent occupying it as his residence for the preceding five years or for periods amounting to five years in the preceding 10 years (section 1(1)(*b*) of the 1967 Act). It is common ground that the tenants of 1-5 Petyt Place satisfied these conditions, except that it is not agreed that J
the tenants of no 3 satisfied the conditions of residence.

The upshot of this is that if after February 1 1978 (the date when Mr Grieves negotiated with Mr Taylor) the net rateable value of no 3 and no 4 Petyt Place had been reduced to £1,500 or less, then the plaintiffs (subject to the residence point in the case of no 3) could have acquired their freeholds at a price computed in accordance with section 1(1A) of the 1967 Act. This price is computed at the time when the tenant gives notice of his desire to enfranchise (see sections 9(1A) and 37(1)(*d*) of the 1967 Act).

It is common ground that a net rateable value of £1,500 is equivalent to a rateable value of £1,834; the latter is thus the crucial figure upon which concentration throughout the case has been focused. K

The duty of a surveyor

In this connection I have been referred to the useful treatment of the topic in *Jackson and Powell's Professional Negligence*, pp 91 *et seq*. The plaintiffs' claims are pleaded both in contract and in tort; it was not suggested that the duty in tort differed from that in contract and I do not distinguish between the two forms of action.

In the area of medical negligence it is well settled by the highest authority that the test of negligence is that laid down by McNair J in *Bolam* v *Friern Hospital Management Committee* [1957] 1 WLR 582, approved in three cases in the House of Lords: *Whitehouse* v *Jordan* [1981] 1 WLR 246, *Maynard* v *West Midlands Regional Health Authority* [1984] 1 WLR 634, and *Sidaway* v *Bethlem Royal Hospital* L
[1985] AC 871. It is to be noted that *Bolam* was concerned with the duty of a doctor to warn his patient of the risks of a certain course of treatment. The present case is at least in part concerned with the duty of a surveyor to advise his clients of the possible benefits of a certain course of action. There is an analogy between a duty to warn of risks and a duty to advise of benefits and it is relevant to note that in *Sidaway* the majority of their Lordships' House held that the test of liability in respect of a doctor's duty to warn his patients of risks inherent in medical treatment was the same as that applicable to diagnosis and treatment itself. I therefore draw no distinction (nor was I urged so to do) between Mr Grieves' duty in relation to his conduct of negotiations with Mr Taylor and his duty with regard to advising the plaintiffs what they should or should not do in the light M
of those negotiations.

I can see no reason for formulating the duty of a surveyor in different terms from that of any other professional man, including a doctor, nor did counsel suggest otherwise. In this I gratefully follow a decision of Watkins J (as he then was) in *Singer & Friedlander Ltd* v *John D Wood & Co* (1977) 243 EG 212. That was a case in which the plaintiffs claimed that they had lost money in reliance upon the negligent valuation of a firm of surveyors, auctioneers, valuers and estate agents. Watkins J applied to the duty of these defendants the formulation of the duty in *Bolam*. I quote two passages from his judgment, at pp 217 and 302 of the report respectively:

Mr Justice McNair said in *Bolam* v *Friern Hospital Management Committee*

[1957] 1 WLR 582, this case being a medical negligence case, that a doctor who had acted in accordance with a practice accepted at the time as proper by a responsible body of medical opinion skilled in the particular form of treatment in question was not guilty of negligence merely because there was a body of competent professional opinion which might adopt a different technique. Therefore, what the experts tell me is standard practice is obviously invaluable to the completion of my task in deciding whether Mr Ross was in breach of duty, or in another word "negligent." Where the views differ, if they do, about what the standard practice is, I must rely upon my judgment for deciding which expert I prefer in this respect, or decide that although the versions of standards differ, providing one or other of them is followed by a valuer, he cannot be criticised for adopting that one.

In another medical negligence case, namely *Vancouver General Hospital* v *McDaniel and Another* (1934) 152 LT 56 at p 57 it was stated:

"Not only do these medical men approve in terms of the appellants' technique, but they affirm as will be observed from the passages cited sup, that the technique challenged by the respondent is in accord with general if not with universal practice today in Canada and the United States. If that be so, it is, in their Lordships' opinion, again difficult to affirm that negligence on the part of the appellants is proved. A defendant charged with negligence can clear his feet if he shows that he has acted in accord with general and approved practice."

That was a case of alleged medical negligence in the treatment of smallpox.

I turn to the second quotation:

I return to *Bolam's* case to quote again from the judgment of McNair J at p 586. Having said what in his consideration was the test of reasonable care in relation to the ordinary man he passed to the test in relation to the ordinary skilled man and said:

"The test is the standard of the ordinary skilled man exercising and professing to have that special skill. A man need not possess the highest expert skill; it is well established law that it is sufficient if he exercises the ordinary skill of an ordinary competent man exercising that particular art."

On the following page, at p 587, quoting from the Scottish case of *Hunter* v *Hanley* and the judgment of Lord President Clyde, he went on:

"In the realm of diagnosis and treatment there is ample scope for genuine difference of opinion and one man clearly is not negligent merely because his conclusion differs from that of other professional men, nor because he has displayed less skill or knowledge than others should have shown. The true test for establishing negligence in diagnosis or treatment on the part of a doctor is whether he has been proved to be guilty of such failure as no doctor of ordinary skill would be guilty of, if acting with ordinary care."

It is these principles which I shall apply in the present case. I desire also to add that in so far as in the present case Mr Grieves is criticised for failing to advance a valuation based on a proper foundation, I bear well in mind the approach of Watkins J in the *Singer & Friedlander* case as set out in his judgment at p 213. I quote two passages from it:

The valuation of land by trained, competent and careful professional men is a task which rarely, if ever, admits of precise conclusion. Often beyond certain well-founded facts so many imponderables confront the valuer that he is obliged to proceed on the basis of assumptions. Therefore, he cannot be faulted for achieving a result which does not admit of some degree of error. Thus, two able and experienced men, each confronted with the same task, might come to different conclusions without any one being justified in saying that either of them has lacked competence and reasonable care, still less integrity, in doing his work.

. . . . As Mr Ross said, valuation is an art, not a science. Pinpoint accuracy in the result is not, therefore, to be expected by he who requests the valuation. There is, as I have said, a permissible margin of error, the "bracket" as I have called it. What can properly be expected from a competent valuer using reasonable skill and care is that his valuation falls within this bracket.

In that case the expert evidence was agreed that the permissible margin of error was generally 10% either side of the right figure, a bracket capable of extension, in exceptional cases, to 15% either way. I accept Mr Hague's submission that the widths of these brackets reflected only the evidence in that particular case and that there is no proposition of law that in valuation cases a valuer is not negligent if his valuation falls within such a bracket. The permissible width of the bracket must be a matter to be decided upon the evidence of any particular case.

The pleadings

The allegations made by the plaintiffs in the two actions are identical, and I need refer only to the amended statement of claim in the McIntyre case. This pleading begins at 1/3. I read paras 11-13 of it.

11 In or about February 1978 the defendants negotiated a settlement on behalf of the plaintiffs of the aforesaid appeal whereby it was negotiated by the defendants on behalf of the plaintiffs that the rateable value of the premises would be reduced to £1,638 with effect from April 1 1973.

On February 3 1973 the defendants wrote to the plaintiffs stating that the figure of £1,638 negotiated as aforesaid "could not be improved upon if we went to the valuation court hearing", and accordingly advised the plaintiffs to confirm their acceptance of the figure which the defendants had negotiated which confirmation the plaintiffs gave to the defendants in accordance with such advice by a letter dated February 6 1978.

12 Accordingly the rateable value of the premises was treated pursuant to the provisions of the General Rate Act 1967 as being £1,638 with effect from April 1 1973.

13 In acting for the plaintiffs and/or negotiating the aforesaid rateable value of £1,638 on behalf of the plaintiffs and/or in advising the plaintiffs to agree the aforesaid rateable value of £1,638 the defendants were in breach of their aforesaid contractual and/or common law duties:

Particulars

(1) The defendants based their negotiations and advice on the assumption and belief that the area from which the gross value of the premises would be calculated was 250 or 228 sq m or thereabouts whereas the appropriate area was 169.3 and 195.5 sq m or thereabouts.

(2) Accordingly the defendants did not measure or did not properly measure the area of the premises. (a) The defendants did not agree or attempt to agree the area of the premises with the valuation officer and/or failed to discover or correct the misapprehension under which the valuation officer laboured, namely that the area of the premises was 228 sq m or thereabouts whereas it was in fact 195.5 sq m or thereabouts. (b) The defendants did not make any end-allowance whereas in all the circumstances they should have made an end-allowance and that allowance should have been 10%.

(3) The defendants valued the ground and first floor at £12.50 per sq m and the third floor at £5 per sq m whereas the proper rate to be applied was £10 per sq m and £2 per sq m respectively.

(4) The defendants did not appreciate or in the alternative did not advise the plaintiffs of the importance of the rateable value being reduced to £1,500 or less by virtue of the provisions of section 1(6) of the Leasehold Reform Act 1967 as amended.

Of the particulars given under para 13, (1) and (2) have been withdrawn by Mr Neuberger and he agrees that particular (2)(a) is not made out if I accept Mr Grieves' evidence that he gave to Mr Taylor his own areas and if Mr Taylor indicated that Mr Grieves' figures were close to his own. It is convenient to state here that I do accept this part of Mr Grieves' evidence, so that particular (2)(a) no longer falls to be considered. The case, therefore, turns upon whether the plaintiffs have made out one or more of particulars (2)(b), (3) or (4).

Although the advice given to the plaintiffs and set out in para 11 of the amended statement of claim is not expressed in the particulars to have been negligent, this allegation is made in the body of para 13 and the case was fought throughout on the basis that, *inter alia*, Mr Grieves was negligent both in advising the plaintiffs to accept a reduced net rateable value of £1,638, and in advising that the figure could not be improved upon if the plaintiffs appealed to the local valuation court.

Further and better particulars of the statement of claim and better particulars of those further and better particulars were subsequently served and later amended. It is not necessary that I should refer to them.

In their defence the defendants admit, *inter alia*, the contents of paras 10, 11 and 12 of the amended statement of claim and they deny breach of duty. They expressly plead that their calculation of the rateable value of the premises was arrived at as set out in para 8 of that pleading, the genesis of which is no doubt the letter from Mr Grieves to Mr Sharp (a later owner of 2 Petyt Place) dated September 26 1969, to be found at 6/7.

The rating history of Petyt Place

On the occasion of the 1963 general rating revaluation 2-5 Petyt Place were given a rateable value of £900 and no 1 was given a value of £1,000.

In October 1968 these values were reduced to £850 and £900 respectively, due to the disadvantage caused by the erection of a block of flats nearby.

In the general rating revaluation of 1973 nos 2-5 were revalued at £2,380 and no 1 at £2,500.

Mr D N Huckle, who has been the valuation officer for Kensington and Chelsea since September 1984, told me from enquiries that he had made that in order to revalue in 1973 the valuation office analysed rents obtained from rent returns and compared them with the 1963 valuation. The office then calculated a factor for each area or street by which the new rateable values were obtained by multiplication. In the case of Petyt Place the factor was 2.8. He described this as a method which was expedient having regard to the limited resources of the valuation office.

Although the factor of 2.8 was applied to the 1963 values, it must, as I understand it, have taken into account those matters which resulted in the 1968 reduction, since it was rental values shortly

A before 1973 which were the basis for the analysis. However this may be, it was not suggested before me that the 2.8 factor was too high having regard to what happened in 1968, although Mr Neuberger for the plaintiffs did suggest that the 1973 values were too high for other reasons.

The 1973 revaluation generated a flood of proposals by aggrieved residents who complained that their rateable values were too high. Among these were the occupants of 1–5 Petyt Place. Mr Pride, the then occupier of 1 Petyt Place, concerted their attack and recommended the instruction of Mr Grieves of the defendants to wage battle (see 5/1). In September 1973 the proposals of the plaintiffs, to which I have already referred, were lodged (3/31,30).

Because the valuation office was overwhelmed with such B proposals they were not able to consider those of the plaintiffs until February 1978.

On the first of that month Mr Grieves negotiated with Mr Taylor reductions in the 1973 rateable values which the occupiers accepted. The revised valuations (which dated back to April 1 1973) were, for nos 2–5, £2,000 and for no 1, £2,150.

At the end of 1978 Mr Sharp, a chartered surveyor with rating experience, became the lessee of no 2. He thought his rateable value was high and in March 1980 he made a proposal for its reduction. In circumstances which I shall have to consider with care the valuation officer (in the person of Mr Taylor) agreed that Mr Sharp's valuation was too high and reduced it to £1,775. The valuation officer then, C very properly of his own motion, made proposals for the reduction of the rateable values of 3, 4 and 5 Petyt Place, the result whereof was that the rateable values of nos 3 and 4 were likewise reduced to £1,775 and that of no 5 to £2,000. Whether any alteration was made in the rateable value of no 1 does not appear.

Petyt Place and the houses

The location of Petyt Place appears from a plan at 11/4. It is a cul-de-sac running approximately east-west about 100 yards north of and parallel to the Chelsea Embankment and it debouches into Old Church Street. Old Church Street runs north-south just to the east of Battersea Bridge.

Between Petyt Place and the Embankment are sunken gardens D surrounded by a wall. The wall cuts off a view of the river from the ground-floor windows of the houses, but such a view is afforded from the first and upper floors.

The fronts of all five houses can be seen in the photograph at 11/3. They are attractive houses in a good residential area. Nos 1 and 5 alone have garages. The garage of no 5 is detached and can be seen in the photograph, facing east. The garage of no 1, the corner house, is integral and faces Old Church Street. Alone of the houses, no 1 has a basement.

The houses have ground, first, second and third floors. The photographs in bundle 11 were taken recently and show the third floor of no 1 with dormer windows. It appeared from the evidence of E Mr McIntyre, which I entirely accept, that the third floor of his house has no windows, being storage space with staircase access. Similarly, he said that there were no external windows to nos 2, 3 and 5 at that level. It did not appear from the evidence when it was that the dormer windows of no 1 were put in, but from such evidence as was given it seemed unlikely that they were there as long ago as April 1973. Mr Grieves told me that he did not think that no 1 had third-floor bedrooms when he looked over it in September 1973 and I approach the case on the basis that at the material time there were no dormers there.

This row of houses was constructed in 1929. Nos 2–5 remain substantially as they were built, but no 1 suffered extensive bomb damage and was rebuilt in 1951.

F The nature of Old Church Street appears from the photographs at 11/23. It is one way north-south and carries a heavy volume of traffic. There is congestion at the junction of Petyt Place due to parking, and vehicles entering Petyt Place have to turn round to emerge. The Embankment, as any Londoner knows, carries continuous and heavy traffic and I entirely accept that the noise so generated reaches the ears of the occupants of the houses with which I am concerned.

Floor areas

Whatever method of valuation for rating is used, the value finally arrived at reflects as a factor at least the superficial floor area of the hereditament. Indeed, one valuation method proceeds on the basis of ascribing a unit price per square metre to the superficial area of each floor, and aggregating the values thus calculated, subject to a final G adjustment for matters independent of price and area.

In the Kensington and Chelsea valuation area there was a standard method of superficial measurement according to which certain areas were omitted (eg where headroom was less than 1.5 metres). The inherent difficulties of measuring with accuracy the dimensions of an oddly-shaped room, and of deciding precisely what areas are to be included and excluded, necessarily lead to variations between the areas arrived at after measurement by equally competent surveyors. The first page of exhibit D2, which sets out the areas in the same premises achieved by five different surveyors or measurers, shows this quite clearly. These measurements relate to 2 and 4 Petyt Place and produce total floor areas of between 182.42 and 200.13 m^2. The lowest end of the bracket is that obtained by Mr Sharp and may, I H think, be rather lower than it should be. When in 1981 he gave his figure to the valuation officer, Mr Taylor had the premises remeasured by a Mr Bellion of the valuation office, whose total area came to 191.7m^2, which is, in my judgment, nearer the mark, and in the event Mr Sharp accepted a reduction in his rateable value based on this area, adjusted in a manner which I shall have to explain in due course.

Leaving aside Mr Sharp's figures on D2, p1, it seems to me that any floor area for no 2 or no 4 which fell within the bracket of 190–200m^2 would be incapable of criticism and that one can do no more than state that the "correct" area is probably about 195m^2 with a tolerance of plus or minus 2½%. J

The floor area of no 1 was measured by the defendants for the purpose of Mr Pride's appeal against the 1963 valuation. The plans and figures are at 9/2A-C. The measurements are in imperial; these areas were transferred by Mr Grieves to his note at 9/5, prepared for the purpose of the meeting with Mr Taylor on February 1 1978, and beside them are the metric equivalents. Excluding the basement, the total is 202m^2; with the basement the total is 213m^2. Although these measurements were not strictly proved nor was any Civil Evidence Act notice served in respect of them, as it could easily have been, there was in the course of the case no suggestion that they were inaccurate and, subject to the necessary tolerance, I accept them.

It is appropriate here to refer to the valuation office file relating to no 1. Of nos 1–5, no 1 was the only house in respect of which, in 1978, K the valuation office had any measurements whatever. These appear upon 10/2-4. From 10/4 it appears that they were scaled from a plan; this is less accurate than measuring on site, particularly in places where the 1.5-metre height restriction may operate. These areas, with their metric equivalents, appear on D1, p 1; the total area, excluding the basement, is 228.64m^2; including the basement, it is 250.48m^2. I am quite satisfied that these, the only recorded areas of any Petyt Place House in the possession of the valuation office in 1978, were significantly inaccurate, as a comparison with the defendants' reference areas (202m^2 without, and 213m^2 with, basement) clearly shows.

It follows that no 1 excluding its basement was on the defendants' measurements marginally greater in area than nos 2–4, being some L 7m^2 in excess of the mean of 195m^2 which I consider appropriate to nos 2–4. It may, of course, be that the difference is in fact smaller than it appears, since the defendants' areas are only very slightly in excess of the highest measured area of no 4, that being the 200.13m^2 arrived at by Mr Hassall. Ignoring the basement of no 1, houses nos 1–4 were substantially similar in floor area.

The floor areas of no 5 did not enter into the case; when it was referenced by the defendants in 1968 it was found to have the same areas as nos 3 and 4; by the time the valuation office referenced it in 1981 its areas were considerably larger than those of the other houses, totalling 216.8m^2. Perhaps there had been alterations in the meantime. M

Methods of valuation for rating

In the course of the evidence I was acquainted with three methods of arriving at the rateable value of residential premises in Kensington and Chelsea. None was put forward as being superior to the others, and a surveyor is perfectly entitled to cross-check by using more than one.

What is quite clear is that the method of valuation required by section 19 of the General Rate Act 1967 (valuation upon the statutory hypothesis) is not used in the case of residential properties in this area. Mr Huckle, a valuation officer of long experience who was, in my judgment, a first-class witness, said quite candidly that after 1973 he would not argue rental values of domestic properties. Neither of

A the expert surveying witnesses called before me so argued. No doubt the existence of a scheme of statutory control of rents is at least in part the reason for putting the statute to one side; when I asked Mr Hague, who has great experience in this field, how all this could be, he told me that really the statutory hypothesis, in the context here to be considered, was a neglected relic of a bygone age. I turn with a sense of wonderment from what Parliament enjoins to what in fact is done on the ground.

The three practical methods are these:

(1) An experienced rating surveyor who knows the property and is familiar with the district may simply say that using his experience the rateable value of the subject premises is about £X. This almost intuitive approach is widely used in other fields, eg in the case of a B valuation of personal injuries.

(2) The surveyor may place the subject properties alongside other properties which in his view are comparable in size, character and location. Then, using the current valuations of his comparables, he will fit the subject property into an appropriate slot. This, too, is a method of valuation widely used in other fields.

(3) The surveyor may attempt an apparently more precise valuation by ascribing to each floor of the subject property a price per m^2, and in the case of each floor multiplying the price by the area. The aggregate of the values thus obtained may then be adjusted up or down by applying an ''end-allowance''; for example, in Kensington and Chelsea if the premises are centrally heated an end-allowance of C 50p per m^2 is added; if the premises are particularly beset by traffic noise or congestion an end-allowance will be deducted, typically (according to Mr Huckle) one of 10%.

When this method is applied it is usual to apply one unit per m^2 to all floors, although floors above first, and basements, merit a lower rateable value per m^2. This is achieved by adjusting basements, and floors above first, according to a factor, reducing their actual floor area to an ''adjusted'' or ''reduced'' floor area. At all material times in Kensington and Chelsea the ground and first were taken to be unity; the second was taken to be 0.8, and the third (if it was a living space) 0.6, and storage areas 0.2; the factor of a basement would depend, Mr Huckle told me, upon the nature of its access; he would have factorised the basement of no 1 at 0.5.

D The manner in which the calculation is done can be seen from the first page of exhibit D1. The utility of working with adjusted floor areas is that changes in the unit price per m^2 for ground and first floors can immediately be translated into changes in rateable value for the premises as a whole.

Garages are sometimes separately rated (the garage to no 5 was separately rated); if rated with the house the garage is usually given a spot value rather than a value based upon the inclusion of its area within the global computation. For the purposes of the present case, garages can be left out of account.

Although this third method has an outward appearance of accuracy due to the mathematics involved, it is to be observed that E there are several occasions upon which value judgments have to be made. I leave aside floor areas, which must usually be capable of agreement on site. However, the final valuation is highly sensitive to fluctuations in the unit price per m^2; if there is no end-allowance a variation of X% in the unit price will produce a corresponding variation of X% in the rateable value. The assessment of the unit price must itself be arrived at by one of the other methods to which I have referred, that is to say by experience or the use of comparables or both. There may be room for argument as to the factors to be used, particularly in the case of storage space. The size of the end-allowance may be fixed by convention (in the case of central heating) or it may be open to debate (as in the case of traffic noise).

I have dealt in some detail with the third method, sometimes F referred to as ''in terms of ground and first'', abbreviated to''ITG1'', because it was that primarily used by Mr Grieves in the present case, and many of the documents are comprehensible only if the nature of the methodology is understood.

Finally under this head I refer to what is called the ''tone of the list''. This phrase enshrines a concept which resists precise definition and it has achieved respectability in rating circles by its inclusion within the heading to section 20 of the General Rate Act 1967, and maybe within the heading of the predecessors to that provision, which speaks of ''valuation according to tone of the list''. As I understood it from the evidence, this arcane rubric connotes no more than the general level of rateable values applicable to a particular street or area within a rating district. Valuation officers, it is said (and I accept), jealously guard the tone of their lists. Any attack upon the G integrity of the tone of the list is vigorously repelled. This is because if the tone of the list is once permitted to slip, the gates are open to a flood of proposals for reduction from other rateable occupiers. Therefore, the valuation officer, whose duty it is to fix a fair valuation rather than the lowest valuation of premises he is considering, much prefers, if he feels that a reduction is justified, to arrive at it by taking into account circumstances special to the hereditament, often through the operation of an end-allowance.

The negotiations of February 1978

In May 1973 Mr Grieves was retained by all the plaintiffs and also by the then occupiers of nos 1, 2 and 5 Petyt Place to act for them in seeking a reduction of their rateable values. H

In September 1973 the interested occupiers met Mr Grieves. Mr Grieves told me, and I of course accept, that he has been in practice for 40 years. Between 1954 and 1982 he was with the defendants and at the material time was a partner in that firm. He has had long experience of rating valuation and negotiated about 8,000 properties and assessed for himself as many again, mostly in the London area. In 1973 he acted for Mr Pride in connection with his revaluation of that year. He then obtained a reduction by agreement. In 1968 he acted for all occupiers of nos 1–5 and secured reductions, to which I have already referred, due to the erection of the nearby flats. This matter is dealt with in bundle 4. In May 1968 Mr Honeybun, an experienced measurer (or ''referencer'') of the defendants measured nos 2–5 and obtained the results set out at 4/9,10. Thereafter, Mr J Grieves was at all times in possession of accurate floor areas for, among other houses, nos 3 and 4 Petyt Place. No 1 Petyt Place had been referenced by the defendants in about 1963 and I have already set out the results of that reference. For the purpose of the 1968 negotiations Mr Grieves inspected either no 2 or no 3, and before 1968 he had been inside no 1 at least three times.

Therefore, by February 1978, as it seems to me, Mr Grieves had at his disposal detailed measurements of floor areas of all the five properties and he was familiar in a general way with the exteriors of all, and with the interiors of no 1 and of one of the other four identical houses.

Mr Taylor, on the other hand, was short of reliable information. K Mr Grieves told me, as I accept, that Mr Taylor described his files as being ''in disarray''. No one who has examined the extracts from them in bundle 10 would cavil at this description. It is necessary to advert to what those files contained.

First, there was a valuation précis sheet (10/1). The evident purpose of this sheet was to summarise such information relating to 1–5 Petyt Place as was relevant to the valuations. As the sheet now appears, it contains much detail concerning the nature of the rooms of each house and the floor areas, including a total of the adjusted floor areas. Not so in 1978. From enquiries made by Mr Huckle he was able to tell me that before 1981 the only entries appearing would have been those in columns 1, 22, upper 22 (850) 23 (2.8), 24, 25 (2380 only), 26 and some of the references in the ''remarks'' column. L

When he gave this evidence, Mr Huckle was in fact referring to an earlier and expurgated version at 10/1, upon which entries relating to nos 1, 2 and 5 had been blanked out. This was because his subpoena related only to nos 3 and 4, and natural delicacy forbade any more liberal disclosure. However, I may with confidence assume that the entries relating to nos 2 and 5 were as sparse as those relating to nos 3 and 4. Whether the entries on 10/1 relating to no 1 were also incomplete is a question not solved by the evidence. Certainly, as I go on to describe, the valuation office was then in possession of the areas set out on 10/1 in their imperial form.

The file of each house included a survey sheet upon which should have appeared the dimensions of each room and its area, together with a plan. M

In the case of no 1 there were measurements and areas for each room, to which I have already referred. The rubber stamp in the top right-hand corner of 10/2 and the like stamp on the survey sheets of the remaining houses were not there in 1978. These stamps have a space in which should be written the adjusted floor area of the premises. None of the stamps relating to any of the houses in fact correctly records the adjusted floor area, and these errors will fall for consideration at a later stage when I consider Mr Sharp's proposal. I have already made explicit my finding that the areas recorded on 10/2-4 were significantly greater than was in fact the case.

The survey sheet in the file relating to no 2 is at 10/9. In 1978 it contained no information whatever as to the area of any part of that house. Upon the sheet is written, followed by a plaintive mark of interrogation, the legend "As no 1". It was simply a useless piece of paper.

The survey sheets relating to the remaining houses in Petyt Place were in like condition.

Such, then, was the documentary information relevant to the valuation of these houses available to Mr Taylor when he met Mr Grieves in February 1978. At this date Mr Taylor had been in the valuation office for a little over six years, having joined as a cadet. He was a main grade valuer, which indicates medium seniority.

Mr Taylor was not called as a witness. So far as the evidence went he was available. I draw no adverse inference against either party from his non-appearance. I decide the case entirely on the evidence I have heard, without speculation as to what Mr Taylor might have said had he been called.

Two important contemporary documents bear upon the content of the discussion between Mr Grieves and Mr Taylor.

Before the meeting Mr Grieves had prepared notes relating to all five houses (9/5-6). Brief additions were made to 9/5 during the meeting; these are the entries to the left of the column of figures. I am satisfied that they were jottings made by Mr Grieves during the meeting as a result of what Mr Taylor said to him.

At the meeting Mr Taylor made brief notes. Their accuracy was in no way proved, but both parties accept that they were written by Mr Taylor and that inferences may be drawn from them as to what occurred at the meeting.

In 1981, in connection with Mr Sharp's proposal, Mr Taylor generated memoranda which refer back to the 1978 meeting. These memoranda are not in my view admissible evidence either as to what took place in 1978 or as to Mr Taylor's state of mind in 1978, and I disregard them in considering the 1978 meeting. In the same category I place a letter from the valuation office dated July 1 1981.

The evidence of Mr Grieves is plainly of crucial importance. Mr Grieves was patently an honest witness doing his best to recollect events impartially. He was the first to admit that upon matters of detail his memory might be unreliable and this is hardly surprising. He was endeavouring to remember one discussion out of very many which took place nine years ago. However, this particular discussion perhaps merited unusual retention in his memory because of the highly exceptional reticence of Mr Taylor, and I am satisfied that Mr Grieves still has an actual recall of the main thrust of the discussion. I am also satisfied that he put before Mr Taylor such contents of his notes, as he says he used, for that purpose, so that I can confidently refer to his notes for the detailed measurements and prices he used. However, once Mr Grieves was left without the assistance of his notes, he was naturally unsure of his ground, so that what I accept as reasonably accurate from his evidence are the broad lines of the discussion and such details as are also to be found in the contemporary notes and furnished by Mr Grieves to Mr Taylor.

Mr Grieves' letter to Mr Sharp of September 26 1979 (6/7) sets out "the basis of the decision I made" with regard to no 2 after meeting Mr Taylor. The floor areas are plainly taken from 9/5, but the floor prices do not agree. In his evidence Mr Grieves explained the price of £7.50 as being the mean of £7 and £8, and he said that the £12.50 came from him. The floor prices for no 2 on 9/5, although not expressed, must be £10, £10, £7 and £5, and so 6/7 is *ex post facto* rationalisation. I do not think that it assists me.

I have spoken of a "discussion" between Mr Grieves and Mr Taylor, but I accept Mr Grieves' evidence that the flow of conversation was almost entirely one way. Mr Grieves put his facts and arguments to Mr Taylor but in return got little inkling of what was in Mr Taylor's mind. As he graphically put it, "It was almost as if I was the valuation officer and he was the appellant".

Mr Grieves' main prong of attack against the valuation of £2,380 was that it took too little account of the traffic noise and congestion suffered by the occupants of all the houses, and he made use of "Points for Discussion" prepared by Mr Price. He thought little of the nuisance caused by tramps using the sunken garden and did not press such an argument.

Mr Grieves told me, and I accept, that he began by giving to Mr Taylor his floor areas for no 3 and no 1, floor by floor. He did not deal separately with 2-5, since they were going to have the same rateable value as no 3, excluding the garage to no 5.

According to Mr Grieves, Mr Taylor said "That's fine, that's the information I have got, the figures I've got here are similar to yours". Although pressed, Mr Taylor refused to say what his figures were but repeated more than once that they were close to Mr Grieves'.

I accept Mr Grieves' evidence as to what Mr Taylor said, and that thereafter the discussion proceeded on the basis of Mr Grieves' floor areas. The conclusion must be that Mr Taylor was incompetent or dishonest or both; if he had calculated the metric equivalent of the areas for no 1 he would at once have realised that there was a significant discrepancy in Mr Grieves' areas; if he had not calculated the metric equivalents it was incompetent of him not to have done so and dishonest of him to have said that his figures were similar to Mr Grieves'. In relation to no 3, for which he had *no* figures, I do not understand how he could honestly have stated that Mr Grieves' figures and his were similar. Mr Grieves told me that Mr Taylor appeared to be stalling and that he, Mr Grieves, was concerned and suspicious. Rightly so, in my judgment.

Having given his areas, Mr Grieves gave his prices. They included 50p per m^2 for central heating. In relation to no 3 he gave floor by floor the prices in the column which begins with £10 on 9/5, together with the values per floor. These in fact total £1,680, a figure which appears to the right-hand side in the middle of an alternative set of prices. This alternative set (which begins with £12 for ground and first floor) he also gave to Mr Taylor. It produces a total of £1,966.

Mr Taylor's note at 10/6 has jottings on the right-hand side "£1,966 less £166, £1,850 GV." The "116" appears to have begun as "196". Now that Mr Grieves has seen 10/6 he thinks that he contended alternatively in respect of no 3 for a unit price of £12 per m^2 and an end-allowance of 10%. 10% of £1,966 is approximately £196, which he considers to be the provenance of the £196, later altered to £116, on 10/6. As he put it, "I think that I would have said 196 (10%) and 1,800 GV. It is my guess as to what I said. I am not sure if I contended for 1,770 or 1,800". £1,770 is, of course, the accurate deduction of £196 from 1966.

In answer to Mr Neuberger, Mr Grieves agreed that in respect of no 3, had he used a 0.2 factor for the third floor he could have justified going in at £1,620 on his £10 ground and first-floor price, or at £1,775 using his £12 ground and first-floor price. A little further on in cross-examination he stated in terms: "In the local valuation court I could have justified £1,775, that would have brought the clients to a net rateable value of £1,451". He added that he did not think they would have got £1,775, "I didn't think the figure was there". Mr Grieves told me that Mr Taylor said that he thought the premises were worth £2,000 gross rental; he asked Mr Taylor what he would go in at and did not get an answer. Still later, when it was suggested that he should have told his clients that if they achieved a net rateable value of less than £1,500 they could enfranchise, Mr Grieves said "I thought we were too far away from £1,500 to broach the subject", and that he did not think there was a prospect of achieving it in the local valuation court.

Mr Taylor's jottings at 10/6 occur opposite unit prices of £12 for ground and first floor, £8 for second and £5 for third, which are indeed the prices put forward by Mr Grieves for no 3 in his alternative formulation and which were never (as a band) the prices he put forward for no 1. It seems plain to me that Mr Taylor was here recording in summary form Mr Grieves' contention for 1966 rateable value in the case of no 3, less an end-allowance, and that the end-allowance being sought was one of 10%. Mr Taylor's end-allowance of £116 was, in fact, one of only 6%.

In both sets of unit prices relating to no 3 at 9/5 Mr Grieves departed from what he knew to be the invariable practice of Kensington and Chelsea to use the factors of 1.0, 0.8 and 0.6, and so on, to which I referred earlier. He agrees that the price for the uninhabitable third floor is an error and that it should have been 0.2 of the ground-floor price, ie £2 or £2.40, as the case may be. He says, and I accept, that Mr Taylor pointed out the error to him at the time. But, of course, he was negotiating, not endeavouring dispassionately to arrive at a fair rateable value. If one corrects the error by using 0.2 for the third floor, and takes the Kensington and Chelsea factor of 0.8 for the second floor, the totals become £1,676 in the case of the £10 ground- and first-floor unit price and £2,005 in the case of the £12 unit price, minimal variations from the totals in fact put forward by Mr Grieves. Therefore, neither the error of putting the third floor in at £5 nor the departure from the Kensington and Chelsea factors made any difference in the result.

Mr Taylor said that he thought that Mr Grieves' valuation was too low, and so Mr Grieves, as I accept, concentrated on the major disadvantages of the locality to which I referred earlier.

A In respect of no 1, Mr Grieves told me, and I accept, that he gave similar alternative sets of figures, as set out at the top of 9/5. Those dependent upon a ground- and first-floor unit price of £10 produced a rateable value of £1,734 without basement and £1,756 with. On the £12 basis the totals were £2,183 without basement and £2,205 with. In the latter case Mr Grieves claimed an end-allowance of 15% (ie £330) to produce £1,875. On Mr Taylor's note at 10/6 appears the calculation "£2,205 less £330 = £1,875", which ties in with Mr Grieves' recollection and may well have prompted it, since the end-allowance appearing on 9/5 is one of 10% only (in conformity with that used in the case of no 3), producing a rateable value of £1,985. Mr Grieves told me that he thought that in the case of no 1 an end-allowance of 15% was justified by virtue of its corner position and the
B traffic noise from Old Church Street. This seems to me to be a tenable argument and I am satisfied that in respect of no 1 Mr Grieves advanced the rateable values of £1,756 at a unit price of £10 without end-allowance and £1,875 at a unit price of £12 with end-allowance.

If one modifies the no 1 figures on 9/5 to take account of the Kensington and Chelsea factors and puts in the third floor at 0.2 a valuation of £1,796 replaces that of £1,756 in respect of the £10 unit price, and a valuation of £2,152 replaces that of £2,085.

Mr Grieves told me that Mr Taylor then considered his figures and resumed the discussion by saying that he was not going to come below £2,000. He thought that Petyt Place was a valuable group of five houses, more so than those in adjoining streets. Mr Grieves tried to
C justify his values, in the knowledge, as he told me, that the bracket of £10-£15 per m² was the Kensington and Chelsea rate for 1973 prime values. Mr Grieves referred to one comparable only, 21 Brompton Square, concerning which his notes are at 9/24. The rateable value of this hereditament was £2,300. Mr Grieves' devaluation of this property to a unit price of £12.50 ground and first floor is, I think, in error, in that, for example, he took the second floor price as 0.8 of £12.50 whereas it should have been £8. A correct calculation would have produced a devaluation of approximately £11.50 for ground and first floor. Where, therefore, as he did, he put forward 21 Brompton Square as an inferior comparable producing £12.50, he was in fact understating his case.

Mr Taylor produced no comparables and refused to budge from
D his £2,000 figure.

On the following day Mr Grieves spoke to Mr Taylor over the telephone and tried once more to achieve a reduction below £2,000, but to no avail. Mr Taylor was adamant.

In the course of his evidence Mr Grieves told me:

> I sat in purdah thinking about it all and decided I would not get better in the local valuation court and had to accept the figure. I considered what the result of the local valuation court might be. I did not expect to do any better. At the local valuation court I would have to give evidence. That would be £12 minus 10% on no 3 to include central heating. I thought I would not get it. I thought the local valuation court would split the difference between them and run down to £2,250. Mr Taylor did not say what he would do if we went to court. I know about the upper limit of the Leasehold Reform Act being £1,500 and I
> E was aware of that in 1978. I thought it was relevant to my clients. There was a possibility of their buying a freehold if it was less than £1,500. Could I reduce the value? I could not see it.

The £12 less 10% referred to in that passage produces a rateable value of approximately £1,800 and a net rateable value of less than £1,500. Mr Grieves was thus telling me that as a conscientious expert surveying witness he could in the local valucation court have advanced a case for his clients which, if accepted, would have brought their net rateable values below the enfranchisement limit, although I entirely accept that his opinion was equally that he would not succeed in doing so.

Following the telephone conversation with Mr Taylor, Mr Grieves wrote to all his clients a letter dated February 3 1978. I quote the last
F paragraph of this important letter:

> I am certain that the change in assessment could not be improved upon if we went to the valuation court hearing. All the arguments which were raised by Mr Price in great detail about the local nuisances have been taken fully into account in making what is a substantial reduction of assessment within the Kensington valuation list for residential accommodation, but I have no hesitation in saying this is now a very reasonable assessment.

It is the advice there given which is said to have constituted a breach of duty by Mr Grieves as set out in paras 11 and 13 of the amended statement of claim in both these actions.

In accordance with this advice the plaintiffs and the occupiers of nos 1, 2 and 5 Petyt Place accepted a reduction to £2,000 rateable value in the case of 2-5 and to £2,150 in the case of no 1. The additional £150 in the case of no 1 is probably referable to its garage. G

In achieving for his clients a reduction of about 16% in their rateable values, it was said by Mr Hague that Mr Grieves had done all that could be expected of a competent rating surveyor, since all the experts agree that a reduction of this order was on the face of it a good achievement; it was unusual to have a reduction of greater than 15%. Whether this is so or not must depend upon whether the existing valuation is itself fair, and I have to remember that the existing valuation was achieved not by any original valuation process but by applying a factor of 2.8 to the then existing valuations. In these circumstances I consider that there is force in Mr Neuberger's submission that in evaluating what has been achieved one should assess the achievement, not in terms relative to its predecessor but in absolute terms, asking whether it represents a fair valuation standing H upon its own feet. This, I think, was also Mr Grieves' approach, for in re-examination he told me that although the factor of 2.8 sounded to him rather high, he disregarded it as material and turned immediately to what *he* thought was the value of the property. The plaintiffs' case is that recourse to the local valuation court would or might have secured a reduction from the existing value of £2,380 of greater than 15%; a reduction to the magic £1,834 is one of £546, or 23%.

Mr Sharp's proposal

It was the opinion of Mr Marr-Johnson, the professional expert surveying witness called on behalf of the plaintiffs, that the reduction achieved by Mr Sharp in 1981 might have been achieved in 1978 and J should have been, and his contention compels an examination of Mr Sharp's proposal. The relevant documents are in bundle 6. Mr Sharp gave evidence which I entirely accept. The suggested new net rateable value of £1,350 was an entirely arbitrary figure. On October 27 1980 Mr Taylor wrote to Mr Sharp and there was an ensuing telephone conversation, during which Mr Taylor said that his figures were based on a floor area of 205m². The figures given by Mr Taylor were recorded at 6/22,23. The floor areas are those of no 1 minus basement taken from 10/2-4 and converted to metric. The areas total 228m².

The total floor area of 205m² spoken of by Mr Taylor is also to be found on a memorandum written by him on January 23 1981. From this it is clear that 205m² was an adjusted floor area. The provenance K of the figure must, I think, have been the Précis Sheet 10/1, upon which the entries for no 1 may have been more complete than those relating to the other houses. This records an adjusted floor area for no 1 of 205m² with a note "excluding basement". The adjusted floor area of 205m² assumes that the third-floor factor was 0.6. Whatever may have been the case in respect of no 1, the appropriate third-floor factor for no 2 was at most 0.2. If 0.2 were taken, the adjusted floor area would be reduced to 195.6m². Mr Huckle describes Mr Taylor's figure as an "absurd error".

In January 1981 Mr Taylor visited Mr Sharp at his house and agreed that something was wrong with the valuation. Thereafter Mr Taylor had no 2 properly referenced; the record of this reference is at L 10/10-12; there is a page missing from the survey sheets. This reference was carried out on January 20 1981 by Mr Bellion. His areas total 191.7m². The related adjusted floor area is 169.3m² according to Mr Huckle and as I accept. This is the area which should have appeared within the rubber stamp on 10/10 in place of 175 m².

At about the same time Mr Bellion referenced nos 3-5 Petyt Place. For no 3 he obtained a total area of 190.7m² and an adjusted area of 168.3m². It is the latter figure which should have appeared in the stamp at 10/19. For no 4 the figures are 191.0m² and 168.5m² respectively and the figure in the stamp is also wrong. For no 5 the figures are 216.8m² and 186.1m² respectively; the figure in the stamp is also wrong. To the referencing of no 5 I have already referred.

Thereafter, Mr Taylor wrote his memorandum of January 23 1981 M which contains the sentence:

> Peter Bellion re-referenced 20/1/81, and adjusted area is 169.3 sq m not 205 (ie 17½% smaller than the areas on which VO/Herring Daw agreed assessment).

Since I have found that Mr Taylor and Mr Grieves agreed in 1978 on the basis of Mr Grieves' figures, Mr Taylor here fell into serious error. From 9/5 it can be seen that in 1978 Mr Grieves was working on an adjusted floor area for no 2 of 168m², that is on an area virtually the same as Mr Bellion's.

Mr Taylor's next step was to devalue no 2 on the basis of 205m²: he obtained a unit price of approximately £10 per m². He then used the houses in Ormonde Gate as comparables — they devalued at £10.05

A per m². By way of end-allowance he added 25p per m² for partial central heating and 25p per m² for double-glazing, arriving at a valuation of £1,786, which he rounded to £1,775.

The utility of Ormonde Gate as a comparable was in dispute between the two expert surveying witnesses. Mr Marr-Johnson thought that it was a useful comparable. He said, however, that there were tall buildings around and he did not know if the houses were modernised in 1973. Mr Hassall's comments appear at 11/25 beneath a photograph of some of the houses. He said in evidence that in 1973 there was much student occupation in the vicinity, that the rooms of the houses were neither well planned nor shaped, and that it was curious that Mr Taylor relied upon these houses for his devaluation. Doing the best I can on exiguous material, I have formed the view that B in 1973 the accommodation and amenities of Ormonde Gate were markedly inferior to those of Petyt Place, so that the devaluation of £10.05 per m² is too low if applied to Petyt Place.

Mr Taylor's valuation of £1,775 was approved by his superior, a Mr Brittan, and agreed to by Mr Sharpe.

In my judgment, the reduction in the rateable values to £1,775 was unjustified and was the result of an incompetent error by Mr Taylor, purportedly supported by an inappropriate comparable. It follows that in determining what could have been achieved for Petyt Place in 1978, where there was no room for such an error, this rateable value of £1,775 is of no assistance. It follows also that the devaluation to £10 per m² unit price, in so far as it was based on an adjusted floor area C of 205 m², is irrelevant.

The liability of Mr Grieves

Mr Neuberger's primary submission was that Mr Grieves was in breach of duty in advising the plaintiffs to agree a net rateable value of £1,638 (see paras 11 and 13 of the amended statement of claim).

Apart from the allegation of faulty advice, the amended statement of claim contains five separate allegations of breach of duty which I have already read. Those under particulars (1), (2) and (2)(a) of para 13 have been disposed of, leaving the three allegations under particulars (2)(b), (3) and (4).

Particulars (2)(b) and (3) relate, as I understand them, both to the valuations which Mr Grieves himself made of the relevant properties D and to the manner in which he advanced his contentions to Mr Taylor.

Particular (4) is a quite different allegation, related to the possibility of enfranchisement. Mr Grieves knew the relevance of the £1,500 limit (see the passage quoted earlier from his evidence-in-chief). Particular (4) thus resolves itself into an allegation of culpable failure to advise, and is conveniently considered together with the principal allegation relating to that failure.

I propose first to deal with particulars (2)(b) and (3). Mr Neuberger did not suggest that Mr Grieves was at fault in not securing Mr Taylor's assent to a reduction to about £1,800 rateable value. I am satisfied that Mr Grieves deployed his figures and arguments with clarity and tenacity, as I have set out earlier in this judgment. One can E take a horse to water but one cannot make it drink; Mr Taylor would not budge from £2,000 and that he would not do so is, in my view, in no way the fault of Mr Grieves.

The allegations in particulars (2)(b) and (3) then resolve themselves into allegations that Mr Grieves valued the premises too highly and, of course, so regarded they impinge upon the allegations of culpable advice, in the sense that if Mr Grieves valued the premises too highly, then this might well explain the advice which he in fact gave.

As to particular (2)(b), it seems to me plain that Mr Grieves, in preparing for his meeting with Mr Taylor, valued the subject premises alternatively, both with and without an end-allowance. He told me that this is what he did, and I accept this. It is indeed the philosophy of his notes at 9/5,6. If what particular (2)(b) means is F that ultimately Mr Grieves justified to himself his valuation on the basis of a unit price without end-allowance, then the allegation adds nothing to particular (3), which is really the nub of this aspect of the case.

In his evidence Mr Grieves was referred to 9/5 and he said that he justified to himself the rateable value of £2,000 by taking £12 or £11.75 per m² for the ground floor. He was then referred to his letter of September 26 1979; he agreed, as I have already stated, that £5 for the third floor was an error and that the £7.50 was probably the mean of £7 and £8. The £12.50 figure was, he said, his. There is no end-allowance on 6/7. Mr Grieves' prices, as he told me, included 50p per m² for central heating.

Much expert valuation evidence was put before me concerning the proper valuation of these properties. This evidence included G reference to a number of comparables, attractively photographed and presented in bundle 11.

I do not consider that the resolution of the issues in this case requires me to make a finding as to what the proper valuation for the subject houses in fact was. It is, in my judgment, sufficient if I limit my inquiry to the question, whether a reasonably competent valuation surveyor would have considered that there was available to him evidence and argument which might have led to the local valuation court reducing the valuation significantly below Mr Taylor's £2,000.

Accordingly, I do not subject the comparables advanced on both sides to analysis. I am the more willing to put them aside since it seems to me that no true comparable was produced. If there is no true H comparable a difficult exercise in value-judgment emerges when one endeavours to assess the pros and cons of a comparable in terms of Petyt Place and then to adjust the rateable value of the comparable accordingly. Furthermore, it is the state of the comparable in April 1973 which is relevant, and one could rarely be sure that the condition of the comparable spoken to by the witnesses was indeed the 1973 condition or that their environment had remained unchanged since.

I therefore take a more direct route. I look to see what is the expert evidence as to the unit price per m² of 3 and 4 Petyt Place. Mr Marr-Johnson thought that an acceptable bracket was £11-£13 per m² or £10 plus, after deduction of end-allowance. This suggests to me that he had in mind the mean of his bracket (£12) as the most likely unit J cost; by implication this included central heating.

Mr Grieves himself went in at £12 to include central heating in the case of no 3, before end-allowance. He said in evidence that in justifying to himself the agreed £2,000 rateable value he took a unit price of £12 or £11.75.

Mr Hassall, the defendants' expert surveying witness, thought that the correct unit price was £12.50 to include central heating and before deduction of any end-allowance.

In the light of this evidence, I have formed the view that a unit price of £12 to include central heating and before deduction of any end-allowance was a perfectly acceptable costing for Petyt Place, and that a valuation reflecting an even lower price might ultimately have been achieved. Nevertheless, I take £12 to include central heating and see K where it leads me.

Mr Hassall's valuation of no 3 is at 2/30; his valuation of no 4 is at 2/31. In each case he has taken a unit price of £12 plus 50p for central heating.

In respect of no 3 his adjusted floor area is 171.92m²; in respect of no 4,175.61m². These areas are, I think, too high. This is because his ground-floor areas are too high. Exhibit D2, p1, enables in the case of no 4 a comparison to be made. Mr Hassall's area of 60.19m² is far above that of any other professional measurement, even leaving out of account Mr Sharp, whose areas as a whole were less than those of the others. I reject Mr Hassall's floor areas and turn elsewhere.

Mr Grieves' floor areas as used by him in negotiation fall, in my L judgment, within an acceptable bracket and I do not see why I should not make use of them. The actual floor area he took for no 3 was 196.4m² and adjusting this by the use of the Kensington and Chelsea factors produces 167.6m². The like figures for no 4 are 198.1 and 169.3m² respectively.

I take these adjusted floor areas and carry out in a revised form the computations carried out by Mr Hassall at 2/30,31. I use a unit price of £12 to include central heating. The calculation of no 3 becomes $167.6 \times 12 = 2{,}011.2$. The calculation for no 4 becomes $169.3 \times 12 = 2{,}031.6$.

At this stage Mr Hassall applied, correctly in my view, an end-allowance "for environmental factors". These included, he said, traffic congestion in the vicinity of the premises and loitering M vagrants. It does not seem, therefore, that his report took into account the noise from Embankment traffic, which I find on the evidence to be a significant detraction from the amenity of Petyt Place compared with locations free from such noise. In his evidence he did, I think, refer to the 5% also in the context of noise.

I am satisfied that an end-allowance of only 5% to reflect traffic congestion in and near Petyt Place, and noise from traffic in the immediate vicinity of Petyt Place and along the Embankment, is far too low. Mr Grieves was perfectly prepared to contend for 10%. Mr Huckle told me that Petyt Place suffered heavily from traffic, and that at the present time an end-allowance of 12½% on that account would be appropriate. The plaintiffs contend for only 10% (see the amended statement of claim para 13(2)(b). Upon the evidence before me, I think the 10% was the least end-allowance that could be justified and that slightly more might ultimately have been achieved. In what follows I use 10%.

Applying an end-allowance of 10% to the valuations of nos 3 and 4 already arrived at, I obtain final valuations of £1,810 in respect of no 3, and of £1,828 in the case of no 4. I emphasise again that these are not valuations made of the properties which purport to be "correct" valuations; they are valuations which seem to me to have been on the cards in 1978, in the sense that a cogent case could have been advanced for them. They are each below the determinator of £1,834, although only marginally so. Moreover, I consider that a reasonably competent valuation surveyor should have appreciated that by the use of the appropriate procedures they might be achieved, although I express no view whatever upon the likelihood of such an achievement.

I return to the pleadings to deal first with the allegations under particulars (2)(b) and (3) of para 13 of the amended statement of claim.

It is true to say, if one looks solely at Mr Grieves' letter of September 26 1979, that he there valued the ground floor and first floor at £12.50 per m^2 and the third floor at £5 per m^2. the price for the third floor was a slip, as I have already recounted. It was *de minimis* and immaterial, and not such, in my judgment, as to constitute a breach of duty on the part of Mr Grieves. It is also true to say that the computation in that letter does not include an end-allowance. However, to my mind the vital question in the case is not *how* Mr Grieves valued the premises, or *how* he later justified his approval of the £2,000 proposed by Mr Taylor, but whether a reasonably competent valuation surveyor would have recommended Mr Taylor's figure to his clients for acceptance in the terms employed by Mr Grieves. I do not, therefore, consider that in so far as Mr Grieves at one time valued the premises as alleged in particulars (2)(b) and (3) he was in breach of duty merely in so doing. This valuation was long after the event and is not congruent with the manner in which, I am satisfied, Mr Grieves advanced his clients' case before Mr Taylor. I think that it was really an *ex post facto* rationalisation which does not reflect the reality of the negotiation.

I therefore come to what seems to me to be the crux of this case, and I deal together with the allegation contained in the body of para 13 of the amended statement of claim relating to the advice given by Mr Grieves to the plaintiffs, and with what remains of that in particular (4), that he did not advise them of the importance of the net rateable value being reduced to £1,500 or less.

Mr Grieves knew well before his meeting with Mr Taylor that his clients were interested in purchasing their freeholds (letter of January 10 1975 from Mr McIntyre to the defendants and Mr Grieves' reply). Mr McIntyre told me, and I accept, that he did not know about a tenant's legal right to purchase the freehold in certain circumstances until Mr Sharp alerted him to that possibility; this must of course have been after Mr Sharp purchased the lease of no 2 in December 1978. In writing his letter of January 10 1975 Mr McIntyre had in mind purchase by agreement, not compulsory purchase. Mrs Johnson did not give evidence, so I do not know what her knowledge or intentions were. I have already quoted a passage from Mr Grieves' evidence which shows that he was well aware both of the £1,500 limit and that it was relevant to his clients.

After his meeting with Mr Taylor, Mr Grieves wrote to his clients, including the plaintiffs, the letter of February 3 1978, the last paragraph of which I read earlier in the course of my judgment. It is in forthright terms: Mr Grieves expressed himself as "certain" that no improvement upon Mr Taylor's figure could be obtained from the local valuation court. It is entirely understandable that in the light of that advice the plaintiffs agreed Mr Taylor's proposal.

Was this uncompromising advice in breach of Mr Grieves' duty to the plaintiffs? I consider that it was. By giving it in the terms he did Mr Grieves effectively deprived his clients of the opportunity of deciding for themselves after appropriate advice and reflection whether or not to appeal. I bear in mind particularly the following matters:

(1) Mr Grieves' own evidence that if the proposals had gone to appeal he could as an honest expert witness have justified a rateable value of £1,775.

(2) My own conclusions upon the evidence that a valuation of little more than £1,800 was on the cards in respect of each of the subject properties.

(3) The patent weakness of Mr Taylor's case as evidenced by his acknowledged lack of proper records and his unwillingness to support his proposal by figures or reasoned argument.

(4) The substantial financial gain to the plaintiffs if they were able compulsorily to enfranchise, compared with the likely costs of an appeal to the local valuation court.

(5) The opinion of the defendants' own expert, Mr Hassall, who in cross-examination gave the following evidence. Mr Neuberger asked him this: "Suppose you have a client who is a tenant under a long lease wanting to buy the freehold and that £1,500 is critical. Suppose you have done calculations which justify you going to court and striving for less than £1,500 but you don't think you can get it?"

Mr Hassall did not at first give germane answers to that important question, but ultimately he was pressed and he said this:

> It would be his duty to tell his clients of the possibility of getting less than £1,500 if he could honestly give evidence to that effect, even if he thought at the end of the day he would not get it.

I can only say that I agree with Mr Hassall that it was in all the circumstances the duty of Mr Grieves to put the plaintiffs in possession of all the facts of which they needed to be cognizant in order for themselves to make an informed decision and also, of course, his duty to tender his own advice upon the situation. Mr Grieves should have made sure that the plaintiffs were well aware of the significance of the vital £1,500 limit. In criticising him for not doing so, I do not suggest that he should have donned the mantle of a lawyer. The £1,500 limit was part of rating surveying expertise and its significance had to be made plain to the plaintiffs if they were to make an informed decision on their rating decision. Mr Grieves failed to advert to that significance and, in my judgment, he fell below a standard of reasonable competence in so failing. Mr Grieves should have advised the plaintiffs, not that he was certain that Mr Taylor's proposal could not be improved upon but that there was at least a reasonable possibility that it might be, and that he, Mr Grieves, could himself advance evidence in argument in support of a case that the net valuation should be reduced below £1,500. It would also, of course, have been Mr Grieves' duty to express the view, which I have no doubt he had, that at the end of the day such a reduction would not be achieved, but had he acted as should have, in my judgment, a competent and reasonable surveyor, the plaintiffs would have been afforded the occasion of making up their own minds. As it was, Mr Grieves virtually withdrew the matter from their consideration. I regret having to make this finding of breach of duty against a respectable and respected professional man who has, I am sure, served his profession and his clients well over a long span of years. There are very few professional men who can truly assert that they have never fallen below the highest standards demanded of them. Usually when there is a lapse no harm is done, and often no one else knows or need know. It is often a matter of chance whether harm results or not and whether, therefore, the lapse eventually becomes the occasion for inquiry and criticism. I hope that Mr Grieves will understand that there is no moral stigma whatever implied in my findings.

Upon the preliminary issue, therefore, there must be judgment for the plaintiffs in each action.

Official Referees' Business

November 19 1987

(Before His Honour Judge James FOX-ANDREWS QC)

GREEN AND ANOTHER v IPSWICH BOROUGH COUNCIL AND ANOTHER

Negligence — Dismissal of allegations of negligence by purchasers of house against local authority and surveyors instructed by building society to carry out mortgage valuation — Bewildering variety of diverse expert opinions as to structural condition of house — House in question was situated in marshy area on made-up ground where some properties had suffered from settlement — An adjoining house had tie bars on the front and side, evidently put there at some time to control further movement — Plaintiffs were young first-time buyers who purchased the subject house for

£12,000 with the aid of a mortgage from the Abbey National Building Society — Soon after the contract had been made, and before they moved in, certain cracks had appeared — They were advised by a structural engineer that the cracking was caused by excessive movement of the gable wall of the adjoining house with the tie bars and they were told that no major repairs should be undertaken until the adjoining house was stabilised — Other experts took a completely different view, advising that the settlement in the adjoining house was slight and had no, or no significant, effect on the plaintiffs' house — This view was taken by employees of the local authority and by the surveyors who carried out the mortgage valuation for the building society — The surveyors' report indicated that the settlement in the plaintiffs' house was slight and would be rectified by such remedial works as were required by the authority; if such works were carried out the house would have a value of £12,000 and would represent suitable security for an advance of £11,400 odd by the society — The surveyors' report, of which the plaintiffs had a copy, was a "Report and Mortgage Valuation" only and, together with other documents given to the plaintiffs, contained a clear disclaimer of liability — The report recommended that the plaintiffs should approach the building control department of the local authority about the necessary remedial works, but the plaintiffs did not take this action — They brought proceedings against the authority and the surveyors — The case against the authority was that its officers had made false and misleading statements that the house was suffering from a slight settlement only and that comparatively minor remedial works were all that was needed — The case against the surveyors was that their report to the building society contained statements to the same effect

A great deal of evidence was given at the trial, analysed in detail by the judge — He identified the major issues of fact which he had to determine, namely, (1) whether the subject house had in fact suffered major subsidence by the time the plaintiffs were actively considering a purchase; (2) was there at that time continuing movement in the adjoining property; (3) was it necessary to stabilise the latter before carrying out works to the subject property? — The judge's answers to all three questions was "no" — "The basic fabric of the house after eighty years remained sound" — No movement in the adjoining house between 1982 and 1985 had taken place — There was no evidence that the adjoining house had caused damage to the subject property — These answers really disposed of the action, but the judge expressed views obiter as to the duties of the authority and the surveyors — The authority owed no duty to the plaintiffs to take reasonable care to prevent damage save to the extent that they had assumed responsibility in their schedule of repair works; that responsibility was the narrow one of advising where it was necessary to rectify a settlement specifically identified in the schedule — The surveyors were under a duty of care to the mortgagees, as was clear from *Smith* v *Bush*, but they had fulfilled that duty by approaching the matter in the way an experienced and skilled surveyor should — The judge expressed the view that if they had failed to exercise reasonable care the disclaimer would have been invalidated by the Unfair Contract Terms Act 1977, but these opinions were academic in view of the findings of fact* — Action dismissed

The following cases are referred to in this report.

Hedley Byrne & Co Ltd v *Heller & Partners Ltd* [1964] AC 465; [1963] 3 WLR 101; [1963] 2 All ER 575; [1963] 1 Lloyd's Rep 485, HL
Smith v *Eric S Bush (a firm)* [1987] 3 WLR 889; [1987] 3 All ER 179; [1987] 1 EGLR 157; (1987) 282 EG 326, CA

*Editor's note: See on this point the decision of the Court of Appeal in *Harris* v *Wyre Forest District Council* at p 132 *ante*.

This was an action by the plaintiffs, Mr and Mrs Green, against Ipswich Borough Council and Spurgeons, chartered surveyors, of Ipswich, alleging breach of duty in respect of the plaintiffs' purchase of the property at 49 Gatacre Road, Ipswich.

John Peppitt QC and Julian Chichester (instructed by Birketts, of Ipswich) appeared on behalf of the plaintiffs; Michael de Navarro and Martin Porter (instructed by Prettys, of Ipswich) represented the first defendants, Ipswich Borough Council; Mark Lomas (instructed by Mills & Reeve Francis) represented the second defendants, Spurgeons.

Giving judgment, JUDGE FOX-ANDREWS QC said: On July 21 1982 Mr and Mrs Green contracted to purchase 49 Gatacre Road, Ipswich, for the sum of £12,000. This was a terraced house, on the south side of the road, built in 1903. It was a modest quality house which was likely to make it particularly attractive to young first-time buyers. Mr and Mrs Green, the plaintiffs in this action, were such buyers. They had married only shortly before the purchase.

The plaintiffs entered into the contract to purchase on the basis of an offer of an advance of £11,484.25 from the Abbey National, who had taken the advice of surveyors called Spurgeons. The plaintiffs wished to make substantial improvements to the house. They hoped to obtain a 75% grant from the council towards the cost of these improvements.

Through no fault of Mr and Mrs Green, what should have been an exciting venture and challenge turned into a nightmare. Less than three weeks after they had contracted to purchase and before they had moved into occupation, they were advised, on August 10 1982, that, because of cracking observed by Mr King and Mr Hook, employees of the council, they should obtain the advice of a structural engineer. The plaintiffs went at once to a Mr Matthews, of Maddocks Lusher & Matthews, chartered civil and structural engineers of Ipswich. Mr Matthews is a member of both the principal professional bodies. He advised in writing in September 1982, sending his report to Mr Green on September 7. His primary conclusion was that the majority of the cracking in no 49 had been caused by excessive movement of the gable wall of the adjoining house no 47, which lay immediately to the east of no 49 and was the end house of that terrace. The gable wall was therefore the eastern wall of the property. Mr Matthews' principal recommendation was that no 47 should be stabilised before major repairs were undertaken to no 49.

A sustained attack on Mr Matthews for reaching that conclusion and for making that recommendation has been made by the council in the course of this trial. The council's own borough engineer, a Mr G E Mann, who was not called as a witness, in an internal report dated January 27 1983, reached and made essentially the same conclusion and recommendation. His conclusion is not evidence, but the fact that he reached that conclusion is evidence.

The problems facing Mr and Mrs Green and indeed the council were exacerbated by the fact that the occupier of no 47 had died intestate. By October 1983 the council had obtained a grant of letters of administration. It was hoped that the Abbey National would be the buyers of no 47. It was then to be improved by an improvement grant provided by the council. For a short period it was to be used jointly by the Abbey National and the council as a show house. It was then to be sold to a particular mortgage applicant of the Abbey National. The council of course had to take and be seen to be taking all reasonable steps to obtain the best price, and for this reason no 47 was put on the market.

On January 26 1984, Abbey National wrote a letter to Spurgeons, in which they said:

> Unfortunately due to the instability of the adjoining property at no 47 causing structural damage to the mortgaged property, our applicants have been unable to occupy it since the date of purchase. Neither have they yet been able to obtain agreement to a local authority grant until no 47 has been sold and stabilised by the purchaser. The latter arrangements are proceeding at the present time.

Spurgeons were asked to make a report, particularly on whether there had been any deterioration to no 49. On February 1 1984, Spurgeons replied, saying:

> It is our opinion that no further significant movement has taken place since our original inspection and whilst further movement cannot be ruled out in view of the state of the property adjoining, it is our opinion that further movement in the immediate future is unlikely providing that the adjoining property is dealt with as required by the local authority.

By the spring of 1984 funds for improvement grants were drying up

A and both the council and the Abbey National were reconsidering their positions. The council became more insistent for a detailed structural engineer's investigation of no 47. The Abbey National were by that time looking for a greater financial contribution from the council in order to reduce their losses.

On August 20 1984, a Dr Wady, who was not called as a witness, a partner of Mr Matthews, made a report on no 47.

In September 1984 the plaintiffs instructed the firm O A Chapman & Son, chartered surveyors and valuers of Norwich and Lowestoft. The partner concerned was Mr Michael W Chapman FRICS and an associate of the Incorporated Association of Architects and Surveyors. He inspected in September and October 1984, on one occasion with Mr Matthews, and took some 38 photographs. He
B reported on October 16 1984. His finding was that there had been extensive subsidence in and movement to no 49. He went on:

> Whilst accentuated by the severe movement in 47 Gatacre Road, there can be no doubt that 49 Gatacre Road and its neighbour at 51 Gatacre Road has subsided of their own accord.

He criticised Spurgeons.

In October 1984 the Abbey National reached the decision not to purchase no 47. On November 13 1984 there was a formal letter before action sent to both defendants on behalf of the plaintiffs. On November 16 1984 Spurgeons wrote to the Abbey National, saying:

> and it is our opinion in any event and notwithstanding anything that the Society has done, that Mr Green should have obtained a full structural survey
> C on receiving a copy of our report.

On the same day a Mr Hill, a partner of Mr Matthews, set up monitoring points on no 49. The reason for this was that at that time there was a possibility of no 47 being demolished and therefore accurate measurements of crack widths might be needed in evidence.

In November 1984 Mr A N Towell FRICS, associate of the Chartered Institute of Arbitrators, a chartered building surveyor, was instructed on behalf of the underwriters of Spurgeons. Before carrying out any inspection of no 49, Mr Towell was instructed on behalf of AS Construction Services Ltd (AS), who were considering purchasing this property, to prepare a scheme for the underpinning of no 47. He concluded that substantial settlement of no 47 could
D have been caused by drainage under the foundations, the mechanism of which I shall have to consider later. He recommended an underpinning scheme, for reasons I shall have to consider later. As regards no 49 he reached an opinion, which he says he never altered, that the settlement of no 49 was slight and that the movement of no 47 had virtually no effect, let alone structural effect, on no 49. He took a number of photographs in December 1984. In reaching his conclusion, he had considered copies of Mr Matthews' report of September 1982 and Mr Mann's report of January 1983.

On January 21 1985 AS contracted to purchase no 47 for £5,800. On January 24 1985 closing orders were served by the council in respect of both 47 and 49. Thereafter, AS speculatively carried out remedial works to no 47, including underpinning of the gable wall
E and by the rear of no 47, in order to stabilise the premises.

On October 17 1985 Mr and Mrs Green commenced proceedings against both defendants. In December 1985, if not earlier, the council were satisfied that all necessary works to no 47 to ensure structural stability had been carried out. The closing order on no 47 was lifted.

In December 1985 Mr Thomas Akroyd was instructed by the council. He is a chartered civil engineer, chartered structural engineer, bachelor of science and technology, Manchester master of science, researcher in soil mechanism and bachelor of law (King's College) London. His assistant, Mr Hooper, carried out an extensive investigation in December 1985, at which time he took a number of photographs. On January 20 1986 Mr Hill took a second set of readings at the monitoring points on no 49. On his estimation the only
F significant change appeared to be that the crack on the front wall had opened up an average approximately 0.8 mm.

By early 1986 the council had approved the plans and specification for structural repairs to no 49, in accordance with plans submitted by Maddocks Lusher & Matthews. No underpinning of the existing parts of no 49 was required. The work to no 49 was then carried out. Mr and Mrs Green were able to get a £7,650 improvement grant, being 75% of £10,200. In about March 1986 the closing order on no 49 was lifted. Finally, nearly four years later the unfortunate Mr and Mrs Green were able to go to live at no 49.

Because of the divergence of views between Mr Matthews and Mr Mann on the one hand and Mr Towell on the other, the underwriters of Spurgeons, on Mr Towell's recommendation, brought in an engineer, Mr Brian Cowley, of Jenkins & Potter, who is a chartered G
engineer FICE, fellow of the Institute of Structural Engineers, member of the Construction Engineers, associate of the Chartered Institute of Arbitrators, FRSA, and a partner of that firm. He gave a report in June 1986, essentially agreeing with Mr Towell. He was not prepared to say that the movement of the gable wall of no 47 and the other walls attached to it would have no effect at all on no 49, but such effect as it would have had would have been insignificant and would in his opinion have in no way affected the structural stability of no 49. In June 1986 Mr Cowley gave a written report containing the views already indicated.

In September 1986 Mr Chapman gave a written report containing the views that I have already indicated. On October 1 1986 Mr Towell likewise gave a written report. On October 7 1986, Mr Akroyd gave H
his written report, in which he damned Mr Matthews' views in trenchant terms. His view was that the settlement of nos 47 and 49 had ceased years ago, that there had been no continuing movement of either house and that the underpinning of no 47 had been unnecessary.

The case against the council as it ultimately emerged is a narrow one. The plaintiffs allege that on June 17 and July 20 1982 building control officers made statements and representations which were false, untrue and misleading, which the council had no or no reasonable grounds for believing to be true. The statements said to be false were (a) those made on June 17, namely , that no 49 was only suffering from slight settlement and that the only remedial work J
required in relation to such settlement was some repointing and mortar filling, and (b) those made on July 20 that no 49 was only suffering from slight settlement and that the only additional remedial work to the repointing and mortar filling would be the provision of steel retaining brackets for the rear stairwell wall, together with the rebuilding of the brickwork over the doorway in the wall, including the provision of a new lintel. It is further alleged that the council were negligent in the carrying out of the inspections on those two dates.

The case against Spurgeons is that in a report and mortgage valuation made by them on June 23 1982 to the Abbey National, Spurgeons made representations which were false, untrue, inaccurate and misleading and which they had no or no reasonable grounds for believing to be true. The representations said to be false K
were that the settlement suffered by no 49 was slight, that it would be rectified by such remedial works as were required by the council, that, provided works listed in the council's schedule of early June 1982 were carried out and no 49 was redecorated, it would have a value of £12,000 and would represent suitable security for an advance of £11,400 odd by the Abbey National and would be readily saleable for £12,000 for the purposes of owner-occupation.

The trial commenced on April 29 of this year. During the course of the trial on April 30 the parties' experts, excluding Mr Cowley, carried out distortion surveys of the party walls of 49/47 and 49/51, together with the plumbing of the front elevation of 47, 49 and 51 (exhibit 4). On May 12 1987 some of the experts investigated the L
position and the nature of the tie bars in no 47 and a remarkably different state of affairs was established from that previously believed to be the case.

I have had the evidence of all the experts already indicated. In addition, I have had the evidence of Mr B V Spurgeon, Mr and Mrs Green and from four employees of the council: Mr Hook, Mr King, Mr Best and Mr Eaton.

Against that background, it is now necessary to examine in detail the evidence. This case depends to a significant extent on the nature of the construction of nos 47 and 49.

Various plans and photographs of nos 47 and 49 have been put in evidence. I should identify the photographs as far as possible in chronological order. Exhibit 1A is a photograph of the front M
elevation of no 49, taken in the summer of 1981. Exhibit 1B is a photograph of no 49, taken in the autumn of 1983. As Appendix D to Mr Chapman's September 1986 report are some 35 photographs, taken in September and October 1984. A description of these photographs is to be found in section 3.3 of his report. Three additional photographs, taken at the same time, form exhibit 12. As Appendix D to Mr Towell's report of October 1 1986 are six photographs taken by him in December 1984. A description is given against each photograph. These I have numbered. On December 18 1985 Mr Hooper of Akroyd took the 17 photographs which are annexed to the Akroyd report dated October 7 1986. Mr Cowley took two photographs in June 1986, but I need not refer to them. Exhibit 9

is a photograph taken by Mr and Mrs Green, which shows the front elevation of no 49 in 1986 after the remedial works so far as that brickwork is concerned was completed. Exhibit 10 is a photograph taken by Mr Hooper on April 30 1987 after remedial works to the brickwork above the dining-room window at the rear of no 49 had been completed. The positions of the camera for the photographs in the reports of Mr Chapman and Mr Akroyd have been helpfully put on a plan, which has not been given an exhibit number. No 49 was a two-storey house, with a two-storey back extension and, beyond that, a single-storey extension. For the purposes of identification I use figure 2 to Mr Akroyd's report, which contains the floor plan of no 49. On the ground floor, room no 1, which is the front room, was also described in the evidence as the sitting-room. Room 2, which is the back room, was also described as the dining-room. Behind room 2 was the kitchen. Although not shown in this plan, behind the kitchen was a larder, fuel store and lavatory. On the first floor, room 3, the front bedroom, was sometimes described as bedroom no 1. Behind room 3 was room 4, also called the back bedroom or bedroom no 2. Room 5 has been variously described as the bathroom or bedroom 3.

The house was built in a marshy area on what was made-up ground. The terrace, in which it was one of five, was built of solid nine-inch brickwork, with a pitched roof. In 1982 the original roofing of the terrace had been replaced with interlocking concrete tiles. The party walls of no 49 with nos 47 and 51 were also of nine-inch brickwork. Access to upstairs was gained by a staircase set in a stairwell of cross walls, running east-west, between the front and back rooms. Access to the staircase from the ground floor was via an opening in the front cross wall of the dining-room at its western end, that is to say close to the party wall with 51. The staircase debouched between the front and back bedrooms on the west side. As figure 2 of Mr Chapman's report shows, there were openings in both cross walls on the first floor at both the eastern and western ends.

Except for the reversed positions of the kitchen and bathroom no 47 was of a similar layout to no 49, as appears from Mr Towell's exhibit C, sketch 1. But of course, as already indicated, it had a gable wall to the east, being the end of the terrace. In 1982 the front of no 47 was already faced with render and the other walls with pebbledash. Certain tie-bar plates were visible from the outside: see, for example, Mr Chapman's photographs 4, 16, 34, 35, exhibit 12 (the photographs showing the flank walls of both nos 47 and 45) and Akroyd's figure 4.

The purlins of the five houses in the terrace ran from east to west. In no 47 they had originally been embedded in the gable wall. The joists ran from north to south. The flooring was supported on the joists. The openings for the doors and windows in the front elevations in nos 47, 49 and 51 can best be seen in the top elevation drawing of sketch 3 of Mr Towell.

Under the Housing Act 1974, Part VII as amended by the Housing Rents and Subsidies Act 1975 and the Housing Act 1980, Ipswich Borough Council were empowered to make certain grants. The reference to sections in this part of the judgment are references to sections of the 1974 Act.

One such grant was an improvement grant under section 56(2)(*a*), being the grant for the improvement of a dwelling by works which encompassed more than the provision of standard amenities (section 56(2)(*b*)). However, the council could not approve an application for such a grant unless they were satisfied that on completion of the particular works the dwelling would attain the required standard. The required standard was defined in section 61(3) as being that the dwelling *inter alia* was in good repair, in conformity with such requirements with respect to construction and physical condition as might for the time being be specified by the Secretary of State. One such requirement was that the dwelling was in a stable structural condition and was likely to provide satisfactory housing accommodation for a period of 30 years. A council was empowered to make an intermediate grant where the dwelling on the completion of the works would provide satisfactory housing accommodation for a period of not less than 10 years.

Under section 57(1), no grant can be paid by the council unless the application is made in accordance with the provisions of Part VII of the Act. By section 57(2), the application for a grant must contain particulars of the works in respect of which a grant is sought (the relevant works) and an estimate of their costs.

Mr Eaton, the council's present improvement officer, has satisfied me that a strict interpretation and application of these provisions could dissuade the very people who most needed the grants from applying. Once in general terms the applicant's works' proposal was known, a council employee from the council's environmental health department (EHO) would be required by the housing department, who handled grant applications, to visit the premises and make an inspection record. Based on that record he would then propose a schedule of improvements and repairs. I am satisfied that there was no statutory obligation on the council to produce such a document. It was produced as a matter of convenience both for the council and the applicant. In the absence of such a schedule, the applicant himself would have had to prepare or cause to be prepared a specification of the works. This in many cases might require the services of a surveyor. From the council's point of view, by producing and serving a schedule they do not have to check a layman's list, which could be a time-consuming business. The council sent a copy of the schedule to the applicants in early June. It was invariable that the copy of the schedule sent to the applicant had attached to it advisory notes.

The building control department was different from that which handled the improvement applications. In section C of the blue bundle was a very detailed improvement grant "flow chart". Like the contents of such documents, it differed significantly from what actually took place. Good sense demanded that whenever a schedule contained any reference to settlement a copy of the application, together with a copy of the schedule, should have been sent to the building control department, who should have opened a file in the matter. Thus when the applicant communicated with the building control department, as he was required to do, that department would — or at least in theory should — have had readily to hand the applicant's file.

However, I am satisfied that it did not ultimately work out in that way and certainly did not work out in that way so far as the plaintiff's application was concerned. From that failure to lay down and follow that system, unfortunate consequences could flow and it will be a matter for consideration in this case as to whether they did and, if so, what the legal consequences are.

Mr Green was unfortunately out of work in the spring of 1982. Mrs Green, however, was working. She had previously been employed by the council and knew or was acquainted with some of the council employees concerned in this action. At an early stage in the hearing it seemed likely that Mrs Green's former position with the council had caused them to treat her application differently from the manner in which applications were ordinarily dealt with. However, I am satisfied that that is not so. The notes of the application form prepared by the council at Red Bundle 10 appear to contemplate that only persons with an interest in the property or a statutory right to be there will be applicable, but question 3 on the previous page appears to contemplate that persons hoping to acquire an interest but who had not yet done so may also apply.

Mrs Green approached the Abbey National in early May 1982 in respect of premises in Harvey Street, Ipswich, in which the plaintiffs were then interested. However, the plaintiffs did not proceed with those premises because bad cracking was noted.

By mid-May Mr and Mrs Green had become interested in 49 Gatacre Road. Mrs Green approached the Abbey National, who gave her encouragement. She wrote a letter to the council's area improvement officer, Mr Best, on or shortly before May 18 (Red 14) and enclosed a two-page document (Red 15 and 16). This had been prepared by Mrs Green without assistance from others. This document listed matters to be done or things to be checked. In respect of both the front and back bedrooms, she had written "check on settlement". In Red Bundle 9 to 11 is the application form, partly completed by Mrs Green on May 18 1982 and partly completed by Mr Green in April 1985 when the stabilisation works to no 47 were under consideration. It is not clear whether the application form in its partially completed state was passed to the council at the same or a different time from the letter and the list to which I have just referred. Mr Best says that at this time he knew that Mrs Green had not bought the property. It is unnecessary to consider in detail what happened within the council to those documents. I am satisfied that neither the letter nor the list was received by the building control officer, at least for a very long time.

The housing department, which embraced the area improvement officers, informed the EHO of the application. Because it was possible that an intermediate rather than an improvement grant might be apposite, two memoranda (Red 17 and 22) were sent by the director of housing to the EHO. The reason for these

A communications is that, as earlier indicated, it is the EHO who carry out an inspection of the property and prepare a schedule of items for improvement and repair. This was done by a Mr Huffey of the EHO, and his documentation appears at Red 20, 21 and 21A. On the basis of this inspection he then produced on June 8 the schedule which appears at Red 26–28, which he sent to the director of housing. It is expressly conceded by the plaintiffs that Mr Huffey did not act wrongly in preparing that schedule in the manner he did. On receipt of the schedule the area improvement officer (within the housing department) checked the schedule for typing errors, added the advisory notes which appear at Red 32 and 33, and sent it, together with a pro forma letter (Red 29) to Mrs Green, shortly after June 8 1982. It was not sent to the building control officer (BCO), which
B came within a department initialled DTS. It is unnecessary to set out in length the contents of the schedule and the advisory notes, but I should refer to certain limited passages:

> The front and rear entrance doors and frames are worn and/or ill-fitting and require overhaul or replacement as necessary to leave in a sound working order (Red 31). All internal doors are worn and/or ill-fitting and require overhaul or replacement as necessary to leave in a sound working order on completion. There is evidence of slight settlement to the front main wall of the property. This matter should be discussed with a Building Control Officer to establish the extent of the works required to remedy the defect (Red 32).
>
> Settlement. If it is necessary to rectify settlement or implement underpinning please consult the Building Control Officer for advice. An estimate for carrying out the works specified, based on his requirements, must
> **C** be submitted as part of the Grant Scheme (Red 33).
>
> Completion of works. You must complete all the repairs required to make your house fit for habitation. These repairs are marked in the schedule. Where other grants are approved the repairs should be completed to a reasonable standard, as listed in the schedule. In some cases the schedule of repairs may be negotiable, but you must agree all adjunctions with the improvement officer before starting work (Red 33).

By the accompanying letters to the plaintiffs they were told that once they had agreed with the improvement officer which works they had to do they should get estimates and submit the estimates to the improvement officer.

The schedule and its notes appear to be clear and straightforward. Where, as here, settlement was raised, it was essential that the
D applicant should contact the BCO. If, however, the applicant wanted to negotiate repairs to a reasonable standard, as distinct from work necessary to ensure *fitness for habitation*, then she had to go to the area improvement officer. Because of the unsatisfactory system then in force, when Mrs Green, as she was required to, spoke on the telephone on June 17 to Mr King, who was then the senior BCO present, as his superior Mr Hook was on holiday, there was nothing in the building control office relating to Mrs Green's application or the inspection and the EHO's schedule. The discussion between Mrs Green and Mr King took place in the morning. Mr King frankly conceded that his recollection of the telephone call was not good. It would be surprising if it were otherwise. Mr King clearly had
E extensive duties to perform. His primary task was to ensure that building regulations and other building control requirements were complied with. This meant inspecting numerous buildings in the course of a month. On any particular day Mr King might have to make a substantial number of inspections. I am satisfied that he received the telephone call just before he set off to make a number of such inspections. I am satisfied that he said he had not got a copy of the schedule. I think it likely that Mrs Green read out the second paragraph at Red 32 that I have just referred to.

A meeting was arranged for a short time later that day at no 49. I am satisfied that Mr King never said that it would be an informal visit. I find it likely that in the course of the telephone conversation Mrs Green told Mr King that she was going to buy the house. I am
F satisfied, though, that nothing was said by Mrs Green to Mr King to indicate that her decision whether or not to purchase was dependent upon Mr King's advice. I think it likely that Mrs Green had a copy of the schedule when Mr King arrived at no 49. The only settlement referred to in the schedule was that to the front main wall of the property. This was the only matter that Mrs Green had to refer to the BCO. It was for Mr King to establish the extent of the works required to remedy that settlement only, so that through him the council could advise. I am satisfied that primarily Mr King directed his attention to the front wall of no 49. I find it likely, however, that he entered no 49, walked through the front and back rooms on the ground floor and went outside to the rear. I find that it is likely that Mrs Green did say that in some places the window frames had come away. I am not
G satisfied Mrs Green ever mentioned no 47 to Mr King or any worries in relation to that house. I am satisfied that Mr King said that the settlement to the front main wall was not serious, that all buildings in that area had some settlement, that it was a nice little house in a quiet road and that all that was required was the repointing of the joints. I am satisfied that he expressed no concern. Whether he was justified in expressing those views is something I shall have to consider later.

The principal reason why I am satisfied that this was not an informal meeting is that, following the visit, Mr King brought a log into existence for no 49, which commences at Red 311. His entry for this visit on June 17 reads:

> Discussed settlement to front of property. Joints opened up in brickwork between living room and front bedroom window. Minor movement joints to
> **H** be repointed.

If this was no more than an informal visit to be followed up by a formal visit, it is likely that no log would then have been brought into existence, or some reference to the fact that a formal visit was to take place would have been recorded in the log. If the council policy had been followed, Mr King's task on returning to his office was to have obtained a copy of the schedule and then to have written a memorandum to the area improvement officer stating the extent of the works required to remedy the settlement defect. Unfortunately, by the time he went on holiday sometime later, he had done nothing. I have had put in evidence as exhibit 13 a statement Mr King made to the council's solicitors on November 6 1985. Save in so far as I have
J made findings that accord with Mr King's statement, I am afraid I must reject it. In particular, I do not accept that Mr King informed the plaintiffs that a further visit might be necessary or that following his visit of June 17 Mr King contacted the EHO. I shall have to consider what, if any, reliance was placed on the statement by Mr King to Mr and Mrs Green.

Mr Green made a formal mortgage application to the Abbey National on June 19 (Red 38–41). Under section 25 of the Building Societies Act 1962 a building society which made an advance on the security of a freehold or leasehold property had to obtain a report from an experienced person as to the value of that property and as to any matters likely to affect its value. Abbey National provided notes for the guidance of mortgage applicants. I am satisfied that Mrs
K Green saw a copy. I should read parts of the notes.

> What are inspection fees for?
>
> The Society is required by law to obtain reports on properties on which mortgage advances are made. For this purpose Surveyors are empowered to inspect each property and report to the Board of Management treating the applicant's business as strictly confidential. When calling at occupied premises the Surveyor may quote your name where necessary.
>
> The Society is able to offer you the choice of two types of report on the property, neither of which is a structural survey. A copy of a *Report and Mortgage Valuation* is available at no extra cost but neither the Society nor the Surveyor or firm of Surveyors will warrant or give any legal assurance as to the accuracy of statements made in the report or the validity of conclusions and opinions expressed or implied therein. As an alternative you are
> **L** recommended to have a copy of a more comprehensive *Report on Condition and Valuation* made available for a fee of £57.50 including VAT which is additional to the normal inspection fee payable. This report which is subject to the Conditions of Engagement (see copy enclosed) will incorporate details of the condition of the property; comment on the purchase price or fair market value and give a mortgage valuation.

The inspection fee for the Report and Mortgage Valuation on a house whose purchase price was £12,000 was £27.13. Mrs Green opted for the Report and Mortgage Valuation.

I should read part of the declaration that she signed:

> I accept that the Society will provide me with a copy of the Report and Mortgage Valuation which the Society will obtain in relation to this application. I understand that the Society is not the agent of the Surveyor or firm of Surveyors and that I am making no agreement with the Surveyor or
> **M** firm of Surveyors. I understand that neither the Society nor the Surveyor or firm of Surveyors will warrant, represent or give any assurance to me that the statements, conclusions and opinions expressed or implied in the Report and Mortgage Valuation will be accurate or valid and the Surveyor's report will be supplied without any acceptance of responsibility on his part to me.

On June 21 Spurgeons were instructed by Abbey National to prepare a Report and Mortgage Valuation. I am satisfied that Spurgeons received the council's schedule (Red 30–32) but not the advisory notes. Abbey National provided the form which Spurgeons were required to complete. The form as completed by Spurgeons on June 23 is at Red 43–46. I should read out some of the contents of the completed form. That part includes both the questions and the answers.

A 10 GENERAL OBSERVATIONS ON STATE OF BUILDINGS INCLUDED IN MORTGAGE VALUATION
The property is situated in area where certainly some settlement has taken place and this property has suffered slight settlement both to the front and rear elevations. The property next door ie no 47 has been tied back to the front and side and we would imagine that the rods were inserted to prevent further movement of that particular property. We note in the Schedule for the Borough of Ipswich the settlement mentioned on the front elevation but there is no mention of the rear movements. Providing the matter is discussed with the Building Control Officer and the necessary remedial works carried out, we are happy with this situation. We further refer to the Schedule from the Local Authority and providing these works are satisfactorily completed as a condition of the granting of the mortgage, together with the interior and exterior redecoration, consider the property will represent a suitable security.
B 11 REPAIRS RECOMMENDED AS A CONDITION OF MORTGAGE
As per Schedule received from the Local Authority together with redecoration of interior and exterior.
13 SUBSIDENCE, SETTLEMENT AND LANDSLIP
Have you any knowledge of significant Subsidence/Settlement/Landslip/ Ground Heave in other properties in the immediate vicinity? If so please comment.
Some evidence of settlement in this property at both front and rear *and also to the adjoining one.* This matter is covered, however, in the works required by the Borough of Ipswich and we are prepared to accept the remedial works they have advised.
14. GENERAL
(A) IS THE PROPERTY READILY SALEABLE FOR THE PURPOSE OF OWNER OCCUPATION AND IS IT LIKELY TO REMAIN SO AT OR ABOUT THE MORTGAGE
C VALUATION?
Will be when the works have been satisfactorily completed.
(B) ANY OTHER IMPORTANT FACTORS?
Completion of necessary works.

In red, in a box at the foot of the form above Spurgeons' certificate, appeared these words incorporated by the Abbey National:

TO THE MORTGAGE APPLICANT(S)
IMPORTANT
1 THIS DOCUMENT IS NOT A MARKET VALUATION. IT IS NOT, AND SHOULD NOT BE TAKEN AS A STRUCTURAL SURVEY. It has been obtained by the Society from a qualified surveyor or firm of surveyors to comply with section 25 of the Building Societies Act 1962.
2 If you are purchasing the property, you will receive a notice that the Society
D does not warrant that the purchase price is reasonable.
3 This is a report to the Society by its Surveyor(s) and neither the Society nor the Surveyor(s) give any warranty, representation or assurance to you that the statements, conclusions and opinions expressed or implied in the document are accurate or valid.
4 The Surveyor(s) has/have made this report without any acceptance or responsibility on his/their part to you.

Spurgeons therefore had clear notice that a copy of their report was likely to be furnished to the mortgage applicants, who were named as Mr and Mrs Green. And this is what occurred. I should read Abbey National's letter of July 1 1982 to Mr and Mrs Green:

To assist the Society in considering your mortgage application a limited report on the property has been obtained from a qualified Surveyor. I have pleasure
E in enclosing a copy of this report for your personal information and hope that you will find this helpful.

This report has been obtained for the information of the Society as mortgagee and as a prospective purchaser you may well require further and amplified information.

Your attention is drawn to the important Notice in the report and I recommend that you seek your own independent professional advice on any points concerning your purchase over which you wish to be satisfied.

In the report Spurgeons had given a gross value for mortgage purposes of £12,000.

The only communication between Spurgeons and the plaintiffs prior to the plaintiffs binding themselves to purchase the property was this copy of Spurgeon's report. I find the report in some respects unsatisfactory. In the course of Mr Spurgeon's examination he had
F observed what he considered to be slight settlement not only to the front of the house but also in three places on the back. These places were, firstly, to the soldier arch over the dining-room window, as appears in Mr Chapman's photo 5 and in Mr Towell's appendix D; secondly and thirdly, in the rear wall of the back extension, as appears in Mr Chapman's photo 8. Fractured brickwork below and above the window opening to the bathroom can be seen. I think that the specific areas of manifestation of slight settlement to the rear should have been identified. The answers to paras 10 and 11 of the report should have made specific mention that the works to be done would also include such additional works as the BCO might require to rectify the slight settlement at the rear albeit the works were of a minor nature. Save for the words "also to the adjoining one", Mr Spurgeon did not answer the question posed in question 13. What is G
abundantly clear from the report is that there is not the slightest suggestion that the plaintiffs should obtain a full structural survey. In these circumstances his comments on November 16 1984 to Abbey National, earlier referred to, were incorrect.

On July 1 Abbey National sent to Mr and Mrs Green a copy of Spurgeons' report, a mortgage offer of £11,484.25, an acceptance and receipt of notices form and a copy of their general mortgage advance conditions (GMA). The offer was on the terms of the GMA conditions. The offer document warned that the GMA conditions should be read carefully.

Para 3 of the GMA provided:

SURVEY SERVICE H
(a) Report and Mortgage Valuation.
This report is made for mortgage purposes only.
(b) Report on Condition and Valuation.
This report is offered as an alternative to 3(a) and gives a more comprehensive opinion on condition and marketability.
In neither case is the Surveyor instructed to carry out a structural survey nor to test the electricity, gas, water or drainage installations. Applicants who are still unsure on any points are recommended to obtain independent specialist advice. In offering a mortgage the Society does not warrant the price to be reasonable nor accepts responsibility for the condition of the property.

I should read three of the special conditions contained in the mortgage offer:

8 You are recommended to arrange execution of the repairs detailed in section
11 of the Report and Mortgage Valuation, enclosed herewith, but no J
retention will be made for the advance.
9 Improvement grants to be advised to Society as soon as possible.
10 Final cost of essential repairs and improvements to be advised to Society as soon as possible.

Mr and Mrs Green signed the acceptance note on July 2 1982 (Red 57).

Mrs Green was questioned about the disclaimer wording appearing in the various documents that I have referred to in this judgment. Her recollection as to whether or not she had read the various passages was vague. If it was necessary for me to make a finding on this I would have held that she had read the various passages. But the relevant findings appear to me to be that the
disclaimers were sufficiently clearly brought to Mrs Green's K
attention or were sufficiently clearly expressed for Mrs Green to be able to understand them. At the material time she was acting both on behalf of herself and on behalf of herself and her husband.

Although the plaintiffs were now moving towards committing themselves to the purchase of no 49, they did not in fact act on the recommendation in clause 10 of Spurgeons' report and make an approach to the BCO about the necessary works in respect of the settlement to the rear of the premises.

While the other matters were going on, the plaintiffs sensibly were having the electrical installations checked by the Eastern Electricity Board, and the timbers and damp-proofing checked by Polysyne
Ltd. The recommendations of these two organisations were made on L
June 25 and 30 respectively (Red 47 and 49). Builders at some stage were called in to give estimates.

The plaintiffs instructed solicitors, Birketts, to act for them on the proposed purchase. They made the usual enquiries before contract. They also employed an architectural and structural designer, a Mr Hillick, who was not called as a witness. He was employed for a kitchen extension that the plaintiffs wanted. I do not consider, however, that his employment has any relevance to this action.

Somewhere between July 14 and 16, Birketts sent a "Report on Contract" (Red 64) to the plaintiffs.
I should read para 3 of that report:

THE CONDITION OF THE PROPERTY
The Vendors give no guarantee as to the condition of the property or its M
services and equipment so that if defects materialise at a later date (whether or not they exist now) you will, generally speaking, have no claim against the vendors. You should therefore take steps by survey or otherwise to check as far as possible that the building and all services are in good order.

Mrs Green, when cross-examined on this paragraph, said:

Overlying all this was the fact that Mr King had said there was nothing wrong with the property. It was this that stuck in my mind. All these warnings were from people who had not seen the property: Mr King had.

By July 16 the plaintiffs had signed their part of the contract and so far as they were concerned they considered that they were committed. On that date, Birketts sent the signed part to the vendor's solicitors.

I come now to the events on July 20.

In their letter of June 30 1982, Polysyne Ltd at para 3 had stated:

It is virtually certain that some areas of plaster have been contaminated by mineral salts brought up into the rising damp. Contaminated plaster is hygroscopic and will remain persistently damp even after the dpc has been installed and the underlying brickwork has dried out. It is therefore essential that the old plaster is hacked off the treated walls to a height of 1 ft above the "tide mark" or to a height of 3 ft where no tide mark is visible, three to five days after the installation of the dpc.

On or shortly before July 20, Mr Green began hacking off the plaster of the cross wall facing into the dining room. He gave as his explanation in chief that he had misread the letter and thought it had to be done before Polysyne put in the damp-proof course. Mrs Green, on the other hand, said that they thought they had found a secret treasure. Mr Green described the matter in this way, in chief:

I moved the plaster after contracts had been exchanged. This was [on] the dining room front wall between two doors. It was over the door and the whole wall area. Entire work was done in half an hour. It was easy to move plaster. Over the top of the door to stairs was a step crack the size of a pencil. This caused concern. The bricks were not well bonded.

Somewhat unsatisfactorily, there is no good photograph showing the brickwork which was exposed on this occasion.

I reject entirely Mr Green's explanation linked to the Polysyne letter. There was no suggestion of rising damp in the cross walls. Mr Green agreed in cross-examination by Mr de Navarro that rising damp had not appeared in the cross walls on the ground floor. Mr Green did not at any time in his evidence say that he was removing plaster up to the heights indicated by Polysyne. In the same cross-examination Mr Green conceded that this work had nothing to do with Polysyne. He gave then as his explanation that he was taking the plaster off because, on tapping it, it had sounded hollow. It is perhaps significant that this visit and the alleged representations now relied on only appeared by way of amendment of the statement of claim on June 17 1986.

I find that the following was the likely course of events. The only person from the council who had visited the premises and with whom the plaintiffs had had contact was Mr King. Mr and Mrs Green wanted to know whether the cost of remedying the fractured brickwork exposed on removing the plaster was recoverable under the improvement grant although not mentioned in the schedule. Believing that they were the owners, naturally they did not want advice as to whether or not they should go ahead with the purchase, but they did want to be advised whether the remedial costs were recoverable. The obvious person to go to was Mr King. Para 11A of the reamended statement of claim is inaccurate when it states that the plaintiffs contacted Mr King by telephone. He was then on holiday. When the plaintiffs tried to get through it is likely they were put on to Mr Hook. I think it likely that the telephone call took place on July 19, and I think it likely that the schedule and other papers had still not come through to the BCO. Mr Hook obtained copies. He would at once have noted that, although Mr King had visited the premises, he had failed to make recommendations as to how the slight settlement at the front should be dealt with. When Mr Hook arrived, he had his attention drawn also to the defects on the rear elevation discovered by Mr Spurgeon and to the cracking of the brickwork exposed by the removal of the plaster. I am satisfied that essentially Mr Hook was saying that the manifestations of settlement were not matters that gave him cause for concern, that they represented only minor subsidence and that if certain works specified by him — which were subsequently set out in a memorandum dated July 20 1982, which appears under the signature of a Mr Ramsdale in Red 71 — were carried out the slight subsidence would be effectively remedied.

I find that these statements were made before contracts were exchanged. There is nothing surprising in the fact that Mrs Green should have thought that contracts had been exchanged. I am satisfied on the evidence of both Mr and Mrs Green that they thought on July 20 that they were the owners of no 49 and the property was theirs to do what they liked with. I am satisfied that on his visit Mr Hook was unaware that Mr and Mrs Green were other than the owners. They did not suggest otherwise. The pulling down of the plaster was an act more consistent with ownership than otherwise.

But the fact remains that, had Mr Hook on July 20 expressed concern as to the stability of no 49, it is likely that Mr and Mrs Green would have at once contacted their solicitor. By good chance the transaction would have been avoided. In these circumstances it appears to me that if these statements were false they may be material.

The only subsequent events not earlier dealt with in this judgment that appear material are the following. On July 23 1982 Mrs Green received an estimate in the sum of £5,426 plus VAT from A E Blasby & Son Ltd on a pricing of a number of paragraphs of the schedule. I am satisfied that it is likely that they were asked to do this shortly before July 19. In their letter they said:

We have not included in the above estimate for Item 3.8 as the work involved to rectify the settlement has not been specified. Day work carried out against this item will be charged at our current daywork rate.

Prior to August 10 no letter from the area improvement officer was sent to Mr and Mrs Green setting out the works that were required to remedy the settlement in the various areas seen by Mr Hook on July 20. By August 10 1982 when Mr Hook and Mr King together visited the premises and advised the plaintiffs to instruct a structural engineer, substantial further plaster had been removed than had been the position when Mr Hook visited on July 20. Subsequently the remainder of the plaster was taken off. The removal of the plaster revealed various cracks and misalignment.

Three primary issues of fact I have to determine are:

(1) Had no 49 suffered major subsidence by June/July 1982?

(2) In June/July 1982 was there continuing movement in no 47?

(3) Was it necessary to stabilise no 47 before carrying out works to make no 49 fit for human habitation?

As to the first issue, only Mr Chapman gives as his opinion that this was so. This is not of course a head-counting exercise. But if the evidence was otherwise in balance, the fact that all those with engineering qualifications were lined up against him would obviously tip the scales. This issue is best approached by a consideration of the cracking and other manifestations of settlement at its worst at any time between June and July 1982 and the time of trial. If, for example, such was the condition of cracking and misalignment and the like of no 49 in December 1985 that by that time a proper assessment was that it had suffered severe subsidence, it will then be necessary to ascertain whether that condition also obtained in June/July 1982.

FRONT ELEVATION

(1) *Front wall in area between western end of cill of upper window and western lintel of lower window* (Chapman photo 3; Akroyd photo 14). Although described as a crack, it is in fact an opening-up. The opening-up has been repointed in cement mortar at some stage. Parts of the repointing had come out. There has been a movement of 49 to the east. I shall have to consider later whether this was due to a settlement of the 47/49 party wall or movement of the gable wall of 47. Across the face of 49 there had been by December 1985 a fall of 18mm. This was significant movement.

(2) *Gaps between western end of lintel of upper window and brickwork* (Akroyd photo 15). This I find only to be minor movement.

(3) *Gap between eastern end of lintel of upper window and brickwork* (Akroyd photo 15). This I find only to be minor movement.

REAR ELEVATION

(4) *Soldier arch over dining-room window* (Chapman photo 5; Towell photo 4; Akroyd photo 4). The difference between these three photographs is that the prop holding part of the arch up in the earlier photograph is absent in the later ones. When looking at Mr Chapman's photograph, Mr Green said that before the prop was put up four bricks were in danger of falling out. I cannot accept that. When the prop was removed the only effect was to allow two bricks to fall very slightly. I accept the defendants' evidence that the cost of remedying this defect would have been very small. I think it likely that this defect was due to the design.

(5) *Brickwork to arch above kitchen window opening* (Chapman photo 6). Such movement as took place here was of a minimal nature.

(6) *Brickwork over back door opening* (Chapman photo 7). Such vertical misalignment as there is was of a very minimal nature.

(7) *Brickwork above and below window opening to bathroom* (Chapman photo 8). There is indication of some settlement here.

(8) *Room No 1 Front Room* (Front walls and/or party walls) (Chapman photos 11, 12 and 13). Despite my earlier belief that these photographs indicated that poor quality bricks had been used, I am satisfied that that was not the case. They were soft bricks, and lime mortar — not cement mortar — was used.

These photographs show that there were some vertical cracks in this room at the junction between the window frames and the brickwork and between the transom light over the front-door opening. I am satisfied that when the plaster was taken off substantial lime mortar will have been removed at the same time. Its effect will have been more marked in other places than the front and party walls of this front room. In appendix E to his report, Mr Towell showed as regards the window frame and door opening of this room the movement that existed in November 1985. The majority of movement was towards no 47 but at the bottom of the door opening the movement was towards no 51. On the distortion survey carried out on April 30 1987, the degree to which racking of the door opening and window frame had taken place is shown in exhibit 4 DS6. The movement was all towards no 47 and downwards and accordingly was worse than in December 1985. This possibly shows the danger of too much reliance being placed on figures such as these when it is recalled that during the whole of this period no 47 had been underpinned. No other cause of further significant subsidence to no 49 has been advanced other than the settlement of no 47.

(9) *Front room north facing staircase wall and party wall 47/49 at junction of party wall and in staircase wall above lintel over opening on west side.* (Chapman photos 9 and 10; Towell photo 5; Akroyd photos 5 and 6). (Reverse side of staircase wall at junction with party wall 47/49: Chapman photo 19). So far as the crack at the party wall junction is concerned, the wall had moved apart more at the top than at the bottom. But I am satisfied by Mr Towell's evidence that the maximum entry of the cross wall (that is to say, the staircase wall into the party wall) was 1in, ie 25mm. Mr Akroyd deals with these cracks at para 3.03 of his report. I am not satisfied that there was any deterioration in either of these cracks between September/October 1984 and December 1985. I am satisfied that even in December 1985 these cracks were all of a minor nature.

DINING ROOM

(10) *Lintel over opening in south facing staircase wall on western side. Plate out of horizontal alignment.* (Chapman photo 17; Akroyd photo 7). I find these were of a very minimal nature.

(11) *Eastern end of southern wall vertical fractures to the left and right-hand side of window.* (Southern wall eastern end. Chapman photos 15 and 16). This is the other side of the wall considered under rear elevation (4) above. I am satisfied that these were of a relatively minor nature.

(12) *Stairwell to first floor. Vertical fracture at junction of partition and 47/49 party wall.* (Chapman photo 20). This was relatively minor.

FRONT BEDROOM

(13) *Front wall and party wall. Vertical fractures in front wall, stepped fracture in brickwork above window opening, stepped fracture to 47/49 party wall.* (Chapman photos 21, 22, 23 and 24; Towell photo 6; Akroyd photos 8 and 9). Towell photo 6 and Akroyd photo 9, taken 14 months apart, show no deterioration. The fracture in the party wall was relatively minor. The fractures to the front wall were significant. Mr Towell in his appendix E showed slight movement of the window frame towards no 47. Again exhibit 4 DS6, produced in April 1987, shows the general distortion.

(14) *North facing staircase wall and party walls 47/49 and 49/51.* Door head between front bedroom and stairwell out of horizontal alignment, stepped fracture in brickwork above the door head, bow to plate supporting ceiling joints. (Chapman photos 25, 26, 27; Akroyd photos 10 and 11). I do not find that any of these was significant.

BACK BEDROOM

(15) *South facing stairwell.* Plaster above both openings supporting ceiling joints bowed. Stepped fracture in brickwork over eastern door opening. Crack in party wall. (Chapman photos 29 and 30; Akroyd photo 12). The bowing was of a minor nature. Although the stepped fracture at first sight appears fairly serious, I am satisfied that it was not. The remedial work required was the taking down of five courses of brickwork over the opening, the installation of a steel lintel and the rebuilding of the brickwork. Mr Akroyd dealt with the crack in the party wall 47/49 in para 3.07 of his report. I am satisfied that this was minor.

(16) *Rear wall. Vertical and stepped fracturing in brickwork to three sides of window opening.* (Chapman photo 28; Akroyd photo 13). A number of the bricks were loose. This was an inherently weak area. There was little to keep these bricks together. The bond had gone. This was of moderate significance.

(17) *Bathroom. Stepped and general fractures in brickwork above window opening.* (Chapman photos 31 and 32). These were relatively minor.

I have considered under these 17 heads the primary areas of subsidence manifestation. In appendix E Mr Towell showed various misalignments in doorway opening, as well as the front door. A plumbing survey of party walls 47/49 and 49/51 was carried out on April 30 1987 and appears as part of exhibit 4. I shall have to consider this survey in greater depth when I turn to issues (2) and (3), but as regards the findings in relation to the effect of subsidence in 49 I consider them to be relatively unimportant. A small number of other manifestations were considered in evidence, but they also, I find, were of no importance.

Reliance was placed on the fact that before removal the plaster was hollow in a number of places and that that was an indication of major subsidence. I reject that contention. Houses of this age and quality may well have hollow plaster without any subsidence having taken place. I have indicated that some of the manifestations were significant, particularly in the front wall of no 49. But overall I find this evidence falls far short of that required to establish that there has been major subsidence of no 49, even in its very worst state. The basic fabric of the house after 80 years remained sound. The answer to this issue is therefore "No".

I turn, therefore, to consider the second and third issues of fact. These require a consideration in particular of no 47 and to a much lesser extent of no 45. No 47, it will be recalled, lies at the eastern end of the terrace of five houses. Then there is a passageway and no 45 is at the western end of the next terrace of five houses. The eastern wall of no 47 is a gable or flank wall. The photographs of the front of no 47 consist of Chapman photos 1, 33 and an additional one taken by him at the same time in September/October 1984, which forms part of exhibit 12; in addition, there are Akroyd photo 3 and Towell photo 1. The flank wall can be seen in Chapman photo 34 in part of exhibit 12. The flank wall, as can be seen, has been finished off in pebbledash. The front wall has been rendered. Superficially it appears that at first-floor level tie bars had been inserted east/west and north/south. The tie bar plates or washers on the front wall had been rendered over after installation.

It will be recalled that at the time of Spurgeons' inspection no 47 was unoccupied and boarded up.

The first suggestion that no 47 had to be stabilised before effective work could be done to 49 came from Mr Matthews in his report, as I have indicated. He reached the conclusion that the majority of cracking in no 49 had been caused by the excessive movement of the gable wall of no 47, which appeared to him to have undergone settlement relative to the party wall. Issues (2) and (3) and the question whether that belief was well justified depend very much on the evidence of the expert engineering witnesses whom I have had before me. I should start with Mr Akroyd. I consider him a very impressive witness. He has an unusually wide range of qualifications. His expertise I find to be considerable. By the end of his evidence, however, I felt bound to treat some of his statements with reserve. It appeared that he was inclined to see everything in stark terms.

Let me illustrate that. It is axiomatic that, if there are manifestations of subsidence, it is difficult to determine whether subsidence is continuing unless and until monitoring over a substantial period has taken place. Mr Akroyd was highly critical of Mr Matthews for having formed the view in September 1982 that the gable wall of no 47 had potential for further movement. An attack on Mr Matthews' competence was vigorously pursued. But it often happens that time simply is not available for the necessary monitoring and engineers do have to reach conclusions on limited evidence as to whether subsidence is continuing. Such an engineer may reach a wrong conclusion, but that does not necessarily mean he is incompetent. Without monitoring, it is equally difficult to reach a conclusion that subsidence which has taken place is not continuing. Yet Mr Akroyd in December 1985 reached a very positive conclusion that subsidence had ceased, without any monitoring evidence to support such an opinion. Mr Akroyd essentially dismissed no 47, however unstable, as being a cause or potential cause of any damage to no 49. As I shall indicate, that view is unrealistic.

A I did not find Mr Matthews a persuasive witness in many respects. Basically the views he expressed lacked an adequate reasoned basis. I did not at any time consider that he was consciously seeking to mislead the court. I thought he was a practical engineer but given often to an enthusiastic rather than a reasoned approach. The criticisms made against Mr Matthews for reaching the conclusion that he did in September 1982 were various, such as failure to dig a trial pit along the flank wall, failure to monitor movement, failure to investigate the true position of the tie bars in no 47, failure to inspect the interior of no 47, failure to plumb the party wall. I am not satisfied that these impugn his competence, but the failure to carry out at least some of these matters necessarily makes the judgment which led to his conclusion somewhat suspect. By the end of his
B evidence he was largely basing his opinion that 47 potentially would damage 49 on the presence of the tie bars, although this emphasis had not appeared in his report.

The last engineering expert was Mr Cowley. I found him an exceptionally reliable witness. He gave reasoned answers to all questions. His reasons stood up to the most searching cross-examination. They accorded with commonsense. They fitted in entirely with the factual evidence such as it was.

I do not refer to the evidence of Mr Chapman or Mr Spurgeon on this issue, because not surprisingly they were unable to make a useful contribution.

The actual evidence relating to no 47 is relatively slight. The
C inference in Mr Matthews' report of September 1982 was that movement in no 47 would continue. But except that reference was made to actual settlement that had taken place, the report was silent. The amount of rotation of the gable wall was 65mm at eaves level. Mr Mann, the council's engineer, gave a written report on January 27 1983. So far as the report contains factual matters, it is agreed that I should accept those findings, save to the extent that other evidence satisfied me otherwise. There was no such other evidence. I need not read all the facts into this judgment but this judgment relies upon them. The principal findings — apart from the fact that the gable wall leaned over — were:

(i) the north-east corner of the front wall had settled significantly;
D (ii) a vertical crack in the pebbledash under the ground-floor window cill;
(iii) a tapered vertical crack on the gable wall at the junction of the rear single-storey extension with the two-storey extension;
(iv) radial cracking in the pebbledash in the rear wall emanating from the rearmost place on the gable wall.

On August 20 1984 Dr Wady, who is, as I indicated, Mr Matthews' partner, made a report on no 47. In so far as it contains factual matters, they are agreed. I should read one extract from it:

> . . . the gable wall . . . appears to have settled by at least 25mm relative to the front and rear elevations and has moved significantly out of plumb. The settlement has caused the front door and window lintels to slope and induced a crack over the first floor window to eaves level. The out of plumb readings
> E (outwards at the top) varied from about 50mm near the front of the gable wall to 75mm near the back. The front and rear walls appear to be approximately plumb. An approximately vertical crack varying from 10mm at roof level to 3mm near ground level in the flank wall indicates that the rear single storey part of the house has settled significantly further than the main part of the flank wall.

In so far as reliance can be placed on distortion surveys, it would appear that little or no additional rotation had occurred since September 1982.

Although from paras 5.3 and 5.4 of Mr Towell's report it appeared that he was accepting that movement of no 47 was taking place requiring no 47 to be underpinned, it became clear in his evidence that in fact his conclusion was that no evidence had been put before the
F court which suggested to him that there was any continuing movement in no 47 at any time since 1982.

The position was the same with Mr Cowley. In his report on p 8, Mr Cowley clearly had been influenced by the conclusions reached by Mr Mann in his report of January 27 1983, which, as I have indicated, are conclusions which are not in evidence before me. Mr Cowley did not criticise the underpinning of no 47 but was unaware of any evidence that there had in fact been continued movement in no 47.

Mr Akroyd in trenchant terms indicated that underpinning of no 47 had been entirely unnecessary because there was no evidence of movement.

There was put before me as exhibit 24 an agreed summary of relevant information in the council's file relating to no 45. I have considered that summary, but I do not find it affords me any G
assistance.

Mr Matthews in his evidence, in support of his proposition that until underpinning no 47 had continued to move, suggested that some evidence of this was continued movement in the front wall and in the soldier arch at the rear of no 49.

On November 24 1984 Mr Matthews' partner, Mr Hill, selected a number of monitoring points on no 49, as appears from sheet III. Readings were taken at that time. The only other reading was on January 20 1986, shortly before remedial works were undertaken at no 49 and after 47 had been underpinned. This showed an average increase of movement of 0.8mm. Mr Matthews sought particularly to derive support from the readings on the front elevation. A H
comparison of all of the two sets of readings fails to provide any reliable overall pattern. All the experts, apart from Mr Matthews, regard this particular exercise as fruitless, since in fact only one subsequent reading had taken place. I consider that criticism is justified. Seasonal and thermal changes are such that only if frequent readings are taken can a pattern be relied on. I have reached the conclusion that on the evidence before me no movement at no 47 between June 1982 and the end of 1985 had taken place.

I turn therefore to consider issue (3). Earlier in this judgment I dealt with the form of construction of the buildings, including the joists and purlins. I am satisfied that the structure that has to be considered is the terrace of five houses. Loads which are imposed at one point may be transmitted through the whole structure. The J
evidence satisfies me that, before the gable wall of no 47 began to settle and lean outwards, the wind load on no 47 would have been approximately 2.7 tonnes. That load would have been transmitted westwards in a number of ways. Primarily the loads would have been transmitted through the 9-in brickwork of the front and rear walls of no 47, through to no 49 and then in ever diminishing form through the whole structure. To a much more limited extent, loads would have been transmitted westwards through the cross walls of no 47 into the party wall of 47/49 and onwards. These internal walls were weak. They were only of $4\frac{1}{2}$-in brickwork, and I am satisfied that the bonding in of these walls into the party wall was very weak. A more satisfactory conveyance of the loads would have been the floor structure. Although it is likely that there was a gap between the K
westward joists in no 47 and the 47/49 party wall, the rigidity of the floor structure would have been likely to operate as a good transmitter of loads. Finally, the roof construction would have transmitted loads to some extent through the structure. Each party wall would have been capable of providing considerable resistance to any loads transmitted to them. The party wall appears to have been well bonded into the front and back walls. While the party wall did not gain much strength from the cross walls, chimney breasts in the party wall on each side of the cross walls would have helped to provide a very strong structure. It seems likely that at some time settlement of the gable wall particularly became substantial. It is unnecessary to speculate more than briefly as to how this occurred. It L
could have been differential movement due to differing subsoils. A more likely explanation is that the foundations were affected when a drain was laid north-south in the passageway between nos 45 and 47 and partly under 47's foundations.

As the gable wall subsided so the load on that wall increased. As it increased it seems likely that the integrity of the fabric of no 47 became affected to some extent so that loads could not be transmitted as previously. I am satisfied, however, that a situation was never reached similar to that at the back of no 37 Boston Road, shown in the photograph in appendix 9 to Mr Towell's report. I think it likely that it was at this time that ties were introduced to no 47. Until the trial had been proceeding for some time all experts were under a false belief as to the true nature of these ties. The true position emerged on May 12 M
1987 and appears in the six Jenkins & Potter drawings forming exhibit 15. There were four east/west ties, all at first-floor level, that is to say about 8 ft above the ground. The first and third ran on the inside faces of the front and rear walls. Some distance short of the party wall they were anchored by a bolt through the brickwork in the manner shown in details A and B. From there, each continued to the party wall as a strap; at the angle there being a bent plate. It was anchored to the westernmost joist of no 47 by coachscrews. The investigation on May 12 established that no crushing of brickwork at the wall connections had taken place. There was no damage to the anchor plate bolt or nut or the strap, bent plate or coachscrews.

The second tie was bolted to the northern side of the starcase wall,

A as shown in detail C. It then continued towards the party wall, again with bent plate and attachment to the joists. One coachscrew appeared to be missing, but as there was no damage as such visible at all I am satisfied that this absent coachscrew is irrelevant. This tie bar was found to be cut at its eastern end. I consider that it was still intact when Dr Wady made his inspection in August 1984; otherwise he was likely to have commented about it. The probabilities point, therefore, to its having been cut when works were done in 1985.

The fourth tie bar at the back of the premises is irrelevant.

At the time the tie bars were installed the front and interior walls would have been restored to their former load-transmitting potential. Consideration was given in the later stages of the trial to the B loading additional to that of wind which would have been imposed by the eccentricity of the gable wall. Mr Akroyd's calculation shows that approximately 0.3 tonne would be added, making a total of 3 tonnes. In broad terms Mr Cowley agreed with this. It is to be appreciated that, even if the eccentric loading were doubled, it would only increase the overall loading on the gable wall by 10%.

Before turning to consider Mr Matthews' theory, I propose to consider the reasoning of Mr Cowley as to why no adverse effect on 49 could occur from loading on the gable wall of no 47. With a 3-tonne loading on the gable wall, then theoretically each tie connection could have had a 1-tonne load imposed upon it. But of course in the real world the load will be transmitted in a number of ways other than through the anchorage points, and so the load on C each connection will be less and probably substantially less. In any event Mr Cowley was of the view that at about half-a-tonne loading the connection would be overloaded, causing damage, which we know does not exist. He accepts that loadings for the gable wall would have been transmitted to 49 and indeed on through the structure, but he simply cannot see the mechanism whereby the loadings would be such as to cause significant damage to 49.

In so far as loads were transmitted to the tie-bar connections, they would in part have been transmitted elsewhere, eg to the walls, but triangular forces would have transmitted much of the load back to the gable wall, as he sketched in exhibit 21. All the known physical manifestations militated against loading on the gable wall ever being a cause of cracking to 49. The cracking in no 47 that would be caused D by such loading was likely to be diagonal. Yet, as the evidence showed, much of the cracking in fact was vertical. The whole pattern of cracking in 49 and the position of its party walls, says Mr Cowley, is strong evidence that no 49 has not been significantly affected by 47. In his two-sheet exhibit 7 he plotted the information obtained on April 30 1987, as shown in exhibit 4. What he regards as so significant is that the maximum distortion shown on the 47/49 party wall is at the front, whereas all agree that it is the rear of no 47 which has suffered the greatest lean and therefore has the greatest load imposed upon it. So a precisely opposite picture is presented to that necessary to support Mr Matthews' theory.

I entirely accept Mr Cowley. It seems immediately to eliminate no E 47 as being in any way causative of the damage to the soldier arch or to the bathroom. The measurements taken on April 30 were of course taken for plastered brickwork. Even if the brickwork had been laid completely in plumb, the plastering could well result in distortion. I am not satisfied that the party wall was out of plumb at any point along its length in 1987 to be significant. I am satisfied that the damage noted in no 49 was due to settlement in 49 typical of settlement in many other houses in this street and nearby streets. A substantial cause of some of the cracks in no 49 was a significant settlement of the 47/49 party wall, which settlement was calculated to cause the subsidence seen and was wholly unrelated to the subsidence of the gable wall.

Mr Akroyd essentially not only agreed with Mr Cowley but went F further. He considered that the connections could accept a loading of about 1.32 tonnes. But I accept Mr Cowley's figure, although, in the absence of any damage, it matters not which figure is correct. He went further than Mr Cowley, in that he was adamant that no load for 47 could be transmitted to 49. This I find to be incorrect. Given the nature of the structure as a whole, it would be extraordinary if some loads were not transmitted. Although I am unable to accept that part of his evidence, I consider that his reasoning why no 49 would not be adversely affected by any movement in no 47 was very persuasive. Except in the respects specifically indicated in this judgment I have no hesitation in accepting his evidence. For Mr Matthews' theory that loads would not only be transmitted to no 49 but would be of such a nature as to damage the fabric of no 49, loads much in excess of those calculated by Mr Cowley and Mr Akroyd would be necessary. G Somewhat late in the day Mr Matthews advanced an ingenious theory as to how such loads could have been created. I trust that Mr Matthews will not regard it as a discourtesy on my part if I do not deal extensively with his theory. In brief, it requires the leaking away under the leading — that is to say the eastern end of the gable wall foundations, causing the point of support to move westwards towards 49, coming as far possibly as the western end of the gable wall foundations. All the visual evidence contradicts this theory. It was based on the merest supposition, which I find to be contrary to all the probabilities. Had loads of the kind necessary been transmitted, I am sure that, as in 37 Boston Road, there would have been on the front and back walls of no 47 major breaches in the brickwork. Further, as the point of support moved westward, the downward H loading would almost certainly have caused significant vertical subsidence of the gable wall. I am left in no doubt on the evidence that Mr Cowley's views are correct and therefore, on this third issue, I find that no 47 never caused damage to no 49 from June/July ownards, nor did it have any potential effect for causing such damage.

My findings of fact essentially dispose of all the issues in the action, but I should deal with certain matters. Special damages were agreed at £7,500, subject to liability. Had the plaintiffs succeeded in establishing negligence, I would have awarded each of them £500 for the physical consequences of the breach. Such consequences would not of course have included for any distress and the like arising from J the legal proceedings.

As to the duty of the council, I am satisfied that the council owed no duty to the plaintiffs to take reasonable care to prevent their suffering damage, save to the extent that they assumed that responsibility in their schedule. Had they not offered to give advice, they would have owed no relevant duty to the plaintiffs. If, for example, Mr Huffey had failed to list in his record or his schedule defects which seriously affected the stability of 49, that would have afforded no cause of action to the plaintiffs against the council. The responsibility that the council assumed was a narrow one, namely, to advise where it was necessary to rectify a settlement specifically identified in the schedule.

The words of Lord Morris of Borth-y-Gest in *Hedley Byrne & Co Ltd* v *Heller & Partners Ltd* [1964] AC 465 at p 502 are apposite: K

> I consider that it follows and that it should now be regarded as settled that if someone possessed of a special skill undertakes, quite irrespective of contract, to apply that skill for the assistance of another person who relies upon such skill, a duty of care will arise. The fact that the service is to be given by means of or by the instrumentality of words can make no difference. Furthermore if in a sphere in which a person is so placed that others could reasonably rely upon his judgment or his skill or upon his ability to make careful inquiry, a person takes it upon himself to give information or advice to, or allows his information or advice to be passed on to, another person who, as he knows or should know, will place reliance upon it, then a duty of care will arise.

I am satisfied that on June 17 1982 Mr King attended at no 49 with the intention of giving and in fact gave advice. But the only advice L that he was required to give and the only advice that in fact he gave was in respect of the slight settlement at the front. I am satisfied that, if Mr King's advice had been that the condition of the front wall was such that either Mr or Mrs Green should instruct a structural engineer or that substantial works were required, the plaintiffs would not have proceeded further. To the extent therefore that Mr King stated that only repointing and the like were required, Mr and Mrs Green relied upon Mr King, even though they subsequently failed to get Blasbys to price item 3.8 on that basis.

The visit of Mr Hook is somewhat anomalous. I am satisfied that Mr Hook attended on July 20 with the purpose of advising how the settlement at the front and the back and the cracking in the cross wall should be rectified. But I do not consider that that was the advice that M Mr Green was seeking or was intending to rely on. What he wanted to know was whether the improvement grant would extend to settlement and cracks, apart from that already listed in item 3.8. If, however, Mr Hook had advised that major subsidence had taken place, I am sure that the result would have been that the deal would have been off. It follows therefore that indirectly the plaintiffs did rely on the council's advice. But in the light of my findings all this is academic.

I turn, therefore, to consider the position of Spurgeons. It assists to consider the state of the law if I had been satisfied on the evidence (1) that Mr and Mrs Green were induced by this report to commit themselves to the purchase and (2) that Spurgeons failed to exercise

A reasonable care and skill in making the report. Some years prior to 1982 the Abbey National had been the first to break away from a long-established practice of not allowing an applicant to see the surveyor's report obtained by the society. In *Smith* v *Eric S Bush* [1987] 3 WLR 889 the Court of Appeal considered a case which had a number of features similar to the present. The plaintiff there had sued a firm of surveyors who had provided the Abbey National with a report and mortgage valuation. This contained an apparently identically worded notice. The mortgage application form signed by the plaintiff had contained essentially the same wording as applied in the present application form. The report and mortgage valuation completed by the surveyor contained a disclaimer notice identical to that in the report completed by Spurgeons. The Court of Appeal held
B that the surveyor was required to make a reasonably careful visual inspection of the property and complete the society's form. The surveyor's report showed nothing seriously wrong with the property, but, as was found, the house was in truth a dangerous structure and unfit for habitation until adequate support for the flues had been provided. It was not in issue in that case that the surveyor owed a duty of care to the plaintiff apart from the disclaimer, and the plaintiff relied on the report and that the surveyor was negligent. The issue in the Court of Appeal was whether the surveyor was protected from liability by the disclaimer. The court held that apart from the Unfair Contract Terms Act 1977 the disclaimer did exempt the surveyor from liability for the consequences of his negligence. In that case it
C was conceded by the surveyor that the words "all the circumstances" in section 2 of the 1977 Act were not limited to considering merely the question whether the notice referred to in the section was sufficiently clearly brought to the attention of the person concerned or was sufficiently clearly expressed to that person to be able to understand them. All three judges held that the disclaimer did not satisfy the requirement of reasonableness. I might just read part of the judgment of Dillon LJ at p 897:

> I cannot see that it is fair or reasonable that a professional surveyor making a mortgage report at the lower end of the property market, when he knows that the would-be-purchaser who is applying for a mortgage on the property has paid the fee for the report, will be supplied with a copy of the report and is likely to rely on the report and so is not likely to instruct any other surveyor,
> D should be able to rely on any general disclaimers, such as those in the present case, unrelated to any special factors affecting the particular property, to exempt him from liability to the purchaser for negligence if it should so happen that he, the surveyor, carries out his visual inspection of the property without due care.

From the wording of the disclaimer notice in the report and E valuation Spurgeons knew that a copy of the report was likely to come into the hands of the applicants before they contracted to purchase the property. I find that there were no special factors regarding either the property or the applicants which satisfied the requirement as to reasonableness. Therefore, on the assumptions above made, Spurgeons would be liable in damages.

But I must now turn away from assumptions to my findings of fact. Although I have criticised Mr Spurgeon for the manner in which he completed the Abbey National form I do not find anything to justify the criticism as to his competence. I find that he approached the matter in the way that an able and experienced surveyor should. He not only ascertained the existence of subsidence at the back but also commented on the position of no 47. He reached proper conclusions F as to the state of the property in all respects. I find it would have been strange if he had recommended that a structural engineer should be instructed and equally surprising if he had raised doubt about the stability of no 49 or no 47. The evidence leaves me in no doubt that Mr Spurgeon behaved in a way that a skilled surveyor should.

It is contended that, even if he was negligent, in any event the Greens have failed to establish they were induced to contract to purchase 49 on the strength of Mr Spurgeon's report. It is unnecessary in this judgment on my finding of the facts to consider interesting questions as to whether it would have been a combination of the council or Spurgeons' statement or one or other of them which induced Mr and Mrs Green to purchase. G

But I should deal with a further matter. Much reliance is placed upon the fact that, although Mr Spurgeon advised the Abbey National that the works set out in the council's schedule should be satisfactorily completed before they granted a mortgage, the Abbey National in fact did not impose such a condition. Mr Spurgeon contemplated that, if his advice had been followed, Mr and Mrs Green would have obtained a bridging loan which later would have been replaced. But Mr and Mrs Green would still have had to purchase the property before they carried out the works. The only difference would have been that the original source of money for the purchase would have been the bank and not the building society. I do not regard that as a significant difference.

On my findings of fact, the regretful result is that the unfortunate H plaintiffs are left to bear their considerable losses.

Judgment was given for the defendants with costs, not to be levied without leave of the court; legal aid taxation of plaintiffs' costs.

For further cases on this subject see p 130

REAL PROPERTY AND CONVEYANCING

Court of Appeal

November 21 1986

(Before Lord Justice SLADE, Lord Justice NEILL and Sir George WALLER)

TUDOR v HAMID AND OTHERS

Vendor and purchaser — Whether money held by vendors' solicitors as stakeholders or as agents for the vendor — Appeal by solicitors from decision of Judge Paul Baker QC, sitting as a judge of the High Court, holding that they received the money (which was the full purchase price, not merely a deposit) in the capacity of stakeholders and must make repayment to the purchaser on the rescission of the contract of sale — There had been a somewhat confused and unhappy history following the purchaser's agreement to take the assignment of the lease of an hotel at Southend-on-Sea from the vendors — The arrangements had been made in some haste and the agreement was an "ineptly drafted and ambiguous document" — The subsequent deed of assignment of the lease was executed by the vendors and was effective to transfer the lease to the purchaser, although it was never in fact executed by her — The purchaser found that the state of the hotel and the clientele were not what they were represented to be and she complained that she had been misled — She issued a writ against the vendors and their solicitors — A few months after the contract of sale the purchaser, under pressure from the landlords in regard to repairs, surrendered the lease — At the trial the purchaser sought to set aside the contract on the ground that she had been induced to enter into it by misrepresentations by the vendors — She claimed repayment of the money paid and she joined the solicitors in the action with a view to enforcing the latter claim — The judge found that false representations had been made by the vendors; he ordered that the contract be set aside and that the money paid should be repaid to the purchaser — He decided that the purchase money had been paid to the solicitors as stakeholders, so that they, not the vendors, were accountable for it — The vendors did not appeal, but the solicitors did, although not challenging the judge's finding of fact as to misrepresentation by the vendors

The appeal therefore centred on the capacity in which the solicitors held the purchase money — It was clear from authorities referred to by the court that if the solicitors had received the money as stakeholders their duty was to hold it until the event occurred which entitled them to release it — If, on the other hand, they received it as the vendors' agents the position was the same as if it had been paid at that instant to the vendors themselves; they were entitled to have it handed over on demand — These rules had been developed in regard to deposits, but they were applicable to the whole purchase money in this case — There were two issues, namely, (1) did the solicitors receive the money as agents for the vendor or as stakeholders pending the occurrence of some future contingency, and (2) if they received it as stakeholders, then, at the date when they released the money had the contingency occurred which entitled them to do so without any authority from the vendors?

The original agreement provided that a deposit of 10 per cent should be held by the solicitors as stakeholders, but the effect of this was rendered questionable by the subsequent arrangement that the purchase money should be paid immediately — However, on the basis of the *contra proferentem* rule, or even on the assumption that the original clause in the agreement should be rejected altogether, the court held that the judge was right in deciding that the solicitors held the money as stakeholders

On the second issue the court held that the contingency which entitled the solicitors to release the money to the order of the vendors occurred on the date when the vendors had performed their obligation to assign the lease of the hotel and its business in accordance with the agreement; and the fact that the deed of assignment was not executed by the purchaser was immaterial — The money was released on or soon after that date and the solicitors were justified in doing so — The appeal was allowed, the judge's order set aside and the action against the solicitors dismissed

The following cases are referred to in this report.

Burt v *Claude Cousins & Co Ltd* [1971] 2 QB 426; [1971] 2 WLR 930; [1971] 2 All ER 611; [1971] EGD 565; (1971) 218 EG 413, CA
Ellis v *Goulton* (1893) 1 QB 350
Potters v *Loppert* [1973] Ch 399; [1973] 2 WLR 469; [1973] 1 All ER 658; (1972) 25 P&CR 82; [1972] EGD 960; 224 EG 1717

This was an appeal by F T Fisher & Lang, solicitors, of Southend-on-Sea, from part of the decision of Judge Paul Baker QC, sitting as a judge of the Chancery Division. The respondent was Mrs Dorothy Anne Tudor, plaintiff in the court below. The proceedings arose from a contract for the assignment of a lease of the property known as The Argyle Hotel, 12 Clifftown Parade, Southend-on-Sea, by the vendors, S Abdul Hamid and Mohamed Nisar Patel, to Mrs Tudor. F T Fisher & Lang were the vendors' solicitors.

G M Godfrey QC and A T C Nicholson (instructed by Reynolds Porter Chamberlain) appeared on behalf of the appellants; G D Conlin (instructed by Jack Bernstein & Co) represented the respondent.

Giving judgment, SLADE LJ said: This is an appeal by a firm of solicitors, F T Fisher & Lang, from part of an order of His Honour Judge Baker QC, sitting as a judge of the Chancery Division, made at the trial of an action on January 25 1985. The plaintiff in the action was Mrs Tudor, the first defendant was Mr Hamid, the second defendant was the appellant firm and the third defendant was Mr Patel.

The proceedings stemmed from a contract dated May 31 1978 by which the plaintiff, Mrs Tudor, agreed for a consideration of £12,500 to take from Mr Hamid and Mr Patel, as vendors (*inter alia*), an assignment of the benefit of a lease of certain property known as The Argyle Hotel, 12 Clifftown Parade, Southend-on-Sea ("the hotel"). The vendors retained the appellants to act for them. Mrs Tudor paid the whole sum of £12,500 to the appellants on May 31 1978. Subsequently she was let into possession and by her solicitors accepted the vendors' title and submitted a proposed form of assignment of the lease before the vendors had executed any assignment in her favour. On July 4 1978 the vendors executed an assignment in the form proposed. Thereupon the appellants in effect released to the vendors the money paid to them by Mrs Tudor.

She then brought proceedings against the vendors, Mr Hamid and Mr Patel, to set aside the agreement on the grounds that it had been procured by fraudulent misrepresentations and for repayment of all the sums paid by her pursuant to the sale (which in fact totalled £17,500). Presumably in the anticipation (which I understand has regrettably proved well founded) that she might not succeed in enforcing any judgment which she might obtain against the vendors, she also joined the appellants as defendants to the action, claiming repayment of the money which had been paid to them. The judge upheld her claim against the vendors and ordered rescission of the contract, so that the vendors became liable to make repayment to Mrs Tudor.

It then became necessary for him to deal with her claim for the sum of £12,500 against the appellants. Her case was that it had been paid to them as stakeholders and not as agents for the vendors, as the appellants themselves claimed. The judge decided this point in her favour. The appellants contend that he was wrong and should have decided that Mrs Tudor paid the £12,500 to them simply as agents for the vendors. They further submitted that, even if this sum had been paid to them as stakeholders and not as agents for the vendors, once the vendors had executed an assurance of the property in favour of the purchaser, the appellants were entitled and bound to pay the £12,500 to the vendors. The learned judge did not accept this submission. The appellants submit that he was wrong not to do so.

Before reverting to the issues which we have to decide on this appeal, I must state the relevant facts. The history of the matter is unusual and by no means straightforward. I derive the facts largely from the learned judge's very careful judgment.

The facts

By a lease of August 21 1963 the hotel was demised to a Mr J C Campbell for a term of 21 years from August 21 1963. The lease was thus due to expire on August 21 1984. It contained a tenant's repairing covenant and a covenant precluding the tenant from assigning the hotel without the written consent of the lessor, such consent not to be unreasonably withheld. Unusually, however, it contained no proviso for forfeiture for breach of this or any other covenant.

In February 1976 Mr Hamid and Mr Patel entered into a partnership agreement to purchase, finance and operate the hotel.

In or about May 1976 they took an assignment of the lease. The appellants acted for them in regard to that assignment. Mr Hamid thereafter remained their client until after the present proceedings began.

On November 14 1977 Mr Hamid decided to offer the lease for sale. Before the end of May 1978 there had been a number of serious inquiries by prospective purchasers. During the course of these inquiries or consequent negotiations the appellants had prepared a draft form of agreement providing for the sale of the hotel and of the hotel business.

At the end of May 1978 Mrs Tudor came on the scene. She was married to an electronics engineer. He had the name Tahbub, but she preferred, for reasons of convenience, to use the name Mrs Tudor. At that time she had three young children and was carrying another. Though the contrary was suggested at the trial, the learned judge found that she had no legal training and had made no study of the law. She and her husband were in fairly urgent need to acquire new accommodation because they had sold their previous house and might shortly find themselves without a roof over their heads. As the judge found, they had no very great financial resources at their disposal and were "singularly ill-equipped for the venture into which they now rushed headlong".

On Friday May 26 1978 Mr Hamid and Mr Tahbub had a discussion in the hotel about the possibility of Mr Tahbub purchasing the lease. Mr Tahbub gave evidence to the effect that in this conversation he required assurances that the landlord was raising no outstanding requirements about repairs to the property and that, if the deal went through, the landlord would agree to him and his wife moving into the hotel on the following Friday, June 2.

That same evening, after telephoning to say that he was coming, Mr Hamid went to Mrs Tudor's home, having equipped himself with a document intended for signature. At that time Mrs Tudor herself had not seen the property. At this meeting a document bearing the date "26.5.78" was signed by Mr Tahbub and Mr Hamid. It was by way of a preliminary agreement for the sale and purchase of the hotel. The learned judge found that it contained certain representations by Mr Hamid and that further oral representations were made by him at this meeting. For the purposes of this appeal, however, nothing turns on the precise nature of these representations. Nor does anything turn on the precise terms of the document of May 26, since it is common ground that on any footing it was superseded by the agreement of May 31. I need refer to only the final paragraph, which provided that "deposit of £5,000 is paid to Mr Hamid and the contract will be exchanged within 7 days".

A sum of £5,000 was duly paid to Mr Hamid at this meeting. At the trial there was a suggestion that as between Mrs Tudor and her husband on the one hand and Mr Hamid on the other hand, it was agreed or contemplated that the £5,000 should be treated as a deposit towards a proposed purchase of an entirely different property in Seaforth Gardens, Southend. The judge, however, did not accept this suggestion. He considered that the payment related only to the proposed purchase of the hotel. He accepted Mrs Tudor's evidence that at the meeting Mr Hamid also said that he would take her and her husband to his solicitors and that his solicitors would do what was necessary to tie matters up.

During the following week Mrs Tudor and her husband made visits to Southend and saw the hotel. They decided to proceed with the purchase, apparently on the basis that the assignment would be taken in her name rather than his.

On May 31 1978 Mr Hamid, as contemplated, took them to see his solicitors, the appellants, or rather their conveyancing partner, Mr Gilbert. The purpose of the meeting was substantially to conclude a contract for the sale and purchase of the hotel and its business which would enable Mrs Tudor to take possession within a very short time. Mr Gilbert, however, was unaware of its purpose until his client, Mr Hamid, was shown in to him. While Mrs Tudor and her husband waited in another room for quite a short time, Mr Hamid explained to Mr Gilbert that he had got a purchaser waiting in Mr Gilbert's office to sign a contract with a view to getting possession almost immediately. The judge found the situation as presented to Mr Gilbert was as follows:

> Both parties (not just Mrs Tudor but both Mr Hamid and Mrs Tudor) were very anxious for this to proceed and were putting pressure on the solicitor. Mr Hamid had been trying to sell this property for some time and he was in current difficulties. His wife was ill and he had proceedings hanging over him. The landlords were being uncooperative and he had had a number of abortive sales which had given him a lot of trouble, and he now had a very firm purchaser who had put up £5,000. It would have seemed a God-sent way, I imagine, to him of getting out of his difficulties. On the other hand Mrs Tudor plainly saw this as a matter of great urgency, to get this accommodation, and over and above that she had paid £5,000 the previous Friday and at the moment had not got anything for it. But I also find that when she went there she was not expecting then and there to have to pay the whole balance. She went there for the further documentation which was necessary to enable her to get possession within the next few days.

The situation which thus confronted Mr Gilbert was a difficult one, bearing in mind that Mrs Tudor had no legal representation. He could have refused to act altogether. However, he did not take that attitude, and the judge considered that he was not to be criticised for trying to do his best to assist his client by at least seeing Mrs Tudor and her husband. Consequently, they were introduced to him. Mr Tahbub, although present throughout, took no part in the ensuing meeting. He was feeling unwell.

At the trial there was a considerable conflict of evidence as to what was said at this meeting on May 31. However, on the basis of the findings of the learned judge, I derive the following picture as to its general course. At the start, Mr Gilbert warned Mrs Tudor that he himself could in no way act on her behalf nor advise her and that she would be most unwise to transact the business without the assistance of a solicitor. He advised her, therefore, to consult a solicitor of her own. Mrs Tudor, however, found this advice unacceptable. She was ready to sign whatever contract was appropriate to enable her to get possession on the immediately following Friday, June 2, which was what she wished. She had been promised by Mr Hamid that contracts would be exchanged that day.

In this context questions then arose as to repairs to the hotel and the landlords' consent. Mr Hamid told those present that the repairs had been done and Mr Gilbert accepted the assurance of his client on this point. References were then discussed. Mrs Tudor gave the names of three referees, a bank and two personal referees.

This point having been dealt with, the discussion moved on to the

A fact that the possession was required on the immediately following Friday. At this point Mr Hamid told Mr Gilbert that the purchase price was to be £12,500. Mr Gilbert was not told about the £5,000 or the document of May 26. He had no knowledge of these matters. So far as he knew, the entire price was to be £12,500 and no more than that (though, as I have already indicated, the judge found that the £5,000 had in fact also been paid on account of the purchase of the hotel).

By this time Mr Gilbert found himself committed to the preparation of a contract and he set about preparing one. The contract which he produced was based on a form in common use in his office for the sale of businesses such as this. Because of Mr Hamid's previous attempts to sell the hotel, this form had already
B been typed up to some extent, in the form of a draft contract for the sale by Mr Hamid and Mr Patel of the lease and business of the hotel.

Having regard to the fact that possession of the hotel was to be taken by Mrs Tudor in two days' time, Mr Gilbert told the parties that he could not allow his client, Mr Hamid, to execute the proposed agreement providing for possession to be taken so soon unless the full sum of £12,500 was paid at once. However, in adapting the available form of draft in order to produce a contract for signature by the vendors and Mrs Tudor, he was clearly under some pressure of time. Otherwise, he would presumably have noticed that a number of the provisions appearing in the form of contract which he produced were in terms not appropriate to a situation in which the entire purchase
C price was to be paid immediately.

Mr Gilbert, however, did direct his attention to one important point. Condition 8(3)(c) of the Law Society's General Conditions of Sale (1973 Revision) provides that in a case where on the sale of a leasehold property the reversioner's licence to assign is requisite, then, if the licence is not granted within three months after the date of the contract, either party may rescind it. As at May 31 1978 it was clear that the parties could not have obtained the reversioner's consent by June 2, when it was intended that Mrs Tudor should take possession. In order to ensure against the ultimate refusal of consent, Mr Gilbert therefore advised Mr Hamid that a clause should be added modifying condition 8. The clause which was added as clause 15, near the end of the agreement, read as follows:

D This Contract is not conditional upon the landlord's consent to the assignment of the Lease being obtained and General Condition No 8 shall be modified accordingly and rescission of the Contract shall not be entertained by either party in the event of the landlord's consent not being forthcoming within the period of three months from the date hereof.

As the learned judge pointed out, clause 15 did nothing to abrogate the vendors' contractual duty to do their best to obtain the landlord's consent to the assignment. However, provided that their best endeavours had been used, it meant that the ultimate failure to obtain such consent would not entitle the purchaser to rescind the contract.

I need refer to only a few of the other provisions of the agreement which Mr Gilbert produced for signature.

E Clause 1 provided in effect that the vendors, Mr Hamid and Mr Patel, would sell and Mrs Tudor would purchase, first, the leasehold hotel; second, the fixtures, fittings and utensils; third, the business, and fourth the stock in trade. Clause 2 provided that the purchase price should be £12,500, apportioned as between the first three items in the manner therein set out and that, in default of agreement, the price of the stock should be determined by a valuation.

Clause 3 provided:

A deposit of 10% of the purchase price shall be paid to Messrs F T Fisher & Lang . . . as Stakeholders on or before the signing hereof.

Clause 4 provided:

The purchase shall be completed on the 2nd day of June 1978 at the offices of the Vendor's Solicitors when the balance of the purchase money together with
F the purchase price of the stock shall be paid.

Clause 13 provided that the sale was subject to the Law Society's Conditions of Sale (1973 Revision) so far as such conditions were not inconsistent with the foregoing provisions of the agreement, but varied the rate of interest payable under condition 16.

Having been given the document and the copies of the lease and the inventory which was to be annexed to it, the parties withdrew from Mr Gilbert's office separately for the specific purpose of allowing Mrs Tudor and her husband to read through the document and see whether they would indeed be willing to sign it. They went to Mr Gilbert's waiting room. They read through the document. The upshot was that Mrs Tudor was prepared to sign. The agreement of May 31 1978 was duly concluded and they paid over a cheque for the full amount of £12,500 to the appellant firm. The judge found that on G
the oral evidence (apart from the documentation) it was not made clear to Mrs Tudor in what capacity the appellants received this sum. It was, however, clear that it was paid in advance of possession and in advance of completion. Later in this judgment I will consider in greater detail the terms upon which this cheque was paid over.

On the following day, June 1 1978, a firm of solicitors, H Kennard & Son, entered the scene, having been instructed to deal with the matter on Mrs Tudor's behalf. On June 1 they wrote to the appellants requesting an abstract of title to enable them to prepare a draft assignment for their approval. They also inquired as to the position regarding the landlord's licence.

When June 2 arrived, no one was nearly ready for a formal completion. However, as had been contemplated, Mrs Tudor took H
possession of the hotel and of the business. This brought into play condition 14 of the General Conditions, which deals with the position where a purchaser is authorised by the vendor before actual completion to take up physical occupation.

Mrs Tudor soon became disturbed by what she found. Suffice it to say that the state of repair of the hotel was not as it had been represented to be. Allegations made by her at the trial as to after-hours drinking and prostitution at the hotel were accepted by the judge, who described the general picture as being that of a "disorderly house".

Meanwhile, on June 5 1978, the appellants sent to Mrs Tudor's solicitors an abstract of title. On June 9 1978 she wrote to Mr Gilbert J
complaining that she and her husband had been misled and that the lease was worth much less than they had paid. She asked for a reduction of the purchase price and that the appellants should find a builder to finish the necessary repairs. Before she wrote this letter she had already telephoned Mr Gilbert. His evidence as to this telephone conversation was summarised by the judge as follows: "She raised the question of £5,000 and I told her I would have to discuss that with Mr Hamid. I said I would not part with the money until I had dealt with the assignment." Either in this or some earlier telephone conversation she apparently offered to surrender the lease, but this offer was never taken up. Mr Gilbert did not reply to her letter of June 9 and on June 13 1978 she wrote again to him, threatening to sue him personally for the return of the money paid to his firm. K

However, in neither of these letters did Mrs Tudor suggest that it was her intention to rescind the contract or to move out of the hotel. On June 14 1978 her solicitors wrote to the appellants enclosing their requisitions on title and, subject thereto, a draft assignment of the lease for their approval. On June 15 1978 the appellants answered this letter enclosing replies to these requisitions together with the draft assignment duly approved. They asked Mrs Tudor's solicitors to request her not to communicate further with the appellants. Thenceforth all correspondence appears to have been confined to the two firms of solicitors.

On June 16 1978 Mrs Tudor's solicitors wrote to the appellants saying: "We are now putting in hand the preparation of the L
engrossment of the Assignment for execution by our client." In saying this, they appear to have been following the conventional conveyancing course. Condition 11(1) of the General Conditions, so far as material, provides:

the draft conveyance prepared by or on behalf of the purchaser shall be delivered at the office of the vendor's solicitors at least ten days before the date fixed for completion and the engrossment (first executed by the purchaser, where requisite) delivered there within four days after return of the approved draft.

However, it would appear that at some time after the dispatch of the last-mentioned letter, Mrs Tudor may have given her solicitors instructions, at least for the time being, to take no further steps to complete the transaction. For whatever reason, her solicitors neither M
sent to the appellants an engrossed assignment nor, so far as the documentation and evidence before us reveals, gave any explanation to the appellants for their failure to do so.

By the beginning of July 1978 the appellants, though not having yet received an engrossed assignment from Mrs Tudor's solicitors, had in their hands a form of assignment of the lease and of the goodwill, which had been approved by her solicitors. Furthermore, by that time the vendors' title to the hotel had been accepted (see condition 10 of the General Conditions) and the vendors had received the full purchase price. The form of assignment submitted by Mrs Tudor's solicitors named the vendors as the parties of the one part and Mrs Tudor as the party of the other part. It also contained an appropriate

A space for Mrs Tudor's signature and the attestation of that signature by a witness. However, it contained no provisions by which Mrs Tudor would bind herself to anything. All the provisions were for her benefit. Clause 1 contained the requisite assignments of the lease and the goodwill. The only other substantive provisions were covenants on the part of the vendors. It had not been possible to obtain the landlord's consent to the assignment, but clause 15 of the agreement of May 31 1978 covered this situation. Presumably with considerations such as these in mind, the appellants took the view that Mrs Tudor's execution of the assignment would not be necessary and that there was nothing to prevent their clients from completing the transaction unilaterally. They therefore themselves prepared an engrossment of the assignment from the approved draft and this was B executed by the two vendors. It bore the date July 4 1978. Though it named Mrs Tudor as a party, it has never been executed by her. The learned judge found that this assignment was effectively delivered as an executed deed by Mr Hamid and Mr Patel. This finding has not been challenged on this appeal. On July 4 1978 Mr Gilbert informed Mrs Tudor's solicitors by telephone, and confirmed by letter, that "we now hold the executed assignment and are proposing now to complete this matter". So far as the evidence and documentation before us shows, Mrs Tudor's solicitors made no protest on receipt of this information.

On July 4 1978, following the sending of this letter, the appellants "distributed" the £12,500 in accordance with a typed distribution C statement. More than £3,000 was paid to a bank to which the vendors had mortgaged the lease. Various sums were retained or distributed on account of costs or disbursements. The statement also provided for the retention by the appellants of a sum of £750 in respect of

> costs on account re proceedings to be undertaken by you in defence of action brought by Mrs Tudor for misrepresentation to include further action to be taken by landlords against you to forfeit lease, counsel's fees, disbursements, VAT and other expenses.

No other proceedings had yet been instituted. After deduction of certain further items, a cheque for the balance of £4,907.43 was paid over by the appellants to the vendors.

On July 18 1978, Mrs Tudor issued a writ endorsed with a D statement of claim against the vendors and the appellants seeking the relief summarised at the start of this judgment. In para 7 of her statement of claim she alleged (*inter alia*) that, in order to induce her to enter into the agreement, Mr Hamid represented to her as follows:

> (a) The annual takings of the said business amounted to £16,500. It was implicit in the said representation that the takings of the said business were lawfully earned and did not result in whole or in part from trade carried on in breach of the licensing laws or through any other improper, immoral or illegal trade.
>
> (b) That the Landlord has given its consent to the assignment of the residue of the lease.
>
> (c) The First Defendant has completed all the necessary repairs and decoration required by the Landlord.
>
> (d) The First Defendant has completed all the necessary structural alterations E and repairs in the ladies and gentlemen's toilets as required by the Health Authorities.

In October 1978 Mrs Tudor, under pressure from the landlords with regard to repairs, voluntarily surrendered to them the lease.

The judgment

At the trial the judge held that the four representations listed in para 7(a) to (d) of the statement of claim had indeed been made by Mr Hamid, were false, were known by him to be false and were relied upon by and were part of the inducement to Mrs Tudor to enter into the agreements both of May 26 and May 31 1978. The relief claimed in respect of the misrepresentations was rescission and repayment of F the sums paid. Mr Patel did not appear at the trial. Mr Hamid appeared in person. The judge therefore did not have the benefit of full legal argument on the vendors' behalf and clearly felt some doubts as to whether rescission was the appropriate remedy, since it could be said that full restitution on Mrs Tudor's part was impossible. However, for the reasons set out by him in his judgment, he concluded that rescission was the appropriate remedy as against the vendors. Substantially the form of relief which he granted against them was an order that the agreement dated May 31 1978 be rescinded and that they should pay to Mrs Tudor the sum of £17,000 and interest, that sum representing £17,500 less a sum of £500, which he assessed as the value of the benefits accrued to Mrs Tudor under the lease.

Neither of the vendors has appealed from his judgment and Mr Godfrey QC, on behalf of the appellant firm, has not sought to G challenge the judge's findings of fact as to the misrepresentations made by Mr Hamid. However, he has submitted, so far as this may be relevant, that rescission was not the appropriate remedy, particularly having regard to the surrender by Mrs Tudor of the lease which had presumably been made on the footing that the assignment of the term had been fully effective to pass the legal estate to her (albeit that the term had expired before the trial).

In relation to the appellants, the gist of the learned judge's conclusions was as follows. He found that the £12,500 had been taken by them "as stakeholders", and they were accordingly subjected to the duties summarised as follows in Cyprian Williams on *The Contract of Sale of Land* in a passage cited with approval by Pennycuick V-C in *Potters* v *Loppert* [1973] Ch 399 at p 409: H

> Where a deposit is paid to any person as stakeholder, he shall hold it to abide the event of the contract, and shall be bound to pay the amount thereof to the vendor upon (but not before) the completion of the sale or the forfeiture of the deposit, or to the purchaser if and when the purchaser shall acquire the right to the return of the deposit; but until the event of the contract shall be decided the stakeholder shall not pay over the deposit to either party without the other party's consent. The stakeholder shall be accountable to the parties accordingly; but he shall not be chargeable with any interest for any time during which he shall rightfully retain the deposit, or be accountable for any profit which he may make, in the way of interest or otherwise, by the use of the deposit money during that time.

In reliance on these rules, as I read his judgment, the learned judge concluded that the appellants as stakeholders *ex hypothesi* had no J authority to release the sum of £12,500 to the vendors until "completion". However, in his view "completion is a consensual act". A little later he said:

> If one gets to a situation where one party refuses to complete then the remedy for the other in my judgment is not to complete unilaterally but seek an order for specific performance. That admittedly was not obtained here. If it could be demonstrated that the vendor would get an order for specific performance without any trouble, say under Order 86, I would be inclined to accept it for the purposes of this case, but to my mind it would be more than doubtful whether on July 4 1978 the vendor could have obtained an order for specific performance. The very question as to whether the contract should be rescinded, as I found was justified, would have been raised as a defence to any such application. It must follow from that, that the stakeholder could not say K that there has been a completion. He would have to take the attitude that until the event of the contract shall be decided (referring again to the passage from *Potters* v *Loppert*), he could not release the stake without the consent of both parties.

Accordingly, since at most there had been no more than a completion by unilateral acts of the vendors and no completion of a consensual nature, he considered that the appellants had no right to release the sum of £12,500. He accordingly ordered that the appellants should pay this sum to Mrs Tudor, but without interest.

The judge rejected an argument submitted in the alternative on behalf of Mrs Tudor, but not repeated in this court, based on constructive trust.

L

The issues

Substantially the following issues fall to be dealt with on this appeal in the light of the arguments presented to us:

(1) Did the appellants receive the £12,500 as agents for the vendors or as stakeholders pending the occurrence of some future contingency or contingencies?

(2) If they received it as stakeholders, then, as at July 4 1978, had some contingency occurred which entitled them to release the £12,500 to the vendors without any further authority from Mrs Tudor?

Issue (1)

In the context of issue (1) Mr Godfrey, on behalf of the appellants, M has placed considerable reliance on the decision of this court in *Ellis* v *Goulton* [1893] 1 QB 350, or rather a dictum of Bowen LJ in that case, neither of which appears to have been cited to Judge Baker. In that case on the sale of certain premises by auction the purchaser had paid a deposit to the vendor's solicitor under the terms of a condition which expressly provided that the solicitor should receive it as agent for and on account of the vendor. Thus, as Lord Esher MR pointed out (at p 352): "by the very terms of the condition the vendee was to pay the money, not to a person who could be agent for him, but to the vendor's agent, who was to receive it for his principal". Subsequently the vendor failed to make out a title to the property, so that, as between him and the purchaser, the purchaser became entitled to have the deposit repaid. However, the purchaser claimed

A return of the deposit not only from the vendor but from the solicitor. This court rejected the claim on grounds summarised by Lord Esher MR as follows (at p 352):

> In the first place, it cannot be said that Jackson was in any respect trustee for the plaintiff; there was nothing in the circumstances to raise any trust. There was no relation of principal and agent, and no bailment as between the plaintiff and Jackson; and I have asked in vain for any authority to shew that there was any fiduciary relation between these parties.

Lord Esher MR went on to point out (*ibid*):

> The cases cited draw the distinction between a stakeholder and an agent for one of the parties; the former is the agent of both, but the latter is the agent of one only, and responsible only to that one.

B Bowen LJ agreed with Lord Esher MR, but added the following general observations (at pp 352-353):

> When a deposit is paid by a purchaser under a contract for the sale of land, the person who makes the payment may enter into an agreement with the vendor that the money shall be held by the recipient as agent for both vendor and purchaser. If this is done, the person who receives it becomes a stakeholder, liable, in certain events, to return the money to the person who paid it. In the absence of such agreement, the money is paid to a person who has not the character of stakeholder; and it follows that, when the money reaches his hands, it is the same thing so far as the person who pays it is concerned as if it had reached the hands of the principal.

In argument before us Mr Conlin, on behalf of Mrs Tudor, submitted that the dictum of Bowen LJ was obiter and that it was too C widely expressed. I accept the former, but not the latter, submission. With respect to Bowen LJ, I think that his dictum accords both with common sense and with principle. No authority has been cited to us which suggests to me that it is ill-founded.

Accordingly, I accept the submission of Mr Godfrey that the relevant rule is that, in the absence of an express or implied agreement to the contrary, money paid by a purchaser of land to the vendor's solicitor is received by the solicitor as agent for the vendor, not as stakeholder. It will therefore be necessary to look to see whether there was an express or implied agreement to the contrary in the present case.

Before doing so, I should say a few words about terminology. The concept of "stakeholders" in the context of contracts for the sale and D purchase of land is well known to conveyancers. Lord Denning MR summarised the position thus in *Burt* v *Claude Cousins & Co Ltd* [1971] 2 QB 426 at pp 435-436:

> If an estate agent or a solicitor, being duly authorised in that behalf, receives a deposit "as stakeholder", he is under a duty to hold it in medio pending the outcome of a future event. . . . Until the event is known, it is his duty to keep it in his own hands: or to put it on deposit at the bank: in which case he is entitled to keep for himself any interest that accrues to it: . . . If the purchaser should become entitled to the return of his deposit, he must sue the estate agent or solicitor for it: . . .. He cannot sue the vendor, because the vendor has never received it, or become entitled to receive it.

On the other hand, as Lord Denning MR pointed out (*ibid* at p 435):

> E If an estate agent or a solicitor, being duly authorised in that behalf, receives a deposit "as agent for the vendor", it is considered in law to be just as if it was paid at that very instant to the vendor himself. The vendor is entitled to have it paid over to him on demand, together with interest from the moment when the estate agent or solicitor received it.

In using the phrase "stakeholder" in this judgment, I use it in the broad sense indicated by Lord Denning MR at the start of the first of these two passages.

For the purpose of ascertaining the terms upon which the £12,500 was paid to the appellants and the capacity in which they received it, the evidence as to what was said by the respective parties to one another before the payment was made on May 31 1978 is clearly of F great potential importance. However, no doubt in the interests of economy, we have been supplied with no transcripts of the oral evidence given at the trial. This included the evidence of Mr Gilbert and Mrs Tudor. Mr Gilbert's oral evidence apparently included a passage which the learned judge quoted in his judgment as follows:

> I told the plaintiff that if she wanted possession I could not allow her in without receiving the full purchase price as solicitors and agents for Messrs Hamid and Patel.

On the basis of the notes of evidence taken by Mr Nicholson, the appellant's counsel at the trial, we were told that Mr Gilbert in cross-examination also said:

> I made clear it was paid to me as solicitor and agent.

The learned judge, without suggesting that Mr Gilbert had been trying to mislead him, did not accept this particular evidence. He G said: "I find on the oral evidence that it was not made clear to Mrs Tudor in what capacity that money was being received."

The oral evidence of Mrs Tudor for her part contained a passage which the learned judge quoted as follows:

> I agreed to give Mr Gilbert a cheque for £12,500. I understood that Mr Gilbert would hold the cheque until a final document was drafted.

However, the notes of her evidence in chief taken by Mr Nicholson at the trial recorded her as going a little further than this, by saying:

> We agreed to make out cheque for £12,500 and gave to Gilbert and I understood [he] would hold it till such time as a final document was drafted and signed by [Hamid] *and myself*. (The emphasis is mine.)

Mr Conlin submitted that Mrs Tudor's evidence constituted H material upon which the learned judge was entitled to find, and did find as a fact, that at the meeting of May 31 1978 the interested parties *expressly* agreed that the appellants would not hand over the £12,500 to the vendors until the assurance of the property had been executed by both the vendors and the purchaser. In the absence of a full transcript of the evidence, it is by no means clear to me that Mrs Tudor, though referring to her own understanding of the position, ever went so far as to assert that Mr Gilbert had made any *express* agreement or representation to this effect. However, it is clear that if she did so assert in her evidence, the judge did not go so far as to accept the assertion. If he had done so, this by itself would have rendered Mrs Tudor's present claim against the appellants more or less unanswerable; and it would not have been necessary for him to J explore the surrounding circumstances in nearly so detailed a manner as he did.

In these circumstances I think that we must proceed on the basis that, before the £12,500 was paid to the appellants, (i) no oral agreement was reached between the parties specifically defining the status or conditions upon which the appellants would hold the £12,500 if and when it was handed over to them, and (ii) no oral representation was made to Mrs Tudor specifically defining such status or conditions.

I now turn to consider what other evidence is available as to the terms upon which the £12,500 was handed over, in the light of a specific finding by the learned judge that the money was taken by the appellants as stakeholders. I begin with the written agreement of May K 31 1978 itself.

No doubt because Mr Gilbert had very little time within which to make all the adaptations to the available form of draft contract, which were necessary and appropriate to cover the peculiar circumstances which were presented to him by his client at the meeting of May 31 1978, the final form of agreement as prepared by him, proffered to Mrs Tudor for execution and signed by the vendors and Mrs Tudor was in a number of respects in an inept form. Clause 3, as has been seen, provided that "a deposit of 10% of the purchase price shall be paid to [the appellants] as stakeholders on or before the signing hereof."

In truth, as everyone present at the meeting appreciated, the L intention of the parties was that the full purchase price should be paid straightaway. Clause 4 provided (*inter alia*) for the purchase to be completed on June 2 and for "the balance of the purchase money" then to be paid. There would, in fact, be no outstanding balance on that date. Furthermore, as the learned judge observed, there was really no chance of there being a completion in the legal sense on June 2 1978 (only two days after the making of the agreement). Clause 7, to which I need not refer in detail, was no less inept, if the agreed date for completion was in fact to be June 2 1978.

The learned judge made these comments in relation to clauses 3 and 4:

> Any lay person reading that might be forgiven for thinking that whatever is paid then and there, whether it is ten per cent or whether it is all the price, is to M be paid to the solicitors as stakeholders. That is the document that was put in front of them for this purpose. It is said that this was a mistake and it should not have been in there, and that I can readily believe. But I am not concerned here to rectify the document. There is no doubt that this was the document which they were sent out to read, and it had those clear clauses in it, and they were admonished to read it carefully before they signed it. They paid the money then and there in order that some time in the future, albeit only a few days away on June 2, they were to have possession, and at any rate at the outset they were paying their money then and not upon completion on June 2. The clear indication is that Fisher & Lang were taking that money as stakeholders.

Mr Godfrey submitted that the learned judge erred in his conclusion that the appellants had received the £12,500 as stakeholders and in the route by which he reached that conclusion. In his submission, it is irrelevant to consider what a lay person reading the form of agreement placed before Mrs Tudor "might be forgiven for thinking". The agreement, he suggested, cannot be read as providing either that the whole purchase price shall be paid to the appellants on or before completion or that they shall receive the purchase price as stakeholders. As a matter of construction, the whole of clause 3 and the reference in clause 4 to the payment of "the balance of the purchase money" has to be rejected as otiose and meaningless in the light of the circumstances subsisting at the time when the agreement was signed. Indeed, he submitted, the very fact that the agreement did contain an express but inapplicable condition providing that the conventional 10% deposit if paid (which it was not) should be held by the vendors' solicitors as stakeholders made it the more difficult to spell out from the documentation an agreement that the full purchase price, if paid in advance of the contract (which it was), should be so held. Assuming that the appellants received the £12,500 as agents for the vendors, he submitted that they were bound to pay the money to the vendors on demand and any action to recover it lies only against the vendors (see *Ellis* v *Goulton* (*supra*)).

Mr Conlin, however, on behalf of Mrs Tudor, submitted that in construing clauses 3 and 4 of the agreement of May 31 1978, so as to make sense of it in the circumstances in which it was signed, it is not necessary or correct to carry the "blue pencil" operation as far as Mr Godfrey suggested. All that is necessary, in his submission, is to reject the meaningless words "A deposit of 10% of" in clause 3 and the reference to payment of "the balance of the purchase money" in clause 4.

Neither side seeks rectification of the agreement of May 31 1978. The court therefore has to make the best sense of it which it can. Each of the two suggested interpretations is a more or less equally possible one. No possible third alternative construction of clauses 3 and 4 occurs to me. The agreement, however, was a document drafted by the appellants and proffered by them to Mrs Tudor in their capacity as the vendors' solicitors. In these circumstances, in my judgment, the rule of construction, *omnia praesumuntur contra proferentem* should be applied. All other things being equal, the alternative construction should be applied which is less favourable to the vendors and the appellants and more favourable to Mrs Tudor, to whom this, in part, ineptly drafted and ambiguous document was proffered. This, I conceive, may have been essentially the thought behind the criticised passage in the judgment beginning with the words "Any lay person reading that might be forgiven for thinking". Whether or not it was, I for my part on balance prefer the construction attributed by Mr Conlin to the document to that attributed by Mr Godfrey. For this reason I conclude that the document itself embodied an agreement to the contrary so as to exclude the application of the general rule expressed by Bowen LJ in *Ellis* v *Goulton* that, in default of agreement to the contrary, money paid by a purchaser to the vendor's solicitor is received by the solicitor as agent for the vendor, not as stakeholder.

However, let it be supposed that any reading of the agreement of May 31 1978 was incorrect and Mr Godfrey was right in submitting that clause 3 should be rejected in toto, so that the document contains no provision which indicates the capacity in which the appellants are to hold the cheque for £12,500 when paid over to them.

On this footing it would be necessary for the court to draw inferences as to the nature of the agreement made in this respect from all the circumstances of the case. Even on this footing, I would still hold that it must have been the parties' intention that the appellants should initially receive the cheque as stakeholders in the broad sense which I have indicated. They cannot initially, as from May 31 1978, have held it as agents for the vendors. At least for a time they must have held it, to use Lord Denning's words, "in medio pending the outcome of a future event". The contrary conclusion (namely that from the beginning they held it as agents for the vendors) would mean that they would immediately have had the right to endorse the cheque over to the vendors, or to pay them its proceeds, and indeed the duty to do so on demand *even before the vendors had delivered up possession of the hotel*. Mr Godfrey submitted that this would not necessarily be an unreasonable intention to impute to the parties, bearing in mind that the vendors were under a contractual obligation to hand over possession as soon as May 31. I cannot agree. In my judgment, on the evidence and findings of the judge, it is plain that this cannot have been the parties' intentions, whatever the construction of the agreement of May 31 1978. I therefore think that he was right to decide issue (1) by holding that the appellants received the £12,500 as stakeholders pending the occurrence of some future contingency.

Issue (2)

I now turn to consider whether as at July 4 1978 a contingency had occurred which entitled the appellants to release the £12,500 to the vendors without any further specific authority from Mrs Tudor. I have already explained that, in my opinion, we must proceed on the basis that no oral agreement was reached between the parties and no representation was made to Mrs Tudor specifically defining the status or conditions upon which the appellants would hold this sum if and when it was handed over to them. The court therefore has to attempt to spell out by a process of implication the contingency or contingencies attached to the appellants' stakeholding.

Four alternative possible relevant contingencies have been canvassed in argument. It has been suggested that, on a proper interpretation of the arrangements made between the parties on May 31 1978, the appellants would have had authority to release the £12,500 either:

(i) on June 2 1978, the agreed date for the delivery up of possession of the hotel to Mrs Tudor (whether or not possession had in fact been given on that date), or

(ii) the date when possession of the hotel was actually delivered up to Mrs Tudor (in the events which happened, also June 2), or

(iii) the date when the vendors had fully performed their obligation to assign the lease of the hotel and its business in accordance with the agreement of May 31 1978, or

(iv) the date when the appropriate assignment of the lease and the goodwill of the business would have been executed by the vendors *and* the purchaser.

Mr Godfrey has sought to support all or any of the first three alternatives. Mr Conlin has sought to support only the fourth.

In order to test which of these alternatives is the correct one, it is necessary to examine the circumstances in which payment of the £12,500 was demanded and made. On the evidence, Mr Gilbert made it clear to Mrs Tudor that the reason why he insisted on this requirement was that the parties intended that she should take possession of the hotel and its business as early as June 2. As the judge pointed out, there was no chance of the conveyancing formalities being concluded by that date. The potential risk to the vendors of allowing Mrs Tudor to take possession of the hotel and its business (particularly the latter) would obviously have been substantial unless the full purchase price had already been paid; Mrs Tudor, having taken possession, might have defaulted and refused to complete by paying the balance. Effective rescission would have been very difficult, particularly in relation to the business, because of the problems of restoring the previous status quo. Mr Gilbert's requirement was clearly designed to mitigate these risks by giving the vendors the benefit of the security of the payment of the full purchase price as soon as Mrs Tudor had taken possession of the hotel and its business. On the other hand, it surely cannot have been intended that the appellants should be free to release the £12,500 to the vendors even before the vendors had given Mrs Tudor possession.

In the circumstances, while I would unhesitatingly reject alternative (i) above, I think that on the evidence and the findings of primary fact made by the learned judge, a fairly strong case could be made out for imputing to the parties the intention that the appellants should have authority to release the £12,500 in contingency (ii) above. However, I do not find it necessary to decide this point because, for reasons which I will state later, I am of the opinion that the proper intention to impute to the parties is that the appellants, as stakeholders, should in any event have authority to release the £12,500 in contingency (iii).

The learned judge, as I have indicated, concluded that alternative (iv) above was the correct one. He did not find that on May 31 1978 any express agreement had been reached or any representation had been made by the appellants to this effect. The route by which he reached his conclusion was a rather different one. In the light of the rules referred to in *Potters* v *Loppert* (*supra*) he appears to have considered that since the £12,500 had been received by the appellants as stakeholders, it necessarily followed that the relevant contingency was "completion of the agreement" and he considered that completion is by its very nature a transaction of consensual nature.

Those rules are in terms directed to the duties of persons who receive a *deposit* as stakeholders in connection with a contract for the sale of land. Subject to the live possibility that this might amount to a penalty clause, I suppose that in theory the parties to such a contract

A might specifically agree that the whole amount of the purchase price was to be paid by way of deposit. At least in the great majority of cases, however, a deposit will represent only a proportion (usually a modest proportion) of the agreed full purchase price, so that in such a case completion necessarily must be a bilateral or "consensual" transaction, since it will involve payment of the balance of purchase price. I have no doubt that the rules referred to in *Potters* v *Loppert* were directed to this common situation. Where in that conventional situation vendor and purchaser agree simply that persons shall hold the deposit "as stakeholders", the natural inference will be that the bilateral transaction of completion will be the relevant contingency which entitles the stakeholders to release the deposit. I am not, however, convinced that the rules referred to in *Potters* v *Loppert* B afford any useful guidance in a case such as the present, where the entire purchase price has been paid and completion will necessitate merely unilateral action on the part of the vendors.

Nor have I derived much assistance from the various authorities which have been cited to us relating to the meaning of the word "completion" in different contexts. The circumstances of the present case are exceptional. After the transactions of May 31 1978 had been concluded, the outstanding obligations as regards completion of the sale and purchase of the lease and goodwill were solely those of the vendors. Though the learned judge briefly observed that "apportionment would appear to be necessary", it has not, I think, been suggested in argument on behalf of Mrs Tudor that C any apportionment as at July 4 1978 would have been likely to result in a balance due from the vendors to her. On the contrary, I infer that any such immediate apportionment, if carried out, would have resulted in the assessment of a further modest sum due from her to the vendors. It was, however, always open to the vendors to waive any rights to insist on immediate apportionment, and to proceed to execute the assignment of the lease without it. If, having made title, they were thereafter to execute a deed of assignment in the proper form otherwise than as an escrow, it would take effect as against them according to its purport even though Mrs Tudor were to be named as a party to the deed and did not herself execute it: see *Halsbury's Laws of England* (4th ed) vol 12, para 1358).

In the circumstances I can see no sufficient reason for imputing to D the parties the intention that the appellants should be obliged to wait until Mrs Tudor herself had executed the assignment before releasing the deposit to the vendors. The agreement of May 31 1978 did not contemplate that she herself would enter into any obligations under the assignment when it came to be executed. She was merely to take thereunder the benefit of (a) the lease and the goodwill of the business; (b) a restrictive covenant on the part of the vendors in the terms set out in clause 10 of the agreement; and (c) a covenant on the part of the vendors to pay debts and outgoings in respect of the hotel and business in accordance with clause 8. The form of the draft assignment subsequently prepared by Mrs Tudor's solicitors implicitly confirmed that such would have been her reasonable E contemplation.

In the course of the hearing before us Mr Conlin sought leave to put in a respondent's notice for the purpose of submitting, in effect, that in any event the appellants were under a duty to retain the £12,500 once they were notified of a substantial dispute between Mrs Tudor and the vendors (as, he submitted, the appellants were notified in June 1978) and only to release the sum thereafter in accordance with the resolution of the dispute. We did not think it right to permit this point to be raised, since it was not raised in the court below and, if it had been, it might well have affected the course of the evidence.

However, when the appellants became stakeholders on May 31 1978, the possibility existed (as in the case of any other stakeholding) that a dispute between the interested parties would arise the F resolution of which might affect their ultimate rights to the stake. Mr Conlin was free to submit, as he did submit, that the very existence of this possibility made it the more reasonable to impute to the parties the intention that the stake should not be released until not only the vendors but also Mrs Tudor herself (by executing the assignment) had signified her consent to such release. I cannot, however, accept this submission. It involves the suggestion that even though the title had been accepted by or on behalf of Mrs Tudor and the vendors had executed an assignment in her favour so as actually to pass title to her, she would still have been in a position (without purporting to rescind the contract) to preclude the appellants from handing over the £12,500 to the vendors by raising some complaint against the vendors in relation to the purchase, however reasonable or unreasonable such complaint might appear to be. I do not think we can properly impute G to the parties an intention to enter into the stakeholding arrangements on this basis.

I therefore conclude that on any footing the arrangements made on May 31 1978 conferred on the appellants the authority to release to the vendors the £12,500 on or after the date when the vendors had fully performed their obligation to execute an assignment of the lease of the hotel and its business in accordance with the agreement of May 31 1978 (ie contingency (iii) above), if not earlier.

In the events that actually happened, the vendors on July 4 1978 fully performed this obligation. The learned judge found as a fact that the assignment of July 4 1978 was effectively delivered as a deed by the vendors. It was not an escrow. It operated to pass the legal estate in the leasehold interest to Mrs Tudor. Thereafter the H appellants were, in my judgment, justified in releasing the £12,500 to the vendors. An event had occurred upon which they had been implicitly authorised to effect the release.

Finally, though these do not constitute grounds for my conclusion, I again draw attention to two points. First, though on July 4 1978 the appellants were aware that Mrs Tudor had made complaints against the vendors, she had never purported or even threatened to rescind the contract, so far as the evidence shows. On the contrary, her letters to Mr Gilbert of the previous month had indicated her intention to proceed with it. Secondly, Mrs Tudor's solicitors had been informed on the telephone on July 4 1978 that the appellants intended to "complete" the matter, but they apparently made no protest or J request to the appellants to postpone the distribution of the money so that they could apply for an interim injunction.

For the reasons stated, I would allow this appeal. I would set aside the order of the learned judge and dismiss the action against the appellants.

NEILL LJ agreed and did not add anything.

Also agreeing, SIR GEORGE WALLER said: Since we are differing from the learned judge, I had prepared a short judgment giving my reasons for allowing the appeal. However, I find the reasons for allowing the appeal so much more fully and elegantly expressed in the judgment of Slade J that I will merely say I entirely agree. K

The appeal was allowed with costs in the Court of Appeal and below, such order not to be enforced save with the leave of the court; legal aid taxation of the respondent's costs ordered.

Queen's Bench Division

July 24 1987

(Before Mr P J CRAWFORD QC, sitting as a deputy judge of the division)

COOPER AND ANOTHER v TAMMS L

Vendor and purchaser — Whether false (but not fraudulent) misrepresentation was made by vendor of flat — Whether, if so, purchasers bought the flat in reliance on the misrepresentation — Whether vendor believed on reasonable grounds that facts represented were true — Claims by purchasers relying on section 2(1) of Misrepresentation Act 1967 and on negligence at common law — Effect of disclaimer of liability — Both the purchasers and the vendor in this case were described by the judge as decent and honest but also young, inexperienced and ignorant — The plaintiff purchasers, in addition, were "touched by a degree of M **naïvety" — The purchasers had obtained a limited survey which found signs of dampness and slight wet rot — The report advised that the firm which had previously carried out damp-proof treatment should be recalled if a guarantee still existed; if not, a specialist firm should be called in to inspect and prescribe treatment — In the end, for various reasons, this course was not pursued before completion — In answer to an additional inquiry before contract as to whether, to the vendor's knowledge, the property had been subject to woodworm, rising damp, dry or other rot or infestation, the vendor answered "No" — This answer was subsequently to be**

the basis of the purchasers' case — Unfortunately, extensive wet rot and some dry rot was later in fact discovered, involving the need for considerable remedial works — After considering the evidence the judge held that the purchasers had failed to establish that they were materially influenced by the vendor's representation — They relied on their surveyors' report and advice received from their solicitors — The surveyors' report made it perfectly clear that there were problems with dampness which required investigation before the purchase was completed, but the purchasers did not pursue the matter, perhaps because they believed that they were protected by a timber company's guarantee obtained by a previous owner, but which turned out to be useless — The judge also held that if the purchasers had been found to have relied on the representation, the vendor had an alternative defence under the concluding part of section 2(1), namely, that she "had reasonable ground to believe and did believe up to the time the contract was made that the facts represented were true" — She had taken steps, with independent advice, to ascertain the true position as to the damp-related problems and had been told by a responsible adviser that the remedial work would cost about £500-£600; she had already agreed to a reduction of £500 in the purchase price (from £27,250 to £26,750) in consequence of the original surveyors' report — As to the allegation of negligence, the judge held that, on the assumption that a duty of care existed, the purchasers had not proved that the vendor had failed to take due care; she had in fact discharged a heavier burden of proof under section 2(1) of the 1967 Act — The judge expressed the opinion that, if the vendor had been found liable, she would not have been protected by a disclaimer clause, which, following Dillon J in *Walker* v *Boyle*, he would have ignored — In case the matter went further, and he was held to be wrong, he assessed damages on the tortious basis of the reasonable loss flowing directly from the breach, with general damages for vexation and inconvenience — The total would come to £6,918.82

The following cases are referred to in this report.

Doyle v *Olby (Ironmongers) Ltd* [1969] 2 QB 158; [1969] 2 WLR 673; [1969] 2 All ER 119, CA
Ford v *White & Co* [1964] 1 WLR 885; [1964] 2 All ER 755; [1964] EGD 283; 190 EG 595
Hedley Byrne & Co Ltd v *Heller & Partners Ltd* [1964] AC 465; [1963] 3 WLR 101; [1963] 2 All ER 575; [1963] 1 Lloyd's Rep 485, HL
Parry v *Cleaver* [1970] AC 1; [1969] 2 WLR 821; [1969] 1 All ER 555; [1969] 1 Lloyd's Rep 183, HL
Perry v *Sidney Phillips & Son* [1982] 1 WLR 1297; [1982] 3 All ER 705; [1982] EGD 412; (1982) 263 EG 888, CA
Philips v *Ward* [1956] 1 WLR 471; [1956] 1 All ER 874, CA
Redpath v *Belfast & County Down Railway* [1947] NI 167
Saunders v *Edwards* [1987] 1 WLR 1116; [1987] 2 All ER 651, CA
Treml v *Ernest W Gibson & Partners* [1984] EGD 922; (1984) 272 EG 68
Walker v *Boyle* [1982] 1 WLR 495; [1982] 1 All ER 634; (1981) 44 P & CR 20; [1982] EGD 1149; 261 EG 1090

This was an action by the plaintiff purchasers, Colin Lester Cooper and Karen Margaret Cooper, against the defendant vendor, Shirley Anne Tamms, complaining that they had bought a ground-floor flat at 2 Strathville Road, Southfields, London SW18, in reliance on a false representation by the defendant as to the condition of the flat.

Alan Bishop (instructed by Evill & Coleman) appeared on behalf of the plaintiffs; Miss Alison Hampton (instructed by Crane & Walton, of Leicester) represented the defendant.

Giving judgment, Mr P J CRAWFORD QC said: This action concerns a contract made in 1983 in relation to the sale and purchase of the ground-floor flat at 2 Strathville Road, Southfields, London SW18. The seller of the flat was the defendant, then Miss Shirley Anne Siggins, now Mrs Tamms. The purchasers were the plaintiffs, Mr Colin Lester Cooper and his then fiancée Karen Margaret Doyle, now Mrs Cooper.

All three of the parties to this action, as I find, were decent, honourable and honest people. They were also, all of them, young, inexperienced and ignorant. It was the first time that the plaintiffs had ever bought any property and the first time that the defendant had ever sold any. In addition to their inexperience and ignorance, the plaintiffs were touched by a degree of naïvety. They were also, however, in my judgment, let down by their professional advisers and led into litigation which careful, competent and conscientious advice at the proper time would have spared them.

The short issue in the case is whether the plaintiffs bought the house in reliance upon a false representation by the defendant vendor that it had not been affected by woodworm, rising damp, dry or other rot or infestation, those being matters referred to in the additional inquiries before contract which had been submitted by the plaintiffs' solicitors to the defendant's solicitor as the transaction proceeded.

From the plaintiffs' point of view, the facts were these. On October 1 1983 they were house-hunting in the Southfields area and came across agents who had this house on their books. They went to see the premises and liked them and made an offer to purchase that same day at the asking price of £27,250. I should say that they were escorted to and shown round the premises by the agents, the defendant at that time having recently married and being away on her honeymoon.

The plaintiffs had no experience of property or conveyancing matters. They knew no solicitors. They had made no preparations for paying the purchase price, apart from having some savings in the Woolwich Equitable Building Society. The following Monday, October 3, they got in touch with the building society and set arrangements in train for a mortgage. Knowing of no solicitors, they were provided with a list by the building society and selected (as I understand it, more or less at random from that list) solicitors who they then instructed as being near at hand. They were also provided by the building society with a list of surveyors locally. The plaintiffs telephoned round the surveyors with a view to getting some kind of survey on the premises.

Mrs Cooper told me that she fixed on a local firm, Frank R Marchant, as being the most economical. She, having telephoned a number of firms, gave that firm instructions. It is possible that that firm was the most economical because they were the same firm of surveyors as had been instructed by the Woolwich Building Society. By a memorandum of October 7 1983 the building society instructed the surveyors, Marchants, to inspect and report on the property.

It is to be observed that there is a typewritten note on the common form memorandum saying: "Applicants require structural survey." In fact they never had a structural survey. It seems to me unlikely that they knew of the difference between one kind of survey and another. The survey that was in fact carried out by Marchants was the Royal Institution of Chartered Surveyors' "Flat Buyers Report and Valuation", which is substantially less thorough, and no doubt also substantially cheaper, than a structural survey.

On October 14 1983, Mr Geoffrey Marchant of that firm inspected the premises and produced the report which is document 18 in the bundle which is before the court. By that time the plaintiffs had instructed their solicitors, Evill & Coleman. They did so by telephone. They did not meet their solicitors until a much later stage when the transaction had already started to go wrong. In the course of the routine processing of the conveyancing, Evill & Coleman, by a letter dated October 10 1983 sent to the defendant's solicitors, Crane & Walton, of Leicester, a form of preliminary inquiries. The preliminary inquiries were mostly in common form, but they incorporated certain additional inquiries to which I will return.

The surveyor's report is dated October 17 1983 and must have been in the possession of the plaintiffs by October 19 1983, that being the date of a letter from the vendor's agents, H V Bragg & Co, to the plaintiffs referring to a reduction in the purchase price of the premises of £500 which was agreed, as I find, after and in consequence of the surveyor's report.

The surveyor's report does not purport to be a full structural survey. It was, and was expressed to be, limited. On p 1 there is this warning note:

> The inspection is to provide a report on the general state of repair of the flat described below. It is not a structural survey but a report by a Chartered Surveyor on those matters expressly set out in this report together with valuation advice. The information set out below must be read in conjunction with the marginal notes.

It also says:

> Your attention is drawn particularly to clauses 4, 5, 6, 7, 8, 9 and 10 which define the scope and limit of the report.

It is noteworthy that by condition 4 it is provided that:

> The surveyor will inspect as much of the surface area as is practicable from ground level externally and internally, will lift loose floorboards and trap

A doors where accessible but he will be under no obligation to raise fixed floorboards which, in any event, are frequently covered with hardboard, or to inspect those areas of the flat that are covered, unexposed or inaccessible.

By condition 7 it is provided that:

Except where the contrary is stated, the parts of the structure and of the woodwork which are covered, unexposed or inaccessible will not be inspected and will be assumed to be sound and in good repair.

In the body of the report it appears that the premises are subject to problems regarding damp. On p 2 there appears this comment against the side heading "Damp-proof course and sub-floor ventilation":

Following breakdown of original DPC we understand that a chemical damp-proof course was inserted through most external walls by a specialist firm and B extra air bricks were fitted all under guarantee. However, the treatment does not appear to have been effective and further remedial work is required. See 17.

Para 17 appears on p 4 where these two relevant paras are set out. At para 17:

There are signs that chemical treatment has been carried out to most of the external walls but on taking readings with a Protimeter moisture meter positive results were obtained in all rooms except the bathroom. Should a guarantee be in existence, the firm which carried out the work should be instructed to re-inspect with a view to undertaking additional treatment which is likely to include the hacking off of damp plaster affected by hygrascopic salts.

C Against para 21, where the side note reads:

Woodworm, dry rot and other timber defects within the flat as revealed by the examination of the structure but excluding those areas of the building which are covered, unexposed or not readily accessible . . .

there is this:

We detected slight wet rot in the flooring to the covered entrance lobby and old standing woodworm infection in some floors and to the stair cupboard panelling. It is possible that treatment may have been carried out in the past and guaranteed, in which event the company which undertook the work should be recalled to ascertain whether any further treatment is required. In any event, the rotted flooring in the communal lobby should be renewed and, if no guarantee is forthcoming, a specialist firm should inspect the property with a view to preparing a report covering treatment of woodworm and wet D rot.

The report concludes by stating:

We consider the current value of the property in its present condition to be £26,750, that is to say £500 less than the asking price and there would appear to be a fairly good demand.

It is noteworthy from the passages that I have read out that Mr Marchant, the surveyor who prepared the report, advised on p 4, first:

Should a guarantee be in existence, the firm which carried out the work should be instructed to re-inspect.

Second

E if no guarantee is forthcoming a specialist firm should inspect the property with a view to preparing a report covering treatment of woodworm and wet rot.

When he gave evidence, Mr Marchant told me that he very quickly detected the presence of damp in the walls of the premises, with his Protimeter, saw that there had been some remedial work carried out at some date in the past, that it was obvious that that work was not effective and that, accordingly, he very soon formed the view that he would advise a specialist firm to be called in to inspect, which indeed he did, in his report. Having taken that view, he himself paid less attention to damp. In particular, he made no attempt to get at the underfloor area of the house and, if his report is read carefully, he makes it plain that that is so.

F The report, in my judgment, is limited in its scope because so much of the premises which is relevant was not in fact inspected.

The plaintiffs received the report and read it, no doubt, with care and noticed that there was a reference to wet rot in the covered entrance lobby and the reference to a guarantee. Miss Doyle (as she then was) telephoned her solicitor. By the attendance note of the solicitor's representative, Mrs Lawrence, dated October 21 1983, Mrs Lawrence noted:

Miss Doyle rang me when she said that she had heard from the building society that the property had been over-valued and there were some queries regarding the damp-proofing guarantees and there was some dry rot. She said that, as a result, the price had been reduced by £500. I said that when I received instructions from the building society, I would deal with the necessary guarantees. I said I was still waiting for replies to Enquiries which might show G these.

One small point which arises on that attendance note is the reference by Mrs Lawrence to dry rot. There is no reference to dry rot in the surveyor's report. Mrs Lawrence told me that what Miss Doyle must have told her was that there was some wet rot and that she should have put "wet rot" in, but that when the original of the attendance note was being dictated she must have used the expression "dry rot" by mistake.

On consideration, I have come to the conclusion that I can accept Mrs Lawrence's evidence in that respect and I do accept it. There was no reason why dry rot should have been mentioned in relation to those premises at that stage. Though I do observe in passing that it displays, to my judgment, a somewhat lackadaisical approach to H conveyancing when the words "dry rot" are written instead of "wet rot", and the implications of the one as against the other appear not to have been considered.

But the fact of the matter is that, accepting Mrs Lawrence's evidence, it was well known to the plaintiffs by that date, October 21, that there was rot in the premises and that there were queries regarding the damp-proofing guarantees.

What the plaintiffs told me was that they, having read the report, noticed that the references to wet rot were limited to the flooring in the covered entrance lobby, which was a communal entrance lobby outside the area of the demise, and, accordingly, felt that that was a matter to be dealt with in conjunction with the occupier of the upper J flat in the same premises and did not represent a particular problem to themselves.

So far as the guarantee was concerned, they understood (though from what source at that stage is not clear) that there was a guarantee given by the firm which had carried out the work to the damp-proof course which covered the other signs of damp which had been found by the surveyor. In that, as it turns out, they were mistaken. The damp-proof course had been installed by specialist contractors, namely the South London Timber Preservation Co Ltd, a number of years before. A guarantee had been given in relation to that work, as it usually is for that class of work, but the extent of the work was very limited. In particular, that company had not been called upon to carry out any work to the wooden flooring of the premises and, as it K turned out, it was the wooden flooring that was substantially the root of the trouble.

Now, the position at or about October 21 was that the plaintiffs had received Mr Marchant's report in which he recommended that the premises be inspected, either under the guarantee (if there was one) or (if there was not) with a view to preparing a report covering the treatment of woodworm and dry rot. In fact, as it turned out, there was no effective guarantee of the premises such as to cover woodworm and dry rot.

Having had the surveyor's report, the plaintiffs returned to the Woolwich Equitable Building Society; and the society, having then received their own report from the same valuer — a much more L abbreviated document — offered an advance by letter of October 24.

The next stage, from the purchasers' point of view, was their receipt of a letter from their solicitor dated October 25 1983, which records that the purchase price of the property had been reduced by £500 and also incorporating this paragraph:

I also enclose copies of the following for your information.
1 Search in the London Borough of Wandsworth.
2 Enquiries with the vendor's replies and copy documents referred to.
3 Lease.
4 Office Copy entries.

The copy documents referred to in subpara 2 included the guarantee which had previously been given in respect of these premises by the South London Timber Preservation Co Ltd, dated August 18 1980, M and appears in the papers at document 2.

Unfortunately, the guarantee by itself is meaningless. It recites that:

Pursuant to the terms of the above mentioned contract this warranty is issued by the South London Timber Preservation Company Limited in respect of the work carried out at the above property by the company's operatives.

The "above-mentioned contract" is identified by number and date and the words "chemical DPC". The guarantee is valueless as a document because the works guaranteed are not more particularly described. In other words, it tells the reader nothing without knowing more about the contract which is identified. If it were accompanied by the report and estimate identified, then no doubt it would be

A explicit. On its own it does not indicate with any particularity what work has been carried out which is the subject of the guarantee.

Now, with the copy documents were the vendor's replies to the inquiries before contract, and it is on these that the plaintiffs rely. The inquiries before contract were in common form. They were then followed by a series of additional inquiries which are special to the plaintiffs' solicitors, containing a great many inquiries of more or less relevance to this class of transaction.

One question (and its subquestions) only is relevant. It is question no 20 and reads as follows:

Has the property been subject to any of the following to the vendor's knowledge?
1 Flooding.
B 2 Structural building or drainage defects.
3 Subsidence or heave.
4 Dampness.
5 Woodworm, rising damp, dry or other rot or infestation.
If so, please supply full details thereof including any treatment or remedial works carried out.

The answer to those questions are as follows:

1 Flooding — no.
2 Structural building or drainage defects — no.
3 Subsidence or heave — no.
4 Dampness — yes.
5 Woodworm, rising damp, dry or other rot or infestation — no.

C Then, in answer to the request to supply full details thereof, there appears this note:

DPC fitted 1980. When I purchased the property £500 retention put on mortgage in case further work necessary. Premises needed to dry out. Subsequent survey carried out by Rentokil and building society surveyor and cleared of damp. £500 retention lifted by building society.

The replies therefore indicated "Yes" as the answer to dampness but "No" as the answer to rising damp, dry or other rot.

The plaintiffs' case is that they relied on those representations in proceeding with the contract. I have no doubt that the plaintiffs did in fact read those answers as they read all the documentation relating to this transaction. They did not receive any help in relation to those answers from their solicitor, who merely sent them the documents D under cover of the letter which I have mentioned.

It is to be observed at this stage that the plaintiffs were aware that the property suffered problems in relation to damp. They were aware or must be taken to be aware that the premises had only partially been inspected. And they knew that their surveyor had advised that there should be an inspection by a specialist firm. This is illustrated by the attendance note from the plaintiffs' solicitors' clerk, Mrs Lawrence, of October 31 1983, where she said:

Attending Mr Cooper

attending, that is, by telephone

when he said that he had now received his mortgage offer and this was subject to the damp-proofing guarantee being in existence. I said that the property E had the benefit of a guarantee and Mr Cooper said that both the building society's surveyor and his own had reported that this was not working.

Interposing, there seems to be some elision there. What was not working was not the guarantee but the work which was supposed to have been guaranteed, namely the damp-proof course. Resuming to what Mrs Lawrence says:

I said that in that event he must go back to Miss Siggins

the vendor

and arrange for the South London Timber Preservation Company Limited to re-inspect and report on why the treatment was not working, although I said that he should be quite prepared for them to say that this was because the replastering had not been done properly and it might be better if he obtained a F report and estimate from another company and renegotiated the price with Miss Siggins. Mr Cooper said that he had to give one month's notice where he was living and asked when he should do this. I said that if everything was in order with the damp-proofing treatment, then he could exchange contracts immediately and complete on December 1.

The following day, November 1, the plaintiffs' solicitors wrote to the defendant's solicitors saying this:

The mortgage offer is subject to the damp-proof guarantee being assigned to our clients on completion. The mortgagee's surveyor also reports that the treatment by the South London Timber Preservation Company does not appear to be working and the property should be re-inspected and the company called in to carry out any further treatment necessary. Our clients say that their own surveyor also reported that the damp-proof course did not appear to be working.

There appears to be some confusion in the mind of Mrs Lawrence G who wrote that letter because, of course, the building society's surveyor and her clients' own surveyor were, of course, the same person. Continuing:

We shall be glad if you will ask your client to arrange for the property to be re-inspected in accordance with the guarantee. Would you also please let us have a copy of the report and estimate to produce to the mortgagees?

Now, that letter was written, it is quite plain, without a full consideration of the surveyor's report of his inspection on October 14, a fortnight before. Indeed, Mrs Lawrence told me that she had never seen the surveyor's report at that stage or at any stage until after this matter had been completed. It seems to me a great pity that she did not see it, because it contained several warning signs to the experienced eye. Further, the letter was written under a H misapprehension because the report said that if there was a specialist guarantee, then the specialists should be recalled, and if there was not, then a specialist should be called in to prepare a report.

What happened here was that a form of guarantee had been forwarded but nobody took the trouble to find out how much the guarantee actually covered, which in fact was not very much.

The letter to which I have just referred met with this reply from the vendor's solicitors, dated November 4 1983:

We note what you say concerning the assignment of the guarantees which you have seen and our client has no objection to this. However, the price of the property has been reduced by £500 to allow some expenditure on your clients' part to perhaps remedy the situation and our client is not prepared to proceed with the South London Timber Preservation Company as you suggest. You J must take whatever action your clients think appropriate following completion. Would you please let us know by return whether your client wishes to proceed?

The substance of that letter was passed on by the plaintiffs' solicitors to their clients by letter dated November 10:

With regard to the assignment of the existing damp-proofing agreement, this will be done but the price of the property has been reduced by £500 to allow for the purpose of allowing you to carry out necessary repairs to the property, and the vendor is not prepared to proceed any further with the South London Timber Preservation Company. You must therefore take whatever appropriate steps are required to put the matter right after completion. The Building Society have agreed that necessary repairs can be carried K

evidently "out" is omitted

within a period of 3 months on your undertaking.

which is a rather less than clear way of putting the position.

Now that letter was an invitation to the solicitors' own clients to complete without having inspection. It is simply a quotation from the vendor's solicitors' letter — "You must therefore take whatever appropriate steps are required to put the matter right after completion". In my judgment, that was fundamentally unsound advice. The matter should have been put right before completion; indeed, before exchange. Unhappily, that was not done.

Now, the position at this stage was that the plaintiffs were in L possession of a surveyor's inspection report recommending in a perfectly explicit way inspection of the premises by a specialist and saying why, namely previous damp-proofing work was not effective and there were signs of wet rot immediateley adjacent to the demise and on the same floor, and making it plain that the subfloor area had not been inspected.

The plaintiffs knew that the premises were damp and knew that their surveyor had advised an inspection. They also had the replies to the inquiries which said, *inter alia*, that there had been a damp problem in the premises and which also said that there was no rising damp, dry or other rot.

The plaintiffs then made some attempt to get South London M Timber Preservation Co — the company which had issued the guarantee — to come back to the premises and inspect. No inspection took place within the week following the letter of November 10 to which I have just referred. The solicitors appear to have been content to allow their clients to organise this on their own.

The conveyancing side of the matter proceeded as if everything was in order, although plainly, in my judgment, it was not. Contracts were exchanged on November 16 with a view to completion on December 1. During the fortnight between contract and completion, it appears that there was some contact between the plaintiffs and South London Timber Preservation Co with a view to their going into the premises and inspecting. In fact they did not. An issue in this case

A is concerned with the extent to which the defendant was concerned in that omission.

Throughout this period (that is to say during the whole period from October 1, when the plaintiffs first saw the flat, to completion) the defendant, the vendor, was not living there. She had in fact got married on September 16 1983, had gone away on honeymoon immediately thereafter and was, therefore, for that reason, away when the plaintiffs first saw the house on October 1. She returned sometime after October 1 and went to live with her husband in his flat, which he had had before marriage, not far away. The defendant had left her furniture and various clothing and personal possessions in the flat but was not living there. The keys, she told me, had been left with the agents in order to sell the premises before she went on
B honeymoon and were there all the time.

Mrs Cooper told me that from about November 8 onwards she tried to have the premises inspected by South London Timber Co and, in order to do that, they had to get access. She telephoned Mrs Tamms, the defendant, saying that they had to get the guarantees checked out. She told me that Mrs Tamms was very pleasant on the telephone and agreed to that, and that a number of arrangements were made for South London Timber Preservation Co to inspect the premises but that Mrs Tamms changed the arrangements so that the inspection could not take place as arranged. What Mrs Cooper told me was that a day or so before the timber company was going in to inspect she received a telephone call from Mrs Tamms saying that she
C was unable to be there on the date and time and they should have to make other arrangements. It happened more than once and the days went by; exchange took place and then completion took place, and it was not until after completion, when the plaintiffs were in occupation, that the inspection took place.

Mrs Tamms, when she gave evidence, told me that she was perfectly happy that there should be an inspection but she was working and had no intention of being present there herself. The keys were with the agents, Bragg & Co, and the plaintiffs could have had access with the assistance of the agents at any time. She tells me she remembers Mrs Cooper telephoning twice or maybe three times; that she had no intention of keeping them out of the property and she herself did not intend to be there when it was inspected.

D Now, the only finding which it is necessary for me to make on these facts is whether Mrs Tamms was deliberately playing out time or making difficulties so as to obstruct an inspection until after completion. I am satisfied that that is not so. There may have been misunderstandings between Mrs Tamms and Mrs Cooper at that time. It may have been that the difficulties occurred through the South London Timber Co, and I bear in mind that Mrs Tamms told me when she gave evidence that when she tried to get South London Timber Co to inspect on one occasion, they said that it would take about five or six weeks before they could do it.

I am satisfied that the fact that there was no inspection before completion is not attributable to any act or default on the part of Mrs
E Tamms. It is, I may add, a great pity, in my judgment, that Mrs Cooper did not tell her solicitors that inspection had not taken place (for whatever reason and whoever's fault, or whether nobody was at fault), and her solicitors ought then to have refused to complete until the inspection had taken place. But that was never done. Possibly, having regard to the letter of November 10 1983 which referred to appropriate steps being taken to put the matter right after completion, it did not seem necessary to the plaintiffs that they should do that.

After completion, South London Timber Preservation Co did carry out their inspection — on December 15 1983. It turns out that the condition of the ground-floor flooring was very poor and a good deal of work was required. There was extensive wet rot and there was
F a relatively small area in the underfloor of dry rot which at that stage had not invaded any of the timbers.

The plaintiffs' claim is that they entered into that contract in reliance on the representation contained in the answers to the preliminary inquiries. The answer to question 20, subquestion 5, of the preliminary inquiries was a representation of fact limited by reference to the vendor's knowledge and it was not true. In fact it was wrong. In order to succeed, the plaintiffs have to satisfy me that they were materially influenced by that representation.

I have to say that the plaintiffs have failed to satisfy me that they were materially influenced by that representation. In my judgment, they relied upon the surveyor's report and the advice which they were given from their solicitors. The surveyor's report made it perfectly
plain that there were problems with dampness which had to be G
investigated, and they should certainly have been investigated before this transaction went any further, and they never were.

I do not think, having heard the evidence of Mr Cooper and Mrs Cooper that the reply to question 20(5) would have made any difference. It would have been taken as simply another illustration of the problem of dampness, which was or should have been under investigation. I am fortified in this by noting that the answer "Yes" to question 4 and "No" to question 5 is, on the face of it, inconsistent, and yet the inconsistency was not picked up by the plaintiffs. They did not ask their solicitors, nor did their solicitors ask of their own volition, for any further information. They knew that the premises, being damp, required work done to put it right. They
hoped — indeed, they assumed — that it could be put right under H
guarantee, but they were not entitled to make that assumption on what they were told. The guarantee did not explain, on its face, what it covered. The report from their surveyor left open whether there was a valid guarantee in existence or not. The report from their surveyor made it plain that if there was not an effective guarantee, then the premises should be inspected by specialists.

But the purchasers blithely went ahead. They did not find out the precise scope of the guarantee by obtaining from the South London Timber Co the contract details of the contract to which it related. If they had done that, they would have found that the guarantee was of very limited scope, even if it were fully effective. And they did not
have the premises inspected. When I say "the purchasers", I do not J
mean to refer to the purchasers individually alone, I refer to the purchasers and their solicitors collectively.

In this case I am satisfied that, as Mrs Cooper told me in examination-in-chief: "We assumed that as we had a guarantee for the damp-proof course, everything was all right. There was no real discussion between us" — that is to say, between herself and her husband. "We assumed that everything was covered. We just assumed that we could get the company back which had done the work and it would be all okay." In fact the guarantee was of very loose scope and turned out to be useless.

Having come to the conclusion they were protected by the guarantee, that may be an explanation why they did not pursue more
actively the inspection. But the emphasis which they seek now to put K
on the replies to the inquiries before contract, in my judgment, is made with the wisdom of hindsight. Those answers raised certainly a question concerning dampness which was not pursued. A surveyor's report raised questions which were not pursued. Their surveyor gave advice which was not followed.

In saying this, as I do, I want to make it clear that I am not criticising the plaintiffs individually or personally, and I do not find that they acted unreasonably. I think that they should have been taken care of in a way in which they were not taken care of.

It follows from the conclusion to which I have come that, in my judgment, the plaintiffs' claim fails. But in view of the evidence and
argument which I have heard over five days raising other matters, I L
ought to indicate my findings on those as well.

The plaintiffs' claim is put under section 2(1) of the Misrepresentation Act 1967 (as amended), and also negligence at common law. By section 2(1) of the Act:

> Where a person has entered into a contract after a misrepresentation has been made to him by another party thereto and as a result thereof he has suffered loss, then, if the person making the misrepresentation would be liable to damages in respect thereof had the misrepresentation been made fraudulently, that person shall be so liable notwithstanding that the misrepresentation was not made fraudulently, unless he proves that he had reasonable ground to believe and did believe up to the time that the contract was made that the facts represented were true.

Now, the primary defence raised by the defendant was that there M
was no reliance on the representation. In my judgment, as I have indicated, that defence succeeds. If it had failed, the defendant would have a second limb under the proviso that concludes subsection (1) — the defendant to prove that she "had reasonable ground to believe and did believe up to the time the contract was made that the facts represented were true". I must turn to and deal with that.

From the defendant's point of view, the position is that she bought the flat in 1981 from developers who had carried out the conversion work. There was dampness in the premises when she bought them. Because of the dampness her mortgagee had insisted on a retention of £500. The defendant knew that some damp-proofing work had been carried out by the South London Timber Co. The defendant, after a time, asked the South London Timber Co to return and inspect the premises with a view to clearing them. She was surprised and disturbed to find that they made a charge of £20 plus VAT for that service, which she thought was unreasonable having regard to the fact that they were being asked to check and verify their own work.

Accordingly, she called in Rentokil (another well-known firm in that field) to give her a free report, which they did. The document containing the report, unhappily, has been lost and apparently cannot now be traced either through the defendant's solicitors or her then mortgagees. The defendant tells me, and I accept, that the report from Rentokil cleared the premises of damp; that she sent the report to her mortgagees, who then lifted the retention and advanced the £500 which had been retained. I accept the defendant's secondary evidence in relation to the contents of that report.

When the defendant put the flat on the market some time before the plaintiffs saw it, she had a potential purchaser who was interested. The purchaser instructed a surveyor. After the survey, the purchaser withdrew. In the light of that, the defendant sought to get the South London Timber Preservation Co back to reinspect the premises. Again, to her annoyance and irritation, she was told that there would be a charge. And she was also told that it would take five or six weeks for them to attend.

Accordingly, the following day she telephoned them again in the guise of a prospective customer and asked them to come and conduct a survey. That had the desired effect because a surveyor, Mr McDonald, came from the South London Timber Preservation Co within a couple of days. This, no doubt, did nothing to improve the standing of the company in her eyes, since the request to come and inspect work of their own was met by the response that it would take five or six weeks but a telephone call inviting them to come to a potential new customer resulted in a call in two days.

Mr McDonald, a director of the company, came in response to her call. He went to the cupboard under the stairs and by cutting and lifting some boards in that cupboard obtained access to the underfloor area. There was a minor issue in this case as to whether Mr McDonald had cut the boards himself or had made use of some pre-existing door. I am quite satisfied that he cut the boards himself, in so far as they needed cutting. When he emerged from the underfloor area, he told the defendant the whole floor needed replacing and all sorts of things were wrong with the place.

Mr McDonald gave evidence to the same effect. He said that he inspected the underfloor area, saw substantial areas of wet rot and one area of dry rot about one and a half m^2.

Mrs Tamms was surprised at this news, having regard to the fact she had, two years earlier, managed to get her own retention released, and was somewhat sceptical about its accuracy, taking the view (mistakenly, as it turned out) that Mr McDonald was simply trying to sell the services of his company and was overestimating the amount of work that required to be done.

If Mrs Tamms had left the matter there, she would, I think, have been open to criticism. But she did not leave it there. What she did was to call in a surveyor friend of her husband's, a Mr Rolls-King, to give her what she could rely on as an unbiased opinion. Mr Rolls-King came to the premises at her request shortly afterwards and he inspected the underfloor area some time in the week before the defendant's wedding. Mr Rolls-King told me that he spent 50 minutes or thereabouts under the floor making a detailed inspection.

Mr Rolls-King told me that he was 42 years of age, he had worked for 23 years in the field as a building surveyor, having been employed by well-known chartered surveyors, and had been for the last 10 years in practice as a consultant surveyor, although he had no formal qualifications. I have, of course, given careful consideration as to the weight which I should place on Mr Rolls-King's evidence, having regard to his unqualified status. But I think I should say that I found him an impressive and accurate witness. In my judgment, he was careful in what he said to the court, and I have no doubt that he was careful in what he said to Mrs Tamms, and I have no doubt that he made a careful inspection. It is to be noted that his inspection lasted for 50 minutes whereas Mr McDonald's lasted for 15 minutes only, or thereabouts. I say that not by way of criticism of Mr McDonald, because it may be that Mr McDonald was enabled to be perfectly satisfied on a quarter of an hour's inspection. I mention it only as an illustration of the thoroughness of Mr Rolls-King.

Mr Rolls-King told me that he, too, had seen the mycelium dry rot growth in the underfloor area of this house. He did not dissent from Mr McDonald's estimate that it was about one and a half m^2 in extent. He also said that he found extensive wet rot; again agreeing with Mr McDonald that the dry rot at that stage was not invasive.

Having made his inspection, he emerged and spoke to Mrs Tamms. He told me that he told her on that occasion that she had damp-related problems in her floorboards. He told her that there was a problem; that, in his opinion (contrary to what had been said by Mr McDonald of the South London Timber Preservation Co), it was not necessary to replace the floors and the cost of the work would be about £500 to £600.

He told me that Mrs Tamms had asked him whether she should do the work herself or whether she should apply a reduction in the sale price of the house. His reply was that the best way was to reduce the price by an amount and let the purchasers get on with the work. He told me that he did not give her any detailed explanation of what was wrong; that he did not use the term "wet rot" or the term "dry rot", because it was his experience that lay persons were unaware of the significance of those terms or the difference between one kind of rot and another; that some persons tended to be unreasonably fearful when either of them was mentioned, and it was his practice not to mention specific kinds of infestation. I accept his evidence that what he said to her was just "damp" or "damp-related problems".

I find that Mrs Tamms was not unreasonable in relying on what Mr Rolls-King had told her. She was, in a certain measure, cynical about South London Timber Preservation Co, for the reasons which I have briefly mentioned. She was sceptical of their advice that the whole floor needed replacing; sceptical because she herself knew perfectly well that it was a new conversion which had been done in about 1980; that she was the first occupant of the flat, having bought it in 1981; that there was no visible sign of damp; and that the building society had released her own retention. And when Mr McDonald told her that the whole floor needed to be replaced at some unspecified but no doubt considerable cost, she was, not unreasonably, in my view, sceptical. And she acted perfectly properly, in my judgment, in asking for an independent second opinion and, having regard to Mr Rolls-King's obvious skill and care in this matter, I think that she was entitled to accept his judgment that there was damp or damp-related problems under the floor, that it would cost about £500 or £600 to put right and that the best course was to make an abatement in the purchase price of about that amount and let the purchaser do whatever work was necessary.

I should say that the next step, so far as she was concerned, was that when she returned from her honeymoon, she found a report from South London Timber Preservation Co indicating that repairs were necessary to an amount of £2,400 including £800 or thereabouts for replastering — a great deal in excess of the figures which she had been given by Mr Rolls-King, which no doubt served to exacerbate her scepticism, having regard to the confidence which she had in Mr Rolls-King; although in this respect I find that the figures put forward by South London Timber Preservation Co, with the advantage of the further material which has been acquired, were not excessive. Mr Rolls-King's figures were on the optimistic side, though to the figure of £500 or £600 I have mentioned has to be added a figure of £400 to £500 in respect of plastering, so that the sum total of Mr Rolls-King's figure was in the order of £1,000. And I think that perhaps he omitted the profit element that a builder would require.

But Mrs Tamms, in my judgment, had reasonable ground to believe that the representation she made in answer 20(5) was true, and I accept her evidence that she did believe it was true. Accordingly, even if the plaintiffs had relied on the representation, in my judgment, the defendant would have established the defence (the burden being on her) under that subsection.

The plaintiffs' claim is also put at negligence in common law on the basis of *Hedley Byrne & Co Ltd* v *Heller & Partners Ltd* [1964] AC 465, namely that the defendant made the representation of fact negligently, that is to say, in breach of the duty of care. Assuming (as, for present purposes, I do assume, though it has not been argued) that Mrs Tamms did have a duty of care to the prospective purchasers, the burden of proving that she failed to take due care would rest on the plaintiffs. I have already found that Mrs Tamms has discharged the heavier burden that lies upon her under section 2(1) of the 1967 Act. And, having done that, the claim in negligence at common law also, in my judgment, fails.

Before passing from this part of the case, I should say that the preliminary inquiries were prefaced by a disclaimer in these terms:

These replies are given on behalf of the proposed vendor and are without

> A responsibility on the part of his solicitors, their partners or employees. They are believed to be correct but the accuracy is not guaranteed. They do not obviate the need to make appropriate searches and enquiries and inspections.

In so far as it is relevant, having regard to my other findings, I do find that such a disclaimer would not have been sufficient to protect the defendant, and I refer in this context to the decision of Dillon J (as he then was) in *Walker* v *Boyle* [1982] 1 All ER 634, where he says:

> My attention has been drawn to certain small print which appears above the replies on the form of preliminary inquiries. This small print reads as follows: "These replies on behalf of the vendor are believed to be correct but accuracy is not guaranteed and they do not obviate the need to make appropriate searches, enquiries and inspections." This may be a warning to a purchaser's solicitor to do what any experienced solicitor would know was his duty and
> B make appropriate searches, inquiries and inspections, but it cannot prevent the answers given from being representations of fact. . . . Any vendor will know that that is the object of asking the questions, and will know that the answers are likely to be relied on. A person who makes a representation of fact cannot negative the representation by words such as those in the small print which I have quoted. I accordingly ignore those words.

And I do the same in this judgment.

Finally, I ought, lest this matter goes any further and lest also it is considered in any other context, to make findings in relation to damages. The plaintiffs' claim is under section 2(1) of the 1967 Act. By that Act, if the person making the representation would be liable to damages in respect thereof had the misrepresentation been made fraudulently, that person shall be so liable, notwithstanding that the
C misrepresentation was not made fraudulently. In this case, the representation — if I had found that there was an effective representation — was not made fraudulently. But the damages are to be the same as if it had been made fraudulently.

The principles in regard to the damages for fraudulent misrepresentation were laid down by the Court of Appeal in the well-known case of *Doyle* v *Olby (Ironmongers) Ltd* [1969] 2 QB 158, where Lord Denning MR at p 167A said:

> On principle the distinction seems to be this: in contract, the defendant has made a promise and broken it. The object of damages is to put the plaintiff in as good a position, as far as money can do it, as if the promise had been performed. In fraud, the defendant has been guilty of a deliberate wrong by inducing the plaintiff to act to his detriment. The object of damages is to
> D compensate the plaintiff for all the loss he has suffered, so far, again, as money can do it. In contract, the damages are limited to what may reasonably be supposed to have been in the contemplation of the parties. In fraud, they are not so limited. The defendant is bound to make reparation for all the actual damages directly flowing from the fraudulent inducement. The person who has been defrauded is entitled to say: "I would not have entered into this bargain at all but for your representation. Owing to your fraud, I have not only lost all the money I paid you, but, what is more, I have been put to a large amount of extra expense as well and suffered this or that extra damages."

Winn LJ at p 168F, put the law this way:

> It appears to me that in a case where there has been a breach of warranty or authority, and still more clearly where there has been a tortious wrong consisting of a fraudulent inducement, the proper starting-point for any court
> E called upon to consider what damages are recoverable by the defrauded person is to compare his position before the representation was made to him with his position after it, brought about by that representation, always bearing in mind that no element in the consequential position can be regarded as attributable loss and damage if it be too remote a consequence: it will be too remote not necessarily because it was not contemplated by the representor, but in any case where the person deceived has not himself behaved with reasonable prudence, reasonable common sense, or can in any true sense be said to have been the author of his own misfortune. The damage that he seeks to recover must have flowed directly from the fraud perpetrated upon him.

In a case of innocent misrepresentation (such as this would be if the plaintiffs succeeded) the same principles apply as in the case of fraudulent misrepresentation. That is to say that the plaintiff is entitled to recover, if he recovers at all, all the loss and damage that
F flows, provided that it is not too remote or, to borrow Lord Denning's phrase, "not only all the lost money but, what is more, the large amount of extra expenses as well and this or that extra damages".

In my judgment, in a case such as this where a misrepresentee has entered into a contract on the strength of misrepresentation and has entered into the premises and then discovered the loss, the damages which flow are the cost of putting him in the position in which he would have been if the misrepresentation had not been made. This includes the cost of putting it right and the cost of consequential damages. The case where such works have had to be carried out is different from those where no such work is required. An example of the latter class is to be found in *Saunders* v *Edwards* [1987] 2 All ER 651, where a flat was purchased on the representation that it had a G
roof terrace, and it had not. Well, of course, there was no question of putting a roof terrace in; the purchaser had to make do with a flat without a roof terrace. And the measure of damages there was held to be the difference in value.

Another well-known case in which the same point is made is *Ford* v *White & Co* [1964] 1 WLR 885, where the land was purchased on the basis that a restriction relating to it was not operative, when in fact it was. The appropriate measure of damage there also was the difference in value. In that case, as it happens, the learned judge found that the difference in value was nil. But those are illustrations of that principle operating when no work has to be done.

In a case such as this where work does have to be done, in my judgment the same principle operates, albeit in a different way, in H
that a reasonable cost for doing the work is recoverable. An illustration of this is to be found in *Perry* v *Sidney Phillips & Son* [1982] 1 WLR 1297. For example, in the judgment of Lord Denning MR at p 1301F:

> We now have to consider how the damages are to be assessed. The cases show up many differences. I need only draw attention to these:
> First, where there is a contract to build a wall or a house, or to do repairs to it, then if the contractor does not do the work or does it badly, the employer is entitled, by way of damages, to recover the reasonable cost of doing such work as is reasonable to make good the breach. The cost is to be assessed at the time when it would be reasonable for the employer to do it, having regard to all the circumstances of the case, including therein any delay due to a denial of liability by the contractor or the financial situation of the employer. . . . J
> Second, where there is a contract by a prospective buyer with a surveyor under which the surveyor agrees to survey a house and make a report on it — and he makes it negligently — and the client buys the house on the faith of the report, then the damages are to be assessed at the time of the breach, according to the difference in price which the buyer would have given if the report had been carefully made from that which he in fact gave owing to the negligence of the surveyor. The surveyor gives no warranty that there are no defects other than those in his report. There is no question of specific performance. The contract has already been performed, albeit negligently. The buyer is not entitled to remedy the defects and charge the cost to the surveyor. He is only entitled to damages for the breach of contract or for negligence.

The Master of the Rolls refers to *Philips* v *Ward* [1956] 1 WLR 471. That is an illustration of the measure of damages in contract against a surveyor. But in that case the cost of repairs never arose, as pointed K
out by Kerr LJ in his judgment at p 1305:

> The plaintiff has sold the house and has changed his life by moving to another job and living in another locality. So the question of doing the repairs, and of the actual cost of the repairs, has never arisen. In these circumstances, this appeal has proceeded on both sides on the basis that the costs of the repairs as such — let alone the question as at what date — did not arise. For myself, I am not in any way expressing any dissent from the approach of the judge, but I would say that I would reserve my view as to whether in a case like this the approach by way of cost of repairs is necessarily right. We have not heard the point argued and it does not arise for decision.

In this case the point *has* been argued and, in my judgment, it flows from the principles in *Doyle* v *Olby (Ironmongers) Ltd* that the L
reasonable loss flowing from the breach is recoverable.

Now, various figures have been put to me and I have to bear in mind that if the plaintiffs had succeeded they would have succeeded on the basis that there had been a representation in relation to the dry rot and — although I regard this as artificial — to some extent wet rot as well because, on any view, the plaintiffs were well aware of the presence of damp.

But, doing the best I can and having regard to the multiplications of the schedule P2 put in front of me, the plaintiffs would, in my judgment, have recovered the sum of £3,364.90 under the second head of item 1 of the schedule, less the sums in respect of the damp-proof course and the plastering, which would result in a figure, as I make it, of £2,482.65, plus £805, £178.05, £140, £37.17 and £250, M
making a total of £3,892.87 under that head.

Under item 2 of the schedule P2, it is agreed that the plaintiffs would have been entitled to recover heads A, B, C and E, which total £1,005.95, plus an arrangement fee of £20, making a total of £1,025.95. In addition, the plaintiffs would have been entitled to general damages for the vexation and inconvenience, and I bear in mind that it was certainly a difficult time for them: the completion of the works, the funding of the works, the obtaining of grants of additional building society moneys took many months. They had to leave the house for six weeks and they had this matter on their minds throughout most of 1984. They lost a holiday when they did a certain amount of decorating themselves. In my judgment the appropriate

A figure would have been £1,000 each, making £2,000 in all.

The total plaintiffs' claim, then, would amount to £6,918.82.

One final question arises and that is the effect, if any, which I ought to give to the fact that part of that work was funded by a grant which the plaintiffs obtained from the local authority. It was a discretionary grant, and in December 1983 it would not have been possible to say whether the grant was obtained or not. In fact it was obtained.

In this connection my attention has been drawn to a decision of Popplewell J of June 27 1984 reported in *Estates Gazette**. In that case the learned judge held that the proper method of assessing damages was the difference in value between what the plaintiff actually paid and what the market value of the house was in its then B condition. Accordingly, the measure of damages had to be assessed as at the date of the breach. There was in that case also, however, a discretionary grant which was obtained, and it was urged on behalf of the defendant that the grant should be taken into account. The plaintiff objected that the grant was payable at another date, that is, not until after the breach; that it was only a small proportion of the actual building costs; and that it was irrelevant on the principles of *Redpath* v *Belfast & County Down Railway* [1947] NI 167, 170, and *Parry* v *Cleaver* [1970] AC 1. Popplewell J expressed himself as being clearly of the view that the sum paid by way of grant does not fall to be deducted, first, because it is irrelevant to a claim where the difference in value is the measure of damage and, second, because there is a C collateral benefit which does not have to be taken into account.

It is to be noted that that was a case where the appropriate measure of damages was the difference in value. Accordingly, the damages had to be assessed as at a particular date. This is not such a case. This is a case where the plaintiffs would be entitled under the principles of *Doyle's* case to the loss flowing from the breach, which includes the direct losses which I have found it unnecessary to describe in detail but the amounts of which are agreed.

But the expenditure which the plaintiffs incurred was in some part met from the local authority grant. Now, in my judgment, this raises different principles from those where the measure of damage is the difference in value. The measure of damage is here the loss that actually is sustained — the actual out-of-pocket loss. The point of D time at which the loss is considered is when the loss has been sustained. In such a case it is possible to say whether some grant has been obtained and whether some part of the work has been funded elsewhere.

In my judgment, that part of the grant which was attributable to work which was carried out in consequence of the defendant's misrepresentation should be taken into account. There has been no evidence as to what that figure is. The work covered by the grant included a number of areas which were not the subject of these proceedings. Accordingly, I am unable to make a finding of the precise figure that should be set off by way of credit against that sum of £6,918.82. However, in view of my findings on the primary issue, E as I have already indicated, the plaintiffs' claim fails.

No order was made in regard to costs except for legal aid taxation in the case of both parties.

Court of Appeal

March 2 1988

(Before Lord Justice LLOYD, Lord Justice NICHOLLS and Sir Roger ORMROD)

JOHNSTON & SONS LTD v HOLLAND

F **Derogation from grant — Scope of principle — Comprehensive and important review of authorities — Application in the case of interference with rights reserved to use the flank wall of a property for advertising purposes — Appeal from decision of county court judge in favour of claim that the rights in question had been frustrated — The respondent to the present appeal (the defendant below) was entitled to a 99-year lease of the property — The appellants, a company with two shareholder-directors, had in the course of time acquired the freehold and also a sublease of the property, G leaving the respondent with the intermediate leasehold interest — The flank wall with its advertisement hoardings faced an open site which became a car park and eventually became vested in the appellant company's shareholders — For a long time no difficulties arose about access from the site to the hoardings and at some stage a passageway was constructed between the wall and the remainder of the site — Later, however, the appellant company objected to ladders being placed on the site, at a point beyond the passageway, for safety in gaining access to the hoardings — When a working platform was placed above the passageway to avoid trespassing on the site the appellant company retaliated by H erecting a massive hoarding close to the passageway thereby obscuring any advertisement on the flank wall — The respondent had granted a 21-year business subtenancy of the property, reserving the use of the flank wall for the purpose of advertisements — This subtenancy became vested in the appellant company**

The present proceedings were commenced in the county court by a claim on behalf of the appellant company for the repayment of rent, arising from a difference as to when an increased rent under the business subtenancy began — The main issue in the proceedings, however, was the counterclaim J by the respondent for damages in respect of interference with the use and enjoyment of the flank wall for advertisement purposes — The county court judge dismissed the company's rent claim but awarded the respondent £4,700 plus interest on her counterclaim — The company appealed

The Court of Appeal considered in the light of the authorities the application of the derogation from grant principle and agreed that the appellant company was in breach of its obligations under that principle when it erected the hoardings which obscured the advertisements on the flank wall and when it allowed advertising contractors to construct and maintain a replacement hoarding — The judge below had K also given judgment against the two shareholder-directors of the company, but this judgment could not be supported as it was based on a matter which was not put to them at the trial; for this reason their appeals succeeded — The respondent had relied on two other causes of action, in addition to derogation from grant — These failed, but the respondent had succeeded on the main ground of derogation and the company's appeal must therefore be dismissed — "An appalling saga", ***per*** **Nicholls LJ**

The following cases are referred to in this report.

L *Aldin* v *Latimer Clark, Muirhead & Co* [1894] 2 Ch 437
Beddington v *Atlee* (1887) 35 ChD 317
Booth v *Alcock* (1873) 8 Ch App 663
British Leyland Motor Corporation Ltd v *Armstrong Patents Co Ltd* [1986] AC 577; [1986] 2 WLR 400; [1986] 1 All ER 850, HL
Browne v *Flower* [1911] 1 Ch 219
Cable v *Bryant* [1908] 1 Ch 259
Davis v *Town Properties Investment Corporation Ltd* [1903] 1 Ch 797
Harmer v *Jumbil (Nigeria) Tin Areas Ltd* [1921] 1 Ch 200
Hollywood Silver Fox Farm Ltd v *Emmett* [1936] 2 KB 468
Lyttelton Times Co Ltd v *Warners Ltd* [1907] AC 476
Molton Builders Ltd v *City of Westminster London Borough Council* (1975) 30 P&CR 182, CA
Myers v *Catterson* (1889) 43 ChD 470
Ricketts v *Enfield Churchwardens* [1909] 1 Ch 544

M This was an appeal against a decision of Judge Harris QC at Westminster County Court. The plaintiffs, Johnston & Son Ltd, appealed against the dismissal by the judge of a claim against the defendant, the present respondent, Miss Ida Kate Holland, for repayment of certain sums paid to her as rent. The judge also found in favour of a counterclaim by Miss Holland against both the company and the two individual shareholder-directors, Sidney Johnston and his wife, Elsie Johnston, in respect of interference with her rights to the use and enjoyment of the flank wall (northern) at 6 Queen's Parade, Green Lanes, Harringay, London N8.

Romie Tager (instructed by Rose & Birn) appeared on behalf of the appellants; Ivan Krolick (instructed by Alan R Pulver & Co)

* Editor's note: *Treml* v *Ernest W Gibson & Partners* [1984] EGD 922; (1984) 272 EG 68.

A represented the respondent.

Giving the first judgment at the invitation of Lloyd LJ, NICHOLLS LJ said: These parties have been engaged in disputes and litigation for over 13 years. The disputes seem to have their origin in an incident which occurred in 1971. A poster hoarding fell off the end wall of 6 Queen's Parade, Green Lanes, Harringay, London N8 (which I shall call "no 6") and damaged some cars standing below. Miss Ida Holland owns a long lease of no 6. The cars belonged to a company called Johnston & Sons Ltd (which I shall call "the company"). The two shareholders and directors of the company were Mr Sidney Johnston and his wife Mrs Elsie Johnston. The company's car sales business was managed by their son Colin. The poster company's insurers were slow in settling the company's claim for damage to the B cars. Since then, as the judge found, Mr Colin Johnston has been almost paranoid in his desire to spite and do down Mrs Holland. Hence this appalling saga.

The appeal is from a decision of His Honour Judge J P Harris QC, sitting in the Westminster County Court on June 8 1987. The judge dismissed a claim brought by the company against Mrs Holland for repayment of sums, totalling £2,812, paid by the company to her as rent, and he found in favour of Miss Holland on a counterclaim brought by her against the company and Mr and Mrs Johnston for damages in respect of interference with her use and enjoyment of the flank wall of no 6 for advertising purposes. He awarded her the sum of £4,700, plus interest. From that decision the company and Mr and C Mrs Johnston have appealed.

The history

I must first refer to the essential history. The facts are a little complicated. No 6 is at the northernmost end of a terraced building. Miss Holland, who is now in her eighties, has been connected with no 6 all her life. In 1907 her father, Henry Holland, carried on a cycle and sports goods business in a shop on the ground floor, and he and his family lived in the residential accommodation above. Initially he seems to have been a subtenant from a Mr Brewis. In 1919, after the first world war, Mr Brewis assigned to Mr Holland the residue of a 99-year lease of no 6 which had been granted on September 10 1898.

The northern flank wall of no 6 overlooked an open site, beyond D which was the Queen's Head public house. At the time when Mr Holland acquired the lease of no 6, the licensee of this public house was a Mr Robert Chattey. He held a lease of the public house plus the adjoining open site for a term of 70 years from Midsummer Day 1885.

From at least 1909 Mr Holland used the northern flank wall of no 6 for advertising purposes. The wall was readily visible to people passing along Green Lanes. On July 23 1919, which was the date on which Mr Holland acquired the lease of no 6 from Mr Brewis, Mr Holland entered into an agreement with Mr Chattey, the licensee of the Queen's Head. By that agreement, which I shall call "the 1919 agreement", Mr Chattey, described as "the Landlord" and Mr Holland, described as "the Tenant" agreed

E that the Landlord shall let and the Tenant shall rent all that space on the South side of the grounds adjoining the "Queen's Head" Hotel, Green Lanes, Harringay in the County of Middlesex on premises belonging to the Tenant including the forecourt thereof for advertising purposes as from the 24th day of June 1919 at the yearly rent of one shilling (1/-) payable on the four usual quarter days in each year the first payment to be made on the 29th day of September 1919.

The agreement provided that the tenancy thereby created was terminable on three months' written notice on either side. The agreement superseded a similar agreement made in 1909 between Mr Chattey and Mr Brewis.

Miss Holland relies on the 1919 agreement as one of her causes of action in the present case. For the moment it is sufficient to note that F the advertisement hoardings fastened on to the flank wall of no 6 projected forward some inches from that wall and overhung the Queen's Head open site of which Mr Chattey had a long lease. The 1919 agreement was intended to regulate the position regarding those hoardings. Although the drafting of the agreement leaves much to be desired, it is tolerably clear that the agreement either operated as a demise to Mr Holland by Mr Chattey of the narrow strip of air space of a few inches forward from the northern flank wall over the whole area of that wall or, if that space were not sufficiently clearly defined to be capable of being the subject-matter of a demise, operated as a grant to Mr Holland of an easement to affix and retain advertisement hoardings on the flank wall. It matters not for the purposes of this action which of these alternatives is the correct one. In either event, by implication there was included a grant of rights of access over Mr G Chattey's land reasonably necessary for maintaining and using the hoardings.

I return to the history. Mr Holland died in 1927. On March 28 1935 his executors, who were his widow, Mrs Sarah Holland, and a Mr Hyman Stark, executed an assent in favour of Mrs Holland of the lease of no 6. By then the Holland family had moved away from no 6. No 6 was sublet, but advertising on the flank wall of no 6, by or on behalf of Mrs Holland, continued.

At some time during the 1930s an alteration was made to no 6. An external staircase giving direct access to the first floor was constructed against the northern wall of no 6. From the foot of this staircase to the front of no 6 a passageway, some 4 ft 6 in wide, was constructed. This was separated from the Queen's Head open site by H a fence. It is not known now who carried out these works — whether Mrs Holland or a subtenant of hers — or what arrangements were made with Mr Chattey or the freeholders of the Queen's Head. Thenceforth the open site was reduced in size by the removal from it of this passageway. I shall refer to the residue of the Queen's Head open site remaining after the removal of this passageway from it as "the car lot" and to this passageway as "the passageway".

The second world war came and went. In 1950 the then lessee of the Queen's Head surrendered the residue of the 1885 lease to the freeholder, the Cannon Brewery Co Ltd.

In 1953 Cannon Brewery let the car lot to Mr Johnston and two partners, who were trading as Atomic Cars, for use by them for J buying and selling secondhand cars. The letting was for three years, and thereafter until determined. Meanwhile advertising on the flank wall of no 6, licensed by Mrs Holland, continued. In 1954 the Borough Billposting Co, which belonged to Odham's Press Ltd, was the licensed advertiser. On September 1 1954 it entered into a written agreement with Cannon Brewery regarding access over the car lot for the purpose of posting advertisements. If ladders up to the hoarding were placed in the passageway they were at too steep an angle for safety. So it was better to place the foot of the ladders on the car lot side of the fence dividing the car lot from the passageway. The 1954 agreement recited the 1919 agreement and that the interest of Mr Chattey in that agreement "is now vested" in Cannon Brewery and that the interest of Mr Holland "is now vested in Mrs Holland his K widow". The agreement provided for Cannon Brewery, "with the consent which has been obtained of its tenants Messrs W A Johnston, S A Johnston and L N Johnston trading as Johnston & Sons", to grant to Odham's

all necessary access over and through the private forecourt of [Cannon Brewery's] property known as the Queen's Head Hotel, Green Lanes, Harringay aforesaid for the purpose of maintaining and posting advertisement boards on the flank wall of the adjoining property known as No 6 Queen's Parade, Harringay . . .

The first of the four conditions then set out was that the agreement should remain in force for one year and thereafter until determined by three months' notice or unless and until the 1919 agreement should L be determined.

Advertising on the flank wall of no 6 continued throughout the 1950s, with access to the hoardings being achieved by ladders resting in the car lot used by Atomic Cars.

In 1960 Mrs Holland granted to her tenant of no 6, a Mr Henry Wade, a new 21-year lease, at a rent of £500 per annum, in place of his then existing lease. The new lease was dated January 6 1960. It is upon the reservation contained in this lease, which I shall call "the 1960 lease", that Miss Holland principally founds her cause of action against Mr and Mrs Johnston and the company. The property demised by the 1960 lease was described as:

ALL THAT lock-up shop messuage and premises with dwelling rooms over situate at and known as 6 Queens Parade, Harringay in the County of M Middlesex and all ways sewers drains lights easements and appurtenances thereto belonging or appertaining . . ..

It is common ground that by this lease Mrs Holland demised to Mr Wade no 6 and also all her estate or interest or rights, whatever precisely they may have been, in and over the passageway. The 1960 lease contained the following reservation:

except and reserving unto the Lessor the use of the exterior of the flank wall of the premises on the north side thereof for the purpose of affixing posters or advertisements thereto and the right to grant licences therefor.

Mrs Holland died soon afterwards, on May 2 1960. Her personal representatives were her daughter Miss Holland and Mr Hyman Stark. On June 7 1960 they made a formal written assent, in favour of

A Miss Holland, of Mrs Holland's interest in no 6 under the old 99-year lease which had been granted in 1898.

On July 24 1961 the brewery company, now known as Ind Coope (London) Ltd, granted a tenancy of the car lot to the company for three years and thereafter until determined by notice. The company had been incorporated on January 25 1960. This tenancy superseded the tenancy of the car lot which the three partners had from 1953. This tenancy is still subsisting. In the following year the company expanded further. On September 10 1962 the company purchased from Mr Wade for £750 the goodwill of Mr Wade's car sales business and the residue of the 1960 lease of no 6. In this way the company became Miss Holland's lessee.

Matters remained thus until, in 1968, the company acquired the B freehold of no 6. Thus the position regarding no 6 then became as follows: the company was the freeholder, Miss Holland was entitled to the residue of the old 99-year headlease, due to expire at midsummer 1997, and the company was the tenant from her under the 21-year lease granted in 1960 which had reserved the advertising rights in favour of the landlord.

The previous litigation

I can leave the conveyancing history there for the moment and turn to the previous litigation between the parties. I have mentioned the incident concerning the damaged cars in 1971. The poster company sought to re-erect its hoarding, but it was prevented by the company from having access to the car lot for this purpose. On October 23 1974 C Miss Holland issued a writ against the company claiming an injunction and damages. Meanwhile, the brewery company, at Mr Colin Johnston's behest, gave notice in August 1974 terminating the 1954 access agreement from, apparently, the end of 1974. The action was tried on February 9 1977 by Sir Douglas Frank QC, sitting as a deputy High Court judge. Before him it seems to have been common ground that, following the termination of the 1954 access agreement, Miss Holland no longer had any right of access over the car lot. Furthermore the company accepted that it had no right to obstruct use of the passageway by Miss Holland for the purpose of obtaining access to the hoardings. Nevertheless, having regard to the course of conduct of the company, the judge granted an injunction. The D injunction was expressed in wide terms, but I do not think that the injunction can have been intended to apply to the company's activities on the car lot. It was intended to deal with the company's activities on the only land over which Miss Holland's rights, as then presented to the court, extended, namely the passageway. The judge also awarded Miss Holland £525 damages, plus interest and costs.

The subsequent history

I continue with the history. The 1960 lease, held by the company as a business tenancy, was due to end at Christmas 1980. In May 1980 Miss Holland gave notice under Part II of the Landlord and Tenant Act 1954 determining the tenancy. She did not object to the grant of a new lease. The company duly served a counternotice and applied to E the county court for a new lease. I shall have to return to that aspect of the relationship between the parties at a later stage.

Meanwhile, after the trial in February 1977, Miss Holland sought to resume advertising on the flank wall of no 6. In July 1982 she entered into a licence with A W May Ltd. A W May agreed to take the site for advertising purposes for five years at £1,000 per annum. The company, and in particular Mr Colin Johnston, were adamant in refusing A W May any access on to the car lot with ladders.

By September 1982 it seemed that A W May had found a solution, by which access could be had to the hoardings without needing to go on to the car lot at all. A working platform was to be erected, above the passageway, against the flank wall. In order to frustrate this the company, acting by Mr Colin Johnston, erected a massive hoarding F of its own, along the edge of the car lot, close against the passageway. The company's hoarding was about 5 ft away from and parallel to the flank wall of no 6. Its size and height were such as substantially to obscure any advertisement which might be affixed to the flank wall of no 6. The company's hoarding did not prevent physical access to that wall from within the passageway, but by obscuring most of the wall any advertisement on it would be effectively hidden from public view. So the wall ceased to be of any use as a site for advertisements. The judge expressed himself as wholly satisfied that the only reason why that hoarding was put up by the company was to try to stop A W May and Miss Holland gaining some advantage from this advertisement site.

Initially the company's hoarding was blank. Then for a few months in the spring of 1983 the company put up some simple G advertisements concerning its own car sales business. In July 1983 professional advertisers erected a replacement hoarding and since then they have displayed posters on a panel measuring 21 ft in width and 10 ft in height. The bottom edge of the panel is some 12 ft or so from the ground.

I can now conclude the conveyancing history by mentioning three matters. First, on August 31 1982 the company, the sitting tenant of the car lot, entered into an agreement to buy the freehold of the car lot. That agreement was completed by a conveyance made on November 17 1982 by Ind Coope (London) Ltd, in favour not of the company but of Mr and Mrs Johnston personally. Second, in September 1985 the freehold of no 6 was transferred by the company to Mr and Mrs Johnston. Third, in May 1987 the landlord and tenant H proceedings were disposed of by a new lease of no 6 being granted by Miss Holland to the company for the residue (save for a nominal reversion) of her interest under the 1898 headlease. The new lease contained a reservation of advertising rights similar to that contained in the 1960 lease.

The present litigation

The present proceedings were started by the company against Miss Holland on January 3 1985. She served a counterclaim against the company and Mr and Mrs Johnston in respect of the obstruction of the advertisement on the flank wall of no 6.

I shall consider the counterclaim first. Miss Holland's claims in respect of the erection of the hoarding on the car lot, and the denial of J access over the car lot, are put forward under three heads. First, her rights under the reservation in the 1960 lease. Second, her rights under the 1919 agreement. Third, nuisance. The judge found in her favour on the first ground. He was also satisfied that the 1919 agreement had not been determined and that until it was ended it still bound the owners of the land to which it related. He rejected the claim in nuisance.

The reservation

First, the reservation in the 1960 lease. For the company and the Johnstons, Mr Tager presented a straightforward argument. He accepted that the reservation of the advertising rights in favour of the K original lessor, Mrs Holland, operated by way of a regrant of those rights by the original lessee, Mr Wade. He also accepted that Mr Wade was under an implied obligation not to derogate from that grant made (or notionally made) by him. Mr Tager submitted, and this is clearly right, that that implied obligation falls to be construed by reference to the circumstances existing at the time of the grant. In 1960 Mr Wade had an interest in no 6 and the passageway, but he had no interest in the car lot. Accordingly, submitted Mr Tager, the implied obligation then assumed by Mr Wade would not have extended to any acts which might be done by him on the car lot. If in the future Mr Wade should ever buy the car lot, or acquire a lease of it, he would be free to build or erect hoardings, or whatever, on the car lot as he might choose, regardless of any impact this might have L on the use of the flank wall of no 6 as an advertising site. Mr Wade's successors stand in Mr Wade's shoes. Hence the company and the Johnstons are not affected in their use and enjoyment of the car lot by the reservation contained in the 1960 lease of no 6.

Mr Tager adopted, as part of his submissions, the following sentence appearing in *Hill and Redman, Law of Landlord and Tenant,* 17th ed, vol 1, at p 206:

> The doctrine
>
> of derogation from grant
>
> is limited by the condition of affairs existing when the lease is made, and as a man cannot by implication grant that which he has not got, or grant easements over adjoining land which does not belong to him, the rule as to derogation M will not extend to property acquired by the grantor after the grant of the lease.

A passage to the same effect appears in *Halsbury's Laws of England*, 4th ed, vol 27, para 334. Mr Tager drew our attention to the decision of the Court of Appeal in Chancery in *Booth* v *Alcock* (1873) 8 Ch App 663. There the lessor of a house, 26 Old Change, also owned an underlease of an adjoining property, 3 Distaff Lane, expiring in 1868. Subsequently he acquired the freehold reversion of 3 Distaff Lane. In 1872 he pulled down 3 Distaff Lane and was proposing to build on the site in a manner which, it was said, would interfere with the access of light to 26 Old Change. The court held that general words in a grant must be restricted to that which the grantor then had power to grant and they will not extend to anything which he might subsequently acquire. Likewise, in *Beddington* v *Atlee* (1887) 35 Ch D 317. That was another rights of light case and Chitty J said (at p 326):

> If the adjoining land is owned by a stranger, of course there is no implied grant.

Mr Tager also reminded us of the classical exposition of the derogation from grant principle given by Parker J in *Browne* v *Flower* [1911] 1 Ch 219. Running through this is the theme that the principle is one which, in appropriate circumstances, restricts the activities of the grantor on land which is retained by him at the time of the grant. For example, at p 224 Parker J said:

> The plaintiffs next relied on the maxim that no one can be allowed to derogate from his own grant. This maxim is generally quoted as explaining certain implications which may arise from the fact that, or the circumstances under which, an owner of land grants or demises part of it, retaining the remainder in his own hands. The real difficulty is in each case to ascertain how far such implications extend.

Later he said (at p 225):

> But the implications usually explained by the maxim that no one can derogate from his own grant do not stop short with easements. Under certain circumstances there will be implied on the part of the grantor or lessor obligations which restrict the user of the land retained by him further than can be explained by the implication of any easement known to the law. Thus, if the grant or demise be made for a particular purpose, the grantor or lessor comes under an obligation not to use the land retained by him in such a way as to render the land granted or demised unfit or materially less fit for the particular purpose for which the grant or demise was made.

In the present case the judge's conclusion on this point was as follows:

> What has to be decided in this counterclaim is what is the extent of this reservation or exception and what terms should be implied therewith. Does it bind the lessees in respect of land which they subsequently acquired? Can they frustrate the effect of the reservation and exception which they agreed to by acts which they have done by erecting, or causing to be erected, this hoarding. I am satisfied, as I have said, on the evidence that the Company initially and thereafter from November 1982, the Directors of the Company, intentionally and maliciously frustrated this reservation and exception. In my judgment the law must be in a sorry state if the plaintiff Company by its directors can get away with this. I distinguish the derogation from the lessor's grant cases as this is not a covenant on the part of the lessor. We are dealing with a reservation or exception in the lease whereby the lessor reserves something to her.
>
> I find that by erecting this hoarding and maintaining it there, the Company by its servants or agents are in breach of the term which must be implied in the said reservation or exception. Otherwise that reservation and exception agreed in 1960 and apparently re-agreed in 1987 would be useless and worthless and could be defeated by the malicious acts of the plaintiff Company and/or its directors.

With great respect to the judge, I am unable to accept the distinction he seems to be drawing. I see no reason to suppose that the obligations arising under the non-derogation from grant principle, in relation to land acquired by the grantor after making the relevant grant, are more onerous in the case of a regrant by a lessee than they are in the case of a grant by a lessor.

The purpose of the reservation

Before addressing Mr Tager's submissions I turn to consider the purpose of the reservation in the 1960 lease.

In 1960 no 6 had two uses. Internally, the building was used as a lock-up shop on the ground floor with living accommodation upstairs. The principal purpose of the 1960 lease was that Mr Wade should have the benefit of that use for 21 years. Externally, the flank wall was used as an advertisement site. The purpose of the reservation was that this use should be retained by Mrs Holland. This use was sufficiently valuable for it to be worthwhile for Mrs Holland, with Mr Wade's agreement and as part of their bargain, expressly to retain this right.

Mr Wade did not own the car lot nor did he have any tenancy of it. Neither did Mrs Holland. She was at risk as to what the present or a future owner or tenant of the car lot might do on that land. He might obscure the flank wall by building on the car lot. He might erect his own advertising hoardings, wherever he wanted on the car lot. But in my view the parties cannot for one moment have contemplated that, whatever a stranger might do, if Mr Wade, lessee of no 6, should himself acquire some estate or interest in the car lot while he was still the lessee of no 6, he would be free wholly to frustrate the purpose of the retention of the advertising rights by putting up his own advertising hoarding immediately in front of Mrs Holland's hoarding. That would be, in substance, to take over the advertising site for himself; *a fortiori*, that he would not erect a blank screen for the sole purpose of obscuring Mrs Holland's advertisements. That would be to make use of the car lot for the sole purpose of defeating the reservation. To my mind it was necessarily implicit in the terms of the 1960 lease and the reservation of the advertising rights that Mr Wade would not himself frustrate the purpose of that reservation by taking either of the steps I have just mentioned.

Of course, depending upon its size and position, the erection of a building on the car lot might equally have the effect of rendering Mrs Holland's wall useless as an advertising site. But it would be going much too far to spell out a common intention that if Mr Wade bought the car lot he would not build on it in such a way as to obstruct the use of the flank wall of no 6 for advertising. However, I do not think that this is inconsistent with Mr Wade not being free to take either of the two steps mentioned above. Constructing a building is one thing. Erecting a blank hoarding for the sole purpose of screening the flank wall, or erecting one's own advertising hoarding for the purpose of effectively taking over the advertising site for oneself, is altogether different.

Derogation from grant

I turn next to Mr Tager's submissions on the law. They are formidable. We have not been referred to any case in which it has been decided, or even suggested, that the non-derogation from grant principle can apply in respect of activities of the grantor on neighbouring land subsequently acquired by him. Despite this, a rigid inflexible limitation excluding such activities by a grantor from the scope of the principle in all circumstances is capable of producing such unattractive results that, given the rationale underlying the principle, I am emboldened to consider whether, indeed, there is any such overriding limitation.

The expression "derogation from grant" conjures up images of parchment and sealing wax, of copperplate handwriting and fusty title deeds. But the principle is not based on some ancient technicality of real property. As Younger LJ observed in *Harmer* v *Jumbil (Nigeria) Tin Areas Ltd* [1921] 1 Ch 200 at pp 225, 226, it is a principle which merely embodies in a legal maxim a rule of common honesty. It was imposed in the interest of fair dealing:

> "A grantor having given a thing with one hand," as Bowen LJ put it in *Birmingham, Dudley & District Banking Co* v *Ross*, "is not to take away the means of enjoying it with the other." "If A lets a plot of land to B," as Lord Loreburn phrases it in *Lyttelton Times Co* v *Warners*, "he may not act so as to frustrate the purpose for which in the contemplation of both parties the land was hired." The rule is clear but the difficulty is, as always, in its application.

As one would expect, the principle applies to all forms of grants. It was applied recently by the House of Lords to the sale of a car by the manufacturer: see *British Leyland Motor Corporation Ltd* v *Armstrong Patents Co Ltd* [1986] AC 577. The principle operates to restrict the future activities of a grantor. One field, and perhaps it was the earliest field, in which the principle finds expression is the grant by implication of easements over land retained by the grantor corresponding to the continuous or apparent quasi-easements enjoyed at the time of the grant over the property retained (see Parker J in *Browne* v *Flower*, at p 225). In such cases there must necessarily be retained land over which the easement being implied is to operate. *Booth* v *Alcock* (1873) 8 Ch App 663 is an example of this.

It is also well established that the principle is not confined to cases of the implication of an easement. For instance, in *Aldin* v *Latimer Clark, Muirhead & Co* [1894] 2 Ch 437 Stirling J applied the principle to protect the access of air to sheds used for drying timber even though the law does not recognise a right by way of easement to the access of air except where such right is enjoyed through a definite aperture or channel. Parker J in *Browne* v *Flower* (at p 226) regarded the implied obligation in such cases as analogous to that which arises under a restrictive covenant. In *Megarry and Wade on the Law of Real Property*, 5th ed, p 849, the view is expressed that in truth the doctrine is an independent rule of law. This approach was approved by Lord Denning MR in *Molton Builders Ltd* v *City of Westminster* (1975) 30 P&CR 182, at p 186. He stated the broad principle thus:

> . . . if one man agrees to confer a particular benefit on another, he must not do anything which substantially deprives the other of the enjoyment of that benefit: because that would be to take away with one hand what is given with the other.

That being the general principle, the next step must be to apply it to a particular factual situation. In a case such as the present, that

A exercise involves identifying what obligations, if any, on the part of the grantor can fairly be regarded as necessarily implicit, having regard to the particular purpose of the transaction when considered in the light of the circumstances subsisting at the time the transaction was entered into.

It is at this point that I find myself unable to see why that exercise must be subject to the overriding limitation contended for by Mr Tager in the present case. The proposition can be tested by taking, by way of illustration, a building adjoining open land and having, in consequence, potential as an advertising site: as in the present case. Assume that the building and the open site are in common ownership. The owner lets the right to put up and maintain a poster on the flank wall of the building for five years, for a lump sum or at annual rent.
B Unless there are further factors which rebut the implication, it seems to me necessarily implicit in such a transaction that the lessor cannot, within the period of the lease, erect on the open site his own advertising hoarding a few feet in front of the wall of the building, obscuring the wall, and thus make it impossible in practice for the wall to be used for the very purpose for which he has granted rights and for which he has been paid or is being paid.

Take, next, the same facts, save that at the time of the grant the lessor does not own the adjoining site, but subsequently, while the lease of the advertising rights is still current, he buys the adjoining site. Can he, in such a case, any more than in the first example, be at liberty to frustrate the whole purpose of the letting by erecting his
C own advertising hoarding hard up against the flank wall of the building, as in the first example? In my view common honesty permits of only one answer to that question.

Of course in considering what is necessarily implicit in a transaction in a case where the grantor owns no other land, very great weight indeed must be given to that factor. It will be a very exceptional case for it to be necessarily implicit in a lease that the activities of a lessor who owns no adjoining land, and has no plans to buy any adjoining land, are to be restricted on the adjoining land should he ever become owner or tenant of that land. Whether it is so implicit or not will depend on all the circumstances, including the purpose of the grant and the nature of the activities sought to be restrained. But if the facts in a given case point clearly to such a
D restriction being implicit, I can see no reason in principle why the law should treat that case differently from one where the lessor already owns the adjoining land at the time of the lease. Why, I ask myself, should the law, as much in such a case as any other, not give effect to what, echoing the words of Bowen LJ in *Myers* v *Catterson* (1889) 43 Ch D 470 at p 481,

> was the obvious intention of the parties, so as to give the transaction between them that minimum of efficacy and value, which, upon any view of the case, it must have been their common intention that it should have?

I have given much thought to the question whether the conclusion I have reached on this point, if it is correct, is one which is still open to any court below the House of Lords. I mention one passage in
E particular from the judgment of Younger LJ in *Harmer* v *Jumbil (Nigeria) Tin Areas Ltd* [1921] 1 Ch 200 at p 226:

> . . . the obligation laid upon the grantor is not unqualified. If it were, that which was imposed in the interest of fair dealing might, in unscrupulous hands, become a justification for oppression, or an instrument of extortion. The obligation therefore must in every case be construed fairly, even strictly, if not narrowly. It must be such as, in view of the surrounding circumstances, was within the reasonable contemplation of the parties at the time when the transaction was entered into, and was at that time within the grantor's power to fulfil. But so limited, the obligation imposed may, I think, be infinitely varied in kind, regard being had to the paramount purpose to serve which it is imposed.

I do not think that the implication of an obligation in the terms I have mentioned in the present case offends against what is said there. An
F obligation on Mr Wade to ensure that the owner or occupier of the car lot would not erect a hoarding immediately in front of the flank wall of no 6 would not be one within his power to fulfil. But an obligation that he would not erect such a hoarding if he should become the owner or tenant of the car lot was one which it would always be within his power to fulfil.

In the end, while I recognise that the conclusion I have reached on the facts in the present case is an application of the derogation from grant principle in novel circumstances, I think that this conclusion accords with basic principle and is not one from which this court is constrained by authority.

For these reasons I reject Mr Tager's submission that because Mr Wade had no interest in the car lot in 1960, he and his successors as tenants under the 1960 lease are untouched by the derogation from G grant principle in relation to anything they may do on the car lot.

Successors in title

As Mr Wade had no interest in the car lot when the 1960 lease was entered into, there can be no question of any obligation undertaken by him operating in the nature of a restrictive covenant over that land so as to bind successive owners of that land as such. But Mr Wade himself remained bound by the obligation so long as he continued as the tenant of no 6. Furthermore, the company as his successor to the lease of no 6, was and is bound by the obligation. The obligation not to derogate from a grant applies to a person deriving title from a grantor in the same way as it applies to the grantor himself (*Cable* v *Bryant* [1908] 1 Ch 259). H

Reliance was placed by Mr Tager on the decision of this court in *Davis* v *Town Properties Investment Corporation Ltd* [1903] 1 Ch 797. The lessor of an office building assigned the reversion to the defendant company. The company bought some adjoining land and erected a new building of such a height that it caused the chimneys in the office building, of which the plaintiff was the original lessee, to smoke. The plaintiff's action for breach of the covenant for quiet enjoyment was dismissed. Collins MR said (at p 804):

> Even if the lessor's covenant could be extended to cover user by him of land subsequently acquired, such an obligation could not pass with the assignment of the reversion to the defendants, who acquired the adjoining land under another title, and have done what they have done under that other title, and not as claiming under the lessor. So far as the covenant was personal or J collateral, it would not run with reversion and bind the assignee. This was decided on the statute of Hen 8 in *Spencer's Case*.

In the present case the obligations of Mr Wade are not collateral. A covenant can "touch and concern" demised property even though nothing has to be done on the demised property. In *Ricketts* v *Enfield Churchwardens* [1909] 1 Ch 544 it was held that a covenant by a lessor not to build on a certain part of the adjoining land touched and concerned the demised property. Likewise here: the obligation on Mr Wade was not to do certain things on the adjoining land which if done would wholly frustrate the purpose for which the reservation was made.

Substance and not form K

I add a footnote on this part of the appeal. Although, in strict law, the reservation in the 1960 lease took effect as a regrant by Mr Wade of the advertisement rights, the conclusion I have reached on the application of the derogation from grant principle is, I should emphasise, not dependent on that highly technical conveyancing notion. I have sought to indicate the broad, commonsense rationale of the principle which bears the title of "derogation from grant". In 1960 Mrs Holland and Mr Wade, I have no doubt, viewed their transaction as a simple lease of the building, with a retention by Mrs Holland of certain rights. I think that it is safe to assume that the notion that Mrs Holland had granted to Mr Wade a lease of the building including the advertising site, and that he had then at once L granted the advertising rights back to her, was not at the forefront of either of their minds on January 6 1960. So be it. The principles underlying the doctrine labelled "derogation from grant" apply as much to grantees as to grantors. The point was covered by Lord Loreburn LC in *Lyttelton Times Co Ltd* v *Warners Ltd* [1907] AC 476. He referred (at p 481) to the "similar and equally binding duty" that may be laid on a grantee.

> The maxim that a grantor cannot derogate from his grant expresses the duty ordinarily laid on a man who sells or leases land. But it does not touch a similar and equally binding duty that may in certain cases be laid on a man who buys or hires land. If A lets a plot to B, he may not act so as to frustrate the purpose for which in the contemplation of both parties the land was hired. So also if B takes a plot from A, he may not act so as to frustrate the purpose for which in M the contemplation of both parties the adjoining plot remaining in A's hands was destined . . . When it is a question of what shall be implied from the contract, it is proper to ascertain what in fact was the purpose, or what were the purposes, to which both intended the land to be put, and, having found that, both should be held to all that was implied in this common intention.

Whichever way one approaches the reservation in the 1960 lease, whether one views it as a regrant by Mr Wade or views it less technically, either way the conclusion is the same.

A breach by the company

It follows from what I have said above that in my view the company was in breach of its obligations under the derogation from grant principle when it erected the hoardings against the passageway in

A 1982 and also when it permitted Mr Gary Johnston to license advertising contractors to construct and maintain a replacement hoarding in 1983. When the company took an assignment of the 1960 lease from Mr Wade in 1962 it already had a tenancy of the car lot. But at that time the end wall was unobscured and the advertisements, licensed by Miss Holland, were being displayed on the end wall. If notice were material, which I do not think it is, the company could have been in no doubt on what was the purpose of the reservation in the lease being acquired by it. The company was content to take the benefit of the lease, knowing of that purpose. Common honesty seems to me to forbid the company at one and the same time taking the benefit of that lease and taking steps to frustrate the purpose for which the reservation was made in favour of the lessor.

B In my view the judge's assessment of damages is unimpeachable. Had the company not erected the hoardings, Miss Holland's deal with A W May would have proceeded to fruition. The 1898 lease of no 6 contained a covenant against making alterations or additions unless the previous licence in writing of the lessor had been obtained. But section 19(2) of the Landlord and Tenant Act 1927 applied to that covenant, and it would have been unreasonable for the company, as freeholder of no 6, to have refused its consent to Miss Holland's contractor A W May fixing a working platform to the wall.

Judgment was given against Mr and Mrs Johnston as well as the company. This appears to have been on the basis of the judge's finding that when Mr and Mrs Johnston had the contract for the C purchase of the car lot made by the company in August 1982 completed in favour of themselves and not the company, that was "a device or endeavour to try and escape any liability on the part of the company" for a breach of the reservation and the injunction granted by Sir Douglas Frank. I understand that, in saying this, the judge was deciding that in truth the beneficial owner of the car lot was the company and not Mr and Mrs Johnston.

There is a procedural difficulty here. At the trial Mr Tager submitted to the judge that no such case had ever been pleaded. The judge decided that there was no necessity to plead such a case. He reached his conclusion on the facts from the answers given in cross-examination by the only witness of substance called for the company and Mr and Mrs Johnston: Mr Colin Johnston. The judge was rightly D determined to get at the true facts and decide the case accordingly, but I do not think that it can have been right to give judgment against Mr and Mrs Johnston on the footing that they were, in truth, holding the car lot merely as agents for the company, without that point having been brought out properly into the open, and alleged, in time for the company and the Johnstons and their advisers to consider the point and lead evidence on this point. In the circumstances I do not think that this conclusion adverse to Mr and Mrs Johnston ought to stand.

I add a cautionary footnote. Mr Krolick voiced the fear that the company might now surrender its tenancy of the car lot to Mr and Mrs Johnston and then disclaim all responsibility for the advertising hoarding, saying that the company no longer has any estate or E interest in the car lot. What the legal position would be in that event was, rightly, not pursued on this appeal. I will say only this. One recoils from the prospect of yet further litigation between the parties. But if the company should follow such a course, the court would be astute to see that the obligations imposed on the company by the derogation from grant principle are not circumvented simply by property being shuffled from the company, wholly owned by Mr and Mrs Johnston, to Mr and Mrs Johnston personally. The law is not so ineffectual that an obligation springing from a rule of common honesty can be turned aside so easily.

The 1919 agreement

The second cause of action relied on by Miss Holland was based on F the 1919 agreement. I can deal with this quite shortly. I need to mention only one of Mr Tager's arguments. The grantor under the 1919 agreement was Mr Chattey. Mr Chattey was not the freeholder of the Queen's Head and the adjoining open site. His lease, granted in 1885, was due to expire at June 24 1955. Thus, although the 1919 agreement was to continue until determined, from its inception the life of the agreement, as a matter of strict law, was confined to midsummer 1955. In 1950 the then lessee surrendered the 1885 lease. This surrender did not, *per se*, prejudice the 1919 agreement nor, conversely, did it operate automatically to extend the life of that agreement.

Of course this is a somewhat technical point and if there were evidence that after 1955 Mrs Holland and the freeholder of the car lot, Cannon Brewery, had continued to act on the assumption that G the 1919 agreement was still subsisting, the court would readily infer that by their conduct those parties had impliedly agreed that the agreement should continue. But, not surprisingly, there is no such evidence here. After the construction of the external staircase in the 1930s the passageway (apparently) ceased to be regarded as part of the Queen's Head open site and became treated as part of no 6. Thereafter there was no need in practice for any agreement between Mrs Holland and Cannon Brewery regarding the fixing of advertisements to the flank wall of no 6. The hoardings were above the passageway. All that was required was, at most, a formal or informal arrangement for Mrs Holland's advertising contractors to be afforded access on to the car lot from time to time with their ladders. H In practical terms nothing was needed beyond a simple access agreement such as the 1954 access agreement.

Mr Krolick relied on the recitals and provisions of the 1954 agreement as showing that Cannon Brewery understood that the 1919 agreement was still in existence, that Mrs Holland was entitled to the benefit of it, and that the 1919 agreement would be continuing beyond June 24 1955. But Mrs Holland was not a party to the 1954 agreement and, given the practical considerations I have mentioned, I do not think that the terms of that agreement, or the continuing enjoyment by Odham's of access arrangements under that agreement until 1971, provide a satisfactory basis for inferring the continuance of the 1919 agreement after 1955, particularly as the 1919 agreement seems to have been wholly forgotton by everyone who had ever J known of it until, over 30 years later, Miss Holland happened to come across the documents among other old papers, shortly before the trial of this action in June 1987. The 1919 agreement was not even resurrected in the previous litigation between the parties, although the 1954 access agreement was prominent in that litigation.

For this reason, which was not canvassed before the judge, I do not think that the 1919 agreement assists Miss Holland in the present litigation. I should add, for good measure, that after Miss Holland found and produced the 1919 agreement, notice was given on behalf of the company and the Johnstons determining it. So, in any event, it is now gone for good.

Nuisance K

I agree with the judge that the claim advanced on the basis of nuisance fails. Mr Krolick relied on the well-known case of *Hollywood Silver Fox Farm Ltd* v *Emmett* [1936] 2 KB 468. In that case the defendant was making noises, which emitted from his property to the plaintiff's for the purpose of annoying and injuring the plaintiff. In the present case nothing is being emitted from the car lot. I am not aware of any case which suggests that, in the absence of interference with an easement or some other proprietary right, the occupier of land may commit a nuisance if he erects a structure wholly on his own land, whether this is done with or without the intention of annoying or injuring his neighbour.

Nor do I think that the absence of planning permission for the erection of the hoardings by the company assists Miss Holland. L Her remedies in respect of any such breach of the planning legislation lie elsewhere.

The company's repayment claim against Miss Holland

I turn now to the company's claim against Miss Holland. This claim relates to part of the sums paid by rent by the company to Miss Holland between September 1982 and the date when the new 1987 lease of no 6 was entered into. In short, what happened was that after the county court proceedings for a new lease were started in 1980, the parties through their solicitors negotiated on the terms of a new lease. On February 2 1982 Miss Holland's solicitors, O H Parsons & Partners, wrote to Rose & Birn, the company's solicitors, seeking M confirmation that the Johnstons would permit access for A W May to place advertisements on the flank wall of no 6, and stating that if that confirmation were forthcoming Miss Holland would agree the terms of the new lease as follows:

(1) A new lease should be for ten years;
(2) There should be rent reviews every four years;
(3) There should be a rental of £1,750 per annum.

In their answer dated March 16 1982 Rose & Birn stated their clients were prepared to agree terms 1-3, but the only confirmation they gave concerning the flank wall was that the terms of the injunction granted by Sir Douglas Frank would be adhered to.

Nearly six months later Parsons wrote again, on September 16.

A They referred to a separate letter written regarding the hoarding and then said this:

> There remains the question of the new Lease. The terms of the Lease have of course been agreed, and my client will be submitting a demand for the rent at the new rate of £1,750 for the next quarter.
> Do you want a new Lease prepared, or do you simply wish a memorandum setting out the agreed terms to be attached to the old Lease?

Miss Holland then sent to Mr Johnston a demand for a quarter's rent at the new rate, and this was paid on September 21. On October 1 Rose & Birn wrote to Parsons, stating that it would be sufficient to attach to the old lease a memorandum of the new terms, and they asked for a draft of such a memorandum.

Having reviewed this and other correspondence, the judge B concluded that the parties' solicitors thought the terms of the new lease had all been agreed and that it was on this basis, together with the willingness of the company to pay the rent at the new rate, that the company started paying rent at the increased rate in September 1982.

The dispute which arose subsequently concerned the date of commencement of the new term. The company's solicitors put forward March 1982, being the date when the terms were agreed: the 10 years would commence then, and the increased rent would be payable from then. Miss Holland's solicitors put forward the date of expiry of the 21-year term of the old lease, namely, Christmas Day 1980. Eventually the parties' solicitors agreed that the correspondence I have referred to did not constitute a binding C contract for the grant of a new lease, because there had been no agreement on the date of commencement of the new, 10-year term. The county court proceedings were then revived, and in May 1987 terms for a new lease were agreed, and a new lease was executed. I need not mention the detailed terms of the new lease. What is in issue on the claim is the difference between the old rent (of £500 per annum) and the amount actually paid (£1,750 per annum) over the period from September 1982 to September 1984.

In this court the company based its claim exclusively on the ground of money paid under a mistake of fact, the mistake being Mr Colin Johnston's belief that the parties had reached a binding agreement for a new lease whereas, as it later transpired, they had not. The judge concluded that the mistake was one of law, as to the construction of D the letters, and not one of fact, and dismissed the company's claim accordingly.

I agree with the judge that this claim fails, although I would prefer to rest my decision on a different ground. The contractual term of the 1960 lease expired at the end of 1980, subject to the operation of Part II of the 1954 Act. The company wanted to obtain a new lease for the residue of Miss Holland's headlease of no 6, which itself was due to expire in 1997. In 1982 the company was ready to pay rent at the increased rate of £1,750 per annum in place of the old rent of £500 per annum fixed as long ago as 1960. When the rent payments were made, in September 1982 and subsequently, the company believed that it had secured agreement for the grant of a new, 10-year lease. It later turned out this was wrong. Had that been the end of the story, and E had the company not succeeded in obtaining a new lease of no 6, there might have been some substance in its claim for repayment.

In the event, however, the company did obtain the new lease it wanted and it did obtain what, when it made the rent payments from 1982 to 1984, it thought it had already agreed. Indeed, it obtained more: it obtained a lease, stretching right through to 1997. And it remained in occupation throughout. Thus the mistake made in 1982 was overtaken by subsequent events. I can see no justification for the company now being entitled to claim repayment of any of the rent it has paid. In my view the claim for restitution fails.

Accordingly, I would allow the appeal by Mr and Mrs Johnston regarding the counterclaim, but dismiss the company's appeal in relation both to the claim and the counterclaim. F

LLOYD LJ and SIR ROGER ORMROD agreed and did not add anything.

The appeal of the first plaintiffs, Johnston & Sons Ltd, in their claim and counterclaim was dismissed. The appeal of the second and third plaintiffs, Mr and Mrs Johnston, was allowed with costs. The defendant, Ida Kate Holland, was ordered to pay the costs of the second and third plaintiffs in the Court of Appeal and below. The plaintiffs' application for leave to appeal to the House of Lords was refused.

For further cases on this subject see p 171